Niall Cooney

497 0521

533-4828

Victoria Hall 613D

Assignment 7 (Homework)

Section 2.3. 10, 14, 20, 30, 34, 40

3.4   4, 8, 26

T or F   39

Linear Algebra
with Applications

Linear Algebra
with Applications

Second Edition

Otto Bretscher
Colby College

Prentice Hall
Upper Saddle River, NJ 07458

Library of Congress Cataloging-in-Publication Data

Bretscher, Otto.
    Linear algebra with applications / Otto Bretscher.—2nd ed.
      p.   cm.
    Includes index.
    ISBN 0-13-019857-9
    1. Algebras, Linear.   I. Title: Linear algebra. II. Title.

QA184 .B73 2001
512′.55—dc21

                       00-052413

Acquisition Editor: *George Lobell*
Editor in Chief: *Sally Yagan*
Vice President/Director of Production and Manufacturing: *David W. Riccardi*
Executive Managing Editor: *Kathleen Schiaparelli*
Senior Managing Editor: *Linda Mihatov Behrens*
Production Editor: *Barbara Mack*
Manufacturing Buyer: *Alan Fischer*
Manufacturing Manager: *Trudy Pisciotti*
Marketing Manager: *Angela Battle*
Marketing Assistant: *Vince Jansen*
Director of Marketing: *John Tweeddale*
Associate Editor, Mathematics/Statistics Media: *Audra J. Walsh*
Editorial Assistant/Supplements Editor: *Gale Epps*
Art Director: *Maureen Eide*
Assistant to the Art Director: *John Christiana*
Interior Designer: *Elizabeth Nemeth*
Cover Designer: *Joseph Sengotta*
Art Editor: *Grace Hazeldine*
Director of Creative Services: *Paul Belfanti*
Cover Photo: *Tomio Ohashi/Itsuko Hasegawa*

Prentice-Hall International (UK) Limited, *London*
Prentice-Hall of Australia Pty. Limited, *Sydney*
Prentice-Hall Canada Inc., *Toronto*
Prentice-Hall Hispanoamericana, S.A., *Mexico*
Prentice-Hall of India Private Limited, *New Delhi*
Prentice-Hall of Japan, Inc., *Tokyo*
Pearson Education Asia Pte. Ltd.
Editora Prentice-Hall do Brasil, Ltda., *Rio de Janeiro*

To my parents
Otto and Margrit Bretscher-Zwicky
with love and gratitude

Contents

9 Linear Differential Equations 396

Continuing Text Features

- *Linear transformations* are introduced early on in the text to make the discussion of matrix operations more meaningful and easier to visualize.
- *Visualization and geometrical interpretation* are emphasized extensively throughout.
- The reader will find an abundance of *thought-provoking* (and occasionally delightful) *problems and exercises.*
- *Abstract concepts* are introduced gradually throughout the text. The major ideas are carefully developed at various levels of generality before the student is introduced to abstract vector spaces.
- *Discrete and continuous dynamical systems* are used as a motivation for eigenvectors, and as a unifying theme thereafter.

New Features in the Second Edition

Students and instructors generally found the first edition to be accurate and well structured. Still, some changes seemed in order.

- Responding to suggestions from users, we have moved the treatment of abstract vector spaces from the end of the text to Chapter 4. This leaves the instructor with more options to develop a syllabus that best fits the students' needs, as illustrated in the diagram of chapter dependencies on p. xiii. After treating the introductory three chapters, the user can turn to abstract vector spaces (Chapter 4), geometry (Chapter 5), or determinants and eigenvalues (Chapters 6 and 7), and then take it from there.
- Over 300 true/false questions have been added to the exercise sections, at the end of most chapters. I have found these to be a valuable tool in guiding students towards conceptual thought, certainly more effective than just performing computations, and often more useful than asking students to prove a theorem.
- There are stronger motivations for many concepts (such as complex eigenvalues).
- There is (even) more visualization (of Cramer's rule, for example).
- In addition, there are hundreds of small editorial changes, offering a hint in a difficult exercise, for example, or choosing a more sensible notation in a theorem.

Free Website

There is an Internet home page for this text that is free to all users, at www.prenhall.com/bretscher. Included are interactive exercises for many chapters and links to many interesting sites related to linear algebra topics.

Preface (with David Steinsaltz)

A police officer on patrol at midnight, so runs an old joke, notices a man crawling about on his hands and knees under a streetlamp. He walks over to investigate, whereupon the man explains in a tired and somewhat slurred voice that he has lost his housekeys. The policeman offers to help, and for the next five minutes he too is searching on his hands and knees. At last he exclaims, "Are you absolutely certain that this is where you dropped the keys?"

"Here? Absolutely not. I dropped them a block down, in the middle of the street."

"Then why the devil have you got me hunting around this lamppost?"

"Because this is where the light is."

It is mathematics, and not just (as Bismarck claimed) politics, that consists in "the art of the possible." Rather than search in the darkness for solutions to problems of pressing interest, we contrive a realm of problems whose interest lies above all in the fact that solutions can conceivably be found.

Perhaps the largest patch of light surrounds the techniques of matrix arithmetic and algebra, and in particular matrix multiplication and row reduction. Here we might begin with Descartes, since it was he who discovered the conceptual meeting-point of geometry and algebra in the identification of Euclidean space with $\mathbb{R}^3$; the techniques and applications proliferated since his day. To organize and clarify those is the role of a modern linear algebra course.

COMPUTERS AND COMPUTATION

An essential issue that needs to be addressed in establishing a mathematical methodology is the role of computation and of computing technology. Are the proper subjects of mathematics algorithms and calculations, or are they grand theories and abstractions that evade the need for computation? If the former, is it important that the students learn to carry out the computations with pencil and paper, or should the algorithm "press the calculator's x^{-1} button" be allowed to substitute for the traditional method of finding an inverse? If the latter, should the abstractions be taught through elaborate notational mechanisms, or through computational examples and graphs?

We seek to take a consistent approach to these questions: algorithms and computations are primary, and precisely for this reason computers are not. Again and again we examine the nitty-gritty of row reduction or matrix multiplication in order to derive new insights. Most of the proofs, whether of rank-nullity theorem, the volume-change formula for determinants, or the spectral theorem for symmetric matrices, are in this way tied to hands-on procedures.

The aim is not just to know how to compute the solution to a problem, but to *imagine* the computations. The student needs to perform enough row reductions by hand to be equipped to follow a line of argument of the form: "If we calculate the reduced row echelon form of such a matrix . . .," and to appreciate in advance what the possible outcomes to a particular computation are.

In applications the solution to a problem is hardly more important than recognizing its range of validity, and appreciating how sensitive it is to perturbations of the input. We emphasize the geometric and qualitative nature of the solutions, notions of approximation, stability, and "typical" matrices. The discussion of Cramer's rule, for instance, underscores the value of closed-form solutions for visualizing a system's behavior, and understanding its dependence from initial conditions.

The availability of computers is, however, neither to be ignored nor regretted. Each student and instructor will have to decide how much practice is needed to be sufficiently familiar with the inner workings of the algorithm. As the explicit computations are being replaced gradually by a theoretical overview of how the algorithm works, the burden of calculation will be taken up by technology, particularly for those wishing to carry out the more numerical and applied exercises.

It is possible to turn your linear algebra course into a more computer oriented or enhanced course by wrapping with this text either *ATLAST Computer Exercises for Linear Algebra, Linear Algebra Labs with MATLAB,* or *Visualizing Linear Algebra with Maple* (see *Solutions Manuals, MATLAB, and Maple*). Each of these supplements goes beyond just using the computer for computational matters. Each takes the standard topics in linear algebra and finds a method of illuminating key ideas visually with the computer. Both have M-files available that can be delivered by the Internet.

EXAMPLES, EXERCISES, APPLICATIONS, AND HISTORY

The exercises and examples are the heart of this book. Our objective is not just to show our readers a "patch of light" where questions may be posed and solved, but to convince them that there is indeed a great deal of useful, interesting material to be found in this area, if they take the time to look around. Consequently, we have included genuine applications of the ideas and methods under discussion to a broad range of sciences: physics, chemistry, biology, economics, and, of course, mathematics itself. Often we have simplified them to sharpen the point, but they use the methods and models of contemporary scientists.

With such a large and varied set of exercises in each section, instructors should have little difficulty in designing a course that is suited to their aims and to the needs of their students. Quite a few straightforward computation problems are offered, of course. Simple (and, in a few cases, not so simple) proofs and derivations are required in some exercises. In many cases, theoretical principles that are discussed at length in more abstract linear algebra courses are here found broken up in bite-size exercises.

The examples make up a significant portion of the text; we have kept abstract exposition to a minimum. It is a matter of taste whether general theories should give rise to specific examples, or be pasted together from them. In a text such as this one, attempting to keep an eye on applications, the latter is clearly preferable: the examples always precede the theorems in this book.

Scattered throughout the mathematical exposition are quite a few names and dates, some historical accounts, and anecdotes as well. Students of mathematics are too rarely shown that the seemingly strange and arbitrary concepts they study are the results of long and hard struggles. It will encourage the readers to know that a mere two centuries ago some of the most brilliant mathematicians were wrestling with problems such as the meaning of dimension or the interpretation of e^{it}, and to realize that the advance of time and understanding actually enables them, with some effort of their own, to see farther than those great minds.

OUTLINE OF THE TEXT

Chapter 1. This chapter provides a careful introduction to the solution of systems of linear equations by Gauss–Jordan elimination. Once the concrete problem is solved, we restate it in terms of matrix formalism, and discuss the geometric properties of the solutions.

Chapter 2. Here we raise the abstraction a notch and reinterpret matrices as **linear transformations.** The reader is introduced to the modern notion of a function, as an arbitrary association between an input and an output, which leads into a discussion of inverses. The traditional method for finding the inverse of a matrix is explained: it fits in naturally as a sort of automated algorithm for Gauss–Jordan elimination.

We define linear transformations primarily in terms of matrices, since that is how they are used; the abstract concept of linearity is presented as an auxiliary notion. Rotations in $\mathbb{R}^2$ are emphasized, both as archetypal, easily visualized examples, and as preparation for future applications.

Chapter 3. We introduce the central concepts of linear algebra: subspaces, image and kernel, linear independence, bases, coordinates, and **dimension,** still firmly fixed in $\mathbb{R}^n$.

Chapter 4. Generalizing the ideas of the preceding chapter and using an abundance of examples, we introduce abstract **vector spaces** (which are called "linear spaces" here, to prevent the confusion some students experience with the term "vector").

Chapter 5. This chapter includes some of the most basic applications. We introduce orthonormal bases and the Gram–Schmidt process, along with the QR factorization. The calculation of correlation coefficients is discussed, and the important technique of least-squares approximations is explained, in a number of different contexts.

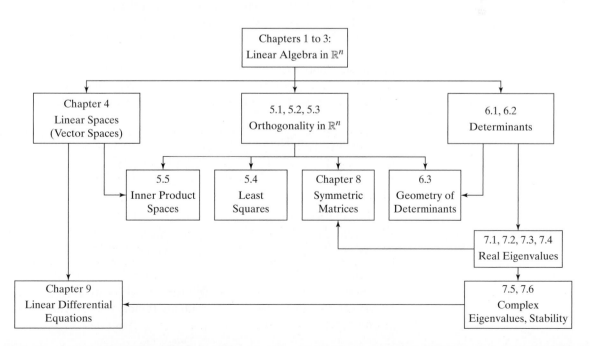

Chapter 6. Our discussion of determinants is algorithmic, based on the counting of "patterns" (a transparent way to deal with permutations). We derive the properties of the determinant from careful analysis of this procedure, and tie it together with Gauss–Jordan elimination. The goal is to prepare for the main application of determinants: the computation of characteristic polynomials.

Chapter 7. This chapter introduces the central application of the latter half of the text: linear **dynamical systems.** We begin with discrete systems, and are naturally led to seek eigenvectors, which characterize the long-term behavior of the system. Qualitative behavior is emphasized, particularly stability conditions. Complex eigenvalues are explained, without apology, and tied into earlier discussions of two-dimensional rotation matrices.

Chapter 8. The ideas and methods of Chapter 7 are applied to geometry. We discuss the spectral theorem for symmetric matrices and its applications to quadratic forms, conic sections, and singular values.

Chapter 9. Here we apply the methods developed for discrete dynamical systems to continuous ones, that is, to systems of first-order linear differential equations. Again, the cases of real and complex eigenvalues are discussed.

SOLUTIONS MANUALS, MATLAB, AND MAPLE

- *Student's Solutions Manual,* with carefully worked solutions to all odd-numbered problems in the text (ISBN 0-13-032856-1)
- *Instructor's Solutions Manual,* with solutions to all the problems in the text (ISBN 0-13-032855-3)
- *ATLAST Computer Exercises for Linear Algebra,* S. Leon, G. Herman, R. Faulkenberry (editors), Prentice Hall, 1997 (ISBN 0-13-270273-8)
- *Linear Algebra Labs with MATLAB,* 2nd ed., D. Hill, D. Zitarelli, Prentice Hall, 1996 (ISBN 0-13-505439-7)
- *Visualizing Linear Algebra with Maple,* Sandra Keith, Prentice Hall, 2001 (ISBN 0-13-041816-1)

ACKNOWLEDGMENTS

I first thank my students and colleagues at Colby College and Harvard University for the key role they have played in developing this text out of a series of rough lecture notes. The following colleagues, who have taught the course with me, have made invaluable contributions:

Persi Diaconis	David Kazhdan
Jordan Ellenberg	Barry Mazur
Matthew Emerton	David Mumford
Edward Frenkel	David Steinsaltz
Fernando Gouvêa	Shlomo Sternberg
Jan Holly	Richard Taylor

I owe special thanks to John Boller, William Calder, and Robert Kaplan for their thoughtful review of the manuscript.

I wish to thank Sylvie Bessette for the careful preparation of the manuscript and Paul Nguyen for his well-drawn figures.

I am grateful to those who have contributed to the book in many ways: Marlisa Smith, Menoo Cung, Srdjan Divac, Robin Gottlieb, Luke Hunsberger, Bridget Neale, Akilesh Palanisamy, Rita Pang, Esther Silberstein, Radhika de Silva, Jonathan Tannenhauser, and Larry Wilson.

I have received valuable feedback from the book's reviewers for the first and second editions:

Loren Argabright, Drexel University

Frank Beatrous, University of Pittsburgh

Tracy Bibelnieks, University of Minnesota

Jeff D. Farmer, University of Northern Colorado

Herman Gollwitzer, Drexel University

David G. Handron, Jr., Carnegie Mellon University

Willy Hereman, Colorado School of Mines

Konrad J. Heuvers, Michigan Technological University

Charles Holmes, Miami University

Michael Kallaher, Washington State University

Daniel King, Oberlin College

Richard Kubelka, San Jose State University

Michael G. Neubauer, California State University–Northridge

Peter C. Patton, University of Pennsylvania

Jeffrey M. Rabin, University of California, San Diego

Daniel B. Shapiro, Ohio State University

David Steinsaltz, Technische Universität Berlin

James A. Wilson, Iowa State University

I also thank my editor, George Lobell, for his encouragement and advice, and Barbara Mack for coordinating book production.

The development of this text has been supported by a grant from the Instructional Innovation Fund of Harvard College.

Otto Bretscher
o_bretsc@colby.edu

Linear Equations

1.1 INTRODUCTION TO LINEAR SYSTEMS

Traditionally, algebra was the art of solving equations and systems of equations. The word *algebra* comes from the Arabic *al-jabr,* which means *restoration* (of broken parts).[1] The term was first used in a mathematical sense by Mohammed al-Khowarizmi (c. 780–850), who worked at the House of Wisdom, an academy established by Caliph al-Ma'mun in Baghdad. Linear algebra, then, is the art of solving systems of linear equations.

The need to solve systems of linear equations frequently arises in mathematics, statistics, physics, astronomy, engineering, computer science, and economics.

Solving systems of linear equations is not conceptually difficult. For small systems, ad hoc methods certainly suffice. Larger systems, however, require more systematic methods. The approach generally used today was beautifully explained 2,000 years ago in a Chinese text, the "Nine Chapters on the Mathematical Art" (Chiu-chang Suan-shu), which contains the following example:[2]

> The yield of one sheaf of inferior grain, two sheaves of medium grain, and three sheaves of superior grain is 39 tou.[3] The yield of one sheaf of inferior grain, three sheaves of medium grain, and two sheaves of superior grain is 34 tou. The yield of three sheaves of inferior grain, two sheaves of medium grain, and one

---

[1] At one time, it was not unusual to see the sign *Algebrista y Sangrador* (bone setter and blood letter) at the entrance of a Spanish barber's shop.

[2] B. L. v.d.Waerden: *Geometry and Algebra in Ancient Civilizations*, Springer-Verlag, Berlin, 1983.

[3] A *tou* is a bronze bowl used as a food container during the middle and late Chou dynasty (c. 900–255 B.C.).

1

sheaf of superior grain is 26 tou. What is the yield of inferior, medium, and superior grain?

In this problem the unknown quantities are the yields of one sheaf of inferior, one sheaf of medium, and one sheaf of superior grain. Let us denote these quantities by x, y, and z, respectively. The problem can then be represented by the following system of linear equations:

$$\left|\begin{array}{l} x + 2y + 3z = 39 \\ x + 3y + 2z = 34 \\ 3x + 2y + z = 26 \end{array}\right|.$$

To solve for x, y, and z, we need to transform this system from the form

$$\left|\begin{array}{l} x + 2y + 3z = 39 \\ x + 3y + 2z = 34 \\ 3x + 2y + z = 26 \end{array}\right| \quad \text{into the form} \quad \left|\begin{array}{l} x = \ldots \\ y = \ldots \\ z = \ldots \end{array}\right|.$$

In other words, we need to eliminate the terms that are off the diagonal, those circled in the following equations, and make the coefficients of the variables along the diagonal equal to 1:

$$x \; + \; \boxed{2y} \; + \; \boxed{3z} = 39$$

$$\boxed{x} \; + \; 3y \; + \; \boxed{2z} = 34$$

$$\boxed{3x} \; + \; \boxed{2y} \; + \; z = 26$$

We can accomplish these goals step by step, one variable at a time. In the past, you may have simplified systems of equations by adding equations to one another or subtracting them. In this system, we can eliminate the variable x from the second equation by subtracting the first equation from the second:

$$\left|\begin{array}{l} x + 2y + 3z = 39 \\ x + 3y + 2z = 34 \\ 3x + 2y + z = 26 \end{array}\right| \quad \begin{array}{c} \longrightarrow \\ -1\text{st equation} \end{array} \quad \left|\begin{array}{l} x + 2y + 3z = 39 \\ y - z = -5 \\ 3x + 2y + z = 26 \end{array}\right|.$$

To eliminate the variable x from the third equation, we subtract the first equation from the third equation three times. We multiply the first equation by 3 to get

$$3x + 6y + 9z = 117 \qquad (3 \times \text{1st equation})$$

and then subtract this result from the third equation:

$$\left|\begin{array}{l} x + 2y + 3z = 39 \\ y - z = -5 \\ 3x + 2y + z = 26 \end{array}\right| \quad \begin{array}{c} \longrightarrow \\ -3 \times \text{1st equation} \end{array} \quad \left|\begin{array}{l} x + 2y + 3z = 39 \\ y - z = -5 \\ -4y - 8z = -91 \end{array}\right|.$$

Similarly, we eliminate the variable y off the diagonal:

$$\left|\begin{array}{l} x + 2y + 3z = 39 \\ y - z = -5 \\ -4y - 8z = -91 \end{array}\right| \quad \begin{array}{c} -2 \times \text{2nd equation} \\ \longrightarrow \\ +4 \times \text{2nd equation} \end{array} \quad \left|\begin{array}{l} x + 5z = 49 \\ y - z = -5 \\ -12z = -111 \end{array}\right|.$$

Before we eliminate the variable z off the diagonal, we make the coefficient of z on the diagonal equal to 1, by dividing the last equation by -12:

$$\begin{vmatrix} x & + & 5z & = & 49 \\ & y & - & z & = & -5 \\ & & & -12z & = & -111 \end{vmatrix} \quad \underset{\div\,(-12)}{\longrightarrow} \quad \begin{vmatrix} x & & +5z & = & 49 \\ & y & - & z & = & -5 \\ & & & z & = & 9.25 \end{vmatrix}.$$

Finally, we eliminate the variable z off the diagonal:

$$\begin{vmatrix} x & & +5z & = & 49 \\ & y & - & z & = & -5 \\ & & & z & = & 9.25 \end{vmatrix} \quad \begin{array}{l} -5 \times \text{last equation} \\ + \text{ last equation} \\ \longrightarrow \end{array} \quad \begin{vmatrix} x & & & = & 2.75 \\ & y & & = & 4.25 \\ & & z & = & 9.25 \end{vmatrix}.$$

The yields of inferior, medium, and superior grain are 2.75, 4.25, and 9.25 tou per sheaf, respectively.

By substituting these values, we can check that $x = 2.75$, $y = 4.25$, $z = 9.25$ is indeed the solution of the system:

$$2.75 + 2 \times 4.25 + 3 \times 9.25 = 39$$
$$2.75 + 3 \times 4.25 + 2 \times 9.25 = 34$$
$$3 \times 2.75 + 2 \times 4.25 + 9.25 = 26$$

Happily, in linear algebra, you are almost always able to check your solutions. It will help you if you get into the habit of checking now.

Geometric Interpretation

How can we interpret this result geometrically? Each of the three equations of the system defines a plane in x–y–z space. The solution set of the system consists of those points (x, y, z) that lie in all three planes (i.e., the intersection of the three planes). Algebraically speaking, the solution set consists of those ordered triples of numbers (x, y, z) that satisfy all three equations simultaneously. Our computations show that the system has only one solution, $(x, y, z) = (2.75, 4.25, 9.25)$. This means that the planes defined by the three equations intersect at the point $(x, y, z) = (2.75, 4.25, 9.25)$, as shown in Figure 1.

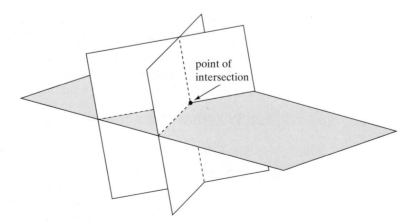

point of intersection

Figure 1 Three planes in space, intersecting at a point.

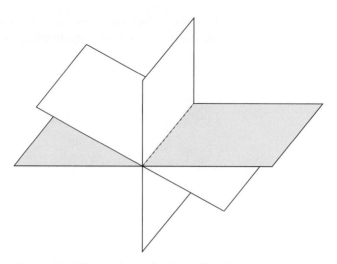

Figure 2(a) Three planes having a line in common.

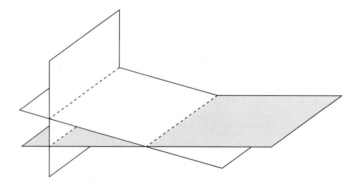

Figure 2(b) Three planes with no common intersection.

While three different planes in space usually intersect at a point, they may have a line in common (see Figure 2(a)) or may not have a common intersection at all, as shown in Figure 2(b). Therefore, a system of three equations with three unknowns may have a unique solution, infinitely many solutions, or no solutions at all.

A System with Infinitely Many Solutions

Next, let's consider a system of linear equations that has infinitely many solutions:

$$\left| \begin{array}{l} 2x + 4y + 6z = 0 \\ 4x + 5y + 6z = 3 \\ 7x + 8y + 9z = 6 \end{array} \right|.$$

We can solve this system using elimination as previously discussed. For simplicity, we label the equations with Roman numerals.

$$
\begin{vmatrix} 2x + 4y + 6z = 0 \\ 4x + 5y + 6z = 3 \\ 7x + 8y + 9z = 6 \end{vmatrix}
\begin{matrix} \div 2 \\ \longrightarrow \\ {} \end{matrix}
\begin{vmatrix} x + 2y + 3z = 0 \\ 4x + 5y + 6z = 3 \\ 7x + 8y + 9z = 6 \end{vmatrix}
\begin{matrix} \longrightarrow \\ -4(\mathrm{I}) \\ -7(\mathrm{I}) \end{matrix}
$$

$$
\begin{vmatrix} x + 2y + 3z = 0 \\ -3y - 6z = 3 \\ -6y - 12z = 6 \end{vmatrix}
\begin{matrix} \longrightarrow \\ \div(-3) \\ {} \end{matrix}
\begin{vmatrix} x + 2y + 3z = 0 \\ y + 2z = -1 \\ -6y - 12z = 6 \end{vmatrix}
\begin{matrix} -2(\mathrm{II}) \\ \longrightarrow \\ +6(\mathrm{II}) \end{matrix}
$$

$$
\begin{vmatrix} x - z = 2 \\ y + 2z = -1 \\ 0 = 0 \end{vmatrix}
\quad \longrightarrow \quad
\begin{vmatrix} x - z = 2 \\ y + 2z = -1 \end{vmatrix}
$$

After omitting the trivial equation $0 = 0$, we have only two equations with three unknowns. The solution set is the intersection of two nonparallel planes in space (i.e., a line). This system has infinitely many solutions.

The two foregoing equations can be written as follows:

$$
\begin{vmatrix} x = z + 2 \\ y = -2z - 1 \end{vmatrix}.
$$

We see that both x and y are determined by z. We can freely choose a value of z, an arbitrary real number; then, the two equations above give us the values of x and y for this choice of z. For example,

- Choose $z = 1$. Then, $x = z + 2 = 3$ and $y = -2z - 1 = -3$. The solution is $(x, y, z) = (3, -3, 1)$.
- Choose $z = 7$. Then, $x = z + 2 = 9$ and $y = -2z - 1 = -15$. The solution is $(x, y, z) = (9, -15, 7)$.

More generally, if we choose $z = t$, an arbitrary real number, we get $x = t + 2$ and $y = -2t - 1$. Therefore, the general solution is

$$
(x, y, z) = (t + 2, -2t - 1, t) = (2, -1, 0) + t(1, -2, 1).
$$

This equation represents a line in space, as shown in Figure 3.

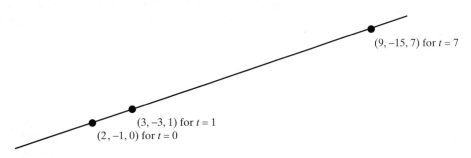

Figure 3 The line $(x, y, z) = (t + 2, -2t - 1, t)$.

A System without Solutions

In the following system, perform the eliminations yourself to obtain the result shown:

$$\begin{vmatrix} x + 2y + 3z = 0 \\ 4x + 5y + 6z = 3 \\ 7x + 8y + 9z = 0 \end{vmatrix} \longrightarrow \begin{vmatrix} x \quad - \quad z = \quad 2 \\ y + 2z = -1 \\ 0 = -6 \end{vmatrix}.$$

Whatever values we choose for x, y, and z, the equation $0 = -6$ cannot be satisfied. This system is *inconsistent;* that is, it has no solutions.

EXERCISES

GOAL Set up and solve systems with as many as three linear equations with three unknowns, and interpret the equations and their solutions geometrically.

In Exercises 1 through 10, find all solutions of the linear systems using elimination as discussed in this section. Then check your solutions.

1. $\begin{vmatrix} x + 2y = 1 \\ 2x + 3y = 1 \end{vmatrix}.$

2. $\begin{vmatrix} 4x + 3y = 2 \\ 7x + 5y = 3 \end{vmatrix}.$

3. $\begin{vmatrix} 2x + 4y = 3 \\ 3x + 6y = 2 \end{vmatrix}.$

4. $\begin{vmatrix} 2x + 4y = 2 \\ 3x + 6y = 3 \end{vmatrix}.$

5. $\begin{vmatrix} 2x + 3y = 0 \\ 4x + 5y = 0 \end{vmatrix}.$

6. $\begin{vmatrix} x + 2y + 3z = \quad 8 \\ x + 3y + 3z = 10 \\ x + 2y + 4z = \quad 9 \end{vmatrix}.$

7. $\begin{vmatrix} x + 2y + 3z = 1 \\ x + 3y + 4z = 3 \\ x + 4y + 5z = 4 \end{vmatrix}.$

8. $\begin{vmatrix} x + 2y + \quad 3z = 0 \\ 4x + 5y + \quad 6z = 0 \\ 7x + 8y + 10z = 0 \end{vmatrix}.$

9. $\begin{vmatrix} x + 2y + 3z = 1 \\ 3x + 2y + \quad z = 1 \\ 7x + 2y - 3z = 1 \end{vmatrix}.$

10. $\begin{vmatrix} x + 2y + \quad 3z = 1 \\ 2x + 4y + \quad 7z = 2 \\ 3x + 7y + 11z = 8 \end{vmatrix}.$

In Exercises 11 through 13, find all solutions of the linear systems. Represent your solutions graphically, as intersections of lines in the x–y-plane.

11. $\begin{vmatrix} x - 2y = \quad 2 \\ 3x + 5y = 17 \end{vmatrix}.$

12. $\begin{vmatrix} x - 2y = 3 \\ 2x - 4y = 6 \end{vmatrix}.$

13. $\begin{vmatrix} x - 2y = 3 \\ 2x - 4y = 8 \end{vmatrix}.$

In Exercises 14 through 16, find all solutions of the linear systems. Describe your solutions in terms of intersecting planes. You need not sketch these planes.

14. $\begin{vmatrix} x + \quad 4y + \quad z = 0 \\ 4x + 13y + \quad 7z = 0 \\ 7x + 22y + 13z = 1 \end{vmatrix}.$

15. $\begin{vmatrix} x + y - \quad z = 0 \\ 4x - y + 5z = 0 \\ 6x + y + 4z = 0 \end{vmatrix}.$

16. $\begin{vmatrix} x + \quad 4y + \quad z = 0 \\ 4x + 13y + \quad 7z = 0 \\ 7x + 22y + 13z = 0 \end{vmatrix}.$

17. Find all solutions of the linear system

$$\begin{vmatrix} x + 2y = a \\ 3x + 5y = b \end{vmatrix},$$

where a and b are arbitrary constants.

18. Find all solutions of the linear system

$$\begin{vmatrix} x + 2y + 3z = a \\ x + 3y + 8z = b \\ x + 2y + 2z = c \end{vmatrix},$$

where a, b, and c are arbitrary constants.

19. Consider a two-commodity market. When the unit prices of the products are P_1 and P_2, the quantities demanded, D_1 and D_2, and the quantities supplied, S_1 and S_2, are given by

$$D_1 = \quad 70 - 2P_1 + P_2, \qquad S_1 = -14 + 3P_1,$$
$$D_2 = 105 + \quad P_1 - P_2, \qquad S_2 = \quad -7 \qquad + 2P_2.$$

 a. What is the relationship between the two commodities? Do they compete, as do Volvos and BMWs, or do they complement one another, as do shirts and ties?

 b. Find the equilibrium prices (i.e., the prices for which supply equals demand), for both products.

20. The Russian-born U.S. economist and Nobel laureate Wassily Leontief (1906–1999) was interested in the following question: What output should each of the industries in an economy produce to satisfy the total demand for all products? Here, we consider a very simple example of input-output analysis, an economy with only two industries, A and B. Assume that the consumer demand for their products is, respectively, 1,000 and 780, in millions of dollars per year.

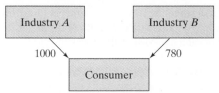

Which outputs a and b (in millions of dollars per year) should the two industries generate to satisfy the demand? You may be tempted to say 1,000 and 780, respectively, but things are not quite as simple as that. We have to take into account the interindustry demand as well. Let us say that industry A produces electricity. Of course, producing almost any product will require electric power. Suppose that industry B needs 10¢ worth of electricity for each $1 of output B produces and that industry A needs 20¢ worth of B's products for each $1 of output A produces. Find the outputs a and b needed to satisfy both consumer and interindustry demand.

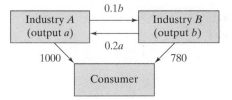

21. Find the outputs a and b needed to satisfy the consumer and interindustry demands given in the following figure (see Exercise 20):

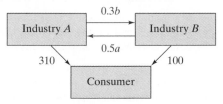

22. Consider the differential equation

$$\frac{d^2x}{dt^2} - \frac{dx}{dt} - x = \cos(t).$$

This equation could describe a forced damped oscillator, as we will see in Chapter 9. We are told that the differential equation has a solution of the form

$$x(t) = a\sin(t) + b\cos(t).$$

Find a and b, and graph the solution.

23. Find all solutions of the system

$$\begin{vmatrix} 7x - y = \lambda x \\ -6x + 8y = \lambda y \end{vmatrix}, \qquad \text{for}$$

a. $\lambda = 5$,   b. $\lambda = 10$,   and   c. $\lambda = 15$.

24. On your next trip to Switzerland, you should take the scenic boat ride from Rheinfall to Rheinau and back. The trip downstream from Rheinfall to Rheinau takes 20 minutes, and the return trip takes 40 minutes; the distance between Rheinfall and Rheinau along the river is 8 kilometers. How fast does the boat travel (relative to the water), and how fast does the river Rhein flow in this area? You may assume both speeds to be constant throughout the journey.

25. Consider the linear system

$$\begin{vmatrix} x + y - z = -2 \\ 3x - 5y + 13z = 18 \\ x - 2y + 5z = k \end{vmatrix},$$

where k is an arbitrary number.

a. For which value(s) of k does this system have one or infinitely many solutions?

b. For each value of k you found in part a, how many solutions does the system have?

c. Find all solutions for each value of k.

26. Consider the linear system

$$\begin{vmatrix} x + y - z = 2 \\ x + 2y + z = 3 \\ x + y + (k^2 - 5)z = k \end{vmatrix},$$

where k is an arbitrary constant. For which choice(s) of k does this system have a unique solution? For which choice(s) of k does the system have infinitely many solutions? For which choice(s) of k is the system inconsistent?

27. Emile and Gertrude are brother and sister. Emile has twice as many sisters as brothers, and Gertrude has just as many brothers as sisters. How many children are there in this family?

28. In a grid of wires, the temperature at exterior mesh points is maintained at constant values (in °C) as shown in the accompanying figure. When the grid is in thermal equilibrium, the temperature T at each interior mesh point is the average of the temperatures at the four adjacent points. For example,

$$T_2 = \frac{T_3 + T_1 + 200 + 0}{4}.$$

Find the temperatures T_1, T_2, and T_3 when the grid is in thermal equilibrium.

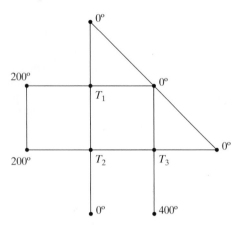

29. Find a polynomial of degree 2 (a polynomial of the form $f(t) = a + bt + ct^2$) whose graph goes through the points $(1, -1)$, $(2, 3)$, and $(3, 13)$. Sketch the graph of this polynomial.

30. Find a polynomial of degree 2 (a polynomial of the form $f(t) = a + bt + ct^2$) whose graph goes through the points $(1, p)$, $(2, q)$, $(3, r)$, where p, q, r are arbitrary constants. Does such a polynomial exist for all choices of p, q, r?

31. Find all points (a, b, c) in space for which the system

$$\begin{vmatrix} x + 2y + 3z = a \\ 4x + 5y + 6z = b \\ 7x + 8y + 9z = c \end{vmatrix}$$

has at least one solution.

32. Linear systems are particularly easy to solve when they are in *triangular* form (i.e., all entries above or below the diagonal are zero).

a. Solve the lower triangular system

$$\begin{vmatrix} x_1 & & & = -3 \\ -3x_1 + x_2 & & & = 14 \\ x_1 + 2x_2 + x_3 & & = 9 \\ -x_1 + 8x_2 - 5x_3 + x_4 & = 33 \end{vmatrix}$$

by forward substitution, finding first x_1, then x_2, then x_3, then x_4.

b. Solve the upper triangular system

$$\begin{vmatrix} x_1 + 2x_2 - x_3 + 4x_4 = -3 \\ x_2 + 3x_3 + 7x_4 = 5 \\ x_3 + 2x_4 = 2 \\ x_4 = 0 \end{vmatrix}.$$

33. Consider the linear system

$$\begin{vmatrix} x + y = 1 \\ x + \dfrac{t}{2}y = t \end{vmatrix},$$

where t is a nonzero constant.

a. Determine the x- and y-intercepts of the lines $x + y = 1$ and $x + (t/2)y = t$; sketch these lines. For which values of the constant t do these lines intersect? For these values of t, the point of intersection (x, y) depends on the choice of the constant t; that is, we can consider x and y as functions of t. Draw rough sketches of these functions.

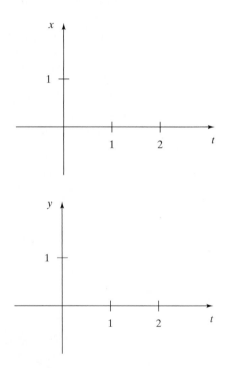

Explain briefly how you found these graphs. Argue geometrically, without solving the system algebraically.

b. Now solve the system algebraically. Verify that the graphs you sketched in part a are compatible with your algebraic solution.

34. "A certain person buys sheep, goats, and hogs, to the number of 100, for 100 crowns; the sheep cost him $\frac{1}{2}$ a crown a-piece; the goats $1\frac{1}{3}$ crown; and the hogs, $3\frac{1}{2}$ crowns. How many had he of each?" (From Leonhard Euler: *Elements of Algebra*, St. Petersburg, 1770.

Translated by Rev. John Hewlett.) Find *all* solutions to this problem.

35. Find a system of linear equations with three unknowns whose solutions are

$$x = 6 + 5t, \qquad y = 4 + 3t, \quad \text{and} \quad z = 2 + t,$$

where t is an arbitrary constant.

36. Boris and Marina are shopping for chocolate bars. Boris observes, "If I add half my money to yours, it will be enough to buy two chocolate bars." Marina naively asks, "If I add half my money to yours, how many can we buy?" Boris replies, "One chocolate bar." How much money did Boris have? (From Yuri Chernyak and Robert Rose: *The Chicken from Minsk*, Basic Books, New York, 1995.)

37. Here is another method to solve a system of linear equations: Solve one of the equations for one of the variables, and substitute the result into the other equations. Repeat this process until you run out of variables or equations. Consider the example discussed early in this section:

$$\begin{vmatrix} x + 2y + 3z = 39 \\ x + 3y + 2z = 34 \\ 3x + 2y + z = 26 \end{vmatrix}.$$

We can solve the first equation for x:

$$x = 39 - 2y - 3z.$$

Then we substitute into the other equations:

$$\begin{vmatrix} (39 - 2y - 3z) + 3y + 2z = 34 \\ 3(39 - 2y - 3z) + 2y + z = 26 \end{vmatrix}.$$

We can simplify:

$$\begin{vmatrix} y - z = -5 \\ -4y - 8z = -91 \end{vmatrix}.$$

Now, $y = z - 5$, so that $-4(z - 5) - 8z = -91$, or

$$-12z = -111.$$

We find that $z = \dfrac{111}{12} = 9.25$. Then,

$$y = z - 5 = 4.25,$$

and

$$x = 39 - 2y - 3z = 2.75.$$

Explain why this method is essentially the same as the method discussed in this section, only the bookkeeping is different.

1.2 MATRICES AND GAUSS–JORDAN ELIMINATION

In practice, systems of linear equations are usually solved by computer. Suppose you want to solve the following system on a computer:

$$\begin{vmatrix} 2x + 8y + 4z = 2 \\ 2x + 5y + z = 5 \\ 4x + 10y - z = 1 \end{vmatrix}.$$

What information about this system would the computer need to solve it? With the right software, all you need to enter is the pattern of coefficients of the variables and the numbers on the right-hand side of the equations:

$$\begin{bmatrix} 2 & 8 & 4 & 2 \\ 2 & 5 & 1 & 5 \\ 4 & 10 & -1 & 1 \end{bmatrix}.$$

All the information about the system is conveniently stored in this array of numbers, called a **matrix**.[4] Since this particular matrix has three rows and four columns, it is called a 3×4 matrix (read "three by four").

---

[4]It appears that the term "matrix" was first used in this sense by the English mathematician J. J. Sylvester, in 1850.

The four columns of the matrix

$$
\begin{array}{c}
\text{The three rows of the matrix} \rightarrow
\end{array}
\begin{bmatrix}
2 & 8 & 4 & 2 \\
2 & 5 & 1 & 5 \\
4 & 10 & -1 & 1
\end{bmatrix}
$$

Note that the first column of this matrix corresponds to the first variable of the system, while the first row corresponds to the first equation.

It is customary to label the entries of a 3×4 matrix A with double subscripts as follows:

$$
A = \begin{bmatrix}
a_{11} & a_{12} & a_{13} & a_{14} \\
a_{21} & a_{22} & a_{23} & a_{24} \\
a_{31} & a_{32} & a_{33} & a_{34}
\end{bmatrix}.
$$

The first subscript refers to the row, and the second to the column: The entry a_{ij} is located in the ith row and the jth column.

Two matrices A and B are equal if they have the same size and if corresponding entries are equal: $a_{ij} = b_{ij}$.

If the number of rows of a matrix A equals the number of columns (A is $n \times n$), then A is called a *square matrix,* and the entries $a_{11}, a_{22}, \ldots, a_{nn}$ form the (main) *diagonal* of A. A square matrix A is called *diagonal* if all its entries off the main diagonal are zero; that is, $a_{ij} = 0$ whenever $i \neq j$. A square matrix A is called *upper triangular* if all its entries below the main diagonal are zero; that is, $a_{ij} = 0$ whenever i exceeds j. *Lower triangular* matrices are defined analogously. A matrix whose entries are all zero is called a *zero matrix* and is denoted by 0 (regardless of its size). Consider the matrices

$$
A = \begin{bmatrix} 1 & 2 & 3 \\ 4 & 5 & 6 \end{bmatrix}, \qquad
B = \begin{bmatrix} 1 & 2 \\ 3 & 4 \end{bmatrix}, \qquad
C = \begin{bmatrix} 2 & 0 & 0 \\ 0 & 3 & 0 \\ 0 & 0 & 0 \end{bmatrix},
$$

$$
D = \begin{bmatrix} 2 & 3 \\ 0 & 4 \end{bmatrix}, \qquad
E = \begin{bmatrix} 5 & 0 & 0 \\ 4 & 0 & 0 \\ 3 & 2 & 1 \end{bmatrix}.
$$

The matrices $B, C, D,$ and E are square, C is diagonal, C and D are upper triangular, and C and E are lower triangular.

Matrices with only one column or row are of particular interest.

A matrix with only one column is called a column vector, or simply a **vector**. The entries of a vector are called its *components*. The set of all column vectors with n components is denoted by $\mathbb{R}^n$.

A matrix with only one row is called a *row vector*.

In this text, the term *vector* refers to column vectors, unless otherwise stated. The reason for our preference for column vectors will become apparent in the next section.

Examples of vectors are

$$\begin{bmatrix} 1 \\ 2 \\ 9 \\ 1 \end{bmatrix},$$

a (column) vector in $\mathbb{R}^4$, and

$$[1 \quad 5 \quad 5 \quad 3 \quad 7],$$

a row vector with five components. Note that the n columns of an $m \times n$ matrix are vectors in $\mathbb{R}^m$.

In previous courses in mathematics or physics, you may have thought about vectors from a more geometric point of view. (See the appendix for a summary of basic facts on vectors.) In this course, it will often be helpful to think about a vector numerically, as a (finite) sequence of numbers, which we will usually write in a column.

In our digital age, information is often transmitted and stored as a sequence of numbers (i.e., as a vector). A sequence of 10 seconds of music on a CD is stored as a vector with 440,000 components. A weather photograph taken by a satellite is digitized and transmitted to Earth as a sequence of numbers.

Consider again the system

$$\begin{vmatrix} 2x + 8y + 4z = 2 \\ 2x + 5y + z = 5 \\ 4x + 10y - z = 1 \end{vmatrix}.$$

Sometimes, we are interested in the matrix

$$\begin{bmatrix} 2 & 8 & 4 \\ 2 & 5 & 1 \\ 4 & 10 & -1 \end{bmatrix},$$

which contains the coefficients of the system, called its *coefficient matrix*.

By contrast, the matrix

$$\begin{bmatrix} 2 & 8 & 4 & 2 \\ 2 & 5 & 1 & 5 \\ 4 & 10 & -1 & 1 \end{bmatrix},$$

which displays all the numerical information contained in the system, is called its *augmented matrix*. For the sake of clarity, we will often indicate the position of the equal signs in the equations by a dotted line:

$$\begin{bmatrix} 2 & 8 & 4 & \vdots & 2 \\ 2 & 5 & 1 & \vdots & 5 \\ 4 & 10 & -1 & \vdots & 1 \end{bmatrix}.$$

Even when you solve a linear system by hand, rather than by computer, it may be more efficient to perform the elimination on the augmented matrix, rather than on the system of equations. The two approaches apply the same concept, but working with the augmented matrix requires less writing, saves time, and is easier to read.

Instead of dividing an *equation* by a scalar,[5] you can divide a *row* by a scalar. Instead of adding a multiple of an equation to another equation, you can add a multiple of a row to another row.

As you perform elimination on the augmented matrix, you should always remember the linear system lurking behind the matrix. To illustrate this method, we perform the elimination both on the augmented matrix and on the linear system it represents:

$$
\begin{bmatrix} 2 & 8 & 4 & \vdots & 2 \\ 2 & 5 & 1 & \vdots & 5 \\ 4 & 10 & -1 & \vdots & 1 \end{bmatrix} \div 2
\qquad
\begin{vmatrix} 2x + & 8y + & 4z = & 2 \\ 2x + & 5y + & z = & 5 \\ 4x + & 10y - & z = & 1 \end{vmatrix} \div 2
$$

$$\downarrow \qquad\qquad\qquad \downarrow$$

$$
\begin{bmatrix} 1 & 4 & 2 & \vdots & 1 \\ 2 & 5 & 1 & \vdots & 5 \\ 4 & 10 & -1 & \vdots & 1 \end{bmatrix} \begin{matrix} \\ -2(\mathrm{I}) \\ -4(\mathrm{I}) \end{matrix}
\qquad
\begin{vmatrix} x + & 4y + & 2z = & 1 \\ 2x + & 5y + & z = & 5 \\ 4x + & 10y - & z = & 1 \end{vmatrix} \begin{matrix} \\ -2(\mathrm{I}) \\ -4(\mathrm{I}) \end{matrix}
$$

$$\downarrow \qquad\qquad\qquad \downarrow$$

$$
\begin{bmatrix} 1 & 4 & 2 & \vdots & 1 \\ 0 & -3 & -3 & \vdots & 3 \\ 0 & -6 & -9 & \vdots & -3 \end{bmatrix} \begin{matrix} \\ \div(-3) \\ \, \end{matrix}
\qquad
\begin{vmatrix} x + & 4y + & 2z = & 1 \\ & -3y - & 3z = & 3 \\ & -6y - & 9z = & -3 \end{vmatrix} \begin{matrix} \\ \div(-3) \\ \, \end{matrix}
$$

$$\downarrow \qquad\qquad\qquad \downarrow$$

$$
\begin{bmatrix} 1 & 4 & 2 & \vdots & 1 \\ 0 & 1 & 1 & \vdots & -1 \\ 0 & -6 & -9 & \vdots & -3 \end{bmatrix} \begin{matrix} -4(\mathrm{II}) \\ \\ +6(\mathrm{II}) \end{matrix}
\qquad
\begin{vmatrix} x + & 4y + & 2z = & 1 \\ & y + & z = & -1 \\ & -6y - & 9z = & -3 \end{vmatrix} \begin{matrix} -4(\mathrm{II}) \\ \\ +6(\mathrm{II}) \end{matrix}
$$

$$\downarrow \qquad\qquad\qquad \downarrow$$

$$
\begin{bmatrix} 1 & 0 & -2 & \vdots & 5 \\ 0 & 1 & 1 & \vdots & -1 \\ 0 & 0 & -3 & \vdots & -9 \end{bmatrix} \begin{matrix} \\ \\ \div(-3) \end{matrix}
\qquad
\begin{vmatrix} x & & - & 2z = & 5 \\ & y + & & z = & -1 \\ & & & -3z = & -9 \end{vmatrix} \begin{matrix} \\ \\ \div(-3) \end{matrix}
$$

$$\downarrow \qquad\qquad\qquad \downarrow$$

$$
\begin{bmatrix} 1 & 0 & -2 & \vdots & 5 \\ 0 & 1 & 1 & \vdots & -1 \\ 0 & 0 & 1 & \vdots & 3 \end{bmatrix} \begin{matrix} +2(\mathrm{III}) \\ -(\mathrm{III}) \\ \, \end{matrix}
\qquad
\begin{vmatrix} x & & - & 2z = & 5 \\ & y + & & z = & -1 \\ & & & z = & 3 \end{vmatrix} \begin{matrix} +2(\mathrm{III}) \\ -(\mathrm{III}) \\ \, \end{matrix}
$$

$$\downarrow \qquad\qquad\qquad \downarrow$$

$$
\begin{bmatrix} 1 & 0 & 0 & \vdots & 11 \\ 0 & 1 & 0 & \vdots & -4 \\ 0 & 0 & 1 & \vdots & 3 \end{bmatrix}
\qquad
\begin{vmatrix} x & & & = & 11 \\ & y & & = & -4 \\ & & z & = & 3 \end{vmatrix}
$$

The solution is often represented as a vector:

$$
\begin{bmatrix} x \\ y \\ z \end{bmatrix} = \begin{bmatrix} 11 \\ -4 \\ 3 \end{bmatrix}.
$$

In this example, the process of elimination works very smoothly. We can eliminate all entries off the diagonal and can make each coefficient on the diagonal equal to 1. The process of elimination works well unless we encounter a zero along the

---

[5] In vector and matrix algebra, the term "scalar" is synonymous with (real) number.

diagonal. These zeros represent missing terms in some equations. The following example illustrates how to solve such a system:

$$\begin{vmatrix} & & x_3 - & x_4 - & x_5 = 4 \\ 2x_1 + 4x_2 + 2x_3 + 4x_4 + 2x_5 = 4 \\ 2x_1 + 4x_2 + 3x_3 + 3x_4 + 3x_5 = 4 \\ 3x_1 + 6x_2 + 6x_3 + 3x_4 + 6x_5 = 6 \end{vmatrix}.$$

The augmented matrix of this system is

$$M = \begin{bmatrix} 0 & 0 & 1 & -1 & -1 & \vdots & 4 \\ 2 & 4 & 2 & 4 & 2 & \vdots & 4 \\ 2 & 4 & 3 & 3 & 3 & \vdots & 4 \\ 3 & 6 & 6 & 3 & 6 & \vdots & 6 \end{bmatrix}.$$

As in the previous examples, we are trying to bring the matrix into diagonal form. To keep track of our work, we will place a cursor in the matrix, as you would on a computer screen. Initially, the cursor is placed at the top position of the first nonzero column of the matrix:

$$\begin{bmatrix} \nearrow 0 & 0 & 1 & -1 & -1 & \vdots & 4 \\ 2 & 4 & 2 & 4 & 2 & \vdots & 4 \\ 2 & 4 & 3 & 3 & 3 & \vdots & 4 \\ 3 & 6 & 6 & 3 & 6 & \vdots & 6 \end{bmatrix}.$$

Our first goal is to make the cursor entry equal to 1. We can accomplish this in two steps as follows:

Step 1 If the cursor entry is 0, swap the cursor row with some row below to make the cursor entry nonzero.[6]

In Step 1, we are merely writing down the equations in a different order. This will certainly not affect the solutions of the system:

$$\begin{bmatrix} \nearrow 0 & 0 & 1 & -1 & -1 & \vdots & 4 \\ 2 & 4 & 2 & 4 & 2 & \vdots & 4 \\ 2 & 4 & 3 & 3 & 3 & \vdots & 4 \\ 3 & 6 & 6 & 3 & 6 & \vdots & 6 \end{bmatrix} \qquad \begin{vmatrix} & & x_3 - & x_4 - & x_5 = 4 \\ 2x_1 + 4x_2 + 2x_3 + 4x_4 + 2x_5 = 4 \\ 2x_1 + 4x_2 + 3x_3 + 3x_4 + 3x_5 = 4 \\ 3x_1 + 6x_2 + 6x_3 + 3x_4 + 6x_5 = 6 \end{vmatrix}$$

$$\downarrow \qquad\qquad\qquad\qquad \downarrow$$

$$\begin{bmatrix} \nearrow 2 & 4 & 2 & 4 & 2 & \vdots & 4 \\ 0 & 0 & 1 & -1 & -1 & \vdots & 4 \\ 2 & 4 & 3 & 3 & 3 & \vdots & 4 \\ 3 & 6 & 6 & 3 & 6 & \vdots & 6 \end{bmatrix} \qquad \begin{vmatrix} 2x_1 + 4x_2 + 2x_3 + 4x_4 + 2x_5 = 4 \\ & & x_3 - & x_4 - & x_5 = 4 \\ 2x_1 + 4x_2 + 3x_3 + 3x_4 + 3x_5 = 4 \\ 3x_1 + 6x_2 + 6x_3 + 3x_4 + 6x_5 = 6 \end{vmatrix}$$

Now we can proceed as in the previous examples.

Step 2 Divide the cursor row by the cursor entry to make the cursor entry equal to 1.

---

[6]To make the process unambiguous, swap the cursor row with the *first* row below that has a nonzero entry in the cursor column.

Step 2 does not change the solutions of the system, because the equation corresponding to the cursor row has the same solutions before and after the operation:

$$\begin{bmatrix} \nearrow 2 & 4 & 2 & 4 & 2 & \vdots & 4 \\ 0 & 0 & 1 & -1 & -1 & \vdots & 4 \\ 2 & 4 & 3 & 3 & 3 & \vdots & 4 \\ 3 & 6 & 6 & 3 & 6 & \vdots & 6 \end{bmatrix} \div 2 \qquad \left| \begin{array}{l} 2x_1 + 4x_2 + 2x_3 + 4x_4 + 2x_5 = 4 \\ \qquad\qquad\quad x_3 - x_4 - x_5 = 4 \\ 2x_1 + 4x_2 + 3x_3 + 3x_4 + 3x_5 = 4 \\ 3x_1 + 6x_2 + 6x_3 + 3x_4 + 6x_5 = 6 \end{array} \right| \div 2$$

$$\downarrow \qquad\qquad\qquad\qquad\qquad \downarrow$$

$$\begin{bmatrix} \nearrow 1 & 2 & 1 & 2 & 1 & \vdots & 2 \\ 0 & 0 & 1 & -1 & -1 & \vdots & 4 \\ 2 & 4 & 3 & 3 & 3 & \vdots & 4 \\ 3 & 6 & 6 & 3 & 6 & \vdots & 6 \end{bmatrix} \qquad \left| \begin{array}{l} x_1 + 2x_2 + x_3 + 2x_4 + x_5 = 2 \\ \qquad\qquad\quad x_3 - x_4 - x_5 = 4 \\ 2x_1 + 4x_2 + 3x_3 + 3x_4 + 3x_5 = 4 \\ 3x_1 + 6x_2 + 6x_3 + 3x_4 + 6x_5 = 6 \end{array} \right|$$

> **Step 3** Eliminate all other entries in the cursor column by subtracting suitable multiples of the cursor row from the other rows.[7]

$$\begin{bmatrix} \nearrow 1 & 2 & 1 & 2 & 1 & \vdots & 2 \\ 0 & 0 & 1 & -1 & -1 & \vdots & 4 \\ 2 & 4 & 3 & 3 & 3 & \vdots & 4 \\ 3 & 6 & 6 & 3 & 6 & \vdots & 6 \end{bmatrix} \begin{array}{l} \\ \\ -2(\mathrm{I}) \\ -3(\mathrm{I}) \end{array} \quad \left| \begin{array}{l} x_1 + 2x_2 + x_3 + 2x_4 + x_5 = 2 \\ \qquad\qquad\quad x_3 - x_4 - x_5 = 4 \\ 2x_1 + 4x_2 + 3x_3 + 3x_4 + 3x_5 = 4 \\ 3x_1 + 6x_2 + 6x_3 + 3x_4 + 6x_5 = 6 \end{array} \right| \begin{array}{l} \\ \\ -2(\mathrm{I}) \\ -3(\mathrm{I}) \end{array}$$

$$\downarrow \qquad\qquad\qquad\qquad\qquad \downarrow$$

$$\begin{bmatrix} \nearrow 1 & 2 & 1 & 2 & 1 & \vdots & 2 \\ 0 & 0 & 1 & -1 & -1 & \vdots & 4 \\ 0 & 0 & 1 & -1 & 1 & \vdots & 0 \\ 0 & 0 & 3 & -3 & 3 & \vdots & 0 \end{bmatrix} \qquad \left| \begin{array}{l} x_1 + 2x_2 + x_3 + 2x_4 + x_5 = 2 \\ \qquad\qquad x_3 - x_4 - x_5 = 4 \\ \qquad\qquad x_3 - x_4 + x_5 = 0 \\ \qquad\quad 3x_3 - 3x_4 + 3x_5 = 0 \end{array} \right|$$

Convince yourself that this operation does not change the solutions of the system. (See Exercise 28.)

Now we have taken care of the first column (the first variable), so we can move the cursor to a new position.

Following the approach taken in Section 1.1, we move the cursor down diagonally (i.e., down one row and over one column):

$$\begin{bmatrix} 1 & 2 & 1 & 2 & 1 & \vdots & 2 \\ 0 & \nearrow 0 & 1 & -1 & -1 & \vdots & 4 \\ 0 & 0 & 1 & -1 & 1 & \vdots & 0 \\ 0 & 0 & 3 & -3 & 3 & \vdots & 0 \end{bmatrix}.$$

For our method to work as before, we need a nonzero cursor entry. Since not only the cursor entry, but also all entries below, are zero, we cannot accomplish this by swapping the cursor row with some row below, as we did in step 1. It would not help us to swap the cursor row with the row above; this would affect the first column of the matrix, which we have already fixed. Thus, we have to give up on the second

---

[7]We may also *add* a multiple of a row, of course. Think of this as subtracting a negative multiple of a row.

column (the second variable); we will move the cursor to the next column:

$$\left[\begin{array}{ccccc:c} 1 & 2 & 1 & 2 & 1 & 2 \\ 0 & 0 & \nearrow 1 & -1 & -1 & 4 \\ 0 & 0 & 1 & -1 & 1 & 0 \\ 0 & 0 & 3 & -3 & 3 & 0 \end{array}\right].$$

Let's summarize.

Step 4 Move the cursor down diagonally (i.e., down one row and over one column). If the new cursor entry and all entries below are zero, move the cursor to the next column (remaining in the same row). Repeat this step if necessary. Then return to step 1.

Here, since the cursor entry is 1, we can proceed directly to step 3 and eliminate all other entries in the cursor column:

$$\left[\begin{array}{ccccc:c} 1 & 2 & 1 & 2 & 1 & 2 \\ 0 & 0 & \nearrow 1 & -1 & -1 & 4 \\ 0 & 0 & 1 & -1 & 1 & 0 \\ 0 & 0 & 3 & -3 & 3 & 0 \end{array}\right] \begin{array}{l} -(\text{II}) \\ \\ -(\text{II}) \\ -3(\text{II}) \end{array}$$

$$\downarrow \text{ Step 3}$$

$$\left[\begin{array}{ccccc:c} 1 & 2 & 0 & 3 & 2 & -2 \\ 0 & 0 & \nearrow 1 & -1 & -1 & 4 \\ 0 & 0 & 0 & 0 & 2 & -4 \\ 0 & 0 & 0 & 0 & 6 & -12 \end{array}\right]$$

$$\downarrow \text{ Step 4}$$

$$\left[\begin{array}{ccccc:c} 1 & 2 & 0 & 3 & 2 & -2 \\ 0 & 0 & 1 & -1 & -1 & 4 \\ 0 & 0 & 0 & 0 & \nearrow 2 & -4 \\ 0 & 0 & 0 & 0 & 6 & -12 \end{array}\right] \begin{array}{l} \\ \\ \div 2 \\ \\ \end{array}$$

$$\downarrow \text{ Step 2}$$

$$\left[\begin{array}{ccccc:c} 1 & 2 & 0 & 3 & 2 & -2 \\ 0 & 0 & 1 & -1 & -1 & 4 \\ 0 & 0 & 0 & 0 & \nearrow 1 & -2 \\ 0 & 0 & 0 & 0 & 6 & -12 \end{array}\right] \begin{array}{l} -2(\text{III}) \\ +(\text{III}) \\ \\ -6(\text{III}) \end{array}$$

$$\downarrow \text{ Step 3}$$

$$E = \left[\begin{array}{ccccc:c} 1 & 2 & 0 & 3 & 0 & 2 \\ 0 & 0 & 1 & -1 & 0 & 2 \\ 0 & 0 & 0 & 0 & \nearrow 1 & -2 \\ 0 & 0 & 0 & 0 & 0 & 0 \end{array}\right]$$

When we try to apply step 4 to this matrix, we run out of columns; the process of row reduction comes to an end. We say that the matrix E is in *reduced row-echelon form*, or *rref* for short. We write

$$E = \text{rref}(M),$$

where M is the augmented matrix of the system.

A matrix is in **reduced row-echelon form** if it satisfies all of the following conditions:

a. If a row has nonzero entries, then the first nonzero entry is 1, called the *leading* 1 in this row.
b. If a column contains a leading 1, then all other entries in that column are zero.
c. If a row contains a leading 1, then each row above contains a leading 1 further to the left.

A matrix E in reduced row-echelon form may contain rows of zeros, as in the example above. By condition c, these rows must appear as the last rows of the matrix.

Convince yourself that the procedure just outlined (repeatedly performing steps 1 through 4) indeed produces a matrix with these three properties.

For emphasis, let's circle the leading 1's in the reduced row-echelon form of the matrix:

$$\left[\begin{array}{ccccc:c} \textcircled{1} & 2 & 0 & 3 & 0 & 2 \\ 0 & 0 & \textcircled{1} & -1 & 0 & 2 \\ 0 & 0 & 0 & 0 & \textcircled{1} & -2 \\ 0 & 0 & 0 & 0 & 0 & 0 \end{array}\right].$$

This matrix represents the following system:

$$
\begin{array}{rcrcrcr}
\textcircled{x_1} + 2x_2 & & & + & 3x_4 & = & 2 \\
& & \textcircled{x_3} & - & x_4 & = & 2 \\
& & & & \textcircled{x_5} & = & -2
\end{array}.
$$

Again, we say that this system is in reduced row-echelon form. The *leading variables* correspond to the leading 1's in the echelon form of the matrix. We also draw the staircase formed by the leading variables. That is where the name echelon form comes from: According to Webster, an echelon is a formation "like a series of steps."

Now we can solve each of the equations above for the leading variable:

$$x_1 = 2 - 2x_2 - 3x_4$$
$$x_3 = 2 + x_4$$
$$x_5 = -2$$

We can freely choose the nonleading variables, $x_2 = s$ and $x_4 = t$, where s and t are arbitrary real numbers. The leading variables are then determined by our choices for s and t; that is, $x_1 = 2 - 2s - 3t$, $x_3 = 2 + t$, and $x_5 = -2$.

This system has infinitely many solutions, namely,

$$x_1 = 2 - 2s - 3t, \quad x_2 = s, \quad x_3 = 2 + t, \quad x_4 = t, \quad x_5 = -2,$$

where s and t are arbitrary real numbers. We can represent the solutions as vectors in $\mathbb{R}^5$:

$$
\begin{bmatrix} x_1 \\ x_2 \\ x_3 \\ x_4 \\ x_5 \end{bmatrix} = \begin{bmatrix} 2 - 2s - 3t \\ s \\ 2 + t \\ t \\ -2 \end{bmatrix} \qquad (s, t \text{ arbitrary}).
$$

We often find it helpful to write this solution as

$$
\begin{bmatrix} x_1 \\ x_2 \\ x_3 \\ x_4 \\ x_5 \end{bmatrix} = \begin{bmatrix} 2 \\ 0 \\ 2 \\ 0 \\ -2 \end{bmatrix} + s \begin{bmatrix} -2 \\ 1 \\ 0 \\ 0 \\ 0 \end{bmatrix} + t \begin{bmatrix} -3 \\ 0 \\ 1 \\ 1 \\ 0 \end{bmatrix}.
$$

For example, if we set $s = t = 0$, we get the particular solution

$$
\begin{bmatrix} x_1 \\ x_2 \\ x_3 \\ x_4 \\ x_5 \end{bmatrix} = \begin{bmatrix} 2 \\ 0 \\ 2 \\ 0 \\ -2 \end{bmatrix}.
$$

Here is a summary of the elimination process just outlined:

Solving systems of linear equations

Write the augmented matrix of the system. Place a cursor in the top entry of the first nonzero column of this matrix.

Step 1   If the cursor entry is zero, swap the cursor row with some row below to make the cursor entry nonzero.

Step 2   Divide the cursor row by the cursor entry.

Step 3   Eliminate all other entries in the cursor column, by subtracting suitable multiples of the cursor row from the other rows.

Step 4   Move the cursor down one row and over one column. If the new cursor entry and all entries below are zero, move the cursor to the next column (remaining in the same row). Repeat the last step if necessary.

Return to step 1.

The process ends when we run out of rows or columns. Then, the matrix is in reduced row-echelon form (rref).

Write down the linear system corresponding to this matrix, and solve each equation in the system for the leading variable. You may choose the nonleading variables freely; the leading variables are then determined by these choices. If the echelon form contains the equation $0 = 1$, then there are no solutions; the system is *inconsistent*.

The operations performed in steps 1, 2, and 3 are called *elementary row operations*: Swap two rows, divide a row by a scalar, or subtract a multiple of a row from another row.

The following is an inconsistent system:

$$\begin{vmatrix} x_1 - 3x_2 & - 5x_4 = -7 \\ 3x_1 - 12x_2 - 2x_3 - 27x_4 = -33 \\ -2x_1 + 10x_2 + 2x_3 + 24x_4 = 29 \\ -x_1 + 6x_2 + x_3 + 14x_4 = 17 \end{vmatrix}.$$

The augmented matrix of the system is

$$\begin{bmatrix} 1 & -3 & 0 & -5 & \vdots & -7 \\ 3 & -12 & -2 & -27 & \vdots & -33 \\ -2 & 10 & 2 & 24 & \vdots & 29 \\ -1 & 6 & 1 & 14 & \vdots & 17 \end{bmatrix}.$$

The reduced row-echelon form for this matrix is

$$\begin{bmatrix} 1 & 0 & 0 & 1 & \vdots & 0 \\ 0 & 1 & 0 & 2 & \vdots & 0 \\ 0 & 0 & 1 & 3 & \vdots & 0 \\ 0 & 0 & 0 & 0 & \vdots & 1 \end{bmatrix}.$$

(We leave it to you to perform the elimination.)

Since the last row of the echelon form represents the equation $0 = 1$, the system is inconsistent.

This method of solving linear systems is sometimes referred to as *Gauss–Jordan elimination*, after the German mathematician Carl Friedrich Gauss (1777–1855; see Figure 1), perhaps the greatest mathematician of modern times, and the German engineer Wilhelm Jordan (1844–1899). Gauss himself called the method *eliminatio vulgaris*. Recall that the Chinese were using this method 2,000 years ago.

Figure 1 Carl Friedrich Gauss appears on an old German 10-mark note. (In fact, this is the *mirror image* of a well-known portrait of Gauss.[8])

---

[8]Reproduced by permission of the German Bundesbank.

How Gauss developed this method is noteworthy. On January 1, 1801, the Sicilian astronomer Giuseppe Piazzi (1746–1826) discovered a planet, which he named "Ceres," in honor of the patron goddess of Sicily. Today, Ceres is called an asteroid or minor planet; it is only about 1,000 km in diameter. The public was very interested in this discovery. At that time, the number of planets in the solar system was still an issue debated by many philosophers and representatives of the Church. Piazzi was able to track the orbit of Ceres for forty days, but then it was lost from view because Piazzi fell ill. Gauss, however, at the age of 24, succeeded in calculating the orbit of Ceres, even though the task seemed hopeless on the basis of a few observations. His computations were so accurate that the German astronomer W. Olbers (1758–1840) located the asteroid on December 31, 1801. In the course of his computations, Gauss had to solve systems of 17 linear equations. In dealing with this problem, Gauss also used the method of least squares, which he had developed around 1794. (See Section 5.4.) Since Gauss at first refused to reveal the methods that led to this amazing accomplishment, some even accused him of sorcery. Gauss later described his methods of orbit computation in his book *Theoria Motus Corporum Coelestium* (1809).

The method of solving a linear system by Gauss–Jordan elimination is called an *algorithm*.[9] An algorithm can be defined as "a finite procedure, written in a fixed symbolic vocabulary, governed by precise instructions, moving in discrete steps, 1, 2, 3, . . . , whose execution requires no insight, cleverness, intuition, intelligence, or perspicuity, and that sooner or later comes to an end" (David Berlinski, *The Advent of the Algorithm: The Idea that Rules the World,* Harcourt Inc., 2000).

Gauss–Jordan elimination is well suited for solving linear systems on a computer, at least in principle. In practice, however, some tricky problems associated with roundoff errors can occur.

Numerical analysis tells us that we can reduce the proliferation of roundoff errors by modifying Gauss–Jordan elimination with partial or complete *pivoting* techniques. Partial pivoting requires us to modify step 1 of the algorithm as follows: Swap the cursor row with a row below to make the cursor entry as large as possible (in absolute value). This swap is performed even if the initial cursor entry is nonzero, as long as there is an entry below with a larger absolute value. In the first example worked in the text, we would start by swapping the first row with the last:

$$\begin{bmatrix} 1 & 2 & 3 & \vdots & 39 \\ 1 & 3 & 2 & \vdots & 34 \\ 3 & 2 & 1 & \vdots & 26 \end{bmatrix} \longrightarrow \begin{bmatrix} 3 & 2 & 1 & \vdots & 26 \\ 1 & 3 & 2 & \vdots & 34 \\ 1 & 2 & 3 & \vdots & 39 \end{bmatrix}.$$

In modifying Gauss–Jordan elimination, an interesting question arises: If we transform a matrix A into a matrix B by a sequence of elementary row operations and if B is in reduced row-echelon form, is it necessarily true that $B = \text{rref}(A)$? Fortunately (and perhaps surprisingly) this is indeed the case.

In this text, we will not utilize this fact, so there is no need to present the somewhat technical proof. If you feel ambitious, try to work out the proof yourself after studying Chapter 3. (See Exercises 3.3.64 through 3.3.67.)

---

[9]The word algorithm is derived from the name of the mathematician al-Khowarizmi, who introduced the term *algebra* into mathematics. (See page 1.)

EXERCISES

GOAL Use Gauss–Jordan elimination to solve linear systems. Do simple problems using paper and pencil, and use technology to solve more complicated problems.

In Exercises 1 through 12, find all solutions of the equations with paper and pencil using Gauss–Jordan elimination. Show all your work. Solve the system in Exercise 8 for the variables $x_1, x_2, x_3, x_4,$ and x_5.

1. $\begin{vmatrix} x + y - 2z = 5 \\ 2x + 3y + 4z = 2 \end{vmatrix}$.

2. $\begin{vmatrix} 3x + 4y - z = 8 \\ 6x + 8y - 2z = 3 \end{vmatrix}$.

3. $x + 2y + 3z = 4$.

4. $\begin{vmatrix} x + y = 1 \\ 2x - y = 5 \\ 3x + 4y = 2 \end{vmatrix}$.

5. $\begin{vmatrix} x_3 + x_4 = 0 \\ x_2 + x_3 \quad\; = 0 \\ x_1 + x_2 \quad\quad\; = 0 \\ x_1 \quad\quad\quad + x_4 = 0 \end{vmatrix}$.

6. $\begin{vmatrix} x_1 - 7x_2 \quad + x_5 = 3 \\ x_3 - 2x_5 = 2 \\ x_4 + x_5 = 1 \end{vmatrix}$.

7. $\begin{vmatrix} x_1 + 2x_2 \quad 2x_4 + 3x_5 = 0 \\ x_3 + 3x_4 + 2x_5 = 0 \\ x_3 + 4x_4 - x_5 = 0 \\ x_5 = 0 \end{vmatrix}$.

8. $\begin{vmatrix} x_2 + 2x_4 + 3x_5 = 0 \\ 4x_4 + 8x_5 = 0 \end{vmatrix}$.

9. $\begin{vmatrix} x_4 + 2x_5 - x_6 = 2 \\ x_1 + 2x_2 \quad\quad + x_5 - x_6 = 0 \\ x_1 + 2x_2 + 2x_3 \quad - x_5 + x_6 = 2 \end{vmatrix}$.

10. $\begin{vmatrix} 4x_1 + 3x_2 + 2x_3 - x_4 = 4 \\ 5x_1 + 4x_2 + 3x_3 - x_4 = 4 \\ -2x_1 - 2x_2 - x_3 + 2x_4 = -3 \\ 11x_1 + 6x_2 + 4x_3 + x_4 = 11 \end{vmatrix}$.

11. $\begin{vmatrix} x_1 + 2x_3 + 4x_4 = -8 \\ x_2 - 3x_3 - x_4 = 6 \\ 3x_1 + 4x_2 - 6x_3 + 8x_4 = 0 \\ - x_2 + 3x_3 + 4x_4 = -12 \end{vmatrix}$.

12. $\begin{vmatrix} 2x_1 - 3x_3 + 7x_5 + 7x_6 = 0 \\ -2x_1 + x_2 + 6x_3 - 6x_5 - 12x_6 = 0 \\ x_2 - 3x_3 + x_5 + 5x_6 = 0 \\ - 2x_2 + x_4 + x_5 + x_6 = 0 \\ 2x_1 + x_2 - 3x_3 + 8x_5 + 7x_6 = 0 \end{vmatrix}$.

Solve the linear systems in Exercises 13 through 17. You may use technology.

13. $\begin{vmatrix} 3x + 11y + 19z = -2 \\ 7x + 23y + 39z = 10 \\ -4x - 3y - 2z = 6 \end{vmatrix}$.

14. $\begin{vmatrix} 3x + 6y + 14z = 22 \\ 7x + 14y + 30z = 46 \\ 4x + 8y + 7z = 6 \end{vmatrix}$.

15. $\begin{vmatrix} 3x + 5y + 3z = 25 \\ 7x + 9y + 19z = 65 \\ -4x + 5y + 11z = 5 \end{vmatrix}$.

16. $\begin{vmatrix} 3x_1 + 6x_2 + 9x_3 + 5x_4 + 25x_5 = 53 \\ 7x_1 + 14x_2 + 21x_3 + 9x_4 + 53x_5 = 105 \\ -4x_1 - 8x_2 - 12x_3 + 5x_4 - 10x_5 = 11 \end{vmatrix}$.

17. $\begin{vmatrix} 2x_1 + 4x_2 + 3x_3 + 5x_4 + 6x_5 = 37 \\ 4x_1 + 8x_2 + 7x_3 + 5x_4 + 2x_5 = 74 \\ -2x_1 - 4x_2 + 3x_3 + 4x_4 - 5x_5 = 20 \\ x_1 + 2x_2 + 2x_3 - x_4 + 2x_5 = 26 \\ 5x_1 - 10x_2 + 4x_3 + 6x_4 + 4x_5 = 24 \end{vmatrix}$.

18. Determine which of the matrices below are in reduced row-echelon form:

a. $\begin{bmatrix} 1 & 2 & 0 & 2 & 0 \\ 0 & 0 & 1 & 3 & 0 \\ 0 & 0 & 1 & 4 & 0 \\ 0 & 0 & 0 & 0 & 1 \end{bmatrix}$
b. $\begin{bmatrix} 0 & 1 & 2 & 0 & 3 \\ 0 & 0 & 0 & 1 & 4 \\ 0 & 0 & 0 & 0 & 0 \end{bmatrix}$

c. $\begin{bmatrix} 1 & 2 & 0 & 3 \\ 0 & 0 & 0 & 0 \\ 0 & 0 & 1 & 2 \end{bmatrix}$
d. $[0 \; 1 \; 2 \; 3 \; 4]$

19. Find all 4×1 matrices in reduced row-echelon form.

20. We say that two $m \times n$ matrices in reduced row-echelon form are of the same type if they contain the same number of leading 1's in the same positions. For example,

$$\begin{bmatrix} ① & 2 & 0 \\ 0 & 0 & ① \end{bmatrix} \text{ and } \begin{bmatrix} ① & 3 & 0 \\ 0 & 0 & ① \end{bmatrix}$$

are of the same type. How many types of 2×2 matrices in reduced row-echelon form are there?

21. How many types of 3×2 matrices in reduced row-echelon form are there? (See Exercise 20.)

22. How many types of 2×3 matrices in reduced row-echelon form are there? (See Exercise 20.)

23. Suppose you apply Gauss–Jordan elimination to a matrix. Explain how you can be sure that the resulting matrix is in reduced row-echelon form.

24. Suppose matrix A is transformed into matrix B by means of an elementary row operation. Is there an elementary row operation that transforms B into A? Explain.

25. Suppose matrix A is transformed into matrix B by a sequence of elementary row operations. Is there a sequence of elementary row operations that transforms B into A? Explain your answer. (See Exercise 24.)

26. Consider an $m \times n$ matrix A. Can you transform $\mathrm{rref}(A)$ into A by a sequence of elementary row operations? (See Exercise 25.)

27. Is there a sequence of elementary row operations that transforms

$$\begin{bmatrix} 1 & 2 & 3 \\ 4 & 5 & 6 \\ 7 & 8 & 9 \end{bmatrix} \quad \text{into} \quad \begin{bmatrix} 1 & 0 & 0 \\ 0 & 1 & 0 \\ 0 & 0 & 0 \end{bmatrix} ?$$

Explain.

28. Suppose you subtract a multiple of an equation in a system from another equation in the system. Explain why the two systems (before and after this operation) have the same solutions.

29. *Balancing a Chemical Reaction.* Consider the chemical reaction

$$a\ NO_2 + b\ H_2O \rightarrow c\ HNO_2 + d\ HNO_3,$$

where a, b, c, and d are unknown positive integers. The reaction must be balanced; that is, the number of atoms of each element must be the same before and after the reaction. For example, because the number of oxygen atoms must remain the same,

$$2a + b = 2c + 3d.$$

While there are many possible choices for a, b, c, and d that balance the reaction, it is customary to use the smallest possible positive integers. Balance this reaction.

30. Find a polynomial of degree 3 (a polynomial of the form $f(t) = a + bt + ct^2 + dt^3$) whose graph goes through the points $(0, 1)$, $(1, 0)$, $(-1, 0)$, and $(2, -15)$. Sketch the graph of this cubic.

31. Find the polynomial of degree 4 whose graph goes through the points $(1, 1)$, $(2, -1)$, $(3, -59)$, $(-1, 5)$, and $(-2, -29)$. Graph this polynomial.

32. *Cubic Splines.* Suppose you are in charge of the design of a roller coaster ride. This simple ride will not make any left or right turns; that is, the track lies in a vertical plane. The accompanying figure shows the ride as viewed from the side. The points (a_i, b_i) are given to

you, and your job is to connect the dots in a reasonably smooth way. Let $a_{i+1} > a_i$.

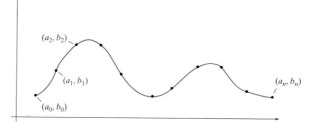

One method often employed in such design problems is the technique of *cubic splines*. We choose $f_i(t)$, a polynomial of degree 3, to define the shape of the ride between (a_{i-1}, b_{i-1}) and (a_i, b_i), for $i = 1, \ldots, n$.

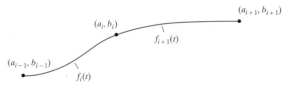

Obviously, it is required that $f_i(a_i) = b_i$ and $f_i(a_{i-1}) = b_{i-1}$, for $i = 1, \ldots, n$. To guarantee a smooth ride at the points (a_i, b_i), we want the first and the second derivatives of f_i and f_{i+1} to agree at these points:

$$f_i'(a_i) = f_{i+1}'(a_i) \quad \text{and}$$
$$f_i''(a_i) = f_{i+1}''(a_i), \quad \text{for } i = 1, \ldots, n-1.$$

Explain the practical significance of these conditions. Explain why, for the convenience of the riders, it is also required that

$$f_1'(a_0) = f_n'(a_n) = 0.$$

Show that satisfying all these conditions amounts to solving a system of linear equations. How many variables are in this system? How many equations? (*Note:* It can be shown that this system has a unique solution.)

33. Find a polynomial $f(t)$ of degree 3 such that $f(1) = 1$, $f(2) = 5$, $f'(1) = 2$, and $f'(2) = 9$, where $f'(t)$ is the derivative of $f(t)$. Graph this polynomial.

34. The *dot product* of two vectors

$$\vec{x} = \begin{bmatrix} x_1 \\ x_2 \\ \vdots \\ x_n \end{bmatrix} \quad \text{and} \quad \vec{y} = \begin{bmatrix} y_1 \\ y_2 \\ \vdots \\ y_n \end{bmatrix}$$

in $\mathbb{R}^n$ is defined by

$$\vec{x} \cdot \vec{y} = x_1 y_1 + x_2 y_2 + \cdots + x_n y_n.$$

Note that the dot product of two vectors is a scalar. We say that the vectors $\vec{x}$ and $\vec{y}$ are *perpendicular* if $\vec{x} \cdot \vec{y} = 0$.

Find all vectors in $\mathbb{R}^3$ perpendicular to

$$\begin{bmatrix} 1 \\ 3 \\ -1 \end{bmatrix}.$$

Draw a sketch.

35. Find all vectors in $\mathbb{R}^4$ that are perpendicular to the three vectors

$$\begin{bmatrix} 1 \\ 1 \\ 1 \\ 1 \end{bmatrix}, \quad \begin{bmatrix} 1 \\ 2 \\ 3 \\ 4 \end{bmatrix}, \quad \begin{bmatrix} 1 \\ 9 \\ 9 \\ 7 \end{bmatrix}.$$

(See Exercise 34.)

36. Find all solutions x_1, x_2, x_3 of the equation

$$\vec{b} = x_1 \vec{v}_1 + x_2 \vec{v}_2 + x_3 \vec{v}_3,$$

where

$$\vec{b} = \begin{bmatrix} -8 \\ -1 \\ 2 \\ 15 \end{bmatrix}, \quad \vec{v}_1 = \begin{bmatrix} 1 \\ 4 \\ 7 \\ 5 \end{bmatrix}, \quad \vec{v}_2 = \begin{bmatrix} 2 \\ 5 \\ 8 \\ 3 \end{bmatrix}, \quad \vec{v}_3 = \begin{bmatrix} 4 \\ 6 \\ 9 \\ 1 \end{bmatrix}.$$

37. For some background on this exercise, see Exercise 1.1.20.

Consider an economy with three industries, I_1, I_2, I_3. What outputs x_1, x_2, x_3 should they produce to satisfy both consumer demand and interindustry demand? The demands put on the three industries are shown in the accompanying figure.

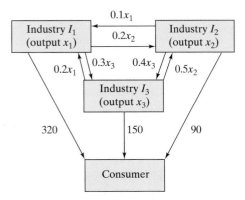

38. If we consider more than three industries in an input–output model, it is cumbersome to represent all the demands in a diagram as in Exercise 37. Suppose we have the industries $I_1, I_2, \ldots, I_n$, with outputs $x_1, x_2, \ldots, x_n$. The *output vector* is

$$\vec{x} = \begin{bmatrix} x_1 \\ x_2 \\ \vdots \\ x_n \end{bmatrix}.$$

The *consumer demand vector* is

$$\vec{b} = \begin{bmatrix} b_1 \\ b_2 \\ \vdots \\ b_n \end{bmatrix},$$

where b_i is the consumer demand on industry I_i. The demand vector for industry I_j is

$$\vec{v}_j = \begin{bmatrix} a_{1j} \\ a_{2j} \\ \vdots \\ a_{nj} \end{bmatrix},$$

where a_{ij} is the demand industry I_j puts on industry I_i, for each \$1 of output industry I_j produces. For example, $a_{32} = 0.5$ means that industry I_2 needs 50¢ worth of products from industry I_3 for each \$1 worth of goods I_2 produces. The coefficient a_{ii} need not be 0: Producing a product may require goods or services from the same industry.

a. Find the four demand vectors for the economy in Exercise 37.

b. What is the meaning in economic terms of $x_j \vec{v}_j$?

c. What is the meaning in economic terms of $x_1 \vec{v}_1 + x_2 \vec{v}_2 + \cdots + x_n \vec{v}_n + \vec{b}$?

d. What is the meaning in economic terms of the equation

$$x_1 \vec{v}_1 + x_2 \vec{v}_2 + \cdots + x_n \vec{v}_n + \vec{b} = \vec{x}\ ?$$

39. Consider the economy of Israel in 1958.[10] The three industries considered here are

$$\begin{array}{ll} I_1 : & \text{agriculture,} \\ I_2 : & \text{manufacturing,} \\ I_3 : & \text{energy.} \end{array}$$

---

[10]W. Leontief: *Input–Output Economics*, Oxford University Press, 1966.

Outputs and demands are measured in millions of Israeli pounds, the currency of Israel at that time. We are told that

$$\vec{b} = \begin{bmatrix} 13.2 \\ 17.6 \\ 1.8 \end{bmatrix}, \quad \vec{v}_1 = \begin{bmatrix} 0.293 \\ 0.014 \\ 0.044 \end{bmatrix},$$

$$\vec{v}_2 = \begin{bmatrix} 0 \\ 0.207 \\ 0.01 \end{bmatrix}, \quad \vec{v}_3 = \begin{bmatrix} 0 \\ 0.017 \\ 0.216 \end{bmatrix}.$$

a. Why do the first components of $\vec{v}_2$ and $\vec{v}_3$ equal 0?

b. Find the outputs x_1, x_2, x_3 required to satisfy demand.

40. Consider some particles in the plane with position vectors $\vec{r}_1, \vec{r}_2, \ldots, \vec{r}_n$ and masses $m_1, m_2, \ldots, m_n$.

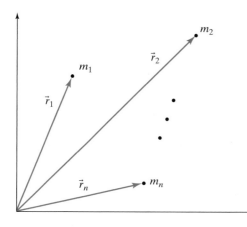

The position vector of the *center of mass* of this system is

$$\vec{r}_{cm} = \frac{1}{M}(m_1\vec{r}_1 + m_2\vec{r}_2 + \cdots + m_n\vec{r}_n),$$

where $M = m_1 + m_2 + \cdots + m_n$.

Consider the triangular plate shown in the accompanying sketch. How must a total mass of 1 kg be distributed among the three vertices of the plate so that the plate can be supported at the point $\begin{bmatrix} 2 \\ 2 \end{bmatrix}$; that is, $\vec{r}_{cm} = \begin{bmatrix} 2 \\ 2 \end{bmatrix}$? Assume that the mass of the plate itself is negligible.

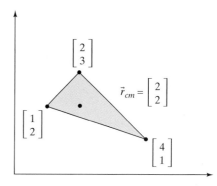

41. The *momentum* $\vec{P}$ of a system of n particles in space with masses $m_1, m_2, \ldots, m_n$ and velocities $\vec{v}_1, \vec{v}_2, \ldots, \vec{v}_n$ is defined as

$$\vec{P} = m_1\vec{v}_1 + m_2\vec{v}_2 + \cdots + m_n\vec{v}_n.$$

Now consider two elementary particles with velocities

$$\vec{v}_1 = \begin{bmatrix} 1 \\ 1 \\ 1 \end{bmatrix} \quad \text{and} \quad \vec{v}_2 = \begin{bmatrix} 4 \\ 7 \\ 10 \end{bmatrix}.$$

The particles collide. After the collision, their respective velocities are observed to be

$$\vec{w}_1 = \begin{bmatrix} 4 \\ 7 \\ 4 \end{bmatrix} \quad \text{and} \quad \vec{w}_2 = \begin{bmatrix} 2 \\ 3 \\ 8 \end{bmatrix}.$$

Assume that the momentum of the system is conserved throughout the collision. What does this experiment tell you about the masses of the two particles? (See the accompanying figure.)

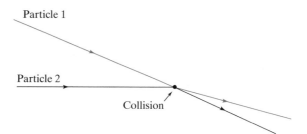

42. The accompanying sketch represents a maze of one-way streets in a city in the United States. The traffic volume through certain blocks during an hour has been measured. Suppose that the vehicles leaving the area during this hour were exactly the same as those entering it.

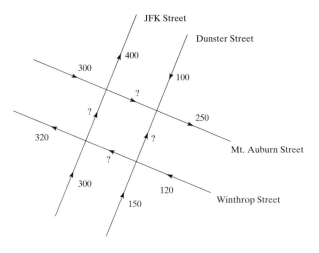

What can you say about the traffic volume at the four locations indicated by a question mark? Can you figure out exactly how much traffic there was on each block? If not, describe one possible scenario. For each of the four locations, find the highest and the lowest possible traffic volume.

43. Let $S(t)$ be the length of the tth day of the year in Bombay, India (measured in hours, from sunrise to sunset). We are given the following values of $S(t)$:

t	$S(t)$
47	11.5
74	12
273	12

For example, $S(47) = 11.5$ means that the time from sunrise to sunset on February 16 is 11 hours and 30 minutes. For locations close to the equator, the function $S(t)$ is well approximated by a trigonometric function of the form

$$S(t) = a + b \cos \left(\frac{2\pi t}{365} \right) + c \sin \left(\frac{2\pi t}{365} \right).$$

(The period is 365 days, or 1 year.) Find this approximation for Bombay, and graph your solution. According to this model, how long is the longest day of the year in Bombay?

1.3 ON THE SOLUTIONS OF LINEAR SYSTEMS

In the last section we discussed how a system of linear equations can be solved by Gauss–Jordan elimination. Now we will investigate what this method tells us about the number of solutions of a linear system. How many solutions can a linear system possibly have? How can we tell whether a system has any solutions at all?

First, we observe that a linear system has no solutions if (and only if) its reduced row-echelon form contains a row of the form

$$\begin{bmatrix} 0 & 0 & 0 & \dots & 0 & \vdots & 1 \end{bmatrix},$$

representing the equation $0 = 1$. In this case we say that the system is *inconsistent*.

If a linear system is *consistent* (i.e., if it does have solutions), how many solutions can it have? The number of solutions depends on whether or not there are nonleading variables. If there is at least one nonleading variable, then there will be infinitely many solutions, since we can assign any value to a nonleading variable. If all variables are leading, on the other hand, there will be only one solution, since we cannot make any choices in assigning values to the variables.

We have shown the following:[11]

---

[11] Starting in this section, we will number the definitions we give and the facts we derive. The nth fact stated in Section $p.q$ is labeled as Fact $p.q.n$.

Fact 1.3.1   **Number of solutions of a linear system**
A linear system has either

- *no solutions* (it is inconsistent),
- *exactly one solution* (if the system is consistent and all variables are leading), or
- *infinitely many solutions* (if the system is consistent and there are nonleading variables).

EXAMPLE 1   The reduced row-echelon forms of the augmented matrices of three systems are given. How many solutions are there in each case?

$$
\text{a. } \left[\begin{array}{ccc:c} 1 & 2 & 0 & 1 \\ 0 & 0 & 1 & 2 \\ 0 & 0 & 0 & 0 \end{array}\right]
\quad
\text{b. } \left[\begin{array}{ccc:c} 1 & 0 & 0 & 1 \\ 0 & 1 & 0 & 2 \\ 0 & 0 & 1 & 3 \end{array}\right]
\quad
\text{c. } \left[\begin{array}{ccc:c} 1 & 2 & 0 & 0 \\ 0 & 0 & 1 & 0 \\ 0 & 0 & 0 & 1 \\ 0 & 0 & 0 & 0 \end{array}\right]
$$

Solution

a. Infinitely many solutions. (The second variable is nonleading.)
b. Exactly one solution. (All variables are leading.)
c. No solutions. (The third row represents the equation $0 = 1$.)   ∎

Example 1 shows that the number of leading 1's in the echelon form tells us about the number of solutions of a linear system. This observation motivates the following definition:

Definition 1.3.2   **Rank**
The *rank* of a matrix A is the number of leading 1's in rref(A).[12]

EXAMPLE 2

$$
\text{rank} \begin{bmatrix} 1 & 2 & 3 \\ 4 & 5 & 6 \\ 7 & 8 & 9 \end{bmatrix} = 2, \text{ since rref} \begin{bmatrix} 1 & 2 & 3 \\ 4 & 5 & 6 \\ 7 & 8 & 9 \end{bmatrix} = \begin{bmatrix} \boxed{1} & 0 & -1 \\ 0 & \boxed{1} & 2 \\ 0 & 0 & 0 \end{bmatrix}.
$$
∎

EXAMPLE 3   Consider a system of m linear equations with n unknowns. Its coefficient matrix A has the size $m \times n$. Show that:

a. rank(A) $\leq m$ and rank(A) $\leq n$.
b. If rank(A) $= m$, then the system is consistent.
c. If rank(A) $= n$, then the system has at most one solution.
d. If rank(A) $< n$, then the system has either infinitely many solutions, or none.

----

[12]This is a preliminary, rather technical definition. In Chapter 3, we will gain a better conceptual understanding of the rank of a matrix.

Solution

a. By the definition of the reduced row-echelon form, there is at most one leading 1 in each row and in each column. If there is a leading 1 in each row, then $\text{rank}(A) = m$; otherwise, $\text{rank}(A) < m$. Likewise for the columns.

b. The equation $\text{rank}(A) = m$ means that there is a leading 1 in each row of the echelon form of A. This implies that the echelon form of the augmented matrix does not contain the row

$$\begin{bmatrix} 0 & 0 & 0 & \ldots & 0 & \vdots & 1 \end{bmatrix}.$$

Therefore, the system must have at least one solution.

c. The equation $\text{rank}(A) = n$ means that there is a leading 1 in each column; that is, all variables are leading. Therefore, either the system is inconsistent or it has a unique solution (by Fact 1.3.1).

d. There are nonleading variables in this case. Therefore, either the system is inconsistent or it has infinitely many solutions (by Fact 1.3.1). ■

EXAMPLE 4   Consider a linear system with fewer equations than unknowns. How many solutions could this system have?

Solution

Suppose there are m equations and n unknowns; we are told that $m < n$. Let A be the coefficient matrix of the system. By Example 3(a),

$$\text{rank}(A) \leq m < n.$$

Thus, the system will have infinitely many solutions or no solutions at all (by Example 3(d)). The key observation is that there are nonleading variables in this case. ■

Fact 1.3.3

A linear system with fewer equations than unknowns has either no solutions or infinitely many solutions.

To illustrate this observation, consider two equations in three variables: Two planes in space either intersect in a line or are parallel (see Figure 1), but they will never intersect at a point! This means that a system of two equations with three unknowns cannot have a unique solution.

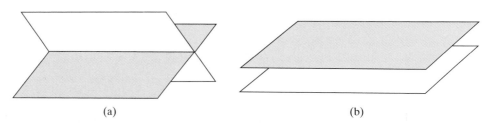

(a)   (b)

Figure 1 (a) Two planes intersect in a line. (b) Two parallel planes.

EXAMPLE 5 Consider a linear system of n equations with n unknowns. When does this system have a unique solution? Use the rank of the coefficient matrix A.

Solution

We observe first that rank$(A) \leq n$, by Example 3(a). If rank$(A) = n$, then the system has a unique solution, by Example 3, parts (b) and (c). However, if rank$(A) < n$, then the system does not have a unique solution, by Example 3(d). ∎

Fact 1.3.4 A linear system of n equations with n unknowns has a unique solution if (and only if) the rank of its coefficient matrix A is n. This means that

$$\text{rref}(A) = \begin{bmatrix} 1 & 0 & 0 & \cdots & 0 \\ 0 & 1 & 0 & \cdots & 0 \\ 0 & 0 & 1 & \cdots & 0 \\ \vdots & \vdots & \vdots & \ddots & \vdots \\ 0 & 0 & 0 & \cdots & 1 \end{bmatrix}.$$

The Vector Form and the Matrix Form of a Linear System

We now introduce some notations that allow us to represent systems of linear equations more succinctly. As a simple example, consider the linear system

$$\begin{vmatrix} 3x + y = 7 \\ x + 2y = 4 \end{vmatrix}.$$

In Section 1.1, we interpreted the solution of this system as the intersection of two lines in the x–y-plane. Here is another interpretation. We can write the system as

$$\begin{bmatrix} 3x + y \\ x + 2y \end{bmatrix} = \begin{bmatrix} 7 \\ 4 \end{bmatrix},$$

or

$$\begin{bmatrix} 3x \\ x \end{bmatrix} + \begin{bmatrix} y \\ 2y \end{bmatrix} = \begin{bmatrix} 7 \\ 4 \end{bmatrix},$$

or

$$x \begin{bmatrix} 3 \\ 1 \end{bmatrix} + y \begin{bmatrix} 1 \\ 2 \end{bmatrix} = \begin{bmatrix} 7 \\ 4 \end{bmatrix}.$$

To solve this system means to write the vector $\begin{bmatrix} 7 \\ 4 \end{bmatrix}$ as the sum of a scalar multiple of $\begin{bmatrix} 3 \\ 1 \end{bmatrix}$ and a scalar multiple of $\begin{bmatrix} 1 \\ 2 \end{bmatrix}$. This problem and its solution can be represented geometrically, as illustrated in Figure 2.

The unique solution of this system is $x = 2$, $y = 1$. The equation

$$x \begin{bmatrix} 3 \\ 1 \end{bmatrix} + y \begin{bmatrix} 1 \\ 2 \end{bmatrix} = \begin{bmatrix} 7 \\ 4 \end{bmatrix}$$

is called the *vector form* of the linear system. Note that the vectors $\begin{bmatrix} 3 \\ 1 \end{bmatrix}$ and $\begin{bmatrix} 1 \\ 2 \end{bmatrix}$ are the columns of its coefficient matrix.

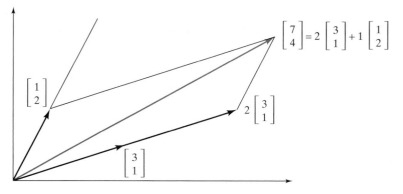

Figure 2

Now consider the general linear system

$$\begin{vmatrix} a_{11}x_1 + a_{12}x_2 + \cdots + a_{1n}x_n = b_1 \\ a_{21}x_1 + a_{22}x_2 + \cdots + a_{2n}x_n = b_2 \\ \vdots \qquad \vdots \qquad \qquad \vdots \qquad \vdots \\ a_{m1}x_1 + a_{m2}x_2 + \cdots + a_{mn}x_n = b_m \end{vmatrix}.$$

We can write

$$\begin{bmatrix} a_{11}x_1 + a_{12}x_2 + \cdots + a_{1n}x_n \\ a_{21}x_1 + a_{22}x_2 + \cdots + a_{2n}x_n \\ \vdots \qquad \vdots \qquad \qquad \vdots \\ a_{m1}x_1 + a_{m2}x_2 + \cdots + a_{mn}x_n \end{bmatrix} = \begin{bmatrix} b_1 \\ b_2 \\ \vdots \\ b_m \end{bmatrix},$$

or

$$x_1 \overset{\vec{v}_1}{\underset{\downarrow}{\begin{bmatrix} a_{11} \\ a_{21} \\ \vdots \\ a_{m1} \end{bmatrix}}} + x_2 \overset{\vec{v}_2}{\underset{\downarrow}{\begin{bmatrix} a_{12} \\ a_{22} \\ \vdots \\ a_{m2} \end{bmatrix}}} + \cdots + x_n \overset{\vec{v}_n}{\underset{\downarrow}{\begin{bmatrix} a_{1n} \\ a_{2n} \\ \vdots \\ a_{mn} \end{bmatrix}}} = \overset{\vec{b}}{\underset{\downarrow}{\begin{bmatrix} b_1 \\ b_2 \\ \vdots \\ b_m \end{bmatrix}}},$$

or, more succinctly,

$$x_1 \vec{v}_1 + x_2 \vec{v}_2 + \cdots + x_n \vec{v}_n = \vec{b}.$$

This is the *vector form* of the linear system. In this context, the following definition is useful:

Definition 1.3.5

Linear combinations

A vector $\vec{b}$ in $\mathbb{R}^m$ is called a *linear combination* of the vectors $\vec{v}_1, \vec{v}_2, \ldots, \vec{v}_n$ in $\mathbb{R}^m$ if there are scalars $x_1, x_2, \ldots, x_n$ such that

$$\vec{b} = x_1 \vec{v}_1 + x_2 \vec{v}_2 + \cdots + x_n \vec{v}_n.$$

Solving a linear system with augmented matrix $[A \,\vdots\, \vec{b}]$ amounts to writing the vector $\vec{b}$ as a linear combination of the column vectors of A.

We now give another definition that permits an even more compact representation of a linear system. We *define* the linear combination

$$x_1 \vec{v}_1 + x_2 \vec{v}_2 + \cdots + x_n \vec{v}_n$$

to be the *product* of the $m \times n$ matrix

$$A = \begin{bmatrix} | & | & & | \\ \vec{v}_1 & \vec{v}_2 & \cdots & \vec{v}_n \\ | & | & & | \end{bmatrix}$$

and the vector

$$\vec{x} = \begin{bmatrix} x_1 \\ x_2 \\ \vdots \\ x_n \end{bmatrix}.$$

Definition 1.3.6

The product $A\vec{x}$

If the column vectors of an $m \times n$ matrix A are $\vec{v}_1, \vec{v}_2, \ldots, \vec{v}_n$, and $\vec{x}$ is a vector in $\mathbb{R}^n$, then the *product* $A\vec{x}$ is defined as

$$A\vec{x} = \begin{bmatrix} | & | & & | \\ \vec{v}_1 & \vec{v}_2 & \cdots & \vec{v}_n \\ | & | & & | \end{bmatrix} \begin{bmatrix} x_1 \\ x_2 \\ \vdots \\ x_n \end{bmatrix} = x_1 \vec{v}_1 + x_2 \vec{v}_2 + \cdots + x_n \vec{v}_n.$$

In words, $A\vec{x}$ is the linear combination of the columns of A with the components of $\vec{x}$ as coefficients.

Note that the product $A\vec{x}$ is defined only if the number of *columns* of A matches the number of *components* of $\vec{x}$. Also, note that the product $A\vec{x}$ is a vector in $\mathbb{R}^m$. This definition allows us to represent the linear system with vector form $x_1 \vec{v}_1 + \cdots + x_n \vec{v}_n = \vec{b}$ as

$$A\vec{x} = \vec{b},$$

the *matrix form* of the linear system.

Here are some examples of matrix products:

EXAMPLE 6

$$\begin{bmatrix} 1 & 0 & -1 \\ 1 & 2 & 3 \end{bmatrix} \begin{bmatrix} 3 \\ 1 \\ 2 \end{bmatrix} = 3 \begin{bmatrix} 1 \\ 1 \end{bmatrix} + 1 \begin{bmatrix} 0 \\ 2 \end{bmatrix} + 2 \begin{bmatrix} -1 \\ 3 \end{bmatrix} = \begin{bmatrix} 1 \\ 11 \end{bmatrix}.$$ ∎

EXAMPLE 7 The product

$$\begin{bmatrix} 1 & 0 & -1 \\ 1 & 2 & 3 \end{bmatrix} \begin{bmatrix} 3 \\ 1 \end{bmatrix}$$

is undefined, because the number of columns of the matrix does not match the number of components of the vector. ∎

EXAMPLE 8   If

$$D = \begin{bmatrix} 1 & 0 & 0 & 0 \\ 0 & 1 & 0 & 0 \\ 0 & 0 & 1 & 0 \\ 0 & 0 & 0 & 1 \end{bmatrix}$$

and $\vec{x}$ is any vector in $\mathbb{R}^4$, find $D\vec{x}$.

Solution

$$D\vec{x} = \begin{bmatrix} 1 & 0 & 0 & 0 \\ 0 & 1 & 0 & 0 \\ 0 & 0 & 1 & 0 \\ 0 & 0 & 0 & 1 \end{bmatrix}\begin{bmatrix} x_1 \\ x_2 \\ x_3 \\ x_4 \end{bmatrix} = x_1\begin{bmatrix} 1 \\ 0 \\ 0 \\ 0 \end{bmatrix} + x_2\begin{bmatrix} 0 \\ 1 \\ 0 \\ 0 \end{bmatrix} + x_3\begin{bmatrix} 0 \\ 0 \\ 1 \\ 0 \end{bmatrix} + x_4\begin{bmatrix} 0 \\ 0 \\ 0 \\ 1 \end{bmatrix} = \begin{bmatrix} x_1 \\ x_2 \\ x_3 \\ x_4 \end{bmatrix}.$$

Thus, $D\vec{x} = \vec{x}$, for all $\vec{x}$ in $\mathbb{R}^4$.   ∎

EXAMPLE 9   Represent the system

$$\begin{vmatrix} 2x_1 - 3x_2 + 5x_3 = 7 \\ 9x_1 + 4x_2 - 6x_3 = 8 \end{vmatrix}$$

in matrix form $A\vec{x} = \vec{b}$.

Solution

Begin by writing the system in vector form:

$$\begin{bmatrix} 2x_1 - 3x_2 + 5x_3 \\ 9x_1 + 4x_2 - 6x_3 \end{bmatrix} = \begin{bmatrix} 7 \\ 8 \end{bmatrix}$$

$$x_1\begin{bmatrix} 2 \\ 9 \end{bmatrix} + x_2\begin{bmatrix} -3 \\ 4 \end{bmatrix} + x_3\begin{bmatrix} 5 \\ -6 \end{bmatrix} = \begin{bmatrix} 7 \\ 8 \end{bmatrix}$$

$$\begin{bmatrix} 2 & -3 & 5 \\ 9 & 4 & -6 \end{bmatrix}\begin{bmatrix} x_1 \\ x_2 \\ x_3 \end{bmatrix} = \begin{bmatrix} 7 \\ 8 \end{bmatrix}$$

$$A\vec{x} = \vec{b},$$

where $A = \begin{bmatrix} 2 & -3 & 5 \\ 9 & 4 & -6 \end{bmatrix}$ and $\vec{b} = \begin{bmatrix} 7 \\ 8 \end{bmatrix}$.   ∎

Here are two important algebraic rules concerning the product $A\vec{x}$:

Fact 1.3.7   For an $m \times n$ matrix A, two vectors $\vec{x}$ and $\vec{y}$ in $\mathbb{R}^n$, and a scalar k,

a. $A(\vec{x} + \vec{y}) = A\vec{x} + A\vec{y}$,

b. $A(k\vec{x}) = k(A\vec{x})$.

We will prove the first equation and leave the verification of the second as Exercise 45:

$$A(\vec{x} + \vec{y}) = \begin{bmatrix} | & | & & | \\ \vec{v}_1 & \vec{v}_2 & \cdots & \vec{v}_n \\ | & | & & | \end{bmatrix} \begin{bmatrix} x_1 + y_1 \\ x_2 + y_2 \\ \vdots \\ x_n + y_n \end{bmatrix}$$

$$= (x_1 + y_1)\vec{v}_1 + (x_2 + y_2)\vec{v}_2 + \cdots + (x_n + y_n)\vec{v}_n$$

$$= x_1\vec{v}_1 + x_2\vec{v}_2 + \cdots + x_n\vec{v}_n + y_1\vec{v}_1 + y_2\vec{v}_2 + \cdots + y_n\vec{v}_n$$

$$= A\vec{x} + A\vec{y}. \qquad \blacktriangle^{13}$$

In Definition 1.3.6, we express the product $A\vec{x}$ in terms of the columns of the matrix A. This product can also be expressed in terms of the rows of A. This characterization involves the *dot product* of two vectors; you may wish to review this concept in the appendix. (See Definition A4.)

Consider the product of an $m \times n$ matrix A with a vector $\vec{x}$ in $\mathbb{R}^n$:

$$A\vec{x} = \begin{bmatrix} a_{11} & a_{12} & \cdots & a_{1n} \\ a_{21} & a_{22} & \cdots & a_{2n} \\ \vdots & \vdots & & \vdots \\ a_{m1} & a_{m2} & \cdots & a_{mn} \end{bmatrix} \begin{bmatrix} x_1 \\ x_2 \\ \vdots \\ x_n \end{bmatrix}$$

$$= x_1 \begin{bmatrix} a_{11} \\ a_{21} \\ \vdots \\ a_{m1} \end{bmatrix} + x_2 \begin{bmatrix} a_{12} \\ a_{22} \\ \vdots \\ a_{m2} \end{bmatrix} + \cdots + x_n \begin{bmatrix} a_{1n} \\ a_{2n} \\ \vdots \\ a_{mn} \end{bmatrix}$$

$$= \begin{bmatrix} a_{11}x_1 + a_{12}x_2 + \cdots + a_{1n}x_n \\ a_{21}x_1 + a_{22}x_2 + \cdots + a_{2n}x_n \\ \vdots & \vdots & \vdots \\ a_{m1}x_1 + a_{m2}x_2 + \cdots + a_{mn}x_n \end{bmatrix}.$$

Note that the first component of $A\vec{x}$ is the dot product of the first row of A with $\vec{x}$, etc.; the ith component of $A\vec{x}$ is the dot product of the ith row of A with $\vec{x}$. Thus, we have the following characterization:

Fact 1.3.8 If $\vec{x}$ is a vector in $\mathbb{R}^n$ and A is an $m \times n$ matrix with *row* vectors $\vec{w}_1, \ldots, \vec{w}_m$, then

$$A\vec{x} = \begin{bmatrix} — & \vec{w}_1 & — \\ — & \vec{w}_2 & — \\ & \vdots & \\ — & \vec{w}_m & — \end{bmatrix} \vec{x} = \begin{bmatrix} \vec{w}_1 \cdot \vec{x} \\ \vec{w}_2 \cdot \vec{x} \\ \vdots \\ \vec{w}_m \cdot \vec{x} \end{bmatrix}.$$

(That is, the ith component of $A\vec{x}$ is the dot product of $\vec{w}_i$ and $\vec{x}$.)

---

[13]The symbol $\blacktriangle$ marks the end of a proof.

EXAMPLE 10

$$\begin{bmatrix} 1 & 0 & -1 \\ 1 & 2 & 3 \end{bmatrix} \begin{bmatrix} 3 \\ 1 \\ 2 \end{bmatrix} = \begin{bmatrix} 1 \cdot 3 & + & 0 \cdot 1 & + & (-1) \cdot 2 \\ 1 \cdot 3 & + & 2 \cdot 1 & + & 3 \cdot 2 \end{bmatrix} = \begin{bmatrix} 1 \\ 11 \end{bmatrix}.$$

(Compare this with Example 6.) ∎

Let us define two other operations involving matrices:

Definition 1.3.9

Sums of matrices

The sum of two matrices of the same size is defined entry by entry:

$$\begin{bmatrix} a_{11} & \cdots & a_{1n} \\ \vdots & & \vdots \\ a_{m1} & \cdots & a_{mn} \end{bmatrix} + \begin{bmatrix} b_{11} & \cdots & b_{1n} \\ \vdots & & \vdots \\ b_{m1} & \cdots & b_{mn} \end{bmatrix} = \begin{bmatrix} a_{11} + b_{11} & \cdots & a_{1n} + b_{1n} \\ \vdots & & \vdots \\ a_{m1} + b_{m1} & \cdots & a_{mn} + b_{mn} \end{bmatrix}.$$

Scalar multiples of matrices

The product of a scalar k with an $m \times n$ matrix is defined entry by entry:

$$k \begin{bmatrix} a_{11} & \cdots & a_{1n} \\ \vdots & & \vdots \\ a_{m1} & \cdots & a_{mn} \end{bmatrix} = \begin{bmatrix} ka_{11} & \cdots & ka_{1n} \\ \vdots & & \vdots \\ ka_{m1} & \cdots & ka_{mn} \end{bmatrix}.$$

Note that these definitions generalize the corresponding operations for vectors. (See the Appendix, Definition A1.)

EXAMPLE 11

$$\begin{bmatrix} 1 & 2 & 3 \\ 4 & 5 & 6 \end{bmatrix} + \begin{bmatrix} 7 & 3 & 1 \\ 5 & 3 & -1 \end{bmatrix} = \begin{bmatrix} 8 & 5 & 4 \\ 9 & 8 & 5 \end{bmatrix}.$$
∎

EXAMPLE 12

$$3 \begin{bmatrix} 2 & 1 \\ -1 & 3 \end{bmatrix} = \begin{bmatrix} 6 & 3 \\ -3 & 9 \end{bmatrix}.$$
∎

We will conclude this section with an example that applies many of the concepts we have introduced.

EXAMPLE 13 Consider an $m \times n$ matrix A with $\operatorname{rank}(A) < m$. Show that there is a vector $\vec{b}$ in $\mathbb{R}^m$ such that the system $A\vec{x} = \vec{b}$ is inconsistent.

Solution

Consider the reduced row-echelon form $E = \operatorname{rref}(A)$. Since the rank of the matrix A is less than the number of its rows, the last row of E does not contain a leading 1 and is therefore zero:

$$E = \operatorname{rref}(A) = \begin{bmatrix} & & & & \\ & & & & \\ & & & & \\ 0\,0\,0 & \cdots & 0\,0 & \end{bmatrix}.$$

We can find a vector $\vec{c}$ in $\mathbb{R}^m$ such that the system $E\vec{x} = \vec{c}$ is inconsistent: Any vector whose last component is nonzero will do. Using this vector $\vec{c}$, how can we find a vector $\vec{b}$ in $\mathbb{R}^m$ such that the system $A\vec{x} = \vec{b}$ is inconsistent?

The key idea is to work backward through Gauss–Jordan elimination. We know that there is a sequence of elementary row operations that transforms A into E. By inverting each of these operations and performing them in reverse order, we can find a sequence of elementary row operations that transforms E into A. (For example, instead of dividing a row by k, we multiply the same row by k.) If we apply the same row operations to the augmented matrix $\begin{bmatrix} E & \vdots & \vec{c} \end{bmatrix}$, we end up with a matrix $\begin{bmatrix} A & \vdots & \vec{b} \end{bmatrix}$ such that the system $A\vec{x} = \vec{b}$ is inconsistent, as desired. (Recall that the system $E\vec{x} = \vec{c}$ is inconsistent, and elementary row operations do not change the set of solutions.) For example, let

$$A = \begin{bmatrix} 0 & 1 & 2 \\ 0 & 2 & 4 \\ 0 & 3 & 6 \\ 1 & 4 & 8 \end{bmatrix}, \quad \text{with} \quad E = \begin{bmatrix} 1 & 0 & 0 \\ 0 & 1 & 2 \\ 0 & 0 & 0 \\ 0 & 0 & 0 \end{bmatrix} \quad \text{and} \quad \vec{c} = \begin{bmatrix} 1 \\ 1 \\ 1 \\ 1 \end{bmatrix}.$$

$$A = \begin{bmatrix} 0 & 1 & 2 \\ 0 & 2 & 4 \\ 0 & 3 & 6 \\ 1 & 4 & 8 \end{bmatrix} \qquad \begin{bmatrix} A & \vdots & \vec{b} \end{bmatrix} = \begin{bmatrix} 0 & 1 & 2 & \vdots & 2 \\ 0 & 2 & 4 & \vdots & 2 \\ 0 & 3 & 6 & \vdots & 4 \\ 1 & 4 & 8 & \vdots & 5 \end{bmatrix}$$

$$\begin{bmatrix} 1 & 4 & 8 \\ 0 & 2 & 4 \\ 0 & 3 & 6 \\ 0 & 1 & 2 \end{bmatrix} \div(2) \qquad \begin{bmatrix} 1 & 4 & 8 & \vdots & 5 \\ 0 & 2 & 4 & \vdots & 2 \\ 0 & 3 & 6 & \vdots & 4 \\ 0 & 1 & 2 & \vdots & 2 \end{bmatrix}$$

$$\begin{bmatrix} 1 & 4 & 8 \\ 0 & 1 & 2 \\ 0 & 3 & 6 \\ 0 & 1 & 2 \end{bmatrix} \begin{matrix} -4(II) \\ \\ -3(II) \\ -(II) \end{matrix} \qquad \begin{bmatrix} 1 & 4 & 8 & \vdots & 5 \\ 0 & 1 & 2 & \vdots & 1 \\ 0 & 3 & 6 & \vdots & 4 \\ 0 & 1 & 2 & \vdots & 2 \end{bmatrix} \cdot 2$$

$$E = \begin{bmatrix} 1 & 0 & 0 \\ 0 & 1 & 2 \\ 0 & 0 & 0 \\ 0 & 0 & 0 \end{bmatrix} \qquad \begin{bmatrix} E & \vdots & \vec{c} \end{bmatrix} = \begin{bmatrix} 1 & 0 & 0 & \vdots & 1 \\ 0 & 1 & 2 & \vdots & 1 \\ 0 & 0 & 0 & \vdots & 1 \\ 0 & 0 & 0 & \vdots & 1 \end{bmatrix} \begin{matrix} +4(II) \\ \\ +3(II) \\ +(II) \end{matrix}$$

The reasoning above shows that the system

$$A\vec{x} = \vec{b}$$

is inconsistent.   ∎

EXERCISES

GOALS Use the reduced row-echelon form of the augmented matrix to find the number of solutions of a linear system. Apply the definition of the rank of a matrix. Compute the product $A\vec{x}$ in terms of the columns or the rows of A. Represent a linear system in vector or in matrix form.

1. The reduced row-echelon forms of the augmented matrices of three systems are given next. How many solutions does each system have?

a. $\begin{bmatrix} 1 & 0 & 2 & \vdots & 0 \\ 0 & 1 & 3 & \vdots & 0 \\ 0 & 0 & 0 & \vdots & 1 \end{bmatrix}$.   b. $\begin{bmatrix} 1 & 0 & \vdots & 5 \\ 0 & 1 & \vdots & 6 \end{bmatrix}$.

c. $\begin{bmatrix} 0 & 1 & 0 & \vdots & 2 \\ 0 & 0 & 1 & \vdots & 3 \end{bmatrix}$.

Find the rank of the matrices in Exercises 2 through 4.

2. $\begin{bmatrix} 1 & 2 & 3 \\ 0 & 1 & 2 \\ 0 & 0 & 1 \end{bmatrix}$.   **3.** $\begin{bmatrix} 1 & 1 & 1 \\ 1 & 1 & 1 \\ 1 & 1 & 1 \end{bmatrix}$.   **4.** $\begin{bmatrix} 1 & 4 & 7 \\ 2 & 5 & 8 \\ 3 & 6 & 9 \end{bmatrix}$.

5. a. Write the system

$$\begin{vmatrix} x + 2y = \ 7 \\ 3x + \ y = 11 \end{vmatrix}$$

in vector form.

b. Use your answer in part (a) to represent the system geometrically. Solve the system and represent the solution geometrically.

6. Consider the vectors $\vec{v}_1$, $\vec{v}_2$, $\vec{v}_3$ in $\mathbb{R}^2$ (sketched in the accompanying figure). Vectors $\vec{v}_1$ and $\vec{v}_2$ are parallel. How many solutions does the system

$$x\vec{v}_1 + y\vec{v}_2 = \vec{v}_3$$

have? Argue geometrically.

7. Consider the vectors $\vec{v}_1$, $\vec{v}_2$, $\vec{v}_3$ in $\mathbb{R}^2$ shown in the accompanying sketch. How many solutions does the system

$$x\vec{v}_1 + y\vec{v}_2 = \vec{v}_3$$

have? Argue geometrically.

8. Consider the vectors $\vec{v}_1$, $\vec{v}_2$, $\vec{v}_3$, $\vec{v}_4$ in $\mathbb{R}^2$ shown in the accompanying sketch. How many solutions does the system

$$x\vec{v}_1 + y\vec{v}_2 + z\vec{v}_3 = \vec{v}_4$$

have? Argue geometrically.

9. Write the system

$$\begin{vmatrix} x + 2y + 3z = 1 \\ 4x + 5y + 6z = 4 \\ 7x + 8y + 9z = 9 \end{vmatrix}$$

in matrix form.

Compute the dot products in Exercises 10 through 12 (if the products are defined).

10. $\begin{bmatrix} 1 \\ 2 \\ 3 \end{bmatrix} \cdot \begin{bmatrix} 1 \\ -2 \\ 1 \end{bmatrix}$ **11.** $\begin{bmatrix} 1 & 9 & 9 & 7 \end{bmatrix} \cdot \begin{bmatrix} 6 \\ 6 \\ 6 \end{bmatrix}$

12. $\begin{bmatrix} 1 & 2 & 3 & 4 \end{bmatrix} \cdot \begin{bmatrix} 5 \\ 6 \\ 7 \\ 8 \end{bmatrix}$

Compute the products $A\vec{x}$ in Exercises 13 through 15 using paper and pencil. In each case, compute the product two ways: in terms of the columns of A (Definition 1.3.6) and in terms of the rows of A (Fact 1.3.8).

13. $\begin{bmatrix} 1 & 2 \\ 3 & 4 \end{bmatrix} \begin{bmatrix} 7 \\ 11 \end{bmatrix}$.   **14.** $\begin{bmatrix} 1 & 2 & 3 \\ 2 & 3 & 4 \end{bmatrix} \begin{bmatrix} -1 \\ 2 \\ 1 \end{bmatrix}$.

15. $\begin{bmatrix} 1 & 2 & 3 & 4 \end{bmatrix} \begin{bmatrix} 5 \\ 6 \\ 7 \\ 8 \end{bmatrix}$.

Compute the products $A\vec{x}$ in Exercises 16 through 19 using paper and pencil (if the products are defined).

16. $\begin{bmatrix} 0 & 1 \\ 3 & 2 \end{bmatrix}\begin{bmatrix} 2 \\ -3 \end{bmatrix}.$

17. $\begin{bmatrix} 1 & 2 & 3 \\ 4 & 5 & 6 \end{bmatrix}\begin{bmatrix} 7 \\ 8 \end{bmatrix}.$

18. $\begin{bmatrix} 1 & 2 \\ 3 & 4 \\ 5 & 6 \end{bmatrix}\begin{bmatrix} 1 \\ 2 \end{bmatrix}.$

19. $\begin{bmatrix} 1 & 1 & -1 \\ -5 & 1 & 1 \\ 1 & -5 & 3 \end{bmatrix}\begin{bmatrix} 1 \\ 2 \\ 3 \end{bmatrix}.$

20. a. Find

$$\begin{bmatrix} 2 & 3 \\ 4 & 5 \\ 6 & 7 \end{bmatrix} + \begin{bmatrix} 7 & 5 \\ 3 & 1 \\ 0 & -1 \end{bmatrix}.$$

  b. Find

$$9\begin{bmatrix} 1 & -1 & 2 \\ 3 & 4 & 5 \end{bmatrix}.$$

21. Use technology to compute the product

$$\begin{bmatrix} 1 & 7 & 8 & 9 \\ 1 & 2 & 9 & 1 \\ 1 & 5 & 1 & 5 \\ 1 & 6 & 4 & 8 \end{bmatrix}\begin{bmatrix} 1 \\ 9 \\ 5 \\ 6 \end{bmatrix}.$$

22. Consider a linear system of three equations with three unknowns. We are told that the system has a unique solution. What does the reduced row-echelon form of the coefficient matrix of this system look like? Explain your answer.

23. Consider a linear system of four equations with three unknowns. We are told that the system has a unique solution. What does the reduced row-echelon form of the coefficient matrix of this system look like? Explain your answer.

24. Let A be a 4×4 matrix, and let $\vec{b}$ and $\vec{c}$ be two vectors in $\mathbb{R}^4$. We are told that the system $A\vec{x} = \vec{b}$ has a unique solution. What can you say about the number of solutions of the system $A\vec{x} = \vec{c}$?

25. Let A be a 4×4 matrix, and let $\vec{b}$ and $\vec{c}$ be two vectors in $\mathbb{R}^4$. We are told that the system $A\vec{x} = \vec{b}$ is inconsistent. What can you say about the number of solutions of the system $A\vec{x} = \vec{c}$?

26. Let A be a 4×3 matrix, and let $\vec{b}$ and $\vec{c}$ be two vectors in $\mathbb{R}^4$. We are told that the system $A\vec{x} = \vec{b}$ has a unique solution. What can you say about the number of solutions of the system $A\vec{x} = \vec{c}$?

27. *True or False?* Justify your answers.
Consider a system $A\vec{x} = \vec{b}$.

  a. If $A\vec{x} = \vec{b}$ is inconsistent, then rref(A) contains a row of zeros.

  b. If rref(A) contains a row of zeros, then $A\vec{x} = \vec{b}$ is inconsistent.

28. *True or False?* Justify your answers.
Suppose the matrix E is in reduced row-echelon form.

  a. If we omit a row of E, then the remaining matrix is in reduced row-echelon form.

  b. If we omit a column of E, then the remaining matrix is in reduced row-echelon form.

29. *True or False?* Justify your answer.
Consider a system $A\vec{x} = \vec{b}$. This system is consistent if (and only if) rank(A) = rank$\left[A \vdots \vec{b}\right]$.

30. If the rank of a 5×3 matrix A is 3, what is rref(A)?

31. If the rank of a 4×4 matrix A is 4, what is rref(A)?

32. *True or False?*
Consider the system $A\vec{x} = \vec{b}$, where A is an $n \times n$ matrix. This system has a unique solution if (and only if) rank(A) = n.

33. Let A be the $n \times n$ matrix with all 1's on the diagonal and all 0's outside the diagonal. What is $A\vec{x}$, where $\vec{x}$ is a vector in $\mathbb{R}^n$?

34. We define the vectors

$$\vec{e}_1 = \begin{bmatrix} 1 \\ 0 \\ 0 \end{bmatrix}, \qquad \vec{e}_2 = \begin{bmatrix} 0 \\ 1 \\ 0 \end{bmatrix}, \qquad \vec{e}_3 = \begin{bmatrix} 0 \\ 0 \\ 1 \end{bmatrix}$$

in $\mathbb{R}^3$.

  a. For

$$A = \begin{bmatrix} a & b & c \\ d & e & f \\ g & h & k \end{bmatrix},$$

  compute $A\vec{e}_1$, $A\vec{e}_2$, and $A\vec{e}_3$.

  b. If B is an $m \times 3$ matrix with columns $\vec{v}_1$, $\vec{v}_2$, and $\vec{v}_3$, what is $B\vec{e}_1$, $B\vec{e}_2$, $B\vec{e}_3$?

35. In $\mathbb{R}^n$, we define

$$\vec{e}_i = \begin{bmatrix} 0 \\ 0 \\ \vdots \\ 1 \\ \vdots \\ 0 \end{bmatrix} \quad \leftarrow i\text{th component.}$$

If A is an $m \times n$ matrix, what is $A\vec{e}_i$?

36. Find a 3×3 matrix A such that

$$A \begin{bmatrix} 1 \\ 0 \\ 0 \end{bmatrix} = \begin{bmatrix} 1 \\ 2 \\ 3 \end{bmatrix}, \quad A \begin{bmatrix} 0 \\ 1 \\ 0 \end{bmatrix} = \begin{bmatrix} 4 \\ 5 \\ 6 \end{bmatrix},$$

and $\quad A \begin{bmatrix} 0 \\ 0 \\ 1 \end{bmatrix} = \begin{bmatrix} 7 \\ 8 \\ 9 \end{bmatrix}.$

37. Find all vectors $\vec{x}$ such that

$$A\vec{x} = \vec{b}, \quad \text{where} \quad A = \begin{bmatrix} 1 & 2 & 0 \\ 0 & 0 & 1 \\ 0 & 0 & 0 \end{bmatrix} \quad \text{and} \quad \vec{b} = \begin{bmatrix} 2 \\ 1 \\ 0 \end{bmatrix}.$$

38. a. Using technology, generate a "random" 3×3 matrix A. (The entries may be either single-digit integers or numbers between 0 and 1, depending on the technology you are using.) Find $\text{rref}(A)$. Repeat this experiment a few times.

   b. What does the reduced row-echelon form of "most" 3×3 matrices look like? Explain.

39. Repeat Exercise 38 for 3×4 matrices.

40. Repeat Exercise 38 for 4×3 matrices.

41. How many solutions do "most" systems of three linear equations with three unknowns have? Explain in terms of your work in Exercise 38.

42. How many solutions do "most" systems of three linear equations with four unknowns have? Explain in terms of your work in Exercise 39.

43. How many solutions do "most" systems of four linear equations with three unknowns have? Explain in terms of your work in Exercise 40.

44. Consider an $m \times n$ matrix A with more rows than columns ($m > n$). Show that there is a vector $\vec{b}$ in $\mathbb{R}^m$ such that the system $A\vec{x} = \vec{b}$ is inconsistent.

45. Consider an $m \times n$ matrix A, a vector $\vec{x}$ in $\mathbb{R}^n$, and a scalar k. Show that

$$A(k\vec{x}) = k(A\vec{x}).$$

46. Find the rank of the matrix

$$\begin{bmatrix} a & b & c \\ 0 & d & e \\ 0 & 0 & f \end{bmatrix},$$

where a, d, and f are nonzero, and b, c, and e are arbitrary numbers.

47. A linear system of the form

$$A\vec{x} = \vec{0}$$

is called *homogeneous*. Justify the following facts:

   a. All homogeneous systems are consistent.

   b. A homogeneous system with fewer equations than unknowns has infinitely many solutions.

   c. If $\vec{x}_1$ and $\vec{x}_2$ are solutions of the homogeneous system $A\vec{x} = \vec{0}$, then $\vec{x}_1 + \vec{x}_2$ is a solution as well.

   d. If $\vec{x}$ is a solution of the homogeneous system $A\vec{x} = \vec{0}$ and k is an arbitrary constant, then $k\vec{x}$ is a solution as well.

48. Consider a solution $\vec{x}_1$ of the linear system $A\vec{x} = \vec{b}$. Justify the facts stated in parts (a) and (b):

   a. If $\vec{x}_h$ is a solution of the system $A\vec{x} = \vec{0}$, then $\vec{x}_1 + \vec{x}_h$ is a solution of the system $A\vec{x} = \vec{b}$.

   b. If $\vec{x}_2$ is another solution of the system $A\vec{x} = \vec{b}$, then $\vec{x}_2 - \vec{x}_1$ is a solution of the system $A\vec{x} = \vec{0}$.

   c. Now suppose A is a 2×2 matrix. A solution vector $\vec{x}_1$ of the system $A\vec{x} = \vec{b}$ is sketched below (in $\mathbb{R}^2$). We are told that the solutions of the system $A\vec{x} = \vec{0}$ form the line shown in the sketch. Draw the line consisting of all solutions of the system $A\vec{x} = \vec{b}$.

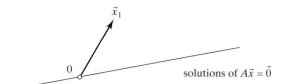

If you are puzzled by the generality of this problem, think about an example first:

$$A = \begin{bmatrix} 1 & 2 \\ 3 & 6 \end{bmatrix}, \quad \vec{b} = \begin{bmatrix} 3 \\ 9 \end{bmatrix}, \quad \text{and} \quad \vec{x}_1 = \begin{bmatrix} 1 \\ 1 \end{bmatrix}.$$

49. Consider the accompanying table. For some linear systems $A\vec{x} = \vec{b}$, you are given either the rank of the coefficient matrix A, or the rank of the augmented matrix $[A \vdots \vec{b}]$. In each case, state whether the system could have no solution, one solution, or infinitely many solutions. There may be more than one possibility for some systems. Justify your answers.

	Number of Equations	Number of Unknowns	Rank of A	Rank of $[A : \vec{b}]$
a.	3	4	—	2
b.	4	3	3	—
c.	4	3	—	4
d.	3	4	3	—

50. Consider a linear system $A\vec{x} = \vec{b}$, where A is a 4×3 matrix. We are told that rank $[A \vdots \vec{b}] = 4$. How many solutions does this system have?

51. Consider an $m \times n$ matrix A, an $r \times s$ matrix B, and a vector $\vec{x}$ in $\mathbb{R}^p$. For which choices of m, n, r, s, p is the product

$$A(B\vec{x})$$

defined?

52. Consider the matrices

$$A = \begin{bmatrix} 1 & 0 \\ 1 & 2 \end{bmatrix} \quad \text{and} \quad B = \begin{bmatrix} 0 & -1 \\ 1 & 0 \end{bmatrix}.$$

Can you find a 2×2 matrix C such that

$$A(B\vec{x}) = C\vec{x},$$

for all vectors $\vec{x}$ in $\mathbb{R}^2$?

53. If A and B are two $m \times n$ matrices, is

$$(A + B)\vec{x} = A\vec{x} + B\vec{x}$$

for all $\vec{x}$ in $\mathbb{R}^n$?

54. Consider two vectors $\vec{v}_1$ and $\vec{v}_2$ in $\mathbb{R}^3$ that are not parallel. Which vectors in $\mathbb{R}^3$ are linear combinations of $\vec{v}_1$ and $\vec{v}_2$? Describe the set of these vectors geometrically. Include a sketch in your answer.

55. Is the vector $\begin{bmatrix} 7 \\ 8 \\ 9 \end{bmatrix}$ a linear combination of $\begin{bmatrix} 1 \\ 2 \\ 3 \end{bmatrix}$ and $\begin{bmatrix} 4 \\ 5 \\ 6 \end{bmatrix}$?

56. Is the vector

$$\begin{bmatrix} 30 \\ -1 \\ 38 \\ 56 \\ 62 \end{bmatrix}$$

a linear combination of

$$\begin{bmatrix} 1 \\ 7 \\ 1 \\ 9 \\ 4 \end{bmatrix}, \quad \begin{bmatrix} 5 \\ 6 \\ 3 \\ 2 \\ 8 \end{bmatrix}, \quad \begin{bmatrix} 9 \\ 2 \\ 3 \\ 5 \\ 2 \end{bmatrix}, \quad \begin{bmatrix} -2 \\ -5 \\ 4 \\ 7 \\ 9 \end{bmatrix}?$$

57. Express the vector $\begin{bmatrix} 7 \\ 11 \end{bmatrix}$ as the sum of a vector on the line $y = 3x$ and a vector on the line $y = x/2$.

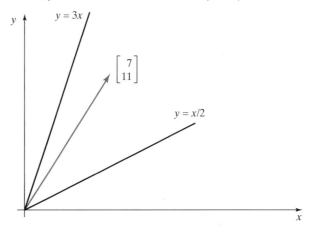

Chapter 1

TRUE OR FALSE?[14]

Determine whether the statements that follow are true or false, and justify your answer.

1. Matrix $\begin{bmatrix} 1 & 2 & 0 \\ 0 & 0 & 1 \\ 0 & 0 & 0 \end{bmatrix}$ is in rref.

---

[14]We will conclude each chapter (except for Chapters 4 and 9) with some True–False questions, over 300 in all. We will start with a group of about 10 straightforward statements that refer directly to definitions and facts given in the chapter. Then there may be some computational exercises, and the remaining ones are more conceptual, calling for independent reasoning. In some chapters, a few of the problems towards the end can be quite challenging. Don't expect a balanced coverage of all the topics; some concepts are much better suited for this kind of questioning than others.

2. A system of four equations in three unknowns is always inconsistent.

3. There is a 3×4 matrix with rank 4.

4. If A is a 3×4 matrix and vector $\vec{v}$ is in $\mathbb{R}^4$, then vector $A\vec{v}$ is in $\mathbb{R}^3$.

5. If the 4×4 matrix A has rank 4, then any linear system with coefficient matrix A will have a unique solution.

6. There exists a system of three linear equations with three unknowns that has exactly three solutions.

7. There is a 5×5 matrix A of rank 4 such that the system $A\vec{x} = \vec{0}$ has only the solution $\vec{x} = \vec{0}$.

8. If matrix A is in rref, then at least one of the entries in each column must be 1.

9. If A is an $n \times n$ matrix and $\vec{x}$ is a vector in $\mathbb{R}^n$, then the product $A\vec{x}$ is a linear combination of the columns of matrix A.

10. If vector $\vec{u}$ is a linear combination of vectors $\vec{v}$ and $\vec{w}$, then we can write $\vec{u} = a\vec{v} + b\vec{w}$ for some scalars a and b.

11. rank $\begin{bmatrix} 2 & 2 & 2 \\ 2 & 2 & 2 \\ 2 & 2 & 2 \end{bmatrix} = 2.$

12. $\begin{bmatrix} 11 & 13 & 15 \\ 17 & 19 & 21 \end{bmatrix} \begin{bmatrix} -1 \\ 3 \\ -1 \end{bmatrix} = \begin{bmatrix} 13 \\ 19 \\ 21 \end{bmatrix}.$

13. There is a matrix A such that $A \begin{bmatrix} -1 \\ 2 \end{bmatrix} = \begin{bmatrix} 3 \\ 5 \\ 7 \end{bmatrix}.$

14. Vector $\begin{bmatrix} 1 \\ 2 \\ 3 \end{bmatrix}$ is a linear combination of vectors $\begin{bmatrix} 4 \\ 5 \\ 6 \end{bmatrix}$ and $\begin{bmatrix} 7 \\ 8 \\ 9 \end{bmatrix}.$

15. The system $\begin{bmatrix} 1 & 2 & 3 \\ 4 & 5 & 6 \\ 0 & 0 & 0 \end{bmatrix} \vec{x} = \begin{bmatrix} 1 \\ 2 \\ 3 \end{bmatrix}$ is inconsistent.

16. There exists a 2×2 matrix A such that $A \begin{bmatrix} 1 \\ 2 \end{bmatrix} = \begin{bmatrix} 3 \\ 4 \end{bmatrix}.$

17. If A is a nonzero matrix of the form $\begin{bmatrix} a & -b \\ b & a \end{bmatrix}$, then the rank of A must be 2.

18. rank $\begin{bmatrix} 1 & 1 & 1 \\ 1 & 2 & 3 \\ 1 & 3 & 6 \end{bmatrix} = 3.$

19. The system $A\vec{x} = \begin{bmatrix} 0 \\ 0 \\ 0 \\ 1 \end{bmatrix}$ is inconsistent for all 4×3 matrices A.

20. There exists a 2×2 matrix A such that $A \begin{bmatrix} 1 \\ 1 \end{bmatrix} = \begin{bmatrix} 1 \\ 2 \end{bmatrix}$ and $A \begin{bmatrix} 2 \\ 2 \end{bmatrix} = \begin{bmatrix} 2 \\ 1 \end{bmatrix}.$

21. There exist scalars a and b such that matrix $\begin{bmatrix} 0 & 1 & a \\ -1 & 0 & b \\ -a & -b & 0 \end{bmatrix}$ has rank 3.

22. If $\vec{v}$ and $\vec{w}$ are vectors in $\mathbb{R}^4$, then $\vec{v}$ must be a linear combination of $\vec{v}$ and $\vec{w}$.

23. If $\vec{u}$, $\vec{v}$, and $\vec{w}$ are nonzero vectors in $\mathbb{R}^2$, then $\vec{w}$ must be a linear combination of $\vec{u}$ and $\vec{v}$.

24. If $\vec{v}$ and $\vec{w}$ are vectors in $\mathbb{R}^4$, then the zero vector in $\mathbb{R}^4$ must be a linear combination of $\vec{v}$ and $\vec{w}$.

25. If A and B are any two 3×3 matrices of rank 2, then A can be transformed into B by means of elementary row operations.

26. If vector $\vec{u}$ is a linear combination of vectors $\vec{v}$ and $\vec{w}$, and $\vec{v}$ is a linear combination of vectors $\vec{p}$, $\vec{q}$, and $\vec{r}$, then $\vec{u}$ is a linear combination of $\vec{p}$, $\vec{q}$, $\vec{r}$, and $\vec{w}$.

27. A system with fewer unknowns than equations must have infinitely many solutions or none.

28. The rank of any upper triangular matrix is the number of nonzero entries on its diagonal.

29. If the system $A\vec{x} = \vec{b}$ has a unique solution, then A must be a square matrix.

30. If A is a 4×3 matrix, then there exists a vector $\vec{b}$ in $\mathbb{R}^4$ such that the system $A\vec{x} = \vec{b}$ is inconsistent.

31. If A is a 4×3 matrix of rank 3 and $A\vec{v} = A\vec{w}$ for two vectors $\vec{v}$ and $\vec{w}$ in $\mathbb{R}^3$, then vectors $\vec{v}$ and $\vec{w}$ must be equal.

32. If A is a 4×4 matrix and the system $A\vec{x} = \begin{bmatrix} 2 \\ 3 \\ 4 \\ 5 \end{bmatrix}$ has a unique solution, then the system $A\vec{x} = \vec{0}$ has only the solution $\vec{x} = \vec{0}$.

33. If vector $\vec{u}$ is a linear combination of vectors $\vec{v}$ and $\vec{w}$, then $\vec{w}$ must be a linear combination of $\vec{u}$ and $\vec{v}$.

34. If $A = \begin{bmatrix} \vec{u} & \vec{v} & \vec{w} \end{bmatrix}$ and rref$(A) = \begin{bmatrix} 1 & 0 & 2 \\ 0 & 1 & 3 \\ 0 & 0 & 0 \end{bmatrix}$, then the equation $\vec{w} = 2\vec{u} + 3\vec{v}$ must hold.

35. If A and B are matrices of the same size, then the formula rank$(A + B) =$ rank$(A) +$ rank(B) must hold.

36. If A and B are any two $n \times n$ matrices of rank n, then A can be transformed into B by means of elementary row operations.

Linear Transformations

2.1 INTRODUCTION TO LINEAR TRANSFORMATIONS AND THEIR INVERSES

Imagine yourself cruising in the Mediterranean as a crew member on a French coast guard boat, looking for spies. Periodically, your boat radios its position to headquarters in Marseille. You have to expect that communications will be intercepted. So, before you broadcast anything, you have to transform the actual position of the boat,

$$\begin{bmatrix} x_1 \\ x_2 \end{bmatrix}$$

(x_1 for Eastern longitude, x_2 for Northern latitude), into an encoded position

$$\begin{bmatrix} y_1 \\ y_2 \end{bmatrix}.$$

You use the following code:

$$\begin{aligned} y_1 &= \ x_1 + 3x_2 \\ y_2 &= 2x_1 + 5x_2. \end{aligned}$$

For example, when the actual position of your boat is 5°E, 42°N, or

$$\vec{x} = \begin{bmatrix} x_1 \\ x_2 \end{bmatrix} = \begin{bmatrix} 5 \\ 42 \end{bmatrix},$$

your encoded position will be

$$\vec{y} = \begin{bmatrix} y_1 \\ y_2 \end{bmatrix} = \begin{bmatrix} 131 \\ 220 \end{bmatrix}.$$

(See Figure 1.)

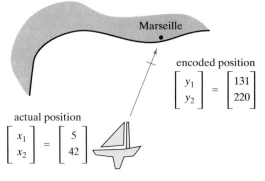

Figure 1

The coding transformation can be represented as

$$\underbrace{\begin{bmatrix} y_1 \\ y_2 \end{bmatrix}}_{\vec{y}} = \begin{bmatrix} x_1 + 3x_2 \\ 2x_1 + 5x_2 \end{bmatrix} = \underbrace{\begin{bmatrix} 1 & 3 \\ 2 & 5 \end{bmatrix}}_{A} \underbrace{\begin{bmatrix} x_1 \\ x_2 \end{bmatrix}}_{\vec{x}},$$

or, more succinctly, as

$$\vec{y} = A\vec{x} \ .$$

The matrix A is called the *coefficient matrix* of the transformation, or simply its matrix.

A transformation of the form

$$\vec{y} = A\vec{x}$$

is called a *linear transformation*. We will discuss this important concept in greater detail later.

As the ship reaches a new position, the sailor on duty at headquarters in Marseille receives the encoded message

$$\vec{b} = \begin{bmatrix} 133 \\ 223 \end{bmatrix}.$$

He must determine the actual position of the boat. He will have to solve the linear system

$$A\vec{x} = \vec{b},$$

or, more explicitly,

$$\begin{vmatrix} x_1 + 3x_2 = 133 \\ 2x_1 + 5x_2 = 223 \end{vmatrix}.$$

Here is his solution. Is it correct?

$$\vec{x} = \begin{bmatrix} x_1 \\ x_2 \end{bmatrix} = \begin{bmatrix} 4 \\ 43 \end{bmatrix}.$$

As the boat travels on and dozens of positions are radioed in, the sailor gets a little tired of solving all those linear systems, and he thinks there must be a general

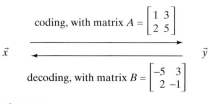

coding, with matrix $A = \begin{bmatrix} 1 & 3 \\ 2 & 5 \end{bmatrix}$

$\vec{x}$ $\vec{y}$

decoding, with matrix $B = \begin{bmatrix} -5 & 3 \\ 2 & -1 \end{bmatrix}$

Figure 2

formula to simplify this task. He wants to solve the system

$$\begin{vmatrix} x_1 + 3x_2 = y_1 \\ 2x_1 + 5x_2 = y_2 \end{vmatrix}$$

when y_1 and y_2 are arbitrary constants, rather than particular numerical values. He is looking for the *decoding transformation*

$$\vec{y} \rightarrow \vec{x},$$

which is the *inverse*[1] of the coding transformation

$$\vec{x} \rightarrow \vec{y}.$$

The method of solution is nothing new. We apply elimination as we have for a linear system with known values y_1 and y_2:

$$\begin{vmatrix} x_1 + 3x_2 = y_1 \\ 2x_1 + 5x_2 = y_2 \end{vmatrix} \quad \begin{matrix} \rightarrow \\ -2(\text{I}) \end{matrix} \quad \begin{vmatrix} x_1 + 3x_2 = y_1 \\ -x_2 = -2y_1 + y_2 \end{vmatrix} \quad \begin{matrix} \rightarrow \\ \div(-1) \end{matrix}$$

$$\begin{vmatrix} x_1 + 3x_2 = y_1 \\ x_2 = 2y_1 - y_2 \end{vmatrix} \quad \begin{matrix} -3(\text{II}) \\ \rightarrow \end{matrix} \quad \begin{vmatrix} x_1 = -5y_1 + 3y_2 \\ x_2 = 2y_1 - y_2 \end{vmatrix}.$$

The formula for the decoding transformation is

$$\begin{aligned} x_1 &= -5y_1 + 3y_2, \\ x_2 &= 2y_1 - y_2, \end{aligned}$$

or

$$\vec{x} = B\vec{y}, \quad \text{where } B = \begin{bmatrix} -5 & 3 \\ 2 & -1 \end{bmatrix}.$$

Note that the decoding transformation is linear and that its coefficient matrix is

$$B = \begin{bmatrix} -5 & 3 \\ 2 & -1 \end{bmatrix}.$$

The relationship between the two matrices A and B is shown in Figure 2.

Since the decoding transformation $\vec{x} = B\vec{y}$ is the inverse of the coding transformation $\vec{y} = A\vec{x}$, we say that the matrix B is the *inverse* of the matrix A. We can write this as $B = A^{-1}$.

Not all linear transformations

$$\begin{bmatrix} x_1 \\ x_2 \end{bmatrix} \rightarrow \begin{bmatrix} y_1 \\ y_2 \end{bmatrix}$$

---

[1] We will discuss the concept of the inverse of a transformation more systematically in Section 2.3.

are invertible. Suppose some ignorant officer chooses the code

$$y_1 = x_1 + 2x_2 \qquad \text{with matrix} \qquad A = \begin{bmatrix} 1 & 2 \\ 2 & 4 \end{bmatrix}$$
$$y_2 = 2x_1 + 4x_2$$

for the French coast guard boats. When the sailor in Marseille has to decode a position, for example,

$$\vec{b} = \begin{bmatrix} 89 \\ 178 \end{bmatrix},$$

he will be chagrined to discover that the system

$$\begin{vmatrix} x_1 + 2x_2 = 89 \\ 2x_1 + 4x_2 = 178 \end{vmatrix}$$

has infinitely many solutions, namely,

$$\begin{bmatrix} x_1 \\ x_2 \end{bmatrix} = \begin{bmatrix} 89 - 2t \\ t \end{bmatrix},$$

where t is an arbitrary number.

Because this system does not have a unique solution, it is impossible to recover the actual position from the encoded position: The coding transformation and the coding matrix A are *noninvertible*. This code is useless!

Now let us discuss the important concept of *linear transformations* in greater detail. Since linear transformations are a special class of functions, it may be helpful to review the concept of a *function* first.

Consider two sets X and Y. A function T from X to Y is a rule that associates with each element x of X a unique element y of Y. The set X is called the *domain* of the function, and Y is its *codomain*. We will sometimes refer to x as the *input* of the function and to y as its *output*. Figure 3 shows an example where domain X and codomain Y are finite.

In precalculus and calculus, you studied functions whose input and output are scalars (i.e., whose domain and codomain are the real numbers $\mathbb{R}$ or subsets of $\mathbb{R}$); for example,

$$y = x^2, \qquad f(x) = e^x, \qquad g(t) = \frac{t^2 - 2}{t - 1}.$$

In multivariable calculus, you may have encountered functions whose input or output were vectors.

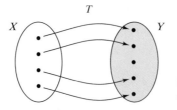

Figure 3 Domain X and codomain Y of a function T.

EXAMPLE 1

$$y = x_1^2 + x_2^2 + x_3^2.$$

This formula defines a function from $\mathbb{R}^3$ to $\mathbb{R}$. The input is the vector $\vec{x} = \begin{bmatrix} x_1 \\ x_2 \\ x_3 \end{bmatrix}$, and the output is the scalar y. ∎

EXAMPLE 2

$$\vec{r} = \begin{bmatrix} \cos(t) \\ \sin(t) \\ t \end{bmatrix}.$$

This formula defines a function from $\mathbb{R}$ to $\mathbb{R}^3$, with input t and output $\vec{r}$. ∎

We now return to the topic of linear transformations.

Definition 2.1.1

> **Linear transformations**
> A function T from $\mathbb{R}^n$ to $\mathbb{R}^m$ is called a *linear transformation* if there is an $m \times n$ matrix A such that
>
> $$T(\vec{x}) = A\vec{x},$$
>
> for all $\vec{x}$ in $\mathbb{R}^n$.

This is a preliminary definition; a more conceptual characterization of linear transformation will follow in Fact 2.2.1.

It is important to note that a linear transformation is a special kind of *function*. The input and the output are both vectors. If we denote the output vector $T(\vec{x})$ by $\vec{y}$, we can write

$$\vec{y} = A\vec{x}.$$

Let us write this equation in terms of its components:

$$\begin{bmatrix} y_1 \\ y_2 \\ \vdots \\ y_m \end{bmatrix} = \begin{bmatrix} a_{11} & a_{12} & \cdots & a_{1n} \\ a_{21} & a_{22} & \cdots & a_{2n} \\ \vdots & \vdots & & \vdots \\ a_{m1} & a_{m2} & \cdots & a_{mn} \end{bmatrix} \begin{bmatrix} x_1 \\ x_2 \\ \vdots \\ x_n \end{bmatrix} = \begin{bmatrix} a_{11}x_1 + a_{12}x_2 + \cdots + a_{1n}x_n \\ a_{21}x_1 + a_{22}x_2 + \cdots + a_{2n}x_n \\ \vdots \qquad \vdots \qquad \qquad \vdots \\ a_{m1}x_1 + a_{m2}x_2 + \cdots + a_{mn}x_n \end{bmatrix},$$

or

$$\begin{aligned} y_1 &= a_{11}x_1 + a_{12}x_2 + \cdots + a_{1n}x_n \\ y_2 &= a_{21}x_1 + a_{22}x_2 + \cdots + a_{2n}x_n \\ \vdots &= \quad \vdots \qquad \vdots \qquad \qquad \vdots \\ y_m &= a_{m1}x_1 + a_{m2}x_2 + \cdots + a_{mn}x_n. \end{aligned}$$

The output variables y_i are linear functions of the input variables x_j. In some branches of mathematics, a first-order function with a constant term, such as $y = 3x_1 - 7x_2 + 5x_3 + 8$, is called linear. Not so in linear algebra: The linear functions of n variables are those of the form $y = c_1x_1 + c_2x_2 + \cdots + c_nx_n$, for some coefficients $c_1, c_2, \ldots, c_n$.

EXAMPLE 3 The linear transformation

$$\begin{aligned} y_1 &= 7x_1 + 3x_2 - 9x_3 + 8x_4 \\ y_2 &= 6x_1 + 2x_2 - 8x_3 + 7x_4 \\ y_3 &= 8x_1 + 4x_2 \qquad\quad\; + 7x_4 \end{aligned}$$

(a function from $\mathbb{R}^4$ to $\mathbb{R}^3$) is represented by the 3×4 matrix

$$A = \begin{bmatrix} 7 & 3 & -9 & 8 \\ 6 & 2 & -8 & 7 \\ 8 & 4 & 0 & 7 \end{bmatrix}.$$ ∎

EXAMPLE 4 The coefficient matrix of the *identity transformation*

$$\begin{aligned} y_1 &= x_1 \\ y_2 &= \quad x_2 \\ &\;\;\vdots \qquad\quad \ddots \\ y_n &= \qquad\quad x_n \end{aligned}$$

(a linear transformation from $\mathbb{R}^n$ to $\mathbb{R}^n$ whose output equals its input) is the $n \times n$ matrix

$$\begin{bmatrix} 1 & 0 & \cdots & 0 \\ 0 & 1 & \cdots & 0 \\ \vdots & \vdots & \ddots & \vdots \\ 0 & 0 & \cdots & 1 \end{bmatrix}.$$

All entries on the main diagonal are 1, and all other entries are 0. This matrix is called the *identity matrix* and is denoted by I_n:

$$I_2 = \begin{bmatrix} 1 & 0 \\ 0 & 1 \end{bmatrix}, \qquad I_3 = \begin{bmatrix} 1 & 0 & 0 \\ 0 & 1 & 0 \\ 0 & 0 & 1 \end{bmatrix}, \qquad \text{etc.}$$ ∎

We have already seen the identity matrix in other contexts. For example, we have shown that a linear system $A\vec{x} = \vec{b}$ of n equations with n unknowns has a unique solution if and only if $\text{rref}(A) = I_n$. (See Fact 1.3.4.)

EXAMPLE 5 Give a geometric interpretation of the linear transformation

$$\vec{y} = A\vec{x}, \quad \text{where } A = \begin{bmatrix} 0 & -1 \\ 1 & 0 \end{bmatrix}.$$

Solution

First, we rewrite the linear transformation to show the components:

$$\begin{bmatrix} y_1 \\ y_2 \end{bmatrix} = \begin{bmatrix} 0 & -1 \\ 1 & 0 \end{bmatrix} \begin{bmatrix} x_1 \\ x_2 \end{bmatrix} = \begin{bmatrix} -x_2 \\ x_1 \end{bmatrix}.$$

Now, consider the geometric relationship between the input vector

$$\vec{x} = \begin{bmatrix} x_1 \\ x_2 \end{bmatrix}$$

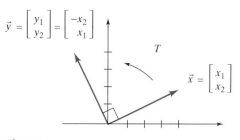

Figure 4

and the corresponding output vector

$$\vec{y} = \begin{bmatrix} y_1 \\ y_2 \end{bmatrix} = \begin{bmatrix} -x_2 \\ x_1 \end{bmatrix}.$$

We observe that the two vectors $\vec{x}$ and $\vec{y}$ have the same length, $\sqrt{x_1^2 + x_2^2}$, and that they are perpendicular to one another (because the dot product equals zero). From the signs of the components, we know that if $\vec{x}$ is in the first quadrant, then $\vec{y}$ will be in the second, as shown in Figure 4.

The output vector $\vec{y}$ is obtained from $\vec{x}$ by *rotating* through an angle of 90° ($\frac{\pi}{2}$ radians) in the counterclockwise direction. Check that the rotation is indeed counterclockwise when $\vec{x}$ is in the second, third, or fourth quadrant. ∎

EXAMPLE 6 Consider the linear transformation $T(\vec{x}) = A\vec{x}$, with

$$A = \begin{bmatrix} 1 & 2 & 3 \\ 4 & 5 & 6 \\ 7 & 8 & 9 \end{bmatrix}.$$

Find

$$T \begin{bmatrix} 1 \\ 0 \\ 0 \end{bmatrix} \quad \text{and} \quad T \begin{bmatrix} 0 \\ 0 \\ 1 \end{bmatrix},$$

where for simplicity we write $T \begin{bmatrix} 1 \\ 0 \\ 0 \end{bmatrix}$ for $T \left(\begin{bmatrix} 1 \\ 0 \\ 0 \end{bmatrix} \right).$

Solution

A straightforward computation shows that

$$T \begin{bmatrix} 1 \\ 0 \\ 0 \end{bmatrix} = \begin{bmatrix} 1 & 2 & 3 \\ 4 & 5 & 6 \\ 7 & 8 & 9 \end{bmatrix} \begin{bmatrix} 1 \\ 0 \\ 0 \end{bmatrix} = \begin{bmatrix} 1 \\ 4 \\ 7 \end{bmatrix}$$

and

$$T \begin{bmatrix} 0 \\ 0 \\ 1 \end{bmatrix} = \begin{bmatrix} 1 & 2 & 3 \\ 4 & 5 & 6 \\ 7 & 8 & 9 \end{bmatrix} \begin{bmatrix} 0 \\ 0 \\ 1 \end{bmatrix} = \begin{bmatrix} 3 \\ 6 \\ 9 \end{bmatrix}.$$

Note that $T\begin{bmatrix} 1 \\ 0 \\ 0 \end{bmatrix}$ is the first column of the matrix A and that $T\begin{bmatrix} 0 \\ 0 \\ 1 \end{bmatrix}$ is its third column. ∎

This observation can be generalized as follows:

Fact 2.1.2 Consider a linear transformation T from $\mathbb{R}^n$ to $\mathbb{R}^m$. Then, the matrix of T is

$$
A = \begin{bmatrix} | & | & & | \\ T(\vec{e}_1) & T(\vec{e}_2) & \cdots & T(\vec{e}_n) \\ | & | & & | \end{bmatrix}, \quad \text{where } \vec{e}_i = \begin{bmatrix} 0 \\ 0 \\ \vdots \\ 1 \\ \vdots \\ 0 \end{bmatrix} \leftarrow i\text{th}.
$$

To justify this result, write

$$
A = \begin{bmatrix} | & | & & | \\ \vec{v}_1 & \vec{v}_2 & \cdots & \vec{v}_n \\ | & | & & | \end{bmatrix}.
$$

Then,

$$
T(\vec{e}_i) = A\vec{e}_i = \begin{bmatrix} | & | & & | & & | \\ \vec{v}_1 & \vec{v}_2 & \cdots & \vec{v}_i & \cdots & \vec{v}_n \\ | & | & & | & & | \end{bmatrix} \begin{bmatrix} 0 \\ 0 \\ \vdots \\ 1 \\ \vdots \\ 0 \end{bmatrix} = \vec{v}_i,
$$

by definition of the product $A\vec{e}_i$.

The vectors $\vec{e}_1, \vec{e}_2, \ldots, \vec{e}_n$ in $\mathbb{R}^n$ are sometimes referred to as the *standard vectors* in $\mathbb{R}^n$. The standard vectors $\vec{e}_1, \vec{e}_2, \vec{e}_3$ in $\mathbb{R}^3$ are often denoted by $\vec{i}, \vec{j}, \vec{k}$.

EXERCISES

GOAL Use the concept of a linear transformation in terms of the formula $\vec{y} = A\vec{x}$, and interpret simple linear transformations geometrically. Find the inverse of a linear transformation from $\mathbb{R}^2$ to $\mathbb{R}^2$ (if it exists). Find the matrix of a linear transformation column by column.

Consider the transformations from $\mathbb{R}^3$ to $\mathbb{R}^3$ defined in

Exercises 1 through 3. Which of these transformations are linear?

1. $\begin{aligned} y_1 &= 2x_2 \\ y_2 &= x_2 + 2 \\ y_3 &= 2x_2 \end{aligned}$

2. $\begin{aligned} y_1 &= 2x_2 \\ y_2 &= 3x_3 \\ y_3 &= x_1 \end{aligned}$

3. $\begin{aligned} y_1 &= x_2 - x_3 \\ y_2 &= x_1 x_3 \\ y_3 &= x_1 - x_2 \end{aligned}$

4. Find the matrix of the linear transformation
$$\begin{aligned} y_1 &= 9x_1 + 3x_2 - 3x_3 \\ y_2 &= 2x_1 - 9x_2 + x_3 \\ y_3 &= 4x_1 - 9x_2 - 2x_3 \\ y_4 &= 5x_1 + x_2 + 5x_3 \end{aligned}$$

5. Consider a linear transformation T from $\mathbb{R}^3$ to $\mathbb{R}^2$, where

$$T \begin{bmatrix} 1 \\ 0 \\ 0 \end{bmatrix} = \begin{bmatrix} 7 \\ 11 \end{bmatrix}, \qquad T \begin{bmatrix} 0 \\ 1 \\ 0 \end{bmatrix} = \begin{bmatrix} 6 \\ 9 \end{bmatrix},$$

and $\quad T \begin{bmatrix} 0 \\ 0 \\ 1 \end{bmatrix} = \begin{bmatrix} -13 \\ 17 \end{bmatrix}.$

Find the matrix A of T.

6. Consider the transformation T from $\mathbb{R}^2$ to $\mathbb{R}^3$ given by

$$T \begin{bmatrix} x_1 \\ x_2 \end{bmatrix} = x_1 \begin{bmatrix} 1 \\ 2 \\ 3 \end{bmatrix} + x_2 \begin{bmatrix} 4 \\ 5 \\ 6 \end{bmatrix}.$$

Is this transformation linear? If so, find its matrix.

7. Suppose $\vec{v}_1, \vec{v}_2, \ldots, \vec{v}_n$ are arbitrary vectors in $\mathbb{R}^m$. Consider the transformation from $\mathbb{R}^n$ to $\mathbb{R}^m$ given by

$$T \begin{bmatrix} x_1 \\ x_2 \\ \vdots \\ x_n \end{bmatrix} = x_1 \vec{v}_1 + x_2 \vec{v}_2 + \cdots + x_n \vec{v}_n.$$

Is this transformation linear? If so, find its matrix A in terms of the vectors $\vec{v}_1, \vec{v}_2, \ldots, \vec{v}_n$.

8. Find the inverse of the linear transformation
$$\begin{aligned} y_1 &= x_1 + 7x_2 \\ y_2 &= 3x_1 + 20x_2 \end{aligned}$$

In Exercises 9 through 12, decide whether the given matrix is invertible. Find the inverse if it exists. In Exercise 12, the constant k is arbitrary.

9. $\begin{bmatrix} 2 & 3 \\ 6 & 9 \end{bmatrix}.$ **10.** $\begin{bmatrix} 1 & 2 \\ 4 & 9 \end{bmatrix}.$ **11.** $\begin{bmatrix} 1 & 2 \\ 3 & 9 \end{bmatrix}.$ **12.** $\begin{bmatrix} 1 & k \\ 0 & 1 \end{bmatrix}.$

13. Prove the following facts:

a. The 2×2 matrix
$$A = \begin{bmatrix} a & b \\ c & d \end{bmatrix}$$
is invertible if and only if $ad - bc \neq 0$. (*Hint:* Consider the cases $a \neq 0$ and $a = 0$ separately.)

b. If
$$\begin{bmatrix} a & b \\ c & d \end{bmatrix}$$

is invertible, then
$$\begin{bmatrix} a & b \\ c & d \end{bmatrix}^{-1} = \frac{1}{ad - bc} \begin{bmatrix} d & -b \\ -c & a \end{bmatrix}.$$
(The formula in part (b) is worth memorizing.)

14. a. For which choices of the constant k is the matrix $\begin{bmatrix} 2 & 3 \\ 5 & k \end{bmatrix}$ invertible?

b. For which choices of the constant k are all entries of $\begin{bmatrix} 2 & 3 \\ 5 & k \end{bmatrix}^{-1}$ integers?

15. For which choices of the constants a and b is the matrix
$$A = \begin{bmatrix} a & -b \\ b & a \end{bmatrix}$$
invertible? What is the inverse in this case?

Give a geometric interpretation of the linear transformations defined by the matrices in Exercises 16 through 23. In each case, decide whether the transformation is invertible. Find the inverse if it exists, and interpret it geometrically.

16. $\begin{bmatrix} 2 & 0 \\ 0 & 2 \end{bmatrix}.$ **17.** $\begin{bmatrix} -1 & 0 \\ 0 & -1 \end{bmatrix}.$ **18.** $\begin{bmatrix} 0.5 & 0 \\ 0 & 0.5 \end{bmatrix}.$

19. $\begin{bmatrix} 1 & 0 \\ 0 & 0 \end{bmatrix}.$ **20.** $\begin{bmatrix} 0 & 1 \\ 1 & 0 \end{bmatrix}.$ **21.** $\begin{bmatrix} 0 & 1 \\ -1 & 0 \end{bmatrix}.$

22. $\begin{bmatrix} 1 & 0 \\ 0 & -1 \end{bmatrix}.$ **23.** $\begin{bmatrix} 0 & 2 \\ -2 & 0 \end{bmatrix}.$

Consider the circular face below. For each of the matrices A in Exercises 24 through 30, draw a sketch showing the effect of the linear transformation $T(\vec{x}) = A\vec{x}$ on this face.

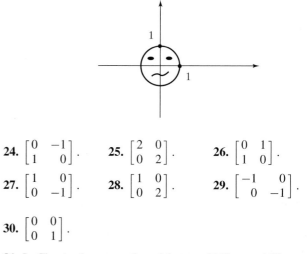

24. $\begin{bmatrix} 0 & -1 \\ 1 & 0 \end{bmatrix}.$ **25.** $\begin{bmatrix} 2 & 0 \\ 0 & 2 \end{bmatrix}.$ **26.** $\begin{bmatrix} 0 & 1 \\ 1 & 0 \end{bmatrix}.$

27. $\begin{bmatrix} 1 & 0 \\ 0 & -1 \end{bmatrix}.$ **28.** $\begin{bmatrix} 1 & 0 \\ 0 & 2 \end{bmatrix}.$ **29.** $\begin{bmatrix} -1 & 0 \\ 0 & -1 \end{bmatrix}.$

30. $\begin{bmatrix} 0 & 0 \\ 0 & 1 \end{bmatrix}.$

31. In Chapter 1, we mentioned that an old German bill shows the mirror image of Gauss' likeness. What linear transformation T can you apply to get the actual picture back?

32. Find an $n \times n$ matrix A such that $A\vec{x} = 3\vec{x}$, for all $\vec{x}$ in $\mathbb{R}^n$.

33. Consider the transformation T from $\mathbb{R}^2$ to $\mathbb{R}^2$ that rotates any vector $\vec{x}$ through an angle of $45°$ in the counterclockwise direction, as shown in the following figure:

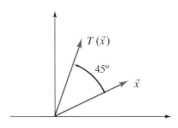

You are told that T is a linear transformation. (This will be shown in the next section.) Find the matrix of T.

34. Consider the transformation T from $\mathbb{R}^2$ to $\mathbb{R}^2$ that rotates any vector $\vec{x}$ through a given angle φ in the counterclockwise direction. (Compare this with Exercise 33.) You are told that T is linear. Find the matrix of T in terms of φ.

35. In the example about the French coast guard in this section, suppose you are a spy watching the boat and listening in on the radio messages from the boat. You collect the following data:

- When the actual position is $\begin{bmatrix} 5 \\ 42 \end{bmatrix}$, they radio $\begin{bmatrix} 89 \\ 52 \end{bmatrix}$.

- When the actual position is $\begin{bmatrix} 6 \\ 41 \end{bmatrix}$, they radio $\begin{bmatrix} 88 \\ 53 \end{bmatrix}$.

Can you crack their code (i.e., find the coding matrix), assuming that the code is linear?

36. Consider a linear transformation T from $\mathbb{R}^n$ to $\mathbb{R}^m$. Using Fact 1.3.7, justify the following equations:

$$T(\vec{v} + \vec{w}) = T(\vec{v}) + T(\vec{w}), \quad \text{for all vectors } \vec{v}, \vec{w} \text{ in } \mathbb{R}^n,$$
$$T(k\vec{v}) = kT(\vec{v}), \quad \text{for all vectors } \vec{v} \text{ in } \mathbb{R}^n \text{ and all scalars } k.$$

37. Consider a linear transformation T from $\mathbb{R}^2$ to $\mathbb{R}^2$. Suppose that $\vec{v}$ and $\vec{w}$ are two arbitrary vectors in $\mathbb{R}^2$ and that $\vec{x}$ is a third vector whose endpoint is on the line segment connecting the endpoints of $\vec{v}$ and $\vec{w}$. Is the endpoint of the vector $T(\vec{x})$ necessarily on the line segment connecting the endpoints of $T(\vec{v})$ and $T(\vec{w})$? Justify your answer.

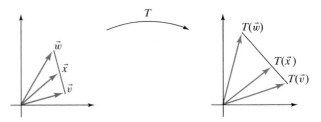

(*Hint:* We can write $\vec{x} = \vec{v} + k(\vec{w} - \vec{v})$, for some scalar k between 0 and 1. Exercise 36 is helpful.)

38. The two column vectors $\vec{v}_1$ and $\vec{v}_2$ of a 2×2 matrix A are shown in the accompanying sketch. Consider the linear transformation $T(\vec{x}) = A\vec{x}$, from $\mathbb{R}^2$ to $\mathbb{R}^2$. Sketch the vector

$$T\begin{bmatrix} 2 \\ -1 \end{bmatrix}.$$

39. Show that if T is a linear transformation from $\mathbb{R}^n$ to $\mathbb{R}^m$, then

$$T\begin{bmatrix} x_1 \\ x_2 \\ \vdots \\ x_n \end{bmatrix} = x_1 T(\vec{e}_1) + x_2 T(\vec{e}_2) + \cdots + x_n T(\vec{e}_n),$$

where $\vec{e}_1, \vec{e}_2, \ldots, \vec{e}_n$ are the standard vectors in $\mathbb{R}^n$.

40. Describe all linear transformations from $\mathbb{R} \, (= \mathbb{R}^1)$ to $\mathbb{R}$. What do their graphs look like?

41. Describe all linear transformations from $\mathbb{R}^2$ to $\mathbb{R} \, (= \mathbb{R}^1)$. What do their graphs look like?

42. When you represent a three-dimensional object graphically in the plane (on paper, the blackboard, or a computer screen), you have to transform spatial coordinates,

$$\begin{bmatrix} x_1 \\ x_2 \\ x_3 \end{bmatrix},$$

into plane coordinates, $\begin{bmatrix} y_1 \\ y_2 \end{bmatrix}$. The simplest choice is a linear transformation, for example, the one given by the matrix

$$\begin{bmatrix} -\frac{1}{2} & 1 & 0 \\ -\frac{1}{2} & 0 & 1 \end{bmatrix}.$$

a. Use this transformation to represent the unit cube with corner points

$$\begin{bmatrix} 0 \\ 0 \\ 0 \end{bmatrix}, \begin{bmatrix} 1 \\ 0 \\ 0 \end{bmatrix}, \begin{bmatrix} 0 \\ 1 \\ 0 \end{bmatrix}, \begin{bmatrix} 0 \\ 0 \\ 1 \end{bmatrix},$$

$$\begin{bmatrix} 1 \\ 1 \\ 0 \end{bmatrix}, \begin{bmatrix} 0 \\ 1 \\ 1 \end{bmatrix}, \begin{bmatrix} 1 \\ 0 \\ 1 \end{bmatrix}, \begin{bmatrix} 1 \\ 1 \\ 1 \end{bmatrix}.$$

Include the images of the x_1, x_2, and x_3 axes in your sketch:

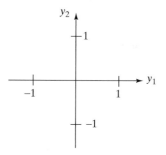

b. Which points

$$\begin{bmatrix} x_1 \\ x_2 \\ x_3 \end{bmatrix}$$

are transformed to $\begin{bmatrix} 0 \\ 0 \end{bmatrix}$? Explain.

43. a. Consider the vector $\vec{v} = \begin{bmatrix} 2 \\ 3 \\ 4 \end{bmatrix}$. Is the transformation $T(\vec{x}) = \vec{v} \cdot \vec{x}$ (the dot product) from $\mathbb{R}^3$ to $\mathbb{R}$ linear? If so, find the matrix of T.

b. Consider an arbitrary vector $\vec{v}$ in $\mathbb{R}^3$. Is the transformation $T(\vec{x}) = \vec{v} \cdot \vec{x}$ linear? If so, find the matrix of T (in terms of the components of $\vec{v}$).

c. Conversely, consider a linear transformation T from $\mathbb{R}^3$ to $\mathbb{R}$. Show that there is a vector $\vec{v}$ in $\mathbb{R}^3$ such that $T(\vec{x}) = \vec{v} \cdot \vec{x}$, for all $\vec{x}$ in $\mathbb{R}^3$.

44. The cross product of two vectors in $\mathbb{R}^3$ is defined by

$$\begin{bmatrix} a_1 \\ a_2 \\ a_3 \end{bmatrix} \times \begin{bmatrix} b_1 \\ b_2 \\ b_3 \end{bmatrix} = \begin{bmatrix} a_2 b_3 - a_3 b_2 \\ a_3 b_1 - a_1 b_3 \\ a_1 b_2 - a_2 b_1 \end{bmatrix}.$$

Consider an arbitrary vector $\vec{v}$ in $\mathbb{R}^3$. Is the transformation $T(\vec{x}) = \vec{v} \times \vec{x}$ from $\mathbb{R}^3$ to $\mathbb{R}^3$ linear? If so, find its matrix in terms of the components of the vector $\vec{v}$.

2.2 LINEAR TRANSFORMATIONS IN GEOMETRY

In the last section, we defined a linear transformation as a function $\vec{y} = T(\vec{x})$ from $\mathbb{R}^n$ to $\mathbb{R}^m$ such that

$$\vec{y} = A\vec{x},$$

for some $m \times n$ matrix A.

EXAMPLE 1   Consider a linear transformation $T(\vec{x}) = A\vec{x}$ from $\mathbb{R}^n$ to $\mathbb{R}^m$.

a. What is the relationship between $T(\vec{v})$, $T(\vec{w})$, and $T(\vec{v} + \vec{w})$, where $\vec{v}$ and $\vec{w}$ are vectors in $\mathbb{R}^n$?

b. What is the relationship between $T(\vec{v})$ and $T(k\vec{v})$, where $\vec{v}$ is a vector in $\mathbb{R}^n$ and k is a scalar?

Solution

a. Applying Fact 1.3.7, we find that

$$T(\vec{v} + \vec{w}) = A(\vec{v} + \vec{w}) = A\vec{v} + A\vec{w} = T(\vec{v}) + T(\vec{w}).$$

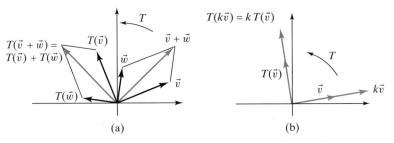

Figure 1 (a) Illustrating the property $T(\vec{v} + \vec{w}) = T(\vec{v}) + T(\vec{w})$.
(b) Illustrating the property $T(k\vec{v}) = kT(\vec{v})$.

In words, the transform of the sum of two vectors equals the sum of the transforms.

b. Again, apply Fact 1.3.7:

$$T(k\vec{v}) = A(k\vec{v}) = kA\vec{v} = kT(\vec{v}).$$

In words: The transform of a scalar multiple of a vector is the scalar multiple of the transform. ∎

Figure 1 illustrates these two properties in the case of the linear transformation T from $\mathbb{R}^2$ to $\mathbb{R}^2$ that rotates a vector through an angle of $90°$ in the counterclockwise direction. (Compare this with Example 2.1.5.)

In Example 1, we saw that a linear transformation satisfies the two equations $T(\vec{v} + \vec{w}) = T(\vec{v}) + T(\vec{w})$ and $T(k\vec{v}) = kT(\vec{v})$. Now we will show that the converse is true as well: Any transformation from $\mathbb{R}^n$ to $\mathbb{R}^m$ that satisfies these two equations is a linear transformation.

Fact 2.2.1

Linear Transformations
A transformation T from $\mathbb{R}^n$ to $\mathbb{R}^m$ is linear[2] if (and only if)

a. $T(\vec{v} + \vec{w}) = T(\vec{v}) + T(\vec{w})$, for all $\vec{v}$, $\vec{w}$ in $\mathbb{R}^n$, and
b. $T(k\vec{v}) = kT(\vec{v})$, for all $\vec{v}$ in $\mathbb{R}^n$ and all scalars k.

Proof   In Example 1, we saw that a linear transformation satisfies the equations in (a) and (b). To prove the converse, consider a transformation T from $\mathbb{R}^n$ to $\mathbb{R}^m$ that satisfies equations (a) and (b). We must show that there is a matrix A such that $T(\vec{x}) = A\vec{x}$, for all $\vec{x}$ in $\mathbb{R}^n$. Let $\vec{e}_1, \dots, \vec{e}_n$ be the standard vectors introduced in Fact 2.1.2.

---

[2]In many texts, a linear transformation is *defined* as a function from $\mathbb{R}^n$ to $\mathbb{R}^m$ with these two properties. The order of presentation does not really matter; think of a linear transformation both as a function T from $\mathbb{R}^n$ to $\mathbb{R}^m$ of the form $T(\vec{x}) = A\vec{x}$ and as a function with the properties (a) and (b) stated in Fact 2.2.1.

$$T(\vec{x}) = T \begin{bmatrix} x_1 \\ x_2 \\ \vdots \\ x_n \end{bmatrix} = T(x_1\vec{e}_1 + x_2\vec{e}_2 + \cdots + x_n\vec{e}_n)$$

$$= T(x_1\vec{e}_1) + T(x_2\vec{e}_2) + \cdots + T(x_n\vec{e}_n) \quad \text{(by property a)}$$
$$= x_1 T(\vec{e}_1) + x_2 T(\vec{e}_2) + \cdots + x_n T(\vec{e}_n) \quad \text{(by property b)}$$

$$= \begin{bmatrix} | & | & & | \\ T(\vec{e}_1) & T(\vec{e}_2) & \cdots & T(\vec{e}_n) \\ | & | & & | \end{bmatrix} \begin{bmatrix} x_1 \\ x_2 \\ \vdots \\ x_n \end{bmatrix} = A\vec{x} \quad \text{(by Definition 1.3.6).}$$

▲

Here is an example illustrating Fact 2.2.1.

EXAMPLE 2   Consider a linear transformation T from $\mathbb{R}^2$ to $\mathbb{R}^2$. The vectors $T(\vec{e}_1)$ and $T(\vec{e}_2)$ are sketched in Figure 2.[3] Sketch the image[4] of the unit square under this transformation.

Solution

The unit square consists of all vectors of the form $\vec{x} = x_1\vec{e}_1 + x_2\vec{e}_2$, where x_1 and x_2 are between 0 and 1. Therefore, the image of the unit square consists of all vectors of the form

$$T(\vec{x}) = T(x_1\vec{e}_1 + x_2\vec{e}_2) = x_1 T(\vec{e}_1) + x_2 T(\vec{e}_2),$$

where x_1 and x_2 are between 0 and 1. The last step follows from Fact 2.2.1.

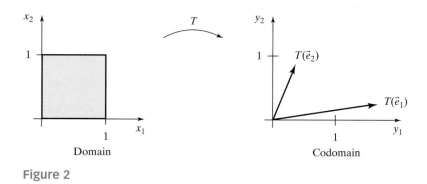

Figure 2

---

[3]Note that there are two slightly different ways to represent a linear transformation from $\mathbb{R}^2$ to $\mathbb{R}^2$ geometrically. Sometimes, we will draw two different planes to represent domain and codomain (as here), and sometimes we will draw the input $\vec{x}$ and the output $\vec{y} = T(\vec{x})$ in the same plane (as in Example 2.1.5). The first representation is less crowded, while the second one clarifies the geometric relationship between $\vec{x}$ and $T(\vec{x})$. (In Example 2.1.5, vectors $\vec{x}$ and $T(\vec{x})$ are perpendicular.)
[4]The *image* of a subset S of the domain consists of the vectors $T(\vec{x})$, for all $\vec{x}$ in S.

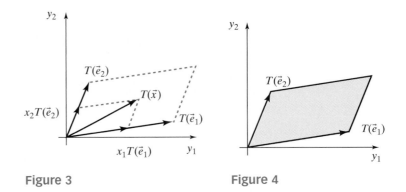

Figure 3   **Figure 4**

One such vector $T(\vec{x})$ is shown in Figure 3. The vector $T(\vec{x})$ is in the shaded parallelogram shown in Figure 4. Conversely, any vector $\vec{b}$ in the shaded parallelogram can be written as

$$\vec{b} = x_1 T(\vec{e}_1) + x_2 T(\vec{e}_2) = T(x_1\vec{e}_1 + x_2\vec{e}_2),$$

for two scalars x_1 and x_2 between 0 and 1. This shows that the *image of the unit square is the parallelogram defined by* $T(\vec{e}_1)$ *and* $T(\vec{e}_2)$. ∎

For generalizations of this example, see Exercises 35 and 36.

EXAMPLE 3   Consider a linear transformation T from $\mathbb{R}^2$ to $\mathbb{R}^2$ such that $T(\vec{v}_1) = \frac{1}{2}\vec{v}_1$ and $T(\vec{v}_2) = 2\vec{v}_2$, for the vectors $\vec{v}_1$ and $\vec{v}_2$ sketched in Figure 5. On the same axes, sketch $T(\vec{x})$, for the given vector $\vec{x}$. Explain your solution.

Solution

Using a parallelogram, we can represent $\vec{x}$ as a linear combination of $\vec{v}_1$ and $\vec{v}_2$, as shown in Figure 6:

$$\vec{x} = c_1\vec{v}_1 + c_2\vec{v}_2.$$

By Fact 2.2.1,

$$T(\vec{x}) = T(c_1\vec{v}_1 + c_2\vec{v}_2) = c_1 T(\vec{v}_1) + c_2 T(\vec{v}_2) = \frac{1}{2}c_1\vec{v}_1 + 2c_2\vec{v}_2.$$

The vector $c_1\vec{v}_1$ is cut in half, and the vector $c_2\vec{v}_2$ is doubled, as shown in Figure 7.

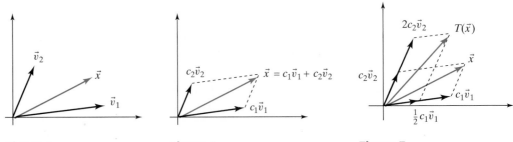

Figure 5   **Figure 6**   **Figure 7**

If you think of the domain $\mathbb{R}^2$ of the transformation T as a rubber sheet, you can imagine that T expands the sheet by a factor of 2 in the $\vec{v}_2$-direction and contracts it by a factor of $\frac{1}{2}$ in the $\vec{v}_1$-direction. ■

Next, we present some classes of linear transformations from $\mathbb{R}^2$ to $\mathbb{R}^2$ that are of interest in geometry.

Rotations

Consider the transformation T from $\mathbb{R}^2$ to $\mathbb{R}^2$ that rotates a vector through an angle α in the counterclockwise direction,[5] as shown in Figure 8. Recall Example 2.1.5, where we studied a rotation through $\alpha = \pi/2$.

EXAMPLE 4   Let T be the counterclockwise rotation through an angle α.

   a. Draw sketches to illustrate that T is a linear transformation.
   b. Find the matrix of T.

Solution

   a. See Figure 1 for the case $\alpha = \pi/2$. For an arbitrary α, we illustrate only the property $T(k\vec{v}) = kT(\vec{v})$, leaving the property $T(\vec{v} + \vec{w}) = T(\vec{v}) + T(\vec{w})$ as an exercise.

   Figure 9 shows that the vectors $T(k\vec{v})$ and $kT(\vec{v})$ are equal; convince yourself that the two vectors have the same length and the same argument.
   b. The matrix of the transformation T is

$$A = \begin{bmatrix} | & | \\ T(\vec{e}_1) & T(\vec{e}_2) \\ | & | \end{bmatrix} = \begin{bmatrix} T\begin{bmatrix} 1 \\ 0 \end{bmatrix} & T\begin{bmatrix} 0 \\ 1 \end{bmatrix} \end{bmatrix},$$

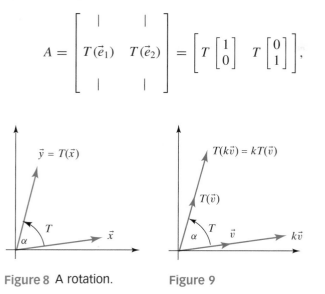

Figure 8 A rotation.   Figure 9

---

[5] It is easiest to define a rotation in terms of polar coordinates. The length of $T(\vec{x})$ equals the length of $\vec{x}$, and the argument (or polar angle) of $T(\vec{x})$ exceeds the argument of $\vec{x}$ by α.

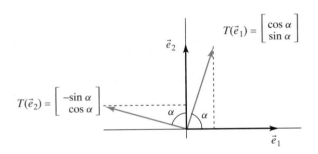

$$T(\vec{e}_2) = \begin{bmatrix} -\sin\alpha \\ \cos\alpha \end{bmatrix}$$

Figure 10

by Fact 2.1.2. To find $T(\vec{e}_1)$ and $T(\vec{e}_2)$, apply basic trigonometry, as shown in Figure 10. Note that the vectors $T(\vec{e}_1)$ and $T(\vec{e}_2)$ both have length 1. ∎

Fact 2.2.2　**Rotations**

The matrix of a counterclockwise *rotation* through an angle α is

$$\begin{bmatrix} \cos\alpha & -\sin\alpha \\ \sin\alpha & \cos\alpha \end{bmatrix}.$$

This formula is important in many applications, from physics to computer graphics; it is worth memorizing.

Here are two special cases:

- The matrix of a counterclockwise rotation through an angle of $\frac{\pi}{2}$ is

$$\begin{bmatrix} \cos\dfrac{\pi}{2} & -\sin\dfrac{\pi}{2} \\ \sin\dfrac{\pi}{2} & \cos\dfrac{\pi}{2} \end{bmatrix} = \begin{bmatrix} 0 & -1 \\ 1 & 0 \end{bmatrix},$$

as in Example 2.1.5.

- The matrix of a counterclockwise rotation through an angle of $\pi/6$ (or 30°) is

$$\begin{bmatrix} \cos\dfrac{\pi}{6} & -\sin\dfrac{\pi}{6} \\ \sin\dfrac{\pi}{6} & \cos\dfrac{\pi}{6} \end{bmatrix} = \begin{bmatrix} \dfrac{\sqrt{3}}{2} & -\dfrac{1}{2} \\ \dfrac{1}{2} & \dfrac{\sqrt{3}}{2} \end{bmatrix}.$$

Rotation–Dilations

EXAMPLE 5　Give a geometric interpretation of the linear transformation

$$T(\vec{x}) = \begin{bmatrix} a & -b \\ b & a \end{bmatrix} \vec{x},$$

where a and b are arbitrary constants.

Solution

Consider the values $T(\vec{e}_1) = \begin{bmatrix} a \\ b \end{bmatrix}$ and $T(\vec{e}_2) = \begin{bmatrix} -b \\ a \end{bmatrix}$ (i.e., the columns of the matrix). Note that $T(\vec{e}_1)$ and $T(\vec{e}_2)$ are perpendicular vectors of the same length, as shown in Figure 11.

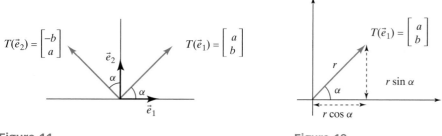

Figure 11   **Figure 12**

We can see that the vectors $T(\vec{e}_1)$ and $T(\vec{e}_2)$ are obtained from $\vec{e}_1$ and $\vec{e}_2$ by first performing a counterclockwise *rotation* through the angle α, where $\tan\alpha = b/a$, and then a *dilation* by the factor $r = \sqrt{a^2 + b^2}$, the length of the vectors $T(\vec{e}_1)$ and $T(\vec{e}_2)$.

To verify that the transformation acts in the same way on all vectors (not just $\vec{e}_1$ and $\vec{e}_2$), write the vector $T(\vec{e}_1) = \begin{bmatrix} a \\ b \end{bmatrix}$ in polar coordinates:

$$\begin{bmatrix} a \\ b \end{bmatrix} = \begin{bmatrix} r\,\cos\alpha \\ r\,\sin\alpha \end{bmatrix}.$$

This is illustrated in Figure 12.

Then,

$$\begin{bmatrix} a & -b \\ b & a \end{bmatrix} = \begin{bmatrix} r\,\cos\alpha & -r\,\sin\alpha \\ r\,\sin\alpha & r\,\cos\alpha \end{bmatrix} = r\begin{bmatrix} \cos\alpha & -\sin\alpha \\ \sin\alpha & \cos\alpha \end{bmatrix}.$$

The matrix $\begin{bmatrix} a & -b \\ b & a \end{bmatrix}$ is a scalar multiple of a rotation matrix. (See Definition 1.3.9 and Fact 2.2.2.)

Therefore,

$$T(\vec{x}) = \begin{bmatrix} a & -b \\ b & a \end{bmatrix}\vec{x} = r\begin{bmatrix} \cos\alpha & -\sin\alpha \\ \sin\alpha & \cos\alpha \end{bmatrix}\vec{x}.$$

A vector $\vec{x}$ is first rotated through an angle α in the counterclockwise direction; the resulting vector is then multiplied by r, which represents a dilation, as shown in Figure 13.

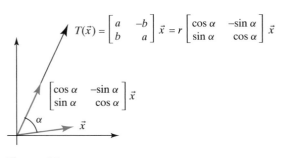

$$T(\vec{x}) = \begin{bmatrix} a & -b \\ b & a \end{bmatrix} \vec{x} = r \begin{bmatrix} \cos\alpha & -\sin\alpha \\ \sin\alpha & \cos\alpha \end{bmatrix} \vec{x}$$

Figure 13   ■

Fact 2.2.3   **Rotation–dilations**

To interpret the linear transformation

$$T(\vec{x}) = \begin{bmatrix} a & -b \\ b & a \end{bmatrix} \vec{x}$$

geometrically, write the vector $\begin{bmatrix} a \\ b \end{bmatrix}$ in polar coordinates: $\begin{bmatrix} a \\ b \end{bmatrix} = \begin{bmatrix} r\cos(\alpha) \\ r\sin(\alpha) \end{bmatrix}$,

where $r = \sqrt{a^2 + b^2}$ and $\tan(\alpha) = \frac{b}{a}$. Then, T is a counterclockwise *rotation* through the angle α followed by a *dilation* by the factor r. We call T a *rotation–dilation*.

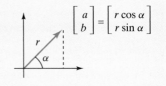

Shears

EXAMPLE 6   Consider the linear transformation

$$\vec{y} = \begin{bmatrix} 1 & \frac{1}{2} \\ 0 & 1 \end{bmatrix} \vec{x}.$$

To understand this transformation geometrically, sketch the image of the unit square.

Solution

By Example 2, the image of the unit square is the parallelogram defined by $T(\vec{e}_1)$ and $T(\vec{e}_2)$, as illustrated in Figure 14. (Compare this with Figure 4.) Note that

$$T(\vec{e}_1) = \begin{bmatrix} 1 & \frac{1}{2} \\ 0 & 1 \end{bmatrix} \begin{bmatrix} 1 \\ 0 \end{bmatrix} = \begin{bmatrix} 1 \\ 0 \end{bmatrix}$$

and

$$T(\vec{e}_2) = \begin{bmatrix} 1 & \frac{1}{2} \\ 0 & 1 \end{bmatrix} \begin{bmatrix} 0 \\ 1 \end{bmatrix} = \begin{bmatrix} \frac{1}{2} \\ 1 \end{bmatrix}.$$

                                                                                 ■

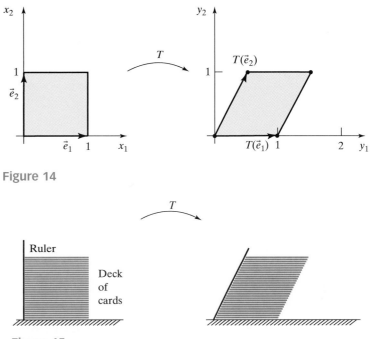

Figure 14

Figure 15

Here is a more tangible description of this transformation. Consider a deck of cards, viewed from the side, as shown in Figure 15. Align a ruler vertically against the deck, as shown. Push the top of the ruler to the right, while holding its bottom in place. (The bottom of the ruler represents the origin.) The higher up a card is, the further it gets pushed to the right, with its elevation unchanged.[6] Compare Figures 14 and 15.

The transformation $T(\vec{x}) = \begin{bmatrix} 1 & \frac{1}{2} \\ 0 & 1 \end{bmatrix} \vec{x}$ is called a *shear* parallel to the x_1-axis.

More generally, we have the following definition:

Definition 2.2.4

> **Shears**
> Let L be a line[7] in $\mathbb{R}^2$. A linear transformation T from $\mathbb{R}^2$ to $\mathbb{R}^2$ is called a *shear parallel to L* if
>
> a. $T(\vec{v}) = \vec{v}$, for all vectors $\vec{v}$ on L, and
> b. $T(\vec{x}) - \vec{x}$ is parallel to L for all vectors $\vec{x}$ in $\mathbb{R}^2$.

Property (a) means that the transformation leaves the line L unchanged, and property (b) says that the tip of any vector $\vec{x}$ is moved parallel to the line L, as illustrated in Figure 16.

---

[6]Two hints for the instructor:
- Use several decks of cards for dramatic effect
- Hold the decks in place with rubber bands to avoid embarrassing and time-consuming accidents.

[7]*Convention:* All lines considered in this text run through the origin unless stated otherwise.

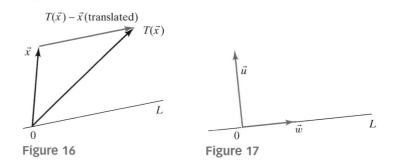

Figure 16   Figure 17

Let us check properties (a) and (b) for the linear transformation

$$T(\vec{x}) = \begin{bmatrix} 1 & \frac{1}{2} \\ 0 & 1 \end{bmatrix} \vec{x}$$

discussed in Example 6, where L is the x_1-axis.

a. $T\begin{bmatrix} x_1 \\ 0 \end{bmatrix} = \begin{bmatrix} 1 & \frac{1}{2} \\ 0 & 1 \end{bmatrix} \begin{bmatrix} x_1 \\ 0 \end{bmatrix} = \begin{bmatrix} x_1 \\ 0 \end{bmatrix}$. The line L remains unchanged.

b. $T\begin{bmatrix} x_1 \\ x_2 \end{bmatrix} - \begin{bmatrix} x_1 \\ x_2 \end{bmatrix} = \begin{bmatrix} 1 & \frac{1}{2} \\ 0 & 1 \end{bmatrix} \begin{bmatrix} x_1 \\ x_2 \end{bmatrix} - \begin{bmatrix} x_1 \\ x_2 \end{bmatrix} = \begin{bmatrix} x_1 + \frac{1}{2}x_2 \\ x_2 \end{bmatrix} - \begin{bmatrix} x_1 \\ x_2 \end{bmatrix} = \begin{bmatrix} \frac{1}{2}x_2 \\ 0 \end{bmatrix}$.

The vector $T(\vec{x}) - \vec{x}$ is parallel to L.

EXAMPLE 7   Consider two (nonzero) perpendicular vectors $\vec{u}$ and $\vec{w}$ in $\mathbb{R}^2$. Show that the transformation

$$T(\vec{x}) = \vec{x} + (\vec{u} \cdot \vec{x})\vec{w}$$

is a shear parallel to the line L spanned by $\vec{w}$. (See Figure 17.)

Solution

We leave it to you to verify that T is a linear transformation. We will check properties a and b of a shear (Definition 2.2.4).

a. If $\vec{v}$ is a vector on L, then

$$T(\vec{v}) = \vec{v} + (\vec{u} \cdot \vec{v})\vec{w} = \vec{v},$$

because $\vec{u} \cdot \vec{v} = 0$ (since the vectors $\vec{u}$ and $\vec{v}$ are perpendicular).

b. For any vector $\vec{x}$ in $\mathbb{R}^2$ the vector

$$T(\vec{x}) - \vec{x} = \vec{x} + (\vec{u} \cdot \vec{x})\vec{w} - \vec{x} = (\vec{u} \cdot \vec{x})\vec{w}$$

is parallel to L, because this vector is a scalar multiple of $\vec{w}$.   ∎

The Scottish scholar d'Arcy Thompson showed how the shapes of related species of plants and animals can often be transformed into one another, using linear as well as nonlinear transformations.[8] In the figure below he uses a shear to transform the shape of one kind of fish into another.

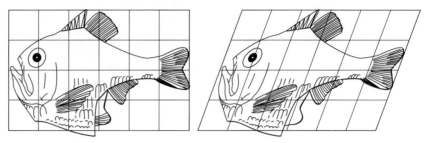

Argyropelecus olfersi.   *Sternoptyx diaphana.*

Projections and Reflections

Consider a line L in $\mathbb{R}^2$. For any vector $\vec{v}$ in $\mathbb{R}^2$, there is a unique vector $\vec{w}$ on L such that $\vec{v} - \vec{w}$ is perpendicular to L, as shown in Figure 18.

This vector $\vec{w}$ is called the *orthogonal projection of $\vec{v}$ onto L*, denoted by $\operatorname{proj}_L \vec{v}$. ("Orthogonal" means "perpendicular.") Intuitively, you can think of $\operatorname{proj}_L \vec{v}$ as the shadow $\vec{v}$ casts on L if we shine a light straight down on L.

How can we generalize the idea of an orthogonal projection to lines in $\mathbb{R}^n$? Let L be a line in $\mathbb{R}^n$ (i.e., the set of all scalar multiples of some nonzero vector $\vec{u}$, which we may choose as a unit vector). For a given vector $\vec{v}$ in $\mathbb{R}^n$, is it possible to find a unique vector $\vec{w}$ on L such that $\vec{v} - \vec{w}$ is perpendicular to L (i.e., perpendicular to $\vec{u}$)? If such a vector $\vec{w}$ exists, it is a scalar multiple $\vec{w} = k\vec{u}$ of $\vec{u}$. See Figure 19. We have to choose k so that $\vec{v} - \vec{w}$ is perpendicular to $\vec{u}$; that is,

$$\vec{u} \cdot (\vec{v} - \vec{w}) = \vec{u} \cdot (\vec{v} - k\vec{u}) = \vec{u} \cdot \vec{v} - k(\vec{u} \cdot \vec{u}) = \vec{u} \cdot \vec{v} - k = 0.$$

We have used the fact that $\vec{u} \cdot \vec{u} = 1$ (since $\vec{u}$ is a unit vector). Therefore, $\vec{v} - k\vec{u}$ is perpendicular to $\vec{u}$ if (and only if) $k = \vec{u} \cdot \vec{v}$.

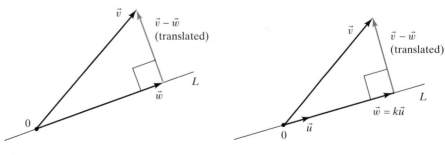

Figure 18 Projecting onto a line.   **Figure 19**

---

[8]Thompson, d'Arcy W., *On Growth and Form* (Cambridge University Press, 1917). P. B. Medawar called this the "finest work of literature in all the annals of science that have been recorded in the English tongue."

Let's summarize.

Fact 2.2.5 **Orthogonal Projections**
- Let L be a line in $\mathbb{R}^n$ consisting of all scalar multiples of some unit vector $\vec{u}$. For any vector $\vec{v}$ in $\mathbb{R}^n$ there is a unique vector $\vec{w}$ on L such that $\vec{v} - \vec{w}$ is perpendicular to L, namely, $\vec{w} = (\vec{u} \cdot \vec{v})\vec{u}$. This vector $\vec{w}$ is called the *orthogonal projection of $\vec{v}$ onto L:*

$$\text{proj}_L \vec{v} = (\vec{u} \cdot \vec{v})\vec{u}.$$

- The transformation $T(\vec{v}) = \text{proj}_L(\vec{v})$ from $\mathbb{R}^n$ to $\mathbb{R}^n$ is linear.

The verification of linearity is straightforward; use Fact 2.2.1. We will check property (b), and leave property (a) as an exercise:

$$\text{proj}_L(k\vec{v}) = (\vec{u} \cdot k\vec{v})\vec{u} = k(\vec{u} \cdot \vec{v})\vec{u} = k\,\text{proj}_L(\vec{v}).$$

Consider again a line L and a vector $\vec{v}$ in $\mathbb{R}^2$. The reflection $\text{ref}_L \vec{v}$ of $\vec{v}$ in the line L is shown in Figure 20. (We are **flipping** vector $\vec{v}$ over the line L.)

Figure 21 illustrates the relationship between projection and reflection:

$$2\,\text{proj}_L \vec{v} = \vec{v} + \text{ref}_L \vec{v}.$$

Now solve for $\text{ref}_L \vec{v}$:

$$\text{ref}_L \vec{v} = 2(\text{proj}_L \vec{v}) - \vec{v} = 2(\vec{u} \cdot \vec{v})\vec{u} - \vec{v},$$

where $\vec{u}$ is a unit vector on L.

We can use the formula $2(\text{proj}_L \vec{v}) - \vec{v}$ to define the reflection in a line in $\mathbb{R}^n$.

Definition 2.2.6 **Reflections**
Let L be a line in $\mathbb{R}^n$. For a vector $\vec{v}$ in $\mathbb{R}^n$, the vector $2(\text{proj}_L \vec{v}) - \vec{v}$ is called the *reflection of $\vec{v}$ in L:*

$$\text{ref}_L \vec{v} = 2(\text{proj}_L \vec{v}) - \vec{v} = 2(\vec{u} \cdot \vec{v})\vec{u} - \vec{v},$$

where $\vec{u}$ is a unit vector on L.

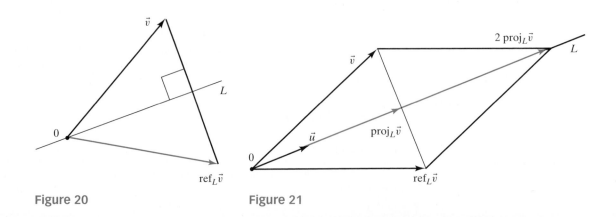

Figure 20   Figure 21

If $\vec{v}$ is perpendicular to L, then $\vec{u} \cdot \vec{v} = 0$, so that

$$\text{ref}_L \vec{v} = 2(\vec{u} \cdot \vec{v})\vec{u} - \vec{v} = \vec{0} - \vec{v} = -\vec{v}.$$

We leave it as an exercise for you to verify that the reflection in a line in $\mathbb{R}^n$ is a linear transformation from $\mathbb{R}^n$ to $\mathbb{R}^n$.

EXERCISES

GOALS Check whether a transformation is linear. Use the matrices of rotations and rotation–dilations. Apply the definitions of shears, projections, and reflections.

1. Sketch the image of the unit square under the linear transformation

$$T(\vec{x}) = \begin{bmatrix} 3 & 1 \\ 1 & 2 \end{bmatrix} \vec{x}.$$

2. Find the matrix of a rotation through an angle of $60°$ in the counterclockwise direction.

3. Consider a linear transformation T from $\mathbb{R}^2$ to $\mathbb{R}^3$. Use $T(\vec{e}_1)$ and $T(\vec{e}_2)$ to describe the image of the unit square geometrically.

4. Interpret the following linear transformation geometrically:

$$T(\vec{x}) = \begin{bmatrix} 1 & -1 \\ 1 & 1 \end{bmatrix} \vec{x}.$$

5. The matrix

$$\begin{bmatrix} -0.8 & -0.6 \\ 0.6 & -0.8 \end{bmatrix}$$

represents a rotation. Find the angle of rotation (in radians).

6. Let L be the line in $\mathbb{R}^3$ that consists of all scalar multiples of the vector $\begin{bmatrix} 2 \\ 1 \\ 2 \end{bmatrix}$. Find the orthogonal projection of the vector $\begin{bmatrix} 1 \\ 1 \\ 1 \end{bmatrix}$ onto L.

7. Let L be the line in $\mathbb{R}^3$ that consists of all scalar multiples of $\begin{bmatrix} 2 \\ 1 \\ 2 \end{bmatrix}$. Find the reflection of the vector $\begin{bmatrix} 1 \\ 1 \\ 1 \end{bmatrix}$ in the line L.

8. Interpret the following linear transformation geometrically:

$$T(\vec{x}) = \begin{bmatrix} 1 & -1 \\ 0 & 1 \end{bmatrix} \vec{x}.$$

9. Interpret the following linear transformation geometrically:

$$T(\vec{x}) = \begin{bmatrix} 1 & 0 \\ 1 & 1 \end{bmatrix} \vec{x}.$$

10. Find the matrix of the orthogonal projection onto the line L in $\mathbb{R}^2$ shown in the following figure:

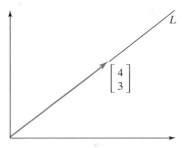

11. Refer to Exercise 10. Find the matrix of the reflection in the line L.

12. Suppose a line L in $\mathbb{R}^2$ contains the unit vector

$$\vec{u} = \begin{bmatrix} u_1 \\ u_2 \end{bmatrix}.$$

Find the matrix A of the linear transformation $T(\vec{x}) = \text{proj}_L \vec{x}$. Give the entries of A in terms of u_1 and u_2.

13. Suppose a line L in $\mathbb{R}^2$ contains the unit vector

$$\vec{u} = \begin{bmatrix} u_1 \\ u_2 \end{bmatrix}.$$

Find the matrix A of the linear transformation $T(\vec{x}) = \text{ref}_L \vec{x}$. Give the entries of A in terms of u_1 and u_2.

14. Suppose a line L in $\mathbb{R}^n$ contains the unit vector

$$\vec{u} = \begin{bmatrix} u_1 \\ u_2 \\ \vdots \\ u_n \end{bmatrix}.$$

a. Find the matrix A of the linear transformation $T(\vec{x}) = \text{proj}_L \vec{x}$. Give the entries of A in terms of the components u_i of $\vec{u}$.

b. What is the sum of the diagonal entries of the matrix A you found in part (a)?

15. Suppose a line L in $\mathbb{R}^n$ contains the unit vector

$$\vec{u} = \begin{bmatrix} u_1 \\ u_2 \\ \vdots \\ u_n \end{bmatrix}.$$

Find the matrix A of the linear transformation $T(\vec{x}) = \text{ref}_L \vec{x}$. Give the entries of A in terms of the components u_i of $\vec{u}$.

16. Let $T(\vec{x}) = \text{ref}_L \vec{x}$ be the reflection in the line L in $\mathbb{R}^2$ shown in the accompanying figure.

a. Draw sketches to illustrate that T is linear.

b. Find the matrix of T in terms of α.

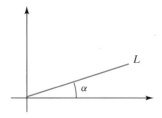

17. Consider the linear transformation

$$T(\vec{x}) = \begin{bmatrix} 1 & 1 \\ 1 & 1 \end{bmatrix} \vec{x}.$$

a. Sketch the image of the unit square under this transformation.

b. Explain how T can be interpreted as a projection followed by a dilation.

18. Interpret the transformation

$$T(\vec{x}) = \begin{bmatrix} 3 & 4 \\ 4 & -3 \end{bmatrix} \vec{x}$$

as a reflection followed by a dilation.

Find the matrices of the linear transformations from $\mathbb{R}^3$ to $\mathbb{R}^3$ given in Exercises 19 through 23. Some of these transformations have not been formally defined in the text. Use common sense. You may assume that all these transformations are linear.

19. The orthogonal projection onto the x–y-plane.

20. The reflection in the x–z-plane.

21. The rotation about the z-axis through an angle of $\pi/2$, counterclockwise as viewed from the positive z-axis.

22. The rotation about the y-axis through an angle α, counterclockwise as viewed from the positive y-axis.

23. The reflection in the plane $y = z$.

24. Consider the linear transformation

$$T(\vec{x}) = \begin{bmatrix} 1 & 0 \\ 2 & 1 \end{bmatrix} \vec{x}.$$

a. Sketch the image of the unit square under this transformation.

b. Show that T is a shear.

c. Find the inverse transformation and describe it geometrically.

25. Find the inverse of the matrix $\begin{bmatrix} 1 & k \\ 0 & 1 \end{bmatrix}$, where k is an arbitrary constant. Interpret your result geometrically.

26. Let

$$T(\vec{x}) = \begin{bmatrix} -1 & 4 \\ -1 & 3 \end{bmatrix} \vec{x}$$

be a linear transformation. Is T a shear? Explain.

27. Consider two linear transformations $\vec{y} = T(\vec{x})$ and $\vec{z} = L(\vec{y})$, where T goes from $\mathbb{R}^n$ to $\mathbb{R}^m$ and L goes from $\mathbb{R}^m$ to $\mathbb{R}^p$. Is the transformation $\vec{z} = L(T(\vec{x}))$ linear as well? (The transformation $\vec{z} = L(T(\vec{x}))$ is called the *composite* of T and L.)

28. Let

$$A = \begin{bmatrix} a & b \\ c & d \end{bmatrix} \quad \text{and} \quad B = \begin{bmatrix} p & q \\ r & s \end{bmatrix}.$$

Find the matrix of the linear transformation $T(\vec{x}) = B(A\vec{x})$. (See Exercise 27.) *Hint:* Find $T(\vec{e}_1)$ and $T(\vec{e}_2)$.

29. Let T and L be transformations from $\mathbb{R}^n$ to $\mathbb{R}^n$. Suppose L is the inverse of T; that is,

$$T(L(\vec{x})) = \vec{x} \quad \text{and} \quad L(T(\vec{x})) = \vec{x},$$

for all $\vec{x}$ in $\mathbb{R}^n$. If T is a linear transformation, is L linear as well? *Hint:* $\vec{x} + \vec{y} = T(L(\vec{x})) + T(L(\vec{y})) = T(L(\vec{x}) + L(\vec{y}))$, because T is linear. Now apply L on both sides.

30. Find a nonzero 2×2 matrix A such that $A\vec{x}$ is parallel to the vector $\begin{bmatrix} 1 \\ 2 \end{bmatrix}$, for all $\vec{x}$ in $\mathbb{R}^2$.

31. Find a nonzero 3×3 matrix A such that $A\vec{x}$ is perpendicular to $\begin{bmatrix} 1 \\ 2 \\ 3 \end{bmatrix}$, for all $\vec{x}$ in $\mathbb{R}^3$.

32. Consider the rotation matrix $D = \begin{bmatrix} \cos\alpha & -\sin\alpha \\ \sin\alpha & \cos\alpha \end{bmatrix}$ and the vector $\vec{v} = \begin{bmatrix} \cos\beta \\ \sin\beta \end{bmatrix}$, where α and β are arbitrary angles.

a. Draw a sketch to explain why $D\vec{v} = \begin{bmatrix} \cos(\alpha + \beta) \\ \sin(\alpha + \beta) \end{bmatrix}$.

b. Compute $D\vec{v}$. Use the result to derive the addition theorems for sine and cosine:

$$\cos(\alpha + \beta) = \dots, \qquad \sin(\alpha + \beta) = \dots.$$

33. Consider two nonparallel lines L_1 and L_2 in $\mathbb{R}^2$. Explain why a vector $\vec{v}$ in $\mathbb{R}^2$ can be expressed uniquely as

$$\vec{v} = \vec{v}_1 + \vec{v}_2,$$

where $\vec{v}_1$ is on L_1 and $\vec{v}_2$ on L_2. Draw a sketch. The transformation $T(\vec{v}) = \vec{v}_1$ is called the *projection onto L_1 along L_2*. Show algebraically that T is linear.

34. One of the five given matrices represents an orthogonal projection onto a line and another represents a reflection in a line. Identify both and briefly justify your choice.

$$A = \frac{1}{3}\begin{bmatrix} 1 & 2 & 2 \\ 2 & 1 & 2 \\ 2 & 2 & 1 \end{bmatrix}, \qquad B = \frac{1}{3}\begin{bmatrix} 1 & 1 & 1 \\ 1 & 1 & 1 \\ 1 & 1 & 1 \end{bmatrix},$$

$$C = \frac{1}{3}\begin{bmatrix} 2 & 1 & 1 \\ 1 & 2 & 1 \\ 1 & 1 & 2 \end{bmatrix}, \qquad D = -\frac{1}{3}\begin{bmatrix} 1 & 2 & 2 \\ 2 & 1 & 2 \\ 2 & 2 & 1 \end{bmatrix},$$

$$E = \frac{1}{3}\begin{bmatrix} -1 & 2 & 2 \\ 2 & -1 & 2 \\ 2 & 2 & -1 \end{bmatrix}.$$

35. Let T be an invertible linear transformation from $\mathbb{R}^2$ to $\mathbb{R}^2$. Let P be a parallelogram in $\mathbb{R}^2$ with one vertex at the origin. Is the image of P a parallelogram as well? Explain. Draw a sketch of the image.

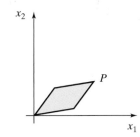

36. Let T be an invertible linear transformation from $\mathbb{R}^2$ to $\mathbb{R}^2$. Let P be a parallelogram in $\mathbb{R}^2$. Is the image of P a parallelogram as well? Explain.

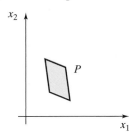

37. Let T be a linear transformation from $\mathbb{R}^2$ to $\mathbb{R}^2$. Three vectors $\vec{v}_1, \vec{v}_2, \vec{w}$ in $\mathbb{R}^2$ and the vectors $T(\vec{v}_1), T(\vec{v}_2)$ are shown in the accompanying figure. Sketch $T(\vec{w})$. Explain your answer.

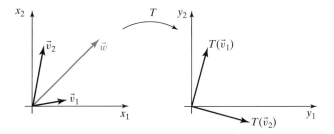

38. Let T be a linear transformation from $\mathbb{R}^2$ to $\mathbb{R}^2$. Let $\vec{v}_1, \vec{v}_2, \vec{w}$ be three vectors in $\mathbb{R}^2$, as shown below. We are told that $T(\vec{v}_1) = \vec{v}_1$ and $T(\vec{v}_2) = 3\vec{v}_2$. On the same axes, sketch $T(\vec{w})$.

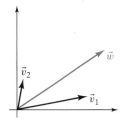

39. Let

$$\vec{v} = \begin{bmatrix} a \\ b \end{bmatrix} \quad \text{and} \quad \vec{w} = \begin{bmatrix} -b \\ a \end{bmatrix}$$

be nonzero perpendicular vectors. Find the matrix A of the shear

$$T(\vec{x}) = \vec{x} + (\vec{v} \cdot \vec{x})\vec{w}.$$

(See Example 7.) Give the entries of A in terms of a and b.

40. Let P and Q be two perpendicular lines in $\mathbb{R}^2$. For a vector $\vec{x}$ in $\mathbb{R}^2$, what is $\text{proj}_P\vec{x} + \text{proj}_Q\vec{x}$? Give your answer in terms of $\vec{x}$. Draw a sketch to justify your answer.

41. Let P and Q be two perpendicular lines in $\mathbb{R}^2$. For a vector $\vec{x}$ in $\mathbb{R}^2$, what is the relationship between $\text{ref}_P\vec{x}$ and $\text{ref}_Q\vec{x}$? Draw a sketch to justify your answer.

42. Let $T(\vec{x}) = \text{proj}_L\vec{x}$ be the projection onto a line in $\mathbb{R}^n$. What is the relationship between $T(\vec{x})$ and $T(T(\vec{x}))$? Justify your answer carefully.

43. Find the inverse of the rotation matrix

$$A = \begin{bmatrix} \cos\alpha & -\sin\alpha \\ \sin\alpha & \cos\alpha \end{bmatrix}.$$

Interpret the linear transformation defined by A^{-1} geometrically. Explain.

44. Find the inverse of the rotation–dilation matrix

$$A = \begin{bmatrix} a & -b \\ b & a \end{bmatrix}.$$

(Assume that A is not the zero matrix.) Interpret the linear transformation defined by A^{-1} geometrically. Explain.

45. Consider two linear transformations T and L from $\mathbb{R}^2$ to $\mathbb{R}^2$. We are told that $T(\vec{v}_1) = L(\vec{v}_1)$ and $T(\vec{v}_2) = L(\vec{v}_2)$ for the vectors $\vec{v}_1$ and $\vec{v}_2$ sketched below. Show that $T(\vec{x}) = L(\vec{x})$, for all vectors $\vec{x}$ in $\mathbb{R}^2$.

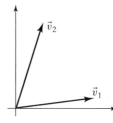

46. Consider a shear T from $\mathbb{R}^2$ to $\mathbb{R}^2$. Show that there are perpendicular vectors $\vec{v}$ and $\vec{w}$ in $\mathbb{R}^2$ such that $T(\vec{x}) = \vec{x} + (\vec{v} \cdot \vec{x})\vec{w}$, for all $\vec{x}$ in $\mathbb{R}^2$.

47. Let T be a linear transformation from $\mathbb{R}^2$ to $\mathbb{R}^2$. Consider the function

$$f(t) = \left(T \begin{bmatrix} \cos(t) \\ \sin(t) \end{bmatrix}\right) \cdot \left(T \begin{bmatrix} -\sin(t) \\ \cos(t) \end{bmatrix}\right),$$

from $\mathbb{R}$ to $\mathbb{R}$. Show the following:

a. The function $f(t)$ is continuous. You may take for granted that the functions $\sin(t)$ and $\cos(t)$ are continuous, and also that sums and products of continuous functions are continuous.

b. $f(\pi/2) = -f(0)$.

c. There is a number c between 0 and $\pi/2$ such that $f(c) = 0$. Use the intermediate value theorem of calculus, which tells us the following: If a function $g(t)$ is continuous for $a \leq t \leq b$, and L is a number between $g(a)$ and $g(b)$, then there is at least one number c between a and b such that $g(c) = L$.

d. There are two perpendicular unit vectors $\vec{v}_1$ and $\vec{v}_2$ in $\mathbb{R}^2$ such that the vectors $T(\vec{v}_1)$ and $T(\vec{v}_2)$ are perpendicular as well. See the accompanying figure. (Compare with Fact 8.3.3 for a generalization.)

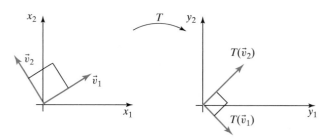

48. Refer to Exercise 47. Consider the linear transformation

$$T(\vec{x}) = \begin{bmatrix} 0 & 4 \\ 5 & -3 \end{bmatrix} \vec{x}.$$

Find the function $f(t)$ defined in Exercise 47, graph it (using technology), and find a number c between 0 and $\pi/2$ such that $f(c) = 0$. Use your answer to find two perpendicular unit vectors $\vec{v}_1$ and $\vec{v}_2$ such that $T(\vec{v}_1)$ and $T(\vec{v}_2)$ are perpendicular. Draw a sketch.

49. Sketch the image of the unit circle under the linear transformation

$$T(\vec{x}) = \begin{bmatrix} 5 & 0 \\ 0 & 2 \end{bmatrix} \vec{x}.$$

50. Let T be an invertible linear transformation from $\mathbb{R}^2$ to $\mathbb{R}^2$. Show that the image of the unit circle is an ellipse centered at the origin.[9] *Hint:* Consider two perpendicular unit vectors $\vec{v}_1$ and $\vec{v}_2$ such that $T(\vec{v}_1)$ and $T(\vec{v}_2)$ are perpendicular. (See Exercise 47(d).) The unit circle consists of all vectors of the form

$$\vec{v} = \cos(t)\vec{v}_1 + \sin(t)\vec{v}_2,$$

where t is a parameter.

---

[9]An ellipse in $\mathbb{R}^2$ centered at the origin may be defined as a curve that can be parametrized as

$$\cos(t)\vec{w}_1 + \sin(t)\vec{w}_2,$$

for two perpendicular vectors $\vec{w}_1$ and $\vec{w}_2$. Suppose the length of $\vec{w}_1$ exceeds the length of $\vec{w}_2$. Then, we call the vectors $\pm\vec{w}_1$ the semimajor axes of the ellipse and $\pm\vec{w}_2$ the semiminor axes.

Convention: All ellipses considered in this text are centered at the origin unless stated otherwise.

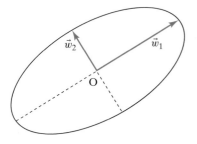

51. Let $\vec{w}_1$ and $\vec{w}_2$ be two nonparallel vectors in $\mathbb{R}^2$. Consider the curve C in $\mathbb{R}^2$ that consists of all vectors of the form $\cos(t)\vec{w}_1 + \sin(t)\vec{w}_2$, where t is a parameter. Show that C is an ellipse. (*Hint:* You can interpret C as the image of the unit circle under a suitable linear transformation; then use Exercise 50.)

52. Consider an invertible linear transformation T from $\mathbb{R}^2$ to $\mathbb{R}^2$. Let C be an ellipse in $\mathbb{R}^2$. Show that the image of C under T is an ellipse as well. (*Hint:* Use the result of Exercise 51.)

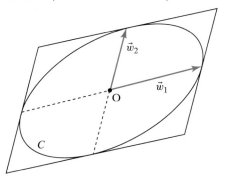

2.3 THE INVERSE OF A LINEAR TRANSFORMATION

Let's first review the concept of an invertible function.

Definition 2.3.1

Invertible functions
A function T from X to Y is called *invertible* if the equation $T(x) = y$ has a unique solution x in X for each y in Y.

Consider the examples in Figure 1, where X and Y are finite sets.

If a function T from X to Y is invertible, then its inverse T^{-1} from Y to X is defined by

$$T^{-1}(y) = \left(\text{the unique } x \text{ in } X \text{ such that } T(x) = y\right).$$

See Figure 2.
Note that

$$T^{-1}(T(x)) = x, \qquad \text{for all } x \text{ in } X,$$
and
$$T(T^{-1}(y)) = y, \qquad \text{for all } y \text{ in } Y.$$

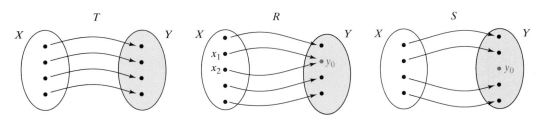

Figure 1 T is invertible. R is not invertible: The equation $R(x) = y_0$ has two solutions, x_1 and x_2. S is not invertible: There is no x such that $S(x) = y_0$.

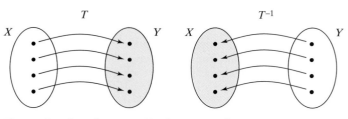

Figure 2 A function T and its inverse T^{-1}.

If a function T is invertible, then so is T^{-1}, and

$$(T^{-1})^{-1} = T.$$

If a function is given by a formula, we may be able to find its inverse by solving the formula for the input variable(s). For example, the inverse of the function

$$y = \frac{x^3 - 1}{5} \qquad \text{(from } \mathbb{R} \text{ to } \mathbb{R}\text{)}$$

is

$$x = \sqrt[3]{5y + 1}.$$

Now consider the case of a *linear transformation* from $\mathbb{R}^n$ to $\mathbb{R}^m$ given by

$$\vec{y} = A\vec{x},$$

where A is an $m \times n$ matrix.

The transformation $\vec{y} = A\vec{x}$ is invertible if the *linear system*

$$A\vec{x} = \vec{y}$$

has a unique solution $\vec{x}$ in $\mathbb{R}^n$ for *all* $\vec{y}$ in $\mathbb{R}^m$. Using techniques developed in Section 1.3, we can determine for which matrices A this is the case. We examine the cases $m < n, m = n$, and $m > n$ separately.

Case 1: $m < n$ ■ The system $A\vec{x} = \vec{y}$ has fewer equations (m) than unknowns (n). Since there must be nonleading variables in this case, the system has either no solutions or infinitely many solutions, for any given $\vec{y}$ in $\mathbb{R}^m$ (Fact 1.3.3). Therefore, the transformation $\vec{y} = A\vec{x}$ is noninvertible.

As an example, consider the case when $m = 2$ and $n = 3$ (i.e., the case of a linear transformation from $\mathbb{R}^3$ to $\mathbb{R}^2$), as shown in Figure 3. Intuitively, we expect that some

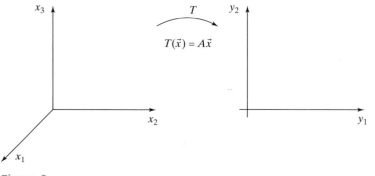

Figure 3

collapsing takes place as we transform $\mathbb{R}^3$ linearly into $\mathbb{R}^2$. In other words, we expect that many points in $\mathbb{R}^3$ get transformed into the same point in $\mathbb{R}^2$. The algebraic reasoning above shows that this is exactly what happens: For some $\vec{y}$ in $\mathbb{R}^2$ (for example, for $\vec{y} = \vec{0}$), the equation $T(\vec{x}) = A\vec{x} = \vec{y}$ will have infinitely many solutions $\vec{x}$ in $\mathbb{R}^3$. See Exercise 2.1.42 for an example.

(Surprisingly, there actually are invertible (but nonlinear) transformations from $\mathbb{R}^3$ to $\mathbb{R}^2$. More generally, there are nonlinear invertible transformations from $\mathbb{R}^n$ to $\mathbb{R}^m$ for any two positive integers n and m.)

Case 2: $m = n$ ■ Here, the number of equations in the system $A\vec{x} = \vec{b}$ matches the number of unknowns. By Fact 1.3.4, the system $A\vec{x} = \vec{y}$ has a unique solution $\vec{x}$ if and only if

$$\text{rref}(A) = \begin{bmatrix} 1 & 0 & 0 & \cdots & 0 \\ 0 & 1 & 0 & \cdots & 0 \\ 0 & 0 & 1 & \cdots & 0 \\ \vdots & \vdots & \vdots & \ddots & \vdots \\ 0 & 0 & 0 & \cdots & 1 \end{bmatrix} = I_n.$$

We conclude that the transformation $\vec{y} = A\vec{x}$ is invertible if and only if $\text{rref}(A) = I_n$, or, equivalently, if $\text{rank}(A) = n$.

Consider the linear transformations from $\mathbb{R}^2$ to $\mathbb{R}^2$ we studied in Section 2.2. You may have seen in the last exercise section that rotations, reflections, and shears are invertible. An orthogonal projection onto a line is noninvertible. (Why?)

Case 3: $m > n$ ■ The transformation $\vec{y} = A\vec{x}$ is noninvertible, because we can find a vector $\vec{y}$ in $\mathbb{R}^m$ such that the system $A\vec{x} = \vec{y}$ is inconsistent. (See Example 1.3.13; note that $\text{rank}(A) \le n < m$ here.)

As an example, consider the case when $m = 3$ and $n = 2$ (i.e., the case of a linear transformation from $\mathbb{R}^2$ to $\mathbb{R}^3$), as shown in Figure 4. Intuitively, we expect that we cannot obtain all points in $\mathbb{R}^3$ by transforming $\mathbb{R}^2$ linearly into $\mathbb{R}^3$. For many vectors $\vec{y}$ in $\mathbb{R}^3$ the equation $T(\vec{x}) = \vec{y}$ will be inconsistent.

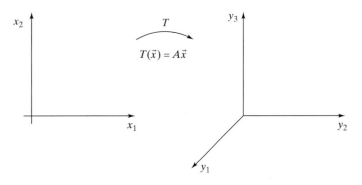

Figure 4

We now introduce some notations and summarize the results we just derived.

Definition 2.3.2

> **Invertible matrices**
> A matrix A is called *invertible* if the linear transformation $\vec{y} = A\vec{x}$ is invertible. The matrix of the inverse transformation[10] is denoted by A^{-1}. If the transformation $\vec{y} = A\vec{x}$ is invertible, its inverse is $\vec{x} = A^{-1}\vec{y}$.

Fact 2.3.3

An $m \times n$ matrix A is *invertible* if and only if

a. A is a square matrix (i.e., $m = n$), and
b. $\operatorname{rref}(A) = I_n$.

EXAMPLE 1   Is the matrix A invertible?

$$A = \begin{bmatrix} 1 & 2 & 3 \\ 4 & 5 & 6 \\ 7 & 8 & 9 \end{bmatrix}.$$

Solution

$$\begin{bmatrix} 1 & 2 & 3 \\ 4 & 5 & 6 \\ 7 & 8 & 9 \end{bmatrix} \begin{array}{c} \to \\ -4(\text{I}) \\ -7(\text{I}) \end{array} \begin{bmatrix} 1 & 2 & 3 \\ 0 & -3 & -6 \\ 0 & -6 & -12 \end{bmatrix} \begin{array}{c} \to \\ \div(-3) \end{array}$$

$$\begin{bmatrix} 1 & 2 & 3 \\ 0 & 1 & 2 \\ 0 & -6 & -12 \end{bmatrix} \begin{array}{c} -2(\text{II}) \\ \to \\ +6(\text{II}) \end{array} \begin{bmatrix} 1 & 0 & -1 \\ 0 & 1 & 2 \\ 0 & 0 & 0 \end{bmatrix}.$$

A fails to be invertible, since $\operatorname{rref}(A) \ne I_3$. ∎

Fact 2.3.4

Let A be an $n \times n$ matrix.

a. Consider a vector $\vec{b}$ in $\mathbb{R}^n$. If A is invertible, then the system $A\vec{x} = \vec{b}$ has the unique solution $\vec{x} = A^{-1}\vec{b}$. If A is noninvertible, then the system $A\vec{x} = \vec{b}$ has infinitely many solutions or none.
b. Consider the special case when $\vec{b} = \vec{0}$. The system $A\vec{x} = \vec{0}$ has $\vec{x} = \vec{0}$ as a solution. If A is invertible, then this is the only solution. If A is noninvertible, then there are infinitely many other solutions.

If a matrix A is invertible, how can we find the inverse matrix A^{-1}? Consider the matrix

$$A = \begin{bmatrix} 1 & 1 & 1 \\ 2 & 3 & 2 \\ 3 & 8 & 2 \end{bmatrix},$$

---

[10]The inverse transformation is linear. (See Exercise 2.2.29.)

or, equivalently, the linear transformation

$$\begin{bmatrix} y_1 \\ y_2 \\ y_3 \end{bmatrix} = \begin{bmatrix} x_1 + x_2 + x_3 \\ 2x_1 + 3x_2 + 2x_3 \\ 3x_1 + 8x_2 + 2x_3 \end{bmatrix}.$$

To find the inverse transformation, we solve this system for the input variables x_1, x_2, and x_3:

$$\left| \begin{array}{l} x_1 + x_2 + x_3 = y_1 \\ 2x_1 + 3x_2 + 2x_3 = \quad\quad y_2 \\ 3x_1 + 8x_2 + 2x_3 = \quad\quad\quad y_3 \end{array} \right| \quad \begin{array}{l} \rightarrow \\ -2(\mathrm{I}) \\ -3(\mathrm{I}) \end{array}$$

$$\left| \begin{array}{l} x_1 + x_2 + x_3 = y_1 \\ \quad\quad x_2 \quad\quad = -2y_1 + y_2 \\ \quad\quad 5x_2 - x_3 = -3y_1 \quad\quad + y_3 \end{array} \right| \quad \begin{array}{l} -(\mathrm{II}) \\ \rightarrow \\ -5(\mathrm{II}) \end{array}$$

$$\left| \begin{array}{l} x_1 \quad\quad + x_3 = \quad 3y_1 - y_2 \\ \quad\quad x_2 \quad\quad = -2y_1 + y_2 \\ \quad\quad\quad - x_3 = \quad 7y_1 - 5y_2 + y_3 \end{array} \right| \quad \begin{array}{l} \rightarrow \\ \\ \div(-1) \end{array}$$

$$\left| \begin{array}{l} x_1 \quad\quad + x_3 = \quad 3y_1 - y_2 \\ \quad\quad x_2 \quad\quad = -2y_1 + y_2 \\ \quad\quad\quad x_3 = -7y_1 + 5y_2 - y_3 \end{array} \right| \quad \begin{array}{l} -(\mathrm{III}) \\ \rightarrow \\ \end{array}$$

$$\left| \begin{array}{l} x_1 \quad\quad\quad = 10y_1 - 6y_2 + y_3 \\ \quad\quad x_2 \quad\quad = -2y_1 + y_2 \\ \quad\quad\quad x_3 = -7y_1 + 5y_2 - y_3 \end{array} \right|.$$

We have found the inverse transformation; its matrix is

$$B = A^{-1} = \begin{bmatrix} 10 & -6 & 1 \\ -2 & 1 & 0 \\ -7 & 5 & -1 \end{bmatrix}.$$

We can write the preceding computations in matrix form:

$$\begin{bmatrix} 1 & 1 & 1 & : & 1 & 0 & 0 \\ 2 & 3 & 2 & : & 0 & 1 & 0 \\ 3 & 8 & 2 & : & 0 & 0 & 1 \end{bmatrix} \begin{array}{l} \rightarrow \\ -2(\mathrm{I}) \\ -3(\mathrm{I}) \end{array} \quad \begin{bmatrix} 1 & 1 & 1 & : & 1 & 0 & 0 \\ 0 & 1 & 0 & : & -2 & 1 & 0 \\ 0 & 5 & -1 & : & -3 & 0 & 1 \end{bmatrix} \begin{array}{l} -(\mathrm{II}) \\ \rightarrow \\ -5(\mathrm{II}) \end{array}$$

$$\begin{bmatrix} 1 & 0 & 1 & : & 3 & -1 & 0 \\ 0 & 1 & 0 & : & -2 & 1 & 0 \\ 0 & 0 & -1 & : & 7 & -5 & 1 \end{bmatrix} \begin{array}{l} \rightarrow \\ \\ \div(-1) \end{array} \quad \begin{bmatrix} 1 & 0 & 1 & : & 3 & -1 & 0 \\ 0 & 1 & 0 & : & -2 & 1 & 0 \\ 0 & 0 & 1 & : & -7 & 5 & -1 \end{bmatrix} \begin{array}{l} -(\mathrm{III}) \\ \rightarrow \\ \end{array}$$

$$\begin{bmatrix} 1 & 0 & 0 & : & 10 & -6 & 1 \\ 0 & 1 & 0 & : & -2 & 1 & 0 \\ 0 & 0 & 1 & : & -7 & 5 & -1 \end{bmatrix}.$$

This process can be described succinctly as follows:

Fact 2.3.5

Finding the inverse of a matrix

To find the *inverse* of an $n \times n$ matrix A, form the $n \times (2n)$ matrix $[A \vdots I_n]$ and compute $\text{rref}[A \vdots I_n]$.

- If $\text{rref}[A \vdots I_n]$ is of the form $[I_n \vdots B]$, then A is invertible, and $A^{-1} = B$.
- If $\text{rref}[A \vdots I_n]$ is of another form (i.e., its left half fails to be I_n), then A is not invertible. (Note that the left half of $\text{rref}[A \vdots I_n]$ is $\text{rref}(A)$.)

The inverse of a 2×2 matrix is particularly easy to find.

Fact 2.3.6

Inverse and determinant of a 2 × 2 matrix

a. The 2×2 matrix

$$A = \begin{bmatrix} a & b \\ c & d \end{bmatrix}$$

is invertible if (and only if) $ad - bc \neq 0$.

Quantity $ad - bc$ is called the *determinant* of A, written $\det(A)$:

$$\det(A) = \det \begin{bmatrix} a & b \\ c & d \end{bmatrix} = ad - bc.$$

b. If

$$A = \begin{bmatrix} a & b \\ c & d \end{bmatrix}$$

is invertible, then

$$\begin{bmatrix} a & b \\ c & d \end{bmatrix}^{-1} = \frac{1}{ad - bc} \begin{bmatrix} d & -b \\ -c & a \end{bmatrix} = \frac{1}{\det(A)} \begin{bmatrix} d & -b \\ -c & a \end{bmatrix}.$$

Compare this with Exercise 2.1.13.

EXERCISES

GOALS Apply the concept of an invertible function. Determine whether a matrix (or a linear transformation) is invertible, and find the inverse if it exists.

Decide whether the matrices in Exercises 1 through 15 are invertible. If they are, find the inverse. Do the computations with paper and pencil. Show all your work.

1. $\begin{bmatrix} 2 & 3 \\ 5 & 8 \end{bmatrix}$.

2. $\begin{bmatrix} 1 & 1 \\ 1 & 1 \end{bmatrix}$.

3. $\begin{bmatrix} 0 & 2 \\ 1 & 1 \end{bmatrix}$.

4. $\begin{bmatrix} 1 & 2 & 1 \\ 1 & 3 & 2 \\ 1 & 0 & 1 \end{bmatrix}$.

5. $\begin{bmatrix} 1 & 2 & 2 \\ 1 & 3 & 1 \\ 1 & 1 & 3 \end{bmatrix}$.

6. $\begin{bmatrix} 1 & 2 & 3 \\ 0 & 1 & 2 \\ 0 & 0 & 1 \end{bmatrix}$.

7. $\begin{bmatrix} 1 & 2 & 3 \\ 0 & 0 & 2 \\ 0 & 0 & 3 \end{bmatrix}$.

8. $\begin{bmatrix} 0 & 0 & 1 \\ 0 & 1 & 0 \\ 1 & 0 & 0 \end{bmatrix}$.

9. $\begin{bmatrix} 1 & 1 & 1 \\ 1 & 1 & 1 \\ 1 & 1 & 1 \end{bmatrix}$.

10. $\begin{bmatrix} 1 & 1 & 1 \\ 1 & 2 & 3 \\ 1 & 3 & 6 \end{bmatrix}$.

11. $\begin{bmatrix} 1 & 0 & 1 \\ 0 & 1 & 0 \\ 0 & 0 & 1 \end{bmatrix}.$
   12. $\begin{bmatrix} 1 & 1 & 2 & 3 \\ 0 & -1 & 0 & 0 \\ 2 & 2 & 5 & 4 \\ 0 & 3 & 0 & 1 \end{bmatrix}.$

13. $\begin{bmatrix} 1 & 0 & 0 & 0 \\ 2 & 1 & 0 & 0 \\ 3 & 2 & 1 & 0 \\ 4 & 3 & 2 & 1 \end{bmatrix}.$
   14. $\begin{bmatrix} 2 & 5 & 0 & 0 \\ 1 & 3 & 0 & 0 \\ 0 & 0 & 1 & 2 \\ 0 & 0 & 2 & 5 \end{bmatrix}.$

15. $\begin{bmatrix} 1 & 2 & 3 & 4 \\ 2 & 4 & 7 & 11 \\ 3 & 7 & 14 & 25 \\ 4 & 11 & 25 & 50 \end{bmatrix}.$

Decide whether the linear transformations in Exercises 16 through 20 are invertible. Find the inverse transformation if it exists. Do the computations with paper and pencil. Show all your work.

16. $\begin{aligned} y_1 &= 3x_1 + 5x_2, \\ y_2 &= 5x_1 + 8x_2. \end{aligned}$
   17. $\begin{aligned} y_1 &= x_1 + 2x_2, \\ y_2 &= 4x_1 + 8x_2. \end{aligned}$

18. $\begin{aligned} y_1 &= x_2, \\ y_2 &= x_3, \\ y_3 &= x_1. \end{aligned}$
   19. $\begin{aligned} y_1 &= x_1 + x_2 + x_3, \\ y_2 &= x_1 + 2x_2 + 3x_3, \\ y_3 &= x_1 + 4x_2 + 9x_3. \end{aligned}$

20. $\begin{aligned} y_1 &= x_1 + 3x_2 + 3x_3, \\ y_2 &= x_1 + 4x_2 + 8x_3, \\ y_3 &= 2x_1 + 7x_2 + 12x_3. \end{aligned}$

Which of the functions f from $\mathbb{R}$ to $\mathbb{R}$ in Exercises 21 through 24 are invertible?

21. $f(x) = x^2.$
   22. $f(x) = 2^x.$

23. $f(x) = x^3 + x.$
   24. $f(x) = x^3 - x.$

Which of the (nonlinear) transformations from $\mathbb{R}^2$ to $\mathbb{R}^2$ in Exercises 25 through 27 are invertible? Find the inverse if it exists.

25. $\begin{bmatrix} y_1 \\ y_2 \end{bmatrix} = \begin{bmatrix} x_1^3 \\ x_2 \end{bmatrix}.$
   26. $\begin{bmatrix} y_1 \\ y_2 \end{bmatrix} = \begin{bmatrix} x_2 \\ x_1^3 + x_2 \end{bmatrix}.$

27. $\begin{bmatrix} y_1 \\ y_2 \end{bmatrix} = \begin{bmatrix} x_1 + x_2 \\ x_1 \cdot x_2 \end{bmatrix}.$

28. Find the inverse of the linear transformation

$$T\begin{bmatrix} x_1 \\ x_2 \\ x_3 \\ x_4 \end{bmatrix} = x_1 \begin{bmatrix} 22 \\ -16 \\ 8 \\ 5 \end{bmatrix} + x_2 \begin{bmatrix} 13 \\ -3 \\ 9 \\ 4 \end{bmatrix} + x_3 \begin{bmatrix} 8 \\ -2 \\ 7 \\ 3 \end{bmatrix} + x_4 \begin{bmatrix} 3 \\ -2 \\ 2 \\ 1 \end{bmatrix}$$

from $\mathbb{R}^4$ to $\mathbb{R}^4$.

29. For which choices of the constant k is the following matrix invertible?

$$\begin{bmatrix} 1 & 1 & 1 \\ 1 & 2 & k \\ 1 & 4 & k^2 \end{bmatrix}.$$

30. For which choices of the constants b and c is the following matrix invertible?

$$\begin{bmatrix} 0 & 1 & b \\ -1 & 0 & c \\ -b & -c & 0 \end{bmatrix}.$$

31. For which choices of the constants a, b, and c is the following matrix invertible?

$$\begin{bmatrix} 0 & a & b \\ -a & 0 & c \\ -b & -c & 0 \end{bmatrix}.$$

32. Find all matrices $\begin{bmatrix} a & b \\ c & d \end{bmatrix}$ such that $ad - bc = 1$ and $A^{-1} = A$.

33. Consider the matrices of the form $A = \begin{bmatrix} a & b \\ b & -a \end{bmatrix}$, where a and b are arbitrary constants. For which choices of a and b is $A^{-1} = A$?

34. Consider the diagonal matrix

$$A = \begin{bmatrix} a & 0 & 0 \\ 0 & b & 0 \\ 0 & 0 & c \end{bmatrix}.$$

a. For which choices of a, b, and c is A invertible? If it is invertible, what is A^{-1}?

b. For which choices of the diagonal elements is a diagonal matrix (of arbitrary size) invertible?

35. Consider the upper triangular 3×3 matrix

$$A = \begin{bmatrix} a & b & c \\ 0 & d & e \\ 0 & 0 & f \end{bmatrix}.$$

a. For which choices of a, b, c, d, e, and f is A invertible?

b. More generally, when is an upper triangular matrix (of arbitrary size) invertible?

c. If an upper triangular matrix is invertible, is its inverse an upper triangular matrix as well?

d. When is a lower triangular matrix invertible?

36. To determine whether a square matrix A is invertible, it is not always necessary to bring it into reduced row-echelon form. Justify the following rule: To determine whether a square matrix A is invertible, reduce it to triangular form (upper or lower), by elementary row operations. A is invertible if (and only if) all entries on the diagonal of this triangular form are nonzero.

37. If A is an invertible matrix and c is a nonzero scalar, is the matrix cA invertible? If so, what is the relationship between A^{-1} and $(cA)^{-1}$?

38. Find A^{-1} for $A = \begin{bmatrix} 1 & k \\ 0 & -1 \end{bmatrix}$.

39. Consider a square matrix that differs from the identity matrix at just one entry, off the diagonal. For example,

$$\begin{bmatrix} 1 & 0 & 0 \\ 0 & 1 & 0 \\ -\frac{1}{2} & 0 & 1 \end{bmatrix}.$$

In general, is a matrix M of this form invertible? If so, what is the M^{-1}?

40. Show that if a square matrix A has two equal columns, then A is not invertible.

41. Which of the following linear transformations T from $\mathbb{R}^3$ to $\mathbb{R}^3$ are invertible? Find the inverse if it exists.

   a.   Reflection in a plane.

   b.   Projection onto a plane.

   c.   Dilation by 5 (i.e., $T(\vec{v}) = 5\vec{v}$, for all vectors $\vec{v}$).

   d.   Rotation about an axis.

42. A square matrix is called a *permutation matrix* if it contains the entry 1 exactly once in each row and in each column, with all other entries being 0. Examples are I_n and

$$\begin{bmatrix} 0 & 0 & 1 \\ 1 & 0 & 0 \\ 0 & 1 & 0 \end{bmatrix}.$$

Are permutation matrices invertible? If so, is the inverse a permutation matrix as well?

43. Consider two invertible $n \times n$ matrices A and B. Is the linear transformation $\vec{y} = A(B\vec{x})$ invertible? If so, what is the inverse? (*Hint:* Solve the equation $\vec{y} = A(B\vec{x})$ first for $B\vec{x}$ and then for $\vec{x}$.)

44. Consider the $n \times n$ matrix M_n which contains all integers $1, 2, 3, \ldots, n^2$ as its entries, written in sequence, column by column; for example,

$$M_4 = \begin{bmatrix} 1 & 5 & 9 & 13 \\ 2 & 6 & 10 & 14 \\ 3 & 7 & 11 & 15 \\ 4 & 8 & 12 & 16 \end{bmatrix}.$$

   a.   Determine the rank of M_4.

   b.   Determine the rank of M_n, for an arbitrary $n \geq 2$.

   c.   For which integers n is M_n invertible?

45. To gauge the complexity of a computational task, mathematicians and computer scientists count the number of elementary operations (additions, subtractions, multiplications, and divisions) required. Since additions and subtractions require very little work compared with multiplications and divisions (think about performing these operations on eight-digit numbers), often only the multiplications and divisions are counted. As an example, we examine the process of inverting a 2×2 matrix by elimination.

$$\begin{bmatrix} a & b & : & 1 & 0 \\ c & d & : & 0 & 1 \end{bmatrix} \qquad \div a, \text{ requires 2 operations:}$$
$$\downarrow \qquad\qquad b/a \text{ and } 1/a$$

$$\begin{bmatrix} 1 & b' & : & e & 0 \\ c & d & : & 0 & 1 \end{bmatrix} \qquad \begin{array}{l} \text{(where } b' = b/a, \text{ and } e = 1/a) \\ -c(I), \text{ requires 2 operations: } cb' \\ \text{and } ce \end{array}$$
$$\downarrow$$

$$\begin{bmatrix} 1 & b' & : & e & 0 \\ 0 & d' & : & g & 1 \end{bmatrix} \qquad \div d', \text{ requires 2 operations}$$
$$\downarrow$$

$$\begin{bmatrix} 1 & b' & : & e & 0 \\ 0 & 1 & : & g' & h \end{bmatrix} \qquad -b'(II), \text{ requires 2 operations}$$
$$\downarrow$$

$$\begin{bmatrix} 1 & 0 & : & e' & f \\ 0 & 1 & : & g' & h \end{bmatrix}$$

The whole process requires 8 operations. Note that we do not count operations with a predictable result, such as $1a$, $0a$, a/a, $0/a$.

   a.   How many operations are required to invert a 3×3 matrix by elimination?

   b.   How many operations are required to invert an $n \times n$ matrix by elimination?

   c.   If it takes a slow hand-held calculator 1 second to invert a 3×3 matrix, how long will it take the same calculator to invert a 12×12 matrix? Assume that the matrices are inverted by Gauss–Jordan elimination and that the duration of the computation depends only on the number of multiplications and divisions involved.

46. Consider the linear system

$$A\vec{x} = \vec{b},$$

where A is an invertible matrix. We can solve this system in two different ways:

• By finding the reduced row-echelon form of the augmented matrix $[A \vdots \vec{b}]$.

• By computing A^{-1} and using the formula $\vec{x} = A^{-1}\vec{b}$.

In general, which approach requires fewer operations? See Exercise 45.

47. Give an example of a noninvertible function f from $\mathbb{R}$ to $\mathbb{R}$ and a number b such that the equation

$$f(x) = b$$

has a unique solution.

48. Give an example of a 3×2 matrix A and a vector $\vec{b}$ in $\mathbb{R}^3$ such that the linear system

$$A\vec{x} = \vec{b}$$

has a unique solution. Why doesn't this example contradict Fact 2.3.4(a)?

49. *Input–Output Analysis.* (This exercise builds on Exercises 1.1.20, 1.2.37, 1.2.38, and 1.2.39.) Consider the industries $J_1, J_2, \ldots, J_n$ in an economy. Suppose the consumer demand vector is $\vec{b}$, the output vector is $\vec{x}$ and the demand vector of the jth industry is $\vec{v}_j$. (The ith component a_{ij} of $\vec{v}_j$ is the demand industry J_j puts on industry J_i, per unit of output of J_j.) As we have seen in Exercise 1.2.38, the output $\vec{x}$ just meets the aggregate demand if

$$\underbrace{x_1\vec{v}_1 + x_2\vec{v}_2 + \cdots + x_n\vec{v}_n + \vec{b}}_{\text{aggregate demand}} = \underbrace{\vec{x}}_{\text{output}}.$$

This equation can be written more succinctly as

$$\begin{bmatrix} | & | & & | \\ \vec{v}_1 & \vec{v}_2 & \cdots & \vec{v}_n \\ | & | & & | \end{bmatrix} \begin{bmatrix} x_1 \\ x_2 \\ \vdots \\ x_n \end{bmatrix} + \vec{b} = \vec{x},$$

or $A\vec{x} + \vec{b} = \vec{x}$. The matrix A is called the *technology matrix* of this economy; its coefficients a_{ij} describe the interindustry demand, which depends on the technology used in the production process. The equation

$$A\vec{x} + \vec{b} = \vec{x}$$

describes a linear system; we can write it in the customary form:

$$\vec{x} - A\vec{x} = \vec{b},$$
$$I_n\vec{x} - A\vec{x} = \vec{b},$$
$$(I_n - A)\vec{x} = \vec{b}.$$

If we want to know the output $\vec{x}$ required to satisfy a given consumer demand $\vec{b}$ (this was our objective in the previous exercises), we can solve this linear system, preferably via the augmented matrix.

In economics, however, we often ask other questions: If $\vec{b}$ *changes,* how will $\vec{x}$ change in response? If the consumer demand on one industry increases by 1 unit, and the consumer demand on the other industries remains unchanged, how will $\vec{x}$ change?[11] If we ask questions like these, we think of the output $\vec{x}$ as a *function* of the consumer demand $\vec{b}$.

If the matrix $(I_n - A)$ is invertible,[12] we can express $\vec{x}$ as a function of $\vec{b}$ (in fact, as a linear transformation):

$$\vec{x} = (I_n - A)^{-1}\vec{b}.$$

a. Consider the example of the economy of Israel in 1958 (discussed in Exercise 1.2.39). Find the technology matrix A, the matrix $(I_n - A)$, and its inverse $(I_n - A)^{-1}$.

b. In the example discussed in part (a), suppose the consumer demand on agriculture (Industry 1) is 1 unit (1 million pounds), and the demands on the other two industries are zero. What output $\vec{x}$ is required in this case? How does your answer relate to the matrix $(I_n - A)^{-1}$?

c. Explain, in terms of economics, why the diagonal elements of the matrix $(I_n - A)^{-1}$ you found in part a must be at least 1.

d. If the consumer demand on manufacturing increases by 1 (from whatever it was), and the consumer demand on the two other industries remains the same, how will the output have to change? How does your answer relate to the matrix $(I_n - A)^{-1}$?

e. Using your answers in parts (a) through (d) as a guide, explain in general (not just for this example) what the columns and the entries of the matrix $(I_n - A)^{-1}$ tell you, in terms of economics. Those who have studied multivariable calculus may wish to consider the partial derivatives

$$\frac{\partial x_i}{\partial b_j}.$$

---

[11] The relevance of questions like these became particularly clear during WWII, when the demand on certain industries suddenly changed dramatically. When U.S. President F. D. Roosevelt asked for 50,000 airplanes to be built, it was easy enough to predict that the country would have to produce more aluminum. Unexpectedly, the demand for copper dramatically increased (why?). A copper shortage then occurred, which was solved by borrowing silver from Fort Knox. People realized that input-output analysis can be effective in modeling and predicting chains of increased demand like this. After WWII, this technique rapidly gained acceptance, and was soon used to model the economies of more than 50 countries.

[12] This will always be the case for a "productive" economy. See Exercise 2.4.75.

50. This exercise refers to Exercise 49(a). Consider the entry $k = a_{11} = 0.293$ of the technology matrix A. Verify that the entry in the first row and the first column of $(I_n - A)^{-1}$ is the value of the geometrical series $1 + k + k^2 + \cdots$. Interpret this observation in terms of economics.

2.4 MATRIX PRODUCTS

Recall the *composite* of two functions: The composite of the functions $y = \sin(x)$ and $z = \cos(y)$ is $z = \cos(\sin(x))$, as illustrated in Figure 1.

Similarly, we can compose two linear transformations.

To understand this concept, let's return to the coding example discussed in Section 2.1. Recall that the position $\vec{x} = \begin{bmatrix} x_1 \\ x_2 \end{bmatrix}$ of your boat is encoded and that you radio the encoded position $\vec{y} = \begin{bmatrix} y_1 \\ y_2 \end{bmatrix}$ to Marseille. The coding transformation is

$$\vec{y} = A\vec{x}, \quad \text{with } A = \begin{bmatrix} 1 & 2 \\ 3 & 5 \end{bmatrix}.$$

In Section 2.1, we left out one detail: Your position is radioed on to Paris, as you would expect in a centrally governed country such as France. Before broadcasting to Paris, the position $\vec{y}$ is again encoded, using the linear transformation

$$\vec{z} = B\vec{y}, \quad \text{with } B = \begin{bmatrix} 6 & 7 \\ 8 & 9 \end{bmatrix}$$

this time, and the sailor in Marseille radios the encoded position $\vec{z}$ to Paris. (See Figure 2.)

We can think of the message $\vec{z}$ received in Paris as a function of the actual position $\vec{x}$ of the boat,

$$\vec{z} = B(A\vec{x}),$$

the composite of the two transformations $\vec{y} = A\vec{x}$ and $\vec{z} = B\vec{y}$. Is this transformation $\vec{z} = T(\vec{x})$ linear, and, if so, what is its matrix? We will show two approaches to these important questions, one using brute force (a), and one using

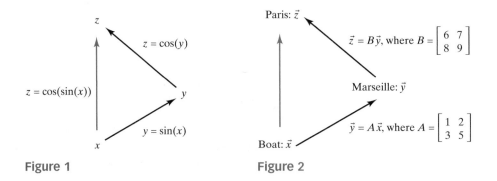

Figure 1   Figure 2

some theory (b).

a. We write the components of the two transformations and substitute.

$$\begin{array}{lll} z_1 = 6y_1 + 7y_2 \\ z_2 = 8y_1 + 9y_2 \end{array} \quad \text{and} \quad \begin{array}{l} y_1 = x_1 + 2x_2 \\ y_2 = 3x_1 + 5x_2 \end{array}$$

so that

$$z_1 = 6(x_1 + 2x_2) + 7(3x_1 + 5x_2) = (6 \cdot 1 + 7 \cdot 3)x_1 + (6 \cdot 2 + 7 \cdot 5)x_2$$

$$= 27x_1 + 47x_2,$$

$$z_2 = 8(x_1 + 2x_2) + 9(3x_1 + 5x_2) = (8 \cdot 1 + 9 \cdot 3)x_1 + (8 \cdot 2 + 9 \cdot 5)x_2$$

$$= 35x_1 + 61x_2.$$

This shows that the composite is indeed linear, with matrix

$$\begin{bmatrix} 6 \cdot 1 + 7 \cdot 3 & 6 \cdot 2 + 7 \cdot 5 \\ 8 \cdot 1 + 9 \cdot 3 & 8 \cdot 2 + 9 \cdot 5 \end{bmatrix} = \begin{bmatrix} 27 & 47 \\ 35 & 61 \end{bmatrix}.$$

b. We can use Fact 1.3.7 to show that the transformation $T(\vec{x}) = B(A\vec{x})$ is linear:

$$T(\vec{v} + \vec{w}) = B(A(\vec{v} + \vec{w})) = B(A\vec{v} + A\vec{w}) = B(A\vec{v}) + B(A\vec{w}) = T(\vec{v}) + T(\vec{w}),$$

$$T(k\vec{v}) = B(A(k\vec{v})) = B(k(A\vec{v})) = k(B(A\vec{v})) = kT(\vec{v}).$$

Once we know that T is linear, we can find its matrix by computing the vectors $T(\vec{e}_1) = B(A\vec{e}_1)$ and $T(\vec{e}_2) = B(A\vec{e}_2)$; the matrix of T is then $[T(\vec{e}_1)\ T(\vec{e}_2)]$, by Fact 2.1.2:

$$T(\vec{e}_1) = B(A\vec{e}_1) = B(\text{first column of } A) = \begin{bmatrix} 6 & 7 \\ 8 & 9 \end{bmatrix} \begin{bmatrix} 1 \\ 3 \end{bmatrix} = \begin{bmatrix} 27 \\ 35 \end{bmatrix},$$

$$T(\vec{e}_2) = B(A\vec{e}_2) = B(\text{second column of } A) = \begin{bmatrix} 6 & 7 \\ 8 & 9 \end{bmatrix} \begin{bmatrix} 2 \\ 5 \end{bmatrix} = \begin{bmatrix} 47 \\ 61 \end{bmatrix}.$$

We find that the matrix of the linear transformation $T(\vec{x}) = B(A\vec{x})$ is

$$\begin{bmatrix} | & | \\ T(\vec{e}_1) & T(\vec{e}_2) \\ | & | \end{bmatrix} = \begin{bmatrix} 27 & 47 \\ 35 & 61 \end{bmatrix}.$$

This result agrees with the result in a, of course.

The matrix of the linear transformation $T(\vec{x}) = B(A\vec{x})$ is called the *product* of the matrices B and A, written as BA. This means that

$$T(\vec{x}) = B(A\vec{x}) = (BA)\vec{x},$$

for all vectors $\vec{x}$ in $\mathbb{R}^2$. (See Figure 3.)

Now let's look at the product of larger matrices. Let B be an $m \times n$ matrix and A an $n \times p$ matrix. These matrices represent linear transformations as shown in Figure 4.

Again, the composite transformation $\vec{z} = B(A\vec{x})$ is linear. (The foregoing justification applies in this more general case as well.) The matrix of the linear transformation $\vec{z} = B(A\vec{x})$ is called the *product* of the matrices B and A, written as BA.

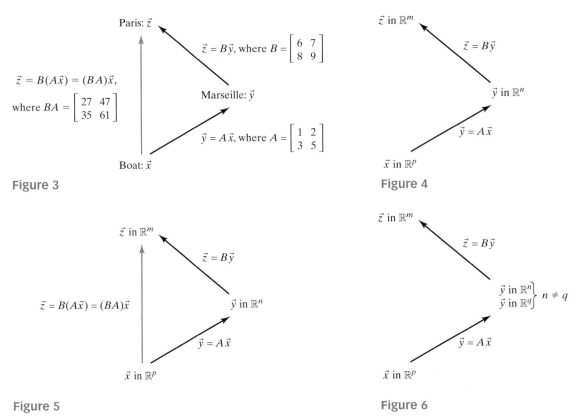

Figure 3

Figure 4

Figure 5

Figure 6

Note that BA is an $m \times p$ matrix (as it represents a linear transformation from $\mathbb{R}^p$ to $\mathbb{R}^m$). As in the case of $\mathbb{R}^2$, the equation

$$\vec{z} = B(A\vec{x}) = (BA)\vec{x}$$

holds for all vectors $\vec{x}$ in $\mathbb{R}^p$, by definition of the product BA. (See Figure 5.)

In the definition of the matrix product BA, the number of columns of B matches the number of rows of A. What happens if these two numbers are different? Suppose B is an $m \times n$ matrix and A is a $q \times p$ matrix, with $n \neq q$.

In this case, the transformations $\vec{z} = B\vec{y}$ and $\vec{y} = A\vec{x}$ cannot be composed, since the codomain of $\vec{y} = A\vec{x}$ is different from the domain of $\vec{z} = B\vec{y}$. (See Figure 6.) To put it more plainly: The output of $\vec{y} = A\vec{x}$ is not an acceptable input for the transformation $\vec{z} = B\vec{y}$. In this case, the matrix product BA is undefined.

Definition 2.4.1

Matrix multiplication
a. Let B be an $m \times n$ matrix and A a $q \times p$ matrix. The product BA is defined if (and only if) $n = q$.
b. If B is an $m \times n$ matrix and A an $n \times p$ matrix, then the product BA is defined as the matrix of the linear transformation $T(\vec{x}) = B(A\vec{x})$. This means that $T(\vec{x}) = B(A\vec{x}) = (BA)\vec{x}$, for all $\vec{x}$ in $\mathbb{R}^p$. The product BA is an $m \times p$ matrix.

Although this definition of matrix multiplication does not give us concrete instructions for computing the product of two numerically given matrices, such instructions can be derived easily from the definition.

As in Definition 2.4.1, let B be an $m \times n$ matrix and A an $n \times p$ matrix. Let's think about the columns of the matrix BA:

$$(i\text{th column of } BA) = (BA)\vec{e}_i$$

$$= B(A\vec{e}_i)$$

$$= B(i\text{th column of } A).$$

If we denote the columns of A by $\vec{v}_1, \vec{v}_2, \ldots, \vec{v}_p$, we can write

$$
BA = B \begin{bmatrix} | & | & & | \\ \vec{v}_1 & \vec{v}_2 & \cdots & \vec{v}_p \\ | & | & & | \end{bmatrix} = \begin{bmatrix} | & | & & | \\ B\vec{v}_1 & B\vec{v}_2 & \cdots & B\vec{v}_p \\ | & | & & | \end{bmatrix}.
$$

Fact 2.4.2   **The matrix product, column by column**

Let B be an $m \times n$ matrix and A an $n \times p$ matrix with columns $\vec{v}_1, \vec{v}_2, \ldots, \vec{v}_p$. Then, the product BA is

$$
BA = B \begin{bmatrix} | & | & & | \\ \vec{v}_1 & \vec{v}_2 & \cdots & \vec{v}_p \\ | & | & & | \end{bmatrix} = \begin{bmatrix} | & | & & | \\ B\vec{v}_1 & B\vec{v}_2 & \cdots & B\vec{v}_p \\ | & | & & | \end{bmatrix}.
$$

To find BA, we can multiply B with the columns of A and combine the resulting vectors.

This is exactly how we computed the product

$$BA = \begin{bmatrix} 6 & 7 \\ 8 & 9 \end{bmatrix} \begin{bmatrix} 1 & 2 \\ 3 & 5 \end{bmatrix} = \begin{bmatrix} 27 & 47 \\ 35 & 61 \end{bmatrix}$$

on page 75, using approach (b).

For practice, let us multiply the same matrices in the reverse order. The first column of AB is $\begin{bmatrix} 1 & 2 \\ 3 & 5 \end{bmatrix} \begin{bmatrix} 6 \\ 8 \end{bmatrix} = \begin{bmatrix} 22 \\ 58 \end{bmatrix}$; the second is $\begin{bmatrix} 1 & 2 \\ 3 & 5 \end{bmatrix} \begin{bmatrix} 7 \\ 9 \end{bmatrix} = \begin{bmatrix} 25 \\ 66 \end{bmatrix}$. Thus,

$$AB = \begin{bmatrix} 1 & 2 \\ 3 & 5 \end{bmatrix} \begin{bmatrix} 6 & 7 \\ 8 & 9 \end{bmatrix} = \begin{bmatrix} 22 & 25 \\ 58 & 66 \end{bmatrix}.$$

Compare the two previous displays to see that $AB \neq BA$: Matrix multiplication is *noncommutative*. This should come as no surprise, because the matrix product represents a composite of transformations. Even for functions of one variable, the order in which we compose matters. Refer to the first example in this section and note that the functions $\cos(\sin(x))$ and $\sin(\cos(x))$ are different.

Fact 2.4.3

Matrix multiplication is *noncommutative*: $AB \neq BA$, in general. However, at times it does happen that $AB = BA$; then, we say that the matrices A and B *commute*.

It is useful to have a formula for the ijth entry of the product BA of an $m \times n$ matrix B and an $n \times p$ matrix A.

Let $\vec{v}_1, \vec{v}_2, \ldots, \vec{v}_p$ be the columns of A. Then, by Fact 2.4.2,

$$BA = B \begin{bmatrix} | & | & & | & & | \\ \vec{v}_1 & \vec{v}_2 & \cdots & \vec{v}_j & \cdots & \vec{v}_p \\ | & | & & | & & | \end{bmatrix} = \begin{bmatrix} | & | & & | & & | \\ B\vec{v}_1 & B\vec{v}_2 & \cdots & B\vec{v}_j & \cdots & B\vec{v}_p \\ | & | & & | & & | \end{bmatrix}.$$

The ijth entry of the product BA is the ith component of the vector $B\vec{v}_j$, which is the dot product of the ith row of B and $\vec{v}_j$, by Fact 1.3.8.

Fact 2.4.4

The matrix product, entry by entry

Let B be an $m \times n$ matrix and A an $n \times p$ matrix. The ijth entry of BA is the dot product of the ith row of B and the jth column of A.

$$BA = \begin{bmatrix} b_{11} & b_{12} & \cdots & b_{1n} \\ b_{21} & b_{22} & \cdots & b_{2n} \\ \vdots & \vdots & \ddots & \vdots \\ b_{i1} & b_{i2} & \cdots & b_{in} \\ \vdots & \vdots & \ddots & \vdots \\ b_{m1} & b_{m2} & \cdots & b_{mn} \end{bmatrix} \begin{bmatrix} a_{11} & a_{12} & \cdots & a_{1j} & \cdots & a_{1p} \\ a_{21} & a_{22} & \cdots & a_{2j} & \cdots & a_{2p} \\ \vdots & \vdots & \ddots & \vdots & \ddots & \vdots \\ a_{n1} & a_{n2} & \cdots & a_{nj} & \cdots & a_{np} \end{bmatrix}$$

is the $m \times p$ matrix whose ijth entry is

$$b_{i1}a_{1j} + b_{i2}a_{2j} + \cdots + b_{in}a_{nj} = \sum_{k=1}^{n} b_{ik}a_{kj}.$$

EXAMPLE 1

$$\begin{bmatrix} 6 & 7 \\ 8 & 9 \end{bmatrix} \begin{bmatrix} 1 & 2 \\ 3 & 5 \end{bmatrix} = \begin{bmatrix} 6 \cdot 1 + 7 \cdot 3 & 6 \cdot 2 + 7 \cdot 5 \\ 8 \cdot 1 + 9 \cdot 3 & 8 \cdot 2 + 9 \cdot 5 \end{bmatrix} = \begin{bmatrix} 27 & 47 \\ 35 & 61 \end{bmatrix}.$$

We have done these computations before. (Where?)   ∎

Matrix Algebra

Next, let's discuss some rules of matrix algebra:

- Consider an invertible $n \times n$ matrix A. By definition of the inverse, A multiplied with its inverse represents the identity transformation.

Fact 2.4.5   For an invertible $n \times n$ matrix A,

$$AA^{-1} = I_n \quad \text{and} \quad A^{-1}A = I_n.$$

- Composing a linear transformation with the identity transformation, on either side, leaves the transformation unchanged.

Fact 2.4.6   For an $m \times n$ matrix A,

$$AI_n = I_m A = A.$$

- If A is an $m \times n$ matrix, B an $n \times p$ matrix, and C a $p \times q$ matrix, what is the relationship between $(AB)C$ and $A(BC)$?

  One way to think about this problem (although perhaps not the most elegant one) is to write C in terms of its columns: $C = [\vec{v}_1 \quad \vec{v}_2 \quad \dots \quad \vec{v}_q]$. Then,

  $$(AB)C = (AB)[\vec{v}_1 \quad \vec{v}_2 \quad \dots \quad \vec{v}_q] = [(AB)\vec{v}_1 \quad (AB)\vec{v}_2 \quad \dots \quad (AB)\vec{v}_q],$$

  and

  $$A(BC) = A[B\vec{v}_1 \quad B\vec{v}_2 \quad \dots \quad B\vec{v}_q] = [A(B\vec{v}_1) \quad A(B\vec{v}_2) \quad \dots \quad A(B\vec{v}_q)].$$

  Since $(AB)\vec{v}_i = A(B\vec{v}_i)$, by definition of the matrix product, we find that $(AB)C = A(BC)$.

Fact 2.4.7   **Matrix multiplication is associative**

$$(AB)C = A(BC).$$

We can write simply ABC for the product $(AB)C = A(BC)$.

A more conceptual proof is based on the fact that the composition of functions is associative. The two linear transformations

$$T(\vec{x}) = ((AB)C)\,\vec{x} \quad \text{and} \quad L(\vec{x}) = (A(BC))\,\vec{x}$$

are identical because, by definition of matrix multiplication,

$$T(\vec{x}) = ((AB)C)\,\vec{x} = (AB)(C\vec{x}) = A(B(C\vec{x}))$$

and

$$L(\vec{x}) = (A(BC))\,\vec{x} = A((BC)\vec{x}) = A(B(C\vec{x})).$$

The domains and codomains of the linear transformations defined by the matrices $A, B, C, BC, AB, A(BC)$, and $(AB)C$ are shown in Figure 7.

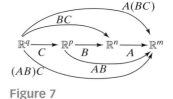

Figure 7

- If A and B are invertible $n \times n$ matrices, is BA invertible as well? If so, what is its inverse?

  To find the inverse of the linear transformation

  $$\vec{y} = BA\vec{x},$$

  we solve the equation for $\vec{x}$ in two steps. First, we multiply both sides of the equation by B^{-1} from the left:

  $$B^{-1}\vec{y} = B^{-1}BA\vec{x} = I_n A\vec{x} = A\vec{x}.$$

  Now, we multiply by A^{-1} from the left:

  $$A^{-1}B^{-1}\vec{y} = A^{-1}A\vec{x} = \vec{x}.$$

  This computation shows that the linear transformation

  $$\vec{y} = BA\vec{x}$$

  is invertible and that its inverse is

  $$\vec{x} = A^{-1}B^{-1}\vec{y}.$$

Fact 2.4.8

The inverse of a product of matrices

If A and B are invertible $n \times n$ matrices, then BA is invertible as well, and

$$(BA)^{-1} = A^{-1}B^{-1}.$$

Pay attention to the order of the matrices.

To verify this result, we can multiply $A^{-1}B^{-1}$ by BA (in either order), and check that the result is I_n:

$$BAA^{-1}B^{-1} = BI_n B^{-1} = BB^{-1} = I_n,$$

$$A^{-1}B^{-1}BA = A^{-1}A = I_n.$$

Everything worked out!

To understand the order of the factors in the formula $(BA)^{-1} = A^{-1}B^{-1}$, think about our French coast guard story again.

To recover the actual position $\vec{x}$ from the doubly encoded position $\vec{z}$, you *first* apply the decoding transformation $\vec{y} = B^{-1}\vec{z}$ and *then* the decoding transformation $\vec{x} = A^{-1}\vec{y}$. The inverse of $\vec{z} = BA\vec{x}$ is therefore $\vec{x} = A^{-1}B^{-1}\vec{z}$, as illustrated in Figure 8.

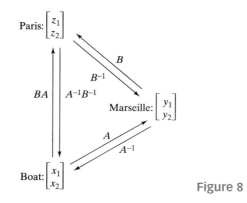

Figure 8

The following result is often useful in finding inverses:

Fact 2.4.9   Let A and B be two $n \times n$ matrices such that

$$BA = I_n.$$

Then,

a. A and B are both invertible,
b. $A^{-1} = B$ and $B^{-1} = A$, and
c. $AB = I_n$.

It follows from the definition of an invertible function that if $AB = I_n$ *and* $BA = I_n$, then A and B are inverses, that is, $A = B^{-1}$ and $B = A^{-1}$. Fact 2.4.9 makes the point that the equation $BA = I_n$ *alone* guarantees that A and B are inverses. Exercises 15 and 79 will illustrate the significance of this claim.

Proof   To demonstrate that A is invertible it suffices to show that the linear system $A\vec{x} = \vec{0}$ has only the solution $\vec{x} = \vec{0}$ (by Fact 2.3.4(b)). If we multiply the equation $A\vec{x} = \vec{0}$ by B from the left, we find that $BA\vec{x} = B\vec{0} = \vec{0}$, so $\vec{x} = \vec{0}$, since $BA = I_n$. Therefore, A is invertible. If we multiply the equation $BA = I_n$ by A^{-1} from the right, we find that $B = A^{-1}$. The matrix B, being the inverse of A, is itself invertible, and $B^{-1} = (A^{-1})^{-1} = A$. (This follows from the definition of an invertible function; see page 66.) Finally, $AB = AA^{-1} = I_n$. ▲

You can use Fact 2.4.9 to check your work when you compute the inverse of a matrix. For example, to check that

$$B = \frac{1}{ad - bc} \begin{bmatrix} d & -b \\ -c & a \end{bmatrix} \quad \text{is the inverse of} \quad A = \begin{bmatrix} a & b \\ c & d \end{bmatrix},$$

it suffices to verify that $BA = I_2$:

$$BA = \frac{1}{ad - bc} \begin{bmatrix} d & -b \\ -c & a \end{bmatrix} \begin{bmatrix} a & b \\ c & d \end{bmatrix} = \frac{1}{ad - bc} \begin{bmatrix} ad - bc & bd - bd \\ ac - ac & ad - bc \end{bmatrix} = I_2.$$

Compare this with Fact 2.3.6.
    Here is another example illustrating Fact 2.4.9.

EXAMPLE 2   Suppose A, B, and C are three $n \times n$ matrices and $ABC = I_n$. Show that B is invertible, and express B^{-1} in terms of A and C.

Solution

Write $ABC = (AB)C = I_n$. We have $C(AB) = I_n$, by Fact 2.4.9(c). Since matrix multiplication is associative, we can write $(CA)B = I_n$. Applying Fact 2.4.9 again, we conclude that B is invertible, and $B^{-1} = CA$.   ∎

Fact 2.4.10
Distributive property for matrices
If A, B are $m \times n$ matrices, and C, D are $n \times p$ matrices, then

$$A(C + D) = AC + AD, \quad \text{and}$$
$$(A + B)C = AC + BC.$$

You will be asked to verify this property in Exercise 63.

Fact 2.4.11
If A is an $m \times n$ matrix, B an $n \times p$ matrix, and k a scalar, then

$$(kA)B = A(kB) = k(AB).$$

You will be asked to verify this property in Exercise 64.

Partitioned Matrices

It is sometimes useful to break a large matrix down into smaller submatrices by slicing it up with horizontal or vertical lines that go all the way through the matrix. For example, we can think of the 4×4 matrix

$$A = \begin{bmatrix} 1 & 2 & 3 & 4 \\ 5 & 6 & 7 & 8 \\ 9 & 8 & 7 & 6 \\ 5 & 4 & 3 & 2 \end{bmatrix}$$

as a 2×2 matrix whose "entries" are four 2×2 matrices:

$$A = \left[\begin{array}{cc|cc} 1 & 2 & 3 & 4 \\ 5 & 6 & 7 & 8 \\ \hline 9 & 8 & 7 & 6 \\ 5 & 4 & 3 & 2 \end{array} \right] = \begin{bmatrix} A_{11} & A_{12} \\ A_{21} & A_{22} \end{bmatrix},$$

with $A_{11} = \begin{bmatrix} 1 & 2 \\ 5 & 6 \end{bmatrix}$, $A_{12} = \begin{bmatrix} 3 & 4 \\ 7 & 8 \end{bmatrix}$, etc.

The submatrices in such a partition need not be of equal size; for example, we could have

$$
B = \begin{bmatrix} 1 & 2 & 3 \\ 4 & 5 & 6 \\ 7 & 8 & 9 \end{bmatrix} = \left[\begin{array}{cc|c} 1 & 2 & 3 \\ 4 & 5 & 6 \\ \hline 7 & 8 & 9 \end{array} \right] = \begin{bmatrix} B_{11} & B_{12} \\ B_{21} & B_{22} \end{bmatrix}.
$$

A useful property of partitioned matrices is the following:

Fact 2.4.12

Multiplying partitioned matrices
Partitioned matrices can be multiplied as though the submatrices were scalars (i.e., using the formula in Fact 2.4.4):

$$
AB = \begin{bmatrix} A_{11} & A_{12} & \cdots & A_{1n} \\ A_{21} & A_{22} & \cdots & A_{2n} \\ \vdots & \vdots & \ddots & \vdots \\ A_{i1} & A_{i2} & \cdots & A_{in} \\ \vdots & \vdots & \ddots & \vdots \\ A_{m1} & A_{m2} & \cdots & A_{mn} \end{bmatrix} \begin{bmatrix} B_{11} & B_{12} & \cdots & B_{1j} & \cdots & B_{1p} \\ B_{21} & B_{22} & \cdots & B_{2j} & \cdots & B_{2p} \\ \vdots & \vdots & \ddots & \vdots & \ddots & \vdots \\ B_{n1} & B_{n2} & \cdots & B_{nj} & \cdots & B_{np} \end{bmatrix}
$$

is the partitioned matrix whose ijth "entry" is the matrix

$$
A_{i1}B_{1j} + A_{i2}B_{2j} + \cdots + A_{in}B_{nj} = \sum_{k=1}^{n} A_{ik}B_{kj},
$$

provided that all the products $A_{ik}B_{kj}$ are defined.

Verifying this fact is left as an exercise. Here is a numerical example.

EXAMPLE 3

$$
\left[\begin{array}{cc|c} 0 & 1 & -1 \\ 1 & 0 & 1 \end{array} \right] \left[\begin{array}{cc|c} 1 & 2 & 3 \\ 4 & 5 & 6 \\ \hline 7 & 8 & 9 \end{array} \right]
$$

$$
= \left[\begin{array}{c|c} \begin{bmatrix} 0 & 1 \\ 1 & 0 \end{bmatrix}\begin{bmatrix} 1 & 2 \\ 4 & 5 \end{bmatrix} + \begin{bmatrix} -1 \\ 1 \end{bmatrix}[7 \quad 8] & \begin{bmatrix} 0 & 1 \\ 1 & 0 \end{bmatrix}\begin{bmatrix} 3 \\ 6 \end{bmatrix} + \begin{bmatrix} -1 \\ 1 \end{bmatrix}[9] \end{array} \right]
$$

$$
= \left[\begin{array}{cc|c} -3 & -3 & -3 \\ 8 & 10 & 12 \end{array} \right].
$$

Compute this product without using a partition, and see whether you find the same result. ∎

In this simple example, using a partition is somewhat pointless; Example 3 merely illustrates Fact 2.4.12. Example 4 shows a more sensible application of the concept of partitioned matrices.

EXAMPLE 4 Let A be a partitioned matrix

$$
A = \begin{bmatrix} A_{11} & A_{12} \\ 0 & A_{22} \end{bmatrix},
$$

where A_{11} is an $n \times n$ matrix, A_{22} is an $m \times m$ matrix, and A_{12} is an $n \times m$ matrix.

a. For which choices of A_{11}, A_{12}, and A_{22} is A invertible?

b. If A is invertible, what is A^{-1} (in terms of A_{11}, A_{12}, A_{22})?

Solution

We are looking for an $(n + m) \times (n + m)$ matrix B such that

$$BA = I_{n+m} = \begin{bmatrix} I_n & 0 \\ 0 & I_m \end{bmatrix}.$$

Let us partition B in the same way as A:

$$B = \begin{bmatrix} B_{11} & B_{12} \\ B_{21} & B_{22} \end{bmatrix},$$

where B_{11} is $n \times n$, B_{22} is $m \times m$, etc. The fact that B is the inverse of A means that

$$\begin{bmatrix} B_{11} & B_{12} \\ B_{21} & B_{22} \end{bmatrix} \begin{bmatrix} A_{11} & A_{12} \\ 0 & A_{22} \end{bmatrix} = \begin{bmatrix} I_n & 0 \\ 0 & I_m \end{bmatrix},$$

or, using Fact 2.4.12,

$$\begin{vmatrix} B_{11}A_{11} = I_n \\ B_{11}A_{12} + B_{12}A_{22} = 0 \\ B_{21}A_{11} = 0 \\ B_{21}A_{12} + B_{22}A_{22} = I_m \end{vmatrix}.$$

We have to solve this system for the submatrices B_{ij}. By Equation 1, A_{11} must be invertible, and $B_{11} = A_{11}^{-1}$. By Equation 3, $B_{21} = 0$. (Multiply by A_{11}^{-1} from the right.) Equation 4 now simplifies to $B_{22}A_{22} = I_m$. Therefore, A_{22} must be invertible, and $B_{22} = A_{22}^{-1}$. Lastly, Equation 2 becomes $A_{11}^{-1}A_{12} + B_{12}A_{22} = 0$, or $B_{12}A_{22} = -A_{11}^{-1}A_{12}$, or $B_{12} = -A_{11}^{-1}A_{12}A_{22}^{-1}$. So,

a. A is invertible if (and only if) both A_{11} and A_{22} are invertible (no condition is imposed on A_{12}).

b. If A is invertible, then its inverse is

$$A^{-1} = \begin{bmatrix} A_{11}^{-1} & -A_{11}^{-1}A_{12}A_{22}^{-1} \\ 0 & A_{22}^{-1} \end{bmatrix}.$$

∎

Verify this result for the following example:

EXAMPLE 5

$$\begin{bmatrix} 1 & 1 & 1 & 2 & 3 \\ 1 & 2 & 4 & 5 & 6 \\ 0 & 0 & 1 & 0 & 0 \\ 0 & 0 & 0 & 1 & 0 \\ 0 & 0 & 0 & 0 & 1 \end{bmatrix}^{-1} = \begin{bmatrix} 2 & -1 & 2 & 1 & 0 \\ -1 & 1 & -3 & -3 & -3 \\ 0 & 0 & 1 & 0 & 0 \\ 0 & 0 & 0 & 1 & 0 \\ 0 & 0 & 0 & 0 & 1 \end{bmatrix}.$$

∎

EXERCISES

GOALS Compute matrix products column by column and entry by entry. Interpret matrix multiplication in terms of the underlying linear transformations. Use the rules of matrix algebra. Multiply partitioned matrices.

If possible, compute the matrix products in Exercises 1 through 13, using paper and pencil.

1. $\begin{bmatrix} 1 & 1 \\ 0 & 1 \end{bmatrix} \begin{bmatrix} 1 & 2 \\ 3 & 4 \end{bmatrix}$

2. $\begin{bmatrix} 1 & -1 \\ -2 & 2 \end{bmatrix} \begin{bmatrix} 7 & 5 \\ 3 & 1 \end{bmatrix}$

3. $\begin{bmatrix} 1 & 2 & 3 \\ 4 & 5 & 6 \end{bmatrix} \begin{bmatrix} 1 & 2 \\ 3 & 4 \end{bmatrix}$

4. $\begin{bmatrix} 1 & -1 \\ 0 & 2 \\ 2 & 1 \end{bmatrix} \begin{bmatrix} 3 & 2 \\ 1 & 0 \end{bmatrix}$

5. $\begin{bmatrix} 1 & 0 \\ 0 & 1 \\ 0 & 0 \end{bmatrix} \begin{bmatrix} a & b \\ c & d \end{bmatrix}$

6. $\begin{bmatrix} a & b \\ c & d \end{bmatrix} \begin{bmatrix} d & -b \\ -c & a \end{bmatrix}$

7. $\begin{bmatrix} 1 & 0 & -1 \\ 0 & 1 & 1 \\ 1 & -1 & -2 \end{bmatrix} \begin{bmatrix} 1 & 2 & 3 \\ 3 & 2 & 1 \\ 2 & 1 & 3 \end{bmatrix}$

8. $\begin{bmatrix} 0 & 1 \\ 0 & 0 \end{bmatrix} \begin{bmatrix} 0 & 1 \\ 0 & 0 \end{bmatrix}$

9. $\begin{bmatrix} 1 & 2 \\ 2 & 4 \end{bmatrix} \begin{bmatrix} -6 & 8 \\ 3 & -4 \end{bmatrix}$

10. $[\, 1 \quad 0 \quad -1 \,] \begin{bmatrix} 1 & 2 \\ 2 & 1 \\ 1 & 1 \end{bmatrix}$

11. $[\, 1 \quad 2 \quad 3 \,] \begin{bmatrix} 3 \\ 2 \\ 1 \end{bmatrix}$

12. $\begin{bmatrix} 1 \\ 2 \\ 3 \end{bmatrix} [\, 1 \quad 2 \quad 3 \,]$

13. $[\, 0 \quad 0 \quad 1 \,] \begin{bmatrix} a & b & c \\ d & e & f \\ g & h & k \end{bmatrix} \begin{bmatrix} 0 \\ 1 \\ 0 \end{bmatrix}$

14. For the matrices

$$A = \begin{bmatrix} 1 & 1 \\ 1 & 1 \end{bmatrix}, \quad B = [\, 1 \quad 2 \quad 3 \,], \quad C = \begin{bmatrix} 1 & 0 & -1 \\ 2 & 1 & 0 \\ 3 & 2 & 1 \end{bmatrix},$$

$$D = \begin{bmatrix} 1 \\ 1 \\ 1 \end{bmatrix}, \quad E = [5],$$

determine which of the 25 matrix products AA, AB, AC, $\ldots$, ED, EE are defined, and compute those that are defined.

15. Compute the matrix product

$$\begin{bmatrix} 1 & -2 & -5 \\ -2 & 5 & 11 \end{bmatrix} \begin{bmatrix} 8 & -1 \\ 1 & 2 \\ 1 & -1 \end{bmatrix}.$$

Explain why the result does not contradict Fact 2.4.9.

For two invertible $n \times n$ matrices A and B, determine which of the formulas stated in Exercises 16 through 25 are necessarily true.

16. $(I_n - A)(I_n + A) = I_n - A^2$.

17. $(A + B)^2 = A^2 + 2AB + B^2$.

18. A^2 is invertible, and $(A^2)^{-1} = (A^{-1})^2$.

19. $A + B$ is invertible, and $(A + B)^{-1} = A^{-1} + B^{-1}$.

20. $(A - B)(A + B) = A^2 - B^2$.

21. $ABB^{-1}A^{-1} = I_n$.

22. $ABA^{-1} = B$.

23. $(ABA^{-1})^3 = AB^3A^{-1}$.

24. $(I_n + A)(I_n + A^{-1}) = 2I_n + A + A^{-1}$.

25. $A^{-1}B$ is invertible, and $(A^{-1}B)^{-1} = B^{-1}A$.

Use the given partitions to compute the products in Exercises 26 and 27. Check your work by computing the same products without using a partition. Show all your work.

26. $\left[\begin{array}{cc|cc} 1 & 0 & 1 & 0 \\ 0 & 1 & 0 & 1 \\ \hline 0 & 0 & 1 & 0 \\ 0 & 0 & 0 & 1 \end{array}\right] \left[\begin{array}{cc|cc} 1 & 2 & 2 & 3 \\ 3 & 4 & 4 & 5 \\ \hline 0 & 0 & 1 & 2 \\ 0 & 0 & 3 & 4 \end{array}\right]$

27. $\left[\begin{array}{cc|c} 1 & 0 & 0 \\ 0 & 1 & 0 \\ \hline 1 & 3 & 4 \end{array}\right] \left[\begin{array}{c|c} 1 & 0 \\ 2 & 0 \\ \hline 3 & 4 \end{array}\right]$

28. Find a nonzero 2×2 matrix A such that $A^2 = \begin{bmatrix} 0 & 0 \\ 0 & 0 \end{bmatrix}$.

29. Find a nonzero 2×2 matrix B such that
$$\begin{bmatrix} 1 & 3 \\ 2 & 6 \end{bmatrix} B = \begin{bmatrix} 0 & 0 \\ 0 & 0 \end{bmatrix}.$$

30. If A is a noninvertible $n \times n$ matrix, can you always find a nonzero $n \times n$ matrix B such that $AB = 0$?

31. For the matrix

$$B = \begin{bmatrix} 1 & 1 & 1 \\ 1 & 2 & 3 \end{bmatrix},$$

find a matrix A such that

$$BA = I_2.$$

How many solutions A does this problem have?

32. Can you find a 3×2 matrix A and a 2×3 matrix B such that the product AB is I_3? *Hint:* Consider any 3×2 matrix A and any 2×3 matrix B. There is a nonzero $\vec{x}$ in $\mathbb{R}^3$ such that $B\vec{x} = \vec{0}$. (Why?) Now think about $AB\vec{x}$.

33. Can you find a 3×2 matrix A and a 2×3 matrix B such that the product AB is invertible? (The hint in Exercise 32 still works.)

34. Consider two $n \times n$ matrices A and B such that the product AB is invertible. Show that the matrices A and B are both invertible. *Hint:* $AB(AB)^{-1} = I_n$ and $(AB)^{-1}AB = I_n$. Use Fact 2.4.9.

35. Consider an $m \times n$ matrix B and an $n \times m$ matrix A such that

$$BA = I_m.$$

a. Find all solutions of the linear system

$$A\vec{x} = \vec{0}.$$

b. Show that the linear system

$$B\vec{x} = \vec{b}$$

is consistent, for all vectors $\vec{b}$ in $\mathbb{R}^m$.

c. What can you say about rank(A)? What about rank(B)?

d. Explain why $m \leq n$.

36. Find all 2×2 matrices X such that $AX = B$, where

$$A = \begin{bmatrix} 1 & 2 \\ 2 & 4 \end{bmatrix}, \quad \text{and} \quad B = \begin{bmatrix} 0 & 0 \\ 0 & 0 \end{bmatrix}.$$

37. Find all 2×2 matrices X that commute with

$$A = \begin{bmatrix} 1 & 0 \\ 0 & 2 \end{bmatrix}.$$

38. Find all 2×2 matrices X that commute with

$$A = \begin{bmatrix} 1 & 2 \\ 0 & 3 \end{bmatrix}.$$

39. Find the 2×2 matrices X that commute with all 2×2 matrices.

40. Consider two 2×2 matrices, A and B. We are told that

$$B^{-1} = \begin{bmatrix} 1 & 2 \\ 3 & 5 \end{bmatrix}, \quad \text{and} \quad (AB)^{-1} = \begin{bmatrix} 1 & 3 \\ 2 & 5 \end{bmatrix}.$$

Find A.

41. Consider the matrix

$$D_\alpha = \begin{bmatrix} \cos\alpha & -\sin\alpha \\ \sin\alpha & \cos\alpha \end{bmatrix}.$$

We know that the linear transformation $T(\vec{x}) = D_\alpha \vec{x}$ is a counterclockwise rotation through an angle α.

a. For two angles, α and β, consider the products $D_\alpha D_\beta$ and $D_\beta D_\alpha$. Arguing geometrically, describe the linear transformations $\vec{y} = D_\alpha D_\beta \vec{x}$ and $\vec{y} = D_\beta D_\alpha \vec{x}$. Are the two transformations the same?

b. Now compute the products $D_\alpha D_\beta$ and $D_\beta D_\alpha$. Do the results make sense in terms of your answer in

part (a)? Recall the trigonometric identities

$$\sin(\alpha \pm \beta) = \sin\alpha \cos\beta \pm \cos\alpha \sin\beta$$
$$\cos(\alpha \pm \beta) = \cos\alpha \cos\beta \mp \sin\alpha \sin\beta.$$

42. Consider the lines P and Q in $\mathbb{R}^2$ sketched below. Consider the linear transformation $T(\vec{x}) = \text{ref}_Q(\text{ref}_P(\vec{x}))$, that is, we first reflect $\vec{x}$ in P and then we reflect the result in Q.

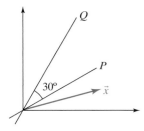

a. For the vector $\vec{x}$ given in the figure, sketch $T(\vec{x})$. What angle do the vectors $\vec{x}$ and $T(\vec{x})$ enclose? What is the relationship between the lengths of $\vec{x}$ and $T(\vec{x})$?

b. Use your answer in part (a) to describe the transformation T geometrically, as a reflection, rotation, shear, or projection.

c. Find the matrix of T.

43. Find a 2×2 matrix $A \neq I_2$ such that $A^3 = I_2$.

44. Find all linear transformations T from $\mathbb{R}^2$ to $\mathbb{R}^2$ such that $T\begin{bmatrix} 1 \\ 2 \end{bmatrix} = \begin{bmatrix} 2 \\ 1 \end{bmatrix}$ and $T\begin{bmatrix} 2 \\ 5 \end{bmatrix} = \begin{bmatrix} 1 \\ 3 \end{bmatrix}$. *Hint:* We are looking for the 2×2 matrices A such that $A\begin{bmatrix} 1 \\ 2 \end{bmatrix} = \begin{bmatrix} 2 \\ 1 \end{bmatrix}$ and $A\begin{bmatrix} 2 \\ 5 \end{bmatrix} = \begin{bmatrix} 1 \\ 3 \end{bmatrix}$. These two equations can be combined to form the matrix equation $A\begin{bmatrix} 1 & 2 \\ 2 & 5 \end{bmatrix} = \begin{bmatrix} 2 & 1 \\ 1 & 3 \end{bmatrix}$.

45. Using the last exercise as a guide, justify the following statement:

Let $\vec{v}_1, \vec{v}_2, \ldots, \vec{v}_n$ be vectors in $\mathbb{R}^n$ such that the matrix

$$S = \begin{bmatrix} | & | & & | \\ \vec{v}_1 & \vec{v}_2 & \cdots & \vec{v}_n \\ | & | & & | \end{bmatrix}$$

is invertible. Let $\vec{w}_1, \vec{w}_2, \ldots, \vec{w}_n$ be arbitrary vectors in $\mathbb{R}^m$. Then there is a unique linear transformation T from $\mathbb{R}^n$ to $\mathbb{R}^m$ such that $T(\vec{v}_i) = \vec{w}_i$, for all $i = 1, \ldots, n$. Find the matrix A of this transformation in terms of S and

$$B = \begin{bmatrix} | & | & & | \\ \vec{w}_1 & \vec{w}_2 & \cdots & \vec{w}_n \\ | & | & & | \end{bmatrix}.$$

46. Find the matrix A of the linear transformation T from $\mathbb{R}^2$ to $\mathbb{R}^3$ with

$$T\begin{bmatrix} 1 \\ 2 \end{bmatrix} = \begin{bmatrix} 7 \\ 5 \\ 3 \end{bmatrix} \quad \text{and} \quad T\begin{bmatrix} 2 \\ 5 \end{bmatrix} = \begin{bmatrix} 1 \\ 2 \\ 3 \end{bmatrix}$$

(compare with Exercise 45).

47. Find the matrix A of the linear transformation T from $\mathbb{R}^2$ to $\mathbb{R}^2$ with

$$T\begin{bmatrix} 3 \\ 1 \end{bmatrix} = 2\begin{bmatrix} 3 \\ 1 \end{bmatrix} \quad \text{and} \quad T\begin{bmatrix} 1 \\ 2 \end{bmatrix} = 3\begin{bmatrix} 1 \\ 2 \end{bmatrix}$$

(compare with Exercise 45).

48. Consider the regular tetrahedron sketched below, whose center is at the origin.

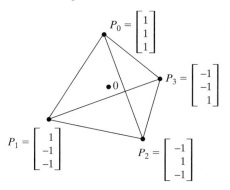

Let T from $\mathbb{R}^3$ to $\mathbb{R}^3$ be the rotation about the axis through the points 0 and P_2 that transforms P_1 into P_3. Find the images of the four corners of the tetrahedron under this transformation.

$$P_0 \xrightarrow{T}$$
$$P_1 \rightarrow P_3$$
$$P_2 \rightarrow$$
$$P_3 \rightarrow$$

Let L from $\mathbb{R}^3$ to $\mathbb{R}^3$ be the reflection in the plane through the points 0, P_0, and P_3. Find the images of the

four corners of the tetrahedron under this transformation.

$$P_0 \xrightarrow{L}$$
$$P_1 \rightarrow$$
$$P_2 \rightarrow$$
$$P_3 \rightarrow$$

Describe the transformations in parts (a) to (c) geometrically.

a. T^{-1}

b. L^{-1}

c. $T^2 = T \circ T$ (the composite of T with itself)

d. Find the images of the four corners under the transformations $T \circ L$ and $L \circ T$. Are the two transformations the same?

	$P_0 \xrightarrow{T \circ L}$		$P_0 \xrightarrow{L \circ T}$
	$P_1 \rightarrow$		$P_1 \rightarrow$
	$P_2 \rightarrow$		$P_2 \rightarrow$
	$P_3 \rightarrow$		$P_3 \rightarrow$

e. Find the images of the four corners under the transformation $L \circ T \circ L$. Describe this transformation geometrically.

49. Find the matrices of the transformations T and L defined in Exercise 48.

50. Consider the matrix

$$E = \begin{bmatrix} 1 & 0 & 0 \\ -3 & 1 & 0 \\ 0 & 0 & 1 \end{bmatrix}$$

and an arbitrary 3×3 matrix

$$A = \begin{bmatrix} a & b & c \\ d & e & f \\ g & h & k \end{bmatrix}.$$

a. Compute EA. Comment on the relationship between A and EA, in terms of the technique of elimination we learned in Section 1.2.

b. Consider the matrix

$$E = \begin{bmatrix} 1 & 0 & 0 \\ 0 & \frac{1}{4} & 0 \\ 0 & 0 & 1 \end{bmatrix}$$

and an arbitrary 3×3 matrix A. Compute EA. Comment on the relationship between A and EA.

c. Can you think of a 3×3 matrix E such that EA is obtained from A by swapping the last two rows (for any 3×3 matrix A)?

d. The matrices of the forms introduced in parts (a), (b), and (c) are called *elementary:* An $n \times n$ matrix E is elementary if it can be obtained from I_n by performing one of the three elementary row operations on I_n. Describe the format of the three types of elementary matrices.

51. Are elementary matrices invertible? If so, is the inverse of an elementary matrix elementary as well? Explain the significance of your answers in terms of elementary row operations.

52. a. Justify the following: If A is an $m \times n$ matrix, then there are elementary $m \times m$ matrices $E_1, E_2, \ldots, E_p$ such that
$$\text{rref}(A) = E_1 E_2 \cdots E_p A.$$

    b. Find such elementary matrices $E_1, E_2, \ldots, E_p$ for
$$A = \begin{bmatrix} 0 & 2 \\ 1 & 3 \end{bmatrix}.$$

53. a. Justify the following: If A is an $m \times n$ matrix, then there is an invertible $m \times m$ matrix S such that
$$\text{rref}(A) = SA.$$

    b. Find such an invertible matrix S for
$$A = \begin{bmatrix} 2 & 4 \\ 4 & 8 \end{bmatrix}.$$

54. a. Justify the following: Any invertible matrix is a product of elementary matrices.

    b. Write $A = \begin{bmatrix} 0 & 2 \\ 1 & 3 \end{bmatrix}$ as a product of elementary matrices.

55. Write all possible forms of elementary 2×2 matrices E. In each case, describe the transformation $\vec{y} = E\vec{x}$ geometrically.

56. Consider an invertible $n \times n$ matrix A and an $n \times n$ matrix B. A certain sequence of elementary row operations transforms A into I_n.

    a. What do you get when you apply the same row operations in the same order to the matrix AB?

    b. What do you get when you apply the same row operations to I_n?

57. Is the product of two lower triangular matrices a lower triangular matrix as well? Explain your answer.

58. Consider the matrix
$$A = \begin{bmatrix} 1 & 2 & 3 \\ 2 & 6 & 7 \\ 2 & 2 & 4 \end{bmatrix}.$$

    a. Find lower triangular elementary matrices $E_1, E_2, \ldots, E_m$ such that the product
$$E_m \cdots E_2 E_1 A$$
    is an upper triangular matrix U. *Hint:* Modify

Gauss–Jordan elimination as follows: In Step 3, eliminate only the entries in the cursor column *below* the leading 1's. This can be accomplished by means of *lower* triangular elementary matrices. Also, you can skip Step 2 (division by the cursor entry).

    b. Find lower triangular elementary matrices $M_1, M_2, \ldots, M_m$ and an upper triangular matrix U such that
$$A = M_1 M_2 \cdots M_m U.$$

    c. Find a lower triangular matrix L and an upper triangular matrix U such that
$$A = LU.$$

    Such a representation of an invertible matrix is called an *LU-factorization*. The method outlined in this exercise to find an LU-factorization can be streamlined somewhat, but we have seen the major ideas. An LU-factorization (as introduced here) does not always exist (see Exercise 60).

    d. Find a lower triangular matrix L with 1's on the diagonal, an upper triangular matrix U with 1's on the diagonal, and a diagonal matrix D such that $A = LDU$. Such a representation of an invertible matrix is called an *LDU-factorization*.

59. Knowing an LU-factorization of a matrix A makes it much easier to solve a linear system
$$A\vec{x} = \vec{b}.$$

Consider the LU-factorization
$$A = \begin{bmatrix} 1 & 2 & -1 & 4 \\ -3 & -5 & 6 & -5 \\ 1 & 4 & 6 & 20 \\ -1 & 6 & 20 & 43 \end{bmatrix}$$
$$= \begin{bmatrix} 1 & 0 & 0 & 0 \\ -3 & 1 & 0 & 0 \\ 1 & 2 & 1 & 0 \\ -1 & 8 & -5 & 1 \end{bmatrix} \begin{bmatrix} 1 & 2 & -1 & 4 \\ 0 & 1 & 3 & 7 \\ 0 & 0 & 1 & 2 \\ 0 & 0 & 0 & 1 \end{bmatrix} = LU.$$

Suppose we have to solve the system $A\vec{x} = LU\vec{x} = \vec{b}$, where
$$\vec{b} = \begin{bmatrix} -3 \\ 14 \\ 9 \\ 33 \end{bmatrix}.$$

    a. Set $\vec{y} = U\vec{x}$, and solve the system $L\vec{y} = \vec{b}$, by forward substitution (finding first y_1, then y_2, etc.) Do this using paper and pencil. Show all your work.

    b. Solve the system $U\vec{x} = \vec{y}$, by back substitution, to find the solution $\vec{x}$ of the system $A\vec{x} = \vec{b}$.

Do this using paper and pencil. Show all your work.

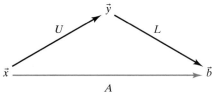

60. Show that the matrix $A = \begin{bmatrix} 0 & 1 \\ 1 & 0 \end{bmatrix}$ cannot be written in the form $A = LU$, where L is lower triangular and U is upper triangular.

61. In this exercise we will examine which invertible $n \times n$ matrices A admit an LU-factorization $A = LU$, as discussed in Exercise 58. The following definition will be useful: For $m = 1, \ldots, n$ the *principal submatrix* $A^{(m)}$ of A is obtained by omitting all rows and columns of A past the mth. For example, the matrix

$$A = \begin{bmatrix} 1 & 2 & 3 \\ 4 & 5 & 6 \\ 7 & 8 & 7 \end{bmatrix}$$

has the principal submatrices

$$A^{(1)} = [1], \quad A^{(2)} = \begin{bmatrix} 1 & 2 \\ 4 & 5 \end{bmatrix}, \quad A^{(3)} = A = \begin{bmatrix} 1 & 2 & 3 \\ 4 & 5 & 6 \\ 7 & 8 & 7 \end{bmatrix}.$$

We will show that an invertible $n \times n$ matrix A admits an LU-factorization $A = LU$ if (and only if) all its principal submatrices are invertible.

  a. Let $A = LU$ be an LU-factorization of an $n \times n$ matrix A. Use partitioned matrices to show that $A^{(m)} = L^{(m)} U^{(m)}$ for $m = 1, \ldots, n$.

  b. Use part (a) to show that if an invertible $n \times n$ matrix A has an LU-factorization, then all its principal submatrices $A^{(m)}$ are invertible.

  c. Consider an $n \times n$ matrix A whose principal submatrices are all invertible. Show that A admits an LU-factorization. *Hint:* By induction, you can assume that $A^{(n-1)}$ has an LU-factorization $A^{(n-1)} = L'U'$. Use partitioned matrices to find an LU-factorization for A. Alternatively, you can explain this result in terms of Gauss–Jordan elimination (if all principal submatrices are invertible, then no row swaps are required).

62. a. Show that if an invertible $n \times n$ matrix A admits an LU-factorization, then it admits an LDU-factorization (see Exercise 58(d)).

  b. Show that if an invertible $n \times n$ matrix A admits an LDU-factorization, then this factorization is unique. *Hint:* Suppose that $A = L_1 D_1 U_1 = L_2 D_2 U_2$. Then

$$U_2 U_1^{-1} = D_2^{-1} L_2^{-1} L_1 D_1 \text{ is diagonal (why?).}$$

Conclude that $U_2 = U_1$.

63. Prove the *distributive laws* for matrices:

$$A(C + D) = AC + AD$$

and

$$(A + B)C = AC + BC.$$

64. Consider an $m \times n$ matrix A, an $n \times p$ matrix B, and a scalar k. Show that

$$(kA)B = A(kB) = k(AB).$$

65. Consider a partitioned matrix

$$A = \begin{bmatrix} A_{11} & 0 \\ 0 & A_{22} \end{bmatrix},$$

where A_{11} and A_{22} are square matrices. For which choices of A_{11} and A_{22} is A invertible? In these cases, what is A^{-1}?

66. Consider a partitioned matrix

$$A = \begin{bmatrix} A_{11} & 0 \\ A_{21} & A_{22} \end{bmatrix},$$

where A_{11} and A_{22} are square matrices. For which choices of A_{11}, A_{21}, and A_{22} is A invertible? In these cases, what is A^{-1}?

67. Consider two matrices A and B whose product AB is defined. Describe the ith row of the product AB in terms of the rows of A and the matrix B.

68. Consider the partitioned matrices

$$A = \begin{bmatrix} k & \vec{v} \\ 0 & B \end{bmatrix} \quad \text{and} \quad S = \begin{bmatrix} 1 & 0 \\ 0 & R \end{bmatrix},$$

where B and R are $n \times n$ matrices (R is invertible), k is a scalar, and $\vec{v}$ is a row vector with n components. Compute $S^{-1} A S$.

69. Consider the partitioned matrix

$$A = \begin{bmatrix} A_{11} & A_{12} & A_{13} \\ 0 & 0 & A_{23} \end{bmatrix},$$

where A_{11} is an invertible matrix. Determine the rank of A in terms of the ranks of the matrices A_{11}, A_{12}, A_{13}, and A_{23}.

70. Consider the partitioned matrix

$$A = \begin{bmatrix} I_n & \vec{v} \\ \vec{w} & 1 \end{bmatrix},$$

where $\vec{v}$ is a vector in $\mathbb{R}^n$, and $\vec{w}$ is a row vector with n components. For which choices of $\vec{v}$ and $\vec{w}$ is A invertible? In these cases, what is A^{-1}?

71. Find all invertible $n \times n$ matrices A such that $A^2 = A$.

72. Find a nonzero $n \times n$ matrix A whose entries are identical such that $A^2 = A$.

$T\left(\begin{bmatrix} R \\ G \\ B \end{bmatrix}\right) = \begin{bmatrix} R \\ G \\ 0 \end{bmatrix}$ $R\begin{bmatrix} 1 \\ 0 \\ 0 \end{bmatrix} + G\begin{bmatrix} 0 \\ 1 \\ 0 \end{bmatrix} + B\begin{bmatrix} 0 \\ 0 \\ 0 \end{bmatrix}$

73. Consider two $n \times n$ matrices A and B whose entries are positive or zero. Suppose that all entries of A are less than or equal to s, and all column sums of B are less than or equal to r (the jth column sum of a matrix is the sum of the entries in its jth column). Show that all entries of the matrix AB are less than or equal to sr.

$\begin{bmatrix} 1 & 0 & 0 \\ 0 & 1 & 0 \\ 0 & 0 & 0 \end{bmatrix}$

74. (This exercise builds on Exercise 73.) Consider an $n \times n$ matrix A whose entries are positive or zero. Suppose that all column sums of A are less than 1. Let r be the largest column sum of A.

   a. Show that the entries of A^m are less than or equal to r^m, for all positive integers m.

   b. Show that

   $$\lim_{m \to \infty} A^m = 0$$

   (meaning that all entries of A^m approach zero).

   c. Show that the infinite series

   $$I_n + A + A^2 + \cdots + A^m + \cdots$$

   converges (entry by entry).

   d. Compute the product

   $$(I_n - A)(I_n + A + A^2 + \cdots + A^m).$$

   Simplify the result. Then let m go to infinity, and thus show that

   $$(I_n - A)^{-1} = I_n + A + A^2 + \cdots + A^m + \cdots.$$

75. (This exercise builds on Exercises 73 and 74 above, as well as Exercise 2.3.49.)

   a. Consider the industries $J_1, \ldots, J_n$ in an economy. We say that industry J_j is *productive* if the jth column sum of the technology matrix A is less than 1. What does this mean in terms of economics?

   b. We say that an economy is productive if all of its industries are productive. Exercise 74 shows that if A is the technology matrix of a productive economy, then the matrix $I_n - A$ is invertible. What does this result tell you about the ability of a productive economy to satisfy any kind of consumer demand?

   c. Interpret the formula

   $$(I_n - A)^{-1} = I_n + A + A^2 + \cdots + A^m + \cdots$$

   derived in Exercise 74(d) in terms of economics.

76. The color of light can be represented in a vector

   $$\begin{bmatrix} R \\ G \\ B \end{bmatrix},$$

where R = amount of red, G = amount of green, and B = amount of blue. The human eye and the brain transform the incoming signal into the signal

$$\begin{bmatrix} I \\ L \\ S \end{bmatrix},$$

where

$$\text{intensity} \quad I = \frac{R + G + B}{3}$$

$$\text{long-wave signal} \quad L = R - G$$

$$\text{short-wave signal} \quad S = B - \frac{R + G}{2}.$$

   a. Find the matrix P representing the transformation from

   $$\begin{bmatrix} R \\ G \\ B \end{bmatrix} \quad \text{to} \quad \begin{bmatrix} I \\ L \\ S \end{bmatrix}.$$

   b. Consider a pair of yellow sunglasses for water sports that cuts out all blue light and passes all red and green light. Find the 3×3 matrix A that represents the transformation incoming light undergoes as it passes through the sunglasses.

   c. Find the matrix for the composite transformation that light undergoes as it first passes through the sunglasses and then the eye.

   d. As you put on the sunglasses, the signal you receive (intensity, long- and short-wave signals) undergoes a transformation. Find the matrix M of this transformation.

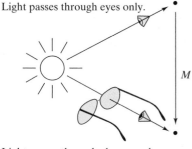

Light passes through eyes only.

M

Light passes through glasses and then through eyes.

77. A village is divided into three mutually exclusive groups called *clans*. Each person in the village belongs to a clan, and this identification is permanent. There are rigid rules concerning marriage: A person from one clan can marry only a person from one other clan. These rules are

encoded in the matrix A below. The fact that the 2-3 entry is 1 indicates that marriage between a man from clan III and a woman from clan II is allowed. The clan of a child is determined by the mother's clan, as indicated by the matrix B. According to this scheme siblings belong to the same clan.

$$
A = \begin{bmatrix} 0 & 1 & 0 \\ 0 & 0 & 1 \\ 1 & 0 & 0 \end{bmatrix} \begin{matrix} \text{I} \\ \text{II} \\ \text{III} \end{matrix} \quad \text{Wife's clan} \qquad B = \begin{bmatrix} 1 & 0 & 0 \\ 0 & 0 & 1 \\ 0 & 1 & 0 \end{bmatrix} \begin{matrix} \text{I} \\ \text{II} \\ \text{III} \end{matrix} \quad \text{Child's clan}
$$

Husband's clan: I II III
Mother's clan: I II III

The identification of a person with clan I can be represented by the vector

$$
\vec{e}_1 = \begin{bmatrix} 1 \\ 0 \\ 0 \end{bmatrix},
$$

and likewise for the two other clans. Matrix A transforms the husband's clan into the wife's clan (if $\vec{x}$ represents the husband's clan, then $A\vec{x}$ represents the wife's clan).

a. Are the matrices A and B invertible? Find the inverses if they exist. What do your answers mean, in practical terms?

b. What is the meaning of B^2, in terms of the rules of the community?

c. What is the meaning of AB and BA, in terms of the rules of the community? Are AB and BA the same?

d. Bueya is a young woman who has many male first cousins, both on her mother's and on her father's sides. The kinship between Bueya and each of her male cousins can be represented by one of the four diagrams below:

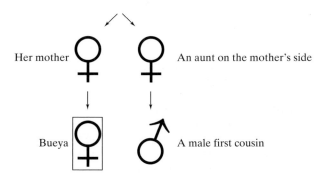

Her mother — An aunt on the mother's side

Bueya — A male first cousin

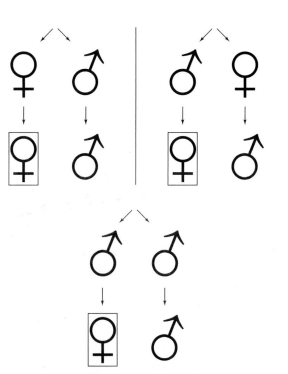

In each of the four cases, find the matrix which gives you the cousin's clan in terms of Bueya's clan.

e. According to the rules of the village, could Bueya marry a first cousin? (We do not know Bueya's clan.)

78. As background to this exercise, see Exercise 2.3.45.

a. If you use Fact 2.4.4, how many multiplications of scalars are necessary to multiply two 2×2 matrices?

b. If you use Fact 2.4.4, how many multiplications are needed to multiply an $m \times n$ and an $n \times p$ matrix?

In 1969, the German mathematician Volker Strassen surprised the mathematical community by showing that two 2×2 matrices can be multiplied with only seven multiplications of numbers. Here is his trick: Suppose you have to find AB for $A = \begin{bmatrix} a & b \\ c & d \end{bmatrix}$ and $B = \begin{bmatrix} p & q \\ r & s \end{bmatrix}$. First compute

$$
\begin{aligned}
h_1 &= (a+d)(p+s) \\
h_2 &= (c+d)p \\
h_3 &= a(q-s) \\
h_4 &= d(r-p) \\
h_5 &= (a+b)s \\
h_6 &= (c-a)(p+q) \\
h_7 &= (b-d)(r+s).
\end{aligned}
$$

Then

$$AB = \begin{bmatrix} h_1 + h_4 - h_5 + h_7 & h_3 + h_5 \\ h_2 + h_4 & h_1 + h_3 - h_2 + h_6 \end{bmatrix}.$$

79. Let $\mathbb{N}$ be the set of all positive integers, 1, 2, 3, ... We define two functions f and g from $\mathbb{N}$ to $\mathbb{N}$:

$$f(x) = 2x, \quad \text{for all } x \text{ in } \mathbb{N}$$
$$g(x) = \begin{cases} x/2 & \text{if } x \text{ is even} \\ (x+1)/2 & \text{if } x \text{ is odd} \end{cases}$$

Find formulas for the composite functions $g(f(x))$ and $f(g(x))$. Is one of them the identity transformation from $\mathbb{N}$ to $\mathbb{N}$? Are the functions f and g invertible?

80. *Geometrical Optics.* Consider a thin bi-convex lens with two spherical faces.

This is a good model for the lens of the human eye and for the lenses used in many optical instruments, such as reading glasses, cameras, microscopes, and telescopes. The line through the centers of the spheres defining the two faces is called the *optical axis* of the lens.

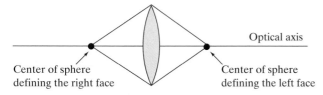

Center of sphere defining the right face

Center of sphere defining the left face

Optical axis

In this exercise, we learn how we can track the path of a ray of light as it passes through the lens, provided that the following conditions are satisfied:

- The ray lies in a plane with the optical axis.
- The angle the ray makes with the optical axis is small.

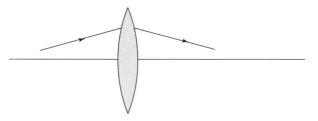

To keep track of the ray, we introduce two *reference planes* perpendicular to the optical axis, to the left and to the right of the lens.

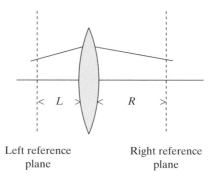

Left reference plane

Right reference plane

We can characterize the incoming ray by its slope m and its intercept x with the left reference plane. Likewise, we characterize the outgoing ray by slope n and intercept y.

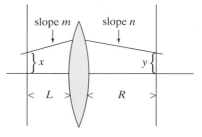

We want to know how the outgoing ray depends on the incoming ray, that is, we are interested in the transformation

$$T: \mathbb{R}^2 \to \mathbb{R}^2; \qquad \begin{bmatrix} x \\ m \end{bmatrix} \to \begin{bmatrix} y \\ n \end{bmatrix}.$$

We will see that T can be approximated by a linear transformation provided that m is small, as we assumed. To study this transformation, we divide the path of the ray into three segments, as shown in the following figure:

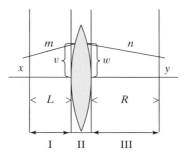

We have introduced two auxiliary reference planes, directly to the left and to the right of the lens. Our transformation $\begin{bmatrix} x \\ m \end{bmatrix} \to \begin{bmatrix} y \\ n \end{bmatrix}$ can now be represented as the composite of three simpler transformations:

$$\begin{bmatrix} x \\ m \end{bmatrix} \rightarrow \begin{bmatrix} v \\ m \end{bmatrix} \rightarrow \begin{bmatrix} w \\ n \end{bmatrix} \rightarrow \begin{bmatrix} y \\ n \end{bmatrix}.$$

From the definition of the slope of a line we get the relations $v = x + Lm$ and $y = w + Rn$.

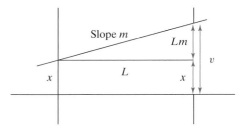

$$\begin{bmatrix} v \\ m \end{bmatrix} = \begin{bmatrix} x + Lm \\ m \end{bmatrix} = \begin{bmatrix} 1 & L \\ 0 & 1 \end{bmatrix} \begin{bmatrix} x \\ m \end{bmatrix}; \quad \begin{bmatrix} y \\ n \end{bmatrix} = \begin{bmatrix} 1 & R \\ 0 & 1 \end{bmatrix} \begin{bmatrix} w \\ n \end{bmatrix}$$

$$\begin{bmatrix} x \\ m \end{bmatrix} \xrightarrow{\begin{bmatrix} 1 & L \\ 0 & 1 \end{bmatrix}} \begin{bmatrix} v \\ m \end{bmatrix} \rightarrow \begin{bmatrix} w \\ n \end{bmatrix} \xrightarrow{\begin{bmatrix} 1 & R \\ 0 & 1 \end{bmatrix}} \begin{bmatrix} y \\ n \end{bmatrix}$$

It would lead us too far into physics to derive a formula for the transformation

$$\begin{bmatrix} v \\ m \end{bmatrix} \rightarrow \begin{bmatrix} w \\ n \end{bmatrix}$$

here.[13] Under the assumptions we have made, the transformation is well approximated by

$$\begin{bmatrix} w \\ n \end{bmatrix} = \begin{bmatrix} 1 & 0 \\ -k & 1 \end{bmatrix} \begin{bmatrix} v \\ m \end{bmatrix},$$

for some positive constant k (this formula implies that $w = v$).

$$\begin{bmatrix} x \\ m \end{bmatrix} \xrightarrow{\begin{bmatrix} 1 & L \\ 0 & 1 \end{bmatrix}} \begin{bmatrix} v \\ m \end{bmatrix} \xrightarrow{\begin{bmatrix} 1 & 0 \\ -k & 1 \end{bmatrix}} \begin{bmatrix} w \\ n \end{bmatrix} \xrightarrow{\begin{bmatrix} 1 & R \\ 0 & 1 \end{bmatrix}} \begin{bmatrix} y \\ n \end{bmatrix}$$

The transformation $\begin{bmatrix} x \\ m \end{bmatrix} \rightarrow \begin{bmatrix} y \\ n \end{bmatrix}$ is represented by the matrix product

$$\begin{bmatrix} 1 & R \\ 0 & 1 \end{bmatrix} \begin{bmatrix} 1 & 0 \\ -k & 1 \end{bmatrix} \begin{bmatrix} 1 & L \\ 0 & 1 \end{bmatrix} = \begin{bmatrix} 1 - Rk & L + R - kLR \\ -k & 1 - kL \end{bmatrix}.$$

a. *Focusing parallel rays.* Consider the lens in the human eye, with the retina as the right reference plane. In an adult, the distance R is about 0.025 meters (about 1 inch). The ciliary muscles allow you to vary the shape of the lens and thus the

lens constant k, within a certain range. What value of k enables you to focus parallel incoming rays, as shown in the figure? This value of k will allow you to see a distant object clearly. (The customary unit of measurement for k is 1 diopter $= \frac{1}{1 \text{ meter}}$.)

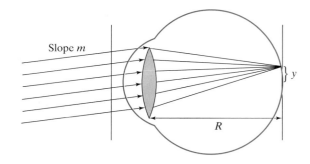

Hint: In terms of the transformation

$$\begin{bmatrix} x \\ m \end{bmatrix} \rightarrow \begin{bmatrix} y \\ n \end{bmatrix},$$

you want y to be independent of x (y must depend on the slope m alone). Explain why $1/k$ is called the *focal length* of the lens.

b. What value of k enables you to read this text from a distance of $L = 0.3$ meters? Consider the following figure (which is not to scale):

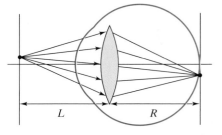

c. *The telescope.* An astronomical telescope consists of two lenses with the same optical axis.

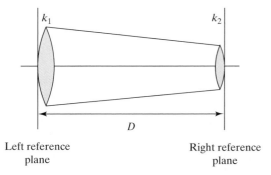

Left reference plane

Right reference plane

---

[13] See, for example, Paul Bamberg and Shlomo Sternberg: *A Course in Mathematics for Students of Physics 1,* Cambridge University Press, 1991.

Find the matrix of the transformation $\begin{bmatrix} x \\ m \end{bmatrix} \rightarrow \begin{bmatrix} y \\ n \end{bmatrix}$, in terms of k_1, k_2, and D. For given values of k_1 and k_2, how do you choose D so that parallel incoming rays are converted into parallel outgoing rays? What is the relationship between D and the focal lengths of the two lenses, $1/k_1$ and $1/k_2$?

C h a p t e r   2

TRUE OR FALSE?

1. The function $T \begin{bmatrix} x \\ y \end{bmatrix} = \begin{bmatrix} x - y \\ y - x \end{bmatrix}$ is a linear transformation.

2. Matrix $\begin{bmatrix} 1/2 & -1/2 \\ 1/2 & 1/2 \end{bmatrix}$ represents a rotation.

3. If A is any invertible $n \times n$ matrix, then $\text{rref}(A) = I_n$.

4. The formula $(A^2)^{-1} = (A^{-1})^2$ holds for all invertible matrices A.

5. The formula $AB = BA$ holds for all $n \times n$ matrices A and B.

6. If $AB = I_n$ for two $n \times n$ matrices A and B, then A must be the inverse of B.

7. If A is a 3×4 matrix and B is a 4×5 matrix, then AB will be a 5×3 matrix.

8. The function $T \begin{bmatrix} x \\ y \end{bmatrix} = \begin{bmatrix} y \\ 1 \end{bmatrix}$ is a linear transformation.

9. The matrix $\begin{bmatrix} 5 & 6 \\ -6 & 5 \end{bmatrix}$ represents a rotation–dilation.

10. If A is any invertible $n \times n$ matrix, then A commutes with A^{-1}.

11. Matrix $\begin{bmatrix} 1 & 2 \\ 3 & 6 \end{bmatrix}$ is invertible.

12. Matrix $\begin{bmatrix} 1 & 1 & 1 \\ 1 & 0 & 1 \\ 1 & 1 & 0 \end{bmatrix}$ is invertible.

13. There is an upper triangular 2×2 matrix A such that $A^2 = \begin{bmatrix} 1 & 1 \\ 0 & 1 \end{bmatrix}$.

14. The function $T \begin{bmatrix} x \\ y \end{bmatrix} = \begin{bmatrix} (y + 1)^2 - (y - 1)^2 \\ (x - 3)^2 - (x + 3)^2 \end{bmatrix}$ is a linear transformation.

15. Matrix $\begin{bmatrix} k & -2 \\ 5 & k - 6 \end{bmatrix}$ is invertible for all real numbers k.

16. There is a real number k such that the matrix $\begin{bmatrix} k - 1 & -2 \\ -4 & k - 3 \end{bmatrix}$ fails to be invertible.

17. There is a real number k such that the matrix $\begin{bmatrix} k - 2 & 3 \\ -3 & k - 2 \end{bmatrix}$ fails to be invertible.

18. Matrix $\begin{bmatrix} -0.6 & 0.8 \\ -0.8 & -0.6 \end{bmatrix}$ represents a rotation.

19. The formula $\det(2A) = 2 \det(A)$ holds for all 2×2 matrices A.

20. There is a matrix A such that
$$\begin{bmatrix} 1 & 2 \\ 3 & 4 \end{bmatrix} A \begin{bmatrix} 5 & 6 \\ 7 & 8 \end{bmatrix} = \begin{bmatrix} 1 & 1 \\ 1 & 1 \end{bmatrix}.$$

21. There is a matrix A such that $A \begin{bmatrix} 1 & 1 \\ 1 & 1 \end{bmatrix} = \begin{bmatrix} 1 & 2 \\ 1 & 2 \end{bmatrix}$.

22. There is a matrix A such that $\begin{bmatrix} 1 & 2 \\ 1 & 2 \end{bmatrix} A = \begin{bmatrix} 1 & 1 \\ 1 & 1 \end{bmatrix}$.

23. Matrix $\begin{bmatrix} -1 & 2 \\ -2 & 3 \end{bmatrix}$ represents a shear.

24. $\begin{bmatrix} 1 & k \\ 0 & 1 \end{bmatrix}^3 = \begin{bmatrix} 1 & 3k \\ 0 & 1 \end{bmatrix}$ for all real numbers k.

25. The matrix product $\begin{bmatrix} a & b \\ c & d \end{bmatrix} \begin{bmatrix} d & -b \\ -c & a \end{bmatrix}$ is always a scalar multiple of I_2.

26. There is a nonzero upper triangular 2×2 matrix A such that $A^2 = \begin{bmatrix} 0 & 0 \\ 0 & 0 \end{bmatrix}$.

27. There is a positive integer n such that $\begin{bmatrix} 0 & -1 \\ 1 & 0 \end{bmatrix}^n = I_2$.

28. There is an invertible 2×2 matrix A such that $A^{-1} = \begin{bmatrix} 1 & 1 \\ 1 & 1 \end{bmatrix}$.

29. There is an invertible $n \times n$ matrix with two identical rows.

30. If $A^2 = I_n$, then matrix A must be invertible.

31. If $A^{17} = I_2$, then A must be I_2.

32. If $A^2 = I_2$, then A must be either I_2 or $-I_2$.

33. If matrix A is invertible, then matrix $5A$ must be invertible as well.

34. If A and B are two 4×3 matrices such then $A\vec{v} = B\vec{v}$ for all vectors $\vec{v}$ in $\mathbb{R}^3$, then matrices A and B must be equal.

35. If matrices A and B commute, then the formula $A^2 B = B A^2$ must hold.

36. If $A^2 = A$ for an invertible $n \times n$ matrix A, then A must be I_n.

37. If matrices A and B are both invertible, then matrix $A + B$ must be invertible as well.

38. The equation $A^2 = A$ holds for all 2×2 matrices A representing an orthogonal projection.

39. If matrix $\begin{bmatrix} a & b & c \\ d & e & f \\ g & h & i \end{bmatrix}$ is invertible, then matrix

$\begin{bmatrix} a & b \\ d & e \end{bmatrix}$ must be invertible as well.

40. If A^2 is invertible, then matrix A itself must be invertible.

41. The equation $A^{-1} = A$ holds for all 2×2 matrices A representing a reflection.

42. The formula $(A\vec{v}) \cdot (A\vec{w}) = \vec{v} \cdot \vec{w}$ holds for all invertible 2×2 matrices A and for all vectors $\vec{v}$ and $\vec{w}$ in $\mathbb{R}^2$.

43. There exist a 2×3 matrix A and a 3×2 matrix B such that $AB = I_2$.

44. There exist a 3×2 matrix A and a 2×3 matrix B such that $AB = I_3$.

45. If $A^2 + 3A + 4I_3 = 0$ for a 3×3 matrix A, then A must be invertible.

46. If A is an $n \times n$ such that $A^2 = 0$, then matrix $I_n + A$ must be invertible.

47. If matrix A represents a shear, then the formula $A^2 - 2A + I_2 = 0$ must hold.

48. If T is any linear transformation from $\mathbb{R}^3$ to $\mathbb{R}^3$, then $T(\vec{v} \times \vec{w}) = T(\vec{v}) \times T(\vec{w})$ for all vectors $\vec{v}$ and $\vec{w}$ in $\mathbb{R}^3$.

49. There is an invertible 10×10 matrix that has 92 ones among its entries.

50. The formula $\text{rref}(AB) = \text{rref}(A)\text{rref}(B)$ holds for all $m \times n$ matrices A and for all $n \times p$ matrices B.

CHAPTER 3

Subspaces of $\mathbb{R}^n$ and Their Dimensions

3.1 IMAGE AND KERNEL OF A LINEAR TRANSFORMATION

You may be familiar with the notion of the *image* of a function.

Definition 3.1.1

Image
The *image* of a function consists of all the values the function takes in its codomain. If f is a function from X to Y, then

$$\text{image}(f) = \{f(x) : x \text{ in } X\}$$
$$= \{y \text{ in } Y : y = f(x), \text{ for some } x \text{ in } X\}.[1]$$

Note that image(f) is a subset of the codomain Y of f.

EXAMPLE 1   Consider the function in Figure 1, whose domain and codomain are finite sets. ■

EXAMPLE 2   The image of the exponential function f from $\mathbb{R}$ to $\mathbb{R}$ given by

$$f(x) = e^x$$

consists of all positive numbers. ■

---

[1] Some authors use the term "range" for what we call the image, while others use the term "range" for what we call the codomain. Make sure to check which definition is used when you encounter the term "range" in a text.

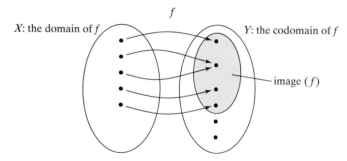

Figure 1

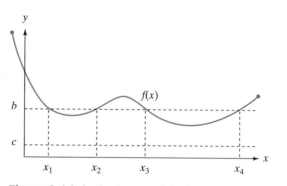

Figure 2 b is in the image of f, since $b = f(x_1) = f(x_2) = f(x_3) = f(x_4)$; c is *not* in the image of f, because there is no x such that $c = f(x)$.

EXAMPLE 3 More generally, the image of a function f from $\mathbb{R}$ to $\mathbb{R}$ consists of all numbers b such that the line $y = b$ intersects the graph of f (possibly more than once), as illustrated in Figure 2. ∎

EXAMPLE 4 The image of the function f from $\mathbb{R}$ to $\mathbb{R}^2$ given by

$$f(t) = \begin{bmatrix} \cos(t) \\ \sin(t) \end{bmatrix}$$

is the unit circle. (See Figure 3.)

   The function f is called a *parametrization* of the unit circle. More generally, a parametrization of a curve C in $\mathbb{R}^2$ is a function g from $\mathbb{R}$ to $\mathbb{R}^2$ whose image is C. ∎

EXAMPLE 5 If the function f from X to Y is invertible, then image$(f) = Y$. For each y in Y, there is one (and only one) x in X such that $y = f(x)$, namely, $x = f^{-1}(y)$. ∎

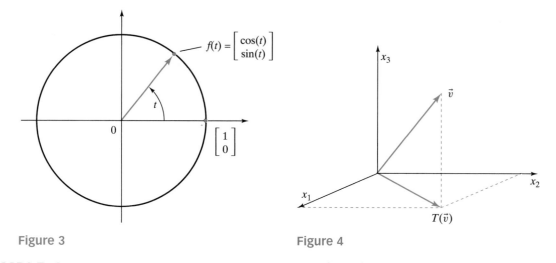

Figure 3   Figure 4

EXAMPLE 6   Consider the linear transformation T from $\mathbb{R}^3$ to $\mathbb{R}^3$ that projects a vector orthogonally into the x_1–x_2-plane, as illustrated in Figure 4.

The image of T is the x_1–x_2-plane in $\mathbb{R}^3$.   ■

EXAMPLE 7   Describe the image of the linear transformation T from $\mathbb{R}^2$ to $\mathbb{R}^2$ given by the matrix

$$A = \begin{bmatrix} 1 & 3 \\ 2 & 6 \end{bmatrix}.$$

Solution

The image of T consists of all values of T; that is, all vectors of the form

$$T\begin{bmatrix} x_1 \\ x_2 \end{bmatrix} = A\begin{bmatrix} x_1 \\ x_2 \end{bmatrix} = \begin{bmatrix} 1 & 3 \\ 2 & 6 \end{bmatrix}\begin{bmatrix} x_1 \\ x_2 \end{bmatrix} = x_1\begin{bmatrix} 1 \\ 2 \end{bmatrix} + x_2\begin{bmatrix} 3 \\ 6 \end{bmatrix}$$

$$= x_1\begin{bmatrix} 1 \\ 2 \end{bmatrix} + 3x_2\begin{bmatrix} 1 \\ 2 \end{bmatrix} = (x_1 + 3x_2)\begin{bmatrix} 1 \\ 2 \end{bmatrix}.$$

Since the vectors $\begin{bmatrix} 1 \\ 2 \end{bmatrix}$ and $\begin{bmatrix} 3 \\ 6 \end{bmatrix}$ are parallel, the image of T is the line of all scalar multiples of the vector $\begin{bmatrix} 1 \\ 2 \end{bmatrix}$, as illustrated in Figure 5.   ■

EXAMPLE 8   Describe the image of the linear transformation T from $\mathbb{R}^2$ to $\mathbb{R}^3$ given by the matrix

$$A = \begin{bmatrix} 1 & 1 \\ 1 & 2 \\ 1 & 3 \end{bmatrix}.$$

Solution

The image of T consists of all vectors of the form

$$T\begin{bmatrix} x_1 \\ x_2 \end{bmatrix} = \begin{bmatrix} 1 & 1 \\ 1 & 2 \\ 1 & 3 \end{bmatrix}\begin{bmatrix} x_1 \\ x_2 \end{bmatrix} = x_1\begin{bmatrix} 1 \\ 1 \\ 1 \end{bmatrix} + x_2\begin{bmatrix} 1 \\ 2 \\ 3 \end{bmatrix},$$

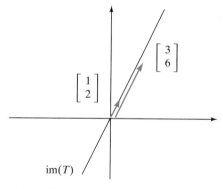

im(T)

Figure 5

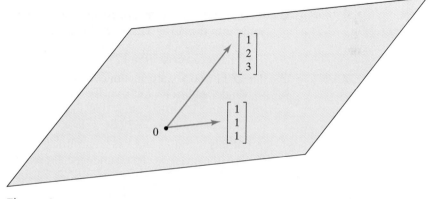

0

Figure 6

that is, all linear combinations of the vectors

$$\begin{bmatrix} 1 \\ 1 \\ 1 \end{bmatrix} \quad \text{and} \quad \begin{bmatrix} 1 \\ 2 \\ 3 \end{bmatrix}.$$

This is the plane "spanned" by the two vectors, in an intuitive, geometric sense. (See Figure 6.) ∎

The observations made in Examples 7 and 8 motivate the following definition.

Definition 3.1.2

Span
Consider the vectors $\vec{v}_1, \vec{v}_2, \ldots, \vec{v}_n$ in $\mathbb{R}^m$. The set of all linear combinations of the vectors $\vec{v}_1, \vec{v}_2, \ldots, \vec{v}_n$ is called their *span:*

$$\text{span}\,(\vec{v}_1, \ldots, \vec{v}_n) = \{c_1\vec{v}_1 + \cdots + c_n\vec{v}_n : c_i \text{ arbitrary scalars}\}$$

Fact 3.1.3   The image of a linear transformation

$$T(\vec{x}) = A\vec{x}$$

is the span of the columns of A.[2] We denote the image of T by im(T) or im(A).

To justify this fact, we write the transformation T in vector form as in Examples 7 and 8:

$$T(\vec{x}) = A\vec{x} = \begin{bmatrix} | & & | \\ \vec{v}_1 & \cdots & \vec{v}_n \\ | & & | \end{bmatrix} \begin{bmatrix} x_1 \\ x_2 \\ \vdots \\ x_n \end{bmatrix} = x_1\vec{v}_1 + x_2\vec{v}_2 + \cdots + x_n\vec{v}_n.$$

This shows that the image of T consists of all linear combinations of the vectors $\vec{v}_i$ (i.e., im(T) is the span of $\vec{v}_1, \vec{v}_2, \ldots, \vec{v}_n$).

Consider a linear transformation T from $\mathbb{R}^n$ to $\mathbb{R}^m$ given by $T(\vec{x}) = A\vec{x}$, for some $m \times n$ matrix A. The image of T has some remarkable properties:

- The zero vector $\vec{0}$ in $\mathbb{R}^m$ is contained in im(T), since $\vec{0} = A\vec{0} = T(\vec{0})$.
- If $\vec{v}_1$ and $\vec{v}_2$ are in im(T), then so is $\vec{v}_1 + \vec{v}_2$. (We say the image is closed under addition.) Let's verify this: Since $\vec{v}_1$ and $\vec{v}_2$ are in im(T), there are vectors $\vec{w}_1$ and $\vec{w}_2$ in $\mathbb{R}^n$ such that $\vec{v}_1 = T(\vec{w}_1)$ and $\vec{v}_2 = T(\vec{w}_2)$. Then, $\vec{v}_1 + \vec{v}_2 = T(\vec{w}_1) + T(\vec{w}_2) = T(\vec{w}_1 + \vec{w}_2)$, so that $\vec{v}_1 + \vec{v}_2$ is in the image as well.
- If $\vec{v}$ is in im(T) and k is an arbitrary scalar, then $k\vec{v}$ is in the image as well. (We say the image is closed under scalar multiplication). Again, let's verify this: The vector $\vec{v}$ can be written as $\vec{v} = T(\vec{w})$ for some $\vec{w}$ in $\mathbb{R}^n$. Then, $k\vec{v} = kT(\vec{w}) = T(k\vec{w})$, so that $k\vec{v}$ is in the image.

Let's summarize:

Fact 3.1.4   **Properties of the image**
The image of a linear transformation T (from $\mathbb{R}^n$ to $\mathbb{R}^m$) has the following properties:

a. The zero vector $\vec{0}$ in $\mathbb{R}^m$ is contained in im(T).
b. The image is closed under addition: If $\vec{v}_1$ and $\vec{v}_2$ are both in im(T), then so is $\vec{v}_1 + \vec{v}_2$.
c. The image is closed under scalar multiplication: If a vector $\vec{v}$ is in im(T) and k is an arbitrary scalar, then $k\vec{v}$ is in the image as well.

It follows from properties (b) and (c) that the image is closed under linear combinations: If some vectors $\vec{v}_1, \vec{v}_2, \ldots, \vec{v}_p$ are in the image and $c_1, c_2, \ldots, c_p$ are arbitrary scalars, then $c_1\vec{v}_1 + c_2\vec{v}_2 + \cdots + c_p\vec{v}_p$ is in the image as well.

---

[2]The image of T is also called the *column space* of A.

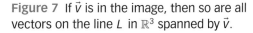

Figure 7 If $\vec{v}$ is in the image, then so are all vectors on the line L in $\mathbb{R}^3$ spanned by $\vec{v}$.

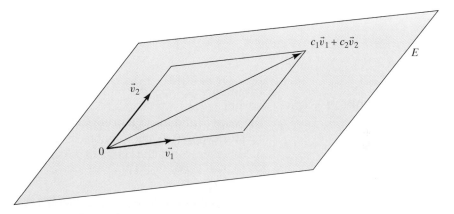

Figure 8 If $\vec{v}_1$ and $\vec{v}_2$ are in the image, then so are all vectors in the plane E in $\mathbb{R}^3$ spanned by $\vec{v}_1$ and $\vec{v}_2$.

If the codomain of the transformation is $\mathbb{R}^3$ (i.e., $m = 3$), we can interpret this property geometrically, as shown in Figures 7 and 8.

EXAMPLE 9   Consider an $n \times n$ matrix A. Show that $\text{im}(A^2)$ is contained in $\text{im}(A)$. (That is, each vector in $\text{im}(A^2)$ is also in $\text{im}(A)$.)

Solution

Consider a vector $\vec{w}$ in $\text{im}(A^2)$. By definition of the image, we can write $\vec{w} = A^2\vec{v} = AA\vec{v}$, for some vector $\vec{v}$ in $\mathbb{R}^n$. We have to show that $\vec{w}$ can be represented as $\vec{w} = A\vec{u}$, for some $\vec{u}$ in $\mathbb{R}^n$. Let $\vec{u} = A\vec{v}$. Then,

$$\vec{w} = AA\vec{v} = A\vec{u}.$$ ∎

The Kernel of a Linear Transformation

When you study functions $y = f(x)$ of one variable, you are often interested in the *zeros* of this function (i.e., the solutions of the equation $f(x) = 0$). For example, the function $y = \sin(x)$ has infinitely many zeros, namely, all integer multiples of π. The zeros of a linear transformation are of interest as well.

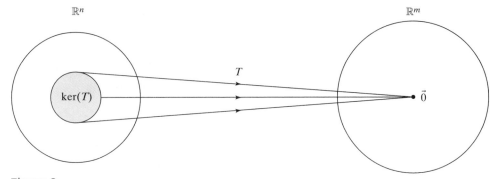

Figure 9

Definition 3.1.5

Kernel
The *kernel*[3] of a linear transformation $T(\vec{x}) = A\vec{x}$ is the set of all zeros of the transformation (i.e., the solutions of the equation $T(\vec{x}) = \vec{0}$). See Figure 9.
   In other words, the kernel of T is the solution set of the linear system

$$A\vec{x} = \vec{0}.$$

We denote the kernel of T by $\ker(T)$ or $\ker(A)$.

For a linear transformation T from $\mathbb{R}^n$ to $\mathbb{R}^m$,

- $\operatorname{im}(T)$ is a subset of the *codomain* $\mathbb{R}^m$ of T, and
- $\ker(T)$ is a subset of the *domain* $\mathbb{R}^n$ of T.

EXAMPLE 10   Consider again the orthogonal projection onto the x_1–x_2-plane, a linear transformation T from $\mathbb{R}^3$ to $\mathbb{R}^3$, as shown in Figure 10.
    The kernel of T consists of all vectors whose orthogonal projection onto the x_1–x_2-plane is $\vec{0}$. These are the vectors on the x_3-axis (the scalar multiples of $\vec{e}_3$). ∎

EXAMPLE 11   Find the kernel of the linear transformation T from $\mathbb{R}^3$ to $\mathbb{R}^2$ given by

$$T(\vec{x}) = \begin{bmatrix} 1 & 1 & 1 \\ 1 & 2 & 3 \end{bmatrix} \vec{x}.$$

Solution

We have to solve the linear system

$$T(\vec{x}) = \begin{bmatrix} 1 & 1 & 1 \\ 1 & 2 & 3 \end{bmatrix} \vec{x} = \vec{0}.$$

---

[3] The kernel of T is also called the *null space* of A.

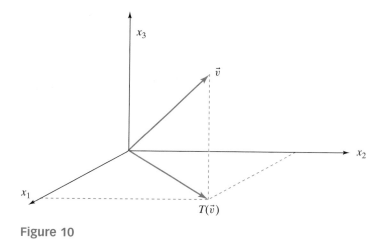

Figure 10

Since we have studied this problem carefully in Section 1.2, we will be brief.

$$\text{rref} \begin{bmatrix} 1 & 1 & 1 & \vdots & 0 \\ 1 & 2 & 3 & \vdots & 0 \end{bmatrix} = \begin{bmatrix} 1 & 0 & -1 & \vdots & 0 \\ 0 & 1 & 2 & \vdots & 0 \end{bmatrix}$$

$$\begin{vmatrix} x_1 & - & x_3 = 0 \\ & x_2 + 2x_3 = 0 \end{vmatrix}$$

$x_3 = t$

$\therefore x_2 = -2t$

$x_1 = t$

$$\begin{bmatrix} x_1 \\ x_2 \\ x_3 \end{bmatrix} = \begin{bmatrix} t \\ -2t \\ t \end{bmatrix} = t \begin{bmatrix} 1 \\ -2 \\ 1 \end{bmatrix}.$$

In the solution, t is an arbitrary constant. The kernel is the line spanned by the vector $\begin{bmatrix} 1 \\ -2 \\ 1 \end{bmatrix}$ in $\mathbb{R}^3$. ∎

Consider a linear transformation $T(\vec{x}) = A\vec{x}$ from $\mathbb{R}^n$ to $\mathbb{R}^m$, where n exceeds m (as in Example 11). There will be nonleading variables for the equation $T(\vec{x}) = A\vec{x} = \vec{0}$; that is, this system has infinitely many solutions. Therefore, the kernel of T consists of infinitely many vectors. This agrees with our intuition: We expect some collapsing to take place as we transform $\mathbb{R}^n$ into $\mathbb{R}^m$.

EXAMPLE 12   Find the kernel of the linear transformation T from $\mathbb{R}^5$ to $\mathbb{R}^4$ given by the matrix

$$A = \begin{bmatrix} 1 & 5 & 4 & 3 & 2 \\ 1 & 6 & 6 & 6 & 6 \\ 1 & 7 & 8 & 10 & 12 \\ 1 & 6 & 6 & 7 & 8 \end{bmatrix}.$$

Solution

We have to solve the linear system $T(\vec{x}) = A\vec{x} = \vec{0}$.

$$\text{rref}(A) = \begin{bmatrix} 1 & 0 & -6 & 0 & 6 \\ 0 & 1 & 2 & 0 & -2 \\ 0 & 0 & 0 & 1 & 2 \\ 0 & 0 & 0 & 0 & 0 \end{bmatrix}.$$

There is no need to write the zeros in the last column of the augmented matrix. The kernel of T consists of the solutions of the system

$$\begin{vmatrix} x_1 & - 6x_3 & + 6x_5 = 0 \\ & x_2 + 2x_3 & - 2x_5 = 0 \\ & & x_4 + 2x_5 = 0 \end{vmatrix}.$$

The solutions are the vectors

$$\vec{x} = \begin{bmatrix} x_1 \\ x_2 \\ x_3 \\ x_4 \\ x_5 \end{bmatrix} = \begin{bmatrix} 6s - 6t \\ -2s + 2t \\ s \\ -2t \\ t \end{bmatrix},$$

where s and t are arbitrary constants.

$$\text{ker}(T) = \left\{ \begin{bmatrix} 6s - 6t \\ -2s + 2t \\ s \\ -2t \\ t \end{bmatrix} : s, t \text{ arbitrary scalars} \right\}.$$

We can write

$$\begin{bmatrix} 6s - 6t \\ -2s + 2t \\ s \\ -2t \\ t \end{bmatrix} = s \begin{bmatrix} 6 \\ -2 \\ 1 \\ 0 \\ 0 \end{bmatrix} + t \begin{bmatrix} -6 \\ 2 \\ 0 \\ -2 \\ 1 \end{bmatrix}.$$

This shows that

$$\text{ker}(T) = \text{span} \left(\begin{bmatrix} 6 \\ -2 \\ 1 \\ 0 \\ 0 \end{bmatrix}, \begin{bmatrix} -6 \\ 2 \\ 0 \\ -2 \\ 1 \end{bmatrix} \right).$$

■

Consider a linear transformation T from $\mathbb{R}^n$ to $\mathbb{R}^m$ given by $T(\vec{x}) = A\vec{x}$, for some $m \times n$ matrix A.

The kernel of T has the following properties:

Fact 3.1.6   **Properties of the kernel**

a. The zero vector $\vec{0}$ in $\mathbb{R}^n$ is contained in ker(T).
b. The kernel is closed under addition.
c. The kernel is closed under scalar multiplication.

Compare these properties of the kernel with the corresponding properties of the image listed in Fact 3.1.4.

The verification of Fact 3.1.6 is straightforward and left as Exercise 49.

Consider an $m \times n$ matrix A. As we just observed in Fact 3.1.6, the kernel of A contains the zero vector in $\mathbb{R}^n$. It is possible that the kernel consists of the zero vector alone, for example, if $A = I_n$. In general, the system $A\vec{x} = \vec{0}$ has only the solution $\vec{x} = \vec{0}$ if (and only if) all the n variables are leading variables, that is, if rank(A) = n. For a square matrix A of size $n \times n$, this condition means that A is invertible.

Fact 3.1.7

a. Consider an $m \times n$ matrix A. Then

$$\ker(A) = \{\vec{0}\}$$

if (and only if) rank(A) = n. (This implies that $n \le m$.)
b. For a *square* matrix A,

$$\ker(A) = \{\vec{0}\}$$

if (and only if) A is invertible.

We conclude this section with a summary that relates many concepts we have introduced thus far.

Summary 3.1.8   Let A be an $n \times n$ matrix. The following statements are equivalent (i.e., they are either all true or all false):

i. A is invertible.
ii. The linear system $A\vec{x} = \vec{b}$ has a unique solution $\vec{x}$, for all $\vec{b}$ in $\mathbb{R}^n$.
iii. rref(A) = I_n.
iv. rank(A) = n.
v. im(A) = $\mathbb{R}^n$.
vi. ker(A) = $\{\vec{0}\}$.

Figure 11 briefly recalls the justification for these equivalences:

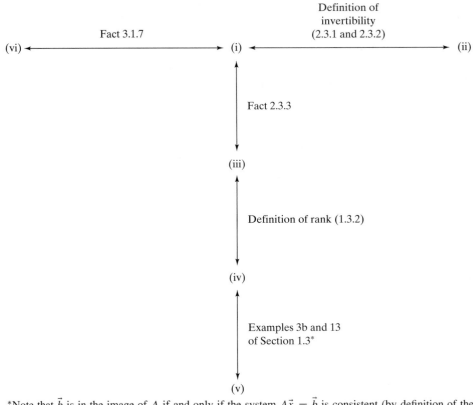

*Note that $\vec{b}$ is in the image of A if and only if the system $A\vec{x} = \vec{b}$ is consistent (by definition of the image).

Figure 11

EXERCISES

GOALS Use the concepts of the image and the kernel of a linear transformation (or a matrix). Express the image and the kernel of any matrix as the span of some vectors. Use kernel and image to determine whether a matrix is invertible.

For each matrix A in Exercises 1 through 13, find vectors that span the kernel of A. Use paper and pencil.

1. $A = \begin{bmatrix} 1 & 2 \\ 3 & 4 \end{bmatrix}$.

2. $A = \begin{bmatrix} 2 & 3 \\ 6 & 9 \end{bmatrix}$.

3. $A = \begin{bmatrix} 0 & 0 \\ 0 & 0 \end{bmatrix}$.

4. $A = \begin{bmatrix} 1 & 2 & 3 \end{bmatrix}$.

5. $A = \begin{bmatrix} 1 & 1 & 1 \\ 1 & 2 & 3 \\ 1 & 3 & 5 \end{bmatrix}$.

6. $A = \begin{bmatrix} 1 & 1 & 1 \\ 1 & 1 & 1 \\ 1 & 1 & 1 \end{bmatrix}$.

7. $A = \begin{bmatrix} 1 & 2 & 3 \\ 1 & 3 & 2 \\ 3 & 2 & 1 \end{bmatrix}$.

8. $A = \begin{bmatrix} 1 & 1 & 1 \\ 1 & 2 & 3 \end{bmatrix}$.

9. $A = \begin{bmatrix} 1 & 1 \\ 1 & 2 \\ 1 & 3 \end{bmatrix}$.

10. $A = \begin{bmatrix} 1 & 2 & 3 & 4 \\ 0 & 1 & 2 & 3 \\ 0 & 0 & 0 & 1 \end{bmatrix}$.

11. $A = \begin{bmatrix} 1 & 0 & 2 & 4 \\ 0 & 1 & -3 & -1 \\ 3 & 4 & -6 & 8 \\ 0 & -1 & 3 & 4 \end{bmatrix}$.

12. $A = \begin{bmatrix} 1 & -1 & -1 & 1 & 1 \\ -1 & 1 & 0 & -2 & 2 \\ 1 & -1 & -2 & 0 & 3 \\ 2 & -2 & -1 & 3 & 4 \end{bmatrix}$.

13. $A = \begin{bmatrix} 1 & 2 & 0 & 0 & 3 & 0 \\ 0 & 0 & 1 & 0 & 2 & 0 \\ 0 & 0 & 0 & 1 & 1 & 0 \\ 0 & 0 & 0 & 0 & 0 & 0 \\ 0 & 0 & 0 & 0 & 0 & 0 \end{bmatrix}.$

For each matrix A in Exercises 14 through 16, find vectors that span the image of A. Give as few vectors as possible. Use paper and pencil.

14. $A = \begin{bmatrix} 1 & 1 \\ 1 & 2 \\ 1 & 3 \\ 1 & 4 \end{bmatrix}.$
    15. $A = \begin{bmatrix} 1 & 1 & 1 & 1 \\ 1 & 2 & 3 & 4 \end{bmatrix}.$

16. $A = \begin{bmatrix} 1 & 2 & 3 \\ 1 & 2 & 3 \\ 1 & 2 & 3 \end{bmatrix}.$

For each matrix A in Exercises 17 through 22, describe the image of the transformation $T(\vec{x}) = A\vec{x}$ geometrically (as a line, plane, etc. in $\mathbb{R}^2$ or $\mathbb{R}^3$).

17. $A = \begin{bmatrix} 1 & 2 \\ 3 & 4 \end{bmatrix}.$
    18. $A = \begin{bmatrix} 1 & 4 \\ 3 & 12 \end{bmatrix}.$

19. $A = \begin{bmatrix} 1 & 2 & 3 & 4 \\ -2 & -4 & -6 & -8 \end{bmatrix}.$

20. $A = \begin{bmatrix} 1 & 1 & 1 \\ 1 & 1 & 1 \\ 1 & 1 & 1 \end{bmatrix}.$
    21. $A = \begin{bmatrix} 4 & 7 & 3 \\ 1 & 9 & 2 \\ 5 & 6 & 8 \end{bmatrix}.$

22. $A = \begin{bmatrix} 2 & 1 & 3 \\ 3 & 4 & 2 \\ 6 & 5 & 7 \end{bmatrix}.$

Describe the images and kernels of the transformations in Exercises 23 through 25 geometrically.

23. Reflection in the line $y = x/3$ in $\mathbb{R}^2$.

24. Orthogonal projection onto the plane $x + 2y + 3z = 0$ in $\mathbb{R}^3$.

25. Rotation through an angle of $\pi/4$ in the counterclockwise direction (in $\mathbb{R}^2$).

26. What is the image of a function f from $\mathbb{R}$ to $\mathbb{R}$ given by
$$f(t) = t^3 + at^2 + bt + c,$$
where a, b, c are arbitrary scalars?

27. Give an example of a noninvertible function f from $\mathbb{R}$ to $\mathbb{R}$ with $\mathrm{im}(f) = \mathbb{R}$.

28. Give an example of a parametrization of the ellipse
$$x^2 + \frac{y^2}{4} = 1$$
in $\mathbb{R}^2$. (See Example 4.)

29. Give an example of a function whose image is the unit sphere
$$x^2 + y^2 + z^2 = 1$$
in $\mathbb{R}^3$.

30. Give an example of a matrix A such that $\mathrm{im}(A)$ is spanned by the vector $\begin{bmatrix} 1 \\ 5 \end{bmatrix}$.

31. Give an example of a matrix A such that $\mathrm{im}(A)$ is the plane with normal vector $\begin{bmatrix} 1 \\ 3 \\ 2 \end{bmatrix}$ in $\mathbb{R}^3$.

32. Give an example of a linear transformation whose image is the line spanned by
$$\begin{bmatrix} 7 \\ 6 \\ 5 \end{bmatrix}$$
in $\mathbb{R}^3$.

33. Give an example of a linear transformation whose kernel is the plane $x + 2y + 3z = 0$ in $\mathbb{R}^3$.

34. Give an example of a linear transformation whose kernel is the line spanned by
$$\begin{bmatrix} -1 \\ 1 \\ 2 \end{bmatrix}$$
in $\mathbb{R}^3$.

35. Consider a nonzero vector $\vec{v}$ in $\mathbb{R}^3$. Arguing geometrically, describe the image and the kernel of the linear transformation T from $\mathbb{R}^3$ to $\mathbb{R}$ given by
$$T(\vec{x}) = \vec{v} \cdot \vec{x}.$$

36. Consider a nonzero vector $\vec{v}$ in $\mathbb{R}^3$. Arguing geometrically, describe the image and the kernel of the linear transformation T from $\mathbb{R}^3$ to $\mathbb{R}^3$ given by
$$T(\vec{x}) = \vec{v} \times \vec{x}.$$

37. For the matrix
$$A = \begin{bmatrix} 0 & 1 & 0 \\ 0 & 0 & 1 \\ 0 & 0 & 0 \end{bmatrix},$$
describe the images and kernels of the matrices A, A^2, and A^3 geometrically.

38. Consider a square matrix A.

  a. What is the relationship between ker(A) and ker(A^2)? More generally, what can you say about ker(A), ker(A^2), ker(A^3), ker(A^4), ...?

  b. What can you say about im(A), im(A^2), im(A^3), ...?

  Hint: Exercise 37 is helpful.

39. Consider an $m \times n$ matrix A and an $n \times p$ matrix B.

  a. What is the relationship between ker(AB) and ker(B)? Are they always equal? Is one of them always contained in the other?

  b. What is the relationship between im(A) and im(AB)?

40. Consider an $m \times n$ matrix A and an $n \times p$ matrix B. If ker(A) = im(B), what can you say about the product AB?

41. Consider the matrix $A = \begin{bmatrix} 0.36 & 0.48 \\ 0.48 & 0.64 \end{bmatrix}$.

  a. Describe ker(A) and im(A) geometrically.

  b. Find A^2. If $\vec{v}$ is in the image of A, what can you say about $A\vec{v}$?

  c. Based on your answers in parts a and b, describe the transformation $T(\vec{x}) = A\vec{x}$ geometrically.

42. Express the image of the matrix

$$A = \begin{bmatrix} 1 & 1 & 1 & 6 \\ 1 & 2 & 3 & 4 \\ 1 & 3 & 5 & 2 \\ 1 & 4 & 7 & 0 \end{bmatrix} \quad 4 \times (4)$$

as the kernel of a matrix B. *Hint:* The image of A consists of all vectors $\vec{y}$ in $\mathbb{R}^4$ such that the system $A\vec{x} = \vec{y}$ is consistent. Write this system more explicitly:

$$\begin{vmatrix} x_1 + & x_2 + & x_3 + 6x_4 = y_1 \\ x_1 + 2x_2 + 3x_3 + 4x_4 = & y_2 \\ x_1 + 3x_2 + 5x_3 + 2x_4 = & y_3 \\ x_1 + 4x_2 + 7x_3 & = y_4 \end{vmatrix}.$$

Now, reduce rows:

$$\begin{vmatrix} x_1 & - & x_3 + 8x_4 = & 4y_3 - 3y_4 \\ & x_2 + 2x_3 - 2x_4 = & -y_3 + y_4 \\ & 0 & = y_1 & -3y_3 + 2y_4 \\ & 0 & = & y_2 - 2y_3 + y_4 \end{vmatrix}.$$

For which vectors $\vec{y}$ is this system consistent? The answer allows you to express im(A) as the kernel of a 2×4 matrix B.

43. Using your work in Exercise 42 as a guide, explain how you can write the image of any matrix A as the kernel of some matrix B.

44. Consider a matrix A, and let $B = $ rref(A).

  a. Is ker(A) necessarily equal to ker(B)? Explain.

  b. Is im(A) necessarily equal to im(B)? Explain.

45. Consider an $m \times n$ matrix A with rank(A) = $r < n$. Explain how you can write ker(A) as the span of $n - r$ vectors.

46. Consider a 3×4 matrix A in reduced row-echelon form. What can you say about the image of A? Describe all cases in terms of rank(A), and draw a sketch for each.

47. Let T be the projection along a line L_1 onto a line L_2. (See Exercise 2.2.33.) Describe the image and the kernel of T geometrically.

48. Consider a 2×2 matrix A with $A^2 = A$.

  a. If $\vec{w}$ is in the image of A, what is the relationship between $\vec{w}$ and $A\vec{w}$?

  b. What can you say about A if rank(A) = 2? What if rank(A) = 0?

  c. If rank(A) = 1, show that the linear transformation $T(\vec{x}) = A\vec{x}$ is the projection onto im(A) along ker(A). (See Exercise 2.2.33.)

49. Verify that the kernel of a linear transformation is closed under addition and scalar multiplication. (See Fact 3.1.6.)

50. Consider a square matrix A with ker(A^2) = ker(A^3). Is ker(A^3) = ker(A^4)? Justify your answer.

51. Consider an $m \times n$ matrix A and an $n \times p$ matrix B such that ker(A) = {$\vec{0}$} and ker(B) = {$\vec{0}$}. Find ker(AB).

52. Consider a $p \times n$ matrix A and a $q \times n$ matrix B, and form the partitioned matrix

$$C = \begin{bmatrix} A \\ B \end{bmatrix}.$$

What is the relationship between ker(A), ker(B), and ker(C)?

53. In Exercises 53 and 54, we will work with the binary digits (or bits) 0 and 1, instead of the real numbers $\mathbb{R}$. Addition and multiplication in this system are defined as usual, except for the rule $1 + 1 = 0$. We denote this number system with $\mathbb{F}_2$, or simply $\mathbb{F}$. The set of all vectors with n components in $\mathbb{F}$ is denoted by $\mathbb{F}^n$; note that $\mathbb{F}^n$ consists of 2^n vectors. (Why?) In information technology, a vector in $\mathbb{F}^8$ is called a *byte*. (A byte is a string of 8 binary digits.)

The basic ideas of linear algebra introduced so far (for the real numbers) apply to $\mathbb{F}$ without modifications.

A *Hamming matrix* with m rows is a matrix that contains all nonzero vectors in $\mathbb{F}^m$ as its columns (in any order). Note that there are $2^m - 1$ columns. Here is an example:

$$H = \begin{bmatrix} 1 & 0 & 0 & 1 & 0 & 1 & 1 \\ 0 & 1 & 0 & 1 & 1 & 0 & 1 \\ 0 & 0 & 1 & 1 & 1 & 1 & 0 \end{bmatrix}, \quad \begin{array}{l} 3 \text{ rows} \\ 2^3 - 1 = 7 \text{ columns.} \end{array}$$

a. Express the kernel of H as the span of four vectors in $\mathbb{F}^7$ of the form

$$\vec{v}_1 = \begin{bmatrix} * \\ * \\ * \\ 1 \\ 0 \\ 0 \\ 0 \end{bmatrix}, \quad \vec{v}_2 = \begin{bmatrix} * \\ * \\ * \\ 0 \\ 1 \\ 0 \\ 0 \end{bmatrix},$$

$$\vec{v}_3 = \begin{bmatrix} * \\ * \\ * \\ 0 \\ 0 \\ 1 \\ 0 \end{bmatrix}, \quad \vec{v}_4 = \begin{bmatrix} * \\ * \\ * \\ 0 \\ 0 \\ 0 \\ 1 \end{bmatrix}.$$

b. Form the 7×4 matrix

$$M = \begin{bmatrix} | & | & | & | \\ \vec{v}_1 & \vec{v}_2 & \vec{v}_3 & \vec{v}_4 \\ | & | & | & | \end{bmatrix}.$$

Explain why $\text{im}(M) = \text{ker}(H)$. If $\vec{x}$ is an arbitrary vector in $\mathbb{F}^4$, what is $H(M\vec{x})$?

54. (See Exercise 53 for some background.) When information is transmitted, there may be some errors in the communication. We present a method of adding extra information to messages so that most errors that occur during transmission can be detected and corrected. Such methods are referred to as *error-correcting codes*. (Compare these with codes whose purpose is to conceal information.) The pictures of man's first landing on the moon (in 1969) were televised just as they had been received and were not very clear, since they contained many errors induced during transmission. On later missions, much clearer error-corrected pictures were obtained.

In computers, information is stored and processed in the form of strings of binary digits, 0 and 1. This stream of binary digits is often broken up into "blocks" of eight binary digits (bytes). For the sake of simplicity, we will work with blocks of only four binary digits (i.e., with vectors in $\mathbb{F}^4$), for example,

$$\dots \mid 1\ 0\ 1\ 1 \mid 1\ 0\ 0\ 1 \mid 1\ 0\ 1\ 0 \mid 1\ 0\ 1\ 1 \mid$$
$$1\ 0\ 0\ 0 \mid \dots.$$

Suppose these vectors in $\mathbb{F}^4$ have to be transmitted from one computer to another, say, from a satellite to ground control in Kourou, French Guiana (the station of the European Space Agency). A vector $\vec{u}$ in $\mathbb{F}^4$ is first transformed into a vector $\vec{v} = M\vec{u}$ in $\mathbb{F}^7$, where M is the matrix you found in Exercise 53. The last four entries of $\vec{v}$ are just the entries of $\vec{u}$; the first three entries of $\vec{v}$ are added to detect errors. The vector $\vec{v}$ is now transmitted to Kourou. We assume that at most one error will occur during transmission; that is, the vector $\vec{w}$ received in Kourou will be either $\vec{v}$ (if no error has occurred) or $\vec{w} = \vec{v} + \vec{e}_i$ (if there is an error in the ith component of the vector).

a. Let H be the Hamming matrix introduced in Exercise 53. How can the computer in Kourou use $H\vec{w}$ to determine whether there was an error in the transmission? If there is no error, what is $H\vec{w}$? If there is an error, how can the computer determine in which component the error was made?

b. Suppose the vector

$$\vec{w} = \begin{bmatrix} 1 \\ 0 \\ 1 \\ 0 \\ 1 \\ 0 \\ 0 \end{bmatrix}$$

is received in Kourou. Determine whether an error was made in the transmission and, if so, correct it. (That is, find $\vec{v}$ and $\vec{u}$.)

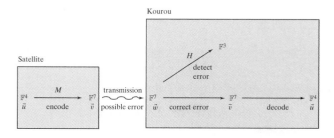

3.2 SUBSPACES OF $\mathbb{R}^n$; BASES AND LINEAR INDEPENDENCE

In the last section, we saw that both the image and the kernel of a linear transformation contain the zero vector (of the codomain and of the domain, respectively), are closed under addition, and are closed under scalar multiplication. Subsets of $\mathbb{R}^n$ that have these three properties are called *subspaces* of the "vector space" $\mathbb{R}^n$.

Definition 3.2.1

> **Subspaces of $\mathbb{R}^n$**
>
> A subset W of $\mathbb{R}^n$ is called a *subspace of* $\mathbb{R}^n$ if it has the following properties:
>
> a. W contains the zero vector in $\mathbb{R}^n$.
> b. W is closed under addition: If $\vec{w}_1$ and $\vec{w}_2$ are both in W, then so is $\vec{w}_1 + \vec{w}_2$.
> c. W is closed under scalar multiplication: If $\vec{w}$ is in W and k is an arbitrary scalar, then $k\vec{w}$ is in W.

Facts 3.1.4 and 3.1.6 tell us the following:

Fact 3.2.2

If T is a linear transformation from $\mathbb{R}^n$ to $\mathbb{R}^m$, then

- $\ker(T)$ is a *subspace* of $\mathbb{R}^n$, and
- $\operatorname{im}(T)$ is a *subspace* of $\mathbb{R}^m$.

The term *space* may seem surprising in this context. We have seen in examples that an image or a kernel can well be a line or a plane, which is not something we ordinarily call a "space." In modern mathematics, the term "space" is applied to structures that are not even remotely similar to the space of our experience (i.e., the space in which we live, which is naively identified with $\mathbb{R}^3$). The term *space* (or vector space) is used to describe $\mathbb{R}^n$ for *any* value of n (not just $n = 3$). Any subset of the space $\mathbb{R}^n$ with the properties just listed is called a subspace of $\mathbb{R}^n$, even if it is not "three-dimensional."

EXAMPLE 1   Is $W = \left\{ \begin{bmatrix} x \\ y \end{bmatrix} \text{ in } \mathbb{R}^2 : x \geq 0, \ y \geq 0 \right\}$ a subspace of $\mathbb{R}^2$?

Solution

Note that W consists of all vectors in the first quadrant, including the positive axes and the origin, as shown in Figure 1.

The answer is no: W contains the zero vector, and it is closed under addition, but it is not closed under multiplication with *negative* scalars. (See Figure 2.)   ∎

EXAMPLE 2   Is $W = \left\{ \begin{bmatrix} x \\ y \end{bmatrix} \text{ in } \mathbb{R}^2 : xy \geq 0 \right\}$ a subspace of $\mathbb{R}^2$?

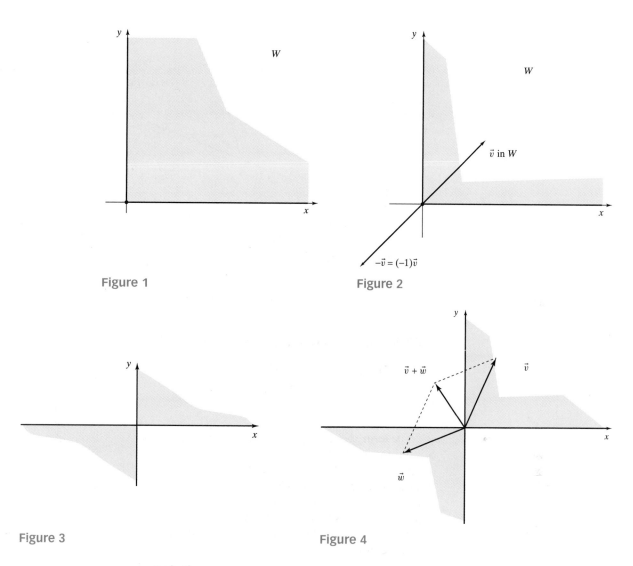

Figure 1

Figure 2

Figure 3

Figure 4

Solution

Note that W consists of all vectors in the first and third quadrant, including the axes. (See Figure 3.) Again, the answer is no: While W contains $\vec{0}$ and is closed under scalar multiplication, it is not closed under addition. (See Figure 4.)   ∎

EXAMPLE 3   Show that the only subspaces of $\mathbb{R}^2$ are $\mathbb{R}^2$ itself, the set $\{\vec{0}\}$, and any of the lines through the origin.

Solution

Suppose W is a subspace of $\mathbb{R}^2$ that is neither the set $\{\vec{0}\}$ nor a line through the origin. We have to show that $W = \mathbb{R}^2$. Pick a nonzero vector $\vec{v}_1$ in W. (We can find such a vector, since $W \neq \{\vec{0}\}$.) The subspace W contains the line L spanned by $\vec{v}_1$, but W does not equal L. Therefore, we can find a vector $\vec{v}_2$ in W that is not on L.

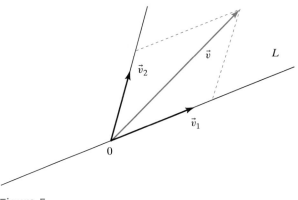

Figure 5

(See Figure 5.) Using a parallelogram, we can express any vector $\vec{v}$ in $\mathbb{R}^2$ as a linear combination of $\vec{v}_1$ and $\vec{v}_2$. Therefore, $\vec{v}$ is contained in W (since W is closed under linear combinations). This shows that $W = \mathbb{R}^2$, as claimed. ∎

Similarly, the only subspaces of $\mathbb{R}^3$ are $\mathbb{R}^3$ itself, the planes through the origin, the lines through the origin, and the set $\{\vec{0}\}$. (See Exercise 5.) Note the hierarchy of subspaces, arranged according to their dimensions. (The concept of dimension will be made precise in the next section.)

	Subspaces of $\mathbb{R}^2$	**Subspaces of $\mathbb{R}^3$**
Dimension 3		$\mathbb{R}^3$
Dimension 2	$\mathbb{R}^2$	planes through $\vec{0}$
Dimension 1	lines through $\vec{0}$	lines through $\vec{0}$
Dimension 0	$\{\vec{0}\}$	$\{\vec{0}\}$

We have seen that both kernel and image of a linear transformation T are subspaces (of the domain and the codomain of T, respectively). Conversely, can we express any subspace V of $\mathbb{R}^n$ as the kernel or the image of a linear transformation (or, equivalently, of a matrix)?

Let us consider some examples. A plane E in $\mathbb{R}^3$ is usually described either by giving a linear equation, such as

$$x_1 + 2x_2 + 3x_3 = 0,$$

or by giving E parametrically, as the span of two vectors, for example,

$$\begin{bmatrix} 1 \\ 1 \\ -1 \end{bmatrix} \quad \text{and} \quad \begin{bmatrix} 1 \\ -2 \\ 1 \end{bmatrix}.$$

In other words, E is described either as

$$\mathrm{ker}[1 \quad 2 \quad 3]$$

or

$$\mathrm{im}\begin{bmatrix} 1 & 1 \\ 1 & -2 \\ -1 & 1 \end{bmatrix}.$$

Similarly, a line L in $\mathbb{R}^3$ may be described either parametrically, as the span of the vector

$$\begin{bmatrix} 3 \\ 2 \\ 1 \end{bmatrix},$$

or by the two linear equations

$$\begin{vmatrix} x_1 - & x_2 - x_3 = 0 \\ x_1 - & 2x_2 + x_3 = 0 \end{vmatrix}.$$

Therefore,

$$L = \operatorname{im} \begin{bmatrix} 3 \\ 2 \\ 1 \end{bmatrix} = \ker \begin{bmatrix} 1 & -1 & -1 \\ 1 & -2 & 1 \end{bmatrix}.$$

A subspace of $\mathbb{R}^n$ is usually presented either as the solution set of a homogeneous linear system (i.e., as a kernel) or as the span of some vectors (i.e., as an image). In Exercise 38, you will see that any subspace of $\mathbb{R}^n$ can be represented as the image of a matrix.

Sometimes, a subspace that has been defined as a kernel must be given as an image, or vice versa. The transition from kernel to image is straightforward: We can represent the solution set of a linear system parametrically by Gauss–Jordan elimination. (See Chapter 1.) A way to write the image of a matrix as a kernel is discussed in Exercises 42 and 43 of Section 3.1.

Bases and Linear Independence

EXAMPLE 4 Consider the matrix

$$A = \begin{bmatrix} 1 & 1 & 2 & 2 \\ 1 & 2 & 2 & 3 \\ 1 & 3 & 2 & 4 \end{bmatrix}.$$

Find vectors $\vec{v}_1, \vec{v}_2, \ldots, \vec{v}_m$ in $\mathbb{R}^3$ that span the image of A. What is the smallest number of vectors needed to span the image of A?

Solution

We know from Fact 3.1.3 that the image of A is spanned by the columns of A,

$$\vec{v}_1 = \begin{bmatrix} 1 \\ 1 \\ 1 \end{bmatrix}, \quad \vec{v}_2 = \begin{bmatrix} 1 \\ 2 \\ 3 \end{bmatrix}, \quad \vec{v}_3 = \begin{bmatrix} 2 \\ 2 \\ 2 \end{bmatrix}, \quad \vec{v}_4 = \begin{bmatrix} 2 \\ 3 \\ 4 \end{bmatrix}.$$

Figure 6 shows that we need only $\vec{v}_1$ and $\vec{v}_2$ to span the image of A. Since $\vec{v}_3 = 2\vec{v}_1$ and $\vec{v}_4 = \vec{v}_1 + \vec{v}_2$, the vectors $\vec{v}_3$ and $\vec{v}_4$ are redundant; that is, they are linear combinations of $\vec{v}_1$ and $\vec{v}_2$:

$$\operatorname{im}(A) = \operatorname{span}(\vec{v}_1, \vec{v}_2, \vec{v}_3, \vec{v}_4) = \operatorname{span}(\vec{v}_1, \vec{v}_2).$$

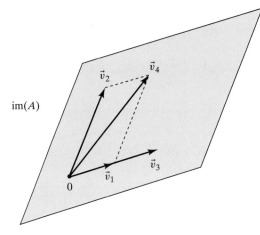

Figure 6

The image of A can be spanned by two vectors, but not by one vector alone. ∎

We often wish to express a subspace V of $\mathbb{R}^n$ as the span of some vectors $\vec{v}_1, \vec{v}_2, \ldots, \vec{v}_m$ in V. It is reasonable to require that none of the $\vec{v}_i$ be a linear combination of the others; otherwise, we might as well omit it. The following definition will make the informal idea of a "redundant vector" more precise.

Definition 3.2.3

Linear independence; basis

Consider a sequence $\vec{v}_1, \ldots, \vec{v}_m$ of vectors in a subspace V of $\mathbb{R}^n$.

The vectors $\vec{v}_1, \ldots, \vec{v}_m$ are called *linearly independent* if none of them is a linear combination of the others.[4] Otherwise, the vectors are called *linearly dependent*.

We say that the vectors $\vec{v}_1, \ldots, \vec{v}_m$ form a *basis* of V if they span V and are linearly independent.[5]

The vectors $\vec{v}_1, \vec{v}_2, \vec{v}_3, \vec{v}_4$ in Example 4 span

$$V = \text{im}(A),$$

but they are linearly dependent, because $\vec{v}_4 = \vec{v}_1 + \vec{v}_2$. Therefore, they do not form a basis of V. The vectors $\vec{v}_1, \vec{v}_2$, on the other hand, do span V and are linearly independent. (Neither of them is a multiple of the other.) Therefore, the vectors $\vec{v}_1, \vec{v}_2$ are a basis of $V = \text{im}(A)$. The vectors $\vec{v}_1, \vec{v}_2$ do not form a basis of $\mathbb{R}^3$, however. While they are linearly independent, they do not span $\mathbb{R}^3$. (They span only a plane.)

---

[4]This definition makes sense if m is at least two. We say that a single vector, $\vec{v}_1$, is linearly independent if $\vec{v}_1 \neq \vec{0}$.

[5]By convention, the empty set Ø is a basis of the space $\{\vec{0}\}$.

The following definition will help us to gain a better conceptual understanding of linear independence.

Definition 3.2.4

> **Linear relations**
>
> Consider the vectors $\vec{v}_1, \vec{v}_2, \ldots, \vec{v}_m$ in $\mathbb{R}^n$. An equation of the form
>
> $$c_1\vec{v}_1 + c_2\vec{v}_2 + \cdots + c_m\vec{v}_m = \vec{0}$$
>
> is called a (linear) *relation* among the vectors $\vec{v}_i$. There is always the trivial relation, with $c_1 = c_2 = \cdots = c_m = 0$. Nontrivial relations may or may not exist among the vectors $\vec{v}_1, \vec{v}_2, \ldots, \vec{v}_m$.

For example, among the vectors introduced in Example 4, we have the nontrivial relation

$$\vec{v}_1 + \vec{v}_2 - \vec{v}_4 = \vec{0},$$

because $\vec{v}_4 = \vec{v}_1 + \vec{v}_2$.

Fact 3.2.5   The vectors $\vec{v}_1, \ldots, \vec{v}_m$ in $\mathbb{R}^n$ are linearly dependent if (and only if) there are nontrivial relations among them.

Proof   • If one of the $\vec{v}_i$ is a linear combination of the others,

$$\vec{v}_i = c_1\vec{v}_1 + \cdots + c_{i-1}\vec{v}_{i-1} + c_{i+1}\vec{v}_{i+1} + \cdots + c_m\vec{v}_m,$$

then we can find a nontrivial relation by subtracting $\vec{v}_i$ from both sides of the equation:

$$c_1\vec{v}_1 + \cdots + c_{i-1}\vec{v}_{i-1} + (-1)\vec{v}_i + c_{i+1}\vec{v}_{i+1} + \cdots + c_m\vec{v}_m = \vec{0}.$$

• Conversely, if there is a nontrivial relation

$$c_1\vec{v}_1 + \cdots + c_i\vec{v}_i + \cdots + c_m\vec{v}_m = \vec{0}, \quad \text{with } c_i \neq 0,$$

then we can solve for $\vec{v}_i$ and thus express $\vec{v}_i$ as a linear combination of the other vectors.   ▲

EXAMPLE 5   Determine whether the following vectors are linearly independent:

$$\begin{bmatrix} 1 \\ 2 \\ 3 \\ 4 \\ 5 \end{bmatrix}, \quad \begin{bmatrix} 6 \\ 7 \\ 8 \\ 9 \\ 10 \end{bmatrix}, \quad \begin{bmatrix} 2 \\ 3 \\ 5 \\ 7 \\ 11 \end{bmatrix}, \quad \begin{bmatrix} 1 \\ 4 \\ 9 \\ 16 \\ 25 \end{bmatrix}.$$

Solution

To find the relations among these vectors, we have to solve the vector equation

$$
c_1 \begin{bmatrix} 1 \\ 2 \\ 3 \\ 4 \\ 5 \end{bmatrix} + c_2 \begin{bmatrix} 6 \\ 7 \\ 8 \\ 9 \\ 10 \end{bmatrix} + c_3 \begin{bmatrix} 2 \\ 3 \\ 5 \\ 7 \\ 11 \end{bmatrix} + c_4 \begin{bmatrix} 1 \\ 4 \\ 9 \\ 16 \\ 25 \end{bmatrix} = \begin{bmatrix} 0 \\ 0 \\ 0 \\ 0 \\ 0 \end{bmatrix},
$$

or the matrix equation

$$
\underbrace{\begin{bmatrix} 1 & 6 & 2 & 1 \\ 2 & 7 & 3 & 4 \\ 3 & 8 & 5 & 9 \\ 4 & 9 & 7 & 16 \\ 5 & 10 & 11 & 25 \end{bmatrix}}_{A} \begin{bmatrix} c_1 \\ c_2 \\ c_3 \\ c_4 \end{bmatrix} = \begin{bmatrix} 0 \\ 0 \\ 0 \\ 0 \\ 0 \end{bmatrix}.
$$

In other words, we have to find the *kernel* of A. To do so, we compute $\mathrm{rref}(A)$. Using technology, we find that

$$
\mathrm{rref}(A) = \begin{bmatrix} 1 & 0 & 0 & 0 \\ 0 & 1 & 0 & 0 \\ 0 & 0 & 1 & 0 \\ 0 & 0 & 0 & 1 \\ 0 & 0 & 0 & 0 \end{bmatrix}.
$$

This shows that the kernel of A is $\{\vec{0}\}$, because there is a leading 1 in each column of $\mathrm{rref}(A)$. There is only the trivial relation among the four vectors, and they are therefore linearly independent.   ■

Since any relation

$$
c_1 \vec{v}_1 + c_2 \vec{v}_2 + \cdots + c_m \vec{v}_m = \vec{0}
$$

can alternatively be written as

$$
\begin{bmatrix} | & | & & | \\ \vec{v}_1 & \vec{v}_2 & \cdots & \vec{v}_m \\ | & | & & | \end{bmatrix} \begin{bmatrix} c_1 \\ c_2 \\ \vdots \\ c_m \end{bmatrix} = \vec{0},
$$

we can generalize the observation made in Example 5.

Fact 3.2.6   The vectors $\vec{v}_1, \ldots, \vec{v}_m$ in $\mathbb{R}^n$ are linearly independent if (and only if)

$$\ker \begin{bmatrix} | & | & & | \\ \vec{v}_1 & \vec{v}_2 & \cdots & \vec{v}_m \\ | & | & & | \end{bmatrix} = \{\vec{0}\},$$

or, equivalently, of

$$\operatorname{rank} \begin{bmatrix} | & | & & | \\ \vec{v}_1 & \vec{v}_2 & \cdots & \vec{v}_m \\ | & | & & | \end{bmatrix} = m.$$

This condition implies that $m \le n$.

We conclude this section with an important alternative characterization of a basis.

If the vectors $\vec{v}_1, \vec{v}_2, \ldots, \vec{v}_m$ are a basis of the subspace V of $\mathbb{R}^n$, then every vector $\vec{v}$ in V can be expressed as a linear combination of the basis vectors (since they span V). This representation is in fact *unique*.

Fact 3.2.7   Consider the vectors $\vec{v}_1, \vec{v}_2, \ldots, \vec{v}_m$ in a subspace V of $\mathbb{R}^n$.
The vectors $\vec{v}_i$ are a basis of V if (and only if) every vector $\vec{v}$ in V can be expressed *uniquely* as a linear combination of the vectors $\vec{v}_i$.

Proof   Suppose the vectors $\vec{v}_i$ form a basis of V, and consider a vector $\vec{v}$ in V. Since the basis vectors span V, the vector $\vec{v}$ can be written as a linear combination of the $\vec{v}_i$. We have to demonstrate that this representation is unique. To do so, we consider two representations of $\vec{v}$, namely,

$$\vec{v} = c_1 \vec{v}_1 + c_2 \vec{v}_2 + \cdots + c_m \vec{v}_m = d_1 \vec{v}_1 + d_2 \vec{v}_2 + \cdots + d_m \vec{v}_m.$$

By subtraction, we find

$$(c_1 - d_1)\vec{v}_1 + (c_2 - d_2)\vec{v}_2 + \cdots + (c_m - d_m)\vec{v}_m = \vec{0},$$

which is a relation among the $\vec{v}_i$. Since the $\vec{v}_i$ are linearly independent, we have $c_i - d_i = 0$, or $c_i = d_i$, for all i: The two representations of $\vec{v}$ are identical, as claimed.

Conversely, suppose that each vector in V can be expressed uniquely as a linear combination of the vectors $\vec{v}_i$. Clearly, the $\vec{v}_i$ span V. The zero vector can be expressed uniquely as a linear combination of the $\vec{v}_i$, namely, as

$$\vec{0} = 0\vec{v}_1 + 0\vec{v}_2 + \cdots + 0\vec{v}_m.$$

But this means that there is only the trivial relation among the $\vec{v}_i$: They are linearly independent (by Fact 3.2.5).   ▲

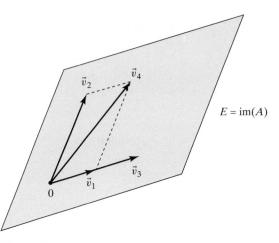

Figure 7

As an example, consider the plane $E = \operatorname{im}(A)$ in $\mathbb{R}^3$ introduced in Example 4. (See Figure 7.)

The vectors $\vec{v}_1$, $\vec{v}_2$, $\vec{v}_3$, $\vec{v}_4$ do not form a basis of E, since every vector in E can be expressed in more than one way as a linear combination of the $\vec{v}_i$. For example, we can write

$$\vec{v}_4 = 1\vec{v}_1 + 1\vec{v}_2 + 0\vec{v}_3 + 0\vec{v}_4,$$

but also

$$\vec{v}_4 = 0\vec{v}_1 + 0\vec{v}_2 + 0\vec{v}_3 + 1\vec{v}_4.$$

Because every vector in E can be expressed *uniquely* as a linear combination of $\vec{v}_1$ and $\vec{v}_2$ alone, the vectors $\vec{v}_1$, $\vec{v}_2$ form a basis of E.

EXERCISES

GOALS Check whether a subset of $\mathbb{R}^n$ is a subspace. Apply the concept of linear independence (in terms of Definition 3.2.3, Fact 3.2.5, and Fact 3.2.6). Apply the concept of a basis, both in terms of Definition 3.2.3 and in terms of Fact 3.2.7.

Which of the sets W in Exercises 1 through 3 are subspaces of $\mathbb{R}^3$?

1. $W = \left\{ \begin{bmatrix} x \\ y \\ z \end{bmatrix} : x + y + z = 1 \right\}.$

2. $W = \left\{ \begin{bmatrix} x \\ y \\ z \end{bmatrix} : x \le y \le z \right\}.$

3. $W = \left\{ \begin{bmatrix} x + 2y + 3z \\ 4x + 5y + 6z \\ 7x + 8y + 9z \end{bmatrix} : x, y, z \text{ arbitrary constants} \right\}.$

4. Consider the vectors $\vec{v}_1, \vec{v}_2, \ldots, \vec{v}_m$ in $\mathbb{R}^n$. Is $\operatorname{span}(\vec{v}_1, \ldots, \vec{v}_m)$ necessarily a subspace of $\mathbb{R}^n$? Justify your answer.

5. Give a geometrical description of all subspaces of $\mathbb{R}^3$. Justify your answer.

6. Consider two subspaces V and W of $\mathbb{R}^n$.

   a. Is the intersection $V \cap W$ necessarily a subspace of $\mathbb{R}^n$?

   b. Is the union $V \cup W$ necessarily a subspace of $\mathbb{R}^n$?

7. Consider a nonempty subset W of $\mathbb{R}^n$ that is closed under addition and under scalar multiplication. Is W necessarily a subspace of $\mathbb{R}^n$? Explain.

8. Find a nontrivial relation among the following vectors:

$$\begin{bmatrix} 1 \\ 2 \end{bmatrix}, \quad \begin{bmatrix} 2 \\ 3 \end{bmatrix}, \quad \begin{bmatrix} 3 \\ 4 \end{bmatrix}.$$

9. Consider the vectors $\vec{v}_1, \vec{v}_2, \ldots, \vec{v}_m$ in $\mathbb{R}^n$, with $\vec{v}_m = \vec{0}$. Are these vectors linearly independent? How can you tell?

In Exercises 10 through 19, use paper and pencil to decide whether the given vectors are linearly independent.

10. $\begin{bmatrix} 2 \\ 1 \end{bmatrix}, \begin{bmatrix} 6 \\ 3 \end{bmatrix}.$

11. $\begin{bmatrix} 7 \\ 11 \end{bmatrix}, \begin{bmatrix} 11 \\ 7 \end{bmatrix}.$

12. $\begin{bmatrix} 7 \\ 11 \end{bmatrix}, \begin{bmatrix} 0 \\ 0 \end{bmatrix}.$

13. $\begin{bmatrix} 1 \\ 2 \end{bmatrix}, \begin{bmatrix} 1 \\ 2 \end{bmatrix}.$

14. $\begin{bmatrix} 1 \\ 0 \\ 0 \end{bmatrix}, \begin{bmatrix} 1 \\ 2 \\ 0 \end{bmatrix}, \begin{bmatrix} 1 \\ 2 \\ 3 \end{bmatrix}.$

15. $\begin{bmatrix} 1 \\ 2 \end{bmatrix}, \begin{bmatrix} 2 \\ 3 \end{bmatrix}, \begin{bmatrix} 3 \\ 4 \end{bmatrix}.$

16. $\begin{bmatrix} 1 \\ 2 \\ 3 \end{bmatrix}, \begin{bmatrix} 4 \\ 5 \\ 6 \end{bmatrix}, \begin{bmatrix} 7 \\ 8 \\ 9 \end{bmatrix}.$

17. $\begin{bmatrix} 1 \\ 1 \\ 1 \end{bmatrix}, \begin{bmatrix} 1 \\ 2 \\ 3 \end{bmatrix}, \begin{bmatrix} 1 \\ 3 \\ 6 \end{bmatrix}.$

18. $\begin{bmatrix} 1 \\ 1 \\ 1 \\ 1 \end{bmatrix}, \begin{bmatrix} 1 \\ 2 \\ 3 \\ 4 \end{bmatrix}, \begin{bmatrix} 1 \\ 4 \\ 7 \\ 10 \end{bmatrix}.$

19. $\begin{bmatrix} 1 \\ 2 \\ 9 \\ 1 \end{bmatrix}, \begin{bmatrix} 1 \\ 5 \\ 1 \\ 5 \end{bmatrix}, \begin{bmatrix} 5 \\ 4 \\ 9 \\ 1 \end{bmatrix}, \begin{bmatrix} 1 \\ 8 \\ 1 \\ 5 \end{bmatrix}, \begin{bmatrix} 1 \\ 1 \\ 1 \\ 1 \end{bmatrix}.$

In Exercises 20 and 21, decide whether the vectors are linearly independent. You may use technology.

20. $\begin{bmatrix} 1 \\ 2 \\ 3 \\ 4 \end{bmatrix}, \begin{bmatrix} 5 \\ 6 \\ 7 \\ 8 \end{bmatrix}, \begin{bmatrix} 9 \\ 10 \\ 11 \\ 12 \end{bmatrix}.$

21. $\begin{bmatrix} 7 \\ 3 \\ 5 \\ 4 \\ 2 \end{bmatrix}, \begin{bmatrix} 2 \\ 5 \\ 3 \\ 9 \\ 7 \end{bmatrix}, \begin{bmatrix} 4 \\ 3 \\ 3 \\ 8 \\ 6 \end{bmatrix}, \begin{bmatrix} 4 \\ 9 \\ 7 \\ 3 \\ 1 \end{bmatrix}.$

22. For which choices of the constants $a, b, c, d, e,$ and f are the following vectors linearly independent? Justify

your answer.

$$\begin{bmatrix} a \\ 0 \\ 0 \\ 0 \end{bmatrix}, \quad \begin{bmatrix} b \\ c \\ 0 \\ 0 \end{bmatrix}, \quad \begin{bmatrix} d \\ e \\ f \\ 0 \end{bmatrix}.$$

23. Consider a subspace V of $\mathbb{R}^n$. We define the *orthogonal complement* $V^\perp$ of V as the set of those vectors $\vec{w}$ in $\mathbb{R}^n$ that are perpendicular to all vectors in V; that is, $\vec{w} \cdot \vec{v} = 0$, for all $\vec{v}$ in V. Show that $V^\perp$ is a subspace of $\mathbb{R}^n$.

24. Consider the line L spanned by $\begin{bmatrix} 1 \\ 2 \\ 3 \end{bmatrix}$ in $\mathbb{R}^3$. Find a basis of $L^\perp$. (See Exercise 23.)

25. Consider the subspace L of $\mathbb{R}^5$ spanned by the given vector. Find a basis of $L^\perp$. (See Exercise 23.)

$$\begin{bmatrix} 1 \\ 2 \\ 3 \\ 4 \\ 5 \end{bmatrix}.$$

26. For which choices of the constants $a, b, \ldots, m$ are the given vectors linearly independent?

$$\begin{bmatrix} a \\ b \\ c \\ d \\ 1 \\ 0 \end{bmatrix}, \begin{bmatrix} e \\ 1 \\ 0 \\ 0 \\ 0 \\ 0 \end{bmatrix}, \begin{bmatrix} f \\ g \\ h \\ i \\ j \\ 1 \end{bmatrix}, \begin{bmatrix} k \\ m \\ 1 \\ 0 \\ 0 \\ 0 \end{bmatrix}.$$

Find a basis of the image of the matrices in Exercises 27 through 33.

27. $\begin{bmatrix} 1 & 1 \\ 1 & 2 \\ 1 & 3 \end{bmatrix}.$

28. $\begin{bmatrix} 1 & 1 & 1 \\ 1 & 2 & 5 \\ 1 & 3 & 7 \end{bmatrix}.$

29. $\begin{bmatrix} 1 & 2 & 3 \\ 4 & 5 & 6 \end{bmatrix}.$

30. $\begin{bmatrix} 0 & 1 & 0 \\ 0 & 0 & 1 \\ 0 & 0 & 0 \end{bmatrix}.$

31. $\begin{bmatrix} 1 & 5 \\ 2 & 6 \\ 3 & 7 \\ 5 & 8 \end{bmatrix}.$

32. $\begin{bmatrix} 1 & 2 \\ 2 & 4 \\ 3 & 6 \\ 4 & 8 \end{bmatrix}.$

33. $\begin{bmatrix} 0 & 1 & 2 & 0 & 3 & 0 \\ 0 & 0 & 0 & 1 & 4 & 0 \\ 0 & 0 & 0 & 0 & 0 & 1 \\ 0 & 0 & 0 & 0 & 0 & 0 \end{bmatrix}.$

34. Consider the 5×4 matrix

$$A = \begin{bmatrix} | & | & | & | \\ \vec{v}_1 & \vec{v}_2 & \vec{v}_3 & \vec{v}_4 \\ | & | & | & | \end{bmatrix}.$$

We are told that the vector $\begin{bmatrix} 1 \\ 2 \\ 3 \\ 4 \end{bmatrix}$ is in the kernel of A.

Write $\vec{v}_4$ as a linear combination of $\vec{v}_1, \vec{v}_2, \vec{v}_3$.

35. Consider the linearly dependent vectors $\vec{v}_1, \vec{v}_2, \ldots, \vec{v}_m$ in $\mathbb{R}^n$, where $\vec{v}_1 \neq \vec{0}$. Show that one of the vectors $\vec{v}_i$ (for $i = 2, \ldots, m$) is a linear combination of the *previous* vectors $\vec{v}_1, \vec{v}_2, \ldots, \vec{v}_{i-1}$.

36. Consider a linear transformation T from $\mathbb{R}^n$ to $\mathbb{R}^p$ and some linearly dependent vectors $\vec{v}_1, \vec{v}_2, \ldots, \vec{v}_m$ in $\mathbb{R}^n$. Are the vectors $T(\vec{v}_1), T(\vec{v}_2), \ldots, T(\vec{v}_m)$ linearly dependent? How can you tell?

37. Consider a linear transformation T from $\mathbb{R}^n$ to $\mathbb{R}^p$ and some linearly independent vectors $\vec{v}_1, \vec{v}_2, \ldots, \vec{v}_m$ in $\mathbb{R}^n$. Are the vectors $T(\vec{v}_1), T(\vec{v}_2), \ldots, T(\vec{v}_m)$ necessarily linearly independent? How can you tell?

38. a. Show that we can find at most n linearly independent vectors in $\mathbb{R}^n$.

   b. Let V be a subspace of $\mathbb{R}^n$. Let m be the largest number of linearly independent vectors we can find in V. (Note that $m \leq n$, by part (a).) Choose linearly independent vectors $\vec{v}_1, \vec{v}_2, \ldots, \vec{v}_m$ in V. Show that the vectors $\vec{v}_1, \vec{v}_2, \ldots, \vec{v}_m$ span V and are therefore a basis of V. This exercise shows that any subspace of $\mathbb{R}^n$ has a basis.
      If you are puzzled, think first about the special case when V is a plane in $\mathbb{R}^3$. What is m in this case?

   c. Show that any subspace V of $\mathbb{R}^n$ can be represented as the image of a matrix.

39. Consider some linearly independent vectors $\vec{v}_1, \vec{v}_2, \ldots, \vec{v}_m$ in $\mathbb{R}^n$ and a vector $\vec{v}$ in $\mathbb{R}^n$ that is not contained in the span of $\vec{v}_1, \vec{v}_2, \ldots, \vec{v}_m$. Are the vectors $\vec{v}_1, \vec{v}_2, \ldots, \vec{v}_m, \vec{v}$ necessarily linearly independent? Justify your answer.

40. Consider an $m \times n$ matrix A and an $n \times p$ matrix B. We are told that the columns of A and the columns of B are linearly independent. Are the columns of the product AB linearly independent as well?

41. Consider an $m \times n$ matrix A and an $n \times m$ matrix B (with $n \neq m$) such that $AB = I_m$. (We say that A is a *left inverse* of B.) Are the columns of B linearly independent? What about the columns of A?

42. Consider some perpendicular unit vectors $\vec{v}_1, \vec{v}_2, \ldots, \vec{v}_m$ in $\mathbb{R}^n$. Show that these vectors are necessarily linearly independent. *Hint:* Form the dot product of $\vec{v}_i$ and both sides of the equation

$$c_1\vec{v}_1 + c_2\vec{v}_2 + \cdots + c_i\vec{v}_i + \cdots + c_m\vec{v}_m = \vec{0}.$$

43. Consider three linearly independent vectors $\vec{v}_1, \vec{v}_2, \vec{v}_3$ in $\mathbb{R}^n$. Are the vectors $\vec{v}_1, \vec{v}_1 + \vec{v}_2, \vec{v}_1 + \vec{v}_2 + \vec{v}_3$ linearly independent as well? How can you tell?

44. Consider the linearly independent vectors $\vec{v}_1, \vec{v}_2, \ldots, \vec{v}_m$ in $\mathbb{R}^n$, and let A be an invertible $m \times m$ matrix. Are the columns of the following matrix linearly independent?

$$\begin{bmatrix} | & | & & | \\ \vec{v}_1 & \vec{v}_2 & \cdots & \vec{v}_m \\ | & | & & | \end{bmatrix} A.$$

45. Are the columns of an invertible matrix linearly independent?

46. Find a basis of the kernel of the matrix

$$\begin{bmatrix} 1 & 2 & 0 & 3 & 5 \\ 0 & 0 & 1 & 4 & 6 \end{bmatrix}.$$

Justify your answer carefully; that is, explain how you know that the vectors you found are linearly independent and span the kernel.

47. Consider three linearly independent vectors $\vec{v}_1, \vec{v}_2, \vec{v}_3$ in $\mathbb{R}^4$. Find

$$\text{rref}\begin{bmatrix} | & | & | \\ \vec{v}_1 & \vec{v}_2 & \vec{v}_3 \\ | & | & | \end{bmatrix}.$$

48. Express the plane E in $\mathbb{R}^3$ with equation $3x_1 + 4x_2 + 5x_3 = 0$ as the kernel of a matrix A and as the image of a matrix B.

49. Express the line L in $\mathbb{R}^3$ spanned by the vector $\begin{bmatrix} 1 \\ 1 \\ 1 \end{bmatrix}$ as the image of a matrix A and as the kernel of a matrix B.

50. Consider two subspaces V and W of $\mathbb{R}^n$. Let $V + W$ be the set of all vectors in $\mathbb{R}^n$ of the form $\vec{v} + \vec{w}$, where $\vec{v}$ is in V and $\vec{w}$ in W. Is $V + W$ necessarily a subspace of $\mathbb{R}^n$?
   If V and W are two distinct lines in $\mathbb{R}^3$, what is $V + W$? Draw a sketch.

51. Consider two subspaces V and W of $\mathbb{R}^n$ whose intersection consists only of the vector $\vec{0}$.

   a. Consider linearly independent vectors $\vec{v}_1, \vec{v}_2, \ldots, \vec{v}_p$ in V and $\vec{w}_1, \vec{w}_2, \ldots, \vec{w}_q$ in W. Explain why the vectors $\vec{v}_1, \vec{v}_2, \ldots, \vec{v}_p, \vec{w}_1, \vec{w}_2, \ldots, \vec{w}_q$ are linearly independent.

   b. Consider a basis $\vec{v}_1, \vec{v}_2, \ldots, \vec{v}_p$ of V and a basis $\vec{w}_1, \vec{w}_2, \ldots, \vec{w}_q$ of W. Explain why $\vec{v}_1, \vec{v}_2, \ldots, \vec{v}_p, \vec{w}_1, \vec{w}_2, \ldots, \vec{w}_q$ is a basis of $V + W$. (See Exercise 50.)

3.3 THE DIMENSION OF A SUBSPACE OF $\mathbb{R}^n$

Consider a plane E in $\mathbb{R}^3$. Using our geometric intuition, we observe that all bases of E consist of two vectors. (Any two nonparallel vectors in E will do; see Figure 1.) One vector is not enough to span E, and three or more vectors are linearly dependent. More generally, all bases of a subspace V of $\mathbb{R}^n$ consist of the same number of vectors. To prove this important fact, we need an auxiliary result.

Fact 3.3.1   Consider vectors $\vec{v}_1, \vec{v}_2, \ldots, \vec{v}_p$ and $\vec{w}_1, \vec{w}_2, \ldots, \vec{w}_q$ in a subspace V of $\mathbb{R}^n$. If the vectors $\vec{v}_i$ are linearly independent, and the vectors $\vec{w}_j$ span V, then $p \leq q$.

For example, let V be a plane in $\mathbb{R}^3$. Our geometric intuition tells us that we can choose *at most* two linearly independent vectors in V, so that $p \leq 2$, and we need *at least* two vectors to span V, so that $2 \leq q$. Therefore, the inequality $p \leq q$ does indeed hold in this case.

Proof   This proof is rather technical and does not really explain the result conceptually. In the next section, when we study coordinate systems, we will gain a more conceptual understanding of this matter.

Since the $\vec{w}_j$ span V, we can express each $\vec{v}_i$ as a linear combination of the vectors $\vec{w}_j$:

$$
\begin{aligned}
\vec{v}_1 &= a_{11}\vec{w}_1 + \cdots + a_{1q}\vec{w}_q, \\
&\ \ \vdots \qquad\quad \vdots \qquad\qquad\quad \vdots \\
\vec{v}_p &= a_{p1}\vec{w}_1 + \cdots + a_{pq}\vec{w}_q.
\end{aligned}
$$

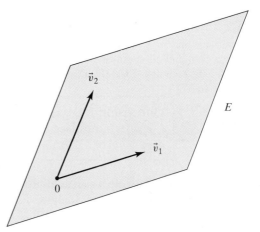

Figure 1 The vectors $\vec{v}_1$, $\vec{v}_2$ form a basis of E.

We write each of these equations in matrix form:

$$
\begin{bmatrix} | & & | \\ \vec{w}_1 & \dots & \vec{w}_q \\ | & & | \end{bmatrix}
\begin{bmatrix} a_{11} \\ \vdots \\ a_{1q} \end{bmatrix} = \vec{v}_1
\quad \dots \quad
\begin{bmatrix} | & & | \\ \vec{w}_1 & \dots & \vec{w}_q \\ | & & | \end{bmatrix}
\begin{bmatrix} a_{p1} \\ \vdots \\ a_{pq} \end{bmatrix} = \vec{v}_p .
$$

We combine all these equations into one matrix equation:

$$
\begin{bmatrix} | & & | \\ \vec{w}_1 & \dots & \vec{w}_q \\ | & & | \end{bmatrix}
\begin{bmatrix} a_{11} & \dots & a_{p1} \\ \vdots & & \vdots \\ a_{1q} & \dots & a_{pq} \end{bmatrix} =
\begin{bmatrix} | & & | \\ \vec{v}_1 & \dots & \vec{v}_p \\ | & & | \end{bmatrix} .
$$

$$MA = N$$

The kernel of A is contained in the kernel of N (if $A\vec{x} = 0$, then $MA\vec{x} = N\vec{x} = \vec{0}$). But the kernel of N is $\{\vec{0}\}$, since the $\vec{v}_i$ are linearly independent (by Fact 3.2.6). Therefore, the kernel of A is $\{\vec{0}\}$ as well. This implies that $\text{rank}(A) = p \leq q$ (by Fact 3.1.7).   ▲

Now we are ready to prove the following:

Fact 3.3.2   All bases of a subspace V of $\mathbb{R}^n$ consist of the same number of vectors.

Proof   Consider two bases $\vec{v}_1, \dots, \vec{v}_p$ and $\vec{w}_1, \dots, \vec{w}_q$ of V. Since the $\vec{v}_i$ are linearly independent and the $\vec{w}_j$ span V, we have $p \leq q$, by Fact 3.3.1. Likewise, since the $\vec{w}_j$ are linearly independent and the $\vec{v}_i$ span V, we have $q \leq p$. Therefore, $p = q$.   ▲

Consider a line L and a plane E in $\mathbb{R}^3$. A basis of L consists of just one vector (any nonzero vector in L will do), while all bases of E consist of two vectors. A basis of $\mathbb{R}^3$ consists of three vectors. (The vectors $\vec{e}_1, \vec{e}_2, \vec{e}_3$ are one possible choice.) In each case, the number of vectors in a basis corresponds to what we intuitively sense is the *dimension* of the subspace.

Definition 3.3.3

> **Dimension**
> Consider a subspace V of $\mathbb{R}^n$. The number of vectors in a basis of V is called the *dimension* of V, denoted by $\dim(V)$.[6]

This algebraic definition of dimension represents a major step in the development of linear algebra: It allows us to conceive of spaces with more than three dimensions.

---

[6]For this definition to make sense, we have to be sure that any subspace of $\mathbb{R}^n$ has a basis. This verification is left as Exercise 3.2.38.

This idea is often poorly understood in popular culture, where some mysticism surrounds higher-dimensional spaces. As the German mathematician Hermann Weyl (1885–1955) said: "We are by no means obliged to seek illumination from the mystic doctrines of spiritists to obtain a clearer vision of multidimensional geometry" (*Raum, Zeit, Materie,* 1918).

The first mathematician who thought about dimension from an algebraic point of view may have been the Frenchman Jean Le Rond d'Alembert (1717–1783). In the article on dimension in the *Encyclopédie,* he writes:

> The way of considering quantities having more than three dimensions is just as right as the other, because letters can always be viewed as representing numbers, whether rational or not. I said above that it was not possible to conceive more than three dimensions. A thoughtful gentleman [*un homme d'esprit*] with whom I am acquainted believes that nevertheless one could view duration as a fourth dimension . . . . This idea may be challenged, but it has, it seems to me, some merit, were it only that of being new [*cette idée peut être contestée, mais elle a, ce me semble, quelque mérite, quand ce ne serait que celui de la nouveauté*]. (*Encyclopédie,* vol. 4, 1754.)

The *homme d'esprit* was no doubt d'Alembert himself. D'Alembert wished to protect himself against being attacked for what appeared as a risky idea.

The idea of dimension was later studied more systematically by the German mathematician Hermann Günther Grassmann (1809–1877), who introduced the concept of a subspace of $\mathbb{R}^n$. Grassmann's methods were only slowly adopted, partly because of his obscure writing. Grassmann expressed his ideas in the book *Die lineare Ausdehnungslehre, ein neuer Zweig der Mathematik* (*The Theory of Linear Extension, a New Branch of Mathematics*), in 1844. Similar work was done at about the same time by the Swiss mathematician Ludwig Schläfli (1814–1895).

Today, dimension is a standard and central tool in mathematics, as well as in physics and statistics. The idea has been applied to certain nonlinear subsets of $\mathbb{R}^n$, called manifolds, thus generalizing the idea of curves and surfaces in $\mathbb{R}^3$.

Let us return to the more mundane: What is the dimension of $\mathbb{R}^n$ itself? Clearly, $\mathbb{R}^n$ ought to have dimension n. This is indeed the case: The vectors $\vec{e}_1, \vec{e}_2, \ldots, \vec{e}_n$ form a basis of $\mathbb{R}^n$ called its *standard basis.*

A plane E in $\mathbb{R}^3$ is two-dimensional. Before, we mentioned that we cannot find more than two linearly independent vectors in E and that we need at least two vectors to span E. If two vectors in E are linearly independent, they form a basis of E. Likewise, if two vectors span E, they form a basis of E. These observations can be generalized as follows:

Fact 3.3.4   Consider a subspace V of $\mathbb{R}^n$ with $\dim(V) = m$.

a. We can find *at most* m linearly independent vectors in V.

b. We need *at least* m vectors to span V.

c. If m vectors in V are linearly independent, then they form a basis of V.

d. If m vectors span V, then they form a basis of V.

The point of parts (c) and (d) is the following: By Definition 3.2.3, some vectors form a basis of a subspace V if they are linearly independent *and* span V. However, when we are dealing with "the right number" of vectors (namely, the dimension of V), it suffices to check only one of the two properties; the other will follow.

Proof   We demonstrate parts (a) and (c), and leave (b) and (d) as Exercises 58 and 59.

Part (a): Consider linearly independent vectors $\vec{v}_1, \vec{v}_2, \ldots, \vec{v}_p$ in V, and choose a basis $\vec{w}_1, \vec{w}_2, \ldots, \vec{w}_m$ of V. Since the $\vec{w}_i$ span V, we have $p \leq m$, as claimed (by Fact 3.3.1).

Part (c): Consider linearly independent vectors $\vec{v}_1, \ldots, \vec{v}_m$ in V. We have to show that the $\vec{v}_i$ span V. Pick a $\vec{v}$ in V. Then, the vectors $\vec{v}_1, \ldots, \vec{v}_m, \vec{v}$ will be linearly dependent, by part (a). Therefore, there is a nontrivial relation

$$c_1 \vec{v}_1 + \cdots + c_m \vec{v}_m + c\vec{v} = 0.$$

We know that $c \neq 0$. (Why?) We can solve the relation for $\vec{v}$ and thus express $\vec{v}$ as a linear combination of the $\vec{v}_i$. We have shown that any vector $\vec{v}$ in V is a linear combination of the $\vec{v}_i$, that is, the $\vec{v}_i$ span V.   ▲

From Section 3.1, we know that the kernel and image of a linear transformation are subspaces of the domain and the codomain of the transformation, respectively. We will now examine how we can find bases of kernel and image, and thus determine their dimensions.

Finding a Basis of the Kernel

EXAMPLE 1   Find a basis of the kernel of the following matrix, and determine the dimension of the kernel:

$$A = \begin{bmatrix} 1 & 2 & 0 & 3 & 0 \\ 2 & 4 & 1 & 9 & 5 \end{bmatrix}.$$

Solution

We have to solve the linear system $A\vec{x} = \vec{0}$. As in Chapter 1, first we find rref(A):

$$A = \begin{bmatrix} 1 & 2 & 0 & 3 & 0 \\ 2 & 4 & 1 & 9 & 5 \end{bmatrix} \begin{matrix} \\ -2(I) \end{matrix} \rightarrow \text{rref}(A) = \begin{bmatrix} 1 & 2 & 0 & 3 & 0 \\ 0 & 0 & 1 & 3 & 5 \end{bmatrix}.$$

This corresponds to the system

$$\begin{vmatrix} x_1 + 2x_2 & + 3x_4 & = 0 \\ & x_3 + 3x_4 + 5x_5 & = 0 \end{vmatrix},$$

with general solution

$$
\begin{bmatrix} x_1 \\ x_2 \\ x_3 \\ x_4 \\ x_5 \end{bmatrix} = \begin{bmatrix} -2s - 3t \\ s \\ -3t - 5r \\ t \\ r \end{bmatrix} = s \underset{\vec{v}_1}{\begin{bmatrix} -2 \\ 1 \\ 0 \\ 0 \\ 0 \end{bmatrix}} + t \underset{\vec{v}_2}{\begin{bmatrix} -3 \\ 0 \\ -3 \\ 1 \\ 0 \end{bmatrix}} + r \underset{\vec{v}_3}{\begin{bmatrix} 0 \\ 0 \\ -5 \\ 0 \\ 1 \end{bmatrix}}.
$$

The three vectors $\vec{v}_1$, $\vec{v}_2$, and $\vec{v}_3$ span ker(A), by construction. To verify that they are also linearly independent, consider the components corresponding to the nonleading variables: the second, fourth, and fifth. As you examine the second components, 1, 0, and 0, respectively, you realize that $\vec{v}_1$ cannot be a linear combination of $\vec{v}_2$ and $\vec{v}_3$. Similarly, we can see that $\vec{v}_2$ isn't a linear combination of $\vec{v}_1$ and $\vec{v}_3$ (likewise for $\vec{v}_3$). We have shown that the vectors $\vec{v}_1$, $\vec{v}_2$, and $\vec{v}_3$ form a basis of the kernel of A. The number of vectors in this basis (i.e., the dimension of ker(A)) is the number of nonleading variables:

$$
\begin{aligned}
\dim(\ker A) &= (\text{number of nonleading variables}) \\
&= (\text{number of columns of } A) - (\text{number of leading variables}) \\
&= (\text{number of columns of } A) - \text{rank}(A) = 5 - 2 = 3.
\end{aligned}
$$ ∎

The method we used in Example 1 applies to all matrices.

Fact 3.3.5   Consider an $m \times n$ matrix A. Then,

$$
\dim(\ker A) = n - \text{rank}(A).
$$

Finding a Basis of the Image

Again, we study an example, making sure that our procedure applies to the general case.

EXAMPLE 2   Find a basis of the image of the linear transformation T from $\mathbb{R}^5$ to $\mathbb{R}^4$ with matrix

$$
A = \begin{bmatrix} 1 & 0 & 1 & 2 & 1 \\ 1 & 0 & 1 & 2 & 2 \\ 2 & 1 & 0 & 1 & 2 \\ 1 & 1 & -1 & -1 & 0 \end{bmatrix},
$$

and thus determine the dimension of the image.

Solution

We know that the columns of A span the image of A (by Fact 3.1.3), but they are linearly dependent in this example. (Why?) To construct a basis of im(A), we could find a relation among the columns of A, express one of the columns as a linear combination of the others, and then omit this vector as redundant. After repeating this procedure a few times, we would be left with linearly independent vectors that still span $\operatorname{im}(A)$; that is, we would have a basis of $\operatorname{im}(A)$. We propose a method for finding a basis of $\operatorname{im}(A)$ that uses the ideas just discussed, but presents them in somewhat streamlined form.

We first find the reduced row-echelon form of A:

$$
A = \begin{array}{c} \begin{array}{ccccc} \vec{v}_1 & \vec{v}_2 & \vec{v}_3 & \vec{v}_4 & \vec{v}_5 \end{array} \\ \begin{bmatrix} 1 & 0 & 1 & 2 & 1 \\ 1 & 0 & 1 & 2 & 2 \\ 2 & 1 & 0 & 1 & 2 \\ 1 & 1 & -1 & -1 & 0 \end{bmatrix} \end{array} \rightarrow E = \operatorname{rref}(A) = \begin{array}{c} \begin{array}{ccccc} \vec{w}_1 & \vec{w}_2 & \vec{w}_3 & \vec{w}_4 & \vec{w}_5 \end{array} \\ \begin{bmatrix} 1 & 0 & 1 & 2 & 0 \\ 0 & 1 & -2 & -3 & 0 \\ 0 & 0 & 0 & 0 & 1 \\ 0 & 0 & 0 & 0 & 0 \end{bmatrix} \end{array}.
$$

We denote the ith columns of A and E by $\vec{v}_i$ and $\vec{w}_i$, respectively. We have to express some of the $\vec{v}_i$ as linear combinations of the other vectors $\vec{v}_j$ to identify redundant vectors. The corresponding problem for the $\vec{w}_i$ is easily solved by inspection:

$$
\vec{w}_3 = \vec{w}_1 - 2\vec{w}_2, \qquad \vec{w}_4 = 2\vec{w}_1 - 3\vec{w}_2.
$$

In general, we can express any column of $\operatorname{rref}(A)$ that does not contain a leading 1 as a linear combination of earlier columns that do contain a leading 1.

It may surprise you that the same relationships hold among the corresponding columns of the matrix A. Verify the following for yourself:

$$
\vec{v}_3 = \vec{v}_1 - 2\vec{v}_2, \qquad \vec{v}_4 = 2\vec{v}_1 - 3\vec{v}_2.
$$

Let us explain why corresponding relations hold among the columns of A and $E = \operatorname{rref}(A)$. Consider a relation $c_1\vec{v}_1 + c_2\vec{v}_2 + \cdots + c_5\vec{v}_5 = \vec{0}$ among the columns of A. This relation can alternatively be written as

$$
A \begin{bmatrix} c_1 \\ c_2 \\ \vdots \\ c_5 \end{bmatrix} = \vec{0}.
$$

Since A and E have the same kernels (the reduced row-echelon form was defined so that the systems $A\vec{x} = \vec{0}$ and $E\vec{x} = \vec{0}$ have the same solutions), it follows that

$$
E \begin{bmatrix} c_1 \\ c_2 \\ \vdots \\ c_5 \end{bmatrix} = \vec{0} \quad \text{or} \quad c_1\vec{w}_1 + c_2\vec{w}_2 + \cdots + c_5\vec{w}_5 = \vec{0}.
$$

Consider the columns $\vec{v}_1$, $\vec{v}_2$, and $\vec{v}_5$ of A corresponding to the columns of E containing the leading 1's. Since $\vec{w}_1$, $\vec{w}_2$, and $\vec{w}_5$ are linearly independent, so are the vectors $\vec{v}_1$, $\vec{v}_2$, and $\vec{v}_5$. The vectors $\vec{v}_1$, $\vec{v}_2$, and $\vec{v}_5$ span the image of A, since any vector $\vec{v}$ in the image of A can be expressed as

$$\vec{v} = c_1\vec{v}_1 + c_2\vec{v}_2 + c_3\vec{v}_3 + c_4\vec{v}_4 + c_5\vec{v}_5 = c_1\vec{v}_1 + c_2\vec{v}_2 + c_3(\vec{v}_1 - 2\vec{v}_2) + c_4(2\vec{v}_1 - 3\vec{v}_2) + c_5\vec{v}_5.$$

That is, $\vec{v}$ can be written as a linear combination of $\vec{v}_1$, $\vec{v}_2$, $\vec{v}_5$ alone.

We have shown that the vectors $\vec{v}_1$, $\vec{v}_2$, and $\vec{v}_5$ form a basis of $\text{im}(A)$, and thus $\dim(\text{im}A) = 3$. ∎

The following definition will be helpful in stating these results more generally:

Definition 3.3.6

A column of a matrix A is called a *pivot column* if the corresponding column of $\text{rref}(A)$ contains a leading 1.

Fact 3.3.7

The pivot columns of a matrix A form a basis of $\text{im}(A)$.

Because the number of pivot columns is the rank of A, by definition, we have the following result:

Fact 3.3.8

For any matrix A,

$$\text{rank}(A) = \dim(\text{im}A).$$

This interpretation of the rank as a dimension is conceptually more satisfactory than the rather technical Definition 1.3.2.

Consider an $m \times n$ matrix A. Facts 3.3.5 and 3.3.8 tell us that

$$\dim(\ker A) = n - \text{rank}(A)$$

and

$$\dim(\text{im}A) = \text{rank}(A).$$

Adding these two equations, we obtain the following:

Fact 3.3.9

Rank-Nullity Theorem

If A is an $m \times n$ matrix, then

$$\dim(\ker A) + \dim(\text{im } A) = n.$$

The dimension of the kernel of matrix A is called the *nullity* of A:

$$\text{nullity}(A) = \dim(\ker A).$$

Using this definition and Fact 3.3.8, we can write

$$\text{nullity}(A) + \text{rank}(A) = n.$$

nullity $+$ rank$(A) = n$

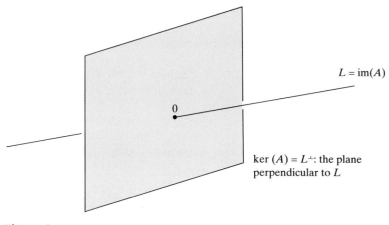

$L = \text{im}(A)$

0

$\ker(A) = L^{\perp}$: the plane
perpendicular to L

Figure 2

For matrices of a given size, the larger the kernel, the smaller the image, and vice versa. We can interpret the formula

$$n - \dim(\ker A) = \dim(\text{im} A)$$

geometrically as follows: Consider a linear transformation

$$T(\vec{x}) = A\vec{x} \quad \text{from } \mathbb{R}^n \text{ to } \mathbb{R}^m.$$

Note that n is the dimension of the domain of transformation T. The quantity $\text{nullity}(A) = \dim(\ker A)$ counts the dimensions that "collapse" as we perform the transformation, and $\text{rank}(A) = \dim(\text{im} A)$ counts the dimensions that "survive" the transformation.

As an example, consider the orthogonal projection onto a line L in $\mathbb{R}^3$. (See Figure 2.)

Here, the dimension n of the domain is 3, two dimensions collapse (the kernel is a plane), and we are left with the 1-dimensional image L:

$$n - \dim(\ker A) = \dim(\text{im} A)$$

$$3 - 2 = 1$$

Bases of $\mathbb{R}^n$

We know that any basis of $\mathbb{R}^n$ consists of n vectors, since we have the standard basis $\vec{e}_1, \ldots, \vec{e}_n$. Conversely, how can we tell whether n given vectors $\vec{v}_1, \vec{v}_2, \ldots, \vec{v}_n$ in $\mathbb{R}^n$ form a basis?

The $\vec{v}_i$ form a basis of $\mathbb{R}^n$ if every vector $\vec{b}$ in $\mathbb{R}^n$ can be written uniquely as a linear combination of the $\vec{v}_i$ (by Fact 3.2.7):

$$\vec{b} = c_1\vec{v}_1 + c_2\vec{v}_2 + \cdots + c_n\vec{v}_n = \begin{bmatrix} | & | & & | \\ \vec{v}_1 & \vec{v}_2 & \cdots & \vec{v}_n \\ | & | & & | \end{bmatrix} \begin{bmatrix} c_1 \\ c_2 \\ \vdots \\ c_n \end{bmatrix}.$$

The *linear system*

$$\begin{bmatrix} | & | & & | \\ \vec{v}_1 & \vec{v}_2 & \cdots & \vec{v}_n \\ | & | & & | \end{bmatrix} \begin{bmatrix} c_1 \\ c_2 \\ \vdots \\ c_n \end{bmatrix} = \vec{b}$$

has a unique solution if (and only if) the $n \times n$ matrix

$$\begin{bmatrix} | & | & & | \\ \vec{v}_1 & \vec{v}_2 & \cdots & \vec{v}_n \\ | & | & & | \end{bmatrix}$$

is invertible.

We have shown the following result:

Fact 3.3.10   **The vectors $\vec{v}_1, \vec{v}_2, \ldots, \vec{v}_n$ in $\mathbb{R}^n$ form a basis of $\mathbb{R}^n$ if (and only if) the matrix**

$$\begin{bmatrix} | & | & & | \\ \vec{v}_1 & \vec{v}_2 & \cdots & \vec{v}_n \\ | & | & & | \end{bmatrix}$$

is invertible.

EXAMPLE 3   Are the following vectors a basis of $\mathbb{R}^4$?

$$\vec{v}_1 = \begin{bmatrix} 1 \\ 2 \\ 9 \\ 1 \end{bmatrix}, \qquad \vec{v}_2 = \begin{bmatrix} 1 \\ 4 \\ 4 \\ 8 \end{bmatrix}, \qquad \vec{v}_3 = \begin{bmatrix} 1 \\ 8 \\ 1 \\ 5 \end{bmatrix}, \qquad \vec{v}_4 = \begin{bmatrix} 1 \\ 9 \\ 7 \\ 3 \end{bmatrix}.$$

Solution

We have to check whether the matrix

$$\begin{bmatrix} 1 & 1 & 1 & 1 \\ 2 & 4 & 8 & 9 \\ 9 & 4 & 1 & 7 \\ 1 & 8 & 5 & 3 \end{bmatrix}$$

is invertible. Using technology, we find that

$$\text{rref} \begin{bmatrix} 1 & 1 & 1 & 1 \\ 2 & 4 & 8 & 9 \\ 9 & 4 & 1 & 7 \\ 1 & 8 & 5 & 3 \end{bmatrix} = I_4.$$

Thus, the vectors $\vec{v}_1, \vec{v}_2, \vec{v}_3, \vec{v}_4$ form a basis of $\mathbb{R}^4$.   ■

Fact 3.3.4 (parts (c) and (d)), applied to $V = \mathbb{R}^n$, and Fact 3.3.10 provide us with new rules for our summary.

Summary 3.3.11   Consider an $n \times n$ matrix

$$A = \begin{bmatrix} | & | & & | \\ \vec{v}_1 & \vec{v}_2 & \cdots & \vec{v}_n \\ | & | & & | \end{bmatrix}.$$

Then, the following statements are equivalent:

  i. A is invertible.
 ii. The linear system $A\vec{x} = \vec{b}$ has a unique solution $\vec{x}$, for all $\vec{b}$ in $\mathbb{R}^n$.
iii. $\text{rref}(A) = I_n$.
 iv. $\text{rank}(A) = n$.
  v. $\text{im}(A) = \mathbb{R}^n$.
 vi. $\text{ker}(A) = \{\vec{0}\}$.
vii. The $\vec{v}_i$ are a basis of $\mathbb{R}^n$.
viii. The $\vec{v}_i$ span $\mathbb{R}^n$.
 ix. The $\vec{v}_i$ are linearly independent.

EXERCISES

GOALS Use the concept of dimension. Find a basis of the kernel and the image of a linear transformation.

In Exercises 1 through 10, find a basis of the kernel of the given matrix, and thus determine the dimension of the kernel. Use paper and pencil.

1. $\begin{bmatrix} 1 & 2 \\ 2 & 4 \end{bmatrix}$.

2. $\begin{bmatrix} 1 & 1 & 1 \\ 1 & 1 & 1 \\ 1 & 1 & 1 \end{bmatrix}$.

3. $\begin{bmatrix} 1 & 3 & 2 \\ 1 & 2 & 3 \\ 1 & 1 & 4 \end{bmatrix}$.

4. $\begin{bmatrix} 1 & 2 \\ 2 & 3 \\ 3 & 4 \end{bmatrix}$.

5. $[\,5 \quad -4 \quad 3\,]$.

6. $\begin{bmatrix} 1 & 3 & 2 \\ 1 & 2 & 3 \\ 1 & 1 & 3 \end{bmatrix}$.

7. $\begin{bmatrix} 0 & 1 & 2 & 0 & 3 \\ 0 & 0 & 0 & 1 & 4 \end{bmatrix}$.

8. $\begin{bmatrix} 1 & 2 & 3 & 2 & 1 \\ 3 & 6 & 9 & 6 & 3 \\ 1 & 2 & 4 & 1 & 2 \\ 2 & 4 & 9 & 1 & 5 \end{bmatrix}$.

9. $[\,1 \quad -1 \quad 1 \quad -1 \quad 1\,]$.

10. $\begin{bmatrix} 1 & 0 & 2 & 4 \\ 0 & 1 & -3 & -1 \\ 3 & 4 & -6 & 8 \\ 0 & -1 & 3 & 4 \end{bmatrix}$.

In Exercises 11 through 20, find a basis of the image of the given matrix and thus determine the dimension of the image. Use paper and pencil.

11. $\begin{bmatrix} 1 & 2 \\ 2 & 4 \end{bmatrix}$.

12. $\begin{bmatrix} 1 & 1 & 1 \\ 1 & 1 & 1 \\ 1 & 1 & 1 \end{bmatrix}$.

13. $\begin{bmatrix} 1 & 3 & 2 \\ 1 & 2 & 3 \\ 1 & 1 & 4 \end{bmatrix}$.

14. $\begin{bmatrix} 1 & 2 \\ 2 & 3 \\ 3 & 4 \end{bmatrix}$.

15. $[\,5 \quad -4 \quad 3\,]$.

16. $\begin{bmatrix} 1 & 3 & 2 \\ 1 & 2 & 3 \\ 1 & 1 & 3 \end{bmatrix}$.

17. $\begin{bmatrix} 1 & 2 & 1 \\ 1 & 2 & 2 \\ 1 & 2 & 3 \\ 1 & 2 & 4 \end{bmatrix}$.

18. $\begin{bmatrix} 4 & 8 & 1 & 1 & 6 \\ 3 & 6 & 1 & 2 & 5 \\ 2 & 4 & 1 & 9 & 10 \\ 1 & 2 & 1 & 1 & 0 \end{bmatrix}$.

19. $\begin{bmatrix} 1 & 0 & 2 & 4 \\ 0 & 1 & -3 & -1 \\ 3 & 4 & -6 & 8 \\ 0 & -1 & 3 & 4 \end{bmatrix}$.

20. $\begin{bmatrix} 1 & 2 & 3 & 2 & 1 \\ 3 & 6 & 9 & 6 & 3 \\ 1 & 2 & 4 & 1 & 2 \\ 2 & 4 & 9 & 1 & 5 \end{bmatrix}$.

For each matrix in Exercises 21 through 23, find a basis of the kernel and a basis of the image and thus determine their dimensions. Use Fact 3.3.9 to check your work.

21. $\begin{bmatrix} 1 & -1 & -1 & 1 & 1 \\ -1 & 1 & 0 & -2 & 2 \\ 1 & -1 & -2 & 0 & 3 \\ 2 & -2 & -1 & 3 & 4 \end{bmatrix}$.

22. $\begin{bmatrix} 1 & 2 & 1 & 2 & 1 \\ 1 & 2 & 2 & 1 & 2 \\ 2 & 4 & 3 & 3 & 3 \\ 0 & 0 & 1 & -1 & -1 \end{bmatrix}$.

23. $\begin{bmatrix} 1 & 1 & 1 & 1 & 1 \\ 1 & 1 & 1 & 1 & 1 \\ 1 & 1 & 1 & 1 & 1 \\ 1 & 1 & 1 & 1 & 1 \\ 1 & 1 & 1 & 1 & 1 \end{bmatrix}$.

24. Find a basis of the subspace of $\mathbb{R}^5$ spanned by the following vectors:

$$\begin{bmatrix} 1 \\ 2 \\ 3 \\ 2 \\ 1 \end{bmatrix}, \begin{bmatrix} 3 \\ 6 \\ 9 \\ 6 \\ 3 \end{bmatrix}, \begin{bmatrix} 3 \\ 2 \\ 4 \\ 1 \\ 2 \end{bmatrix}, \begin{bmatrix} 8 \\ 4 \\ 9 \\ 1 \\ 5 \end{bmatrix}, \begin{bmatrix} 0 \\ 4 \\ 5 \\ 5 \\ 1 \end{bmatrix}.$$

25. Pick a basis of $\mathbb{R}^5$ from the following vectors (if possible):

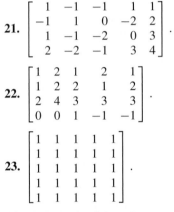

26. Consider the matrix

$$A = \begin{bmatrix} 0 & 1 & 2 & 0 & 3 \\ 0 & 0 & 0 & 1 & 4 \end{bmatrix}.$$

Find a matrix B such that $\ker(A) = \operatorname{im}(B)$.

27. Determine whether the following vectors form a basis of $\mathbb{R}^4$:

$$\begin{bmatrix} 1 \\ 1 \\ 1 \\ 1 \end{bmatrix}, \begin{bmatrix} 1 \\ -1 \\ 1 \\ -1 \end{bmatrix}, \begin{bmatrix} 1 \\ 2 \\ 4 \\ 8 \end{bmatrix}, \begin{bmatrix} 1 \\ -2 \\ 4 \\ -8 \end{bmatrix}.$$

28. For which choice(s) of the constant k do the vectors below form a basis of $\mathbb{R}^4$?

$$\begin{bmatrix} 1 \\ 0 \\ 0 \\ 2 \end{bmatrix}, \begin{bmatrix} 0 \\ 1 \\ 0 \\ 3 \end{bmatrix}, \begin{bmatrix} 0 \\ 0 \\ 1 \\ 4 \end{bmatrix}, \begin{bmatrix} 2 \\ 3 \\ 4 \\ k \end{bmatrix}.$$

29. Find a basis of the subspace of $\mathbb{R}^3$ defined by the equation

$$2x_1 + 3x_2 + x_3 = 0.$$

30. Find a basis of the subspace of $\mathbb{R}^4$ defined by the equation

$$2x_1 - x_2 + 2x_3 + 4x_4 = 0.$$

31. Let V be the subspace of $\mathbb{R}^4$ defined by the equation

$$x_1 - x_2 + 2x_3 + 4x_4 = 0.$$

Find a linear transformation T from $\mathbb{R}^3$ to $\mathbb{R}^4$ such that $\ker(T) = \{\vec{0}\}$ and $\operatorname{im}(T) = V$. Describe T by its matrix A.

32. Find a basis of the subspace of $\mathbb{R}^4$ that consists of all vectors perpendicular to both

$$\begin{bmatrix} 1 \\ 0 \\ -1 \\ 1 \end{bmatrix} \quad \text{and} \quad \begin{bmatrix} 0 \\ 1 \\ 2 \\ 3 \end{bmatrix}.$$

33. A subspace V of $\mathbb{R}^n$ is called a *hyperplane* if V is defined by the homogeneous linear equation

$$c_1 x_1 + c_2 x_2 + \cdots + c_n x_n = 0,$$

where at least one of the coefficients c_i is nonzero. What is the dimension of a hyperplane in $\mathbb{R}^n$? Justify your answer carefully. What is a hyperplane in $\mathbb{R}^2$? What is it in $\mathbb{R}^3$?

34. Consider a subspace V in $\mathbb{R}^n$ that is defined by m homogeneous linear equations:

$$\begin{vmatrix} a_{11}x_1 + a_{12}x_2 + \cdots + a_{1n}x_n = 0 \\ a_{21}x_1 + a_{22}x_2 + \cdots + a_{2n}x_n = 0 \\ \vdots \qquad \vdots \qquad\qquad \vdots \quad \vdots \\ a_{m1}x_1 + a_{m2}x_2 + \cdots + a_{mn}x_n = 0 \end{vmatrix}.$$

What is the relationship between the dimension of V and the quantity $n - m$? State your answer as an inequality. Explain carefully.

35. Consider a nonzero vector $\vec{v}$ in $\mathbb{R}^n$. What is the dimension of the space of all vectors in $\mathbb{R}^n$ that are perpendicular to $\vec{v}$?

36. Can you find a 3×3 matrix A such that $\operatorname{im}(A) = \ker(A)$? Explain.

37. Give an example of a 4×5 matrix A with $\dim(\ker A) = 3$.

38. a. Consider a linear transformation T from $\mathbb{R}^5$ to $\mathbb{R}^3$. What are the possible values of $\dim(\ker T)$? Explain.

  b. Consider a linear transformation T from $\mathbb{R}^4$ to $\mathbb{R}^7$. What are the possible values of $\dim(\operatorname{im} T)$? Explain.

39. We are told that a certain 5×5 matrix A can be written as

$$A = BC,$$

where B is a 5×4 matrix and C is 4×5. Explain how you know that A is not invertible.

40. Consider two subspaces V and W of $\mathbb{R}^n$, where V is contained in W. Explain why $\dim(V) \leq \dim(W)$. (This statement seems intuitively rather obvious. Still, we cannot rely on our intuition when dealing with $\mathbb{R}^n$.)

41. Consider two subspaces V and W of $\mathbb{R}^n$, where V is contained in W. In Exercise 40 we learned that $\dim(V) \leq \dim(W)$. Show that if $\dim(V) = \dim(W)$, then $V = W$.

42. Consider a subspace V of $\mathbb{R}^n$ with $\dim(V) = n$. Explain why $V = \mathbb{R}^n$.

43. Consider two subspaces V and W of $\mathbb{R}^n$, with $V \cap W = \{\vec{0}\}$. What is the relationship between $\dim(V)$, $\dim(W)$, and $\dim(V + W)$? (For the definition of $V + W$, see Exercise 3.2.50; Exercise 3.2.51 also is helpful.)

44. Two subspaces V and W of $\mathbb{R}^n$ are called *complements* if any vector $\vec{x}$ in $\mathbb{R}^n$ can be expressed uniquely as $\vec{x} = \vec{v} + \vec{w}$, where $\vec{v}$ is in V and $\vec{w}$ is in W. Show that V and W are complements if (and only if) $V \cap W = \{\vec{0}\}$ and $\dim(V) + \dim(W) = n$.

45. Consider linearly independent vectors $\vec{v}_1, \vec{v}_2, \ldots, \vec{v}_p$ in a subspace V of $\mathbb{R}^n$ and vectors $\vec{w}_1, \vec{w}_2, \ldots, \vec{w}_q$ that span V. Show that there is a basis of V that consists of *all* the $\vec{v}_i$ and *some* of the $\vec{w}_j$. *Hint:* Find a basis of the image of the matrix

$$A = \begin{bmatrix} | & & | & | & & | \\ \vec{v}_1 & \cdots & \vec{v}_p & \vec{w}_1 & \cdots & \vec{w}_q \\ | & & | & | & & | \end{bmatrix}.$$

46. Use Exercise 45 to construct a basis of $\mathbb{R}^4$ that consists of the vectors

$$\begin{bmatrix} 1 \\ 2 \\ 3 \\ 4 \end{bmatrix}, \quad \begin{bmatrix} 1 \\ 4 \\ 6 \\ 8 \end{bmatrix},$$

and some of the vectors $\vec{e}_1, \vec{e}_2, \vec{e}_3,$ and $\vec{e}_4$ in $\mathbb{R}^4$.

47. Consider two subspaces V and W of $\mathbb{R}^n$. Show that

$$\dim(V) + \dim(W) = \dim(V \cap W) + \dim(V + W).$$

(For the definition of $V + W$, see Exercise 3.2.50.) *Hint:* Pick a basis $\vec{u}_1, \vec{u}_2, \ldots, \vec{u}_m$ of $V \cap W$. Using Exercise 45, construct bases $\vec{u}_1, \vec{u}_2, \ldots, \vec{u}_m, \vec{v}_1, \vec{v}_2, \ldots, \vec{v}_p$ of V and $\vec{u}_1, \vec{u}_2, \ldots, \vec{u}_m, \vec{w}_1, \vec{w}_2, \ldots, \vec{w}_q$ of W. Show that $\vec{u}_1, \vec{u}_2, \ldots, \vec{u}_m, \vec{v}_1, \vec{v}_2, \ldots, \vec{v}_p, \vec{w}_1, \vec{w}_2, \ldots, \vec{w}_q$ is a basis of $V + W$. Demonstrating linear independence is somewhat challenging.

48. Use Exercise 47 to answer the following question: If V and W are subspaces of $\mathbb{R}^{10}$, with $\dim(V) = 6$ and $\dim(W) = 7$, what are the possible dimensions of $V \cap W$?

In Exercises 49 through 52, we will study the row space of a matrix. The row space of an $m \times n$ matrix A is defined as the span of the row vectors of A (i.e., the set of their linear combinations). For example, the row space of the matrix

$$\begin{bmatrix} 1 & 2 & 3 & 4 \\ 1 & 1 & 1 & 1 \\ 2 & 2 & 2 & 3 \end{bmatrix}$$

it is the set of all row vectors of the form

$$a\begin{bmatrix} 1 & 2 & 3 & 4 \end{bmatrix} + b\begin{bmatrix} 1 & 1 & 1 & 1 \end{bmatrix} + c\begin{bmatrix} 2 & 2 & 2 & 3 \end{bmatrix}.$$

49. Find a basis of the row space of the matrix

$$E = \begin{bmatrix} 0 & 1 & 0 & 2 & 0 \\ 0 & 0 & 1 & 3 & 0 \\ 0 & 0 & 0 & 0 & 1 \\ 0 & 0 & 0 & 0 & 0 \end{bmatrix}.$$

50. Consider an $m \times n$ matrix E in reduced row-echelon form. Using your work in Exercise 49 as a guide, explain how you can find a basis of the row space of E. What is the relationship between the dimension of the row space and the rank of E?

51. Consider an arbitrary $m \times n$ matrix A.

    a. What is the relationship between the row spaces of A and $E = \mathrm{rref}(A)$? *Hint:* Examine how the row space is affected by elementary row operations.

    b. What is the relationship between the dimension of the row space of A and the rank of A?

52. Find a basis of the row space of the matrix

$$A = \begin{bmatrix} 1 & 1 & 1 & 1 \\ 2 & 2 & 2 & 2 \\ 1 & 2 & 3 & 4 \\ 1 & 3 & 5 & 7 \end{bmatrix}.$$

53. Consider an $n \times n$ matrix A. Show that there are scalars $c_0, c_1, \ldots, c_n$ (not all zero) such that the matrix $c_0 I_n + c_1 A + c_2 A^2 + \cdots + c_n A^n$ is noninvertible. *Hint:* Pick an arbitrary nonzero vector $\vec{v}$ in $\mathbb{R}^n$. Then, the $n + 1$ vectors $\vec{v}, A\vec{v}, A^2\vec{v}, \ldots, A^n\vec{v}$ will be linearly dependent. (Much more is true: There are scalars c_0, $c_1, \ldots, c_n$ such that $c_0 I_n + c_1 A + c_2 A^2 + \cdots + c_n A^n = 0$. You are not asked to demonstrate this fact here.)

54. Consider the matrix

$$A = \begin{bmatrix} 1 & -2 \\ 2 & 1 \end{bmatrix}.$$

Find scalars c_0, c_1, c_2 (not all zero) such that the matrix $c_0 I_2 + c_1 A + c_2 A^2$ is noninvertible. (See Exercise 53.)

55. Consider an $m \times n$ matrix A. Show that the rank of A is m if (and only if) A has an invertible $m \times m$ submatrix (i.e., a matrix obtained by deleting $n - m$ columns of A).

56. An $n \times n$ matrix A is called *nilpotent* if $A^m = 0$ for some positive integer m. Examples are triangular matrices whose entries on the diagonal are all 0. Consider a nilpotent $n \times n$ matrix A, and choose the smallest number m such that $A^m = 0$. Pick a vector $\vec{v}$ in $\mathbb{R}^n$ such that $A^{m-1}\vec{v} \neq \vec{0}$. Show that the vectors $\vec{v}, A\vec{v}, A^2\vec{v}, \ldots, A^{m-1}\vec{v}$ are linearly independent. *Hint:* Consider a relation $c_0\vec{v} + c_1 A\vec{v} + c_2 A^2\vec{v} + \cdots + c_{m-1} A^{m-1}\vec{v} = \vec{0}$. Multiply both sides of the equation with A^{m-1} to show that $c_0 = 0$. Next, show that $c_1 = 0$, etc.

57. Consider a nilpotent $n \times n$ matrix A. Use the result demonstrated in Exercise 56 to show that $A^n = 0$.

58. Explain why you need at least m vectors to span a space of dimension m. (See Fact 3.3.4(b).)

59. Prove Fact 3.3.4(d): If m vectors span an m-dimensional space, they form a basis of the space.

60. If a 3×3 matrix A represents the orthogonal projection onto a plane in $\mathbb{R}^3$, what is rank (A)?

61. Consider a 4×2 matrix A and a 2×5 matrix B.

    a. What are the possible dimensions of the *kernel* of AB?

    b. What are the possible dimensions of the *image* of AB?

62. Consider two $m \times n$ matrices A and B. What can you say about the relationship between the quantitites $\mathrm{rank}(A)$, $\mathrm{rank}(B)$, and $\mathrm{rank}(A + B)$?

63. Consider an $m \times n$ matrix A and an $n \times p$ matrix B.

    a. What can you say about the relationship between $\mathrm{rank}(A)$ and $\mathrm{rank}(AB)$?

    b. What can you say about the relationship between $\mathrm{rank}(B)$ and $\mathrm{rank}(AB)$?

64. Consider the matrices

$$A = \begin{bmatrix} 1 & 0 & 2 & 0 & 4 & 0 \\ 0 & 1 & 3 & 0 & 5 & 0 \\ 0 & 0 & 0 & 1 & 6 & 0 \\ 0 & 0 & 0 & 0 & 0 & 1 \end{bmatrix} \text{ and } B = \begin{bmatrix} 1 & 0 & 2 & 0 & 4 & 0 \\ 0 & 1 & 3 & 0 & 5 & 0 \\ 0 & 0 & 0 & 1 & 7 & 0 \\ 0 & 0 & 0 & 0 & 0 & 1 \end{bmatrix}.$$

Show that the kernels of matrices A and B are different. *Hint:* Think about ways to write the fifth column as a linear combination of the previous columns.[7]

65. Consider the matrices

$$A = \begin{bmatrix} 1 & 0 & 2 & 0 & 4 & 0 \\ 0 & 1 & 3 & 0 & 5 & 0 \\ 0 & 0 & 0 & 1 & 6 & 0 \\ 0 & 0 & 0 & 0 & 0 & 1 \end{bmatrix} \text{ and } B = \begin{bmatrix} 1 & 0 & 2 & 0 & 0 & 4 \\ 0 & 1 & 3 & 0 & 0 & 5 \\ 0 & 0 & 0 & 1 & 0 & 6 \\ 0 & 0 & 0 & 0 & 1 & 7 \end{bmatrix}.$$

Show that the kernels of matrices A and B are different. *Hint:* Think about ways to write the fifth column as a linear combination of the previous columns.[7]

---

[7]In Exercises 64 through 66, it helps to remember that the nonzero vectors in the kernel of a matrix correspond to the nontrivial relations among its columns, which, in turn, correspond to representations of one column as a linear combination of the other columns. Thus, to show that two matrices A and B of the same size have different kernels, it suffices to give a representation of a column of A in terms of the other columns of A such that there is no corresponding representation for matrix B. For example, the third column of A might be the sum of the first two columns of A, but the third column of B fails to be the sum of the first two columns of B.

66. Let A and B be two different matrices of the same size, both in reduced row-echelon form. Show that the kernels of A and B are different. *Hint:* Focus on the first column in which the two matrices differ, say, the kth column, and think about ways to write that column as a linear combination of the previous columns. You will have to examine two cases: One of the two matrices has a leading 1 in the kth column, as in Example 65, or neither of them does, as in Example 64.[7]

67. Suppose a matrix A in reduced row-echelon form can be obtained from a matrix M by a sequence of elementary row operations. Show that $A = \text{rref}(M)$. *Hint:* Both A and $\text{rref}(M)$ are in reduced row-echelon form, and they have the same kernel. Exercise 66 is helpful.

3.4 COORDINATES

Coordinates are one of the "great ideas" of mathematics. René Descartes (1596–1650) is credited with having introduced them, in an appendix to his treatise *Discours de la Méthode* (Leyden, 1637). Myth has it that the idea came to him as he was laying on his back in bed one lazy Sunday morning, watching a fly on the ceiling above him. It occurred to him that he could describe the position of the fly by giving its distance from two walls.[8]

We have used Cartesian coordinates in the x–y-plane and in x–y–z-space throughout Chapters 1 through 3, without much fanfare. In this section, we will discuss coordinates more systematically.

EXAMPLE 1   Let V be the plane in $\mathbb{R}^3$ with equation $x_1 + 2x_2 + 3x_3 = 0$, a two-dimensional subspace of $\mathbb{R}^3$. We can describe a vector in this plane by its spatial (3D) coordinates; for example, vector

$$\vec{x} = \begin{bmatrix} 5 \\ -1 \\ -1 \end{bmatrix}$$

is in plane V. However, it may be more convenient to introduce a *plane* coordinate system in V, thus making plane V into a copy of $\mathbb{R}^2$.

To define the direction of the axes, we can consider any two vectors in plane V that aren't parallel, for example, vectors

$$\vec{v}_1 = \begin{bmatrix} 1 \\ 1 \\ -1 \end{bmatrix} \quad \text{and} \quad \vec{v}_2 = \begin{bmatrix} 1 \\ -2 \\ 1 \end{bmatrix}.$$

See Figure 1, where we label the new axes c_1 and c_2. The figure also shows the coordinate grid defined by vectors $\vec{v}_1$ and $\vec{v}_2$.

Don't be alarmed by the fact that the axes aren't perpendicular; Cartesian coordinates work just as well with oblique axes. Note that the $c_1 - c_2$ coordinates of vectors $\vec{v}_1$ and $\vec{v}_2$ are $\begin{bmatrix} c_1 \\ c_2 \end{bmatrix} = \begin{bmatrix} 1 \\ 0 \end{bmatrix}$ and $\begin{bmatrix} 0 \\ 1 \end{bmatrix}$, respectively.

How can we find the $c_1 - c_2$ coordinates of a vector $\vec{x}$ in plane V, for example, of $\vec{x} = \begin{bmatrix} 5 \\ -1 \\ -1 \end{bmatrix}$? As Figure 2 suggests, we need to find the scalars c_1 and c_2

---

[8]For a light-hearted account, see Julie Glass' *The Fly on the Ceiling: A Math Myth* (Random House, New York, 1998); the booklet makes a great gift for kids in Grades 1–3.

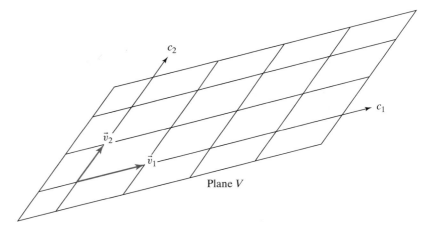

Figure 1

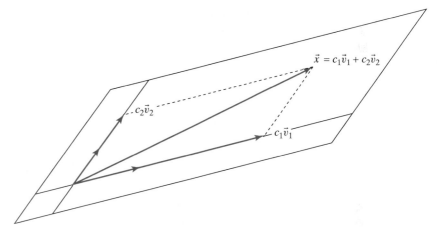

Figure 2

such that

$$\vec{x} = c_1\vec{v}_1 + c_2\vec{v}_2, \quad \text{or} \quad \begin{bmatrix} 5 \\ -1 \\ -1 \end{bmatrix} = c_1 \begin{bmatrix} 1 \\ 1 \\ -1 \end{bmatrix} + c_2 \begin{bmatrix} 1 \\ -2 \\ 1 \end{bmatrix}.$$

A routine computation shows that $c_1 = 3$ and $c_2 = 2$, so that the $c_1 - c_2$ coordinates of $\vec{x}$ are

$$\begin{bmatrix} c_1 \\ c_2 \end{bmatrix} = \begin{bmatrix} 3 \\ 2 \end{bmatrix}.$$

See Figure 3.

The following notation can be helpful when discussing coordinates (although it is a bit heavy). Let's denote the basis $\vec{v}_1, \vec{v}_2$ of V by $\mathfrak{B}$ (Fraktur B). Then, the coordinate vector of $\vec{x}$ with respect to $\mathfrak{B}$ is denoted by $[\vec{x}]_{\mathfrak{B}}$:

$$\text{If} \quad \vec{x} = \begin{bmatrix} 5 \\ -1 \\ -1 \end{bmatrix}, \quad \text{then} \quad [\vec{x}]_{\mathfrak{B}} = \begin{bmatrix} 3 \\ 2 \end{bmatrix}.$$

∎

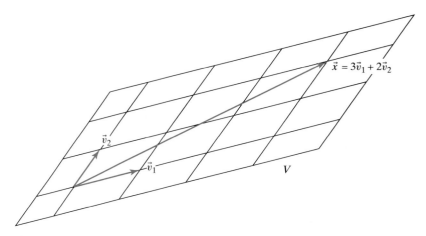

Figure 3

Let's generalize the ideas introduced in Example 1.

Definition 3.4.1

Coordinates in a subspace of $\mathbb{R}^n$
Consider a basis $\mathfrak{B}$ of a subspace V of $\mathbb{R}^n$, consisting of vectors $\vec{v}_1, \vec{v}_2, \ldots, \vec{v}_m$. Any vector $\vec{x}$ in V can be written uniquely as

$$\vec{x} = c_1\vec{v}_1 + c_2\vec{v}_2 + \cdots + c_m\vec{v}_m.$$

The scalars $c_1, c_2, \ldots, c_m$ are called the $\mathfrak{B}$-coordinates of $\vec{x}$, and the vector

$$\begin{bmatrix} c_1 \\ c_2 \\ \ldots \\ c_m \end{bmatrix}$$

is called the $\mathfrak{B}$-coordinate vector of $\vec{x}$, denoted by $[\vec{x}]_{\mathfrak{B}}$.
  Note that

$$\vec{x} = S[\vec{x}]_{\mathfrak{B}}, \quad \text{where} \quad S = \begin{bmatrix} | & | & & | \\ \vec{v}_1 & \vec{v}_2 & \ldots & \vec{v}_m \\ | & | & & | \end{bmatrix}, \quad \text{an } n \times m \text{ matrix.}$$

The last equation, $\vec{x} = S[\vec{x}]_{\mathfrak{B}}$, follows directly from the definition of coordinates:

$$\vec{x} = c_1\vec{v}_1 + c_2\vec{v}_2 + \cdots + c_m\vec{v}_m = \begin{bmatrix} | & | & & | \\ \vec{v}_1 & \vec{v}_2 & \ldots & \vec{v}_m \\ | & | & & | \end{bmatrix} \begin{bmatrix} c_1 \\ c_2 \\ \ldots \\ c_m \end{bmatrix} = S[\vec{x}]_{\mathfrak{B}}.$$

In Example 1, we considered the case where

$$\vec{x} = \begin{bmatrix} 5 \\ -1 \\ -1 \end{bmatrix}, \quad [\vec{x}]_{\mathfrak{B}} = \begin{bmatrix} 3 \\ 2 \end{bmatrix}, \quad \text{and} \quad S = \begin{bmatrix} 1 & 1 \\ 1 & -2 \\ -1 & 1 \end{bmatrix}.$$

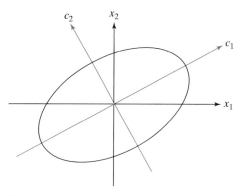

Figure 4

You can verify that

$$\vec{x} = S[\vec{x}]_{\mathcal{B}}, \quad \text{or} \quad \begin{bmatrix} 5 \\ -1 \\ -1 \end{bmatrix} = \begin{bmatrix} 1 & 1 \\ 1 & -2 \\ -1 & 1 \end{bmatrix} \begin{bmatrix} 3 \\ 2 \end{bmatrix}.$$

We leave it to the reader to verify the *linearity* of coordinates (see Exercise 25):

$$[\vec{x} + \vec{y}]_{\mathcal{B}} = [\vec{x}]_{\mathcal{B}} + [\vec{y}]_{\mathcal{B}}, \quad \text{for all vectors } \vec{x} \text{ and } \vec{y} \text{ on } V, \text{ and}$$

$$[k\vec{x}]_{\mathcal{B}} = k[\vec{x}]_{\mathcal{B}}, \qquad \text{for all } \vec{x} \text{ on } V \text{ and for all scalars } k.$$

As an important special case of Definition 3.4.1, consider the case when $V = \mathbb{R}^n$. It is often useful to work with bases of $\mathbb{R}^n$ other than the standard basis, $\vec{e}_1, \vec{e}_2, \ldots, \vec{e}_n$. When dealing with the ellipse in Figure 4, the $c_1 - c_2$ axes aligned with the principal axes are preferable to the standard $x_1 - x_2$ axes.

EXAMPLE 2 Consider the basis $\mathcal{B}$ of $\mathbb{R}^2$ consisting of vectors $\vec{v}_1 = \begin{bmatrix} 3 \\ 1 \end{bmatrix}$ and $\vec{v}_2 = \begin{bmatrix} -1 \\ 3 \end{bmatrix}$.

a. If $\vec{x} = \begin{bmatrix} 10 \\ 10 \end{bmatrix}$, find $[\vec{x}]_{\mathcal{B}}$.

b. If $[\vec{x}]_{\mathcal{B}} = \begin{bmatrix} 2 \\ -1 \end{bmatrix}$, find $\vec{x}$.

Solution

a. To find the coordinates of vector $\vec{x}$, we need to write $\vec{x}$ as a linear combination of the basis vectors:

$$\vec{x} = c_1 \vec{v}_1 + c_2 \vec{v}_2 \quad \text{or} \quad \begin{bmatrix} 10 \\ 10 \end{bmatrix} = c_1 \begin{bmatrix} 3 \\ 1 \end{bmatrix} + c_2 \begin{bmatrix} -1 \\ 3 \end{bmatrix}.$$

The solution is $c_1 = 4$, $c_2 = 2$, so that $[\vec{x}]_{\mathcal{B}} = \begin{bmatrix} 4 \\ 2 \end{bmatrix}$.

Alternatively, we can solve the equation $\vec{x} = S[\vec{x}]_{\mathcal{B}}$ for $[\vec{x}]_{\mathcal{B}}$:

$$[\vec{x}]_{\mathcal{B}} = S^{-1}\vec{x} = \begin{bmatrix} 3 & -1 \\ 1 & 3 \end{bmatrix}^{-1} \begin{bmatrix} 10 \\ 10 \end{bmatrix} = \frac{1}{10} \begin{bmatrix} 3 & 1 \\ -1 & 3 \end{bmatrix} \begin{bmatrix} 10 \\ 10 \end{bmatrix} = \begin{bmatrix} 4 \\ 2 \end{bmatrix}.$$

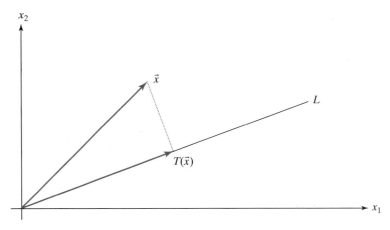

Figure 5

b. By definition of coordinates, $[\vec{x}]_{\mathcal{B}} = \begin{bmatrix} 2 \\ -1 \end{bmatrix}$ means that

$$\vec{x} = 2\vec{v}_1 + (-1)\vec{v}_2 = 2\begin{bmatrix} 3 \\ 1 \end{bmatrix} + (-1)\begin{bmatrix} -1 \\ 3 \end{bmatrix} = \begin{bmatrix} 7 \\ -1 \end{bmatrix}.$$

Alternatively, use the formula

$$\vec{x} = S[\vec{x}]_{\mathcal{B}} = \begin{bmatrix} 3 & -1 \\ 1 & 3 \end{bmatrix}\begin{bmatrix} 2 \\ -1 \end{bmatrix} = \begin{bmatrix} 7 \\ -1 \end{bmatrix}.$$ ∎

We will now go a step further and see how we can express a linear transformation in coordinates.

EXAMPLE 3 Let L be the line in $\mathbb{R}^2$ spanned by vector $\begin{bmatrix} 3 \\ 1 \end{bmatrix}$. Let T be the linear transformation from $\mathbb{R}^2$ to $\mathbb{R}^2$ that projects any vector orthogonally onto line L, as shown in Figure 5. It will facilitate the study of T to introduce a coordinate system where L is one of the axes (say, the c_1-axis), with the c_2-axis perpendicular to L. If we use this coordinate system, then T transforms vector $\begin{bmatrix} c_1 \\ c_2 \end{bmatrix}$ into $\begin{bmatrix} c_1 \\ 0 \end{bmatrix}$. In $c_1 - c_2$ coordinates, T is given by the matrix $B = \begin{bmatrix} 1 & 0 \\ 0 & 0 \end{bmatrix}$, since $\begin{bmatrix} c_1 \\ 0 \end{bmatrix} = \begin{bmatrix} 1 & 0 \\ 0 & 0 \end{bmatrix}\begin{bmatrix} c_1 \\ c_2 \end{bmatrix}$.

Let's make these ideas more precise. We start by introducing a basis $\mathcal{B}$ of $\mathbb{R}^2$ with vector $\vec{v}_1$ on line L and vector $\vec{v}_2$ perpendicular to L, for example, $\vec{v}_1 = \begin{bmatrix} 3 \\ 1 \end{bmatrix}$ and $\vec{v}_2 = \begin{bmatrix} -1 \\ 3 \end{bmatrix}$. If $[\vec{x}]_{\mathcal{B}} = \begin{bmatrix} c_1 \\ c_2 \end{bmatrix}$, then $[T(\vec{x})]_{\mathcal{B}} = \begin{bmatrix} c_1 \\ 0 \end{bmatrix}$. The matrix $B = \begin{bmatrix} 1 & 0 \\ 0 & 0 \end{bmatrix}$ that transforms $[\vec{x}]_{\mathcal{B}}$ into $[T(\vec{x})]_{\mathcal{B}}$ is called the $\mathcal{B}$-matrix of T:

$$[T(\vec{x})]_{\mathcal{B}} = B[\vec{x}]_{\mathcal{B}}.$$

See Figure 6. ∎

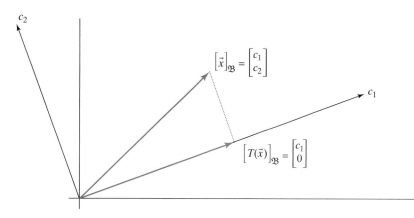

Figure 6

Let's generalize the ideas of Example 3.

Definition 3.4.2

> **The matrix of a linear transformation**
> Consider a linear transformation T from $\mathbb{R}^n$ to $\mathbb{R}^n$ and a basis $\mathcal{B}$ of $\mathbb{R}^n$. The $n \times n$ matrix B that transforms $[\vec{x}]_{\mathcal{B}}$ into $[T(\vec{x})]_{\mathcal{B}}$ is called the $\mathcal{B}$-matrix of T:
>
> $$[T(\vec{x})]_{\mathcal{B}} = B[\vec{x}]_{\mathcal{B}},$$
>
> for all $\vec{x}$ in $\mathbb{R}^n$.

How can we construct the $\mathcal{B}$-matrix B of a linear transformation T column by column? If

$$\vec{x} = c_1 \vec{v}_1 + \cdots + c_n \vec{v}_n,$$

then

$$T(\vec{x}) = c_1 T(\vec{v}_1) + \cdots + c_n T(\vec{v}_n),$$

by linearity of T, and

$$[T(\vec{x})]_{\mathcal{B}} = c_1 [T(\vec{v}_1)]_{\mathcal{B}} + \cdots + c_n [T(\vec{v}_n)]_{\mathcal{B}} = \underbrace{\left[[T(\vec{v}_1)]_{\mathcal{B}} \quad \cdots \quad [T(\vec{v}_n)]_{\mathcal{B}} \right]}_{B} [\vec{x}]_{\mathcal{B}}.$$

Fact 3.4.3

The columns of the matrix of a linear transformation
Consider a linear transformation T from $\mathbb{R}^n$ to $\mathbb{R}^n$ and a basis $\mathcal{B}$ of $\mathbb{R}^n$ consisting of vectors $\vec{v}_1, \vec{v}_2, \ldots, \vec{v}_n$. Then, the $\mathcal{B}$-matrix of T is

$$B = \left[[T(\vec{v}_1)]_{\mathcal{B}} \quad [T(\vec{v}_2)]_{\mathcal{B}} \quad \cdots \quad [T(\vec{v}_n)]_{\mathcal{B}} \right];$$

that is, the columns of B are the $\mathcal{B}$-coordinate vectors of $T(\vec{v}_1), T(\vec{v}_2), \ldots, T(\vec{v}_n)$.

In Example 3, we have

$$B = \left[[T(\vec{v}_1)]_{\mathcal{B}} \quad [T(\vec{v}_2)]_{\mathcal{B}} \right] = \left[[\vec{v}_1]_{\mathcal{B}} \quad [\vec{0}]_{\mathcal{B}} \right] = \begin{bmatrix} 1 & 0 \\ 0 & 0 \end{bmatrix}.$$

EXAMPLE 4   Consider two perpendicular unit vectors $\vec{v}_1$ and $\vec{v}_2$ in $\mathbb{R}^3$. Form the basis $\vec{v}_1$, $\vec{v}_2$, $\vec{v}_3 = \vec{v}_1 \times \vec{v}_2$ of $\mathbb{R}^3$; let's denote this basis by $\mathfrak{B}$. Find the $\mathfrak{B}$-matrix B of the linear transformation $T(\vec{x}) = \vec{v}_1 \times \vec{x}$.

Solution

Use Fact 3.4.3 to construct B column by column:

$$B = \begin{bmatrix} [T(\vec{v}_1)]_{\mathfrak{B}} & [T(\vec{v}_2)]_{\mathfrak{B}} & [T(\vec{v}_3)]_{\mathfrak{B}} \end{bmatrix} = \begin{bmatrix} [\vec{v}_1 \times \vec{v}_1]_{\mathfrak{B}} & [\vec{v}_1 \times \vec{v}_2]_{\mathfrak{B}} & [\vec{v}_1 \times \vec{v}_3]_{\mathfrak{B}} \end{bmatrix}$$

$$= \begin{bmatrix} [\vec{0}]_{\mathfrak{B}} & [\vec{v}_3]_{\mathfrak{B}} & [-\vec{v}_2]_{\mathfrak{B}} \end{bmatrix}$$

$$= \begin{bmatrix} 0 & 0 & 0 \\ 0 & 0 & -1 \\ 0 & 1 & 0 \end{bmatrix} \quad \text{(draw a sketch to see that } \vec{v}_1 \times \vec{v}_3 = -\vec{v}_2\text{)}.$$

Can you interpret this transformation geometrically, in terms of a projection and a rotation? ∎

EXAMPLE 5   Let T be the linear transformation from $\mathbb{R}^2$ to $\mathbb{R}^2$ that projects any vector orthogonally onto the line L spanned by $\begin{bmatrix} 3 \\ 1 \end{bmatrix}$. In Example 3, we found that the matrix of T with respect to the basis $\mathfrak{B}$ consisting of $\begin{bmatrix} 3 \\ 1 \end{bmatrix}$ and $\begin{bmatrix} -1 \\ 3 \end{bmatrix}$ is

$$B = \begin{bmatrix} 1 & 0 \\ 0 & 0 \end{bmatrix}.$$

What is the relationship between B and the *standard matrix* A of T (such that $T(\vec{x}) = A\vec{x}$)? We introduced the standard matrix of a linear transformation back in Section 2.1; alternatively, we can think of A as the matrix of T with respect to the standard basis, in the sense of Definition 3.4.2. (Think about it!)

Solution

Recall from Definition 3.4.1 that

$$\vec{x} = S[\vec{x}]_{\mathfrak{B}}, \quad \text{where } S = \begin{bmatrix} 3 & -1 \\ 1 & 3 \end{bmatrix},$$

and consider the following diagram:

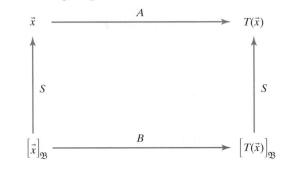

Note that $T(\vec{x}) = AS[\vec{x}]_{\mathfrak{B}}$ and also $T(\vec{x}) = SB[\vec{x}]_{\mathfrak{B}}$, so that $AS[\vec{x}]_{\mathfrak{B}} = SB[\vec{x}]_{\mathfrak{B}}$ for all $\vec{x}$. Thus,

$$AS = SB, \quad B = S^{-1}AS, \quad \text{and} \quad A = SBS^{-1}.$$

Now we can find the standard matrix A of T:

$$A = SBS^{-1} = \begin{bmatrix} 3 & -1 \\ 1 & 3 \end{bmatrix} \begin{bmatrix} 1 & 0 \\ 0 & 0 \end{bmatrix} \left(\frac{1}{10} \begin{bmatrix} 3 & 1 \\ -1 & 3 \end{bmatrix} \right) = \begin{bmatrix} 0.9 & 0.3 \\ 0.3 & 0.1 \end{bmatrix}.$$

Alternatively, we could use Fact 2.2.5 to construct matrix A. The point here was to explore the relationship between matrices A and B. ∎

Fact 3.4.4

Standard matrix versus $\mathcal{B}$-matrix of a linear transformation
Consider a linear transformation T from $\mathbb{R}^n$ to $\mathbb{R}^n$ and a basis $\mathcal{B}$ of $\mathbb{R}^n$ consisting of vectors $\vec{v}_1, \vec{v}_2, \ldots, \vec{v}_n$. Let B be the $\mathcal{B}$-matrix of T and let A be the standard matrix of T (such that $T(\vec{x}) = A\vec{x}$). Then,

$$AS = SB, \quad B = S^{-1}AS, \quad \text{and} \quad A = SBS^{-1}, \quad \text{where} \quad S = \begin{bmatrix} | & | & & | \\ \vec{v}_1 & \vec{v}_2 & \cdots & \vec{v}_n \\ | & | & & | \end{bmatrix}.$$

The formulas in Fact 3.4.4 motivate the following definition.

Definition 3.4.5

Similar matrices
Consider two $n \times n$ matrices A and B. We say that A is *similar* to B if there is an invertible matrix S such that

$$AS = SB, \quad \text{or} \quad B = S^{-1}AS.$$

Two matrices are similar if they represent the same linear transformation with respect to different bases.

EXAMPLE 6 Is matrix $A = \begin{bmatrix} 1 & 2 \\ 4 & 3 \end{bmatrix}$ similar to $B = \begin{bmatrix} 5 & 0 \\ 0 & -1 \end{bmatrix}$?

Solution

At this early stage of the course, we have to tackle this problem with "brute force," using Definition 3.4.5. In Chapter 7, we will develop tools that allow a more conceptual approach.

We are looking for a matrix $S = \begin{bmatrix} x & y \\ z & t \end{bmatrix}$ such that $AS = SB$, or

$$\begin{bmatrix} x + 2z & y + 2t \\ 4x + 3z & 4y + 3t \end{bmatrix} = \begin{bmatrix} 5x & -y \\ 5z & -t \end{bmatrix}.$$

These equations simplify to

$$z = 2x, \quad t = -y,$$

so that any invertible matrix of the form

$$S = \begin{bmatrix} x & y \\ 2x & -y \end{bmatrix}$$

does the job. Note that $\det(S) = -3xy$. Matrix S is invertible if $\det(S) \neq 0$ (i.e., if neither x nor y is zero). We can let $x = y = 1$, for example, so that $S = \begin{bmatrix} 1 & 1 \\ 2 & -1 \end{bmatrix}$.

Thus matrices A and B are indeed similar.   ∎

EXAMPLE 7   Show that if matrix A is similar to B, then its power A^t is similar to B^t for all positive integers t. (That is, A^2 is similar to B^2, A^3 is similar to B^3, etc.)

Solution

We know that $B = S^{-1}AS$ for some invertible matrix S. Now,

$$B^t = \underbrace{(S^{-1}AS)(S^{-1}AS)\cdots(S^{-1}AS)(S^{-1}AS)}_{t \text{ times}} = S^{-1}A^t S,$$

proving our claims. Note the cancellation of many terms of the form SS^{-1}.   ∎

We conclude this section with some noteworthy facts about similar matrices.

Fact 3.4.6   **Similarity is an equivalence relation**

a. An $n \times n$ matrix A is similar to itself (Reflexivity).
b. If A is similar to B, then B is similar to A (Symmetry).
c. If A is similar to B and B is similar to C, then A is similar to C (Transitivity).

We will prove transitivity, leaving reflexivity and symmetry as Exercise 39.
    The assumptions of part (c) tell us that there are invertible matrices P and Q such that $B = P^{-1}AP$ and $C = Q^{-1}BQ$. Then,

$$C = Q^{-1}BQ = Q^{-1}(P^{-1}AP)Q = (Q^{-1}P^{-1})A(PQ) = (PQ)^{-1}A(PQ) = S^{-1}AS,$$

where $S = PQ$. We have shown that A is similar to C.   ▲

EXERCISES

GOALS Use the concept of coordinates. Apply the definition of the matrix of a linear transformation with respect to a basis. Relate this matrix to the standard matrix of the transformation. Find the matrix of a linear transformation (with respect to any basis) column by column. Use the concept of similarity.

1. Consider the plane $x_1 + x_2 + x_3 = 0$ with basis $\mathfrak{B}$ consisting of vectors $\begin{bmatrix} 1 \\ -1 \\ 0 \end{bmatrix}$ and $\begin{bmatrix} 0 \\ 1 \\ -1 \end{bmatrix}$. Find $[\vec{x}]_{\mathfrak{B}}$ for

$$\vec{x} = \begin{bmatrix} 3 \\ 1 \\ -4 \end{bmatrix}.$$

2. Consider the plane $x_1 + 2x_2 + x_3 = 0$ with basis $\mathfrak{B}$ consisting of vectors $\begin{bmatrix} -1 \\ 0 \\ 1 \end{bmatrix}$ and $\begin{bmatrix} -2 \\ 1 \\ 0 \end{bmatrix}$. Find $[\vec{x}]_{\mathfrak{B}}$ for

$$\vec{x} = \begin{bmatrix} -5 \\ 1 \\ 3 \end{bmatrix}.$$

3. Consider the plane $2x_1 - 3x_2 + 4x_3 = 0$ with basis $\mathfrak{B}$ consisting of vectors $\begin{bmatrix} 1 \\ 2 \\ 1 \end{bmatrix}$ and $\begin{bmatrix} -3 \\ 2 \\ 3 \end{bmatrix}$. Find $[\vec{x}]_{\mathfrak{B}}$ for

$$\vec{x} = \begin{bmatrix} -1 \\ 2 \\ 2 \end{bmatrix}.$$

4. Consider the plane $2x_1 - 3x_2 + 4x_3 = 0$ with basis $\mathfrak{B}$ consisting of vectors $\begin{bmatrix} 8 \\ 4 \\ -1 \end{bmatrix}$ and $\begin{bmatrix} 5 \\ 2 \\ -1 \end{bmatrix}$. Find $[\vec{x}]_{\mathfrak{B}}$ for $\vec{x} = \begin{bmatrix} 1 \\ -2 \\ -2 \end{bmatrix}$.

5. Consider the plane $x_1 + 2x_2 + x_3 = 0$ with basis $\mathfrak{B}$ consisting of vectors $\begin{bmatrix} -1 \\ 0 \\ 1 \end{bmatrix}$ and $\begin{bmatrix} -2 \\ 1 \\ 0 \end{bmatrix}$. If $[\vec{x}]_{\mathfrak{B}} = \begin{bmatrix} 2 \\ -3 \end{bmatrix}$, find $\vec{x}$.

6. Consider the plane $2x_1 - 3x_2 + 4x_3 = 0$ with basis $\mathfrak{B}$ consisting of vectors $\begin{bmatrix} 8 \\ 4 \\ -1 \end{bmatrix}$ and $\begin{bmatrix} 5 \\ 2 \\ -1 \end{bmatrix}$. If $[\vec{x}]_{\mathfrak{B}} = \begin{bmatrix} 2 \\ -1 \end{bmatrix}$, find $\vec{x}$.

7. Consider the plane $2x_1 - 3x_2 + 4x_3 = 0$. Find a basis $\mathfrak{B}$ of this plane such that $[\vec{x}]_{\mathfrak{B}} = \begin{bmatrix} 2 \\ 3 \end{bmatrix}$ for $\vec{x} = \begin{bmatrix} 2 \\ 0 \\ -1 \end{bmatrix}$.

8. Consider the plane $x_1 + 2x_2 + x_3 = 0$. Find a basis $\mathfrak{B}$ of this plane such that $[\vec{x}]_{\mathfrak{B}} = \begin{bmatrix} 2 \\ -1 \end{bmatrix}$ for $\vec{x} = \begin{bmatrix} 1 \\ -1 \\ 1 \end{bmatrix}$.

9. Find the coordinate vector of $\begin{bmatrix} 7 \\ 16 \end{bmatrix}$ with respect to the basis $\begin{bmatrix} 2 \\ 5 \end{bmatrix}, \begin{bmatrix} 5 \\ 12 \end{bmatrix}$ of $\mathbb{R}^2$.

10. Find the coordinate vector of $\begin{bmatrix} -4 \\ 4 \end{bmatrix}$ with respect to the basis $\begin{bmatrix} 1 \\ 2 \end{bmatrix}, \begin{bmatrix} 5 \\ 6 \end{bmatrix}$ of $\mathbb{R}^2$.

11. Find the coordinate vector of $\begin{bmatrix} 1 \\ 1 \\ 1 \end{bmatrix}$ with respect to the basis $\begin{bmatrix} 1 \\ 2 \\ 3 \end{bmatrix}, \begin{bmatrix} 0 \\ 1 \\ 2 \end{bmatrix}, \begin{bmatrix} 0 \\ 0 \\ 1 \end{bmatrix}$ of $\mathbb{R}^3$.

12. Find the coordinate vector of $\begin{bmatrix} 3 \\ 7 \\ 13 \end{bmatrix}$ with respect to the basis $\begin{bmatrix} 1 \\ 1 \\ 1 \end{bmatrix}, \begin{bmatrix} 0 \\ 1 \\ 1 \end{bmatrix}, \begin{bmatrix} 0 \\ 0 \\ 1 \end{bmatrix}$ of $\mathbb{R}^3$.

13. Find the matrix of the linear transformation
$$T(\vec{x}) = \begin{bmatrix} 1 & 2 \\ 3 & 4 \end{bmatrix} \vec{x}$$
with respect to the basis $\begin{bmatrix} 1 \\ 1 \end{bmatrix}, \begin{bmatrix} 1 \\ 2 \end{bmatrix}$.

14. Find the matrix of the linear transformation
$$T(\vec{x}) = \begin{bmatrix} 7 & -1 \\ -6 & 8 \end{bmatrix} \vec{x}$$
with respect to the basis $\begin{bmatrix} 1 \\ 2 \end{bmatrix}, \begin{bmatrix} -1 \\ 3 \end{bmatrix}$.

15. Let T from $\mathbb{R}^2$ to $\mathbb{R}^2$ be the orthogonal projection onto the line spanned by $\begin{bmatrix} 1 \\ 3 \end{bmatrix}$.

   a. Find the matrix of T with respect to the basis $\begin{bmatrix} 1 \\ 3 \end{bmatrix}$, $\begin{bmatrix} -3 \\ 1 \end{bmatrix}$. Draw a sketch.

   b. Use your answer in part (a) to find the standard matrix of T.

16. Let T from $\mathbb{R}^3$ to $\mathbb{R}^3$ be the reflection in the plane given by the equation
$$x_1 + 2x_2 + 3x_3 = 0.$$

   a. Find the matrix B of this transformation with respect to the basis
$$\begin{bmatrix} 1 \\ 1 \\ -1 \end{bmatrix}, \begin{bmatrix} -1 \\ 2 \\ -1 \end{bmatrix}, \begin{bmatrix} 1 \\ 2 \\ 3 \end{bmatrix}.$$

   Draw a sketch.

   b. Use your answer in part (a) to find the standard matrix A of T. Feel free to use technology.

17. Consider the linear transformation T from $\mathbb{R}^2$ to $\mathbb{R}^2$ with standard matrix $\begin{bmatrix} 1 & 9 \\ 9 & 4 \end{bmatrix}$. Find the matrix of this transformation with respect to the basis $\begin{bmatrix} 3 \\ 5 \end{bmatrix}, \begin{bmatrix} 5 \\ 8 \end{bmatrix}$.

18. Consider a linear transformation T from $\mathbb{R}^2$ to $\mathbb{R}^2$. We are told that the matrix of T with respect to the basis $\begin{bmatrix} 3 \\ 5 \end{bmatrix}, \begin{bmatrix} 5 \\ 8 \end{bmatrix}$ is $\begin{bmatrix} 1 & 9 \\ 9 & 7 \end{bmatrix}$. Find the standard matrix of T.

19. Consider a linear transformation T from $\mathbb{R}^2$ to $\mathbb{R}^2$. We are told that the matrix of T with respect to the basis $\begin{bmatrix} 0 \\ 1 \end{bmatrix}, \begin{bmatrix} 1 \\ 0 \end{bmatrix}$ is $\begin{bmatrix} a & b \\ c & d \end{bmatrix}$. Find the standard matrix of T in terms of $a, b, c,$ and d.

20. Consider the basis $\mathfrak{B}$ of $\mathbb{R}^2$ consisting of the vectors $\begin{bmatrix} 3 \\ 1 \end{bmatrix}$ and $\begin{bmatrix} -1 \\ 1 \end{bmatrix}$. Find $[\vec{x}]_{\mathfrak{B}}$ for $\vec{x} = \begin{bmatrix} 5 \\ 3 \end{bmatrix}$. Illustrate the result with a sketch.

21. Redo Exercise 20 for $\vec{x} = \begin{bmatrix} 2 \\ 0 \end{bmatrix}$.

22. In the accompanying figure, sketch the vector $\vec{x}$ with $[\vec{x}]_{\mathcal{B}} = \begin{bmatrix} -1 \\ 2 \end{bmatrix}$, where $\mathcal{B}$ is the basis of $\mathbb{R}^2$ consisting of the vectors $\vec{v}, \vec{w}$.

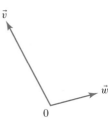

23. Consider the vectors $\vec{u}$, $\vec{v}$, and $\vec{w}$ sketched in the accompanying figure. Find the coordinate vector of $\vec{w}$ with respect to the basis $\vec{u}, \vec{v}$.

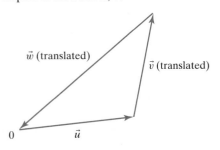

24. Given a hexagonal tiling of the plane, such as you might find on a kitchen floor, consider the basis $\mathcal{B}$ of $\mathbb{R}^2$ consisting of the vectors $\vec{v}, \vec{w}$ in the following sketch:

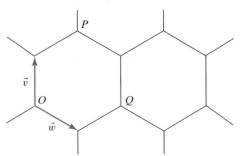

a. Find the coordinate vectors $[\overrightarrow{OP}]_{\mathcal{B}}$ and $[\overrightarrow{OQ}]_{\mathcal{B}}$.

b. We are told that $[\overrightarrow{OR}]_{\mathcal{B}} = \begin{bmatrix} 3 \\ 2 \end{bmatrix}$. Sketch the point R. Is R a vertex or a center of a tile?

c. We are told that $[\overrightarrow{OS}]_{\mathcal{B}} = \begin{bmatrix} 17 \\ 13 \end{bmatrix}$. Is S a center or a vertex of a tile?

25. If $\mathcal{B}$ is a basis of a subspace V of $\mathbb{R}^n$, show that the equations $[\vec{x} + \vec{y}]_{\mathcal{B}} = [\vec{x}]_{\mathcal{B}} + [\vec{y}]_{\mathcal{B}}$ and $[k\vec{x}]_{\mathcal{B}} = k[\vec{x}]_{\mathcal{B}}$ hold for all vectors $\vec{x}$ and $\vec{y}$ in V and for all scalars k.

26. If $\mathcal{B}$ is a basis of $\mathbb{R}^n$, is the transformation T from $\mathbb{R}^n$ to $\mathbb{R}^n$ given by

$$T(\vec{x}) = [\vec{x}]_{\mathcal{B}}$$

linear? Justify your answer.

27. Consider the basis $\mathcal{B}$ of $\mathbb{R}^2$ consisting of the vectors $\begin{bmatrix} 1 \\ 2 \end{bmatrix}$ and $\begin{bmatrix} 3 \\ 4 \end{bmatrix}$. We are told that $[\vec{x}]_{\mathcal{B}} = \begin{bmatrix} 7 \\ 11 \end{bmatrix}$ for a certain vector $\vec{x}$ in $\mathbb{R}^2$. Find $\vec{x}$.

28. Let $\mathcal{B}$ be the basis of $\mathbb{R}^n$ consisting of the vectors $\vec{v}_1, \vec{v}_2, \ldots, \vec{v}_n$, and let $\mathcal{T}$ be some other basis of $\mathbb{R}^n$. Is

$$[\vec{v}_1]_{\mathcal{T}}, \quad [\vec{v}_2]_{\mathcal{T}}, \quad \ldots, \quad [\vec{v}_n]_{\mathcal{T}}$$

a basis of $\mathbb{R}^n$ as well? Explain.

29. Consider the basis $\mathcal{B}$ of $\mathbb{R}^2$ consisting of the vectors $\begin{bmatrix} 1 \\ 1 \end{bmatrix}$ and $\begin{bmatrix} 1 \\ 2 \end{bmatrix}$, and let $\mathfrak{R}$ be the basis consisting of $\begin{bmatrix} 1 \\ 2 \end{bmatrix}$, $\begin{bmatrix} 3 \\ 4 \end{bmatrix}$. Find a matrix P such that

$$[\vec{x}]_{\mathfrak{R}} = P[\vec{x}]_{\mathcal{B}},$$

for all $\vec{x}$ in $\mathbb{R}^2$.

30. Find a basis $\mathcal{B}$ of $\mathbb{R}^2$ such that

$$\begin{bmatrix} 1 \\ 2 \end{bmatrix}_{\mathcal{B}} = \begin{bmatrix} 3 \\ 5 \end{bmatrix} \quad \text{and} \quad \begin{bmatrix} 3 \\ 4 \end{bmatrix}_{\mathcal{B}} = \begin{bmatrix} 2 \\ 3 \end{bmatrix}.$$

31. Consider two perpendicular unit vectors $\vec{v}_1$ and $\vec{v}_2$ in $\mathbb{R}^3$, and form the basis $\mathcal{B}$ consisting of the vectors $\vec{v}_1, \vec{v}_2, \vec{v}_3 = \vec{v}_1 \times \vec{v}_2$. Find the matrix B of the linear transformation

$$T(\vec{x}) = \vec{x} \times \vec{v}_2$$

with respect to the basis $\mathcal{B}$.

32. Consider a 3×3 matrix A and a vector $\vec{v}$ in $\mathbb{R}^3$ such that $A^3\vec{v} = \vec{0}$, but $A^2\vec{v} \neq \vec{0}$.

a. Show that the vectors $A^2\vec{v}$, $A\vec{v}$, $\vec{v}$ form a basis of $\mathbb{R}^3$. *Hint:* It suffices to show linear independence. Consider a relation $c_1 A^2\vec{v} + c_2 A\vec{v} + c_3\vec{v} = \vec{0}$ and multiply by A^2 to show that $c_3 = 0$.

b. Find the matrix of the transformation $T(\vec{x}) = A\vec{x}$ with respect to the basis $A^2\vec{v}, A\vec{v}, \vec{v}$.

33. Is matrix $\begin{bmatrix} 2 & 0 \\ 0 & 3 \end{bmatrix}$ similar to matrix $\begin{bmatrix} 2 & 1 \\ 0 & 3 \end{bmatrix}$?

34. Is matrix $\begin{bmatrix} 1 & 0 \\ 0 & -1 \end{bmatrix}$ similar to matrix $\begin{bmatrix} 0 & 1 \\ 1 & 0 \end{bmatrix}$?

35. Find a basis $\mathcal{B}$ of $\mathbb{R}^2$ such that the $\mathcal{B}$-matrix of the linear transformation

$$T(\vec{x}) = \begin{bmatrix} -5 & -9 \\ 4 & 7 \end{bmatrix} \vec{x} \quad \text{is} \quad B = \begin{bmatrix} 1 & 1 \\ 0 & 1 \end{bmatrix}.$$

36. Find a basis $\mathcal{B}$ of $\mathbb{R}^2$ such that the $\mathcal{B}$-matrix of the linear transformation

$$T(\vec{x}) = \begin{bmatrix} 1 & 2 \\ 4 & 3 \end{bmatrix} \vec{x} \quad \text{is} \quad B = \begin{bmatrix} 5 & 0 \\ 0 & -1 \end{bmatrix}.$$

37. Is matrix $\begin{bmatrix} p & -q \\ q & p \end{bmatrix}$ similar to matrix $\begin{bmatrix} p & q \\ -q & p \end{bmatrix}$ for all p and q?

38. Is matrix $\begin{bmatrix} a & b \\ c & d \end{bmatrix}$ similar to matrix $\begin{bmatrix} a & c \\ b & d \end{bmatrix}$ for all a, b, c, d?

39. Prove parts (a) and (b) of Fact 3.4.6.

40. Show that if $T(\vec{x}) = A\vec{x}$ represents a shear in $\mathbb{R}^2$ (where $A \neq I_2$), then A is similar to $\begin{bmatrix} 1 & 1 \\ 0 & 1 \end{bmatrix}$. Compare this with Exercise 35.

41. If $c \neq 0$, find the matrix of the linear transformation $T(\vec{x}) = \begin{bmatrix} a & b \\ c & d \end{bmatrix} \vec{x}$ with respect to basis $\begin{bmatrix} 1 \\ 0 \end{bmatrix}, \begin{bmatrix} a \\ c \end{bmatrix}$.

42. Find an invertible 2×2 matrix S such that

$$S^{-1} \begin{bmatrix} 1 & 2 \\ 3 & 4 \end{bmatrix} S \text{ is of the form } \begin{bmatrix} 0 & b \\ 1 & d \end{bmatrix}.$$ See Exercise 41.

43. If A is a 2×2 matrix such that $A \begin{bmatrix} 1 \\ 2 \end{bmatrix} = \begin{bmatrix} 3 \\ 6 \end{bmatrix}$ and $A \begin{bmatrix} 2 \\ 1 \end{bmatrix} = \begin{bmatrix} -2 \\ -1 \end{bmatrix}$, show that A is similar to a diagonal matrix D. Find an invertible S such that

$$S^{-1} A S = D.$$

44. Is there a basis $\mathcal{B}$ of $\mathbb{R}^2$ such that $\mathcal{B}$-matrix B of the linear transformation $T(\vec{x}) = \begin{bmatrix} 0 & -1 \\ 1 & 0 \end{bmatrix} \vec{x}$ is upper triangular? *Hint:* Think about the first column of B.

45. Suppose that matrix A is similar to B, with $B = S^{-1}AS$.
  a. Show that if $\vec{x}$ is in ker(B), then $S\vec{x}$ is in ker(A).
  b. Show that nullity(A) = nullity(B). *Hint:* If $\vec{v}_1, \vec{v}_2, \ldots, \vec{v}_p$ is a basis of ker(B), then the vectors $S\vec{v}_1, S\vec{v}_2, \ldots, S\vec{v}_p$ in ker(A) are linearly independent. Now reverse the roles of A and B.

46. If A is similar to B, what is the relationship between rank(A) and rank(B)? See Exercise 45.

47. Let L be the line in $\mathbb{R}^3$ spanned by the vector
$$\vec{v} = \begin{bmatrix} 0.6 \\ 0.8 \\ 0 \end{bmatrix}.$$ Let T from $\mathbb{R}^3$ to $\mathbb{R}^3$ be the rotation about this line through an angle of $\pi/2$, in the direction indicated in the accompanying sketch. Find the matrix A such that $T(x) = A\vec{x}$.

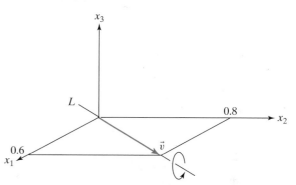

48. Consider the regular tetrahedron in the accompanying sketch whose center is at the origin.

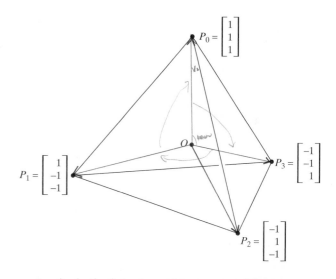

Let $\vec{v}_0, \vec{v}_1, \vec{v}_2, \vec{v}_3$ be the position vectors of the four vertices of the tetrahedron: $\vec{v}_0 = \overrightarrow{OP}_0, \ldots, \vec{v}_3 = \overrightarrow{OP}_3$.
  a. Find the sum $\vec{v}_0 + \vec{v}_1 + \vec{v}_2 + \vec{v}_3$.
  b. Find the coordinate vector of $\vec{v}_0$ with respect to the basis $\vec{v}_1, \vec{v}_2, \vec{v}_3$.
  c. Let T be the linear transformation with $T(\vec{v}_0) = \vec{v}_3$, $T(\vec{v}_3) = \vec{v}_1$, and $T(\vec{v}_1) = \vec{v}_0$. What is $T(\vec{v}_2)$? Describe the transformation T geometrically (as a reflection, rotation, projection, or whatever). Find the

matrix B of T with respect to the basis $\vec{v}_1, \vec{v}_2, \vec{v}_3$. What is B^3? Explain.

49. Find the matrix B of the rotation $T(\vec{x}) = \begin{bmatrix} 0 & -1 \\ 1 & 0 \end{bmatrix} \vec{x}$

with respect to the basis $\begin{bmatrix} 0 \\ 1 \end{bmatrix}, \begin{bmatrix} -1 \\ 0 \end{bmatrix}$. Interpret your answer geometrically.

50. If t is any real number, what is the matrix B of the linear transformation

$$T(\vec{x}) = \begin{bmatrix} \cos(t) & -\sin(t) \\ \sin(t) & \cos(t) \end{bmatrix} \vec{x}$$

with respect to basis $\begin{bmatrix} \cos(t) \\ \sin(t) \end{bmatrix}, \begin{bmatrix} -\sin(t) \\ \cos(t) \end{bmatrix}$? Interpret your answer geometrically.

51. Consider a linear transformation $T(\vec{x}) = A\vec{x}$ from $\mathbb{R}^n$ to $\mathbb{R}^n$. Let B be the matrix of T with respect to the basis $\vec{e}_n, \vec{e}_{n-1}, \ldots, \vec{e}_2, \vec{e}_1$ of $\mathbb{R}^n$. Describe the entries of B in terms of the entries of A.

52. This problem refers to Leontief's input–output model, first discussed in the Exercises 1.1.20 and 1.2.37. Consider three industries I_1, I_2, I_3, each of which produces only one good, with unit prices $p_1 = 2$, $p_2 = 5$, $p_3 = 10$ (in dollars), respectively. Let the three

products be good 1, good 2, and good 3. Let

$$A = \begin{bmatrix} a_{11} & a_{12} & a_{13} \\ a_{21} & a_{22} & a_{23} \\ a_{31} & a_{32} & a_{33} \end{bmatrix} = \begin{bmatrix} 0.3 & 0.2 & 0.1 \\ 0.1 & 0.3 & 0.3 \\ 0.2 & 0.2 & 0.1 \end{bmatrix}$$

be the matrix that lists the interindustry demand in terms of dollar amounts. The entry a_{ij} tells us how many dollars' worth of good i are required to produce one dollar's worth of good j. Alternatively, the interindustry demand can be measured in units of goods by means of the matrix

$$B = \begin{bmatrix} b_{11} & b_{12} & b_{13} \\ b_{21} & b_{22} & b_{23} \\ b_{31} & b_{32} & b_{33} \end{bmatrix},$$

where b_{ij} tells us how many units of good i are required to produce one unit of good j. Find the matrix B for the economy discussed here. Also write an equation relating the three matrices A, B, and S, where

$$S = \begin{bmatrix} 2 & 0 & 0 \\ 0 & 5 & 0 \\ 0 & 0 & 10 \end{bmatrix}$$

is the diagonal matrix listing the unit prices on the diagonal. Justify your answer carefully.

Chapter 3

TRUE OR FALSE?

1. The image of a 3×4 matrix is a subspace of $\mathbb{R}^4$.

2. The span of vectors $\vec{v}_1, \vec{v}_2, \ldots, \vec{v}_n$ consists of all linear combinations of vectors $\vec{v}_1, \vec{v}_2, \ldots, \vec{v}_n$.

3. If $\vec{v}_1, \vec{v}_2, \ldots, \vec{v}_n$ are linearly independent vectors in $\mathbb{R}^n$, then they must form a basis of $\mathbb{R}^n$.

4. There is a 5×4 matrix whose image consists of all of $\mathbb{R}^5$.

5. The kernel of any invertible matrix consists of the zero vector only.

6. The identity matrix I_n is similar to all invertible $n \times n$ matrices.

7. If $2\vec{u} + 3\vec{v} + 4\vec{w} = 5\vec{u} + 6\vec{v} + 7\vec{w}$, then vectors $\vec{u}, \vec{v}, \vec{w}$ must be linearly dependent.

8. The column vectors of a 5×4 matrix must be linearly dependent.

9. If $\vec{v}_1, \vec{v}_2, \ldots, \vec{v}_n$ and $\vec{w}_1, \vec{w}_2, \ldots, \vec{w}_m$ are any two bases of a subspace V of $\mathbb{R}^{10}$, then n must equal m.

10. If A is a 5×6 matrix of rank 4, then the nullity of A is 1.

11. If the kernel of a matrix A consists of the zero vector only, then the column vectors of A must be linearly independent.

12. If the image of an $n \times n$ matrix A is all of $\mathbb{R}^n$, then A must be invertible.

13. If vectors $\vec{v}_1, \vec{v}_2, \ldots, \vec{v}_n$ span $\mathbb{R}^4$, then n must be equal to 4.

14. If vectors $\vec{u}, \vec{v}$, and $\vec{w}$ are in a subspace V of $\mathbb{R}^n$, then vector $2\vec{u} - 3\vec{v} + 4\vec{w}$ must be in V as well.

15. If matrix A is similar to matrix B, and B is similar to C, then C must be similar to A.

16. If a subspace V of $\mathbb{R}^n$ contains none of the standard vectors $\vec{e}_1, \vec{e}_2, \ldots, \vec{e}_n$, then V consists of the zero vector only.

17. If vectors $\vec{v}_1, \vec{v}_2, \vec{v}_3, \vec{v}_4$ are linearly independent, then vectors $\vec{v}_1, \vec{v}_2, \vec{v}_3$ must be linearly independent as well.

18. The vectors of the form $\begin{bmatrix} a \\ b \\ 0 \\ a \end{bmatrix}$ (where a and b are arbitrary real numbers) form a subspace of $\mathbb{R}^4$.

19. Matrix $\begin{bmatrix} 1 & 0 \\ 0 & -1 \end{bmatrix}$ is similar to $\begin{bmatrix} 0 & 1 \\ 1 & 0 \end{bmatrix}$.

20. Vectors $\begin{bmatrix} 1 \\ 0 \\ 0 \end{bmatrix}$, $\begin{bmatrix} 2 \\ 1 \\ 0 \end{bmatrix}$, $\begin{bmatrix} 3 \\ 2 \\ 1 \end{bmatrix}$ form a basis of $\mathbb{R}^3$.

21. Matrix $\begin{bmatrix} 0 & 1 \\ 0 & 0 \end{bmatrix}$ is similar to $\begin{bmatrix} 0 & 0 \\ 0 & 1 \end{bmatrix}$.

22. Vectors $\begin{bmatrix} 1 \\ 2 \\ 3 \\ 4 \end{bmatrix}$, $\begin{bmatrix} 5 \\ 6 \\ 7 \\ 8 \end{bmatrix}$, $\begin{bmatrix} 9 \\ 8 \\ 7 \\ 6 \end{bmatrix}$, $\begin{bmatrix} 5 \\ 4 \\ 3 \\ 2 \end{bmatrix}$, $\begin{bmatrix} 1 \\ 0 \\ -1 \\ -2 \end{bmatrix}$ are linearly independent.

23. If a subspace V of $\mathbb{R}^3$ contains the standard vectors $\vec{e}_1, \vec{e}_2, \vec{e}_3$, then V must be $\mathbb{R}^3$.

24. If a 2×2 matrix P represents the orthogonal projection onto a line in $\mathbb{R}^2$, then P must be similar to matrix $\begin{bmatrix} 1 & 0 \\ 0 & 0 \end{bmatrix}$.

25. If A and B are $n \times n$ matrices, and vector $\vec{v}$ is in the kernel of both A and B, then $\vec{v}$ must be in the kernel of matrix AB as well.

26. If two nonzero vectors are linearly dependent, then each of them is a scalar multiple of the other.

27. If $\vec{v}_1, \vec{v}_2, \vec{v}_3$ are any three vectors in $\mathbb{R}^3$, then there must be a linear transformation T from $\mathbb{R}^3$ to $\mathbb{R}^3$ such that $T(\vec{v}_1) = \vec{e}_1$, $T(\vec{v}_2) = \vec{e}_2$, and $T(\vec{v}_3) = \vec{e}_3$.

28. If vectors $\vec{u}, \vec{v}, \vec{w}$ are linearly dependent, then vector $\vec{w}$ must be a linear combination of $\vec{u}$ and $\vec{v}$.

29. If A and B are invertible $n \times n$ matrices, then AB is similar to BA.

30. If A is an invertible $n \times n$ matrix, then the kernels of A and A^{-1} must be equal.

31. If V is any three-dimensional subspace of $\mathbb{R}^5$, then V has infinitely many bases.

32. Matrix I_n is similar to $2I_n$.

33. If $AB = 0$ for two 2×2 matrices A and B, then BA must be the zero matrix as well.

34. If A and B are $n \times n$ matrices, and vector $\vec{v}$ is in the image of both A and B, then $\vec{v}$ must be in the image of matrix $A + B$ as well.

35. If V and W are subspaces of $\mathbb{R}^n$, then their union $V \cup W$ must be a subspace of $\mathbb{R}^n$ as well.

36. If the kernel of a 5×4 matrix A consists of the zero vector only and if $A\vec{v} = A\vec{w}$ for two vectors $\vec{v}$ and $\vec{w}$ in $\mathbb{R}^4$, then vectors $\vec{v}$ and $\vec{w}$ must be equal.

37. If $\vec{v}_1, \vec{v}_2, \ldots, \vec{v}_n$ and $\vec{w}_1, \vec{w}_2, \ldots, \vec{w}_n$ are two bases of $\mathbb{R}^n$, then there is a linear transformation T from $\mathbb{R}^n$ to $\mathbb{R}^n$ such that $T(\vec{v}_1) = \vec{w}_1$, $T(\vec{v}_2) = \vec{w}_2$, ..., $T(\vec{v}_n) = \vec{w}_n$.

38. If matrix A represents a rotation through $\pi/2$ and matrix B a rotation through $\pi/4$, then A is similar to B.

39. $\mathbb{R}^2$ is a subspace of $\mathbb{R}^3$.

40. If an $n \times n$ matrix A is similar to matrix B, then $A + 7I_n$ must be similar to $B + 7I_n$.

41. There is a 2×2 matrix A such that $\operatorname{im}(A) = \ker(A)$.

42. If two $n \times n$ matrices A and B have the same rank, then they must be similar.

43. If A is similar to B, and A is invertible, then B must be invertible as well.

44. If $A^2 = 0$ for a 10×10 matrix A, then the inequality $\operatorname{rank}(A) \le 5$ must hold.

45. For every subspace V of $\mathbb{R}^3$ there is a 3×3 matrix A such that $V = \operatorname{im}(A)$.

46. There is a nonzero 2×2 matrix A that is similar to $2A$.

47. If the 2×2 matrix R represents the reflection across a line in $\mathbb{R}^2$, then R must be similar to matrix $\begin{bmatrix} 0 & 1 \\ 1 & 0 \end{bmatrix}$.

48. If A is similar to B, then there is one *and only one* invertible matrix S such that $S^{-1}AS = B$.

49. If the kernel of a 5×4 matrix A consists of the zero vector alone, and if $AB = AC$ for two 4×5 matrices B and C, then matrices B and C must be equal.

50. If A is any $n \times n$ matrix such that $A^2 = A$, then the image of A and the kernel of A have only the zero vector in common.

51. There is a 2×2 matrix A such that $A^2 \ne 0$ and $A^3 = 0$.

4

Linear Spaces

4.1 INTRODUCTION TO LINEAR SPACES

Thus far in this course, we have applied the "language of linear algebra" to vectors in $\mathbb{R}^n$. Some of the key words of this language are linear combination, linear transformation, kernel, image, subspace, span, linear independence, basis, dimension, and coordinates. Note that all these concepts can be defined in terms of sums and scalar multiples of vectors. In this chapter, we will see that it can be both natural and useful to apply this language to other mathematical objects, such as functions, matrices, equations, or infinite sequences. Indeed, linear algebra provides a unifying language used throughout modern mathematics and physics.

Here is an introductory example:

EXAMPLE 1 Consider the differential equation[1] (DE)

$$f''(x) + f(x) = 0, \quad \text{or} \quad f''(x) = -f(x).$$

We are asked to find all functions $f(x)$ whose second derivative is the negative of the function itself. Recalling the derivative rules from your introductory calculus class, you will (hopefully) note that

$$\sin(x) \quad \text{and} \quad \cos(x)$$

are solutions of this DE.

Can you find any other solutions?

---

[1] A differential equation is an equation involving derivatives of an unknown function. No previous knowledge of DEs is expected here.

Note that the solution set of this DE is closed under addition and under scalar multiplication. If $f_1(x)$ and $f_2(x)$ are solutions, then so is $f(x) = f_1(x) + f_2(x)$, since

$$f''(x) = f_1''(x) + f_2''(x) = -f_1(x) - f_2(x) = -f(x).$$

Likewise, if $f_1(x)$ is a solution and k is any scalar, then $f(x) = kf_1(x)$ is a solution of the DE as well. (Verify this!)

It follows that all "linear combinations"[2]

$$f(x) = c_1 \sin(x) + c_2 \cos(x)$$

are solutions of this DE. It can be shown that all solutions are of this form; we will omit the proof here, since our focus is not on calculus.

Let $F(\mathbb{R}, \mathbb{R})$ be the set of all functions from $\mathbb{R}$ to $\mathbb{R}$. Since the solution set V of our DE is closed under addition and scalar multiplication, we can say that V is a "subspace" of $F(\mathbb{R}, \mathbb{R})$.

How many solutions does this differential equation have? There are infinitely many solutions, of course, but we can use the language of linear algebra to give a more precise answer. The functions $\sin(x)$ and $\cos(x)$ form a "basis" of the "solution space" V, so that the "dimension" of V is 2.

In summary, the solutions of our DE form a two-dimensional subspace of $F(\mathbb{R}, \mathbb{R})$, with basis $\sin(x)$ and $\cos(x)$. ∎

We will now make the informal ideas presented in Example 1 more precise.

Note again that all the basic concepts of linear algebra can be defined in terms of sums and scalar multiples. Whenever we are dealing with a set (such as $F(\mathbb{R}, \mathbb{R})$ in Example 1) whose elements can be added and multiplied by scalars, subject to certain rules, then we can apply the language of linear algebra just as we do for vectors in $\mathbb{R}^n$. These "certain rules" are spelled out in Definition 4.1.1. (Compare this definition with the rules of vector algebra listed in Appendix A.2.)

Definition 4.1.1

> **Linear spaces**
>
> A *linear space*[3] V is a set endowed with a rule for addition (if f and g are in V, then so is $f + g$) and a rule for scalar multiplication (if f is in V and k in $\mathbb{R}$, then kf is in V) such that these operations satisfy the following eight rules[4] (for all f, g, h in V and all c, k in $\mathbb{R}$):
>
> 1. $(f + g) + h = f + (g + h)$
> 2. $f + g = g + f$
> 3. There is a *neutral element n* in V such that $f + n = f$, for all f in V. This n is unique and denoted by 0.
>
> *(continued)*

---

[2]We are cautious here and use quotes, since the term *linear combination* has been officially defined for vectors in $\mathbb{R}^n$ only.

[3]The term *vector space* is more commonly used in English (but it's *espace linéaire* in French). We prefer the term linear space to avoid the confusion that some students experience with the term *vector* in this abstract sense.

[4]These axioms were established by the Italian mathematician Giuseppe Peano (1858–1932) in his *Calcolo Geometrico* of 1888. Peano calls V a "linear system."

Definition 4.1.1

> **Linear spaces (*continued*)**
>
> 4. For each f in V there is a g in V such that $f + g = 0$. This g is unique and denoted by $(-f)$
> 5. $k(f + g) = kf + kg$
> 6. $(c + k)f = cf + kf$
> 7. $c(kf) = (ck)f$
> 8. $1f = f$

This definition contains a lot of fine print. In brief, a linear space is a set with two reasonably defined operations, addition and scalar multiplication, that allow us to form linear combinations. All the other basic concepts of linear algebra in turn rest on the concept of a linear combination.

EXAMPLE 2   In $\mathbb{R}^n$, the prototype linear space, the neutral element is the zero vector, $\vec{0}$.   ∎

Probably the most important examples of linear spaces, besides $\mathbb{R}^n$, are *spaces of functions*.

EXAMPLE 3   Let $F(\mathbb{R}, \mathbb{R})$ be the set of all functions from $\mathbb{R}$ to $\mathbb{R}$ (see Example 1), with the operations

$$(f + g)(x) = f(x) + g(x)$$
$$\text{and}$$
$$(kf)(x) = kf(x).$$

Then, $F(\mathbb{R}, \mathbb{R})$ is a linear space. The neutral element is the zero function, $f(x) = 0$ for all x.   ∎

EXAMPLE 4   If addition and scalar multiplication are given as in Definition 1.3.9, then $\mathbb{R}^{m \times n}$, the set of all $m \times n$ matrices, is a linear space. The neutral element is the zero matrix, whose entries are all zero.   ∎

EXAMPLE 5   The set of all infinite sequences of real numbers is a linear space, where addition and scalar multiplication are defined term by term:

$$(x_0, x_1, x_2, \ldots) + (y_0, y_1, y_2, \ldots) = (x_0 + y_0, x_1 + y_1, x_2 + y_2, \ldots)$$
$$k(x_0, x_1, x_2, \ldots) = (kx_0, kx_1, kx_2, \ldots).$$

The neutral element is the sequence

$$(0, 0, 0, \ldots).$$ ∎

EXAMPLE 6   The linear equations in three unknowns,

$$ax + by + cz = d,$$

where a, b, c, and d are constants, form a linear space.

The operations (addition and scalar multiplication) are familiar from the process of Gaussian elimination discussed in Chapter 1. The neutral element is the equation $0 = 0$ (with $a = b = c = d = 0$). ∎

EXAMPLE 7 Consider the plane with a point designated as the origin, O, but without a coordinate system (the coordinate-free plane). A *geometric vector* $\vec{v}$ in this plane is an arrow (a directed line segment) with its tail at the origin, as shown in Figure 1. The sum $\vec{v} + \vec{w}$ of vectors $\vec{v}$ and $\vec{w}$ is defined by means of a parallelogram, as illustrated in Figure 2. If k is a positive scalar, then vector $k\vec{v}$ points in the same direction as $\vec{v}$, but $k\vec{v}$ is k times as long as $\vec{v}$; see Figure 3. If k is negative, then $k\vec{v}$ points in the opposite direction, and it is $|k|$ times as long as $\vec{v}$; see Figure 4. The geometric vectors in the plane with these operations forms a linear space. The neutral element is the zero vector $\vec{0}$, with tail and head at the origin.

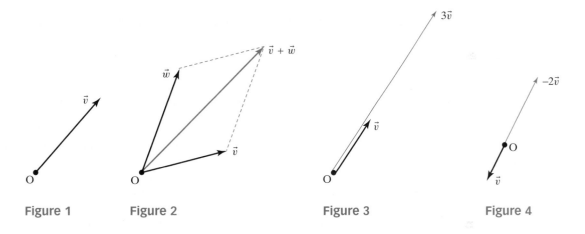

Figure 1   Figure 2   Figure 3   Figure 4

By introducing a *coordinate system,* we can identify the plane of geometric vectors with $\mathbb{R}^2$; this was the great idea of Descartes' *Analytic Geometry.* In Section 4.3, we will study this idea more systematically. ∎

EXAMPLE 8 Let $\mathbb{C}$ be the set of the *complex numbers.* We trust that you have at least a fleeting acquaintance with complex numbers. Without attempting a definition, we recall that a complex number can be expressed as $z = a + bi$, where a and b are real numbers. Addition of complex numbers is defined in a natural way, by the rule

$$(a + ib) + (c + id) = (a + c) + i(b + d).$$

If k is a real scalar, we define

$$k(a + ib) = ka + i(kb).$$

There is also a (less natural) rule for the multiplication of complex numbers, but we are not concerned with this operation here.

The complex numbers $\mathbb{C}$ with the two operations just given form a linear space; the neutral element is the complex number $0 = 0 + 0i$. ∎

We say that an element f of a linear space is a *linear combination* of the elements $f_1, f_2, \ldots, f_n$ if

$$f = c_1 f_1 + c_2 f_2 + \cdots + c_n f_n$$

for some scalars $c_1, c_2, \ldots, c_n$.

EXAMPLE 9   Let $A = \begin{bmatrix} 0 & 1 \\ 2 & 3 \end{bmatrix}$. Show that $A^2 = \begin{bmatrix} 2 & 3 \\ 6 & 11 \end{bmatrix}$ is a linear combination of A and I_2.

Solution

We have to find scalars c_1 and c_2 such that

$$A^2 = c_1 A + c_2 I_2,$$

or

$$\begin{bmatrix} 2 & 3 \\ 6 & 11 \end{bmatrix} = c_1 \begin{bmatrix} 0 & 1 \\ 2 & 3 \end{bmatrix} + c_2 \begin{bmatrix} 1 & 0 \\ 0 & 1 \end{bmatrix}.$$

In this simple example, we can see by inspection that $c_1 = 3$ and $c_2 = 2$. We could do this problem more systematically and solve a system of four linear equations in two unknowns. ∎

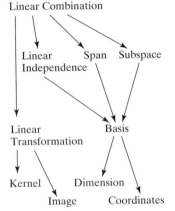

Since the basic notions of linear algebra (initially introduced for $\mathbb{R}^n$) are defined in terms of linear combinations, we can now generalize these notions without modifications. A short version of the rest of this chapter would say that the concepts of linear transformation, kernel, image, linear independence, span, subspace, basis, dimension, and coordinates can be defined for a linear space in just the same way as for $\mathbb{R}^n$. See the diagram in the margin, illustrating the logical dependencies between the key concepts of linear algebra introduced thus far.

What follows is the long version, with many examples.

Definition 4.1.2

> **Subspaces**
> A subset W of a linear space V is called a *subspace* of V if
>
> a. W contains the neutral element 0 of V.
> b. W is closed under addition (if f and g are in W, then so is $f + g$).
> c. W is closed under scalar multiplication (if f is in W and k is a scalar, then kf is in W).
>
> We can summarize parts (b) and (c) by saying that W is closed under linear combinations.

Note that a subspace W of a linear space V is a linear space in its own right (Why do the eight rules listed in Definition 4.1.1 hold for W?).

EXAMPLE 10   Show that the polynomials of degree ≤ 2, of the form $f(x) = a + bx + cx^2$, are a subspace W of the space $F(\mathbb{R}, \mathbb{R})$ of all functions from $\mathbb{R}$ to $\mathbb{R}$.

Solution

a. W contains the neutral element of $F(\mathbb{R}, \mathbb{R})$, the zero function $f(x) = 0$. Indeed, we can write $f(x) = 0 + 0x + 0x^2$.

b. W is closed under addition: If two polynomials $f(x) = a + bx + cx^2$ and $g(x) = p + qx + rx^2$ are in W, then their sum $f(x) + g(x) = (a + p) + (b + q)x + (c + r)x^2$ is in W as well, since $f(x) + g(x)$ is a polynomial of degree ≤ 2.

c. W is closed under scalar multiplication: If $f(x) = a + bx + cx^2$ is a polynomial in W and k is a constant, then $kf(x) = ka + (kb)x + (kc)x^2$ is in W as well. ∎

EXAMPLE 11 Show that the differentiable functions form a subspace W of $F(\mathbb{R}, \mathbb{R})$.

Solution

a. The zero function $f(x) = 0$ is differentiable, with $f'(x) = 0$.

b. W is closed under addition: You learned in your introductory calculus class that the sum of two differentiable functions $f(x)$ and $g(x)$ is differentiable, with $(f(x) + g(x))' = f'(x) + g'(x)$.

c. W is closed under scalar multiplication, since any scalar multiple of a differentiable function is differentiable as well. ∎

In the next example, we will build upon Examples 10 and 11.

EXAMPLE 12 Here are three more subspaces of $F(\mathbb{R}, \mathbb{R})$:

a. C^∞, the smooth functions, that is, functions we can differentiate as many times as we want. This subspace contains all polynomials, exponential functions, $\sin(x)$, and $\cos(x)$, for example.

b. P, the set of all polynomials.

c. P_n, the set of all polynomials of degree $\leq n$. ∎

EXAMPLE 13 Show that the matrices B that commute with $A = \begin{bmatrix} 0 & 1 \\ 2 & 3 \end{bmatrix}$ form a subspace of $\mathbb{R}^{2\times 2}$.

Solution

a. The zero matrix 0 commutes with A, since $A0 = 0A = 0$.

b. If matrices B_1 and B_2 commute with A, then so does matrix $B = B_1 + B_2$, since

$$BA = (B_1 + B_2)A = B_1A + B_2A = AB_1 + AB_2 = A(B_1 + B_2) = AB.$$

c. If B commutes with A, then so does kB, since

$$(kB)A = k(BA) = k(AB) = A(kB).$$

Note that we have not used the special form of A. We have indeed shown that the $n \times n$ matrices B that commute with any given $n \times n$ matrix A form a subspace of $\mathbb{R}^{n \times n}$. ∎

EXAMPLE 14 Consider the set W of all noninvertible 2×2 matrices. Is W a subspace of $\mathbb{R}^{2 \times 2}$?

Solution

The example below shows that W isn't closed under addition:

$$\begin{bmatrix} 1 & 0 \\ 0 & 0 \end{bmatrix} + \begin{bmatrix} 0 & 0 \\ 0 & 1 \end{bmatrix} = \begin{bmatrix} 1 & 0 \\ 0 & 1 \end{bmatrix}$$

$$\underset{\text{in } W}{\nwarrow \qquad \nearrow} \qquad \underset{\text{not in } W}{\uparrow}$$

Therefore, W fails to be a subspace of $\mathbb{R}^{2 \times 2}$. ∎

Next, we will generalize the notions of span, linear independence, basis, coordinates, and dimension.

Definition 4.1.3

> **Span, linear independence, basis, coordinates**
> Consider the elements $f_1, f_2, \ldots, f_n$ of a linear space V.
>
> a. We say that $f_1, f_2, \ldots, f_n$ *span V* if every f in V can be expressed as a linear combination of $f_1, f_2, \ldots, f_n$.
> b. We say that $f_1, f_2, \ldots, f_n$ are *(linearly) independent* if the equation
>
> $$c_1 f_1 + c_2 f_2 + \cdots + c_n f_n = 0$$
>
> has only the trivial solution
>
> $$c_1 = c_2 = \cdots = c_n = 0.$$
>
> c. We say that elements $f_1, f_2, \ldots, f_n$ are a *basis* of V if they span V and are independent. This means that every f in V can be written uniquely as a linear combination
>
> $$f = c_1 f_1 + \cdots + c_n f_n.$$
>
> The coefficients $c_1, c_2, \ldots, c_n$ are called the *coordinates* of f with respect to the basis $f_1, \ldots, f_n$.

Fact 4.1.4

Dimension
If a linear space V has a basis with n elements, then all other bases of V consist of n elements as well. We say that n is the *dimension* of V:

$$\dim(V) = n.$$

We defer the proof of Fact 4.1.4 to Section 4.3.

EXAMPLE 15   Find a basis of $V = \mathbb{R}^{2\times 2}$ and thus determine $\dim(V)$.

Solution

We can write any 2×2 matrix $\begin{bmatrix} a & b \\ c & d \end{bmatrix}$ as

$$\begin{bmatrix} a & b \\ c & d \end{bmatrix} = a \begin{bmatrix} 1 & 0 \\ 0 & 0 \end{bmatrix} + b \begin{bmatrix} 0 & 1 \\ 0 & 0 \end{bmatrix} + c \begin{bmatrix} 0 & 0 \\ 1 & 0 \end{bmatrix} + d \begin{bmatrix} 0 & 0 \\ 0 & 1 \end{bmatrix}.$$

This shows that matrices $\begin{bmatrix} 1 & 0 \\ 0 & 0 \end{bmatrix}, \begin{bmatrix} 0 & 1 \\ 0 & 0 \end{bmatrix}, \begin{bmatrix} 0 & 0 \\ 1 & 0 \end{bmatrix}, \begin{bmatrix} 0 & 0 \\ 0 & 1 \end{bmatrix}$ span $V = \mathbb{R}^{2\times 2}$. The four matrices are also independent: None of them is a linear combination of the others, since each has a 1 in a position where the three others have a 0. This shows that matrices $\begin{bmatrix} 1 & 0 \\ 0 & 0 \end{bmatrix}, \begin{bmatrix} 0 & 1 \\ 0 & 0 \end{bmatrix}, \begin{bmatrix} 0 & 0 \\ 1 & 0 \end{bmatrix}, \begin{bmatrix} 0 & 0 \\ 0 & 1 \end{bmatrix}$ form a basis, called the *standard basis* of $\mathbb{R}^{2\times 2}$, so that $\dim(V) = 4$. ∎

EXAMPLE 16   Find a basis of P_2, the space of all polynomials of degree ≤ 2, and thus determine the dimension of P_2.

Solution

We can write any polynomial $f(x)$ of degree ≤ 2 uniquely[5] as

$$f(x) = a + bx + cx^2 = a \cdot 1 + b \cdot x + c \cdot x^2.$$

(In a sense, $f(x)$ is already written this way.) This shows that the monomials $1, x,$ and x^2 form a basis, called the *standard basis* of P_2, so that $\dim(P_2) = 3$. ∎

Using Examples 15 and 16 as a guide, we can present the following strategy for finding a basis of a linear space V.

> **Finding a basis of a linear space V**
> a. Write down a typical element of V in terms of some arbitrary constants.
> b. Using the arbitrary constants as coefficients, express your typical element as a linear combination.
> c. Verify that the elements of V in this linear combination are linearly independent; then, they form a basis of V.

This is the way we found a basis of the kernel of a matrix; see Example 1 of Section 3.3.

EXAMPLE 17   Find a basis of the space V of all matrices B that commute with $A = \begin{bmatrix} 0 & 1 \\ 2 & 3 \end{bmatrix}$. (See Example 13.)

---

[5]Note that $a = f(0), b = f'(0),$ and $c = \dfrac{1}{2} f''(0).$

Solution

We need to find all matrices $B = \begin{bmatrix} a & b \\ c & d \end{bmatrix}$ such that $\begin{bmatrix} a & b \\ c & d \end{bmatrix} \begin{bmatrix} 0 & 1 \\ 2 & 3 \end{bmatrix} = \begin{bmatrix} 0 & 1 \\ 2 & 3 \end{bmatrix} \begin{bmatrix} a & b \\ c & d \end{bmatrix}$. The entries of B must satisfy the linear equations

$$2b = c, \quad a + 3b = d, \quad 2d = 2a + 3c, \quad c + 3d = 2b + 3d.$$

The last two equations are redundant, so that a typical matrix B in V is of the form

$$B = \begin{bmatrix} a & b \\ 2b & a + 3b \end{bmatrix} = a \begin{bmatrix} 1 & 0 \\ 0 & 1 \end{bmatrix} + b \begin{bmatrix} 0 & 1 \\ 2 & 3 \end{bmatrix} = aI_2 + bA.$$

The matrices I_2 and A form a basis of V, so that $\dim(V) = 2$.   ∎

In the introductory example of this section, we stated that the solutions of the differential equation

$$f''(x) + f(x) = 0$$

form a two-dimensional subspace of C^∞.

We can generalize this result as follows:

Fact 4.1.5   **Linear differential equations**

The solutions of the DE

$$f''(x) + af'(x) + bf(x) = 0 \quad \text{(where } a \text{ and } b \text{ are constants)}$$

form a two-dimensional subspace of the space C^∞ of smooth functions.
   More generally, the solutions of the DE

$$f^{(n)}(x) + a_{n-1}f^{(n-1)}(x) + \cdots + a_1 f'(x) + a_0 f(x) = 0$$

(where the a_i are constants)

form an n-dimensional subspace of C^∞. A DE of this form is called an nth-order linear differential equation.

Second-order linear DEs are frequently used to model oscillatory phenomena in physics. Simple examples are damped harmonic motion and LC circuits.
   Think about how cumbersome it would be to state the second part of Fact 4.1.5 without using the language of linear algebra. This may convince you that it can be both natural and useful to apply the language of linear algebra to functions.
   Fact 4.1.5 will be proven in Section 9.3.

EXAMPLE 18   Find all solutions of the DE

$$f''(x) + f'(x) - 6f(x) = 0.$$

(*Hint:* Find all exponential functions $f(x) = e^{kx}$ that solve the DE.)

Solution

An exponential function $f(x) = e^{kx}$ solves the DE if

$$k^2 e^{kx} + k e^{kx} - 6 e^{kx} = (k^2 + k - 6)e^{kx} = (k + 3)(k - 2)e^{kx} = 0.$$

This is the case for $k = 2$ and for $k = -3$. Thus, e^{2x} and e^{-3x} are solutions of the DE. (Check this!) Fact 4.1.5 tells us that the solution space V is two-dimensional. Thus, the two exponential functions e^{2x} and e^{-3x} form a basis of V, and all solutions are of the form

$$f(x) = c_1 e^{2x} + c_2 e^{-3x}.$$ ∎

EXAMPLE 19   Let $f_1, f_2, \ldots, f_n$ be polynomials. Explain why these polynomials do not span the space P of all polynomials.

Solution

Let N be the maximum of the degrees of these n polynomials. Then, all linear combinations of $f_1, f_2, \ldots, f_n$ are in P_N, the space of the polynomials of degree $\leq N$. Any polynomial of higher degree, such as $f(x) = x^{N+1}$, will not be in the span of $f_1, f_2, \ldots, f_n$. ∎

Example 19 implies that the space P of all polynomials does not have a finite basis $f_1, f_2, \ldots, f_n$.

Definition 4.1.6

> **Finite-dimensional linear spaces**
> A linear space V is called *finite-dimensional* if it has a (finite) basis $f_1, f_2, \ldots, f_n$, so that we can define its dimension $\dim(V) = n$. (See Definition 4.1.4.) Otherwise, the space is called *infinite-dimensional*.[6]

As we have just seen, the space P of all polynomials is infinite-dimensional. Take another look at the linear spaces introduced in Examples 1 through 8 of this section and see which of them are finite-dimensional.

In this introductory course, the focus will be on finite-dimensional spaces.

EXERCISES

GOALS Find a basis of a linear space and thus determine its dimension. Examine whether a subset of a linear space is a subspace.

Which of the subsets of P_2 given in Exercises 1 through 5 are subspaces of P_2? Find a basis for those that are subspaces.

1. $\{p(t): p(0) = 2\}$.

2. $\{p(t): p(2) = 0\}$.

3. $\{p(t): p'(1) = p(2)\}$. (p' is the derivative.)

4. $\{p(t): \int_0^1 p(t)\, dt = 0\}$.

5. $\{p(t): p(-t) = -p(t), \text{ for all } t\}$.

Which of the subsets of $\mathbb{R}^{3 \times 3}$ given in Exercises 6 through 11 are subspaces of $\mathbb{R}^{3 \times 3}$?

6. The invertible 3×3 matrices.

7. The diagonal 3×3 matrices.

8. The upper triangular 3×3 matrices.

9. The 3×3 matrices whose entries are all greater or equal than zero.

10. The 3×3 matrices A such that vector $\begin{bmatrix} 1 \\ 2 \\ 3 \end{bmatrix}$ is in the kernel of A.

---

[6]More advanced texts introduce the concept of an *infinite basis*.

11. The 3×3 matrices in reduced row-echelon form.

Let V be the space of all infinite sequences of real numbers. (See Example 5.) Which of the subsets of V given in Exercises 12 through 15 are subspaces of V?

12. The arithmetic sequences (i.e., sequences of the form $(a, a + k, a + 2k, a + 3k, \ldots)$, for some constants a and k).

13. The geometric sequences (i.e., sequences of the form $(a, ar, ar^2, ar^3, \ldots)$, for some constants a and r).

14. The sequences $(x_0, x_1, \ldots)$ that converge to zero (i.e., $\lim\limits_{n \to \infty} x_n = 0$).

15. The square-summable sequences $(x_0, x_1, \ldots)$, i.e., those for which $\sum\limits_{i=0}^{\infty} x_i^2$ converges.

Find a basis for each of the spaces in Exercises 16 through 31, and determine its dimension.

16. $\mathbb{R}^{3 \times 2}$.

17. $\mathbb{R}^{m \times n}$.

18. P_n.

19. The real linear space $\mathbb{C}^2$.

20. The space of all matrices $A = \begin{bmatrix} a & b \\ c & d \end{bmatrix}$ in $\mathbb{R}^{2 \times 2}$ such that $a + d = 0$.

21. The space of all diagonal 2×2 matrices.

22. The space of all diagonal $n \times n$ matrices.

23. The space of all lower triangular 2×2 matrices.

24. The space of all upper triangular 3×3 matrices.

25. The space of all polynomials $f(t)$ in P_2 such that $f(1) = 0$.

26. The space of all polynomials $f(t)$ in P_3 such that $f(1) = 0$ and $\int_{-1}^{1} f(t) \, dt = 0$.

27. The space of all 2×2 matrices A that commute with $B = \begin{bmatrix} 1 & 0 \\ 0 & 2 \end{bmatrix}$.

28. The space of all 2×2 matrices A that commute with $B = \begin{bmatrix} 1 & 1 \\ 0 & 1 \end{bmatrix}$.

29. The space of all 2×2 matrices A such that $A \begin{bmatrix} 1 & 1 \\ 1 & 1 \end{bmatrix} = \begin{bmatrix} 0 & 0 \\ 0 & 0 \end{bmatrix}$.

30. The space of all 2×2 matrices A such that $\begin{bmatrix} 1 & 2 \\ 3 & 6 \end{bmatrix} A = \begin{bmatrix} 0 & 0 \\ 0 & 0 \end{bmatrix}$.

31. The space of all 2×2 matrices S such that $\begin{bmatrix} 0 & 1 \\ 1 & 0 \end{bmatrix} S = S \begin{bmatrix} 1 & 0 \\ 0 & -1 \end{bmatrix}$.

32. In the linear space of infinite sequences, consider the subspace W of arithmetic sequences (see Exercise 12). Find a basis for W, and thus determine the dimension of W.

33. A function $f(t)$ from $\mathbb{R}$ to $\mathbb{R}$ is called even if $f(-t) = f(t)$, for all t in $\mathbb{R}$, and odd if $f(-t) = -f(t)$, for all t. Are the even functions a subspace of $F(\mathbb{R}, \mathbb{R})$, the space of all functions from $\mathbb{R}$ to $\mathbb{R}$? What about the odd functions? Justify your answers carefully.

34. Find a basis of each of the following linear spaces, and thus determine their dimensions. (See Exercise 33.)

   a. $\{f$ in P_4: f is even$\}$.

   b. $\{f$ in P_4: f is odd$\}$.

35. Let $L(\mathbb{R}^n, \mathbb{R}^m)$ be the set of all linear transformations from $\mathbb{R}^n$ to $\mathbb{R}^m$. Is $L(\mathbb{R}^n, \mathbb{R}^m)$ a subspace of $F(\mathbb{R}^n, \mathbb{R}^m)$, the space of all functions from $\mathbb{R}^n$ to $\mathbb{R}^m$? Justify your answer carefully.

36. Find all the solutions of the differential equation $f''(x) + 8f'(x) - 20f(x) = 0$.

37. Find all the solutions of the differential equation $f''(x) - 7f'(x) + 12f(x) = 0$.

38. Make up a second-order linear DE whose solution space is spanned by the functions e^{-x} and e^{-5x}.

39. Show that if W is an infinite-dimensional subspace of a linear space V, then V itself must be infinite-dimensional.

40. Show that the space $F(\mathbb{R}, \mathbb{R})$ of all functions from $\mathbb{R}$ to $\mathbb{R}$ is infinite-dimensional.

41. Show that the space of infinite sequences of real numbers is infinite-dimensional.

4.2 LINEAR TRANSFORMATIONS AND ISOMORPHISMS

In this section, we will define the concepts of a linear transformation, image, kernel, rank, and nullity in the context of linear spaces.

Definition 4.2.1

> **Linear transformations, image, kernel, rank, nullity**
>
> Consider two linear spaces V and W. A function T from V to W is called a *linear transformation* if
>
> $$T(f + g) = T(f) + T(g) \quad \text{and} \quad T(kf) = kT(f)$$
>
> for all elements f and g of V and for all scalars k.
>
> For a linear transformation T from V to W, we let
>
> $$\text{im}(T) = \{T(f) : f \text{ in } V\}$$
>
> and
>
> $$\ker(T) = \{f \text{ in } V : T(f) = 0\}$$
>
> Note that $\text{im}(T)$ is a subspace of co-domain W and that $\ker(T)$ is a subspace of domain V.
>
> If the image of T is finite-dimensional, then $\dim(\text{im } T)$ is called the *rank* of T, and if the kernel of T is finite-dimensional, then $\dim(\ker T)$ is the *nullity* of T.
>
> If V is finite-dimensional, then the rank-nullity theorem holds (see Fact 3.3.9):
>
> $$\dim(V) = \text{rank}(T) + \text{nullity}(T) = \dim(\text{im } T) + \dim(\ker T)$$

The proof of the last assertion will follow from our work in the next section.

EXAMPLE 1   Consider the transformation $D(f) = f'$ from C^∞ to C^∞. It follows from the rules of calculus that D is a linear transformation:

$$D(f + g) = (f + g)' = f' + g' \quad \text{equals} \quad D(f) + D(g) = f' + g' \quad \text{and}$$
$$D(kf) = (kf)' = kf' \quad \text{equals} \quad kD(f) = kf'.$$

Here f and g are smooth functions, and k is a constant.

What is the *kernel* of D? This kernel consists of all smooth functions f such that $D(f) = f' = 0$. As you may recall from calculus, these are the constant functions $f(x) = k$. Therefore, the kernel of D is one-dimensional; the function $f(x) = 1$ is a basis. The nullity of D is 1.

What about the *image* of D? The image consists of all smooth functions g such that $g = D(f) = f'$ for some function f in C^∞ (i.e., all smooth functions g that have a smooth antiderivative f). The fundamental theorem of calculus tells us that all smooth functions (in fact, all continuous functions) have an antiderivative. We can conclude that

$$\text{im}(D) = C^\infty.$$ ∎

EXAMPLE 2   Let $C[0, 1]$ be the linear space of all continuous functions from the closed interval $[0, 1]$ to $\mathbb{R}$. We define the transformation

$$I(f) = \int_0^1 f(x)\, dx \quad \text{from } C[0, 1] \text{ to } \mathbb{R}.$$

We adopt the simplified notation $I(f) = \int_0^1 f$. To check that I is linear, we apply basic rules of integration:

$$I(f + g) = \int_0^1 (f + g) = \int_0^1 f + \int_0^1 g \quad \text{equals} \quad I(f) + I(g) = \int_0^1 f + \int_0^1 g \quad \text{and}$$

$$I(kf) = \int_0^1 (kf) = k \int_0^1 f \quad \text{equals} \quad kI(f) = k \int_0^1 f$$

What is the image of I? The image of I consists of all real numbers b such that

$$b = I(f) = \int_0^1 f,$$

for some continuous function f. One of many possible choices for f is the constant function $f(x) = b$. Therefore,

$$\text{im}(I) = \mathbb{R}, \quad \text{and} \quad \text{rank}(I) = 1.$$

We leave it to the reader to think about the kernel of I. ∎

EXAMPLE 3   Let V be the space of all infinite sequences of real numbers. Consider the transformation

$$T(x_0, x_1, x_2, \ldots) = (x_1, x_2, x_3, \ldots)$$

from V to V. (We drop the first term, x_0, of the sequence.)

    a. Show that T is a linear transformation.
    b. Find the kernel of T.
    c. Is the sequence $(1, 2, 3, \ldots)$ in the image of T?
    d. Find the image of T.

Solutions

    a. $T\big((x_0, x_1, x_2, \ldots) + (y_0, y_1, y_2, \ldots)\big) = T(x_0 + y_0, x_1 + y_1, x_2 + y_2, \ldots)$
        $= (x_1 + y_1, x_2 + y_2, x_3 + y_3, \ldots) \quad \text{equals}$

      $T(x_0, x_1, x_2, \ldots) + T(y_0, y_1, y_2, \ldots) = (x_1, x_2, x_3, \ldots) + (y_1, y_2, y_3, \ldots)$
        $= (x_1 + y_1, x_2 + y_2, x_3 + y_3, \ldots).$

    We leave it to the reader to verify the second property of a linear transformation.

    b. The kernel consists of everything that is transformed to zero, that is, all sequences $(x_0, x_1, x_2, \ldots)$ such that

$$T(x_0, x_1, x_2, \ldots) = (x_1, x_2, x_3, \ldots) = (0, 0, 0, \ldots).$$

    This means that entries $x_1, x_2, x_3, \ldots$ all have to be zero, while x_0 is arbitrary. Thus, $\ker(T)$ consists of all sequences of the form $(x_0, 0, 0, \ldots)$, where x_0 is arbitrary. The kernel of T is one-dimensional, with basis $(1, 0, 0, 0, \ldots)$. The nullity of T is 1.

    c. We need to find a sequence $(x_0, x_1, x_2, \ldots)$ such that

$$T(x_0, x_1, x_2, \ldots) = (x_1, x_2, x_3, \ldots) = (1, 2, 3, \ldots).$$

It is required that $x_1 = 1$, $x_2 = 2$, $x_3 = 3, \ldots$, and we can choose any value for x_0, for example, $x_0 = 0$. Thus,

$$(1, 2, 3, \ldots) = T(0, 1, 2, 3, \ldots)$$

is indeed in the image of T.

d. Mimicking our solution in part (c), we can write any sequence $(b_0, b_1, b_2, \ldots)$ as

$$(b_0, b_1, b_2, \ldots) = T(0, b_0, b_1, b_2, \ldots),$$

so that $\operatorname{im}(T) = V$. ∎

EXAMPLE 4   Consider the transformation

$$T \begin{bmatrix} a \\ b \\ c \\ d \end{bmatrix} = \begin{bmatrix} a & b \\ c & d \end{bmatrix} \qquad \text{from } \mathbb{R}^4 \text{ to } \mathbb{R}^{2 \times 2}.$$

We are told that T is a linear transformation. Show that transformation T is invertible.

Solution

The most direct way to show that a function is invertible is to find its inverse. We can see that

$$T^{-1} \begin{bmatrix} a & b \\ c & d \end{bmatrix} = \begin{bmatrix} a \\ b \\ c \\ d \end{bmatrix}.$$

There is not that much going on here: The elements of both $\mathbb{R}^4$ and $\mathbb{R}^{2 \times 2}$ are described by four scalars a, b, c, and d. The linear transformation T merely rearranges these scalars (and T^{-1} puts them back into their original places in $\mathbb{R}^4$).

$$\begin{bmatrix} a \\ b \\ c \\ d \end{bmatrix} \xrightarrow[\;\;T^{-1}\;\;]{\;\;T\;\;} \begin{bmatrix} a & b \\ c & d \end{bmatrix}$$

The linear spaces $\mathbb{R}^4$ and $\mathbb{R}^{2 \times 2}$ have essentially the *same structure*. We say that the linear spaces $\mathbb{R}^4$ and $\mathbb{R}^{2 \times 2}$ are *isomorphic* (from Greek ἴσος, isos, same, and μορφή, morphe, structure). The invertible linear transformation T is called an *isomorphism*. ∎

Definition 4.2.2

> **Isomorphisms and isomorphic spaces**
> An invertible linear transformation is called an *isomorphism*. We say that the linear spaces V and W are *isomorphic* if there is an isomorphism from V to W.

EXAMPLE 5   Show that the transformation

$$T(A) = S^{-1}AS \quad \text{from } \mathbb{R}^{2 \times 2} \text{ to } \mathbb{R}^{2 \times 2}$$

is an isomorphism, where $S = \begin{bmatrix} 1 & 2 \\ 3 & 4 \end{bmatrix}$.

Solution

We need to show that T is a linear transformation, and that T is invertible.
Let's think about the linearity of T first:

$$T(M + N) = S^{-1}(M + N)S = S^{-1}(MS + NS) = S^{-1}MS + S^{-1}NS \quad \text{equals}$$
$$T(M) + T(N) = S^{-1}MS + S^{-1}NS,$$

and

$$T(kA) = S^{-1}(kA)S = k(S^{-1}AS) \quad \text{equals} \quad kT(A) = k(S^{-1}AS).$$

The most direct way to show that a function is invertible is to exhibit the inverse. Here we need to solve the equation $B = S^{-1}AS$ for input A. The solution is $A = SBS^{-1}$. Thus, the inverse transformation is

$$T^{-1}(B) = SBS^{-1}. \qquad \blacksquare$$

Next, we state some noteworthy facts concerning isomorphisms.

Fact 4.2.3   **Properties of isomorphisms**

a. If T is an isomorphism, then so is T^{-1}.
b. A linear transformation T from V to W is an isomorphism if (and only if) $\ker(T) = \{0\}$ and $\operatorname{im}(T) = W$.
c. Consider an isomorphism T from V to W. If $f_1, f_2, \ldots, f_n$ is a basis of V, then $T(f_1), T(f_2), \ldots, T(f_n)$ is a basis of W.
d. If V and W are isomorphic and $\dim(V) = n$, then $\dim(W) = n$.

Part (d) should come as no surprise, since isomorphic linear spaces have the same structure.

Proof   a. We must show that T^{-1} is linear. Consider two elements f and g of the codomain of T (i.e., the domain of T^{-1}). Then,

$$T^{-1}(f + g) = T^{-1}\big(TT^{-1}(f) + TT^{-1}(g)\big)$$
$$= T^{-1}\Big(T\big(T^{-1}(f) + T^{-1}(g)\big)\Big) \qquad \text{(since } T \text{ is linear)}$$
$$= T^{-1}(f) + T^{-1}(g).$$

In a similar way, you can show that $T^{-1}(kf) = kT^{-1}(f)$, for all f in the codomain of T and all scalars k.

b. Suppose first that T is an isomorphism. To find the kernel of T, we have to solve the equation $T(f) = 0$. Applying T^{-1} on both sides, we find that $f = T^{-1}(0) = 0$, so that $\ker(T) = \{0\}$, as claimed (compare with Exercise 42). Any g in W can be written as $g = T(T^{-1}(g))$, so that $\text{im}(T) = W$.

Conversely, suppose that $\ker(T) = \{0\}$ and $\text{im}(T) = W$. We have to show that the equation $T(f) = g$ has a unique solution f for any g in W (by Definition 2.3.1). There is at least one such solution f, since $\text{im}(T) = W$. Consider two solutions, f_1 and f_2: $T(f_1) = T(f_2) = g$. Then, $0 = T(f_1) - T(f_2) = T(f_1 - f_2)$, so that $f_1 - f_2$ is in the kernel of T. Since $\ker(T) = \{0\}$, we have $f_1 - f_2 = 0$ and $f_1 = f_2$, as claimed.

c. We will show first that the $T(f_i)$ span W. For any g in W, we can write $T^{-1}(g) = c_1 f_1 + \cdots + c_n f_n$, because the f_i span V. Applying T on both sides and using linearity, we find that $g = c_1 T(f_1) + \cdots + c_n T(f_n)$, as claimed.

To show the linear independence of the $T(f_i)$, consider a relation $c_1 T(f_1) + \cdots + c_n T(f_n) = 0$ or $T(c_1 f_1 + \cdots + c_n f_n) = 0$. Since the kernel of T is 0, we have $c_1 f_1 + \cdots + c_n f_n = 0$. Then, the c_i are zero since the f_i are linearly independent.

d. Follows from part (c). ▲

EXAMPLE 6   We are told that the transformation

$$B = T(A) = \begin{bmatrix} 1 & 2 \\ 3 & 4 \end{bmatrix} A - A \begin{bmatrix} 1 & 2 \\ 3 & 4 \end{bmatrix}$$

from $\mathbb{R}^{2\times 2}$ to $\mathbb{R}^{2\times 2}$ is linear. Is T an isomorphism?

Solution

We need to examine whether transformation T is invertible. First we try to solve the equation

$$\begin{bmatrix} 1 & 2 \\ 3 & 4 \end{bmatrix} A - A \begin{bmatrix} 1 & 2 \\ 3 & 4 \end{bmatrix} = B$$

for input A, as in Examples 4 and 5. However, the fact that matrix multiplication is non-commutative gets in the way, and we are unable to solve for A. (Try it yourself!) If we cannot write the inverse of a linear transformation directly, the criteria established in Fact 4.2.3b can be useful: Let's think about kernel and image of transformation T. The kernel of T consists of all 2×2 matrices A such that

$$T(A) = \begin{bmatrix} 1 & 2 \\ 3 & 4 \end{bmatrix} A - A \begin{bmatrix} 1 & 2 \\ 3 & 4 \end{bmatrix} = \begin{bmatrix} 0 & 0 \\ 0 & 0 \end{bmatrix}, \quad \text{or,} \quad \begin{bmatrix} 1 & 2 \\ 3 & 4 \end{bmatrix} A = A \begin{bmatrix} 1 & 2 \\ 3 & 4 \end{bmatrix},$$

that is, the matrices that commute with $\begin{bmatrix} 1 & 2 \\ 3 & 4 \end{bmatrix}$. We don't really need to find this kernel; we just want to know whether there are nonzero matrices in the kernel. We see that matrices $A = I_2$ and $A = \begin{bmatrix} 1 & 2 \\ 3 & 4 \end{bmatrix}$ are in the kernel, for example, so that the kernel consists of more than just the zero matrix. Thus, T fails to be an isomorphism. ∎

EXERCISES

GOALS Examine whether a transformation is linear. Find image and kernel of a linear transformation. Examine whether a linear transformation is an isomorphism.

Find out which of the transformations in Exercises 1 through 24 are linear. For those that are linear, determine whether they are isomorphisms.

1. $T(A) = A + I_2$ from $\mathbb{R}^{2 \times 2}$ to $\mathbb{R}^{2 \times 2}$.

2. $T(A) = 7A$ from $\mathbb{R}^{2 \times 2}$ to $\mathbb{R}^{2 \times 2}$.

3. $T(A) =$ (sum of the diagonal entries of A) from $\mathbb{R}^{2 \times 2}$ to $\mathbb{R}$.

4. $T(A) = \det(A)$ from $\mathbb{R}^{2 \times 2}$ to $\mathbb{R}$.

5. $T(A) = A^2$ from $\mathbb{R}^{2 \times 2}$ to $\mathbb{R}^{2 \times 2}$.

6. $T(A) = \begin{bmatrix} 1 & 2 \\ 3 & 6 \end{bmatrix} A$ from $\mathbb{R}^{2 \times 2}$ to $\mathbb{R}^{2 \times 2}$.

7. $T(A) = S^{-1} A S$ from $\mathbb{R}^{2 \times 2}$ to $\mathbb{R}^{2 \times 2}$, where $S = \begin{bmatrix} 3 & 4 \\ 5 & 6 \end{bmatrix}$.

8. $T(c) = cA$ from $\mathbb{R}$ to $\mathbb{R}^{2 \times 2}$, where $A = \begin{bmatrix} 2 & 3 \\ 4 & 5 \end{bmatrix}$.

9. $T(A) = A \begin{bmatrix} 1 & 2 \\ 0 & 1 \end{bmatrix} - \begin{bmatrix} 1 & 2 \\ 0 & 1 \end{bmatrix} A$ from $\mathbb{R}^{2 \times 2}$ to $\mathbb{R}^{2 \times 2}$.

10. $T(x + iy) = x - iy$ from $\mathbb{C}$ to $\mathbb{C}$.

11. $T(x + iy) = y + ix$ from $\mathbb{C}$ to $\mathbb{C}$.

12. $T(f) = \int_{-2}^{3} f(t)\, dt$ from P_2 to $\mathbb{R}$.

13. $T(f) = f'' + 4f'$ from P_2 to P_2.

14. $T\big(f(t)\big) = f(-t)$ from P_2 to P_2, that is, $T(a + bt + ct^2) = a - bt + ct^2$.

15. $T\big(f(t)\big) = f(2t)$ from P_2 to P_2, that is, $T(a + bt + ct^2) = a + 2bt + 4ct^2$.

16. $T\big(f(t)\big) = t\big(f'(t)\big)$ from P_2 to P_2.

17. $T(x_0, x_1, x_2, x_3, x_4, \ldots) = (x_0, x_2, x_4, \ldots)$ from the space of infinite sequences into itself (we are dropping every other term).

18. $T(x_0, x_1, x_2, \ldots) = (0, x_0, x_1, x_2, \ldots)$ from the space of infinite sequences into itself.

19. $T(f) = f'' - 5f' + 6f$ from C^∞ to C^∞.

20. $T(f) = f'' + 2f' + f$ from C^∞ to C^∞.

21. $T\big(f(t)\big) = f(7)$ from P_2 to $\mathbb{R}$.

22. $T\big(f(t)\big) = \begin{bmatrix} f(7) \\ f(11) \end{bmatrix}$ from P_2 to $\mathbb{R}^2$.

23. $T\big(f(t)\big) = t\big(f(t)\big)$ from P to P.

24. $T\big(f(t)\big) = f'(t)$ from P to P.

25. Find kernel and nullity of the transformation in Exercise 9.

26. Find kernel and nullity of the transformation in Exercise 6.

27. Find image, rank, kernel, and nullity of the transformation in Exercise 13.

28. Find image, rank, kernel, and nullity of the transformation in Exercise 12.

29. Find image and kernel of the transformation in Exercise 17.

30. Find image, rank, kernel, and nullity of the transformation in Exercise 16.

31. Find kernel and nullity of the transformation in Exercise 19.

32. Find image and kernel of the transformation in Exercise 18.

33. Find image, rank, kernel, and nullity of the transformation in Exercise 21.

34. Find image, rank, kernel, and nullity of the transformation in Exercise 22.

35. Find image and kernel of the transformation in Exercise 23.

36. Find image and kernel of the transformation in Exercise 24.

37. Define an isomorphism from P_3 to $\mathbb{R}^3$, if you can.

38. Define an isomorphism from P_3 to $\mathbb{R}^{2 \times 2}$, if you can.

39. We will define a transformation T from $\mathbb{R}^{m \times n}$ to $F(\mathbb{R}^n, \mathbb{R}^m)$; recall that $F(\mathbb{R}^n, \mathbb{R}^m)$ is the space of all functions from $\mathbb{R}^n$ to $\mathbb{R}^m$. For a matrix A in $\mathbb{R}^{m \times n}$, the value $T(A)$ will be a function from $\mathbb{R}^n$ to $\mathbb{R}^m$; thus we need to define $\big(T(A)\big)(\vec{v})$ for a vector $\vec{v}$ in $\mathbb{R}^n$. We let

$$\big(T(A)\big)(\vec{v}) = A\vec{v}.$$

   a. Show that T is a linear transformation.

   b. Find the kernel of T.

   c. Show that the image of T is the space $L(\mathbb{R}^n, \mathbb{R}^m)$ of all linear transformation from $\mathbb{R}^n$ to $\mathbb{R}^m$. (See Exercise 35 of Section 4.1.)

   d. Find the dimension of $L(\mathbb{R}^n, \mathbb{R}^m)$.

40. Find kernel and nullity of the linear transformation $T(f) = f - f'$ from C^∞ to C^∞.

41. Show that if 0 is the neutral element of a linear space V, then $0 + 0 = 0$ and $k0 = 0$, for all scalars k.

42. Show that if T is a linear transformation from V to W, then $T(0_V) = 0_W$, where 0_V and 0_W are the neutral elements of V and W, respectively.

43. If T is a linear transformation from V to W and L is a linear transformation from W to U, is the composite transformation $L \circ T$ from V to U linear? How can you tell? If T and L are isomorphisms, is $L \circ T$ an isomorphism as well?

44. Let $\mathbb{R}^+$ be the set of positive real numbers. On $\mathbb{R}^+$ we define the "exotic" operations

$$x \oplus y = xy \quad \text{(usual multiplication)}$$

and

$$k \odot x = x^k.$$

a. Show that $\mathbb{R}^+$ with these operations is a linear space; find a basis of this space.

b. Show that $T(x) = \ln(x)$ is a linear transformation from $\mathbb{R}^+$ to $\mathbb{R}$, where $\mathbb{R}$ is endowed with the ordinary operations. Is T an isomorphism?

45. Is it possible to define "exotic" operations on $\mathbb{R}^2$, so that $\dim(\mathbb{R}^2) = 1$?

46. Let X be the set of all students in your linear algebra class. Can you define operations on X that make X into a real linear space? Explain.

4.3 COORDINATES IN A LINEAR SPACE

By introducing coordinates, Descartes transformed the plane into $\mathbb{R}^2$, the set of all pairs $\begin{bmatrix} x \\ y \end{bmatrix}$ of real numbers. (See Section 3.4.) We can generalize his idea: By introducing coordinates, we can transform any n-dimensional linear space into $\mathbb{R}^n$.

Definition 4.3.1

> **Coordinates in a linear space**
> Consider a linear space V with a basis $\mathcal{B}$ consisting of $f_1, f_2, \ldots, f_n$. Then any element f of V can be written uniquely as
>
> $$f = c_1 f_1 + c_2 f_2 + \cdots + c_n f_n,$$
>
> for some scalars $c_1, c_2, \ldots, c_n$. These scalars are called the $\mathcal{B}$-*coordinates* of f, and the vector
>
> $$\begin{bmatrix} c_1 \\ c_2 \\ \cdot \\ \cdot \\ c_n \end{bmatrix}$$
>
> is called the $\mathcal{B}$-*coordinate vector* of f, denoted by $[f]_\mathcal{B}$.
>   The $\mathcal{B}$-*coordinate transformation* $T(f) = [f]_\mathcal{B}$ from V to $\mathbb{R}^n$ is an isomorphism (i.e., an invertible linear transformation). Thus, V is isomorphic to $\mathbb{R}^n$; the linear spaces V and $\mathbb{R}^n$ have the same structure.

Let's reiterate the main point: **Any n-dimensional linear space V is isomorphic to $\mathbb{R}^n$.** This means that we don't need a separate theory for finite-dimensional spaces. By introducing coordinates, we can transform an n-dimensional space into $\mathbb{R}^n$ and then apply the techniques of Chapters 1 through 3. (Infinite-dimensional linear spaces, on the other hand, are largely beyond the reach of the methods of elementary linear algebra.)

EXAMPLE 1 Choose a basis of P_2 and thus transform P_2 into $\mathbb{R}^n$, for an appropriate n.

Solution

To keep things simple, we will work with the basis $\mathfrak{B}$ consisting of 1, x, and x^2, called the *standard basis* of P_2. Then the $\mathfrak{B}$-coordinates of

$$f(x) = a + bx + cx^2 = a \cdot 1 + b \cdot x + c \cdot x^2$$

are the scalars a, b, and c, and the $\mathfrak{B}$-coordinate vector is $f(x)$ is

$$[f(x)]_{\mathfrak{B}} = \begin{bmatrix} a \\ b \\ c \end{bmatrix}.$$

By introducing coordinates, we transform P_2 into $\mathbb{R}^3$. Thus, the three-dimensional space P_2 is isomorphic to $\mathbb{R}^3$. ∎

EXAMPLE 2 Let V be the linear space of upper-triangular 2×2 matrices (that is, matrices of the form $\begin{bmatrix} a & b \\ 0 & c \end{bmatrix}$). Choose a basis of V and thus transform V into $\mathbb{R}^n$, for an appropriate n.

Solution

The equation

$$\begin{bmatrix} a & b \\ 0 & c \end{bmatrix} = a \begin{bmatrix} 1 & 0 \\ 0 & 0 \end{bmatrix} + b \begin{bmatrix} 0 & 1 \\ 0 & 0 \end{bmatrix} + c \begin{bmatrix} 0 & 0 \\ 0 & 1 \end{bmatrix}$$

produces the basis $\mathfrak{B}$ consisting of matrices $\begin{bmatrix} 1 & 0 \\ 0 & 0 \end{bmatrix}$, $\begin{bmatrix} 0 & 1 \\ 0 & 0 \end{bmatrix}$, and $\begin{bmatrix} 0 & 0 \\ 0 & 1 \end{bmatrix}$.

The $\mathfrak{B}$-coordinate vector of $\begin{bmatrix} a & b \\ 0 & c \end{bmatrix}$ is

$$\begin{bmatrix} a & b \\ 0 & c \end{bmatrix}_{\mathfrak{B}} = \begin{bmatrix} a \\ b \\ c \end{bmatrix}.$$

By introducing coordinates, we transform V into $\mathbb{R}^3$. Thus, the three-dimensional space V is isomorphic to $\mathbb{R}^3$. ∎

EXAMPLE 3 Do the polynomials $f_1(x) = 1 + 2x + 3x^2$, $f_2(x) = 4 + 5x + 6x^2$, $f_3(x) = 7 + 8x + 10x^2$ form a basis of P_2?

Solution

Since P_2 is isomorphic to $\mathbb{R}^3$ (by Example 1), we can use a coordinate transformation (with respect to, say, the standard basis 1, x, x^2), to make this into a problem concerning $\mathbb{R}^3$. The three given polynomials form a basis of P_2 if (and only if) the coordinate vectors

$$\vec{v}_1 = \begin{bmatrix} 1 \\ 2 \\ 3 \end{bmatrix}, \quad \vec{v}_2 = \begin{bmatrix} 4 \\ 5 \\ 6 \end{bmatrix}, \quad \vec{v}_3 = \begin{bmatrix} 7 \\ 8 \\ 10 \end{bmatrix}$$

form a basis of $\mathbb{R}^3$ (by Fact 4.2.3(c)). This in turn is the case if the matrix

$$A = \begin{bmatrix} 1 & 4 & 7 \\ 2 & 5 & 8 \\ 3 & 6 & 10 \end{bmatrix}$$

is invertible (by Fact 3.3.10).

Now,

$$\text{rref}(A) = I_3.$$

Thus, matrix A is invertible, so that vectors $\vec{v}_1, \vec{v}_2, \vec{v}_3$ form a basis of $\mathbb{R}^3$, and polynomials $f_1(x), f_2(x), f_3(x)$ form a basis of P_2. ■

Example 3 illustrates how we can transform a problem concerning an n-dimensional linear space into a problem concerning $\mathbb{R}^n$ and then use matrix techniques.

We are now able to prove that any two bases of a linear space consist of the same number of elements (Fact 4.1.4). Suppose two bases of a linear space V are given: basis $\mathfrak{A}$, consisting of $f_1, f_2, \ldots, f_n$ and basis $\mathfrak{B}$ with m elements. We need to show that $m = n$. Let's find the coordinate vectors of the elements of one basis with respect to the other basis. Consider the vectors $[f_1]_\mathfrak{B}, [f_2]_\mathfrak{B}, \ldots, [f_n]_\mathfrak{B}$, for example. These n vectors form a basis of $\mathbb{R}^m$, since the $\mathfrak{B}$-coordinate transformation is an isomorphism from V to $\mathbb{R}^m$. (See Fact 4.2.3(c).) Since all bases of $\mathbb{R}^m$ consist of m elements, by Fact 3.3.2, we have $m = n$, as claimed.

The Matrix of a Linear Transformation

Next we will examine how we can write a linear transformation in coordinates.

EXAMPLE 4   Consider the linear transformation

$$T(f) = f' + f'' \quad \text{from } P_2 \text{ to } P_2.$$

Since P_2 is isomorphic to $\mathbb{R}^3$, this is essentially a linear transformation from $\mathbb{R}^3$ to $\mathbb{R}^3$, given by a 3×3 matrix B. Let's see how we can find this matrix.

We can write transformation T more explicitly as

$$T(a + bx + cx^2) = (b + 2cx) + 2c = (b + 2c) + 2cx.$$

Next let's write the input and the output of T in coordinates with respect to the standard basis $\mathfrak{B}$ of P_2 consisting of $1, x, x^2$:

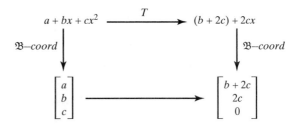

Written in $\mathcal{B}$-coordinates, transformation T takes

$$\begin{bmatrix} a \\ b \\ c \end{bmatrix} \quad \text{into} \quad \begin{bmatrix} b+2c \\ 2c \\ 0 \end{bmatrix} = \begin{bmatrix} 0 & 1 & 2 \\ 0 & 0 & 2 \\ 0 & 0 & 0 \end{bmatrix} \begin{bmatrix} a \\ b \\ c \end{bmatrix}.$$

The matrix

$$B = \begin{bmatrix} 0 & 1 & 2 \\ 0 & 0 & 2 \\ 0 & 0 & 0 \end{bmatrix}$$

is called the $\mathcal{B}$-*matrix of* T. It describes the transformation T if input and output are written in $\mathcal{B}$-coordinates.

Let us summarize our work in a diagram:

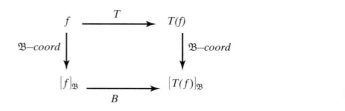

Definition 4.3.2

> **$\mathcal{B}$-Matrix of a linear transformation**
> Consider a linear transformation T from V to V, where V is an n-dimensional linear space. Let $\mathcal{B}$ be a basis of V. Then, there is an $n \times n$ matrix B that transforms $[f]_{\mathcal{B}}$ into $[T(f)]_{\mathcal{B}}$, called the $\mathcal{B}$-*matrix of* T.[7]
>
> $$[T(f)]_{\mathcal{B}} = B[f]_{\mathcal{B}}$$

(Compare this with Definition 3.4.2.)

We can write matrix B column by column (compare this with Fact 3.4.3). Suppose basis $\mathcal{B}$ consists of $f_1, f_2, \ldots, f_n$.

If

$$f = c_1 f_1 + c_2 f_2 + \cdots + c_n f_n,$$

then

$$T(f) = c_1 T(f_1) + c_2 T(f_2) + \cdots + c_n T(f_n),$$

and

$$[T(f)]_{\mathcal{B}} = c_1[T(f_1)]_{\mathcal{B}} + \cdots + c_n[T(f_n)]_{\mathcal{B}} = \begin{bmatrix} [T(f_1)]_{\mathcal{B}} & \cdots & [T(f_n)]_{\mathcal{B}} \end{bmatrix} \begin{bmatrix} c_1 \\ \cdots \\ c_n \end{bmatrix}$$

$$= \begin{bmatrix} [T(f_1)]_{\mathcal{B}} & \cdots & [T(f_n)]_{\mathcal{B}} \end{bmatrix} [f]_{\mathcal{B}}.$$

Note that we have used the linearity of T and of the $\mathcal{B}$-coordinate transformation. The last equation gives the matrix B that transforms $[f]_{\mathcal{B}}$ into $[T(f)]_{\mathcal{B}}$.

---

[7]Some authors denote this matrix by $[T]_{\mathcal{B}}$.

Fact 4.3.3   **The columns of the $\mathcal{B}$-matrix of a linear transformation**
Consider a linear transformation T from V to V, and let B be the matrix of T with respect to a basis $\mathcal{B}$ of V consisting of $f_1, \ldots, f_n$.
Then

$$B = \left[[T(f_1)]_\mathcal{B} \quad \cdots \quad [T(f_n)]_\mathcal{B} \right].$$

That is, the columns of B are the $\mathcal{B}$-coordinate vectors of the transforms of the basis elements.

EXAMPLE 5 Use Fact 4.3.3 to find the matrix B of the linear transformation

$$T(f) = f' + f'' \quad \text{from } P_2 \text{ to } P_2$$

with respect to the standard basis $\mathcal{B}$. (See Example 4.)

Solution

$$B = \left[[T(1)]_\mathcal{B} \quad [T(x)]_\mathcal{B} \quad [T(x^2)]_\mathcal{B} \right] = \left[[0]_\mathcal{B} \quad [1]_\mathcal{B} \quad [2+2x]_\mathcal{B} \right] = \begin{bmatrix} 0 & 1 & 2 \\ 0 & 0 & 2 \\ 0 & 0 & 0 \end{bmatrix}$$

∎

A problem concerning a linear transformation T can often be done by solving the corresponding problem for the matrix B of T with respect to some basis $\mathcal{B}$. We can use this technique to find image and kernel of T, to determine whether T is an isomorphism (this is the case if B is invertible), or to solve an equation $T(f) = g$ for f if g is given.

EXAMPLE 6 Let V be the linear space of all functions of the form $f(x) = a\cos(x) + b\sin(x)$, a subspace of C^∞. Consider the linear transformation

$$T(f) = f'' - 2f' - 3f$$

from V to V.

a. Find the matrix B of T with respect to the basis $\mathcal{B}$ consisting of functions $\cos(x)$ and $\sin(x)$.
b. Is T an isomorphism?
c. How many solutions f in V does the differential equation

$$f''(x) - 2f'(x) - 3f(x) = \cos(x)$$

have?

Solution

a. Using Fact 4.3.3, we find that

$$B = \left[[T(\cos x)]_\mathcal{B} \quad [T(\sin x)]_\mathcal{B} \right]$$

$$= \left[[-4\cos(x) + 2\sin(x)]_\mathcal{B} \quad [-2\cos(x) - 4\sin(x)]_\mathcal{B} \right] = \begin{bmatrix} -4 & -2 \\ 2 & -4 \end{bmatrix}.$$

Matrix B represents a rotation–dilation.

b. Matrix B is invertible, since $\det(B) = ad - bc = 16 + 4 = 20 \neq 0$. Thus, transformation T is invertible as well, so that it is an isomorphism. (We were told that T is linear.)

c. The fact that T is invertible means that the differential equation

$$T(f) = f'' - 2f' - 3f = g$$

has a unique solution f in V for all g in V, namely, $f = T^{-1}(g)$. In particular, this is the case for $g(x) = \cos(x)$, so that the given differential equation has a unique solution in V.   ∎

EXAMPLE 7   Consider the function

$$T(M) = \begin{bmatrix} 0 & 1 \\ 0 & 0 \end{bmatrix} M - M \begin{bmatrix} 0 & 1 \\ 0 & 0 \end{bmatrix}$$

from $\mathbb{R}^{2\times 2}$ to $\mathbb{R}^{2\times 2}$. We are told that T is a linear transformation.

a. Find the matrix B of T with respect to the standard basis $\mathcal{B}$ of $\mathbb{R}^{2\times 2}$.
b. Find image and kernel of B.
c. Find image and kernel of T.
d. Find rank and nullity of transformation T.

Solution

a. We can construct B column by column using Fact 4.3.3, or we can use the definition, $[T(M)]_{\mathcal{B}} = B[M]_{\mathcal{B}}$. For the sake of variety, let us use the second approach here. We have

$$T(M) = T\begin{bmatrix} a & b \\ c & d \end{bmatrix} = \begin{bmatrix} 0 & 1 \\ 0 & 0 \end{bmatrix}\begin{bmatrix} a & b \\ c & d \end{bmatrix} - \begin{bmatrix} a & b \\ c & d \end{bmatrix}\begin{bmatrix} 0 & 1 \\ 0 & 0 \end{bmatrix}$$

$$= \begin{bmatrix} c & d \\ 0 & 0 \end{bmatrix} - \begin{bmatrix} 0 & a \\ 0 & c \end{bmatrix} = \begin{bmatrix} c & d-a \\ 0 & -c \end{bmatrix}.$$

Now we write input and output in $\mathcal{B}$-coordinates:

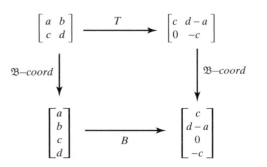

We can see that

$$B = \begin{bmatrix} 0 & 0 & 1 & 0 \\ -1 & 0 & 0 & 1 \\ 0 & 0 & 0 & 0 \\ 0 & 0 & -1 & 0 \end{bmatrix}.$$

b. To find image and kernel of matrix B, we compute rref(B) first:

$$\text{rref}(B) = \begin{bmatrix} 1 & 0 & 0 & -1 \\ 0 & 0 & 1 & 0 \\ 0 & 0 & 0 & 0 \\ 0 & 0 & 0 & 0 \end{bmatrix}.$$

Using the techniques of Section 3.3, we find that

$$\begin{bmatrix} 0 \\ -1 \\ 0 \\ 0 \end{bmatrix}, \begin{bmatrix} 1 \\ 0 \\ 0 \\ -1 \end{bmatrix} \text{ is a basis of im}(B) \quad \text{and} \quad \begin{bmatrix} 1 \\ 0 \\ 0 \\ 1 \end{bmatrix}, \begin{bmatrix} 0 \\ 1 \\ 0 \\ 0 \end{bmatrix} \text{ is a basis of ker}(B).$$

c. We need to transform the vectors we found in part (b) back into $\mathbb{R}^{2\times2}$, domain and co-domain of transformation T:

$$\begin{bmatrix} 0 & -1 \\ 0 & 0 \end{bmatrix}, \begin{bmatrix} 1 & 0 \\ 0 & -1 \end{bmatrix} \text{ is a basis of im}(T)$$

and

$$\begin{bmatrix} 1 & 0 \\ 0 & 1 \end{bmatrix} = I_2, \begin{bmatrix} 0 & 1 \\ 0 & 0 \end{bmatrix} \text{ is a basis of ker}(T).$$

Apply transformation T to the last two matrices to check that they are indeed in ker(T).

d. rank(T) = dim(im$T$) = 2 and nullity(T) = dim(ker$T$) = 2. ∎

It is sometimes necessary to consider the relationship between the matrices of a linear transformation with respect to two different bases.

Fact 4.3.4

The matrices of T with respect to different bases

Suppose that $\mathfrak{A}$ and $\mathfrak{B}$ are two bases of a linear space V and that T is a linear transformation from V to V.

a. There is an invertible matrix S such that $[f]_{\mathfrak{A}} = S[f]_{\mathfrak{B}}$ for all f in V.

b. Let A and B be the $\mathfrak{A}$- and the $\mathfrak{B}$-matrix of T, respectively. Then matrix A is *similar* to B. In fact, $B = S^{-1}AS$ for the matrix S from part (a).

Note that in Fact 3.4.4 we considered the special case where $V = \mathbb{R}^n$ and $\mathfrak{A}$ is the standard basis of $\mathbb{R}^n$.

Proof of Fact 4.3.4:

a. Suppose basis $\mathcal{B}$ consists of $f_1, f_2, \ldots, f_n$. If

$$f = c_1 f_1 + c_2 f_2 + \cdots + c_n f_n,$$

then

$$[f]_{\mathfrak{A}} = [c_1 f_1 + \cdots + c_n f_n]_{\mathfrak{A}} = c_1 [f_1]_{\mathfrak{A}} + \cdots + c_n [f_n]_{\mathfrak{A}}$$

$$= \left[\begin{array}{ccc} [f_1]_{\mathfrak{A}} & \cdots & [f_n]_{\mathfrak{A}} \end{array} \right] \begin{bmatrix} c_1 \\ \cdots \\ c_n \end{bmatrix} = \underbrace{\left[\begin{array}{ccc} [f_1]_{\mathfrak{A}} & \cdots & [f_n]_{\mathfrak{A}} \end{array} \right]}_{S} [f]_{\mathcal{B}}.$$

b. Consider the following diagram:

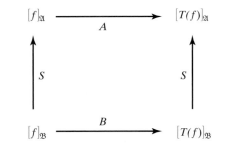

Performing a "diagram chase," as in Example 5 of Section 3.4, we see that

$$AS = SB \quad \text{or} \quad B = S^{-1} A S.$$

We can summarize our work in parts (a) and (b) in one big diagram:

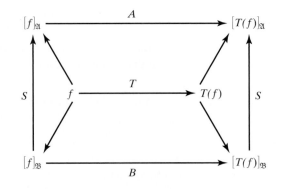

EXAMPLE 8   Let V be the linear space spanned by functions e^x and e^{-x}. Consider the linear transformation $D(f) = f'$ from V to V:

a. Find the matrix A of D with respect to basis $\mathfrak{A}$ consisting of e^x and e^{-x}.

b. Find the matrix B of D with respect to basis $\mathfrak{B}$ consisting of $\frac{1}{2}(e^x + e^{-x})$ and $\frac{1}{2}(e^x - e^{-x})$. (These two functions are called the *hyperbolic cosine*, $\cosh(x)$, and the *hyperbolic sine*, $\sinh(x)$, respectively.)

c. Using the proof of Fact 4.3.4 as a guide, construct a matrix S such that $B = S^{-1}AS$, showing that matrix A is similar to B.

Solution

We can use Fact 4.3.3 to construct matrices A and B column by column:

a. $A = \begin{bmatrix} [D(e^x)]_{\mathfrak{A}} & [D(e^{-x})]_{\mathfrak{A}} \end{bmatrix} = \begin{bmatrix} [e^x]_{\mathfrak{A}} & [-e^{-x}]_{\mathfrak{A}} \end{bmatrix} = \begin{bmatrix} 1 & 0 \\ 0 & -1 \end{bmatrix}$

b. $B = \begin{bmatrix} [D(\frac{1}{2}(e^x + e^{-x}))]_{\mathfrak{B}} & [D(\frac{1}{2}(e^x - e^{-x}))]_{\mathfrak{B}} \end{bmatrix}$

$= \begin{bmatrix} [\frac{1}{2}(e^x - e^{-x})]_{\mathfrak{B}} & [\frac{1}{2}(e^x + e^{-x})]_{\mathfrak{B}} \end{bmatrix} = \begin{bmatrix} 0 & 1 \\ 1 & 0 \end{bmatrix}$

c. Following the proof of Fact 4.3.4, part (a), we find that

$$S = \begin{bmatrix} [\frac{1}{2}(e^x + e^{-x})]_{\mathfrak{A}} & [\frac{1}{2}(e^x - e^{-x})]_{\mathfrak{A}} \end{bmatrix} = \begin{bmatrix} \frac{1}{2} & \frac{1}{2} \\ \frac{1}{2} & -\frac{1}{2} \end{bmatrix} = \frac{1}{2}\begin{bmatrix} 1 & 1 \\ 1 & -1 \end{bmatrix}.$$

Now we can verify that

$$AS = SB = \frac{1}{2}\begin{bmatrix} 1 & 1 \\ -1 & 1 \end{bmatrix},$$

so that $B = S^{-1}AS$, as claimed. ∎

EXERCISES

GOALS Use the concept of coordinates. Find the matrix of a linear transformation. Use this matrix to find image and kernel of a transformation.

1. Are the polynomials $f(t) = 7 + 3t + t^2$, $g(t) = 9 + 9t + 4t^2$, and $h(t) = 3 + 2t + t^2$ linearly independent?

2. Are the matrices

$$\begin{bmatrix} 1 & 1 \\ 1 & 1 \end{bmatrix}, \quad \begin{bmatrix} 1 & 2 \\ 3 & 4 \end{bmatrix}, \quad \begin{bmatrix} 2 & 3 \\ 5 & 7 \end{bmatrix}, \quad \begin{bmatrix} 1 & 4 \\ 6 & 8 \end{bmatrix}$$

linearly independent?

3. Do the polynomials $f(t) = 1 + 2t + 9t^2 + t^3$, $g(t) = 1 + 7t + 7t^3$, $h(t) = 1 + 8t + t^2 + 5t^3$, and $k(t) = 1 + 8t + 4t^2 + 8t^3$ form a basis of P_3?

4. Consider the polynomials $f(t) = t + 1$ and $g(t) = (t + 2)(t + k)$, where k is an arbitrary constant. For which values of the constant k are the three polynomials $f(t), tf(t)$, and $g(t)$ a basis of P_2?

In Exercises 5 through 33, find the matrix of the given linear transformation with respect to the given basis. If no basis is given, use the standard basis: $1, t, t^2$ for P_2,

$$\begin{bmatrix} 1 & 0 \\ 0 & 0 \end{bmatrix}, \begin{bmatrix} 0 & 1 \\ 0 & 0 \end{bmatrix}, \begin{bmatrix} 0 & 0 \\ 1 & 0 \end{bmatrix}, \begin{bmatrix} 0 & 0 \\ 0 & 1 \end{bmatrix} \text{ for } \mathbb{R}^{2\times 2}, \text{ and } 1, i \text{ for}$$

$\mathbb{C}$. *For the space $U^{2\times 2}$ of upper triangular 2×2 matrices use the basis* $\begin{bmatrix} 1 & 0 \\ 0 & 0 \end{bmatrix}, \begin{bmatrix} 0 & 1 \\ 0 & 0 \end{bmatrix}, \begin{bmatrix} 0 & 0 \\ 0 & 1 \end{bmatrix}$ *unless another basis is given. In each exercise, determine whether T is an isomorphism.*

5. $T(M) = \begin{bmatrix} 1 & 2 \\ 0 & 3 \end{bmatrix} M$ from $U^{2\times 2}$ to $U^{2\times 2}$.

6. $T(M) = \begin{bmatrix} 1 & 2 \\ 0 & 3 \end{bmatrix} M$ from $U^{2\times 2}$ to $U^{2\times 2}$, with respect to

the basis $\begin{bmatrix} 1 & 0 \\ 0 & 0 \end{bmatrix}, \begin{bmatrix} 0 & 1 \\ 0 & 0 \end{bmatrix}, \begin{bmatrix} 0 & 1 \\ 0 & 1 \end{bmatrix}$.

7. $T(M) = M \begin{bmatrix} 1 & 2 \\ 0 & 1 \end{bmatrix} - \begin{bmatrix} 1 & 2 \\ 0 & 1 \end{bmatrix} M$ from $U^{2\times2}$ to $U^{2\times2}$,

with respect to the basis $\begin{bmatrix} 1 & 0 \\ 0 & 1 \end{bmatrix}, \begin{bmatrix} 0 & 1 \\ 0 & 0 \end{bmatrix}, \begin{bmatrix} 1 & 0 \\ 0 & -1 \end{bmatrix}$.

8. $T(M) = M \begin{bmatrix} 1 & 2 \\ 0 & 1 \end{bmatrix} - \begin{bmatrix} 1 & 2 \\ 0 & 1 \end{bmatrix} M$ from $U^{2\times2}$ to $U^{2\times2}$.

9. $T(M) = \begin{bmatrix} 1 & 0 \\ 0 & 2 \end{bmatrix}^{-1} M \begin{bmatrix} 1 & 0 \\ 0 & 2 \end{bmatrix}$ from $U^{2\times2}$ to $U^{2\times2}$.

10. $T(M) = \begin{bmatrix} 1 & 2 \\ 0 & 3 \end{bmatrix}^{-1} M \begin{bmatrix} 1 & 2 \\ 0 & 3 \end{bmatrix}$ from $U^{2\times2}$ to $U^{2\times2}$.

11. $T(M) = \begin{bmatrix} 1 & 2 \\ 0 & 3 \end{bmatrix}^{-1} M \begin{bmatrix} 1 & 2 \\ 0 & 3 \end{bmatrix}$ from $U^{2\times2}$ to $U^{2\times2}$,

with respect to the basis $\begin{bmatrix} 1 & -1 \\ 0 & 0 \end{bmatrix}, \begin{bmatrix} 0 & 1 \\ 0 & 1 \end{bmatrix}, \begin{bmatrix} 0 & 1 \\ 0 & 0 \end{bmatrix}$.

12. $T(M) = M \begin{bmatrix} 2 & 0 \\ 0 & 3 \end{bmatrix}$ from $\mathbb{R}^{2\times2}$ to $\mathbb{R}^{2\times2}$.

13. $T(M) = \begin{bmatrix} 1 & 1 \\ 2 & 2 \end{bmatrix} M$ from $\mathbb{R}^{2\times2}$ to $\mathbb{R}^{2\times2}$.

14. $T(M) = \begin{bmatrix} 1 & 1 \\ 2 & 2 \end{bmatrix} M$ from $\mathbb{R}^{2\times2}$ to $\mathbb{R}^{2\times2}$, with respect to

the basis $\begin{bmatrix} 1 & 0 \\ -1 & 0 \end{bmatrix}, \begin{bmatrix} 0 & 1 \\ 0 & -1 \end{bmatrix}, \begin{bmatrix} 1 & 0 \\ 2 & 0 \end{bmatrix}, \begin{bmatrix} 0 & 1 \\ 0 & 2 \end{bmatrix}$.

15. $T(x + iy) = x - iy$ from $\mathbb{C}$ to $\mathbb{C}$.

16. $T(x + iy) = x - iy$ from $\mathbb{C}$ to $\mathbb{C}$, with respect to the basis $1 + i, 1 - i$.

17. $T(z) = iz$ from $\mathbb{C}$ to $\mathbb{C}$.

18. $T(z) = (2 + 3i)z$ from $\mathbb{C}$ to $\mathbb{C}$.

19. $T(z) = (p + iq)z$ from $\mathbb{C}$ to $\mathbb{C}$, where p and q are arbitrary real numbers.

20. $T(f) = f'$ from P_2 to P_2.

21. $T(f) = f' - 3f$ from P_2 to P_2.

22. $T(f) = f'' + 4f'$ from P_2 to P_2.

23. $T(f(t)) = f(3)$ from P_2 to P_2.

24. $T(f(t)) = f(3)$ from P_2 to P_2, with respect to the basis $1, t - 3, (t - 3)^2$.

25. $T(f(t)) = f(-t)$ from P_2 to P_2.

26. $T(f(t)) = f(2t)$ from P_2 to P_2.

27. $T(f(t)) = f(2t - 1)$ from P_2 to P_2.

28. $T(f(t)) = f(2t - 1)$ from P_2 to P_2, with respect to the basis $1, t - 1, (t - 1)^2$.

29. $T(f(t)) = \int_0^2 f(t)dt$ from P_2 to P_2.

30. $T(f(t)) = \dfrac{f(t + h) - f(t)}{h}$ from P_2 to P_2, where h is a nonzero constant. Interpret transformation T geometrically.

31. $T(f(t)) = \dfrac{f(t + h) - f(t - h)}{2h}$ from P_2 to P_2, where h is a nonzero constant. Interpret transformation T geometrically.

32. $T(f(t)) = f(1) + f'(1)(t - 1)$ from P_2 to P_2. Interpret transformation T geometrically.

33. $T(f(t)) = f(1) + f'(1)(t - 1)$ from P_2 to P_2, with respect to the basis $1, t - 1, (t - 1)^2$.

34. Find image, rank, and kernel of the transformation in Exercise 7.

35. Find image, rank, and kernel of the transformation in Exercise 13.

36. Find image, rank, and kernel of the transformation in Exercise 20.

37. Find image, rank, and kernel of the transformation in Exercise 21.

38. Find image, rank, and kernel of the transformation in Exercise 22.

39. Find image, rank, and kernel of the transformation in Exercise 23.

40. Find image, rank, and kernel of the transformation in Exercise 32.

41. If matrix A is the answer to Exercise 5 and B is the answer to Exercise 6, find an invertible matrix S such that $B = S^{-1}AS$.

42. If matrix A is the answer to Exercise 8 and B is the answer to Exercise 7, find an invertible matrix S such that $B = S^{-1}AS$.

43. If matrix A is the answer to Exercise 10 and B is the answer to Exercise 11, find an invertible matrix S such that $B = S^{-1}AS$.

44. If matrix A is the answer to Exercise 13 and B is the answer to Exercise 14, find an invertible matrix S such that $B = S^{-1}AS$.

45. If matrix A is the answer to Exercise 15 and B is the answer to Exercise 16, find an invertible matrix S such that $B = S^{-1}AS$.

46. If matrix A is the answer to Exercise 23 and B is the answer to Exercise 24, find an invertible matrix S such that $B = S^{-1}AS$.

47. If matrix A is the answer to Exercise 27 and B is the answer to Exercise 28, find an invertible matrix S such that $B = S^{-1}AS$.

In Exercises 48 through 53, let V be the space spanned by the two functions $\cos(t)$ and $\sin(t)$. In each exercise, find the matrix of the given transformation T with respect to the basis $\cos(t), \sin(t)$, and determine whether T is an isomorphism.

48. $T(f) = f'$

49. $T(f) = f'' + 2f' + 3f$

50. $T(f) = f'' + af' + bf$, where a and b are arbitrary real numbers.

51. $T(f(t)) = f(t - \pi/2)$.

52. $T(f(t)) = f(t - \pi/4)$.

53. $T(f(t)) = f(t - \delta)$, where δ is an arbitrary real number. *Hint:* Use the addition theorems for sine and cosine.

In Exercises 54 through 58, let V be the plane with equation $x_1 + 2x_2 + 3x_3 = 0$ in $\mathbb{R}^3$. In each exercise, find the matrix B of the given transformation T from V to V, with respect to the

basis $\begin{bmatrix} 1 \\ 1 \\ -1 \end{bmatrix}, \begin{bmatrix} 5 \\ -4 \\ 1 \end{bmatrix}$. *Note that domain and codomain of T*

are restricted to the plane V, so that B will be a 2×2 matrix.

54. The orthogonal projection onto the line spanned by

vector $\begin{bmatrix} 1 \\ 1 \\ -1 \end{bmatrix}$.

55. The orthogonal projection onto the line spanned by

vector $\begin{bmatrix} 1 \\ -2 \\ 1 \end{bmatrix}$.

56. $T(\vec{x}) = \begin{bmatrix} 1 \\ 2 \\ 3 \end{bmatrix} \times \vec{x}$.

57. $T(\vec{x}) = \begin{bmatrix} -2 & -3 & 1 \\ 1 & 0 & -2 \\ 0 & 1 & 1 \end{bmatrix} \vec{x}$.

58. $T(\vec{x}) = \left(\vec{x} \cdot \begin{bmatrix} 1 \\ 1 \\ -1 \end{bmatrix} \right) \begin{bmatrix} 1 \\ 1 \\ -1 \end{bmatrix}$.

59. Consider a linear transformation T from V to V with $\ker(T) = \{0\}$. If V is finite-dimensional, then T is an isomorphism, since the matrix of T will be invertible. Show that this is not necessarily the case if V is infinite-dimensional: Give an example of a linear transformation T from P to P with $\ker(T) = \{0\}$ that is not an isomorphism. (Recall that P is the space of all polynomials.)

60. Let V be the linear space of all functions in two variables of the form $q(x_1, x_2) = ax_1^2 + bx_1x_2 + cx_2^2$. Consider the linear transformation

$$T(f) = \frac{\partial f}{\partial x_1} x_2 - \frac{\partial f}{\partial x_2} x_1$$

from V to V.

a. Find the matrix of T with respect to the basis x_1^2, x_1x_2, x_2^2.

b. Find bases of kernel and image of T.

61. Let V be the linear space of all functions of the form

$$f(t) = c_1 \cos(t) + c_2 \sin(t) + c_3 t \cos(t) + c_4 t \sin(t).$$

Consider the linear transformation T from V to V given by

$$T(f) = f'' + f.$$

a. Find the matrix of T with respect to the basis $\cos(t), \sin(t), t \cos(t), t \sin(t)$.

b. Find all solutions f in W of the differential equation

$$T(f) = f'' + f = \cos(t).$$

Graph your solution(s). (The differential equation $f'' + f = \cos(t)$ describes a forced undamped oscillator. In this example, we observe the phenomenon of *resonance*.)

62. Consider the linear space V of all infinite sequences of real numbers. We define the subset W of L consisting of all sequences $(x_0, x_1, x_2, \ldots)$ such that $x_{n+2} = x_{n+1} + 6x_n$ for all $n \geq 0$.

a. Show that W is a *subspace* of V.

b. Determine the dimension of W.

c. Does W contain any geometric sequences of the form $(1, c, c^2, c^3, \ldots)$, for some constant c? Find all such sequences in W.

d. Can you find a basis of W consisting of geometric sequences?

e. Consider the sequence in W whose first two terms are $x_0 = 0, x_1 = 1$. Find x_2, x_3, x_4. Find a closed formula for the nth term x_n of this sequence. *Hint:* Write this sequence as a linear combination of the sequences you found in part (d).

63. Consider a basis $f_1, \ldots, f_n$ of P_{n-1}. Let $a_1, \ldots, a_n$ be distinct real numbers. Consider the $n \times n$ matrix M whose ijth entry is $f_j(a_i)$. Show that the matrix M is invertible. *Hint:* If the vector

$$\begin{bmatrix} c_1 \\ c_2 \\ \vdots \\ c_n \end{bmatrix}$$

is in the kernel of M, then the polynomial $f = c_1 f_1 + \cdots + c_n f_n$ in P_{n-1} vanishes at $a_1, \ldots, a_n$; therefore, $f = 0$.

64. Consider two finite-dimensional linear spaces V and W. If V and W are isomorphic, then they have the same dimension (by Fact 4.2.3d). Conversely, if V and W have the same dimension, are they necessarily isomorphic? Justify your answer carefully.

65. Let $a_1, \ldots, a_n$ be distinct real numbers. Show that there are "weights" $w_1, \ldots, w_n$ such that

$$\int_{-1}^{1} f(t) \, dt = \sum_{i=1}^{n} w_i f(a_i),$$

for all polynomials $f(t)$ in P_{n-1}. *Hint:* It suffices to prove the claim for a basis $f_1, \ldots, f_n$ of P_{n-1}. Exercise 63 is helpful.

66. Find the weights w_1, w_2, w_3 in Exercise 65 for $a_1 = -1, a_2 = 0, a_3 = 1$. (Compare this with Simpson's rule in calculus.)

5

Orthogonality and Least Squares

5.1 ORTHONORMAL BASES AND ORTHOGONAL PROJECTIONS

Not all bases are created equal. When working in a plane E in $\mathbb{R}^3$, for example, it is particularly convenient to use a basis $\vec{v}_1$, $\vec{v}_2$ consisting of two perpendicular unit vectors. (See Figure 1.)

In this section, we will develop some vocabulary to deal with such bases. Then, we will use a basis consisting of perpendicular unit vectors to study the orthogonal projection onto a subspace. In the next section, we will construct such a basis of a subspace of $\mathbb{R}^n$.

First, we review some basic concepts.

Definition 5.1.1

> **Orthogonality, length, unit vectors**
>
> a. Two vectors $\vec{v}$ and $\vec{w}$ in $\mathbb{R}^n$ are called *perpendicular* or *orthogonal*[1] if $\vec{v} \cdot \vec{w} = 0$.
> b. The *length* (or magnitude or norm) of a vector $\vec{v}$ in $\mathbb{R}^n$ is $\|\vec{v}\| = \sqrt{\vec{v} \cdot \vec{v}}$.
> c. A vector $\vec{u}$ in $\mathbb{R}^n$ is called a *unit vector* if its length is 1, (i.e., $\|\vec{u}\| = 1$, or $\vec{u} \cdot \vec{u} = 1$).

If $\vec{v}$ is a nonzero vector in $\mathbb{R}^n$, then

$$\vec{u} = \frac{1}{\|\vec{v}\|} \vec{v}$$

is a unit vector. (See Exercise 25(b).)

---

[1] The two terms are synonymous: "Perpendicular" comes from Latin and "orthogonal" from Greek.

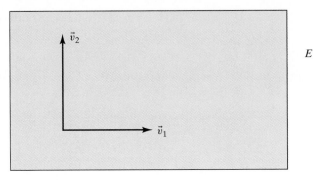

Figure 1

Definition 5.1.2

Orthonormal vectors
The vectors $\vec{v}_1, \vec{v}_2, \ldots, \vec{v}_m$ in $\mathbb{R}^n$ are called *orthonormal* if they are all unit vectors and orthogonal to one another:

$$\vec{v}_i \cdot \vec{v}_j = \begin{cases} 1 & \text{if } i = j, \\ 0 & \text{if } i \neq j. \end{cases}$$

EXAMPLE 1   The vectors $\vec{e}_1, \vec{e}_2, \ldots, \vec{e}_n$ in $\mathbb{R}^n$ are orthonormal.   ∎

EXAMPLE 2   For any scalar α, the vectors $\begin{bmatrix} \cos \alpha \\ \sin \alpha \end{bmatrix}, \begin{bmatrix} -\sin \alpha \\ \cos \alpha \end{bmatrix}$ are orthonormal. (See Figure 2.)   ∎

EXAMPLE 3   The vectors

$$\vec{v}_1 = \begin{bmatrix} 1/2 \\ 1/2 \\ 1/2 \\ 1/2 \end{bmatrix}, \qquad \vec{v}_2 = \begin{bmatrix} 1/2 \\ 1/2 \\ -1/2 \\ -1/2 \end{bmatrix}, \qquad \vec{v}_3 = \begin{bmatrix} 1/2 \\ -1/2 \\ 1/2 \\ -1/2 \end{bmatrix}$$

in $\mathbb{R}^4$ are orthonormal. (Verify this.) Can you find a vector $\vec{v}_4$ in $\mathbb{R}^4$ such that all the vectors $\vec{v}_1, \vec{v}_2, \vec{v}_3, \vec{v}_4$ are orthonormal. (See Exercise 16.)   ∎

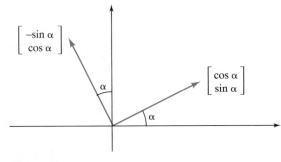

Figure 2

The following properties of orthonormal vectors are often useful:

Fact 5.1.3

a. Orthonormal vectors are linearly independent.

b. Orthonormal vectors $\vec{v}_1, \ldots, \vec{v}_n$ in $\mathbb{R}^n$ form a basis of $\mathbb{R}^n$.

Proof a. Consider a relation

$$c_1\vec{v}_1 + c_2\vec{v}_2 + \cdots + c_i\vec{v}_i + \cdots + c_m\vec{v}_m = \vec{0}$$

among the orthonormal vectors $\vec{v}_1, \vec{v}_2, \ldots, \vec{v}_m$ in $\mathbb{R}^n$. Let us form the dot product of each side of this equation with $\vec{v}_i$:

$$(c_1\vec{v}_1 + c_2\vec{v}_2 + \cdots + c_i\vec{v}_i + \cdots + c_m\vec{v}_m) \cdot \vec{v}_i = \vec{0} \cdot \vec{v}_i = 0.$$

Because the dot product is distributive (see Fact A5 in the Appendix),

$$c_1(\vec{v}_1 \cdot \vec{v}_i) + c_2(\vec{v}_2 \cdot \vec{v}_i) + \cdots + c_i(\vec{v}_i \cdot \vec{v}_i) + \cdots + c_m(\vec{v}_m \cdot \vec{v}_i) = 0.$$

We know that $\vec{v}_i \cdot \vec{v}_i = 1$, and all other dot products are zero. Therefore, $c_i = 0$. Since this holds for all $i = 1, \ldots, m$, it follows that the vectors $\vec{v}_i$ are linearly independent.

b. This follows from part (a) and Summary 3.3.11. (Any n linearly independent vectors in $\mathbb{R}^n$ form a basis of $\mathbb{R}^n$.) ▲

Definition 5.1.4

Orthogonal complement

Consider a subspace V of $\mathbb{R}^n$. The *orthogonal complement* $V^\perp$ of V is the set of those vectors $\vec{x}$ in $\mathbb{R}^n$ that are orthogonal to all vectors in V:

$$V^\perp = \{\vec{x} \text{ in } \mathbb{R}^n : \vec{v} \cdot \vec{x} = 0, \text{ for all } \vec{v} \text{ in } V\}.$$

If we have a basis $\vec{v}_1, \vec{v}_2, \ldots, \vec{v}_m$ of V, then $V^\perp$ is the set of those vectors $\vec{x}$ in $\mathbb{R}^n$ that are orthogonal to all the $\vec{v}_i$ (See Exercise 22):

$$V^\perp = \{\vec{x} \text{ in } \mathbb{R}^n : \vec{v}_i \cdot \vec{x} = 0, \quad \text{for } i = 1, \ldots, m\}.$$

For example, a vector $\vec{x}$ in $\mathbb{R}^3$ is orthogonal to all vectors in a plane E if (and only if) it is orthogonal to two vectors $\vec{v}_1, \vec{v}_2$ that form a basis of E. (See Figure 3.)

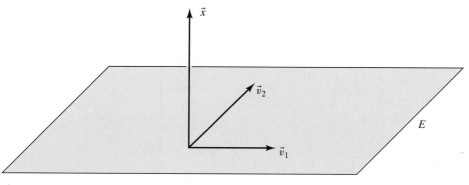

Figure 3

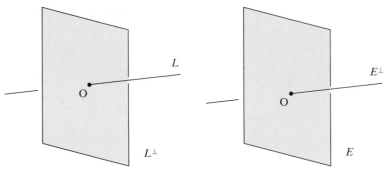

Figure 4

In Figure 4, we sketch the orthogonal complements of a line L and of a plane E in $\mathbb{R}^3$. Note that both $L^\perp$ and $E^\perp$ are subspaces of $\mathbb{R}^3$.

Fact 5.1.5 If V is a subspace of $\mathbb{R}^n$, then its orthogonal complement $V^\perp$ is a subspace of $\mathbb{R}^n$ as well.

Proof   We will verify that $V^\perp$ is closed under scalar multiplication and leave the verification of the two other properties as Exercise 23. Consider a vector $\vec{w}$ in $V^\perp$ and a scalar k. We have to show that $k\vec{w}$ is orthogonal to all vectors $\vec{v}$ in V. Pick an arbitrary vector $\vec{v}$ in V. Then, $(k\vec{w}) \cdot \vec{v} = k(\vec{w} \cdot \vec{v}) = 0$, as claimed.   ▲

Orthogonal Projections

As mentioned at the beginning of this section, it is often convenient to work with an *orthonormal basis* of a subspace V of $\mathbb{R}^n$. (An orthonormal basis is a basis of V that consists of orthonormal vectors.) In the next section, we will see how such an orthonormal basis of a subspace can be constructed. Assuming that we have an orthonormal basis of a subspace V of $\mathbb{R}^n$, we will now examine how we can find the orthogonal projection of a vector in $\mathbb{R}^n$ onto V.

Before we deal with the general case, let us briefly review the case of a one-dimensional subspace V of $\mathbb{R}^n$. In this case, an orthonormal basis of V consists of a single unit vector, $\vec{v}_1$. In Section 2.2, we found that for any vector $\vec{x}$ in $\mathbb{R}^n$, there is a unique vector $\vec{w}$ in V such that $\vec{x} - \vec{w}$ is perpendicular to V. (See Figure 5.)

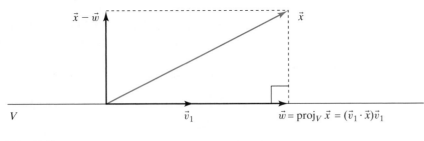

Figure 5

This vector $\vec{w}$ is called the *orthogonal projection* of $\vec{x}$ onto V, denoted by $\text{proj}_V \vec{x}$. In Fact 2.2.5, we stated the formula

$$\text{proj}_V \vec{x} = (\vec{v}_1 \cdot \vec{x})\vec{v}_1.$$

Now consider a subspace V of $\mathbb{R}^n$ with arbitrary dimension m. Suppose we have an orthonormal basis $\vec{v}_1, \vec{v}_2, \ldots, \vec{v}_m$ of V. Consider a vector $\vec{x}$ in $\mathbb{R}^n$. Is it still possible to find a vector $\vec{w}$ in V such that $\vec{x} - \vec{w}$ is in $V^\perp$? If such a $\vec{w}$ exists, we can write

$$\vec{w} = c_1\vec{v}_1 + c_2\vec{v}_2 + \cdots + c_m\vec{v}_m,$$

for some scalars c_i (since $\vec{w}$ is in V). It is required that

$$\vec{x} - \vec{w} = \vec{x} - c_1\vec{v}_1 - c_2\vec{v}_2 - \cdots - c_m\vec{v}_m$$

be perpendicular to V; that is, vector $\vec{x} - \vec{w}$ must be perpendicular to all the vectors $\vec{v}_i$:

$$\begin{aligned}
\vec{v}_i \cdot (\vec{x} - \vec{w}) &= \vec{v}_i \cdot (\vec{x} - c_1\vec{v}_1 - \cdots - c_i\vec{v}_i - \cdots - c_m\vec{v}_m) \\
&= \vec{v}_i \cdot \vec{x} - c_1(\vec{v}_i \cdot \vec{v}_1) - \cdots - c_i(\vec{v}_i \cdot \vec{v}_i) - \cdots - c_m(\vec{v}_i \cdot \vec{v}_m) \\
&= \vec{v}_i \cdot \vec{x} - c_i = 0.
\end{aligned}$$

This equation holds if (and only if) $c_i = \vec{v}_i \cdot \vec{x}$.

We have shown that there is a unique $\vec{w}$ in V such that $\vec{x} - \vec{w}$ is in $V^\perp$, namely,

$$\vec{w} = (\vec{v}_1 \cdot \vec{x})\vec{v}_1 + \cdots + (\vec{v}_m \cdot \vec{x})\vec{v}_m.$$

Let us summarize.

Fact 5.1.6

Orthogonal projection

Consider a subspace V of $\mathbb{R}^n$ with orthonormal basis $\vec{v}_1, \vec{v}_2, \ldots, \vec{v}_m$. For any vector $\vec{x}$ in $\mathbb{R}^n$, there is a unique vector $\vec{w}$ in V such that $\vec{x} - \vec{w}$ is in $V^\perp$.

This vector $\vec{w}$ is called the *orthogonal projection* of $\vec{x}$ onto V, denoted by $\text{proj}_V \vec{x}$. We have the formula

$$\text{proj}_V \vec{x} = (\vec{v}_1 \cdot \vec{x})\vec{v}_1 + \cdots + (\vec{v}_m \cdot \vec{x})\vec{v}_m.$$

The transformation $T(\vec{x}) = \text{proj}_V \vec{x}$ from $\mathbb{R}^n$ to $\mathbb{R}^n$ is linear.

We leave the verification of the last assertion as Exercise 24.

Note that $\text{proj}_V \vec{x}$ is the sum of all vectors $(\vec{v}_i \cdot \vec{x})\vec{v}_i$, representing the orthogonal projections of $\vec{x}$ onto the lines spanned by the vectors $\vec{v}_i$. This may be a good way to memorize the formula for $\text{proj}_V \vec{x}$.

For example, projecting a vector in $\mathbb{R}^3$ orthogonally onto the x_1–x_2-plane amounts to the same as projecting it onto the x_1-axis, then onto the x_2-axis, and then adding the resultant vectors. (See Figure 6.)

EXAMPLE 4   Consider the subspace $V = \text{im}(A)$ of $\mathbb{R}^4$, where

$$A = \begin{bmatrix} 1 & 1 \\ 1 & -1 \\ 1 & -1 \\ 1 & 1 \end{bmatrix}.$$

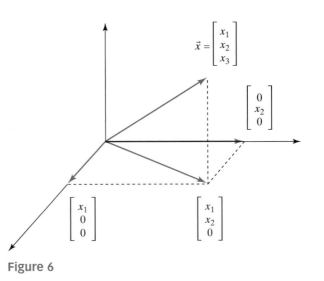

Figure 6

Find $\text{proj}_V \vec{x}$, for

$$\vec{x} = \begin{bmatrix} 1 \\ 3 \\ 1 \\ 7 \end{bmatrix}.$$

Solution

The two columns of A form a basis of V. Since they happen to be orthogonal, we can construct an orthonormal basis of V merely by dividing these two vectors by their length (2 for both vectors):

$$\vec{v}_1 = \begin{bmatrix} 1/2 \\ 1/2 \\ 1/2 \\ 1/2 \end{bmatrix}, \qquad \vec{v}_2 = \begin{bmatrix} 1/2 \\ -1/2 \\ -1/2 \\ 1/2 \end{bmatrix}.$$

Then,

$$\text{proj}_V \vec{x} = (\vec{v}_1 \cdot \vec{x})\vec{v}_1 + (\vec{v}_2 \cdot \vec{x})\vec{v}_2 = 6\vec{v}_1 + 2\vec{v}_2 = \begin{bmatrix} 3 \\ 3 \\ 3 \\ 3 \end{bmatrix} + \begin{bmatrix} 1 \\ -1 \\ -1 \\ 1 \end{bmatrix} = \begin{bmatrix} 4 \\ 2 \\ 2 \\ 4 \end{bmatrix}.$$

To check this answer, verify that $\vec{x} - \text{proj}_V \vec{x}$ is perpendicular to both $\vec{v}_1$ and $\vec{v}_2$.   ∎

What happens when we apply Fact 5.1.6 to the subspace $V = \mathbb{R}^n$ of $\mathbb{R}^n$ with orthonormal basis $\vec{v}_1, \vec{v}_2, \ldots, \vec{v}_n$? Clearly, $\text{proj}_V \vec{x} = \vec{x}$, for all $\vec{x}$ in $\mathbb{R}^n$. Therefore,

$$\vec{x} = (\vec{v}_1 \cdot \vec{x})\vec{v}_1 + \cdots + (\vec{v}_n \cdot \vec{x})\vec{v}_n,$$

for all $\vec{x}$ in $\mathbb{R}^n$.

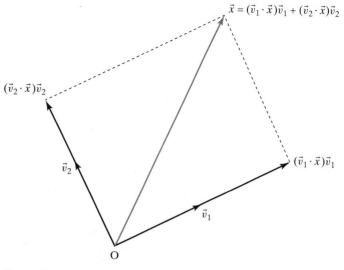

$$\vec{x} = (\vec{v}_1 \cdot \vec{x})\vec{v}_1 + (\vec{v}_2 \cdot \vec{x})\vec{v}_2$$

$(\vec{v}_2 \cdot \vec{x})\vec{v}_2$

$\vec{v}_2$

$(\vec{v}_1 \cdot \vec{x})\vec{v}_1$

$\vec{v}_1$

O

Figure 7

Fact 5.1.7

Consider an orthonormal basis $\vec{v}_1, \ldots, \vec{v}_n$ of $\mathbb{R}^n$. Then,

$$\vec{x} = (\vec{v}_1 \cdot \vec{x})\vec{v}_1 + \cdots + (\vec{v}_n \cdot \vec{x})\vec{v}_n,$$

for all $\vec{x}$ in $\mathbb{R}^n$.

This means that if you project $\vec{x}$ onto all the lines spanned by the basis vectors $\vec{v}_i$ and add the resultant vectors, you get the vector $\vec{x}$ back. Figure 7 illustrates this in the case $n = 2$.

What is the practical significance of Fact 5.1.7? Whenever we have a basis $\vec{v}_1, \ldots, \vec{v}_n$ of $\mathbb{R}^n$, any vector $\vec{x}$ in $\mathbb{R}^n$ can be expressed uniquely as a linear combination of the $\vec{v}_i$, by Fact 3.2.7:

$$\vec{x} = c_1\vec{v}_1 + c_2\vec{v}_2 + \cdots + c_n\vec{v}_n.$$

To find the coordinates c_i, we generally need to solve a linear system, which may involve a fair amount of computation. However, if the basis $\vec{v}_1, \ldots, \vec{v}_n$ is orthonormal, we can find the c_i much more easily, using the formula

$$c_i = \vec{v}_i \cdot \vec{x}.$$

EXAMPLE 5   By using paper and pencil, express the vector $\vec{x} = \begin{bmatrix} 1 \\ 2 \\ 3 \end{bmatrix}$ as a linear combination of

$$\vec{v}_1 = \frac{1}{3}\begin{bmatrix} 2 \\ 2 \\ 1 \end{bmatrix}, \qquad \vec{v}_2 = \frac{1}{3}\begin{bmatrix} 1 \\ -2 \\ 2 \end{bmatrix}, \qquad \vec{v}_3 = \frac{1}{3}\begin{bmatrix} -2 \\ 1 \\ 2 \end{bmatrix}.$$

Solution

Since $\vec{v}_1, \vec{v}_2, \vec{v}_3$ is an orthonormal basis of $\mathbb{R}^3$, we have

$$\vec{x} = (\vec{v}_1 \cdot \vec{x})\vec{v}_1 + (\vec{v}_2 \cdot \vec{x})\vec{v}_2 + (\vec{v}_3 \cdot \vec{x})\vec{v}_3 = 3\vec{v}_1 + \vec{v}_2 + 2\vec{v}_3.$$ ■

From Pythagoras to Cauchy

EXAMPLE 6   Consider a line L in $\mathbb{R}^3$ and a vector $\vec{x}$ in $\mathbb{R}^3$. What can you say about the relationship between the lengths of the vectors $\vec{x}$ and $\text{proj}_L \vec{x}$?

Solution

Applying the Pythagorean theorem to the shaded right triangle in Figure 8, we find that $\|\text{proj}_L \vec{x}\| \le \|\vec{x}\|$. The statement is an equality if (and only if) $\vec{x}$ is on L. ■

Does this inequality hold in higher dimensional cases? We have to examine whether the Pythagorean theorem holds in $\mathbb{R}^n$.

Fact 5.1.8   **Pythagorean theorem**
Consider two vectors $\vec{x}$ and $\vec{y}$ in $\mathbb{R}^n$. The equation

$$\|\vec{x} + \vec{y}\|^2 = \|\vec{x}\|^2 + \|\vec{y}\|^2$$

holds if (and only if) $\vec{x}$ and $\vec{y}$ are orthogonal. (See Figure 9.)

Proof   The verification is straightforward:

$$\|\vec{x} + \vec{y}\|^2 = (\vec{x} + \vec{y}) \cdot (\vec{x} + \vec{y}) = \vec{x} \cdot \vec{x} + 2(\vec{x} \cdot \vec{y}) + \vec{y} \cdot \vec{y} = \|\vec{x}\|^2 + 2(\vec{x} \cdot \vec{y}) + \|\vec{y}\|^2$$

$$= \|\vec{x}\|^2 + \|\vec{y}\|^2 \qquad \text{if (and only if) } \vec{x} \cdot \vec{y} = 0. \qquad \blacktriangle$$

Now we can generalize Example 6.

Fact 5.1.9   Consider a subspace V of $\mathbb{R}^n$ and a vector $\vec{x}$ in $\mathbb{R}^n$. Then,

$$\|\text{proj}_V \vec{x}\| \le \|\vec{x}\|.$$

The statement is an equality if (and only if) $\vec{x}$ is in V.

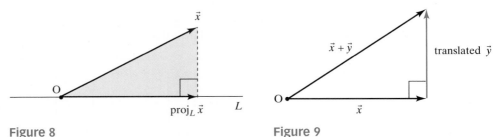

Figure 8   Figure 9

Proof   We can write $\vec{x} = \text{proj}_V \vec{x} + (\vec{x} - \text{proj}_V \vec{x})$ and apply the Pythagorean theorem (see Figure 10):

$$\|\vec{x}\|^2 = \|\text{proj}_V \vec{x}\|^2 + \|\vec{x} - \text{proj}_V \vec{x}\|^2.$$

It follows that $\|\text{proj}_V \vec{x}\| \leq \|\vec{x}\|$, as claimed.   ▲

For example, let V be a one-dimensional subspace of $\mathbb{R}^n$ spanned by a (nonzero) vector $\vec{y}$. We introduce the unit vector

$$\vec{u} = \frac{1}{\|\vec{y}\|} \vec{y}$$

in V. (See Figure 11.)
We know that

$$\text{proj}_V \vec{x} = (\vec{u} \cdot \vec{x})\vec{u} = \frac{1}{\|\vec{y}\|^2} (\vec{y} \cdot \vec{x})\vec{y},$$

for any $\vec{x}$ in $\mathbb{R}^n$. Fact 5.1.9 tells us that

$$\|\vec{x}\| \geq \|\text{proj}_V \vec{x}\| = \left\| \frac{1}{\|\vec{y}\|^2} (\vec{y} \cdot \vec{x})\vec{y} \right\| = \frac{1}{\|\vec{y}\|^2} |\vec{y} \cdot \vec{x}| \, \|\vec{y}\|.$$

To justify the last step, note that $\|k\vec{v}\| = |k| \, \|\vec{v}\|$, for all vectors $\vec{v}$ in $\mathbb{R}^n$ and all scalars k. (See Exercise 25(a).) We conclude that

$$\frac{|\vec{x} \cdot \vec{y}|}{\|\vec{y}\|} \leq \|\vec{x}\|.$$

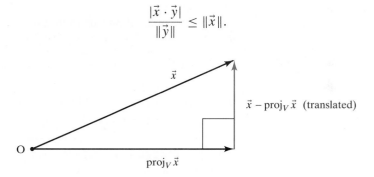

Figure 10

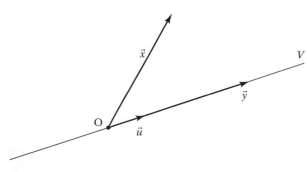

Figure 11

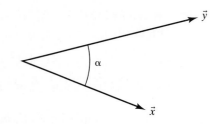

Figure 12

Multiplying both sides of this inequality by $\|\vec{y}\|$, we find the following useful result:

Fact 5.1.10

Cauchy–Schwarz inequality[2]
If $\vec{x}$ and $\vec{y}$ are vectors in $\mathbb{R}^n$, then

$$|\vec{x} \cdot \vec{y}| \leq \|\vec{x}\| \, \|\vec{y}\|.$$

This statement is an equality if (and only if) $\vec{x}$ and $\vec{y}$ are parallel.

Consider two nonzero vectors $\vec{x}$ and $\vec{y}$ in $\mathbb{R}^3$. You may know an expression for the dot product $\vec{x} \cdot \vec{y}$ in terms of the angle α between the two vectors (see Figure 12):

$$\vec{x} \cdot \vec{y} = \|\vec{x}\| \, \|\vec{y}\| \cos \alpha.$$

This formula allows us to find the angle between two nonzero vectors $\vec{x}$ and $\vec{y}$ in $\mathbb{R}^3$:

$$\cos \alpha = \frac{\vec{x} \cdot \vec{y}}{\|\vec{x}\| \, \|\vec{y}\|} \qquad \text{or} \qquad \alpha = \arccos \frac{\vec{x} \cdot \vec{y}}{\|\vec{x}\| \, \|\vec{y}\|}.$$

In $\mathbb{R}^n$, where we have no intuitive notion of an angle between two vectors, we can use this formula to *define* the angle:

Definition 5.1.11

Angle between two vectors
Consider two nonzero vectors $\vec{x}$ and $\vec{y}$ in $\mathbb{R}^n$. The angle α between these vectors is defined as

$$\alpha = \arccos \frac{\vec{x} \cdot \vec{y}}{\|\vec{x}\| \, \|\vec{y}\|}.$$

Note that α is between 0 and π, by definition of the inverse cosine function.

We have to make sure that

$$\arccos \frac{\vec{x} \cdot \vec{y}}{\|\vec{x}\| \, \|\vec{y}\|}$$

---

[2]Named after the French mathematician Augustin-Louis Cauchy (1789–1857) and the German mathematician Hermann Amandus Schwarz (1843–1921).

is defined; that is,

$$\frac{\vec{x} \cdot \vec{y}}{\|\vec{x}\| \|\vec{y}\|}$$

is between -1 and 1, or, equivalently,

$$\left| \frac{\vec{x} \cdot \vec{y}}{\|\vec{x}\| \|\vec{y}\|} \right| = \frac{|\vec{x} \cdot \vec{y}|}{\|\vec{x}\| \|\vec{y}\|} \leq 1.$$

But this follows from the Cauchy–Schwarz inequality $|\vec{x} \cdot \vec{y}| \leq \|\vec{x}\| \|\vec{y}\|$.

EXAMPLE 7   Find the angle between the vectors

$$\vec{x} = \begin{bmatrix} 1 \\ 0 \\ 0 \\ 0 \end{bmatrix} \quad \text{and} \quad \vec{y} = \begin{bmatrix} 1 \\ 1 \\ 1 \\ 1 \end{bmatrix}.$$

Solution

$$\alpha = \arccos \frac{\vec{x} \cdot \vec{y}}{\|\vec{x}\| \|\vec{y}\|} = \arccos \frac{1}{1 \cdot 2} = \frac{\pi}{3}$$ ■

Here is an application to statistics of some concepts introduced in this section.

Correlation

Consider the meat consumption (in grams per day per person) and incidence of colon cancer (per 100,000 women per year) in various industrialized countries:

Country	Meat Consumption	Cancer Rate
Japan	26	7.5
Finland	101	9.8
Israel	124	16.4
Great Britain	205	23.3
United States	284	34
Mean	148	18.2

Can we detect a positive or negative *correlation*[3] between meat consumption and cancer rate? Does a country with high meat consumption have high cancer rates, and vice versa? By "high," we mean "above average," of course. A quick look at the data shows such a *positive correlation:* In Great Britain and the United States, both meat consumption and cancer rate are above average. In the three other countries, they are below average. This positive correlation becomes more apparent

---

[3]We are using the term "correlation" in a colloquial, *qualitative* sense. Our goal is to *quantify* this term.

when we list the preceding data as *deviations from the mean* (above or below the average):

Country	Meat Consumption (Deviation from Mean)	Cancer Rate (Deviation from Mean)
Japan	−122	−10.7
Finland	−47	−8.4
Israel	−24	−1.8
Great Britain	57	5.1
United States	136	15.8

Perhaps even more informative is a *scatter plot* of the deviation data. (See Figure 13.)

A positive correlation is indicated when most of the data points (in our case, all of them) are located in the first and third quadrant.

To process these data numerically, it is convenient to represent the deviation for both characteristics (meat consumption and cancer rate) as vectors in $\mathbb{R}^5$:

$$\vec{x} = \begin{bmatrix} -122 \\ -47 \\ -24 \\ 57 \\ 136 \end{bmatrix}, \qquad \vec{y} = \begin{bmatrix} -10.7 \\ -8.4 \\ -1.8 \\ 5.1 \\ 15.8 \end{bmatrix}$$

We will call these two vectors the *deviation vectors* of the two characteristics.

In the case of a positive correlation, most of the corresponding entries x_i, y_i of the deviation vectors have the same sign (both positive or both negative). In our example, this is the case for all entries. This means that the product $x_i y_i$ will be

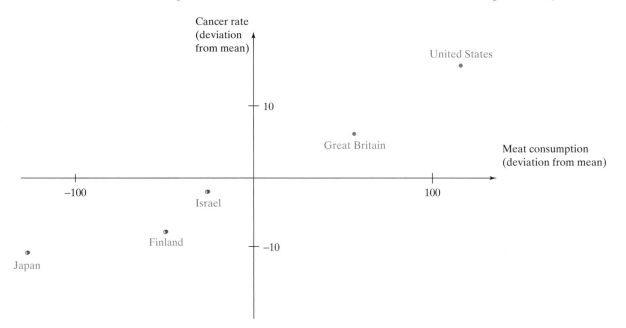

Figure 13

positive most of the time; hence, the sum of all these products will be positive. But this sum is simply the dot product of the two deviation vectors.

Still using the term "correlation" in a colloquial sense, we conclude the following:

> Consider two characteristics of a population, with deviation vectors $\vec{x}$ and $\vec{y}$. There is a *positive correlation* between the two characteristics if (and only if) $\vec{x} \cdot \vec{y} > 0$.

A positive correlation between the characteristics means that the angle α between the deviation vectors is less than 90°. (See Figure 14.)

We can use the cosine of the angle α between $\vec{x}$ and $\vec{y}$ as a quantitative measure for the correlation between the two characteristics.

Definition 5.1.12

Correlation coefficient
The *correlation coefficient* r between two characteristics of a population is the cosine of the angle α between the deviation vectors $\vec{x}$ and $\vec{y}$ for the two characteristics:

$$r = \cos(\alpha) = \frac{\vec{x} \cdot \vec{y}}{\|\vec{x}\| \, \|\vec{y}\|}$$

In the case of meat consumption and cancer, we find that

$$r \approx \frac{4182.9}{198.53 \cdot 21.539} \approx 0.9782.$$

The angle between the two deviation vectors is $\arccos(r) \approx 0.21$ (radians) $\approx 12°$.

Note that the length of the deviation vectors is irrelevant for the correlation: If we had measured the cancer rate per 1,000,000 women (instead of 100,000), the vector $\vec{y}$ would be 10 times longer, but the correlation would be the same.

The correlation coefficient r is always between -1 and 1; the cases when $r = 1$ (representing a perfect positive correlation) and $r = -1$ (perfect negative correlation) are of particular interest. (See Figure 15.) In both cases, the data points (x_i, y_i) will be on the straight line $y = mx$. (See Figure 16.)

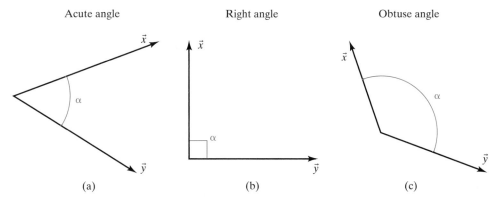

Figure 14 (a) Positive correlation: $\vec{x} \cdot \vec{y} > 0$. (b) No correlation: $\vec{x} \cdot \vec{y} = 0$. (c) Negative correlation: $\vec{x} \cdot \vec{y} < 0$.

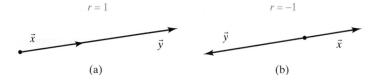

(a)   (b)

Figure 15 (a) $\vec{y} = m\vec{x}$, for positive m. (b) $\vec{y} = m\vec{x}$, for negative m.

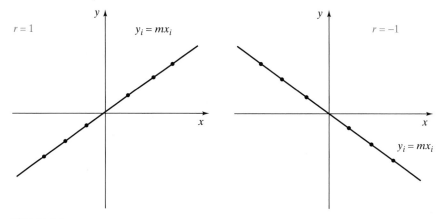

Figure 16

Note that even a strong positive correlation (an r close to 1) does not necessarily imply a causal relationship. Based only on the work we did above, we cannot conclude that high meat consumption causes colon cancer. Take a statistics course to learn more about these important issues!

EXERCISES

GOALS Apply the basic concepts of geometry in $\mathbb{R}^n$: length, angles, orthogonality. Use the idea of an orthogonal projection onto a subspace. Find this projection if an orthonormal basis of the subspace is given.

Find the length of each of the vectors $\vec{v}$ in Exercises 1 through 3.

1. $\vec{v} = \begin{bmatrix} 7 \\ 11 \end{bmatrix}$ **2.** $\vec{v} = \begin{bmatrix} 2 \\ 3 \\ 4 \end{bmatrix}$ **3.** $\vec{v} = \begin{bmatrix} 2 \\ 3 \\ 4 \\ 5 \end{bmatrix}$

Find the angle α between each of the pairs of vectors $\vec{u}$ and $\vec{v}$ in Exercises 4 through 6.

4. $\vec{u} = \begin{bmatrix} 1 \\ 1 \end{bmatrix}$, $\vec{v} = \begin{bmatrix} 7 \\ 11 \end{bmatrix}$ **5.** $\vec{u} = \begin{bmatrix} 1 \\ 2 \\ 3 \end{bmatrix}$, $\vec{v} = \begin{bmatrix} 2 \\ 3 \\ 4 \end{bmatrix}$

6. $\vec{u} = \begin{bmatrix} 1 \\ -1 \\ 2 \\ -2 \end{bmatrix}$, $\vec{v} = \begin{bmatrix} 2 \\ 3 \\ 4 \\ 5 \end{bmatrix}$

For each pair of vectors $\vec{u}$, $\vec{v}$ listed in Exercises 7 through 9, determine whether the angle α between $\vec{u}$ and $\vec{v}$ is acute, obtuse, or right.

7. $\vec{u} = \begin{bmatrix} 2 \\ -3 \end{bmatrix}$, $\vec{v} = \begin{bmatrix} 5 \\ 4 \end{bmatrix}$

8. $\vec{u} = \begin{bmatrix} 2 \\ 3 \\ 4 \end{bmatrix}$, $\vec{v} = \begin{bmatrix} 2 \\ -8 \\ 5 \end{bmatrix}$

9. $\vec{u} = \begin{bmatrix} 1 \\ -1 \\ 1 \\ -1 \end{bmatrix}$, $\vec{v} = \begin{bmatrix} 3 \\ 4 \\ 5 \\ 3 \end{bmatrix}$

10. For which choice(s) of the constant k are the vectors

$$\vec{u} = \begin{bmatrix} 2 \\ 3 \\ 4 \end{bmatrix} \quad \text{and} \quad \vec{v} = \begin{bmatrix} 1 \\ k \\ 1 \end{bmatrix}$$

perpendicular?

11. Consider the vectors

$$\vec{u} = \begin{bmatrix} 1 \\ 1 \\ \vdots \\ 1 \end{bmatrix}, \quad \vec{v} = \begin{bmatrix} 1 \\ 0 \\ \vdots \\ 0 \end{bmatrix} \quad \text{in } \mathbb{R}^n.$$

a. For $n = 2, 3, 4$, find the angle between $\vec{u}$ and $\vec{v}$. For $n = 2$ and 3, represent the vectors graphically.

b. Find the limit of this angle as n approaches infinity.

12. Give an algebraic proof for the *triangle inequality*

$$\|\vec{v} + \vec{w}\| \le \|\vec{v}\| + \|\vec{w}\|.$$

Draw a sketch. (*Hint:* Expand $\|\vec{v} + \vec{w}\|^2 = (\vec{v} + \vec{w}) \cdot (\vec{v} + \vec{w})$. Then use the Cauchy–Schwarz inequality.)

13. *Leg traction.* The accompanying figure shows how a leg may be stretched by a pulley line for therapeutic purposes. We denote by $\vec{F}_1$ the vertical force of the weight. The string of the pulley line has the same tension everywhere; hence, the forces $\vec{F}_2$ and $\vec{F}_3$ have the same magnitude as $\vec{F}_1$. Assume that the magnitude of each force is 10 pounds. Find the angle α so that the magnitude of the force exerted on the leg is 16 pounds. Round your answer to the nearest degree. (Adapted from E. Batschelet: *Introduction to Mathematics for Life Scientists,* Springer, 1979).

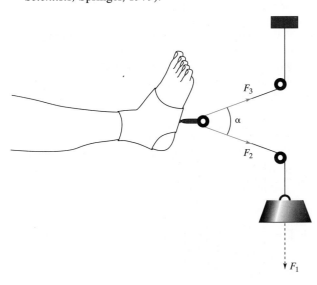

14. *Leonardo da Vinci and the resolution of forces.* Leonardo (1452–1519) asked himself how the weight of a body, supported by two strings of different length, is apportioned between the two strings.

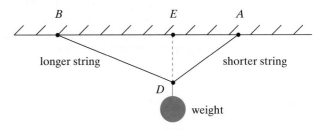

Three forces are acting at the point D: the tensions $\vec{F}_1$ and $\vec{F}_2$ in the strings and the weight $\vec{W}$. Leonardo believed that

$$\frac{\|\vec{F}_1\|}{\|\vec{F}_2\|} = \frac{\overline{EA}}{\overline{EB}}.$$

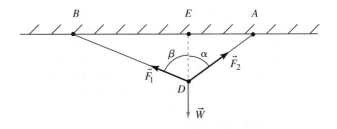

Was he right? (Source: *Les Manuscrits de Léonard de Vinci,* published by Ravaisson-Mollien, Paris, 1890.) *Hint:* Resolve $\vec{F}_1$ into a horizontal and a vertical component; do the same for $\vec{F}_2$. Since the system is at rest, the equation $\vec{F}_1 + \vec{F}_2 + \vec{W} = \vec{0}$ holds. Express the ratios

$$\frac{\|\vec{F}_1\|}{\|\vec{F}_2\|} \quad \text{and} \quad \frac{\overline{EA}}{\overline{EB}}$$

in terms of α and β, using trigonometric functions, and compare the results.

15. Consider the vector

$$\vec{v} = \begin{bmatrix} 1 \\ 2 \\ 3 \\ 4 \end{bmatrix} \quad \text{in } \mathbb{R}^4.$$

Find a basis of the subspace of $\mathbb{R}^4$ consisting of all vectors perpendicular to $\vec{v}$.

16. Consider the vectors

$$\vec{v}_1 = \begin{bmatrix} 1/2 \\ 1/2 \\ 1/2 \\ 1/2 \end{bmatrix}, \quad \vec{v}_2 = \begin{bmatrix} 1/2 \\ 1/2 \\ -1/2 \\ -1/2 \end{bmatrix}, \quad \vec{v}_3 = \begin{bmatrix} 1/2 \\ -1/2 \\ 1/2 \\ -1/2 \end{bmatrix}$$

in $\mathbb{R}^4$. Can you find a vector $\vec{v}_4$ in $\mathbb{R}^4$ such that the vectors $\vec{v}_1, \vec{v}_2, \vec{v}_3, \vec{v}_4$ are orthonormal? If so, how many such vectors are there?

17. Find a basis for $W^{\perp}$, where

$$W = \text{span} \left(\begin{bmatrix} 1 \\ 2 \\ 3 \\ 4 \end{bmatrix}, \begin{bmatrix} 5 \\ 6 \\ 7 \\ 8 \end{bmatrix} \right).$$

18. Here is an infinite-dimensional version of Euclidean space: In the space of all infinite sequences, consider the subspace ℓ_2 of square-summable sequences (i.e., those sequences $(x_1, x_2, \ldots)$ for which the infinite series $x_1^2 + x_2^2 + \cdots$ converges). For $\vec{x}$ and $\vec{y}$ in ℓ_2, we define

$$\|\vec{x}\| = \sqrt{x_1^2 + x_2^2 + \cdots}, \quad \vec{x} \cdot \vec{y} = x_1 y_1 + x_2 y_2 + \cdots.$$

(Why does the series $x_1 y_1 + x_2 y_2 + \cdots$ converge?)

a. Check that $\vec{x} = (1, \frac{1}{2}, \frac{1}{4}, \frac{1}{8}, \frac{1}{16}, \ldots)$ is in ℓ_2, and find $\|\vec{x}\|$. Recall the formula for the geometric series: $1 + a + a^2 + a^3 + \cdots = 1/(1-a)$, if $-1 < a < 1$.

b. Find the angle between $(1, 0, 0, \ldots)$ and $(1, \frac{1}{2}, \frac{1}{4}, \frac{1}{8}, \ldots)$.

c. Give an example of a sequence $(x_1, x_2, \ldots)$ that converges to 0 (i.e., $\lim_{n \to \infty} x_n = 0$), but does not belong to ℓ_2.

d. Let L be the subspace of ℓ_2 spanned by $(1, \frac{1}{2}, \frac{1}{4}, \frac{1}{8}, \ldots)$. Find the orthogonal projection of $(1, 0, 0, \ldots)$ onto L.

The Hilbert space ℓ_2 was initially used mostly in physics: Werner Heisenberg's formulation of quantum mechanics is in terms of ℓ_2. Today, this space is used in many other applications, including economics. (See, for example, the work of the economist Andreu Mas-Colell of the University of Barcelona.)

19. For a line L in $\mathbb{R}^2$, draw a sketch to interpret the following transformations geometrically:

a. $T(\vec{x}) = \vec{x} - \text{proj}_L \vec{x}$.

b. $T(\vec{x}) = \vec{x} - 2\,\text{proj}_L \vec{x}$.

c. $T(\vec{x}) = 2\,\text{proj}_L \vec{x} - \vec{x}$.

20. Refer to Figure 13 of this section. The *least-squares line* for these data is the line $y = mx$ that fits the data best, in

that the sum of the squares of the vertical distances between the line and the data points is minimal. We want to minimize the sum

$$(mx_1 - y_1)^2 + (mx_2 - y_2)^2 + \cdots + (mx_5 - y_5)^2.$$

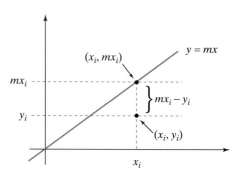

In vector notation, to minimize the sum means to find the scalar m such that

$$\|m\vec{x} - \vec{y}\|^2$$

is minimal. Arguing geometrically, explain how you can find m. Use the accompanying sketch, which is not drawn to scale.

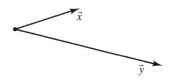

Find m numerically, and explain the relationship between m and the correlation coefficient r. You may find the following information helpful:

$$\vec{x} \cdot \vec{y} = 4182.9, \quad \|\vec{x}\| \approx 198.53, \quad \|\vec{y}\| \approx 21.539.$$

To check whether your solution m is reasonable, draw the line $y = mx$ in Figure 13. (A more thorough discussion of least-squares approximations will follow in Section 5.4.)

21. Find scalars a, b, c, d, e, f, g such that the vectors

$$\begin{bmatrix} a \\ d \\ f \end{bmatrix}, \quad \begin{bmatrix} b \\ 1 \\ g \end{bmatrix}, \quad \begin{bmatrix} c \\ e \\ 1/2 \end{bmatrix}$$

are orthonormal.

22. Consider a basis $\vec{v}_1, \vec{v}_2, \ldots, \vec{v}_m$ of a subspace V of $\mathbb{R}^n$. Show that

$$V^{\perp} = \{\vec{x} \text{ in } \mathbb{R}^n : \vec{v}_i \cdot \vec{x} = 0, \quad \text{for all } i = 1, \ldots, m\}.$$

23. Complete the proof of Fact 5.1.5: The orthogonal complement $V^{\perp}$ of a subspace V of $\mathbb{R}^n$ is a subspace of $\mathbb{R}^n$ as well.

24. Complete the proof of Fact 5.1.6: Orthogonal projections are linear transformations.

25. a. Consider a vector $\vec{v}$ in $\mathbb{R}^n$, and a scalar k. Show that

$$\|k\vec{v}\| = |k| \, \|\vec{v}\|.$$

b. Show that if $\vec{v}$ is a nonzero vector in $\mathbb{R}^n$, then
$$\vec{u} = \frac{1}{\|\vec{v}\|}\vec{v} \text{ is a unit vector.}$$

26. Find the orthogonal projection of $\begin{bmatrix} 49 \\ 49 \\ 49 \end{bmatrix}$ onto the

subspace of $\mathbb{R}^3$ spanned by

$$\begin{bmatrix} 2 \\ 3 \\ 6 \end{bmatrix} \quad \text{and} \quad \begin{bmatrix} 3 \\ -6 \\ 2 \end{bmatrix}.$$

27. Find the orthogonal projection of $9\vec{e}_1$ onto the subspace of $\mathbb{R}^4$ spanned by

$$\begin{bmatrix} 2 \\ 2 \\ 1 \\ 0 \end{bmatrix} \quad \text{and} \quad \begin{bmatrix} -2 \\ 2 \\ 0 \\ 1 \end{bmatrix}.$$

28. Find the orthogonal projection of

$$\begin{bmatrix} 1 \\ 0 \\ 0 \\ 0 \end{bmatrix}$$

onto the subspace of $\mathbb{R}^4$ spanned by

$$\begin{bmatrix} 1 \\ 1 \\ 1 \\ 1 \end{bmatrix}, \quad \begin{bmatrix} 1 \\ 1 \\ -1 \\ -1 \end{bmatrix}, \quad \begin{bmatrix} 1 \\ -1 \\ -1 \\ 1 \end{bmatrix}.$$

29. Consider the orthonormal vectors $\vec{v}_1, \vec{v}_2, \vec{v}_3, \vec{v}_4, \vec{v}_5$ in $\mathbb{R}^{10}$. Find the length of the vector

$$\vec{x} = 7\vec{v}_1 - 3\vec{v}_2 + 2\vec{v}_3 + \vec{v}_4 - \vec{v}_5.$$

30. Consider a subspace V of $\mathbb{R}^n$ and a vector $\vec{x}$ in $\mathbb{R}^n$. Let $\vec{y} = \text{proj}_V \vec{x}$. What is the relationship between the

following quantities?

$$\|\vec{y}\|^2 \quad \text{and} \quad \vec{y} \cdot \vec{x}$$

31. Consider the orthonormal vectors $\vec{v}_1, \vec{v}_2, \ldots, \vec{v}_m$ in $\mathbb{R}^n$, and an arbitrary vector $\vec{x}$ in $\mathbb{R}^n$. What is the relationship between the following two quantities?

$$p = (\vec{v}_1 \cdot \vec{x})^2 + (\vec{v}_2 \cdot \vec{x})^2 + \cdots + (\vec{v}_m \cdot \vec{x})^2 \quad \text{and} \quad \|\vec{x}\|^2$$

When are the two quantities equal?

32. Consider two vectors $\vec{v}_1$ and $\vec{v}_2$ in $\mathbb{R}^n$. Form the matrix

$$G = \begin{bmatrix} \vec{v}_1 \cdot \vec{v}_1 & \vec{v}_1 \cdot \vec{v}_2 \\ \vec{v}_2 \cdot \vec{v}_1 & \vec{v}_2 \cdot \vec{v}_2 \end{bmatrix}.$$

For which choices of $\vec{v}_1$ and $\vec{v}_2$ is the matrix G invertible?

33. Among all the vectors in $\mathbb{R}^n$ whose components add up to 1, find the vector of minimal length. In the case $n = 2$, explain your solution geometrically.

34. Among all the unit vectors in $\mathbb{R}^n$, find the one for which the sum of the components is maximal. In the case $n = 2$, explain your answer geometrically, in terms of the unit circle and the level curves of the function $x_1 + x_2$.

35. Among all the unit vectors $\vec{u} = \begin{bmatrix} x \\ y \\ z \end{bmatrix}$ in $\mathbb{R}^3$, find the one

for which the sum $x + 2y + 3z$ is minimal.

36. There are three exams in your linear algebra class, and you theorize that your score in each exam (out of 100) will be numerically equal to the number of hours you study for that exam. The three exams count 20%, 30%, and 50%, respectively, towards the final grade. If your (modest) goal is to score 76% in the course, how many hours a, b, and c should you study for each of the three exams to minimize quantity $a^2 + b^2 + c^2$? This quadratic model reflects the fact that it may be four times as painful to study for 10 hours than for just 5 hours.

37. Consider a plane E in $\mathbb{R}^3$ with orthonormal basis $\vec{v}_1, \vec{v}_2$. Let $\vec{x}$ be a vector in $\mathbb{R}^3$. Find a formula for the reflection $R(\vec{x})$ of $\vec{x}$ in the plane E.

38. Consider three unit vectors $\vec{v}_1, \vec{v}_2$, and $\vec{v}_3$ in $\mathbb{R}^n$. We are told that $\vec{v}_1 \cdot \vec{v}_2 = \vec{v}_1 \cdot \vec{v}_3 = 1/2$. What are the possible values of $\vec{v}_2 \cdot \vec{v}_3$? What could the angle between the vectors $\vec{v}_2$ and $\vec{v}_3$ be? Give examples; draw sketches for the cases $n = 2$ and $n = 3$.

39. Can you find a line L in $\mathbb{R}^n$ and a vector $\vec{x}$ in $\mathbb{R}^n$ such that

$$\vec{x} \cdot \text{proj}_L \vec{x}$$

is negative? Explain, arguing algebraically.

5.2 GRAM–SCHMIDT PROCESS AND *QR* FACTORIZATION

In the previous section, we saw that it is sometimes useful to have an orthonormal basis of a subspace of $\mathbb{R}^n$. Now we will show how to construct such a basis. We present an algorithm that allows us to convert any basis $\vec{v}_1, \vec{v}_2, \ldots, \vec{v}_m$ of a subspace V of $\mathbb{R}^n$ into an *orthonormal* basis $\vec{w}_1, \vec{w}_2, \ldots, \vec{w}_m$ of V.

Let us first think about simple cases. If V is a line with basis $\vec{v}_1$, we can find an orthonormal basis $\vec{w}_1$ simply by dividing $\vec{v}_1$ by its length:

$$\vec{w}_1 = \frac{1}{\|\vec{v}_1\|}\vec{v}_1.$$

When V is a plane with basis $\vec{v}_1, \vec{v}_2$, we first divide the vector $\vec{v}_1$ by its length to get a unit vector

$$\vec{w}_1 = \frac{1}{\|\vec{v}_1\|}\vec{v}_1.$$

(See Figure 1.)

Now comes the crucial step: We have to find a vector in V orthogonal to $\vec{w}_1$. (Initially, we will not insist that this vector be a unit vector.) Recalling our work on orthogonal projections, we realize that $\vec{v}_2 - \text{proj}_L \vec{v}_2 = \vec{v}_2 - (\vec{w}_1 \cdot \vec{v}_2)\vec{w}_1$ is a natural choice, where L is the line spanned by $\vec{w}_1$. (See Figure 2.)

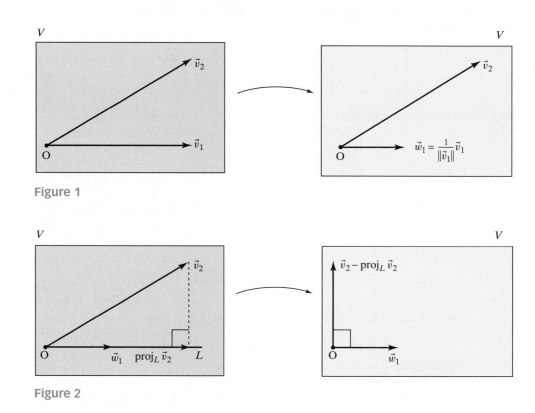

Figure 1

Figure 2

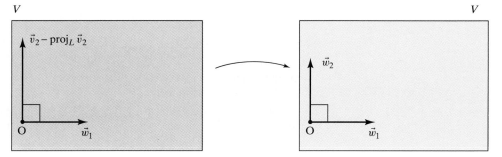

Figure 3

The last step is easy: We divide the vector $\vec{v}_2 - \text{proj}_L \vec{v}_2$ by its length to get the second vector $\vec{w}_2$ of an orthonormal basis. (See Figure 3.)

$$\vec{w}_2 = \frac{1}{\|\vec{v}_2 - \text{proj}_L \vec{v}_2\|}(\vec{v}_2 - \text{proj}_L \vec{v}_2)$$

EXAMPLE 1 Find an orthonormal basis of the subspace

$$V = \text{span}\left(\begin{bmatrix} 1 \\ 1 \\ 1 \\ 1 \end{bmatrix}, \begin{bmatrix} 1 \\ 9 \\ 9 \\ 1 \end{bmatrix}\right)$$

of $\mathbb{R}^4$, with basis

$$\vec{v}_1 = \begin{bmatrix} 1 \\ 1 \\ 1 \\ 1 \end{bmatrix}, \qquad \vec{v}_2 = \begin{bmatrix} 1 \\ 9 \\ 9 \\ 1 \end{bmatrix}.$$

Solution

Using the terminology just introduced, we find the following results:

$$\|\vec{v}_1\| = \sqrt{1^2 + 1^2 + 1^2 + 1^2} = 2,$$

$$\vec{w}_1 = \frac{1}{\|\vec{v}_1\|}\vec{v}_1 = \frac{1}{2}\begin{bmatrix} 1 \\ 1 \\ 1 \\ 1 \end{bmatrix} = \begin{bmatrix} 1/2 \\ 1/2 \\ 1/2 \\ 1/2 \end{bmatrix},$$

$$\vec{w}_1 \cdot \vec{v}_2 = \begin{bmatrix} 1/2 \\ 1/2 \\ 1/2 \\ 1/2 \end{bmatrix} \cdot \begin{bmatrix} 1 \\ 9 \\ 9 \\ 1 \end{bmatrix} = 10,$$

$$\vec{v}_2 - \text{proj}_L \vec{v}_2 = \vec{v}_2 - (\vec{w}_1 \cdot \vec{v}_2)\vec{w}_1 = \begin{bmatrix} 1 \\ 9 \\ 9 \\ 1 \end{bmatrix} - 10\begin{bmatrix} 1/2 \\ 1/2 \\ 1/2 \\ 1/2 \end{bmatrix} = \begin{bmatrix} -4 \\ 4 \\ 4 \\ -4 \end{bmatrix},$$

$$\|\vec{v}_2 - \text{proj}_L \vec{v}_2\| = \sqrt{4 \cdot 16} = 8,$$

and

$$\vec{w}_2 = \frac{1}{\|\vec{v}_2 - \text{proj}_L \vec{v}_2\|}(\vec{v}_2 - \text{proj}_L \vec{v}_2) = \frac{1}{8}\begin{bmatrix} -4 \\ 4 \\ 4 \\ -4 \end{bmatrix} = \begin{bmatrix} -1/2 \\ 1/2 \\ 1/2 \\ -1/2 \end{bmatrix}.$$

We have found an orthonormal basis of V:

$$\vec{w}_1 = \begin{bmatrix} 1/2 \\ 1/2 \\ 1/2 \\ 1/2 \end{bmatrix}, \qquad \vec{w}_2 = \begin{bmatrix} -1/2 \\ 1/2 \\ 1/2 \\ -1/2 \end{bmatrix}$$

We can represent the preceding computations more succinctly in matrix form. Let's solve the equations defining $\vec{w}_1$ and $\vec{w}_2$,

$$\vec{w}_1 = \frac{1}{\|\vec{v}_1\|}\vec{v}_1 \qquad \text{and} \qquad \vec{w}_2 = \frac{1}{\|\vec{v}_2 - \text{proj}_L \vec{v}_2\|}(\vec{v}_2 - \text{proj}_L \vec{v}_2),$$

for vectors $\vec{v}_1$ and $\vec{v}_2$:

$$\vec{v}_1 = \|\vec{v}_1\|\vec{w}_1, \qquad \text{and} \qquad \vec{v}_2 = \text{proj}_L \vec{v}_2 + \|\vec{v}_2 - \text{proj}_L \vec{v}_2\|\vec{w}_2$$
$$= (\vec{w}_1 \cdot \vec{v}_2)\vec{w}_1 + \|\vec{v}_2 - \text{proj}_L \vec{v}_2\|\vec{w}_2.$$

We can write the last two equations in matrix form:

$$\begin{bmatrix} \vec{v}_1 & \vec{v}_2 \end{bmatrix} = \underbrace{\begin{bmatrix} \vec{w}_1 & \vec{w}_2 \end{bmatrix}}_{Q} \underbrace{\begin{bmatrix} \|\vec{v}_1\| & \vec{w}_1 \cdot \vec{v}_2 \\ 0 & \|\vec{v}_2 - \text{proj}_L \vec{v}_2\| \end{bmatrix}}_{R}$$

It is customary to denote the matrices on the right-hand side by Q and R, respectively. Note that we have written the 4×2 matrix with columns $\vec{v}_1$ and $\vec{v}_2$ as the product of the 4×2 matrix Q with orthonormal columns and the upper triangular 2×2 matrix R with positive entries on the diagonal. Matrix Q stores the orthonormal basis $\vec{w}_1, \vec{w}_2$ we constructed, and matrix R gives the relationship between the "old" basis $\vec{v}_1, \vec{v}_2$, and the "new" basis $\vec{w}_1, \vec{w}_2$ of V.

Let's plug in the numbers (note that we computed all the entries of matrix R in the process of finding $\vec{w}_1$ and $\vec{w}_2$):

$$\begin{bmatrix} 1 & 1 \\ 1 & 9 \\ 1 & 9 \\ 1 & 1 \end{bmatrix} = \underbrace{\begin{bmatrix} 1/2 & -1/2 \\ 1/2 & 1/2 \\ 1/2 & 1/2 \\ 1/2 & -1/2 \end{bmatrix}}_{Q} \underbrace{\begin{bmatrix} 2 & 10 \\ 0 & 8 \end{bmatrix}}_{R} \qquad \blacksquare$$

Now that we know how to find an orthonormal basis of a plane, how would we proceed in the case of a three-dimensional subspace V of $\mathbb{R}^n$ with basis $\vec{v}_1, \vec{v}_2, \vec{v}_3$? We can first find an orthonormal basis $\vec{w}_1, \vec{w}_2$ of the plane $E = \text{span}(\vec{v}_1, \vec{v}_2)$, as illustrated in Example 1. Then, we consider the vector $\vec{v}_3 - \text{proj}_E \vec{v}_3$ and divide it by its length to get $\vec{w}_3$, as shown in Figure 4. (How do we know that $\vec{v}_3 - \text{proj}_E \vec{v}_3$ is nonzero?)

Recall from Fact 5.1.6 that

$$\vec{v}_3 - \text{proj}_E \vec{v}_3 = \vec{v}_3 - (\vec{w}_1 \cdot \vec{v}_3)\vec{w}_1 - (\vec{w}_2 \cdot \vec{v}_3)\vec{w}_2.$$

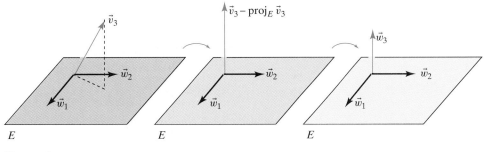

Figure 4

Using the same method, we can construct an orthonormal basis of any subspace of $\mathbb{R}^n$. Unfortunately, the notation gets a bit heavy in the general case.

Algorithm 5.2.1

The Gram–Schmidt process[4]

Consider a subspace V of $\mathbb{R}^n$ with basis $\vec{v}_1, \vec{v}_2, \ldots, \vec{v}_m$. We wish to construct an *orthonormal* basis $\vec{w}_1, \vec{w}_2, \ldots, \vec{w}_m$ of V.

Let $\vec{w}_1 = (1/\|\vec{v}_1\|)\vec{v}_1$. As we define $\vec{w}_j$ for $j = 2, 3, \ldots, m$, we may assume that an orthonormal basis $\vec{w}_1, \vec{w}_2, \ldots, \vec{w}_{j-1}$ of $V_{j-1} = \text{span}\,(\vec{v}_1, \vec{v}_2, \ldots, \vec{v}_{j-1})$ has already been constructed. Let

$$\vec{w}_j = \frac{1}{\|\vec{v}_j - \text{proj}_{V_{j-1}}\vec{v}_j\|}(\vec{v}_j - \text{proj}_{V_{j-1}}\vec{v}_j).$$

Note that

$$\text{proj}_{V_{j-1}}\vec{v}_j = (\vec{w}_1 \cdot \vec{v}_j)\vec{w}_1 + (\vec{w}_2 \cdot \vec{v}_j)\vec{w}_2 + \cdots + (\vec{w}_{j-1} \cdot \vec{v}_j)\vec{w}_{j-1},$$

by Fact 5.1.6.

If you are confused by these formulas, go back to the cases where V is a two- or three-dimensional space.

The *QR* Factorization

The Gram–Schmidt process can be presented succinctly in matrix form, as illustrated in Example 1. Using the terminology introduced in Algorithm 5.2.1, we can write

$$\vec{v}_1 = \|\vec{v}_1\|\vec{w}_1$$

and

$$\begin{aligned}
\vec{v}_j &= \text{proj}_{V_{j-1}}\vec{v}_j + \|\vec{v}_j - \text{proj}_{V_{j-1}}\vec{v}_j\|\vec{w}_j \\
&= (\vec{w}_1 \cdot \vec{v}_j)\vec{w}_1 + \cdots + (\vec{w}_{j-1} \cdot \vec{v}_j)\vec{w}_{j-1} + \|\vec{v}_j - \text{proj}_{V_{j-1}}\vec{v}_j\|\vec{w}_j
\end{aligned}$$

$$(\text{for } j = 2, 3, \ldots, m).$$

---

[4]Named after the Danish actuary Jörgen Gram (1850–1916) and the German mathematician Erhardt Schmidt (1876–1959).

Let $r_{11} = \|\vec{v}_1\|$, $r_{jj} = \|\vec{v}_j - \text{proj}_{V_{j-1}} \vec{v}_j\|$ (for $j = 2, 3, \ldots, m$), and $r_{ij} = \vec{w}_i \cdot \vec{v}_j$, (for $i < j$). Then,

$$\vec{v}_1 = r_{11}\vec{w}_1$$
$$\vec{v}_2 = r_{12}\vec{w}_1 + r_{22}\vec{w}_2$$
$$\vdots$$
$$\vec{v}_m = r_{1m}\vec{w}_1 + r_{2m}\vec{w}_2 + \cdots + r_{mm}\vec{w}_m.$$

We can write these equations in matrix form:

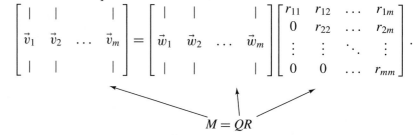

$$M = QR$$

Note that M is an $n \times m$ matrix with linearly independent columns, Q is an $n \times m$ matrix with orthonormal columns, and R is an upper triangular $m \times m$ matrix with positive entries on the diagonal.

Fact 5.2.2

_QR_ factorization

Consider an $n \times m$ matrix M with linearly independent columns $\vec{v}_1, \ldots, \vec{v}_m$. Then there is an $n \times m$ matrix Q whose columns $\vec{w}_1, \vec{w}_2, \ldots, \vec{w}_m$ are orthonormal and an upper triangular $m \times m$ matrix R with positive diagonal entries such that

$$M = QR.$$

This representation is unique. Furthermore,

$$r_{11} = \|\vec{v}_1\|, \quad r_{jj} = \|\vec{v}_j - \text{proj}_{V_{j-1}} \vec{v}_j\| \text{ (for } j > 1), \quad \text{and} \quad r_{ij} = \vec{w}_i \cdot \vec{v}_j \text{ (for } i < j),$$

where $V_{j-1} = \text{span}(\vec{v}_1, \vec{v}_2, \ldots, \vec{v}_{j-1})$.

Take another look at Example 1, where $L = V_1 = \text{span}(\vec{v}_1)$.

The verification of the uniqueness of the QR factorization is left as Exercise 5.3.31. To find the QR factorization of a matrix M, we perform the Gram–Schmidt process on the columns of M, constructing Q and R column by column. No extra computations are required: All the information necessary to build Q and R is provided by the Gram–Schmidt process. QR factorization is an effective way to organize and record the work performed in the Gram–Schmidt process; it is useful for many computational and theoretical purposes.

EXAMPLE 2 Find the QR factorization of the shear matrix $M = \begin{bmatrix} 1 & 0 \\ 1 & 1 \end{bmatrix}$.

Solution

Here

$$\vec{v}_1 = \begin{bmatrix} 1 \\ 1 \end{bmatrix}, \qquad \vec{v}_2 = \begin{bmatrix} 0 \\ 1 \end{bmatrix}.$$

As in Example 1, the QR factorization of M will have the form

$$M = \begin{bmatrix} \vec{v}_1 & \vec{v}_2 \end{bmatrix} = \underbrace{\begin{bmatrix} \vec{w}_1 & \vec{w}_2 \end{bmatrix}}_{Q} \underbrace{\begin{bmatrix} \|\vec{v}_1\| & \vec{w}_1 \cdot \vec{v}_2 \\ 0 & \|\vec{v}_2 - \mathrm{proj}_{V_1}\vec{v}_2\| \end{bmatrix}}_{R}.$$

We will compute the columns of Q and the entries of R step by step:

$$r_{11} = \|\vec{v}_1\| = \sqrt{2}$$

$$\vec{w}_1 = \frac{1}{\|\vec{v}_1\|}\vec{v}_1 = \frac{1}{\sqrt{2}}\begin{bmatrix} 1 \\ 1 \end{bmatrix}$$

$$r_{12} = \vec{w}_1 \cdot \vec{v}_2 = \frac{1}{\sqrt{2}}\begin{bmatrix} 1 \\ 1 \end{bmatrix} \cdot \begin{bmatrix} 0 \\ 1 \end{bmatrix} = \frac{1}{\sqrt{2}}$$

$$\vec{v}_2 - \mathrm{proj}_{V_1}\vec{v}_2 = \vec{v}_2 - (\vec{w}_1 \cdot \vec{v}_2)\vec{w}_1 = \begin{bmatrix} 0 \\ 1 \end{bmatrix} - \frac{1}{\sqrt{2}}\frac{1}{\sqrt{2}}\begin{bmatrix} 1 \\ 1 \end{bmatrix} = \begin{bmatrix} -1/2 \\ 1/2 \end{bmatrix}$$

$$r_{22} = \|\vec{v}_2 - \mathrm{proj}_{V_1}\vec{v}_2\| = \sqrt{\frac{1}{4}+\frac{1}{4}} = \frac{1}{\sqrt{2}}$$

$$\vec{w}_2 = \frac{1}{\|\vec{v}_2 - \mathrm{proj}_{V_1}\vec{v}_2\|}(\vec{v}_2 - \mathrm{proj}_{V_1}\vec{v}_2) = \sqrt{2}\begin{bmatrix} -1/2 \\ 1/2 \end{bmatrix} = \frac{1}{\sqrt{2}}\begin{bmatrix} -1 \\ 1 \end{bmatrix}$$

Now,

$$\begin{bmatrix} 1 & 0 \\ 1 & 1 \end{bmatrix} = M = QR = \begin{bmatrix} \vec{w}_1 & \vec{w}_2 \end{bmatrix}\begin{bmatrix} r_{11} & r_{12} \\ 0 & r_{22} \end{bmatrix} = \underbrace{\left(\frac{1}{\sqrt{2}}\begin{bmatrix} 1 & -1 \\ 1 & 1 \end{bmatrix}\right)}_{Q}\underbrace{\left(\frac{1}{\sqrt{2}}\begin{bmatrix} 2 & 1 \\ 0 & 1 \end{bmatrix}\right)}_{R}.$$

Draw pictures analogous to Figures 1 through 3 to illustrate these computations! ■

EXERCISES

GOALS Perform the Gram–Schmidt process, and thus find the QR factorization of a matrix.

Using paper and pencil, perform the Gram–Schmidt process on the sequences of vectors given in Exercises 1 through 14.

1. $\begin{bmatrix} 2 \\ 1 \\ -2 \end{bmatrix}$

2. $\begin{bmatrix} 6 \\ 3 \\ 2 \end{bmatrix}$, $\begin{bmatrix} 2 \\ -6 \\ 3 \end{bmatrix}$

3. $\begin{bmatrix} 4 \\ 0 \\ 3 \end{bmatrix}$, $\begin{bmatrix} 25 \\ 0 \\ -25 \end{bmatrix}$

4. $\begin{bmatrix} 4 \\ 0 \\ 3 \end{bmatrix}$, $\begin{bmatrix} 25 \\ 0 \\ -25 \end{bmatrix}$, $\begin{bmatrix} 0 \\ -2 \\ 0 \end{bmatrix}$

5. $\begin{bmatrix} 2 \\ 2 \\ 1 \end{bmatrix}$, $\begin{bmatrix} 1 \\ 1 \\ 5 \end{bmatrix}$

6. $\begin{bmatrix} 2 \\ 0 \\ 0 \end{bmatrix}$, $\begin{bmatrix} 3 \\ 4 \\ 0 \end{bmatrix}$, $\begin{bmatrix} 5 \\ 6 \\ 7 \end{bmatrix}$

7. $\begin{bmatrix} 2 \\ 2 \\ 1 \end{bmatrix}$, $\begin{bmatrix} -2 \\ 1 \\ 2 \end{bmatrix}$, $\begin{bmatrix} 18 \\ 0 \\ 0 \end{bmatrix}$

8. $\begin{bmatrix} 5 \\ 4 \\ 2 \\ 2 \end{bmatrix}$, $\begin{bmatrix} 3 \\ 6 \\ 7 \\ -2 \end{bmatrix}$

9. $\begin{bmatrix} 1 \\ 1 \\ 1 \\ 1 \end{bmatrix}$, $\begin{bmatrix} 1 \\ 9 \\ -5 \\ 3 \end{bmatrix}$

10. $\begin{bmatrix} 1 \\ 1 \\ 1 \\ 1 \end{bmatrix}$, $\begin{bmatrix} 6 \\ 4 \\ 6 \\ 4 \end{bmatrix}$

11. $\begin{bmatrix} 4 \\ 0 \\ 0 \\ 3 \end{bmatrix}$, $\begin{bmatrix} 5 \\ 2 \\ 14 \\ 10 \end{bmatrix}$

12. $\begin{bmatrix} 2 \\ 3 \\ 0 \\ 6 \end{bmatrix}$, $\begin{bmatrix} 4 \\ 4 \\ 2 \\ 13 \end{bmatrix}$

13. $\begin{bmatrix} 1 \\ 1 \\ 1 \\ 1 \end{bmatrix}$, $\begin{bmatrix} 1 \\ 0 \\ 0 \\ 1 \end{bmatrix}$, $\begin{bmatrix} 0 \\ 2 \\ 1 \\ -1 \end{bmatrix}$

14. $\begin{bmatrix} 1 \\ 7 \\ 1 \\ 7 \end{bmatrix}$, $\begin{bmatrix} 0 \\ 7 \\ 2 \\ 7 \end{bmatrix}$, $\begin{bmatrix} 1 \\ 8 \\ 1 \\ 6 \end{bmatrix}$

Using paper and pencil, find the QR factorizations of the matrices in Exercises 15 through 28. (Compare with Exercises 1 through 14.)

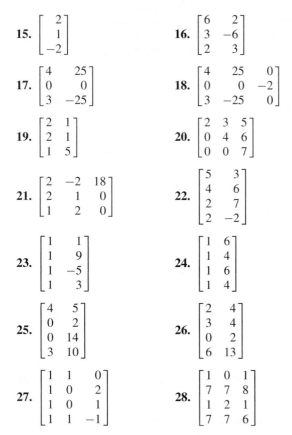

15. $\begin{bmatrix} 2 \\ 1 \\ -2 \end{bmatrix}$

16. $\begin{bmatrix} 6 & 2 \\ 3 & -6 \\ 2 & 3 \end{bmatrix}$

17. $\begin{bmatrix} 4 & 25 \\ 0 & 0 \\ 3 & -25 \end{bmatrix}$

18. $\begin{bmatrix} 4 & 25 & 0 \\ 0 & 0 & -2 \\ 3 & -25 & 0 \end{bmatrix}$

19. $\begin{bmatrix} 2 & 1 \\ 2 & 1 \\ 1 & 5 \end{bmatrix}$

20. $\begin{bmatrix} 2 & 3 & 5 \\ 0 & 4 & 6 \\ 0 & 0 & 7 \end{bmatrix}$

21. $\begin{bmatrix} 2 & -2 & 18 \\ 2 & 1 & 0 \\ 1 & 2 & 0 \end{bmatrix}$

22. $\begin{bmatrix} 5 & 3 \\ 4 & 6 \\ 2 & 7 \\ 2 & -2 \end{bmatrix}$

23. $\begin{bmatrix} 1 & 1 \\ 1 & 9 \\ 1 & -5 \\ 1 & 3 \end{bmatrix}$

24. $\begin{bmatrix} 1 & 6 \\ 1 & 4 \\ 1 & 6 \\ 1 & 4 \end{bmatrix}$

25. $\begin{bmatrix} 4 & 5 \\ 0 & 2 \\ 0 & 14 \\ 3 & 10 \end{bmatrix}$

26. $\begin{bmatrix} 2 & 4 \\ 3 & 4 \\ 0 & 2 \\ 6 & 13 \end{bmatrix}$

27. $\begin{bmatrix} 1 & 1 & 0 \\ 1 & 0 & 2 \\ 1 & 0 & 1 \\ 1 & 1 & -1 \end{bmatrix}$

28. $\begin{bmatrix} 1 & 0 & 1 \\ 7 & 7 & 8 \\ 1 & 2 & 1 \\ 7 & 7 & 6 \end{bmatrix}$

29. Perform the Gram–Schmidt process on the following basis of $\mathbb{R}^2$:
$$\vec{v}_1 = \begin{bmatrix} -3 \\ 4 \end{bmatrix}, \qquad \vec{v}_2 = \begin{bmatrix} 1 \\ 7 \end{bmatrix}$$
Illustrate your work with sketches, as in Figures 1 through 3 of this section.

30. Consider two linearly independent vectors $\vec{v}_1 = \begin{bmatrix} a \\ b \end{bmatrix}$ and $\vec{v}_2 = \begin{bmatrix} c \\ d \end{bmatrix}$ in $\mathbb{R}^2$. Draw sketches (as in Figures 1 through 3 of this section) to illustrate the Gram–Schmidt process for $\vec{v}_1$, $\vec{v}_2$. You need not perform the process algebraically.

31. Perform the Gram–Schmidt process on the following basis of $\mathbb{R}^3$:
$$\vec{v}_1 = \begin{bmatrix} a \\ 0 \\ 0 \end{bmatrix}, \qquad \vec{v}_2 = \begin{bmatrix} b \\ c \\ 0 \end{bmatrix}, \qquad \vec{v}_3 = \begin{bmatrix} d \\ e \\ f \end{bmatrix}$$

Here, a, c, and f are positive constants, and the other constants are arbitrary. Illustrate your work with a sketch, as in Figure 4 of this section.

32. Find an orthonormal basis of the plane
$$x_1 + x_2 + x_3 = 0.$$

33. Find an orthonormal basis of the kernel of the matrix
$$A = \begin{bmatrix} 1 & 1 & 1 & 1 \\ 1 & -1 & -1 & 1 \end{bmatrix}.$$

34. Find an orthonormal basis of the kernel of the matrix
$$A = \begin{bmatrix} 1 & 1 & 1 & 1 \\ 1 & 2 & 3 & 4 \end{bmatrix}.$$

35. Find an orthonormal basis of the image of the matrix
$$A = \begin{bmatrix} 1 & 2 & 1 \\ 2 & 1 & 1 \\ 2 & -2 & 0 \end{bmatrix}.$$

36. Consider the matrix
$$M = \frac{1}{2} \begin{bmatrix} 1 & 1 & 1 \\ 1 & -1 & -1 \\ 1 & -1 & 1 \\ 1 & 1 & -1 \end{bmatrix} \begin{bmatrix} 2 & 3 & 5 \\ 0 & -4 & 6 \\ 0 & 0 & 7 \end{bmatrix}.$$
Find the QR factorization of M.

37. Consider the matrix
$$M = \frac{1}{2} \begin{bmatrix} 1 & 1 & 1 & 1 \\ 1 & -1 & -1 & 1 \\ 1 & -1 & 1 & -1 \\ 1 & 1 & -1 & -1 \end{bmatrix} \begin{bmatrix} 3 & 4 \\ 0 & 5 \\ 0 & 0 \\ 0 & 0 \end{bmatrix}.$$
Find the QR factorization of M.

38. Find the QR factorization of
$$A = \begin{bmatrix} 0 & -3 & 0 \\ 0 & 0 & 0 \\ 2 & 0 & 0 \\ 0 & 0 & 4 \end{bmatrix}.$$

39. Find an orthonormal basis $\vec{w}_1$, $\vec{w}_2$, $\vec{w}_3$ of $\mathbb{R}^3$ such that
$$\text{span}(\vec{w}_1) = \text{span}\left(\begin{bmatrix} 1 \\ 2 \\ 3 \end{bmatrix} \right)$$
and
$$\text{span}(\vec{w}_1, \vec{w}_2) = \text{span}\left(\begin{bmatrix} 1 \\ 2 \\ 3 \end{bmatrix}, \begin{bmatrix} 1 \\ 1 \\ -1 \end{bmatrix} \right).$$

40. Consider an invertible $n \times n$ matrix A whose columns are orthogonal, but not necessarily orthonormal. What does the QR factorization of A look like?

41. Consider an invertible upper triangular $n \times n$ matrix A. What does the QR factorization of A look like?

42. The two column vectors $\vec{v}_1$ and $\vec{v}_2$ of a 2×2 matrix A are shown in the accompanying figure. Let $A = QR$ be the QR factorization of A. Represent the diagonal entries r_{11} and r_{22} of R as lengths in the figure. Interpret the product $r_{11}r_{22}$ as an area.

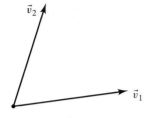

43. Consider a partitioned matrix

$$A = [A_1 \quad A_2]$$

with linearly independent columns. (A_1 is an $n \times m_1$ matrix, and A_2 is $n \times m_2$.) Suppose you know the QR factorization of A. Explain how this allows you to find the QR factorization of A_1.

44. Consider an $n \times m$ matrix A with rank$(A) < m$. Is it always possible to write

$$A = QR,$$

where Q is an $n \times m$ matrix with orthonormal columns and R is upper triangular? Explain.

45. Consider an $n \times m$ matrix A with rank$(A) = m$. Is it always possible to write A as

$$A = QL,$$

where Q is an $n \times m$ matrix with orthonormal columns and L is a lower triangular $m \times m$ matrix with positive diagonal entries? Explain.

46. Several of my students have observed over the years that what we are doing in Sections 5.1 and 5.2 is circular. In Section 5.1, we derive a formula for the orthogonal projection onto a subspace V, assuming that we have an orthonormal basis of V already (see Fact 5.1.6); in this section, we use just that projection formula to *construct* an orthonormal basis (the Gram–Schmidt process, Algorithm 5.2.1). What exactly is going on here?

5.3 ORTHOGONAL TRANSFORMATIONS AND ORTHOGONAL MATRICES

In geometry, we are particularly interested in those linear transformations that preserve the length of vectors.

Definition 5.3.1

> **Orthogonal transformations and orthogonal matrices**
> A linear transformation T from $\mathbb{R}^n$ to $\mathbb{R}^n$ is called *orthogonal* if it preserves the length of vectors:
>
> $$\|T(\vec{x})\| = \|\vec{x}\|, \quad \text{for all } \vec{x} \text{ in } \mathbb{R}^n.$$
>
> If $T(\vec{x}) = A\vec{x}$ is an orthogonal transformation, we say that A is an *orthogonal matrix*.

EXAMPLE 1 The rotation

$$T(\vec{x}) = \begin{bmatrix} \cos\phi & -\sin\phi \\ \sin\phi & \cos\phi \end{bmatrix} \vec{x}$$

is an orthogonal transformation from $\mathbb{R}^2$ to $\mathbb{R}^2$, and

$$A = \begin{bmatrix} \cos\phi & -\sin\phi \\ \sin\phi & \cos\phi \end{bmatrix}$$

is an orthogonal matrix, for all angles ϕ. ∎

EXAMPLE 2 Consider a subspace V of $\mathbb{R}^n$. For a vector $\vec{x}$ in $\mathbb{R}^n$, the vector $R(\vec{x}) = 2\operatorname{proj}_V \vec{x} - \vec{x}$ is called the reflection of $\vec{x}$ in V. (Compare this with Definition 2.2.6; see Figure 1). Show that reflections are orthogonal transformations.

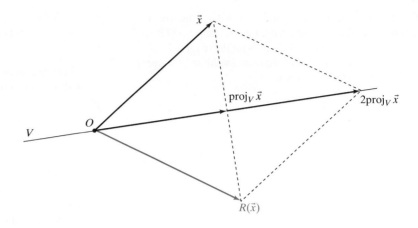

Figure 1

Solution

We can write $R(\vec{x}) = \text{proj}_V \vec{x} + (\text{proj}_V \vec{x} - \vec{x})$ and $\vec{x} = \text{proj}_V \vec{x} + (\vec{x} - \text{proj}_V \vec{x})$. By the Pythagorean theorem, we have

$$\|R(\vec{x})\|^2 = \|\text{proj}_V \vec{x}\|^2 + \|\text{proj}_V \vec{x} - \vec{x}\|^2$$
$$= \|\text{proj}_V \vec{x}\|^2 + \|\vec{x} - \text{proj}_V \vec{x}\|^2 = \|\vec{x}\|^2. \qquad \blacksquare$$

As the name suggests, orthogonal transformations preserve right angles. In fact, orthogonal transformations preserve all angles. (See Exercise 3.)

Fact 5.3.2   **Orthogonal transformations preserve orthogonality**
Consider an orthogonal transformation T from $\mathbb{R}^n$ to $\mathbb{R}^n$. If the vectors $\vec{v}$ and $\vec{w}$ in $\mathbb{R}^n$ are orthogonal, then so are $T(\vec{v})$ and $T(\vec{w})$.

Proof   By the theorem of Pythagoras, we have to show that

$$\|T(\vec{v}) + T(\vec{w})\|^2 = \|T(\vec{v})\|^2 + \|T(\vec{w})\|^2.$$

Let's see:

$$\begin{aligned}
\|T(\vec{v}) + T(\vec{w})\|^2 &= \|T(\vec{v} + \vec{w})\|^2 & (T \text{ is linear}) \\
&= \|\vec{v} + \vec{w}\|^2 & (T \text{ is orthogonal}) \\
&= \|\vec{v}\|^2 + \|\vec{w}\|^2 & (\vec{v} \text{ and } \vec{w} \text{ are orthogonal}) \\
&= \|T(\vec{v})\|^2 + \|T(\vec{w})\|^2 & (T \text{ is orthogonal}). \qquad \blacktriangle
\end{aligned}$$

Fact 5.3.2 is perhaps better explained with a sketch. (See Figure 2.)
The two shaded triangles are congruent, because corresponding sides are the same length (since T preserves length). Since D_1 is a right triangle, so is D_2.

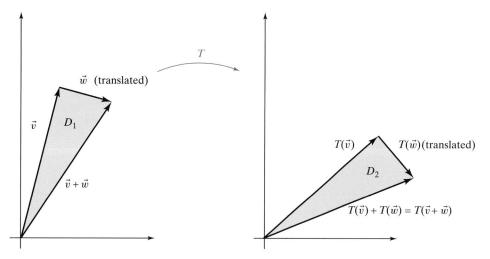

Figure 2

Here is an alternative characterization of orthogonal transformations:

Fact 5.3.3

Orthogonal transformations and orthonormal bases

a. A linear transformation T from $\mathbb{R}^n$ to $\mathbb{R}^n$ is orthogonal if (and only if) the vectors $T(\vec{e}_1), T(\vec{e}_2), \ldots, T(\vec{e}_n)$ form an orthonormal basis of $\mathbb{R}^n$.
b. An $n \times n$ matrix A is orthogonal if (and only if) its columns form an orthonormal basis of $\mathbb{R}^n$.

Figure 3 illustrates part (a) for a linear transformation from $\mathbb{R}^2$ to $\mathbb{R}^2$.

Proof   We prove part (a); part (b) then follows from Fact 2.1.2. If T is orthogonal, then, by definition, the $T(\vec{e}_i)$ are unit vectors, and by Fact 5.3.2, they are orthogonal. Conversely, suppose the $T(\vec{e}_i)$ form an orthonormal basis. Consider a vector

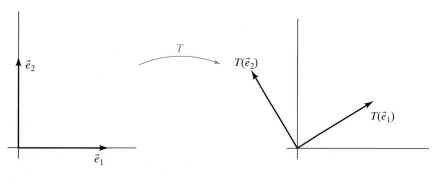

Figure 3

$\vec{x} = x_1\vec{e}_1 + x_2\vec{e}_2 + \cdots + x_n\vec{e}_n$ in $\mathbb{R}^n$. Then,

$$
\begin{aligned}
\|T(\vec{x})\|^2 &= \|x_1 T(\vec{e}_1) + x_2 T(\vec{e}_2) + \cdots + x_n T(\vec{e}_n)\|^2 \\
&= \|x_1 T(\vec{e}_1)\|^2 + \|x_2 T(\vec{e}_2)\|^2 + \cdots + \|x_n T(\vec{e}_n)\|^2 \qquad \text{(by Pythagoras)} \\
&= x_1^2 + x_2^2 + \cdots + x_n^2 \\
&= \|\vec{x}\|^2.
\end{aligned}
$$

▲

Warning: A matrix with orthogonal columns need not be an orthogonal matrix. As an example, consider the matrix $A = \begin{bmatrix} 4 & -3 \\ 3 & 4 \end{bmatrix}$.

EXAMPLE 3   Show that the matrix A is orthogonal:

$$
A = \frac{1}{2} \begin{bmatrix} 1 & -1 & -1 & -1 \\ 1 & -1 & 1 & 1 \\ 1 & 1 & -1 & 1 \\ 1 & 1 & 1 & -1 \end{bmatrix}.
$$

Solution

Check that the columns of A form an orthonormal basis of $\mathbb{R}^4$.   ■

Here are some algebraic properties of orthogonal matrices.

Fact 5.3.4   **Products and inverses of orthogonal matrices**

a. The product AB of two orthogonal $n \times n$ matrices A and B is orthogonal.
b. The inverse A^{-1} of an orthogonal $n \times n$ matrix A is orthogonal.

Proof   In part (a), the linear transformation $T(\vec{x}) = AB\vec{x}$ preserves length, because $\|T(\vec{x})\| = \|A(B\vec{x})\| = \|B\vec{x}\| = \|\vec{x}\|$. In part (b), the linear transformation $T(\vec{x}) = A^{-1}\vec{x}$ preserves length, because $\|A^{-1}\vec{x}\| = \|A(A^{-1}\vec{x})\| = \|\vec{x}\|$. Figure 4 illustrates property (a).   ▲

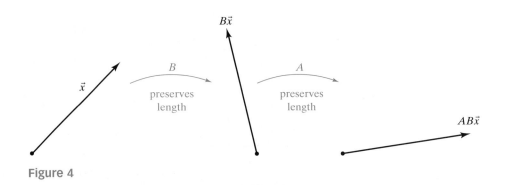

Figure 4

The Transpose of a Matrix

EXAMPLE 4 Consider the orthogonal matrix

$$A = \frac{1}{7}\begin{bmatrix} 2 & 6 & 3 \\ 3 & 2 & -6 \\ 6 & -3 & 2 \end{bmatrix}.$$

Form another 3×3 matrix B whose ijth entry is the jith entry of A:

$$B = \frac{1}{7}\begin{bmatrix} 2 & 3 & 6 \\ 6 & 2 & -3 \\ 3 & -6 & 2 \end{bmatrix}$$

Note that the rows of B correspond to the columns of A.
   Compute BA, and explain the result.

Solution

$$BA = \frac{1}{49}\begin{bmatrix} 2 & 3 & 6 \\ 6 & 2 & -3 \\ 3 & -6 & 2 \end{bmatrix}\begin{bmatrix} 2 & 6 & 3 \\ 3 & 2 & -6 \\ 6 & -3 & 2 \end{bmatrix} = \frac{1}{49}\begin{bmatrix} 49 & 0 & 0 \\ 0 & 49 & 0 \\ 0 & 0 & 49 \end{bmatrix} = I_3$$

This result is no coincidence: The ijth entry of BA is the dot product of the ith row of B and the jth column of A. By definition of B, this is just the dot product of the ith column of A and the jth column of A. Since A is orthogonal, this product is 1 if $i = j$ and 0 otherwise. ∎

   Let us generalize:

Definition 5.3.5

> **The transpose of a matrix; symmetric and skew-symmetric matrices**
> Consider an $m \times n$ matrix A.
>   The *transpose* A^T of A is the $n \times m$ matrix whose ijth entry is the jith entry of A: The roles of rows and columns are reversed.
>   We say that a square matrix A is *symmetric* if $A^T = A$, and A is called *skew-symmetric* if $A^T = -A$.

EXAMPLE 5 If $A = \begin{bmatrix} 1 & 2 & 3 \\ 9 & 7 & 5 \end{bmatrix}$, then $A^T = \begin{bmatrix} 1 & 9 \\ 2 & 7 \\ 3 & 5 \end{bmatrix}$. ∎

EXAMPLE 6 The *symmetric* 2×2 matrices are those of the form $A = \begin{bmatrix} a & b \\ b & c \end{bmatrix}$, for example, $A = \begin{bmatrix} 1 & 2 \\ 2 & 3 \end{bmatrix}$. The symmetric 2×2 matrices form a three-dimensional subspace of $\mathbb{R}^{2\times 2}$, with basis $\begin{bmatrix} 1 & 0 \\ 0 & 0 \end{bmatrix}, \begin{bmatrix} 0 & 1 \\ 1 & 0 \end{bmatrix}, \begin{bmatrix} 0 & 0 \\ 0 & 1 \end{bmatrix}.$

The *skew-symmetric* 2×2 matrices are those of the form $A = \begin{bmatrix} 0 & b \\ -b & 0 \end{bmatrix}$, for example, $A = \begin{bmatrix} 0 & 2 \\ -2 & 0 \end{bmatrix}$. These form a one-dimensional space with basis $\begin{bmatrix} 0 & 1 \\ -1 & 0 \end{bmatrix}$. ∎

Note that the transpose of a (column) vector $\vec{v}$ is a row vector: If

$$\vec{v} = \begin{bmatrix} 1 \\ 2 \\ 3 \end{bmatrix}, \quad \text{then} \quad \vec{v}^T = [1 \quad 2 \quad 3].$$

The transpose gives us a convenient way to express the dot product of two (column) vectors as a matrix product.

Fact 5.3.6   If $\vec{v}$ and $\vec{w}$ are two (column) vectors in $\mathbb{R}^n$, then
$$\vec{v} \cdot \vec{w} \;=\; \vec{v}^T \vec{w}.$$
dot product $\quad$ matrix product

For example,

$$\begin{bmatrix} 1 \\ 2 \\ 3 \end{bmatrix} \cdot \begin{bmatrix} 1 \\ -1 \\ 1 \end{bmatrix} = [1 \quad 2 \quad 3] \begin{bmatrix} 1 \\ -1 \\ 1 \end{bmatrix} = 2.$$

Now we can succinctly state the observation made in Example 4.

Fact 5.3.7   Consider an $n \times n$ matrix A. The matrix A is orthogonal if (and only if) $A^T A = I_n$ or, equivalently, if $A^{-1} = A^T$.

Proof   To justify this fact, write A in terms of its columns:

$$A = \begin{bmatrix} | & | & & | \\ \vec{v}_1 & \vec{v}_2 & \cdots & \vec{v}_n \\ | & | & & | \end{bmatrix}.$$

Then,

$$A^T A = \begin{bmatrix} - & \vec{v}_1^T & - \\ - & \vec{v}_2^T & - \\ & \vdots & \\ - & \vec{v}_n^T & - \end{bmatrix} \begin{bmatrix} | & | & & | \\ \vec{v}_1 & \vec{v}_2 & \cdots & \vec{v}_n \\ | & | & & | \end{bmatrix} = \begin{bmatrix} \vec{v}_1 \cdot \vec{v}_1 & \vec{v}_1 \cdot \vec{v}_2 & \cdots & \vec{v}_1 \cdot \vec{v}_n \\ \vec{v}_2 \cdot \vec{v}_1 & \vec{v}_2 \cdot \vec{v}_2 & \cdots & \vec{v}_2 \cdot \vec{v}_n \\ \vdots & \vdots & \ddots & \vdots \\ \vec{v}_n \cdot \vec{v}_1 & \vec{v}_n \cdot \vec{v}_2 & \cdots & \vec{v}_n \cdot \vec{v}_n \end{bmatrix}.$$

By Fact 5.3.3(b) this product is I_n if (and only if) A is orthogonal.   ▲

Later in this text, we will frequently work with matrices of the form $A^T A$. It is helpful to think of $A^T A$ as a table displaying the dot products $\vec{v}_i \cdot \vec{v}_j$ among the columns of A, as shown above.

We summarize the various characterizations we have found of orthogonal matrices.

Summary 5.3.8 **Orthogonal matrices**

Consider an $n \times n$ matrix A. Then, the following statements are equivalent:

i. A is an orthogonal matrix.
ii. The transformation $L(\vec{x}) = A\vec{x}$ preserves length, that is, $\|A\vec{x}\| = \|\vec{x}\|$ for all $\vec{x}$ in $\mathbb{R}^n$.
iii. The columns of A form an orthonormal basis of $\mathbb{R}^n$.
iv. $A^T A = I_n$.
v. $A^{-1} = A^T$.

Here are some algebraic properties of transposes:

Fact 5.3.9 **Properties of the transpose**

a. If A is an $m \times n$ matrix and B an $n \times p$ matrix, then
$$(AB)^T = B^T A^T.$$
Note the order of the factors.

b. If an $n \times n$ matrix A is invertible, then so is A^T, and
$$(A^T)^{-1} = (A^{-1})^T.$$

c. For any matrix A,
$$\operatorname{rank}(A) = \operatorname{rank}(A^T).$$

Proof   a. Compare entries:
$$ij\text{th entry of } (AB)^T = ji\text{th entry of } AB$$
$$= (j\text{th row of } A) \cdot (i\text{th column of } B)$$
$$ij\text{th entry of } B^T A^T = (i\text{th row of } B^T) \cdot (j\text{th column of } A^T)$$
$$= (i\text{th column of } B) \cdot (j\text{th row of } A)$$

b. We know that
$$AA^{-1} = I_n.$$
Transposing both sides and using part (a), we find that
$$(AA^{-1})^T = (A^{-1})^T A^T = I_n.$$

By Fact 2.4.9, it follows that

$$(A^{-1})^T = (A^T)^{-1}.$$

c. Consider the row space of A (i.e., the span of the rows of A). It is not hard to show that the dimension of this space is $\text{rank}(A)$ (see Exercises 49–52 in Section 3.3):

$$\text{rank}(A^T) = \text{dimension of the span of the columns of } A^T$$
$$= \text{dimension of the span of the rows of } A$$
$$= \text{rank}(A) \qquad \blacktriangle$$

The Matrix of an Orthogonal Projection

The transpose allows us to write a formula for the matrix of an orthogonal projection. Consider first the orthogonal projection

$$\text{proj}_L \vec{x} = (\vec{v}_1 \cdot \vec{x})\vec{v}_1$$

onto a line L in $\mathbb{R}^n$, where $\vec{v}_1$ is a unit vector in L. If we view the vector $\vec{v}_1$ as an $n \times 1$ matrix and the scalar $\vec{v}_1 \cdot \vec{x}$ as a 1×1 matrix, we can write

$$\text{proj}_L \vec{x} = \vec{v}_1 (\vec{v}_1 \cdot \vec{x})$$
$$= \vec{v}_1 \vec{v}_1^T \vec{x}$$
$$= M\vec{x},$$

where $M = \vec{v}_1 \vec{v}_1^T$. Note that $\vec{v}_1$ is an $n \times 1$ matrix and $\vec{v}_1^T$ is $1 \times n$, so that M is $n \times n$, as expected.

More generally, consider the projection

$$\text{proj}_V \vec{x} = (\vec{v}_1 \cdot \vec{x})\vec{v}_1 + \cdots + (\vec{v}_m \cdot \vec{x})\vec{v}_m$$

onto a subspace V of $\mathbb{R}^n$ with orthonormal basis $\vec{v}_1, \ldots, \vec{v}_m$. We can write

$$\text{proj}_V \vec{x} = \vec{v}_1 \vec{v}_1^T \vec{x} + \cdots + \vec{v}_m \vec{v}_m^T \vec{x}$$
$$= \left(\vec{v}_1 \vec{v}_1^T + \cdots + \vec{v}_m \vec{v}_m^T \right) \vec{x}$$
$$= \begin{bmatrix} | & & | \\ \vec{v}_1 & \cdots & \vec{v}_m \\ | & & | \end{bmatrix} \begin{bmatrix} - & \vec{v}_1^T & - \\ & \vdots & \\ - & \vec{v}_m^T & - \end{bmatrix} \vec{x}.$$

We have shown the following result:

Fact 5.3.10

The matrix of an orthogonal projection

Consider a subspace V of $\mathbb{R}^n$ with orthonormal basis $\vec{v}_1, \vec{v}_2, \ldots, \vec{v}_m$. The matrix of the orthogonal projection onto V is

$$AA^T, \quad \text{where} \quad A = \begin{bmatrix} | & | & & | \\ \vec{v}_1 & \vec{v}_2 & \cdots & \vec{v}_m \\ | & | & & | \end{bmatrix}.$$

Pay attention to the order of the factors (AA^T as opposed to $A^T A$).

EXAMPLE 7 Find the matrix of the orthogonal projection onto the subspace of $\mathbb{R}^4$ spanned by

$$\vec{v}_1 = \frac{1}{2} \begin{bmatrix} 1 \\ 1 \\ 1 \\ 1 \end{bmatrix}, \qquad \vec{v}_2 = \frac{1}{2} \begin{bmatrix} 1 \\ -1 \\ -1 \\ 1 \end{bmatrix}.$$

Solution

Note that the vectors $\vec{v}_1$ and $\vec{v}_2$ are orthonormal. Therefore, the matrix is

$$AA^T = \frac{1}{4} \begin{bmatrix} 1 & 1 \\ 1 & -1 \\ 1 & -1 \\ 1 & 1 \end{bmatrix} \begin{bmatrix} 1 & 1 & 1 & 1 \\ 1 & -1 & -1 & 1 \end{bmatrix} = \frac{1}{2} \begin{bmatrix} 1 & 0 & 0 & 1 \\ 0 & 1 & 1 & 0 \\ 0 & 1 & 1 & 0 \\ 1 & 0 & 0 & 1 \end{bmatrix}. \quad ■$$

EXERCISES

GOALS Use the various characterizations of orthogonal transformations and orthogonal matrices. Find the matrix of an orthogonal projection. Use the properties of the transpose.

1. Consider an $m \times n$ matrix A, a vector $\vec{v}$ in $\mathbb{R}^n$, and a vector $\vec{w}$ in $\mathbb{R}^m$. Show that

$$(A\vec{v}) \cdot \vec{w} = \vec{v} \cdot (A^T \vec{w}).$$

2. Consider an orthogonal transformation L from $\mathbb{R}^n$ to $\mathbb{R}^n$. Show that L preserves the dot product:

$$\vec{v} \cdot \vec{w} = L(\vec{v}) \cdot L(\vec{w}),$$

for all $\vec{v}$ and $\vec{w}$ in $\mathbb{R}^n$.

3. Show that an orthogonal transformation L from $\mathbb{R}^n$ to $\mathbb{R}^n$ preserves angles: The angle between two nonzero vectors $\vec{v}$ and $\vec{w}$ in $\mathbb{R}^n$ equals the angle between $L(\vec{v})$ and $L(\vec{w})$. Conversely, is any linear transformation that preserves angles orthogonal?

4. Consider a linear transformation L from $\mathbb{R}^n$ to $\mathbb{R}^m$ that preserves length. What can you say about the kernel of L? What is the dimension of the image? What can you say about the relationship between n and m? If A is the matrix of L, what can you say about the columns of A? What is $A^T A$? What about AA^T? Illustrate your answers with an example where $n = 2$ and $m = 3$.

5. If a matrix A is orthogonal, is A^2 orthogonal as well?

6. If a matrix A is orthogonal, is A^T orthogonal as well?

7. Are the *rows* of an orthogonal matrix A orthonormal?

8. a. Consider an $n \times m$ matrix A such that $A^T A = I_m$. Is it necessarily true that $AA^T = I_n$? Explain.

   b. Consider an $n \times n$ matrix A such that $A^T A = I_n$. Is it necessarily true that $AA^T = I_n$? Explain.

9. Find all orthogonal 2×2 matrices.

10. Find all orthogonal 3×3 matrices of the form

$$\begin{bmatrix} a & b & 0 \\ c & d & 1 \\ e & f & 0 \end{bmatrix}.$$

11. Find an orthogonal transformation T from $\mathbb{R}^3$ to $\mathbb{R}^3$ such that

$$T \begin{bmatrix} 2/3 \\ 2/3 \\ 1/3 \end{bmatrix} = \begin{bmatrix} 0 \\ 0 \\ 1 \end{bmatrix}.$$

12. Find an orthogonal matrix of the form

$$\begin{bmatrix} 2/3 & 1/\sqrt{2} & a \\ 2/3 & -1/\sqrt{2} & b \\ 1/3 & 0 & c \end{bmatrix}.$$

13. Is there an orthogonal transformation T from $\mathbb{R}^3$ to $\mathbb{R}^3$ such that

$$T \begin{bmatrix} 2 \\ 3 \\ 0 \end{bmatrix} = \begin{bmatrix} 3 \\ 0 \\ 2 \end{bmatrix} \quad \text{and} \quad T \begin{bmatrix} -3 \\ 2 \\ 0 \end{bmatrix} = \begin{bmatrix} 2 \\ -3 \\ 0 \end{bmatrix} ?$$

14. If A is any $m \times n$ matrix, show that the matrix $A^T A$ is symmetric. What about AA^T?

15. If two $n \times n$ matrices A and B are symmetric, is AB necessarily symmetric?

16. If an $n \times n$ matrix A is symmetric, is A^2 necessarily symmetric?

17. If an invertible $n \times n$ matrix A is symmetric, is A^{-1} necessarily symmetric?

18. a. Give an example of a (nonzero) skew-symmetric 3×3 matrix A and compute A^2.

    b. If an $n \times n$ matrix A is skew-symmetric, is matrix A^2 necessarily skew-symmetric as well? Or is A^2 necessarily symmetric?

19. Consider a line L in $\mathbb{R}^n$, spanned by a unit vector

$$\vec{v} = \begin{bmatrix} v_1 \\ v_2 \\ \vdots \\ v_n \end{bmatrix}.$$

Consider the matrix A of the orthogonal projection onto L. Describe the ijth entry of A, in terms of the components v_i of $\vec{v}$.

20. Consider the subspace W of $\mathbb{R}^4$ spanned by the vectors

$$\vec{v}_1 = \begin{bmatrix} 1 \\ 1 \\ 1 \\ 1 \end{bmatrix} \quad \text{and} \quad \vec{v}_2 = \begin{bmatrix} 1 \\ 9 \\ -5 \\ 3 \end{bmatrix}.$$

Find the matrix of the orthogonal projection onto W.

21. Find the matrix A of the orthogonal projection onto the line in $\mathbb{R}^n$ spanned by the vector

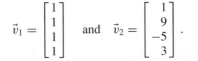

all n components are 1.

22. Let A be the matrix of an orthogonal projection. Find A^2 in two ways:

    a. Geometrically. (Consider what happens when you apply an orthogonal projection twice.)

    b. By computation, using the formula given in Fact 5.3.10.

23. Consider a unit vector $\vec{u}$ in $\mathbb{R}^3$. We define the matrices

$$A = 2\vec{u}\vec{u}^T - I_3 \quad \text{and} \quad B = I_3 - 2\vec{u}\vec{u}^T.$$

Describe the linear transformations defined by these matrices geometrically.

24. Consider an $m \times n$ matrix A. Find

$$\dim\left(\operatorname{im}(A)\right) + \dim\left(\ker(A^T)\right),$$

in terms of m and n.

25. For which $m \times n$ matrices A does the equation

$$\dim\left(\ker(A)\right) = \dim\left(\ker(A^T)\right)$$

hold? Explain.

26. Consider a QR factorization

$$M = QR.$$

Show that

$$R = Q^T M.$$

27. If $A = QR$ is a QR factorization, what is the relationship between $A^T A$ and $R^T R$?

28. Consider an invertible $n \times n$ matrix A. Can you write A as $A = LQ$, where L is a *lower* triangular matrix and Q is orthogonal? *Hint:* Consider the QR factorization of A^T.

29. Consider an invertible $n \times n$ matrix A. Can you write $A = RQ$, where R is an upper triangular matrix and Q is orthogonal?

30. a. Find all $n \times n$ matrices that are both orthogonal and upper triangular, with positive diagonal entries.

    b. Show that the QR factorization of an invertible $n \times n$ matrix is unique. *Hint:* If $A = Q_1 R_1 = Q_2 R_2$, then the matrix $Q_2^{-1} Q_1 = R_2 R_1^{-1}$ is both orthogonal and upper triangular, with positive diagonal entries.

31. a. Consider the matrix product $Q_1 = Q_2 S$, where both Q_1 and Q_2 are $n \times m$ matrices with orthonormal columns. Show that S is an orthogonal matrix. *Hint:* Compute $Q_1^T Q_1 = (Q_2 S)^T Q_2 S$. Note that $Q_1^T Q_1 = Q_2^T Q_2 = I_m$.

   b. Show that the QR factorization of an $n \times m$ matrix M is unique. *Hint:* If $M = Q_1 R_1 = Q_2 R_2$, then $Q_1 = Q_2 R_2 R_1^{-1}$. Now use part (a) and Exercise 30(a).

32. Find a basis of the space V of all symmetric 3×3 matrices, and thus determine the dimension of V.

33. Find a basis of the space V of all skew-symmetric 3×3 matrices, and thus determine the dimension of V.

34. Find the dimension of the space of all skew-symmetric $n \times n$ matrices.

35. Find the dimension of the space of all symmetric $n \times n$ matrices.

36. Is the transformation $L(A) = A^T$ from $\mathbb{R}^{2 \times 3}$ to $\mathbb{R}^{3 \times 2}$ linear? Is L an isomorphism?

37. Is the transformation $L(A) = A^T$ from $\mathbb{R}^{m \times n}$ to $\mathbb{R}^{n \times m}$ linear? Is L an isomorphism?

38. Find image and kernel of the linear transformation $L(A) = A + A^T$ from $\mathbb{R}^{n \times n}$ to $\mathbb{R}^{n \times n}$. *Hint:* Think about symmetric and skew-symmetric matrices.

39. Find the image and kernel of the linear transformation $L(A) = A - A^T$ from $\mathbb{R}^{n \times n}$ to $\mathbb{R}^{n \times n}$. *Hint:* Think about symmetric and skew-symmetric matrices.

40. Find the matrix of the linear transformation $L(A) = A^T$ from $\mathbb{R}^{2 \times 2}$ to $\mathbb{R}^{2 \times 2}$ with respect to the basis
$$\begin{bmatrix} 1 & 0 \\ 0 & 0 \end{bmatrix}, \begin{bmatrix} 0 & 0 \\ 0 & 1 \end{bmatrix}, \begin{bmatrix} 0 & 1 \\ 1 & 0 \end{bmatrix}, \begin{bmatrix} 0 & 1 \\ -1 & 0 \end{bmatrix}.$$

41. Find the matrix of the linear transformation $L(A) = A - A^T$ from $\mathbb{R}^{2 \times 2}$ to $\mathbb{R}^{2 \times 2}$ with respect to the basis
$$\begin{bmatrix} 1 & 0 \\ 0 & 0 \end{bmatrix}, \begin{bmatrix} 0 & 0 \\ 0 & 1 \end{bmatrix}, \begin{bmatrix} 0 & 1 \\ 1 & 0 \end{bmatrix}, \begin{bmatrix} 0 & 1 \\ -1 & 0 \end{bmatrix}.$$

42. Consider the matrix
$$A = \begin{bmatrix} 1 & 1 & -1 \\ 3 & 2 & -5 \\ 2 & 2 & 0 \end{bmatrix}$$

with *LDU* factorization
$$A = \begin{bmatrix} 1 & 0 & 0 \\ 3 & 1 & 0 \\ 2 & 0 & 1 \end{bmatrix} \begin{bmatrix} 1 & 0 & 0 \\ 0 & -1 & 0 \\ 0 & 0 & 2 \end{bmatrix} \begin{bmatrix} 1 & 1 & -1 \\ 0 & 1 & 2 \\ 0 & 0 & 1 \end{bmatrix}.$$

Find the *LDU* factorization of A^T. (See Exercise 2.4.58(d).)

43. Consider a symmetric invertible $n \times n$ matrix A which admits an *LDU* factorization $A = LDU$. (See Exercises 58, 61, and 62 of Section 2.4.) Recall that this factorization is unique. (See Exercise 2.4.62.) Show that $U = L^T$. (This is sometimes called the LDL^T *factorization* of a symmetric matrix A.)

44. This exercise shows one way to define the *quaternions*, discovered in 1843 by the Irish mathematician Sir W. R. Hamilton (1805–1865). Consider the set H of all 4×4 matrices M of the form
$$M = \begin{bmatrix} p & -q & -r & -s \\ q & p & s & -r \\ r & -s & p & q \\ s & r & -q & p \end{bmatrix},$$

where p, q, r, s are arbitrary real numbers. We can write M more succinctly in partitioned form as
$$M = \begin{bmatrix} A & -B^T \\ B & A^T \end{bmatrix},$$

where A and B are rotation–dilation matrices.

   a. Show that H is closed under addition: If M and N are in H, then so is $M + N$.

   b. Show that H is closed under scalar multiplication: If M is in H and k is an arbitrary scalar, then kM is in H.

   c. Parts (a) and (b) show that H is a subspace of the linear space $\mathbb{R}^{4 \times 4}$. Find a basis of H, and thus determine the dimension of H.

   d. Show that H is closed under multiplication: If M and N are in H, then so is MN.

   e. Show that if M is in H, then so is M^T.

   f. For a matrix M in H, compute $M^T M$.

   g. Which matrices M in H are invertible? If a matrix M in H is invertible, is M^{-1} necessarily in H as well?

   h. If M and N are in H, does the equation $MN = NM$ always hold?

5.4 LEAST SQUARES AND DATA FITTING

In this section, we will present an important application of the ideas introduced in this chapter. First, we take another look at orthogonal complements and orthogonal projections.

Another Characterization of Orthogonal Complements

Consider a subspace $V = \operatorname{im}(A)$ of $\mathbb{R}^n$, where $A = \begin{bmatrix} \vec{v}_1 & \vec{v}_2 & \cdots & \vec{v}_m \end{bmatrix}$. Then,

$$
\begin{aligned}
V^\perp &= \{\vec{x} \text{ in } \mathbb{R}^n \colon \vec{v} \cdot \vec{x} = 0, \text{ for all } \vec{v} \text{ in } V\} \\
&= \{\vec{x} \text{ in } \mathbb{R}^n \colon \vec{v}_i \cdot \vec{x} = 0, \text{ for } i = 1, \ldots, m\} \\
&= \{\vec{x} \text{ in } \mathbb{R}^n \colon \vec{v}_i^T \vec{x} = 0, \text{ for } i = 1, \ldots, m\}.
\end{aligned}
$$

In other words, $V^\perp$ is the kernel of the matrix

$$
A^T = \begin{bmatrix} - & \vec{v}_1^T & - \\ - & \vec{v}_2^T & - \\ & \vdots & \\ - & \vec{v}_m^T & - \end{bmatrix}.
$$

Fact 5.4.1   For any matrix A,

$$
(\operatorname{im} A)^\perp = \ker(A^T).
$$

Here is a very simple example: consider the line

$$
V = \operatorname{im} \begin{bmatrix} 1 \\ 2 \\ 3 \end{bmatrix}.
$$

Then,

$$
V^\perp = \ker \begin{bmatrix} 1 & 2 & 3 \end{bmatrix}
$$

is the plane with equation $x_1 + 2x_2 + 3x_3 = 0$. (See Figure 1.)

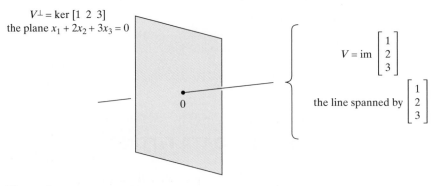

Figure 1

Here are some properties of the orthogonal complement that are obvious for subspaces of $\mathbb{R}^2$ and $\mathbb{R}^3$:

Fact 5.4.2 Consider a subspace V of $\mathbb{R}^n$. Then,

a. $\dim(V) + \dim(V^{\perp}) = n$,
b. $(V^{\perp})^{\perp} = V$,
c. $V \cap V^{\perp} = \{\vec{0}\}$.

Proof a. Let $T(\vec{x}) = \text{proj}_V\vec{x}$ be the orthogonal projection onto V. Note that $\text{im}(T) = V$ and $\ker(T) = V^{\perp}$. Fact 3.3.9 tells us that $n = \dim(\ker T) + \dim(\text{im } T) = \dim(V) + \dim(V^{\perp})$.

b. First observe that $V \subseteq (V^{\perp})^{\perp}$, since a vector in V is orthogonal to every vector in $V^{\perp}$ (by definition of $V^{\perp}$). Furthermore, the dimensions of the two spaces are equal, by part (a):

$$\dim(V^{\perp})^{\perp} = n - \dim(V^{\perp})$$
$$= n - (n - \dim(V))$$
$$= \dim(V).$$

It follows that the two spaces are equal. (See Exercise 3.3.41.)

c. If $\vec{x}$ is in V and in $V^{\perp}$, then $\vec{x}$ is orthogonal to itself; that is, $\vec{x} \cdot \vec{x} = \|\vec{x}\|^2 = 0$, and thus $\vec{x} = \vec{0}$. ▲

The following somewhat technical result will be useful later:

Fact 5.4.3 a. If A is an $m \times n$ matrix, then
$$\ker(A) = \ker(A^T A).$$

b. If A is an $m \times n$ matrix with $\ker(A) = \{\vec{0}\}$, then $A^T A$ is invertible.

Proof a. Clearly, the kernel of A is contained in the kernel of $A^T A$. Conversely, consider a vector $\vec{x}$ in the kernel of $A^T A$, so that $A^T A\vec{x} = \vec{0}$. Then, $A\vec{x}$ is in the image of A and in the kernel of A^T. Since $\ker(A^T)$ is the orthogonal complement of $\text{im}(A)$ by Fact 5.4.1, the vector $A\vec{x}$ is $\vec{0}$ by Fact 5.4.2(c), that is, $\vec{x}$ is in the kernel of A.

b. Note that $A^T A$ is an $n \times n$ matrix. By part (a), $\ker(A^T A) = \{\vec{0}\}$, and $A^T A$ is therefore invertible. (See Summary 3.3.11.) ▲

An Alternative Characterization of Orthogonal Projections

Fact 5.4.4 Consider a vector $\vec{x}$ in $\mathbb{R}^n$ and a subspace V of $\mathbb{R}^n$. Then, the orthogonal projection $\text{proj}_V\vec{x}$ is the vector in V *closest* to $\vec{x}$, in that
$$\|\vec{x} - \text{proj}_V\vec{x}\| < \|\vec{x} - \vec{v}\|,$$
for all $\vec{v}$ in V different from $\text{proj}_V\vec{x}$.

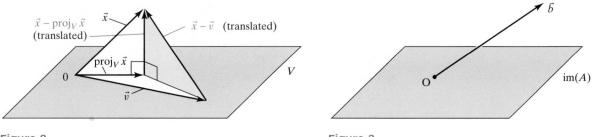

Figure 2   **Figure 3**

To justify this fact, apply the Pythagorean theorem to the shaded right triangle in Figure 2.

Least-Squares Approximations

Consider an *inconsistent* linear system $A\vec{x} = \vec{b}$. The fact that this system is inconsistent means that the vector $\vec{b}$ is not in the image of A. (See Figure 3.)

Although this system cannot be solved, we might be interested in finding a good approximate solution. We can try to find a vector $\vec{x}^*$ such that $A\vec{x}^*$ is "as close as possible" to $\vec{b}$. In other words, we try to minimize the *error* $\|\vec{b} - A\vec{x}\|$.

Definition 5.4.5

> **Least-squares solution**
> Consider a linear system
> $$A\vec{x} = \vec{b},$$
> where A is an $m \times n$ matrix. A vector $\vec{x}^*$ in $\mathbb{R}^n$ is called a *least-squares solution* of this system if $\|\vec{b} - A\vec{x}^*\| \leq \|\vec{b} - A\vec{x}\|$ for all $\vec{x}$ in $\mathbb{R}^n$.

See Figure 4.

The term "least-squares solution" reflects the fact that we are minimizing the sum of the squares of the components of the vector $\vec{b} - A\vec{x}$.

If the system $A\vec{x} = \vec{b}$ happens to be consistent, then the least-squares solutions are its exact solutions: The error $\|\vec{b} - A\vec{x}\|$ is zero.

How can we find the least-squares solutions of a linear system $A\vec{x} = \vec{b}$? Consider the following string of equivalent statements:

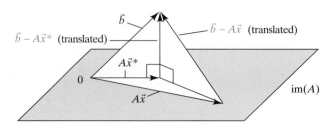

Figure 4

The vector $\vec{x}^*$ is a least-squares solution
of the system $A\vec{x} = \vec{b}$.

$\updownarrow$ Def. 5.4.5

$\|\vec{b} - A\vec{x}^*\| \leq \|\vec{b} - A\vec{x}\|$ for all $\vec{x}$ in $\mathbb{R}^n$.

$\updownarrow$ Fact 5.4.4

$A\vec{x}^* = \text{proj}_V \vec{b}$, where $V = \text{im}(A)$

$\updownarrow$ Facts 5.1.6 and 5.4.1

$\vec{b} - A\vec{x}^*$ is in $V^\perp = (\text{im } A)^\perp = \ker(A^T)$

$\updownarrow$

$A^T(\vec{b} - A\vec{x}^*) = \vec{0}$

$\updownarrow$

$A^T A\vec{x}^* = A^T\vec{b}$

Take another look at Figures 2 and 4.

Fact 5.4.6

The normal equation

The least-squares solutions of the system
$$A\vec{x} = \vec{b}$$
are the exact solutions of the (consistent) system
$$A^T A\vec{x} = A^T\vec{b}.$$
The system $A^T A\vec{x} = A^T\vec{b}$ is called the *normal equation* of $A\vec{x} = \vec{b}$.

The case when $\ker(A) = \{\vec{0}\}$ is of particular importance. Then, the matrix $A^T A$ is invertible (by Fact 5.4.3), and we can give a closed formula for the least-squares solution.

Fact 5.4.7

If $\ker(A) = \{\vec{0}\}$, then the linear system
$$A\vec{x} = \vec{b}$$
has the unique least-squares solution
$$\vec{x}^* = (A^T A)^{-1} A^T\vec{b}.$$

From a computational point of view, it may be more efficient to solve the normal equation $A^T A\vec{x} = A^T\vec{b}$ by Gauss–Jordan elimination, rather than by using Fact 5.4.7.

EXAMPLE 1   Use Fact 5.4.7 to find the least-squares solution $\vec{x}^*$ of the system

$$A\vec{x} = \vec{b}, \quad \text{where} \quad A = \begin{bmatrix} 1 & 1 \\ 1 & 2 \\ 1 & 3 \end{bmatrix} \quad \text{and} \quad \vec{b} = \begin{bmatrix} 0 \\ 0 \\ 6 \end{bmatrix}.$$

What is the geometric relationship between $A\vec{x}^*$ and $\vec{b}$?

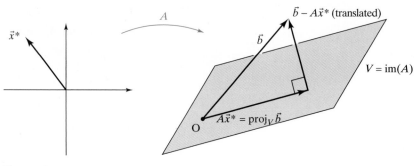

Figure 5

Solution

We compute

$$\vec{x}^* = (A^T A)^{-1} A^T \vec{b} = \begin{bmatrix} -4 \\ 3 \end{bmatrix} \quad \text{and} \quad A\vec{x}^* = \begin{bmatrix} -1 \\ 2 \\ 5 \end{bmatrix}.$$

Recall that $A\vec{x}^*$ is the orthogonal projection of $\vec{b}$ onto the image of A. Check that

$$\vec{b} - A\vec{x}^* = \begin{bmatrix} 1 \\ -2 \\ 1 \end{bmatrix}$$

is indeed perpendicular to the two column vectors of A. (See Figure 5.) ∎

If $\vec{x}^*$ is a least-squares solution of the system $A\vec{x} = \vec{b}$, then $A\vec{x}^*$ is the orthogonal projection of $\vec{b}$ onto im(A). We can use this fact to find a new formula for orthogonal projections. (Compare this with Facts 5.1.6 and 5.3.10.) Consider a subspace V of $\mathbb{R}^n$ and a vector $\vec{b}$ in $\mathbb{R}^n$. Choose a basis $\vec{v}_1, \ldots, \vec{v}_m$ of V, and form the matrix $A = [\vec{v}_1 \quad \ldots \quad \vec{v}_m]$. Note that ker($A$) = $\{\vec{0}\}$, since the columns of A are linearly independent. The least-squares solution of the system $A\vec{x} = \vec{b}$ is $\vec{x}^* = (A^T A)^{-1} A^T \vec{b}$. Thus, the orthogonal projection of $\vec{b}$ onto V is $\text{proj}_V \vec{b} = A\vec{x}^* = A(A^T A)^{-1} A^T \vec{b}$.

Fact 5.4.8

The matrix of an orthogonal projection
Consider a subspace V of $\mathbb{R}^n$ with basis $\vec{v}_1, \vec{v}_2, \ldots, \vec{v}_m$. Let

$$A = \begin{bmatrix} \vec{v}_1 & \vec{v}_2 & \ldots & \vec{v}_m \end{bmatrix}.$$

Then the matrix of the orthogonal projection onto V is

$$A(A^T A)^{-1} A^T.$$

We are not required to find an *orthonormal* basis of V here. If the vectors $\vec{v}_i$ happen to be orthonormal, then $A^T A = I_m$ and the formula simplifies to AA^T. (See Fact 5.3.10.)

EXAMPLE 2   Find the matrix of the orthogonal projection onto the subspace of $\mathbb{R}^4$ spanned by the vectors

$$\begin{bmatrix} 1 \\ 1 \\ 1 \\ 1 \end{bmatrix} \quad \text{and} \quad \begin{bmatrix} 1 \\ 2 \\ 3 \\ 4 \end{bmatrix}.$$

Solution

Let

$$A = \begin{bmatrix} 1 & 1 \\ 1 & 2 \\ 1 & 3 \\ 1 & 4 \end{bmatrix},$$

and compute

$$A(A^T A)^{-1} A^T = \frac{1}{10} \begin{bmatrix} 7 & 4 & 1 & -2 \\ 4 & 3 & 2 & 1 \\ 1 & 2 & 3 & 4 \\ -2 & 1 & 4 & 7 \end{bmatrix}. \qquad \blacksquare$$

Data Fitting

Scientists are often interested in fitting a function of a certain type to data they have gathered. The functions considered could be linear, polynomial, rational, trigonometric, or exponential. The equations we have to solve as we fit data are frequently linear. (See Exercises 29 and 30 of Section 1.1, and Exercises 30 through 33 of Section 1.2.)

EXAMPLE 3   Find a cubic polynomial whose graph passes through the points $(1, 3), (-1, 13), (2, 1), (-2, 33)$.

Solution

We are looking for a function

$$f(t) = c_0 + c_1 t + c_2 t^2 + c_3 t^3$$

such that $f(1) = 3$, $f(-1) = 13$, $f(2) = 1$, $f(-2) = 33$; that is, we have to solve the linear system

$$\begin{vmatrix} c_0 + & c_1 + & c_2 + & c_3 = & 3 \\ c_0 - & c_1 + & c_2 - & c_3 = & 13 \\ c_0 + 2c_1 + & 4c_2 + 8c_3 = & 1 \\ c_0 - 2c_1 + & 4c_2 - 8c_3 = & 33 \end{vmatrix}.$$

This linear system has the unique solution

$$\begin{bmatrix} c_0 \\ c_1 \\ c_2 \\ c_3 \end{bmatrix} = \begin{bmatrix} 5 \\ -4 \\ 3 \\ -1 \end{bmatrix}.$$

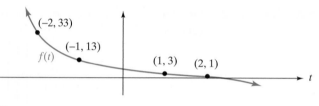

Figure 6

Thus, the cubic polynomial whose graph passes through the four given data points is $f(t) = 5 - 4t + 3t^2 - t^3$, as shown in Figure 6. ∎

Frequently, a data-fitting problem leads to a linear system with more equations than variables. (This happens when the number of data points exceeds the number of parameters in the function we seek.) Such a system is usually inconsistent, and we will look for the least-squares solution(s).

EXAMPLE 4 Fit a quadratic function to the four data points $(a_1, b_1) = (-1, 8)$, $(a_2, b_2) = (0, 8)$, $(a_3, b_3) = (1, 4)$, and $(a_4, b_4) = (2, 16)$.

Solution

We are looking for a function $f(t) = c_0 + c_1 t + c_2 t^2$ such that

$$
\begin{vmatrix} f(a_1) = b_1 \\ f(a_2) = b_2 \\ f(a_3) = b_3 \\ f(a_4) = b_4 \end{vmatrix} \quad \text{or} \quad \begin{vmatrix} c_0 - c_1 + c_2 = 8 \\ c_0 = 8 \\ c_0 + c_1 + c_2 = 4 \\ c_0 + 2c_1 + 4c_2 = 16 \end{vmatrix} \quad \text{or} \quad A \begin{bmatrix} c_0 \\ c_1 \\ c_2 \end{bmatrix} = \vec{b},
$$

where

$$
A = \begin{bmatrix} 1 & -1 & 1 \\ 1 & 0 & 0 \\ 1 & 1 & 1 \\ 1 & 2 & 4 \end{bmatrix} \quad \text{and} \quad \vec{b} = \begin{bmatrix} 8 \\ 8 \\ 4 \\ 16 \end{bmatrix}.
$$

We have four equations, corresponding to the four data points, but only three unknowns, the three coefficients of a quadratic polynomial. Check that this system is indeed inconsistent. The least-squares solution is

$$
\vec{c}^* = \begin{bmatrix} c_0^* \\ c_1^* \\ c_2^* \end{bmatrix} = (A^T A)^{-1} A^T \vec{b} = \begin{bmatrix} 5 \\ -1 \\ 3 \end{bmatrix}.
$$

The least-squares approximation is $f^*(t) = 5 - t + 3t^2$, as shown in Figure 7.
This quadratic function $f^*(t)$ fits the data points best, in that the vector

$$
A\vec{c}^* = \begin{bmatrix} f^*(a_1) \\ f^*(a_2) \\ f^*(a_3) \\ f^*(a_4) \end{bmatrix}
$$

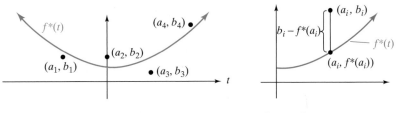

Figure 7

Figure 8

is as close as possible to

$$\vec{b} = \begin{bmatrix} b_1 \\ b_2 \\ b_3 \\ b_4 \end{bmatrix}.$$

This means that

$$\|\vec{b} - A\vec{c}^*\|^2 = \left(b_1 - f^*(a_1)\right)^2 + \left(b_2 - f^*(a_2)\right)^2 + \left(b_3 - f^*(a_3)\right)^2 + \left(b_4 - f^*(a_4)\right)^2$$

is minimal: The sum of the squares of the vertical distances between graph and data points is minimal. (See Figure 8.) ∎

EXAMPLE 5   Find the linear function $c_0 + c_1 t$ that best fits the data points (a_1, b_1), (a_2, b_2), ..., (a_n, b_n), using least squares. Assume that $a_1 \neq a_2$.

Solution

We attempt to solve the system

$$\begin{vmatrix} c_0 + c_1 a_1 = b_1 \\ c_0 + c_1 a_2 = b_2 \\ \vdots \quad \vdots \quad \vdots \\ c_0 + c_1 a_n = b_n \end{vmatrix},$$

or

$$\begin{bmatrix} 1 & a_1 \\ 1 & a_2 \\ \vdots & \vdots \\ 1 & a_n \end{bmatrix} \begin{bmatrix} c_0 \\ c_1 \end{bmatrix} = \begin{bmatrix} b_1 \\ b_2 \\ \vdots \\ b_n \end{bmatrix},$$

or

$$A \begin{bmatrix} c_0 \\ c_1 \end{bmatrix} = \vec{b}.$$

Note that $\text{rank}(A) = 2$, since $a_1 \neq a_2$.

The least-squares solution is

$$\begin{bmatrix} c_0^* \\ c_1^* \end{bmatrix} = (A^T A)^{-1} A^T \vec{b} = \left(\begin{bmatrix} 1 & \cdots & 1 \\ a_1 & \cdots & a_n \end{bmatrix} \begin{bmatrix} 1 & a_1 \\ \vdots & \vdots \\ 1 & a_n \end{bmatrix} \right)^{-1} \begin{bmatrix} 1 & \cdots & 1 \\ a_1 & \cdots & a_n \end{bmatrix} \begin{bmatrix} b_1 \\ \vdots \\ b_n \end{bmatrix}$$

$$= \begin{bmatrix} n & \sum_i a_i \\ \sum_i a_i & \sum_i a_i^2 \end{bmatrix}^{-1} \begin{bmatrix} \sum_i b_i \\ \sum_i a_i b_i \end{bmatrix} \qquad \text{(where } \sum_i \text{ refers to the sum for } i = 1, \ldots, n)$$

$$= \frac{1}{n \left(\sum_i a_i^2 \right) - \left(\sum_i a_i \right)^2} \begin{bmatrix} \sum_i a_i^2 & -\sum_i a_i \\ -\sum_i a_i & n \end{bmatrix} \begin{bmatrix} \sum_i b_i \\ \sum_i a_i b_i \end{bmatrix}.$$

We have found that

$$c_0^* = \frac{\left(\sum_i a_i^2 \right) \left(\sum_i b_i \right) - \left(\sum_i a_i \right) \left(\sum_i a_i b_i \right)}{n \left(\sum_i a_i^2 \right) - \left(\sum_i a_i \right)^2}, \qquad c_1^* = \frac{n \left(\sum_i a_i b_i \right) - \left(\sum_i a_i \right) \left(\sum_i b_i \right)}{n \left(\sum_i a_i^2 \right) - \left(\sum_i a_i \right)^2}.$$

These formulas are well known to statisticians. There is no need to memorize them. ∎

We conclude this section with an example for multivariate data fitting.

EXAMPLE 6 In the accompanying table, we list the scores of five students in the three exams given in a class.

	h: Hour Exam	m: Midterm Exam	f: Final Exam
Gabriel	76	48	43
Kanya	92	92	90
Jessica	68	82	64
Janelle	86	68	69
Wynn	54	70	50

Find the function of the form $f = c_0 + c_1 h + c_2 m$ that best fits these data, using least squares. What score f does your formula predict for Marlisa, another student, whose scores in the first two exams were $h = 92$ and $m = 72$?

Solution

We attempt to solve the system

$$\begin{vmatrix} c_0 + 76 c_1 + 48 c_2 = 43 \\ c_0 + 92 c_1 + 92 c_2 = 90 \\ c_0 + 68 c_1 + 82 c_2 = 64 \\ c_0 + 86 c_1 + 68 c_2 = 69 \\ c_0 + 54 c_1 + 70 c_2 = 50 \end{vmatrix}.$$

The least-squares solution is

$$\begin{bmatrix} c_0^* \\ c_1^* \\ c_2^* \end{bmatrix} = (A^T A)^{-1} A^T \vec{b} \approx \begin{bmatrix} -42.4 \\ 0.639 \\ 0.799 \end{bmatrix}.$$

The function which gives the best fit is approximately

$$f = -42.4 + 0.639h + 0.799m.$$

This formula predicts the score

$$f = -42.4 + 0.639 \cdot 92 + 0.799 \cdot 72 \approx 74$$

for Marlisa.   ■

EXERCISES

GOALS Use the formula $(\text{im } A)^\perp = \ker(A^T)$. Apply the characterization of $\text{proj}_V \vec{x}$ as the vector in V "closest to $\vec{x}$." Find the least-squares solutions of a linear system $A\vec{x} = \vec{b}$ using the normal equation $A^T A\vec{x} = A^T \vec{b}$.

1. Consider the subspace $\text{im}(A)$ of $\mathbb{R}^2$, where

$$A = \begin{bmatrix} 2 & 4 \\ 3 & 6 \end{bmatrix}.$$

Find a basis of $\ker(A^T)$, and draw a sketch illustrating the formula

$$(\text{im } A)^\perp = \ker(A^T)$$

in this case.

2. Consider the subspace $\text{im}(A)$ of $\mathbb{R}^3$, where

$$A = \begin{bmatrix} 1 & 1 \\ 1 & 2 \\ 1 & 3 \end{bmatrix}.$$

Find a basis of $\ker(A^T)$, and draw a sketch illustrating the formula $(\text{im } A)^\perp = \ker(A^T)$ in this case.

3. Consider a subspace V of $\mathbb{R}^n$. Let $\vec{v}_1, \ldots, \vec{v}_p$ be a basis of V and $\vec{w}_1, \ldots, \vec{w}_q$ a basis of $V^\perp$. Is $\vec{v}_1, \ldots, \vec{v}_p$, $\vec{w}_1, \ldots, \vec{w}_q$ a basis of $\mathbb{R}^n$? Explain.

4. Let A be an $m \times n$ matrix. Is the formula

$$(\ker A)^\perp = \text{im}(A^T)$$

necessarily true? Explain.

5. Let V be the solution space of the linear system

$$\begin{vmatrix} x_1 + x_2 + x_3 + x_4 = 0 \\ x_1 + 2x_2 + 5x_3 + 4x_4 = 0 \end{vmatrix}.$$

Find a basis of $V^\perp$.

6. If A is an $m \times n$ matrix, is the formula

$$\text{im}(A) = \text{im}(AA^T)$$

necessarily true? Explain.

7. Consider a symmetric $n \times n$ matrix A. What is the relationship between $\text{im}(A)$ and $\ker(A)$?

8. Consider a linear transformation $L(\vec{x}) = A\vec{x}$ from $\mathbb{R}^n$ to $\mathbb{R}^m$, with $\ker(L) = \{\vec{0}\}$. The *pseudo-inverse* L^+ of L is the transformation from $\mathbb{R}^m$ to $\mathbb{R}^n$ given by

$$L^+(\vec{y}) = (\text{the least-squares solution of } L(\vec{x}) = \vec{y}).$$

a. Show that the transformation L^+ is linear. Find the matrix A^+ of L^+, in terms of the matrix A of L.

b. If L is invertible, what is the relationship between L^+ and L^{-1}?

c. What is $L^+(L(\vec{x}))$, for $\vec{x}$ in $\mathbb{R}^n$?

d. What is $L(L^+(\vec{y}))$, for $\vec{y}$ in $\mathbb{R}^m$?

e. Find L^+ for the linear transformation

$$L(\vec{x}) = \begin{bmatrix} 1 & 0 \\ 0 & 1 \\ 0 & 0 \end{bmatrix} \vec{x}.$$

9. Consider the linear system $A\vec{x} = \vec{b}$, where

$$A = \begin{bmatrix} 1 & 3 \\ 2 & 6 \end{bmatrix} \quad \text{and} \quad \vec{b} = \begin{bmatrix} 10 \\ 20 \end{bmatrix}.$$

a. Draw a sketch showing the following subsets of $\mathbb{R}^2$:
   • the kernel of A, and $(\ker A)^\perp$.
   • the image of A^T.
   • the solution set S of the system $A\vec{x} = \vec{b}$.

b. What relationship do you observe between $\ker(A)$ and $\text{im}(A^T)$? Explain.

c. What relationship do you observe between ker(A) and S? Explain.

d. Find the unique vector $\vec{x}_0$ in the intersection of S and (ker A)$^{\perp}$. Show $\vec{x}_0$ on your sketch.

e. What can you say about the length of $\vec{x}_0$ compared with the length of all other vectors in S?

10. Consider a consistent system $A\vec{x} = \vec{b}$.

a. Show that this system has a solution $\vec{x}_0$ in (ker A)$^{\perp}$. *Hint:* An arbitrary solution $\vec{x}$ of the system can be written as $\vec{x} = \vec{x}_h + \vec{x}_0$, where $\vec{x}_h$ is in ker(A) and $\vec{x}_0$ is in (ker A)$^{\perp}$.

b. Show that the system $A\vec{x} = \vec{b}$ has only one solution in (ker A)$^{\perp}$. *Hint:* If $\vec{x}_0$ and $\vec{x}_1$ are two solutions in (ker A)$^{\perp}$, think about $\vec{x}_1 - \vec{x}_0$.

c. If $\vec{x}_0$ is the solution in (ker A)$^{\perp}$ and $\vec{x}_1$ is another solution of the system $A\vec{x} = \vec{b}$, show that $\|\vec{x}_0\| < \|\vec{x}_1\|$. The vector $\vec{x}_0$ is called the *minimal solution* of the linear system $A\vec{x} = \vec{b}$.

11. Consider a linear transformation $L(\vec{x}) = A\vec{x}$ from $\mathbb{R}^n$ to $\mathbb{R}^m$, where rank(A) = m. The *pseudo-inverse* L^+ of L is the transformation from $\mathbb{R}^m$ to $\mathbb{R}^n$ given by

$$L^+(\vec{y}) = \left(\text{the minimal solution of the system } L(\vec{x}) = \vec{y}\right).$$

(See Exercise 10.)

a. Show that the transformation L^+ is linear.

b. What is $L(L^+(\vec{y}))$, for $\vec{y}$ in $\mathbb{R}^m$?

c. What is $L^+(L(\vec{x}))$, for $\vec{x}$ in $\mathbb{R}^n$?

d. Determine image and kernel of L^+.

e. Find L^+ for the linear transformation

$$L(\vec{x}) = \begin{bmatrix} 1 & 0 & 0 \\ 0 & 1 & 0 \end{bmatrix} \vec{x}.$$

12. Using Exercise 10 as a guide, define the term "minimal least-squares solution" of a linear system. Explain why the minimal least-squares solution $\vec{x}^*$ of a linear system $A\vec{x} = \vec{b}$ is in (ker A)$^{\perp}$.

13. Consider a linear transformation $L(\vec{x}) = A\vec{x}$ from $\mathbb{R}^n$ to $\mathbb{R}^m$. The pseudo-inverse L^+ of L is the transformation from $\mathbb{R}^m$ to $\mathbb{R}^n$ given by

$$L^+(\vec{y}) = \left(\text{the minimal least-squares solution} \right.$$
$$\left. \text{of the system } L(\vec{x}) = \vec{y}\right).$$

(See Exercises 8, 11, and 12 for special cases.)

a. Show that the transformation L^+ is linear.

b. What is $L^+\left(L(\vec{x})\right)$, for $\vec{x}$ in $\mathbb{R}^n$?

c. What is $L\left(L^+(\vec{y})\right)$, for $\vec{y}$ in $\mathbb{R}^m$?

d. Determine image and kernel of L^+ (in terms of im(A^T) and ker(A^T)).

e. Find L^+ for the linear transformation

$$L(\vec{x}) = \begin{bmatrix} 2 & 0 & 0 \\ 0 & 0 & 0 \end{bmatrix} \vec{x}.$$

14. In the accompanying figure, we show the kernel and the image of a linear transformation L from $\mathbb{R}^2$ to $\mathbb{R}^2$, together with some vectors $\vec{v}_1, \vec{w}_1, \vec{w}_2, \vec{w}_3$. We are told that $L(\vec{v}_1) = \vec{w}_1$. For $i = 1, 2, 3$, find the vectors $L^+(\vec{w}_i)$, where L^+ is the pseudo-inverse of L defined in Exercise 13. Show your solutions in the figure, and explain how you found them.

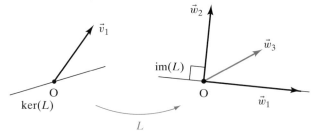

15. Consider an $m \times n$ matrix A with ker(A) = $\{\vec{0}\}$. Show that there is an $n \times m$ matrix B such that $BA = I_n$. *Hint:* $A^T A$ is invertible.

16. Use the formula (im A)$^{\perp}$ = ker(A^T) to prove the equation

$$\text{rank}(A) = \text{rank}(A^T).$$

17. Does the equation

$$\text{rank}(A) = \text{rank}(A^T A)$$

hold for all $m \times n$ matrices A? Explain.

18. Does the equation

$$\text{rank}(A^T A) = \text{rank}(AA^T)$$

hold for all $m \times n$ matrices A? Explain. *Hint:* Exercise 17 is useful.

19. Find the least-squares solution $\vec{x}^*$ of the system

$$A\vec{x} = \vec{b}, \quad \text{where} \quad A = \begin{bmatrix} 1 & 0 \\ 0 & 1 \\ 0 & 0 \end{bmatrix} \quad \text{and} \quad \vec{b} = \begin{bmatrix} 1 \\ 1 \\ 1 \end{bmatrix}.$$

Use paper and pencil. Draw a sketch showing the vector $\vec{b}$, the image of A, the vector $A\vec{x}^*$, and the vector $\vec{b} - A\vec{x}^*$.

20. By using paper and pencil, find the least-squares solution $\vec{x}^*$ of the system

$$A\vec{x} = \vec{b}, \quad \text{where} \quad A = \begin{bmatrix} 1 & 1 \\ 1 & 0 \\ 0 & 1 \end{bmatrix} \quad \text{and} \quad \vec{b} = \begin{bmatrix} 3 \\ 3 \\ 3 \end{bmatrix}.$$

Verify that the vector $\vec{b} - A\vec{x}^*$ is perpendicular to the image of A.

21. Find the least-squares solution $\vec{x}^*$ of the system

$$A\vec{x} = \vec{b}, \quad \text{where} \quad A = \begin{bmatrix} 6 & 9 \\ 3 & 8 \\ 2 & 10 \end{bmatrix} \quad \text{and} \quad \vec{b} = \begin{bmatrix} 0 \\ 49 \\ 0 \end{bmatrix}.$$

Determine the error $\|\vec{b} - A\vec{x}^*\|$.

22. Find the least-squares solution $\vec{x}^*$ of the system

$$A\vec{x} = \vec{b}, \quad \text{where} \quad A = \begin{bmatrix} 3 & 2 \\ 5 & 3 \\ 4 & 5 \end{bmatrix} \quad \text{and} \quad \vec{b} = \begin{bmatrix} 5 \\ 9 \\ 2 \end{bmatrix}.$$

Determine the error $\|\vec{b} - A\vec{x}^*\|$.

23. Find the least-squares solution $\vec{x}^*$ of the system

$$A\vec{x} = \vec{b}, \quad \text{where} \quad A = \begin{bmatrix} 1 & 1 \\ 2 & 8 \\ 1 & 5 \end{bmatrix} \quad \text{and} \quad \vec{b} = \begin{bmatrix} 1 \\ -2 \\ 3 \end{bmatrix}.$$

Explain.

24. Find the least-squares solution $\vec{x}^*$ of the system

$$A\vec{x} = \vec{b}, \quad \text{where} \quad A = \begin{bmatrix} 1 \\ 2 \\ 3 \end{bmatrix} \quad \text{and} \quad \vec{b} = \begin{bmatrix} 3 \\ 2 \\ 7 \end{bmatrix}.$$

Draw a sketch showing the vector $\vec{b}$, the image of A, the vector $A\vec{x}^*$, and the vector $\vec{b} - A\vec{x}^*$.

25. Find the least-squares solutions $\vec{x}^*$ of the system $A\vec{x} = \vec{b}$, where

$$A = \begin{bmatrix} 1 & 3 \\ 2 & 6 \end{bmatrix} \quad \text{and} \quad \vec{b} = \begin{bmatrix} 5 \\ 0 \end{bmatrix}.$$

Use only paper and pencil. Draw a sketch.

26. Find the least-squares solutions $\vec{x}^*$ of the system $A\vec{x} = \vec{b}$, where

$$A = \begin{bmatrix} 1 & 2 & 3 \\ 4 & 5 & 6 \\ 7 & 8 & 9 \end{bmatrix} \quad \text{and} \quad \vec{b} = \begin{bmatrix} 1 \\ 0 \\ 0 \end{bmatrix}.$$

27. Consider an inconsistent linear system $A\vec{x} = \vec{b}$, where A is a 3×2 matrix. We are told that the least-squares solution of this system is $\vec{x}^* = \begin{bmatrix} 7 \\ 11 \end{bmatrix}$. Consider an orthogonal 3×3 matrix S. Find the least-squares solution(s) of the system $SA\vec{x} = S\vec{b}$.

28. Consider an orthonormal basis $\vec{v}_1, \vec{v}_2, \ldots, \vec{v}_n$ in $\mathbb{R}^n$. Find the least-squares solution(s) of the system

$$A\vec{x} = \vec{v}_n,$$

where

$$A = \begin{bmatrix} | & | & & | \\ \vec{v}_1 & \vec{v}_2 & \cdots & \vec{v}_{n-1} \\ | & | & & | \end{bmatrix}.$$

29. Find the least-squares solution of the system

$$A\vec{x} = \vec{b}, \quad \text{where} \quad A = \begin{bmatrix} 1 & 1 \\ 10^{-10} & 0 \\ 0 & 10^{-10} \end{bmatrix} \quad \text{and}$$

$$\vec{b} = \begin{bmatrix} 1 \\ 10^{-10} \\ 10^{-10} \end{bmatrix}.$$

Describe and explain the difficulties you may encounter if you use technology. Then find the solution using paper and pencil.

30. Fit a linear function of the form $f(t) = c_0 + c_1 t$ to the data points $(0, 0)$, $(0, 1)$, $(1, 1)$, using least squares. Use only paper and pencil. Sketch your solution, and explain why it makes sense.

31. Fit a linear function of the form $f(t) = c_0 + c_1 t$ to the data points $(0, 3)$, $(1, 3)$, $(1, 6)$, using least squares. Sketch the solution.

32. Fit a quadratic polynomial to the data points $(0, 27)$, $(1, 0)$, $(2, 0)$, $(3, 0)$, using least squares. Sketch the solution.

33. Find the trigonometric function of the form $f(t) = c_0 + c_1 \sin(t) + c_2 \cos(t)$ that best fits the data points $(0, 0)$, $(1, 1)$, $(2, 2)$, $(3, 3)$, using least squares. Sketch the solution together with the function $g(t) = t$.

34. Find the function of the form

$$f(t) = c_0 + c_1 \sin(t) + c_2 \cos(t) + c_3 \sin(2t) + c_4 \cos(2t)$$

that best fits the data points $(0, 0)$, $(0.5, 0.5)$, $(1, 1)$, $(1.5, 1.5)$, $(2, 2)$, $(2.5, 2.5)$, $(3, 3)$, using least squares. Sketch the solution, together with the function $g(t) = t$.

35. Suppose you wish to fit a function of the form

$$f(t) = c + p \sin(t) + q \cos(t)$$

to a given continuous function $g(t)$ on the closed interval from 0 to 2π. One approach is to choose n equally spaced points a_i between 0 and 2π ($a_i = i \cdot (2\pi/n)$, for $i = 1, \ldots, n$, say). We can fit a function

$$f_n(t) = c_n + p_n \sin(t) + q_n \cos(t)$$

to the data points $(a_i, g(a_i))$, for $i = 1, \ldots, n$. Now examine what happens to the coefficients c_n, p_n, q_n of

$f_n(t)$ as n approaches infinity:

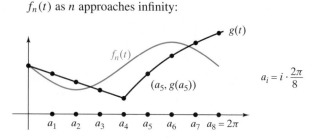

To find $f_n(t)$, we make an attempt to solve the equations

$$f_n(a_i) = g(a_i), \quad \text{for} \quad i = 1, \ldots, n,$$

or

$$\begin{vmatrix} c_n + p_n \sin(a_1) + q_n \cos(a_1) = g(a_1) \\ c_n + p_n \sin(a_2) + q_n \cos(a_2) = g(a_2) \\ \vdots \qquad\qquad \vdots \\ c_n + p_n \sin(a_n) + q_n \cos(a_n) = g(a_n) \end{vmatrix},$$

or

$$A_n \begin{bmatrix} c_n \\ p_n \\ q_n \end{bmatrix} = \vec{b}_n,$$

where

$$A_n = \begin{bmatrix} 1 & \sin(a_1) & \cos(a_1) \\ 1 & \sin(a_2) & \cos(a_2) \\ \vdots & \vdots & \vdots \\ 1 & \sin(a_n) & \cos(a_n) \end{bmatrix}, \quad \vec{b}_n = \begin{bmatrix} g(a_1) \\ g(a_2) \\ \vdots \\ g(a_n) \end{bmatrix}.$$

a. Find the entries of the matrix $A_n^T A_n$ and the components of the vector $A_n^T \vec{b}_n$.

b. Find

$$\lim_{n \to \infty} \left(\frac{2\pi}{n} A_n^T A_n \right) \quad \text{and} \quad \lim_{n \to \infty} \left(\frac{2\pi}{n} A_n^T \vec{b} \right).$$

Hint: Interpret the entries of the matrix $(2\pi/n) A_n^T A_n$ and the components of the vector $(2\pi/n) A^T \vec{b}$ as Riemann sums. Then the limits are the corresponding Riemann integrals. Evaluate as many integrals as you can. Note that

$$\lim_{n \to \infty} \left(\frac{2\pi}{n} A_n^T A_n \right)$$

is a diagonal matrix.

c. Find

$$\lim_{n \to \infty} \begin{bmatrix} c_n \\ p_n \\ q_n \end{bmatrix} = \lim_{n \to \infty} \left(A_n^T A_n \right)^{-1} A_n^T \vec{b}_n$$

$$= \lim_{n \to \infty} \left[\left(\frac{2\pi}{n} A_n^T A_n \right)^{-1} \left(\frac{2\pi}{n} A_n^T \vec{b}_n \right) \right]$$

$$= \left[\lim_{n \to \infty} \left(\frac{2\pi}{n} A_n^T A_n \right) \right]^{-1} \lim_{n \to \infty} \left(\frac{2\pi}{n} A_n^T \vec{b}_n \right).$$

The resulting vector $\begin{bmatrix} c \\ p \\ q \end{bmatrix}$ gives you the coefficient of the desired function

$$f(t) = \lim_{n \to \infty} f_n(t).$$

Write $f(t)$. The function $f(t)$ is called the first *Fourier approximation* of $g(t)$. The Fourier approximation satisfies a "continuous" least-squares condition, an idea we will make more precise in the next section.

36. Let $S(t)$ be the number of daylight hours on the tth day of the year 2001 in Rome, Italy. We are given the following data for $S(t)$:

Day	t	$S(t)$
January 28	28	10
March 17	77	12
May 3	124	14
June 16	168	15

We wish to fit a trigonometric function of the form

$$f(t) = a + b \sin \left(\frac{2\pi}{365} t \right) + c \cos \left(\frac{2\pi}{365} t \right)$$

to these data. Find the best approximation of this form, using least squares.

37. The accompanying table lists several commercial airplanes, the year they were introduced, and the number of displays in the cockpit.

Plane	Year t	Displays d
Douglas DC-3	'35	35
Lockheed Constellation	'46	46
Boeing 707	'59	77
Concorde	'69	133

a. Fit a linear function of the form $\log(d) = c_0 + c_1 t$ to the data points $(t_i, \log(d_i))$, using least squares.

b. Use your answer in part (a) to fit an exponential function $d = ka^t$ to the data points (t_i, d_i).

c. The Airbus A320 was introduced in 1988. Based on your answer in part (b), how many displays do you expect in the cockpit of this plane? (There are 93 displays in the cockpit of an Airbus A320. Explain.)

38. In the accompanying table, we list the height h, the gender g, and the weight w of some young adults.

Height h (in Inches above 5 Feet)	Gender g (1 = "Female," 0 = "Male")	Weight w (in Pounds)
2	1	110
12	0	180
5	1	120
11	1	160
6	0	160

Fit a function of the form

$$w = c_0 + c_1 h + c_2 g$$

to these data, using least squares. Before you do the computations, think about the signs of c_1 and c_2. What signs would you expect if these data were representative of the general population? Why? What is the sign of c_0? What is the practical significance of c_0?

39. In the accompanying table, we list the estimated number g of genes and the estimated number z of cell types for various organisms.

Organism	Number of Genes, g	Number of Cell Types, z
Humans	600,000	250
Annelid worms	200,000	60
Jellyfish	60,000	25
Sponges	10,000	12
Yeasts	2,500	5

a. Fit a function of the form $\log(z) = c_0 + c_1 \log(g)$ to the data points $(\log(g_i), \log(z_i))$, using least squares.

b. Use your answer in part (a) to fit a power function $z = kg^n$ to the data points (g_i, z_i).

c. Using the theory of self-regulatory systems, scientists developed a model that predicts that z is a square-root

function of g (i.e., $a = k\sqrt{g}$, for some constant k). Is your answer in part (b) reasonably close to this form?

40. Consider the data in the following table:

Planet	a Mean Distance from the Sun (in Astronomical Units)	D Period of Revolution (in Earth Years)
Mercury	0.387	0.241
Earth	1	1
Jupiter	5.20	11.86
Uranus	19.18	84.0
Pluto	39.53	248.5

Use the methods discussed in Exercise 39 to fit a power function of the form $D = ka^n$ to these data. Explain, in terms of Kepler's laws of planetary motion. Explain why the constant k is close to 1.

41. In the accompanying table, we list the public debt D of the United States (in billions of dollars), in the year t (as of September 30).

t	'70	'75	'80	'85	'90	'95
D	370	533	908	1,823	3,233	4,974

a. Fit a linear function of the form $\log(D) = c_0 + c_1 t$ to the data points $(t_i, \log(D_i))$, using least squares. Use the result to fit an exponential function to the data points (t_i, D_i).

b. What debt does your formula in part (a) predict for the year 2000?

c. On Sept 30, 2000, the debt was 5,674 billion dollars. What happened?

42. If A is any matrix, show that the linear transformation $L(\vec{x}) = A\vec{x}$ from $\text{im}(A^T)$ to $\text{im}(A)$ is an isomorphism. This provides yet another proof of the formula $\text{rank}(A) = \text{rank}(A^T)$.

5.5 INNER PRODUCT SPACES

Let's take a look back at what we have done thus far in this text. In Chapters 1 through 3, we studied the basic concepts of linear algebra in the concrete context of $\mathbb{R}^n$. Recall that these concepts are all defined in terms of two operations: addition and scalar multiplication. In Chapter 4, we saw that it can be both natural and useful to apply the language of linear algebra to objects other than vectors in $\mathbb{R}^n$, for example, to functions. We introduced the term *linear space* (or *vector space*) for a set that behaves like $\mathbb{R}^n$ as far as addition and scalar multiplication are concerned.

In this chapter, a new operation for vectors in $\mathbb{R}^n$ entered the picture: the *dot product*. In Sections 5.1 through 5.4, we studied concepts that are defined in terms of the dot product, the most important of them being the *length* of vectors and *orthogonality* of vectors. In this section, we will see that in can be useful to define a product analogous to the dot product in a linear space other than $\mathbb{R}^n$. These generalized dot products are called *inner products*. Once we have an inner product in a linear space, we can define length and orthogonality in that space just as in $\mathbb{R}^n$, and we can generalize all the key ideas and facts of Sections 5.1 through 5.4.

Definition 5.5.1

> **Inner products and inner product spaces**
> An *inner product* in a linear space V is a rule that assigns a real scalar (denoted by $\langle f, g \rangle$) to any pair f, g of elements of V, such that the following properties hold for all f, g, h in V, and all c in $\mathbb{R}$:
>
> a. $\langle f, g \rangle = \langle g, f \rangle$.
> b. $\langle f + g, h \rangle = \langle f, h \rangle + \langle g, h \rangle$.
> c. $\langle cf, g \rangle = c \langle f, g \rangle$.
> d. $\langle f, f \rangle > 0$, for all nonzero f in V.
>
> A linear space endowed with an inner product is called an *inner product space*.

Compare these rules with those for the dot product in $\mathbb{R}^n$, listed in the Appendix, Fact A.5. Roughly speaking, an inner product space behaves like $\mathbb{R}^n$ as far as addition, scalar multiplication, and the dot product are concerned.

EXAMPLE 1   Consider the linear space $C[a, b]$ consisting of all continuous functions whose domain is the closed interval $[a, b]$, where $a < b$. See Figure 1.

For functions f and g in $C[a, b]$, we define

$$\langle f, g \rangle = \int_a^b f(t) g(t) \, dt.$$

The verification of the first three axioms for an inner product is straightforward. For

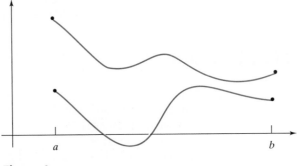

Figure 1

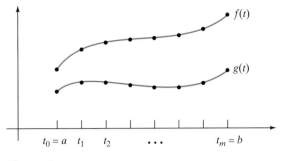

Figure 2

example,

$$\langle f, g \rangle = \int_a^b f(t)g(t) \, dt = \int_a^b g(t)f(t) \, dt = \langle g, f \rangle.$$

The verification of the last axiom requires a bit of calculus. We leave it as Exercise 1.

Recall that the Riemann integral $\int_a^b f(t)g(t) \, dt$ is the limit of the Riemann sum $\sum_{i=1}^m f(t_k)g(t_k) \, \Delta t$, where the t_k can be chosen as equally spaced points in the interval $[a, b]$. See Figure 2.

Then,

$$\langle f, g \rangle = \int_a^b f(t)g(t) \, dt \approx \sum_{k=1}^m f(t_k)g(t_k) \, \Delta t = \left(\begin{bmatrix} f(t_1) \\ f(t_2) \\ \vdots \\ f(t_m) \end{bmatrix} \cdot \begin{bmatrix} g(t_1) \\ g(t_2) \\ \vdots \\ g(t_m) \end{bmatrix} \right) \Delta t$$

for large m.

This approximation shows that the inner product $\langle f, g \rangle = \int_a^b f(t)g(t) \, dt$ for functions is a continuous version of the dot product: The more subdivisions you choose, the better the dot product on the right will approximate the inner product $\langle f, g \rangle$. ■

EXAMPLE 2   Let ℓ_2 be the space of all "square-summable" infinite sequences, i.e., sequences

$$\vec{x} = (x_0, x_1, x_2, \ldots, x_n, \ldots)$$

such that $\sum_{i=0}^{\infty} x_i^2 = x_0^2 + x_1^2 + \cdots$ converges. In this space we can define the inner product

$$\langle \vec{x}, \vec{y} \rangle = \sum_{i=0}^{\infty} x_i y_i = x_0 y_0 + x_1 y_1 + \cdots.$$

(Show that this series converges.) The verification of the axioms is straightforward. Compare this with Exercises 4.1.15 and 5.1.18. ■

EXAMPLE 3   The *trace* of a square matrix is the sum of its diagonal entries. For example,
$$\text{trace} \begin{bmatrix} 1 & 2 \\ 3 & 4 \end{bmatrix} = 1 + 4 = 5.$$

In $\mathbb{R}^{m \times n}$ we can define the inner product

$$\langle A, B \rangle = \text{trace}(A^T B).$$

We will verify the first and the fourth axioms.

$$\langle A, B \rangle = \text{trace}(A^T B) = \text{trace}\left((A^T B)^T\right) = \text{trace}(B^T A) = \langle B, A \rangle.$$

To check that $\langle A, A \rangle > 0$ for nonzero A, write A in terms of its columns:

$$A = \begin{bmatrix} | & | & & | \\ \vec{v}_1 & \vec{v}_2 & \cdots & \vec{v}_n \\ | & | & & | \end{bmatrix}$$

$$\langle A, A \rangle = \text{trace}(A^T A) = \text{trace}\left(\begin{bmatrix} - & \vec{v}_1^T & - \\ - & \vec{v}_2^T & - \\ & \vdots & \\ - & \vec{v}_n^T & - \end{bmatrix} \begin{bmatrix} | & | & & | \\ \vec{v}_1 & \vec{v}_2 & \cdots & \vec{v}_n \\ | & | & & | \end{bmatrix} \right)$$

$$= \text{trace}\left(\begin{bmatrix} \|\vec{v}_1\|^2 & \cdots & \cdots \\ \cdots & \|\vec{v}_2\|^2 & \cdots \\ \vdots & \vdots & \ddots & \vdots \\ & & \cdots & \|\vec{v}_n\|^2 \end{bmatrix} \right) = \|\vec{v}_1\|^2 + \|\vec{v}_2\|^2 + \cdots + \|\vec{v}_n\|^2$$

If A is nonzero, then at least one of the $\vec{v}_i$ is nonzero, so that the sum $\|\vec{v}_1\|^2 + \|\vec{v}_2\|^2 + \cdots + \|\vec{v}_n\|^2$ is positive, as desired. ■

We can introduce the basic concepts of geometry for an inner product space exactly as we did in $\mathbb{R}^n$ for the dot product.

Definition 5.5.2

Norm, orthogonality

The *norm* (or magnitude) of an element f of an inner product space is

$$\|f\| = \sqrt{\langle f, f \rangle}.$$

Two elements f and g of an inner product space are called *orthogonal* (or perpendicular) if

$$\langle f, g \rangle = 0.$$

We can define the *distance* between two elements of an inner product space as the norm of their difference:

$$\text{dist}(f, g) = \|f - g\|.$$

Consider the space $C[a, b]$, with the inner product defined in Example 1.

In physics, the quantity $\|f\|^2$ can often be interpreted as *energy*. For example, it describes the acoustic energy of a periodic sound wave $f(t)$ and the elastic potential energy of a uniform string with vertical displacement $f(x)$. (See Figure 3.) The quantity $\|f\|^2$ may also measure thermal or electric energy.

EXAMPLE 4 In the inner product space $C[0, 1]$ with $\langle f, g \rangle = \int_0^1 f(t)g(t)\, dt$, find $\|f\|$ for $f(t) = t^2$.

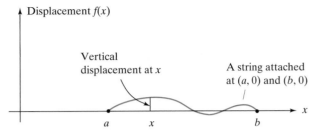

Figure 3

Solution

$$\|f\| = \sqrt{\langle f, f \rangle} = \sqrt{\int_0^1 t^4 \, dt} = \sqrt{\frac{1}{5}} \qquad \blacksquare$$

EXAMPLE 5 Show that $f(t) = \sin(t)$ and $g(t) = \cos(t)$ are perpendicular in the inner product space $C[0, 2\pi]$ with $\langle f, g \rangle = \int_0^{2\pi} f(t)g(t) \, dt$.

Solution

$$\langle f, g \rangle = \int_0^{2\pi} \sin(t) \cos(t) \, dt = \left[\frac{1}{2} \sin^2(t) \right] \Big|_0^{2\pi} = 0 \qquad \blacksquare$$

EXAMPLE 6 Find the distance between $f(t) = t$ and $g(t) = 1$ in $C[0, 1]$.

Solution

$$\text{dist}\langle f, g \rangle = \sqrt{\int_0^1 (t - 1)^2 \, dt} = \sqrt{\left[\frac{1}{3}(t - 1)^3 \right] \Big|_0^1} = \frac{1}{\sqrt{3}} \qquad \blacksquare$$

The results and procedures discussed for the dot product generalize to arbitrary inner product spaces. For example, the Pythagorean theorem holds; the Gram–Schmidt process can be used to construct an orthonormal basis of a (finite-dimensional) inner product space; and the Cauchy–Schwarz inequality tells us that $|\langle f, g \rangle| \leq \|f\| \|g\|$, for two elements f and g of an inner product space.

Orthogonal Projections

In an inner product space V, consider a subspace W with orthonormal basis g_1, ..., g_m. The orthogonal projection $\text{proj}_W f$ of an element f of V onto W is defined as the unique element of W such that $f - \text{proj}_W f$ is orthogonal to W. As in the case of the dot product in $\mathbb{R}^n$, the orthogonal projection is given by the formula below.

Fact 5.5.3   **Orthogonal projection**

If $g_1, \ldots, g_m$ is an orthonormal basis of a subspace W of an inner product space V, then

$$\text{proj}_W f = \langle g_1, f \rangle g_1 + \cdots + \langle g_m, f \rangle g_m,$$

for all f in V.

(Verify this by checking that $\langle f - \text{proj}_W f, g_i \rangle = 0$ for $i = 1, \ldots, m$.)

We may think of $\text{proj}_W f$ as the element of W closest to f. In other words, if we choose another element h of W, then the distance between f and h will exceed the distance between f and $\text{proj}_W f$.

As an example, consider a subspace W of $C[a, b]$, with the inner product introduced in Example 1. Then, $\text{proj}_W f$ is the function g in W that is closest to f, in the sense that

$$\text{dist}(f, g) = \|f - g\| = \sqrt{\int_a^b \left(f(t) - g(t) \right)^2 dt}$$

is least.

The requirement that

$$\int_a^b \left(f(t) - g(t) \right)^2 dt$$

be minimal is a *continuous least-squares condition,* as opposed to the discrete least-squares conditions we discussed in Section 5.4. We can use the discrete least-squares condition to fit a function g of a certain type to some data points (a_k, b_k), while the continuous least-squares condition can be used to fit a function g of a certain type to a given function f. (Functions of a certain type are frequently polynomials of a certain degree or trigonometric functions of a certain form.) See Figures 4(a) and 4(b).

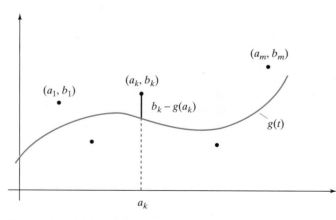

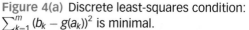

Figure 4(a) Discrete least-squares condition: $\sum_{k=1}^{m} (b_k - g(a_k))^2$ is minimal.

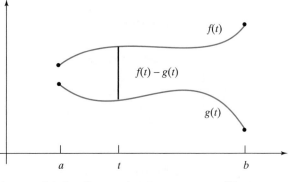

Figure 4(b) Continuous least-squares condition: $\int_a^b (f(t) - g(t))^2 \, dt$ is minimal.

We can think of the continuous least-squares condition as a limiting case of a discrete least-squares condition by writing

$$\int_a^b \left(f(t) - g(t) \right)^2 \, dt = \lim_{m \to \infty} \sum_{k=1}^m \left(f(t_k) - g(t_k) \right)^2 \, \Delta t.$$

EXAMPLE 7   Find the linear function of the form $g(t) = a + bt$ that best fits the function $f(t) = e^t$ over the interval from -1 to 1, in a continuous least-squares sense.

Solution

We need to find $\operatorname{proj}_{P_1} f$. We first find an orthonormal basis of P_1 for the given inner product; then, we will use Fact 5.5.3. In general, we have to use the Gram–Schmidt process to find an orthonormal basis of an inner product space. Because the two functions $1, t$ in the standard basis of P_1 are already orthogonal, or

$$\langle 1, t \rangle = \int_{-1}^1 t \, dt = 0,$$

we merely need to divide each function by its norm:

$$\|1\| = \sqrt{\int_{-1}^1 1 \, dt} = \sqrt{2} \quad \text{and} \quad \|t\| = \sqrt{\int_{-1}^1 t^2 \, dt} = \sqrt{\frac{2}{3}}.$$

An orthonormal basis of P_1 is

$$\frac{1}{\sqrt{2}} 1 \quad \text{and} \quad \sqrt{\frac{3}{2}} t.$$

Now,

$$\operatorname{proj}_{P_1} f = \frac{1}{2} \langle 1, f \rangle 1 + \frac{3}{2} \langle t, f \rangle t$$

$$= \frac{1}{2}(e - e^{-1}) + 3e^{-1}t. \qquad \text{(We omit the straightforward computations.)}$$

See Figure 5.   ∎

What follows is one of the major applications of this theory.

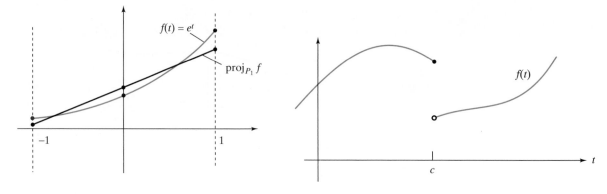

Figure 5

Figure 6 $f(t)$ has a jump-discontinuity at $t = c$.

Fourier Analysis[5]

In the space $C[-\pi, \pi]$, we introduce an inner product that is a slight modification of the definition given in Example 1:

$$\langle f, g \rangle = \frac{1}{\pi} \int_{-\pi}^{\pi} f(t)g(t) \, dt.$$

The factor $1/\pi$ is introduced to facilitate the computations. Convince yourself that this is indeed an inner product. (Compare with Exercise 7.)

More generally, we can consider this inner product in the space of all *piecewise continuous functions* defined in the interval $[-\pi, \pi]$. These are functions $f(t)$ that are continuous except for a finite number of *jump-discontinuities* (i.e., points c where the one-sided limits $\lim_{t \to c^-} f(t)$ and $\lim_{t \to c^+} f(t)$ both exist, but are not equal). Also, it is required that $f(c)$ equal one of the two one-sided limits. Let us consider the piecewise continuous functions with $f(c) = \lim_{t \to c^-} f(t)$. See Figure 6.

For a positive integer n, consider the subspace T_n of $C[-\pi, \pi]$ that is defined as the span of the functions $1, \sin(t), \cos(t), \sin(2t), \cos(2t), \ldots, \sin(nt), \cos(nt)$. The space T_n consists of all functions of the form

$$f(t) = a + b_1 \sin(t) + c_1 \cos(t) + \cdots + b_n \sin(nt) + c_n \cos(nt),$$

called *trigonometric polynomials* of order $\leq n$.

From calculus, you may recall the *Euler identities:*

$$\int_{-\pi}^{\pi} \sin(pt) \cos(mt) \, dt = 0, \qquad \text{for integers } p, m,$$

$$\int_{-\pi}^{\pi} \sin(pt) \sin(mt) \, dt = 0, \qquad \text{for distinct integers } p, m,$$

$$\int_{-\pi}^{\pi} \cos(pt) \cos(mt) \, dt = 0, \qquad \text{for distinct integers } p, m.$$

---

[5]Named after the French mathematician Jean-Baptiste-Joseph Fourier (1768–1830), who developed the subject in his *Théorie analytique de la chaleur* (1822), where he investigated the conduction of heat in very thin sheets of metal. Baron Fourier was also an Egyptologist and government administrator; he accompanied Napoleon on his expedition to Egypt in 1798.

These equations tell us that the functions $1, \sin(t), \cos(t), \ldots, \sin(nt), \cos(nt)$ are orthogonal to one another (and therefore linearly independent).

Another of Euler's identities tells us that

$$\int_{-\pi}^{\pi} \sin^2(mt)\, dt = \int_{-\pi}^{\pi} \cos^2(mt)\, dt = \pi,$$

for positive integers m. This means that the functions $\sin(t), \cos(t), \ldots, \sin(nt), \cos(nt)$ all have norm 1 with respect to the given inner product. This is why we chose the inner product as we did, with the factor $\frac{1}{\pi}$.

The norm of the function $f(t) = 1$ is

$$\|f\| = \sqrt{\frac{1}{\pi} \int_{-\pi}^{\pi} 1\, dt} = \sqrt{2};$$

therefore,

$$g(t) = \frac{f(t)}{\|f(t)\|} = \frac{1}{\sqrt{2}}$$

is a function of norm 1.

Fact 5.5.4

Let T_n be the space of all trigonometric polynomials of order $\leq n$, with the inner product

$$\langle f, g \rangle = \frac{1}{\pi} \int_{-\pi}^{\pi} f(t)g(t)\, dt.$$

Then, the functions

$$\frac{1}{\sqrt{2}}, \sin(t), \cos(t), \sin(2t), \cos(2t), \ldots, \sin(nt), \cos(nt)$$

form an orthonormal basis of T_n.

For a piecewise continuous function f, we can consider

$$f_n = \text{proj}_{T_n} f.$$

As discussed after Fact 5.5.3, f_n is the trigonometric polynomial in T_n that best approximates f, in the sense that

$$\text{dist}(f, f_n) < \text{dist}(f, g),$$

for all other g in T_n.

We can use Facts 5.5.3 and 5.5.4 to find a formula for $f_n = \text{proj}_{T_n} f$.

Fact 5.5.5

Fourier coefficients

If f is a piecewise continuous function defined on the interval $[-\pi, \pi]$, then its best approximation f_n in T_n is

$$f_n(t) = \text{proj}_{T_n} f(t) = a_0 \frac{1}{\sqrt{2}} + b_1 \sin(t) + c_1 \cos(t) + \cdots$$

$$+ b_n \sin(nt) + c_n \cos(nt),$$

where

$$b_k = \langle f(t), \sin(kt) \rangle = \frac{1}{\pi} \int_{-\pi}^{\pi} f(t) \sin(kt)\, dt,$$

$$c_k = \langle f(t), \cos(kt) \rangle = \frac{1}{\pi} \int_{-\pi}^{\pi} f(t) \cos(kt)\, dt,$$

$$a_0 = \left\langle f(t), \frac{1}{\sqrt{2}} \right\rangle = \frac{1}{\sqrt{2\pi}} \int_{-\pi}^{\pi} f(t)\, dt.$$

The b_k, the c_k, and a_0 are called the *Fourier coefficients* of the function f. The function

$$f_n(t) = a_0 \frac{1}{\sqrt{2}} + b_1 \sin(t) + c_1 \cos(t) + \cdots + b_n \sin(nt) + c_n \cos(nt)$$

is called the *n*th-order *Fourier approximation* of f.

Note that the constant term, written somewhat awkwardly, is

$$a_0 \frac{1}{\sqrt{2}} = \frac{1}{2\pi} \int_{-\pi}^{\pi} f(t)\, dt,$$

which is the average value of the function f between $-\pi$ and π. It makes sense that the best way to approximate $f(t)$ by a constant function is to take the average value of $f(t)$.

The function $b_k \sin(kt) + c_k \cos(kt)$ is called the kth *harmonic* of $f(t)$. Using elementary trigonometry, we can write the harmonics alternatively as

$$b_k \sin(kt) + c_k \cos(kt) = A_k \sin\big(k(t - \delta_k)\big),$$

where $A_k = \sqrt{b_k^2 + c_k^2}$ is the *amplitude* of the harmonic and δ_k is the *phase shift*.

Consider the sound generated by a vibrating string, such as in a piano or on a violin. Let $f(t)$ be the air pressure at your eardrum as a function of time t. (The function $f(t)$ is measured as a deviation from the normal atmospheric pressure.) In this case, the harmonics have a simple physical interpretation: They correspond to the various sinusoidal modes at which the string can vibrate. See Figure 7.

The *fundamental frequency* (corresponding to the vibration shown at the bottom of the figure) gives us the *first harmonic* of $f(t)$, while the *overtones* (with frequencies that are integer multiples of the fundamental frequency) give us the other terms of the harmonic series. The quality of a tone is in part determined by the relative amplitudes of the harmonics. When you play concert A (440 Hz) on a piano, the first harmonic is much more prominent than the higher ones, but the same tone played on a violin gives prominence to higher harmonics (especially the fifth). See

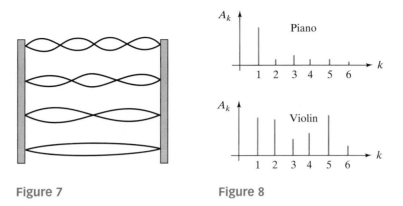

Figure 7   Figure 8

Figure 8. Similar considerations apply to wind instruments; they have a vibrating column of air instead of a vibrating string.

The human ear cannot hear tones whose frequencies exceed 20,000 Hz. We pick up only finitely many harmonics of a tone. What we hear is the projection of $f(t)$ onto a certain T_n.

EXAMPLE 8   Find the Fourier coefficients for the function $f(t) = t$ on the interval $-\pi \le t \le \pi$:

$$b_k = \langle f, \sin(kt) \rangle = \frac{1}{\pi} \int_{-\pi}^{\pi} \sin(kt) t \, dt$$

$$= \frac{1}{\pi} \left\{ -\left[\frac{1}{k} \cos(kt) t \right] \Big|_{-\pi}^{\pi} + \frac{1}{k} \int_{-\pi}^{\pi} \cos(kt) \, dt \right\} \qquad \text{(integration by parts)}$$

$$= \begin{cases} -\dfrac{2}{k} & \text{if } k \text{ is even} \\[2mm] \dfrac{2}{k} & \text{if } k \text{ is odd} \end{cases}$$

$$= (-1)^{k+1} \frac{2}{k}.$$

All c_k and a_0 are zero, since the integrands are odd functions.

The first few Fourier polynomials are

$$f_1 = 2\sin(t),$$
$$f_2 = 2\sin(t) - \sin(2t),$$
$$f_3 = 2\sin(t) - \sin(2t) + \frac{2}{3}\sin(3t),$$
$$f_4 = 2\sin(t) - \sin(2t) + \frac{2}{3}\sin(3t) - \frac{1}{2}\sin(4t).$$

See Figure 9.   ∎

How do the errors $\| f - f_n \|$ and $\| f - f_{n+1} \|$ of the nth and the $(n+1)$st Fourier approximation compare? We hope that f_{n+1} will be a better approximation than f_n, or at least no worse:

$$\| f - f_{n+1} \| \le \| f - f_n \|$$

This is indeed the case, by definition: f_n is a polynomial in T_{n+1}, since T_n is contained

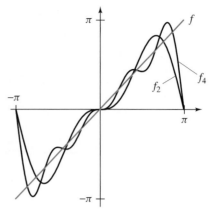

Figure 9

in T_{n+1}, and

$$\|f - f_{n+1}\| \le \|f - g\|,$$

for all g in T_{n+1}, in particular for $g = f_n$. In other words, as n goes to infinity, the error $\|f - f_n\|$ becomes smaller and smaller (or at least not larger). Using somewhat advanced calculus, one can show that this error approaches zero:

$$\lim_{n \to \infty} \|f - f_n\| = 0.$$

What does this tell us about $\lim_{n \to \infty} \|f_n\|$? By the theorem of Pythagoras, we have

$$\|f - f_n\|^2 + \|f_n\|^2 = \|f\|^2.$$

As n goes to infinity, the first summand, $\|f - f_n\|^2$, approaches 0, so that

$$\lim_{n \to \infty} \|f_n\| = \|f\|.$$

We have an expansion of f_n in terms of an orthonormal basis

$$f_n = a_0 \frac{1}{\sqrt{2}} + b_1 \sin(t) + c_1 \cos(t) + \cdots + b_n \sin(nt) + c_n \cos(nt),$$

where the b_k, the c_k, and a_0 are the Fourier coefficients. We can express $\|f_n\|$ in terms of these Fourier coefficients, using the Pythagorean theorem:

$$\|f_n\|^2 = a_0^2 + b_1^2 + c_1^2 + \cdots + b_n^2 + c_n^2.$$

Combining the last two "shaded" equations, we get the following identity:

Fact 5.5.6

$$a_0^2 + b_1^2 + c_1^2 + \cdots + b_n^2 + c_n^2 + \cdots = \|f\|^2$$

The infinite series of the squares of the Fourier coefficients of a piecewise continuous function f converges to $\|f\|^2$.

For the function $f(t)$ studied in Example 8, this means that

$$4 + \frac{4}{4} + \frac{4}{9} + \cdots + \frac{4}{n^2} + \cdots = \frac{1}{\pi} \int_{-\pi}^{\pi} t^2 \, dt = \frac{2}{3} \pi^2,$$

or

$$\sum_{n=1}^{\infty} \frac{1}{n^2} = 1 + \frac{1}{4} + \frac{1}{9} + \frac{1}{16} + \cdots = \frac{\pi^2}{6},$$

an equation discovered by Euler.

Fact 5.5.6 has a physical interpretation when $\|f\|^2$ represents energy. For example, if $f(x)$ is the displacement of a vibrating string, then $b_k^2 + c_k^2$ represents the energy of the kth harmonic, and Fact 5.5.6 tells us that the total energy $\|f\|^2$ is the sum of the energies of the harmonics.

There is an interesting application of Fourier analysis in quantum mechanics. In the 1920s, quantum mechanics was presented in two quite distinct forms: Werner Heisenberg's matrix mechanics and Erwin Schrödinger's wave mechanics. Schrödinger (1887–1961) later showed that the two theories are mathematically equivalent: They use isomorphic inner product spaces. Heisenberg works with the space ℓ_2 introduced in Example 2, while Schrödinger works with a function space related to $C[-\pi, \pi]$. The isomorphism from Schrödinger's space to ℓ_2 is established by taking Fourier coefficients. (See Exercise 13.)

EXERCISES

GOALS Use the idea of an inner product, and apply the basic results derived earlier for the dot product in $\mathbb{R}^n$ to inner product spaces.

1. In $C[a, b]$, define the product

$$\langle f, g \rangle = \int_a^b f(t)g(t) \, dt.$$

Show that this product satisfies the property

$$\langle f, f \rangle > 0$$

for all nonzero f.

2. Does the equation

$$\langle f, g + h \rangle = \langle f, g \rangle + \langle f, h \rangle$$

hold for all elements f, g, h of an inner product space? Explain.

3. Consider a matrix S in $\mathbb{R}^{n \times n}$. In $\mathbb{R}^n$, define the product

$$\langle \vec{x}, \vec{y} \rangle = (S\vec{x})^T S\vec{y}.$$

a. For which choices of S is this an inner product?
b. For which choices of S is $\langle \vec{x}, \vec{y} \rangle = \vec{x} \cdot \vec{y}$ (the dot product)?

4. In $\mathbb{R}^{m \times n}$, consider the inner product

$$\langle A, B \rangle = \text{trace}(A^T B)$$

defined in Example 3.

a. Find a formula for this inner product in $\mathbb{R}^{m \times 1} = \mathbb{R}^m$.
b. Find a formula for this inner product in $\mathbb{R}^{1 \times n}$ (i.e., the space of row vectors with n components).

5. Is $\langle\!\langle A, B \rangle\!\rangle = \text{trace}(AB^T)$ an inner product in $\mathbb{R}^{m \times n}$? (The notation $\langle\!\langle A, B \rangle\!\rangle$ is chosen to distinguish this product from the one considered in Example 3 and Exercise 4.)

6. a. Consider an $m \times n$ matrix P and an $n \times m$ matrix Q. Show that

$$\text{trace}(PQ) = \text{trace}(QP).$$

b. Compare the following two inner products in $\mathbb{R}^{m \times n}$:

$$\langle A, B \rangle = \text{trace}(A^T B),$$

and

$$\langle\!\langle A, B \rangle\!\rangle = \text{trace}(AB^T).$$

(See Example 3 and Exercises 4 and 5.)

7. Consider an inner product $\langle v, w \rangle$ in a space V, and a scalar k. For which choices of k is

$$\langle\!\langle v, w \rangle\!\rangle = k \langle v, w \rangle$$

an inner product?

8. Consider an inner product $\langle v, w \rangle$ in a space V. Let w be a fixed element of V. Is the transformation $T(v) = \langle v, w \rangle$ from V to $\mathbb{R}$ linear? What is its image? Give a geometric interpretation of its kernel.

9. Recall that a function $f(t)$ from $\mathbb{R}$ to $\mathbb{R}$ is called

$$\text{even if } f(-t) = f(t), \quad \text{for all } t,$$

and

$$\text{odd if } f(-t) = -f(t), \quad \text{for all } t.$$

Show that if $f(x)$ is an odd continuous function and $g(x)$ is an even continuous function, then functions $f(x)$ and $g(x)$ are orthogonal in the space $C[-1, 1]$ with the inner product defined in Example 1.

10. Consider the space P_2 with inner product

$$\langle f, g \rangle = \frac{1}{2} \int_{-1}^{1} f(t)g(t) \, dt.$$

Find an *orthonormal* basis of the space of all functions in P_2 orthogonal to $f(t) = t$.

11. The angle between two nonzero elements v and w of an inner product space is defined as

$$\sphericalangle(v, w) = \arccos \frac{\langle v, w \rangle}{\|v\| \, \|w\|}.$$

In the space $C[-\pi, \pi]$ with inner product

$$\langle f, g \rangle = \frac{1}{\pi} \int_{-\pi}^{\pi} f(t)g(t) \, dt,$$

find the angle between $f(t) = \cos(t)$ and $g(t) = \cos(t + \delta)$, where $0 \le \delta \le \pi$. *Hint:* Use the formula $\cos(t + \delta) = \cos(t)\cos(\delta) - \sin(t)\sin(\delta)$.

12. Find all Fourier coefficients of the absolute value function

$$f(t) = |t|.$$

13. For a function f in $C[-\pi, \pi]$ (with the inner product defined on page 232), consider the sequence of all its Fourier coefficients,

$$(a_0, b_1, c_1, b_2, c_2, \ldots, b_n, c_n, \ldots).$$

Is this infinite sequence in ℓ_2? If so, what is the relationship between

$$\|f\| \quad \text{(the norm in } C[-\pi, \pi])$$

and

$$\|(a_0, b_1, c_1, b_2, c_2, \ldots)\| \quad \text{(the norm in } \ell_2)?$$

The inner product space ℓ_2 was introduced in Example 2.

14. Which of the following is an inner product in P_2? Explain.

a. $\langle f, g \rangle = f(1)g(1) + f(2)g(2)$.

b. $\langle\!\langle f, g \rangle\!\rangle = f(1)g(1) + f(2)g(2) + f(3)g(3)$.

15. For which values of the constants b, c, and d is the following an inner product in $\mathbb{R}^2$?

$$\left\langle \begin{bmatrix} x_1 \\ x_2 \end{bmatrix}, \begin{bmatrix} y_1 \\ y_2 \end{bmatrix} \right\rangle = x_1 y_1 + b x_1 y_2 + c x_2 y_1 + d x_2 y_2.$$

Hint: Be prepared to complete a square.

16. a. Find an *orthonormal* basis of the space P_1 with inner product

$$\langle f, g \rangle = \int_0^1 f(t)g(t) \, dt.$$

b. Find the linear polynomial $g(t) = a + bt$ that best approximates the function $f(t) = t^2$ in the interval $[0, 1]$ in the (continuous) least-squares sense. Draw a sketch.

17. Consider a linear space V. For which linear transformations T from V to $\mathbb{R}^n$ is

$$\langle v, w \rangle = T(v) \cdot T(w)$$

$\nwarrow$ dot product

an inner product in V?

18. Consider an orthonormal basis $\mathfrak{B}$ of the inner product space V. For an element f of V, what is the relationship between $\|f\|$ and $\|[f]_\mathfrak{B}\|$ (the norm in $\mathbb{R}^n$ defined by the dot product)?

19. For which 2×2 matrices A is

$$\langle \vec{v}, \vec{w} \rangle = \vec{v}^T A \vec{w}$$

an inner product in $\mathbb{R}^2$? *Hint:* Be prepared to complete a square.

20. Consider the inner product

$$\langle \vec{v}, \vec{w} \rangle = \vec{v}^T \begin{bmatrix} 1 & 2 \\ 2 & 8 \end{bmatrix} \vec{w}$$

in $\mathbb{R}^2$. (See Exercise 19.)

a. Find all vectors in $\mathbb{R}^2$ perpendicular to $\begin{bmatrix} 1 \\ 0 \end{bmatrix}$ with respect to this inner product.

b. Find an orthonormal basis of $\mathbb{R}^2$ with respect to this inner product.

21. If $\|\vec{v}\|$ denotes the standard norm in $\mathbb{R}^n$, does the formula

$$\langle \vec{v}, \vec{w} \rangle = \|\vec{v} + \vec{w}\|^2 - \|\vec{v}\|^2 - \|\vec{w}\|^2$$

define an inner product in $\mathbb{R}^n$?

22. If $f(t)$ is a continuous function, what is the relationship between

$$\int_0^1 \left(f(t) \right)^2 dt \quad \text{and} \quad \left(\int_0^1 f(t)\, dt \right)^2 ?$$

Hint: Use the Cauchy–Schwarz inequality.

23. In the space P_1 of the polynomials of degree ≤ 1, we define the inner product

$$\langle f, g \rangle = \frac{1}{2}\left(f(0)g(0) + f(1)g(1) \right).$$

Find an orthonormal basis for this inner product space.

24. Consider the linear space P of all polynomials, with inner product

$$\langle f, g \rangle = \int_0^1 f(t)g(t)\, dt.$$

For three polynomials f, g, and h we are given the following inner products:

$\langle \cdot \rangle$	f	g	h
f	4	0	8
g	0	1	3
h	8	3	50

For example, $\langle f, f \rangle = 4$ and $\langle g, h \rangle = \langle h, g \rangle = 3$.

a. Find $\langle f, g + h \rangle$.

b. Find $\|g + h\|$.

c. Find $\mathrm{proj}_E h$, where $E = \mathrm{span}(f, g)$. Express your solution as linear combinations of f and g.

d. Find an orthonormal basis of $\mathrm{span}(f, g, h)$. Express the functions in your basis as linear combinations of f, g, and h.

25. Find the norm $\|\vec{x}\|$ of

$$\vec{x} = \left(1, \frac{1}{2}, \frac{1}{3}, \ldots, \frac{1}{n}, \ldots \right) \quad \text{in } \ell_2.$$

(ℓ_2 is defined in Example 2.)

26. Find the Fourier coefficients of the piecewise continuous function

$$f(t) = \begin{cases} -1 & \text{if } t \le 0 \\ 1 & \text{if } t > 0. \end{cases}$$

Sketch the graphs of the first few Fourier polynomials.

27. Find the Fourier coefficients of the piecewise continuous function

$$f(t) = \begin{cases} 0 & \text{if } t \le 0 \\ 1 & \text{if } t > 0. \end{cases}$$

28. Apply Fact 5.5.6 to your answer in Exercise 26.

29. Apply Fact 5.5.6 to your answer in Exercise 27.

30. Consider an ellipse E in $\mathbb{R}^2$ whose center is the origin. Show that there is an inner product $\langle \cdot, \cdot \rangle$ in $\mathbb{R}^2$ such that E consists of all vectors $\vec{x}$ with $\|\vec{x}\| = 1$, where the norm is taken with respect to the inner product $\langle \cdot, \cdot \rangle$.

31. *Gaussian Integration.* In an introductory calculus course, you may have seen approximation formulas for integrals of the form

$$\int_a^b f(t)\, dt \approx \sum_{i=i}^{n} w_i f(a_i),$$

where the a_i are equally spaced points in the interval (a, b), and the w_i are certain "weights" (Riemann sums, trapezoidal sums, and Simpson's rule). Gauss has shown that, with the same computational effort, we can get better approximations if we drop the requirement that the a_i be equally spaced. Next, we discuss his approach.

Consider the space P_n with the inner product

$$\langle f, g \rangle = \int_{-1}^1 f(t)g(t)\, dt.$$

Let $f_0, f_1, \ldots, f_n$ be an orthonormal basis of this space, with $\mathrm{degree}(f_k) = k$. (To construct such a basis, apply the Gram–Schmidt process to the standard basis 1, $t, \ldots, t^n$.) It can be shown that f_n has n distinct roots $a_1, a_2, \ldots, a_n$ in the interval $(-1, 1)$. We can find "weights" $w_1, w_2, \ldots, w_n$ such that

$$\int_{-1}^1 f(t)\, dt = \sum_{i=1}^{n} w_i f(a_1),$$

for all polynomials of degree less than n. (See Exercise 4.3.65.) In fact, much more is true: This formula holds for all polynomials $f(t)$ of degree less than $2n$.

You are not asked to prove the foregoing assertions for arbitrary n, but work out the case $n = 2$: Find a_1, a_2 and w_1, w_2, and show that the formula

$$\int_{-1}^1 f(t)\, dt = w_1 f(a_1) + w_2 f(a_2)$$

holds for all cubic polynomials.

Chapter 5

TRUE OR FALSE?

1. If matrix A is orthogonal, then matrix A^2 must be orthogonal as well.

2. The equation $(AB)^T = A^T B^T$ holds for all $n \times n$ matrices A and B.

3. If A and B are symmetric $n \times n$ matrices, then $A + B$ must be symmetric as well.

4. If matrices A and S are orthogonal, then $S^{-1}AS$ is orthogonal as well.

5. All nonzero symmetric matrices are invertible.

6. If A is an $n \times n$ matrix such that $AA^T = I_n$, then A must be an orthogonal matrix.

7. If $\vec{v}$ is a unit vector in $\mathbb{R}^n$, and $L = \text{span}(\vec{v})$, then $\text{proj}_L(\vec{x}) = (\vec{x} \cdot \vec{v})\vec{x}$ for all vectors $\vec{x}$ in $\mathbb{R}^n$.

8. If A is a symmetric matrix, then $7A$ must be symmetric as well.

9. If T is a linear transformation from $\mathbb{R}^n$ to $\mathbb{R}^n$ such that $T(\vec{e}_1), T(\vec{e}_2), \ldots, T(\vec{e}_n)$ are all unit vectors, then T must be an orthogonal transformation.

10. If A is an invertible matrix, then the equation $(A^T)^{-1} = (A^{-1})^T$ must hold.

11. If A and B are symmetric $n \times n$ matrices, then $ABBA$ must be symmetric as well.

12. If matrices A and B commute, then matrices A^T and B^T must commute as well.

13. There is a subspace V of $\mathbb{R}^5$ such that $\dim(V) = \dim(V^\perp)$, where $V^\perp$ denotes the orthogonal complement of V.

14. Every invertible matrix A can be expressed as the product of an orthogonal matrix and an upper triangular matrix.

15. If $\vec{x}$ and $\vec{y}$ are two vectors in $\mathbb{R}^n$, then the equation $\|\vec{x} + \vec{y}\|^2 = \|\vec{x}\|^2 + \|\vec{y}\|^2$ must hold.

16. If A is an $n \times n$ matrix such that $\|A\vec{u}\| = 1$ for all unit vectors $\vec{u}$, then A must be an orthogonal matrix.

17. If matrix A is orthogonal, then A^T must be orthogonal as well.

18. If A and B are symmetric $n \times n$ matrices, then AB must be symmetric as well.

19. If V is a subspace of $\mathbb{R}^n$ and $\vec{x}$ is a vector in $\mathbb{R}^n$, then the inequality $\vec{x} \cdot (\text{proj}_V \vec{x}) \geq 0$ must hold.

20. If A is any matrix with $\ker(A) = \{\vec{0}\}$, then the matrix AA^T represents the orthogonal projection onto the image of A.

21. The entries of an orthogonal matrix are all less than or equal to 1.

22. For every nonzero subspace of $\mathbb{R}^n$ there is an orthonormal basis.

23. $\begin{bmatrix} 3 & -4 \\ 4 & 3 \end{bmatrix}$ is an orthogonal matrix.

24. If V is a subspace of $\mathbb{R}^n$ and $\vec{x}$ is a vector in $\mathbb{R}^n$, then vector $\text{proj}_V \vec{x}$ must be orthogonal to vector $\vec{x} - \text{proj}_V \vec{x}$.

25. If A and B are orthogonal 2×2 matrices, then $AB = BA$.

26. If A is a symmetric matrix, vector $\vec{v}$ is in the image of A, and $\vec{w}$ is in the kernel of A, then the equation $\vec{v} \cdot \vec{w} = 0$ must hold.

27. The formula $\ker(A) = \ker(A^T A)$ holds for all matrices A.

28. If $A^T A = AA^T$ for an $n \times n$ matrix A, then A must be orthogonal.

29. If A is any symmetric 2×2 matrix, then there must be a real number x such that matrix $A - xI_2$ fails to be invertible.

30. If A is any matrix, then matrix $\frac{1}{2}(A - A^T)$ is skew-symmetric.

31. If A is an invertible matrix such that $A^{-1} = A$, then A must be orthogonal.

32. If the entries of two vectors $\vec{v}$ and $\vec{w}$ in $\mathbb{R}^n$ are all positive, then $\vec{v}$ and $\vec{w}$ must enclose an acute angle.

33. The formula $(\ker B)^\perp = \text{im}(B^T)$ holds for all matrices B.

34. The matrix $A^T A$ is symmetric for all matrices A.

35. If matrix A is similar to B and A is orthogonal, then B must be orthogonal as well.

36. The formula $\text{Im}(B) = \text{Im}(B^T B)$ holds for all square matrices B.

37. If matrix A is symmetric and matrix S is orthogonal, then matrix $S^{-1}AS$ must be symmetric.

38. If A is a square matrix such that $A^T A = AA^T$, then $\ker(A) = \ker(A^T)$.

39. There are orthogonal 2×2 matrices A and B such that $A + B$ is orthogonal as well.

40. If $\|A\vec{x}\| \leq \|\vec{x}\|$ for all $\vec{x}$ in $\mathbb{R}^n$, then A must represent the orthogonal projection onto a subspace V of $\mathbb{R}^n$.

41. Any square matrix can be written as the sum of a symmetric and a skew-symmetric matrix.

42. If $x_1, x_2, \ldots, x_n$ are any real numbers, then the inequality $\left(\sum_{k=1}^{n} x_k \right)^2 \leq n \sum_{k=1}^{n} \left(x_k^2 \right)$ must hold.

43. If $AA^T = A^2$ for a 2×2 matrix A, then A must be symmetric.

Determinants

6.1 INTRODUCTION TO DETERMINANTS

In Chapter 2, we found a criterion for the invertibility of a 2×2 matrix: The matrix

$$A = \begin{bmatrix} a & b \\ c & d \end{bmatrix}$$

is invertible if (and only if)

$$ad - bc \neq 0,$$

by Fact 2.3.6. The quantity $ad - bc$ is called the *determinant* of the matrix A, denoted by $\det(A)$. If the matrix A is invertible, then its inverse can be expressed in terms of the determinant:

$$A^{-1} = \frac{1}{ad - bc} \begin{bmatrix} d & -b \\ -c & a \end{bmatrix} = \frac{1}{\det(A)} \begin{bmatrix} d & -b \\ -c & a \end{bmatrix}.$$

It is natural to ask whether the concept of a determinant can be generalized to square matrices of arbitrary size. Can we assign a number $\det(A)$ to any square matrix A (expressed in terms of the entries of A), such that A is invertible if (and only if) $\det(A) \neq 0$?

The Determinant of a 3×3 Matrix

Let

$$A = \begin{bmatrix} a_{11} & a_{12} & a_{13} \\ a_{21} & a_{22} & a_{23} \\ a_{31} & a_{32} & a_{33} \end{bmatrix}.$$

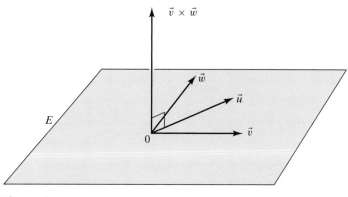

Figure 1

Let us denote the three columns $\vec{u}$, $\vec{v}$, and $\vec{w}$. The matrix A is *not* invertible if the three vectors $\vec{u}$, $\vec{v}$, and $\vec{w}$ are contained in a plane E. (This means that the image of A is not all of $\mathbb{R}^3$.)

In this case, the vector $\vec{u}$ is perpendicular to the cross product $\vec{v} \times \vec{w}$; that is,

$$\vec{u} \cdot (\vec{v} \times \vec{w}) = 0.$$

(Compare this result with Fact A.10(a) in the Appendix.) See Figure 1.

What does the condition $\vec{u} \cdot (\vec{v} \times \vec{w}) = 0$ mean in terms of the entries a_{ij} of the matrix A?

$$\vec{u} \cdot (\vec{v} \times \vec{w}) = \begin{bmatrix} a_{11} \\ a_{21} \\ a_{31} \end{bmatrix} \cdot \left(\begin{bmatrix} a_{12} \\ a_{22} \\ a_{32} \end{bmatrix} \times \begin{bmatrix} a_{13} \\ a_{23} \\ a_{33} \end{bmatrix} \right) = \begin{bmatrix} a_{11} \\ a_{21} \\ a_{31} \end{bmatrix} \cdot \begin{bmatrix} a_{22}a_{33} - a_{32}a_{23} \\ a_{32}a_{13} - a_{12}a_{33} \\ a_{12}a_{23} - a_{22}a_{13} \end{bmatrix}$$

$$= a_{11}(a_{22}a_{33} - a_{32}a_{23}) + a_{21}(a_{32}a_{13} - a_{12}a_{33}) + a_{31}(a_{12}a_{23} - a_{22}a_{13})$$

$$= a_{11}a_{22}a_{33} - a_{11}a_{32}a_{23} + a_{21}a_{32}a_{13} - a_{21}a_{12}a_{33} + a_{31}a_{12}a_{23} - a_{31}a_{22}a_{13}.$$

We define the last expression as the determinant of A. The preceding work shows that the 3×3 matrix A is invertible if (and only if) its determinant is nonzero.

Here is a memory aid for the determinant of a 3×3 matrix:

Fact 6.1.1

Sarrus's rule[1]

To find the determinant of a 3×3 matrix A, write the first two columns of A to the right of A. Then, multiply the entries along the six diagonals:

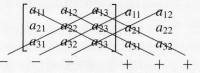

Add or subtract these diagonal products as shown in the diagram:

$$\det(A) = a_{11}a_{22}a_{33} + a_{12}a_{23}a_{31} + a_{13}a_{21}a_{32}$$
$$- a_{13}a_{22}a_{31} - a_{11}a_{23}a_{32} - a_{12}a_{21}a_{33}$$

---

[1] Stated by Pierre Frédéric Sarrus (1798–1861) of Strasbourg, *c.* 1820.

EXAMPLE 1 Find the determinant of

$$A = \begin{bmatrix} 1 & 2 & 3 \\ 4 & 5 & 6 \\ 7 & 8 & 10 \end{bmatrix}.$$

Solution

$$\det(A) = 1 \cdot 5 \cdot 10 + 2 \cdot 6 \cdot 7 + 3 \cdot 4 \cdot 8 - 3 \cdot 5 \cdot 7 - 1 \cdot 6 \cdot 8 - 2 \cdot 4 \cdot 10 = -3.$$

This matrix is invertible. ∎

EXAMPLE 2 Find the determinant of the upper triangular matrix

$$A = \begin{bmatrix} a & b & c \\ 0 & d & e \\ 0 & 0 & f \end{bmatrix}.$$

Solution

We find that $\det(A) = adf$, because all other contributions in Sarrus's formula are zero. The determinant of an upper (or lower) triangular 3×3 matrix is the product of its diagonal entries. ∎

EXAMPLE 3 For which choices of the scalar λ is the matrix

$$A = \begin{bmatrix} 1 - \lambda & 1 & 1 \\ 1 & 1 - \lambda & 1 \\ 1 & 1 & 1 - \lambda \end{bmatrix}$$

invertible?

Solution

$$\det(A) = (1 - \lambda)^3 + 1 + 1 - (1 - \lambda) - (1 - \lambda) - (1 - \lambda)$$
$$= 1 - 3\lambda + 3\lambda^2 - \lambda^3 + 2 - 3 + 3\lambda = -\lambda^3 + 3\lambda^2$$
$$= \lambda^2(3 - \lambda).$$

The determinant is 0 if $\lambda = 0$ or $\lambda = 3$. The matrix A is invertible if λ is neither 0 nor 3. ∎

The Determinant of an $n \times n$ Matrix

We may be tempted to define the determinant of an $n \times n$ matrix by generalizing Sarrus's rule. (See Fact 6.1.1.) For a 4×4 matrix, a naive generalization of Sarrus's rule produces the expression

$$a_{11}a_{22}a_{33}a_{44} + \cdots + a_{14}a_{21}a_{32}a_{43} - a_{14}a_{23}a_{32}a_{41} - \cdots - a_{13}a_{22}a_{31}a_{44}.$$

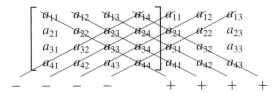

For example, for the *invertible* matrix

$$A = \begin{bmatrix} 1 & 0 & 0 & 0 \\ 0 & 1 & 0 & 0 \\ 0 & 0 & 0 & 1 \\ 0 & 0 & 1 & 0 \end{bmatrix},$$

the expression given by a generalization of Sarrus's rule is 0. This shows that we cannot define the determinant by generalizing Sarrus's rule in this way: recall that we want the determinant of an invertible matrix to be nonzero.

We have to look for a more subtle structure in the formula

$$\det(A) = a_{11}a_{22}a_{33} + a_{12}a_{23}a_{31} + a_{13}a_{21}a_{32}$$
$$- a_{13}a_{22}a_{31} - a_{11}a_{23}a_{32} - a_{12}a_{21}a_{33}$$

for the determinant of a 3×3 matrix. Note that each of the six terms in this expression is a product of three factors involving exactly one entry from each row and each column of the matrix:

For lack of a better word, we call such a choice of a number in each row and column of a square matrix a *pattern* in the matrix.[2] The simplest pattern is the *diagonal pattern,* where we choose all numbers a_{ii} on the main diagonal. For you chess players, a pattern in an 8×8 matrix corresponds to placing 8 rooks on a chessboard so that none of them can attack another.

How many patterns are there in an $n \times n$ matrix? Let us see how we can construct a pattern column by column. In the first column we have n choices. For each of these, we then have $n - 1$ choices left in the second column. Therefore, we have $n(n - 1)$ choices for the numbers in the first two columns. For each of these, there are $n - 2$ choices in the third column, and so on. When we come to the last column, we have no choice, because there is only one row left. We conclude that there are $n(n - 1)(n - 2) \cdots 3 \cdot 2 \cdot 1$ patterns in an $n \times n$ matrix. The quantity $1 \cdot 2 \cdot 3 \cdots (n - 2) \cdot (n - 1) \cdot n$ is called $n!$ (read "n factorial").

We observed that the determinant of a 3×3 matrix is obtained by adding up the products associated with some of the patterns in the matrix and then subtracting the products associated with some other patterns. We note first that the product $a_{11}a_{22}a_{33}$ associated with the diagonal pattern is added. To find out whether the product associated with another pattern is added or subtracted, we have to determine how much this pattern deviates from the diagonal pattern. More specifically, we say that two numbers in a pattern are *inverted* if one of them is to the right and above the other. Let's indicate the inversions for each of the six patterns in a 3×3 matrix.

---

[2]This theory is usually phrased in the language of **permutations**. Here we attempt a less technical presentation, without sacrificing content.

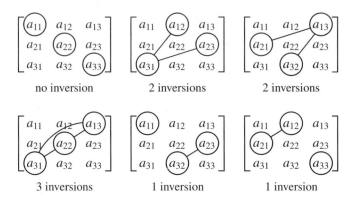

Note that the diagonal pattern is the only pattern without any inversions.

When we compute the determinant of a 3×3 matrix, the product associated with a pattern is added if the number of inversions in the pattern is even and is subtracted if this number is odd. Using this observation as a guide, we now define the determinant of a larger square matrix.

Definition 6.1.2

> **Determinant**
>
> A *pattern* in an $n \times n$ matrix is a way to choose n entries of the matrix so that there is one chosen entry in each row and one in each column of the matrix.
>
> Two numbers in a pattern are *inverted* if one of them is located to the right and above the other in the matrix.
>
> We obtain the *determinant*[3] of an $n \times n$ matrix A by summing the products associated with all patterns with an even number of inversions and subtracting the products associated with all patterns with an odd number of inversions.

A more conceptual definition of the determinant will be discussed in Exercise 6.2.41.

EXAMPLE 4   Count the number of inversions in the following pattern:

$$\begin{bmatrix} 1 & ② & 3 & 4 & 5 & 6 \\ 7 & 8 & 9 & ⑧ & 7 & 6 \\ 5 & 4 & 3 & 2 & 1 & ② \\ ③ & 4 & 5 & 6 & 7 & 8 \\ 9 & 8 & 7 & 6 & ⑤ & 4 \\ 3 & 2 & ① & 2 & 3 & 4 \end{bmatrix}$$

---

[3] It appears that determinants were first considered by the Japanese mathematician Seki Kowa (1642–1708). Seki may have known that the determinant of an $n \times n$ matrix has $n!$ terms and that rows and columns are interchangeable. (See Fact 6.2.1.) The French mathematician Alexandre-Théophile Vandermonde (1735–1796) was the first to give a coherent and systematic exposition of the theory of determinants. Throughout the 19th century, determinants were considered the ultimate tool in linear algebra, used extensively by Cauchy, Jacobi, Kronecker, and others. Recently, determinants have gone somewhat out of fashion, and some people would like to see them eliminated altogether from linear algebra. (See, for example, Sheldon Axler's article "Down with Determinants" in *The American Mathe-*

Solution

There are seven inversions, as indicated by the bars:

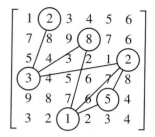

EXAMPLE 5   Apply Definition 6.1.2 to a 2×2 matrix, and verify that the result agrees with the formula given in Fact 2.3.6.

Solution

There are two patterns in the 2×2 matrix $A = \begin{bmatrix} a & b \\ c & d \end{bmatrix}$:

no inversions   one inversion

Therefore, $\det(A) = ad - bc$.   ∎

EXAMPLE 6   Find $\det(A)$ for

$$A = \begin{bmatrix} 2 & 0 & 0 & 0 & 0 \\ 0 & 3 & 0 & 0 & 0 \\ 0 & 0 & 5 & 0 & 0 \\ 0 & 0 & 0 & 7 & 0 \\ 0 & 0 & 0 & 0 & 9 \end{bmatrix}.$$

Solution

The diagonal pattern (with no inversions) makes the contribution $2 \cdot 3 \cdot 5 \cdot 7 \cdot 9 = 1890$. All other patterns contain at least one zero and will therefore make no contribution toward the determinant. We can conclude that

$$\det(A) = 2 \cdot 3 \cdot 5 \cdot 7 \cdot 9 = 1890.$$ ∎

More generally, we observe that the determinant of a diagonal matrix is the product of the diagonal entries.

―――――――
matical Monthly, February 1995, where we read: "This paper will show how linear algebra can be done better without determinants." Read it and see what you think.)

EXAMPLE 7   Find det(A) for

$$A = \begin{bmatrix} 0 & 2 & 0 & 0 & 0 & 0 \\ 0 & 0 & 0 & 8 & 0 & 0 \\ 0 & 0 & 0 & 0 & 0 & 2 \\ 3 & 0 & 0 & 0 & 0 & 0 \\ 0 & 0 & 0 & 0 & 5 & 0 \\ 0 & 0 & 1 & 0 & 0 & 0 \end{bmatrix}.$$

Solution

As in Example 6, only one pattern makes a nonzero contribution toward the determinant:

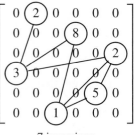

7 inversions

Thus, det(A) = $-3 \cdot 2 \cdot 1 \cdot 8 \cdot 5 \cdot 2 = -480$.   ∎

EXAMPLE 8   Find det(A) for

$$A = \begin{bmatrix} 6 & 0 & 1 & 0 & 0 \\ 9 & 3 & 2 & 3 & 7 \\ 8 & 0 & 3 & 2 & 9 \\ 0 & 0 & 4 & 0 & 0 \\ 5 & 0 & 5 & 0 & 1 \end{bmatrix}.$$

Solution

Again, only one pattern makes a nonzero contribution, but this fact is not as obvious here as it was in Examples 6 and 7. In the second column, we must choose the second component, 3. Then, in the fourth column, we must choose the third component, 2. Next, think about the last column; and so on:

$$\begin{bmatrix} 6 & 0 & 1 & 0 & 0 \\ 9 & 3 & 2 & 3 & 7 \\ 8 & 0 & 3 & 2 & 9 \\ 0 & 0 & 4 & 0 & 0 \\ 5 & 0 & 5 & 0 & 1 \end{bmatrix}$$

1 inversion

det(A) = $-6 \cdot 3 \cdot 4 \cdot 2 \cdot 1 = -144$.   ∎

EXAMPLE 9   Find det(A) for

$$A = \begin{bmatrix} 1 & 2 & 3 & 4 & 5 \\ 0 & 2 & 3 & 4 & 5 \\ 0 & 0 & 3 & 4 & 5 \\ 0 & 0 & 0 & 4 & 5 \\ 0 & 0 & 0 & 0 & 5 \end{bmatrix}.$$

Solution

Note that A is an upper triangular matrix. To make a nonzero contribution, a pattern must contain the first component of the first column, then the second component of the second column, and so on. Thus, only the diagonal pattern makes a nonzero contribution. We conclude that

$$\det(A) = \text{product of the diagonal entries} = 1 \cdot 2 \cdot 3 \cdot 4 \cdot 5 = 120. \quad \blacksquare$$

We can generalize this result:

Fact 6.1.3   The determinant of an (upper or lower) triangular matrix is the product of the diagonal entries of the matrix.

EXAMPLE 10   Find det(A) for

$$A = \begin{bmatrix} a & b & c & d \\ e & f & g & h \\ 0 & 0 & k & m \\ 0 & 0 & 0 & n \end{bmatrix}.$$

Solution

Most of the $4! = 24$ patterns in this matrix contain zeros. To get a pattern that could be nonzero, we have to choose the entry n in the last row and the k in the third row. In the first two rows, we are then left with two choices:

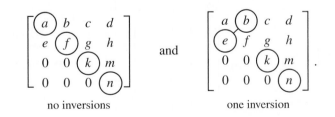

and

no inversions   one inversion

Thus, $\det(A) = afkn - ebkn = (af - eb)kn$. Note that the first factor is the determinant of the matrix $\begin{bmatrix} a & b \\ e & f \end{bmatrix}$. $\blacksquare$

EXERCISES

GOALS Use patterns and inversions to find the determinant of an $n \times n$ matrix. Use determinants to check the invertibility of 2×2 and 3×3 matrices.

Find the determinants of the matrices in Exercises 1 through 20.

1. $\begin{bmatrix} 1 & 2 \\ 3 & 4 \end{bmatrix}$.

2. $\begin{bmatrix} 7 & 11 \\ 7 & 11 \end{bmatrix}$.

3. $\begin{bmatrix} 1 & 2 & 3 \\ 0 & 2 & 3 \\ 0 & 0 & 3 \end{bmatrix}$.

4. $\begin{bmatrix} 2 & 0 & 0 \\ 3 & 2 & 0 \\ 4 & 3 & 2 \end{bmatrix}$.

5. $\begin{bmatrix} 0 & 0 & 2 \\ 0 & 3 & 0 \\ 4 & 0 & 0 \end{bmatrix}$.

6. $\begin{bmatrix} 4 & 3 & 2 \\ 4 & 3 & 0 \\ 4 & 0 & 0 \end{bmatrix}$.

7. $\begin{bmatrix} 1 & 2 & 3 \\ 4 & 5 & 6 \\ 7 & 8 & 9 \end{bmatrix}$.

8. $\begin{bmatrix} 1 & 2 & 3 \\ 4 & 5 & 6 \\ 7 & 8 & 7 \end{bmatrix}$.

9. $\begin{bmatrix} 7 & 8 & 9 \\ 0 & 0 & 0 \\ 5 & 6 & 7 \end{bmatrix}$.

10. $\begin{bmatrix} 7 & 8 & 9 \\ 7 & 8 & 9 \\ 4 & 5 & 6 \end{bmatrix}$.

11. $\begin{bmatrix} 1 & 1 & 1 \\ 1 & 1 & 1 \\ 1 & 1 & 1 \end{bmatrix}$.

12. $\begin{bmatrix} a & b & c \\ a & b & c \\ d & e & f \end{bmatrix}$.

13. $\begin{bmatrix} 2 & 3 & 4 & 5 \\ 0 & 6 & 7 & 8 \\ 0 & 0 & 3 & 2 \\ 0 & 0 & 0 & 1 \end{bmatrix}$.

14. $\begin{bmatrix} 0 & 0 & 0 & 1 \\ 0 & 0 & 1 & 0 \\ 0 & 1 & 0 & 0 \\ 1 & 0 & 0 & 0 \end{bmatrix}$.

15. $\begin{bmatrix} 0 & 2 & 3 & 4 \\ 1 & 2 & 3 & 4 \\ 0 & 0 & 0 & 4 \\ 0 & 0 & 3 & 4 \end{bmatrix}$.

16. $\begin{bmatrix} 0 & 0 & 2 & 3 & 1 \\ 0 & 0 & 0 & 2 & 2 \\ 0 & 9 & 7 & 9 & 3 \\ 0 & 0 & 0 & 0 & 5 \\ 3 & 4 & 5 & 8 & 5 \end{bmatrix}$.

17. $\begin{bmatrix} 0 & 0 & 1 & 0 & 0 \\ 0 & 0 & 0 & 1 & 0 \\ 1 & 0 & 0 & 0 & 0 \\ 0 & 0 & 0 & 0 & 1 \\ 0 & 1 & 0 & 0 & 0 \end{bmatrix}$.

18. $\begin{bmatrix} a & b & c & d & e \\ 0 & 0 & 0 & 0 & 0 \\ f & g & h & i & j \\ k & l & m & n & p \\ q & r & s & t & u \end{bmatrix}$.

19. $\begin{bmatrix} a & b & c \\ 0 & p & q \\ 0 & r & s \end{bmatrix}$.

20. $\begin{bmatrix} 1 & 1 & 1 & 1 & 1 & 1 \\ 1 & 1 & 1 & 1 & 1 & 1 \\ 1 & 1 & 1 & 1 & 1 & 1 \\ 1 & 1 & 1 & 1 & 1 & 1 \\ 1 & 1 & 1 & 1 & 1 & 1 \\ 1 & 1 & 1 & 1 & 1 & 1 \end{bmatrix}$.

Use the determinant to find out which matrices in Exercises 21 through 25 are invertible.

21. $\begin{bmatrix} 7 & 6 \\ 9 & 8 \end{bmatrix}$.

22. $\begin{bmatrix} 3 & 2 \\ 6 & 4 \end{bmatrix}$.

23. $\begin{bmatrix} a & b & c \\ 0 & b & c \\ 0 & 0 & c \end{bmatrix}$.

24. $\begin{bmatrix} 1 & 2 & 3 \\ 4 & 5 & 6 \\ 7 & 8 & 9 \end{bmatrix}$.

25. $\begin{bmatrix} 1 & 2 & 3 \\ 4 & 5 & 6 \\ 7 & 8 & 7 \end{bmatrix}$.

For the matrices A in Exercises 26 through 32, find all (real) numbers λ such that the matrix $\lambda I_n - A$ fails to be invertible.

26. $\begin{bmatrix} 1 & 3 \\ 0 & 3 \end{bmatrix}$.

27. $\begin{bmatrix} 4 & 2 \\ 2 & 7 \end{bmatrix}$.

28. $\begin{bmatrix} 2 & -3 \\ 3 & 2 \end{bmatrix}$.

29. $\begin{bmatrix} 3 & 1 \\ -4 & -1 \end{bmatrix}$.

30. $\begin{bmatrix} 1 & 0 & 0 \\ 1 & 2 & 0 \\ 1 & 2 & 3 \end{bmatrix}$.

31. $\begin{bmatrix} 0 & 0 & 1 \\ 0 & 0 & 1 \\ 1 & 1 & 1 \end{bmatrix}$.

32. $\begin{bmatrix} 0 & 1 & 0 \\ 0 & 0 & 1 \\ 0 & 8 & -2 \end{bmatrix}$.

33. If a matrix A has a row of zeros or a column of zeros, what can you say about its determinant?

34. A square matrix is called a *permutation matrix* if each row and each column contains exactly one entry 1, with all other entries being 0. Examples are

$$I_n, \quad \begin{bmatrix} 0 & 1 \\ 1 & 0 \end{bmatrix}, \quad \begin{bmatrix} 0 & 1 & 0 \\ 0 & 0 & 1 \\ 1 & 0 & 0 \end{bmatrix}.$$

What are the possible values of the determinant of a permutation matrix? Explain.

35. Find the determinant of the 100×100 matrix A:

$$A = \begin{bmatrix} 0 & 0 & \dots & 0 & 1 \\ 0 & 0 & \dots & 1 & 0 \\ \vdots & \vdots & & \vdots & \vdots \\ 0 & 1 & \dots & 0 & 0 \\ 1 & 0 & \dots & 0 & 0 \end{bmatrix}.$$

The ijth entry of A is 1 if $i + j = 101$; all other entries are 0.

36. Find the determinant of the $n \times n$ matrix

$$A = \begin{bmatrix} 0 & 0 & \cdots & 0 & 1 \\ 0 & 0 & \cdots & 1 & 0 \\ \vdots & \vdots & & \vdots & \vdots \\ 0 & 1 & \cdots & 0 & 0 \\ 1 & 0 & \cdots & 0 & 0 \end{bmatrix}.$$

The ijth entry of A is 1 if $i + j = n + 1$; all other entries are 0. Give your answer in terms of the remainder you get when you divide n by 4.

37. For 2×2 matrices A, B, C, form the partitioned 4×4 matrix

$$M = \begin{bmatrix} A & B \\ 0 & C \end{bmatrix}.$$

Express $\det(M)$ in terms of the determinants of A, B, and C.

38. For 2×2 matrices A, B, C, D, form the partitioned 4×4 matrix

$$M = \begin{bmatrix} A & B \\ C & D \end{bmatrix}.$$

Is it necessarily true that $\det(M) = \det(A) \det(D) - \det(B) \det(C)$?

39. Explain why any pattern in a matrix A, other than the diagonal pattern, contains at least one entry below the diagonal and at least one entry above the diagonal.

40. For which choices of α is the matrix

$$A = \begin{bmatrix} \cos \alpha & 1 & -\sin \alpha \\ 0 & 2 & 0 \\ \sin \alpha & 3 & \cos \alpha \end{bmatrix}$$

invertible?

41. For which choices of x is the matrix

$$A = \begin{bmatrix} 1 & 1 & x \\ 1 & x & x \\ x & x & x \end{bmatrix}$$

invertible?

In Exercises 42 through 45, let A be an arbitrary 2×2 matrix with $\det(A) = k$.

42. If B is obtained from A by multiplying the first row by 3, what is $\det(B)$?

43. If C is obtained by swapping the two rows of A, what is $\det(C)$?

44. If D is obtained from A by adding 6 times the second row to the first, what is $\det(D)$?

45. What is $\det(A^T)$?

46. Consider two vectors $\vec{v}$ and $\vec{w}$ in $\mathbb{R}^3$. Form the matrix

$$A = \begin{bmatrix} \vec{v} \times \vec{w} & \vec{v} & \vec{w} \end{bmatrix}.$$

Express $\det(A)$ in terms of $\|\vec{v} \times \vec{w}\|$. For which choices of $\vec{v}$ and $\vec{w}$ is A invertible?

47. If A is an $n \times n$ matrix, what is the relationship between $\det(A)$ and $\det(-A)$?

48. If A is an $n \times n$ matrix and k is a constant, what is the relationship between $\det(A)$ and $\det(kA)$?

49. If A is an invertible 2×2 matrix, what is the relationship between $\det(A)$ and $\det(A^{-1})$?

50. Does the following matrix have an LU factorization (see Exercises 2.4.58 and 2.4.61)?

$$A = \begin{bmatrix} 7 & 4 & 2 \\ 5 & 3 & 1 \\ 3 & 1 & 3 \end{bmatrix}$$

51. Find the determinant of the $(2n) \times (2n)$ matrix

$$A = \begin{bmatrix} 0 & I_n \\ I_n & 0 \end{bmatrix}.$$

52. Is the determinant of the matrix

$$A = \begin{bmatrix} 1 & 1000 & 2 & 3 & 4 \\ 5 & 6 & 7 & 1000 & 8 \\ 1000 & 9 & 8 & 7 & 6 \\ 5 & 4 & 3 & 2 & 1000 \\ 1 & 2 & 1000 & 3 & 4 \end{bmatrix}$$

positive or negative? How can you tell? Do not use technology.

6.2 PROPERTIES OF THE DETERMINANT

The main goal of this section is to show that a square matrix of any size is invertible if (and only if) its determinant is nonzero. As we work toward this goal, we will discuss a number of other remarkable properties of the determinant.

The Determinant of the Transpose[4]

EXAMPLE 1 Let

$$
A = \begin{bmatrix}
1 & 2 & 3 & 4 & 5 \\
6 & 7 & 8 & 9 & 8 \\
7 & 6 & 5 & 4 & 3 \\
2 & 1 & 2 & 3 & 4 \\
5 & 6 & 7 & 8 & 9
\end{bmatrix}.
$$

Express $\det(A^T)$ in terms of $\det(A)$. You need not compute $\det(A)$.

Solution

For each pattern in A, we can consider the corresponding (transposed) pattern in A^T; for example,

$$
A = \begin{bmatrix}
1 & 2 & ③ & 4 & 5 \\
⑥ & 7 & 8 & 9 & 8 \\
7 & 6 & 5 & 4 & ③ \\
2 & ① & 2 & 3 & 4 \\
5 & 6 & 7 & ⑧ & 9
\end{bmatrix}, \qquad
A^T = \begin{bmatrix}
1 & ⑥ & 7 & 2 & 5 \\
2 & 7 & 6 & ① & 6 \\
③ & 8 & 5 & 2 & 7 \\
4 & 9 & 4 & 3 & ⑧ \\
5 & 8 & ③ & 4 & 9
\end{bmatrix}.
$$

The two patterns (in A and A^T) involve the same numbers, and they contain the same number of inversions, but the role of the two numbers in each inversion is reversed. Therefore, the two patterns make the same contributions to the respective determinants. Since these observations apply to all patterns of A, we can conclude that $\det(A^T) = \det(A)$. ■

Since we have not used any special properties of the matrix A in Example 1, we can state more generally:

Fact 6.2.1

Determinant of the transpose
If A is a square matrix, then

$$
\det(A^T) = \det(A).
$$

This symmetry property will prove very useful. Any property of the determinant expressed in terms of the rows holds for the columns as well, and vice versa.

Linearity Properties of the Determinant

The function $T(A) = \det(A)$ from $\mathbb{R}^{n \times n}$ to $\mathbb{R}$ is nonlinear (if $n > 1$). Still, the determinant has some noteworthy linearity properties.

---

[4]If you skipped Chapter 5, read Definition 5.3.5.

EXAMPLE 2   Consider the transformation

$$T \begin{bmatrix} x_1 \\ x_2 \\ x_3 \\ x_4 \end{bmatrix} = \det \begin{bmatrix} 1 & 2 & x_1 & 3 \\ 4 & 5 & x_2 & 6 \\ 7 & 6 & x_3 & 5 \\ 4 & 3 & x_4 & 1 \end{bmatrix}$$

from $\mathbb{R}^4$ to $\mathbb{R}$. Is this transformation linear?

Solution

Each pattern in the matrix

$$A = \begin{bmatrix} 1 & 2 & x_1 & 3 \\ 4 & 5 & x_2 & 6 \\ 7 & 6 & x_3 & 5 \\ 4 & 3 & x_4 & 1 \end{bmatrix}$$

involves three numbers and one of the variables x_i; that is, the product associated with a pattern has the form $k x_i$, for some constant k. The diagonal pattern, for example, makes the contribution $5x_3$. The determinant itself, obtained by adding and subtracting such products, has the form

$$\det(A) = c_1 x_1 + c_2 x_2 + c_3 x_3 + c_4 x_4,$$

for some constants c_i. Therefore, the transformation T is indeed linear, by Definition 2.1.1. (What is the matrix of T?)   ■

We can generalize:

Fact 6.2.2

Linearity of the determinant in the columns
Consider the vectors $\vec{v}_1, \ldots, \vec{v}_{j-1}, \vec{v}_{j+1}, \ldots, \vec{v}_n$ in $\mathbb{R}^n$. Then, the transformation

$$T(\vec{x}) = \det \begin{bmatrix} | & & | & | & | & & | \\ \vec{v}_1 & \cdots & \vec{v}_{j-1} & \vec{x} & \vec{v}_{j+1} & \cdots & \vec{v}_n \\ | & & | & | & | & & | \end{bmatrix}$$

from $\mathbb{R}^n$ to $\mathbb{R}$ is linear. This property is called *linearity of the determinant in the jth column*. Likewise, the determinant is linear in the rows.

For example, the transformation

$$T \begin{bmatrix} x_1 \\ x_2 \\ x_3 \\ x_4 \end{bmatrix} = \det \begin{bmatrix} 1 & 9 & 6 & 9 \\ 1 & 2 & 9 & 1 \\ x_1 & x_2 & x_3 & x_4 \\ 1 & 2 & 3 & 4 \end{bmatrix}$$

is linear (by linearity in the third row).

Consider the linear transformation

$$
T(\vec{x}) = \det
\begin{bmatrix}
| & & | & | & | & & | \\
\vec{v}_1 & \dots & \vec{v}_{j-1} & \vec{x} & \vec{v}_{j+1} & \dots & \vec{v}_n \\
| & & | & | & | & & |
\end{bmatrix}
$$

introduced in Fact 6.2.2.

Using the alternative characterization of a linear transformation (see Fact 2.2.1), we infer that

$$
T(\vec{x} + \vec{y}) = T(\vec{x}) + T(\vec{y}) \quad \text{and} \quad T(k\vec{x}) = kT(\vec{x}),
$$

for all vectors $\vec{x}$ and $\vec{y}$ in $\mathbb{R}^n$ and all scalars k.

We can write these results more explicitly:

Fact 6.2.3

Linearity of the determinant in the columns

a. If three $n \times n$ matrices A, B, C are the same, except for the jth column, and the jth column of C is the sum of the jth columns of A and B, then $\det(C) = \det(A) + \det(B)$:

$$
\underbrace{\det
\begin{bmatrix}
| & & | & & | \\
\vec{v}_1 & \dots & \vec{x} + \vec{y} & \dots & \vec{v}_n \\
| & & | & & |
\end{bmatrix}}_{C}
$$

$$
= \underbrace{\det
\begin{bmatrix}
| & & | & & | \\
\vec{v}_1 & \dots & \vec{x} & \dots & \vec{v}_n \\
| & & | & & |
\end{bmatrix}}_{A}
+ \underbrace{\det
\begin{bmatrix}
| & & | & & | \\
\vec{v}_1 & \dots & \vec{y} & \dots & \vec{v}_n \\
| & & | & & |
\end{bmatrix}}_{B}
$$

b. If two $n \times n$ matrices A and B are the same, except for the jth column, and the jth column of B is k times the jth column of A, then $\det(B) = k \det(A)$:

$$
\underbrace{\det
\begin{bmatrix}
| & & | & & | \\
\vec{v}_1 & \dots & k\vec{x} & \dots & \vec{v}_n \\
| & & | & & |
\end{bmatrix}}_{B}
= k \underbrace{\det
\begin{bmatrix}
| & & | & & | \\
\vec{v}_1 & \dots & \vec{x} & \dots & \vec{v}_n \\
| & & | & & |
\end{bmatrix}}_{A}
$$

The same holds for the rows. For example,

$$
\det
\begin{bmatrix}
1 & 2 & 3 \\
x_1 + y_1 & x_2 + y_2 & x_3 + y_3 \\
4 & 5 & 6
\end{bmatrix}
= \det
\begin{bmatrix}
1 & 2 & 3 \\
x_1 & x_2 & x_3 \\
4 & 5 & 6
\end{bmatrix}
+ \det
\begin{bmatrix}
1 & 2 & 3 \\
y_1 & y_2 & y_3 \\
4 & 5 & 6
\end{bmatrix}.
$$

Determinants and Gauss–Jordan Elimination

Suppose you have to find the determinant of a 20×20 matrix. Since there are $20! \approx 2 \cdot 10^{18}$ patterns in this matrix, you would have to perform more than 10^{19} multiplications to compute the determinant using Definition 6.1.2. Even if a computer performed 1 billion multiplications a second, it would still take over 1,000 years to carry out these computations. Clearly, we have to look for more efficient ways to compute the determinant.

So far in this text, we have found Gauss–Jordan elimination to be a powerful tool for solving numerical problems in linear algebra. If we could understand what happens to the determinant of a matrix as we row-reduce it, we could use Gauss–Jordan elimination to compute determinants as well. We have to understand what happens to the determinant of a matrix as we perform the three elementary row operations: (a) dividing a row by a scalar, (b) swapping two rows, and (c) adding a multiple of a row to another row. See Exercises 42 through 44 of Section 6.1.

a. If

$$A = \begin{bmatrix} - & \vec{v}_1 & - \\ & \vdots & \\ - & \vec{v}_i & - \\ & \vdots & \\ - & \vec{v}_n & - \end{bmatrix} \quad \text{and} \quad B = \begin{bmatrix} - & \vec{v}_1 & - \\ & \vdots & \\ - & \vec{v}_i / k & - \\ & \vdots & \\ - & \vec{v}_n & - \end{bmatrix},$$

then $\det(B) = (1/k) \det(A)$, by linearity in the ith row.

b. Refer to Example 3.

EXAMPLE 3   Consider the matrices

$$A = \begin{bmatrix} 1 & 2 & 3 & 4 & 5 \\ 6 & 7 & 8 & 9 & 8 \\ 7 & 6 & 5 & 4 & 3 \\ 2 & 1 & 2 & 3 & 4 \\ 5 & 6 & 7 & 8 & 9 \end{bmatrix} \quad \text{and} \quad B = \begin{bmatrix} 6 & 7 & 8 & 9 & 8 \\ 1 & 2 & 3 & 4 & 5 \\ 7 & 6 & 5 & 4 & 3 \\ 2 & 1 & 2 & 3 & 4 \\ 5 & 6 & 7 & 8 & 9 \end{bmatrix}.$$

Note that B is obtained from A by swapping the first two rows. Express $\det(B)$ in terms of $\det(A)$.

Solution

For each pattern in A, we can consider the corresponding pattern in B; for example,

$$A = \begin{bmatrix} 1 & 2 & ③ & 4 & 5 \\ ⑥ & 7 & 8 & 9 & 8 \\ 7 & 6 & 5 & 4 & ③ \\ 2 & ① & 2 & 3 & 4 \\ 5 & 6 & 7 & ⑧ & 9 \end{bmatrix} \quad \text{and} \quad B = \begin{bmatrix} ⑥ & 7 & 8 & 9 & 8 \\ 1 & 2 & ③ & 4 & 5 \\ 7 & 6 & 5 & 4 & ③ \\ 2 & ① & 2 & 3 & 4 \\ 5 & 6 & 7 & ⑧ & 9 \end{bmatrix}.$$

The two patterns (in A and B) involve the same numbers, but the number of inversions for the pattern in B is 1 less than for A. (We lose the inversion formed by the numbers in the first two rows of A.) This implies that the respective contributions the two patterns make to the determinants of A and B have the same absolute value, but opposite signs. Since these remarks apply to all patterns of A, we can conclude that

$$\det(B) = -\det(A).$$

(If the numbers in the first two rows of A do not form an inversion, then an inversion is created in B.) ■

If the matrix B is obtained from A by swapping any two rows (rather than the first two), keeping track of the number of inversions gets a little trickier. However, it is still true that the number of inversions changes by an odd number for each pattern, so that $\det(B) = -\det(A)$, as in Example 3. (Think about it.)

EXAMPLE 4   If a matrix A has two equal rows, what can you say about $\det(A)$?

Solution

Swap the two equal rows and call the resulting matrix B. Since we have swapped two *equal* rows, we have $B = A$. Now

$$\det(A) = \det(B) = -\det(A),$$

so that

$$\det(A) = 0.$$ ■

c. Finally, what happens to the determinant if we add k times the ith row to the jth row?

$$A = \begin{bmatrix} \vdots \\ - \ \vec{v}_i \ - \\ \vdots \\ - \ \vec{v}_j \ - \\ \vdots \end{bmatrix} \longrightarrow B = \begin{bmatrix} \vdots \\ - \ \vec{v}_i \ - \\ \vdots \\ - \ \vec{v}_j + k\vec{v}_i \ - \\ \vdots \end{bmatrix}$$

By linearity in the jth row, we find that

$$\det(B) = \det \begin{bmatrix} \vdots \\ - \ \vec{v}_i \ - \\ \vdots \\ - \ \vec{v}_j \ - \\ \vdots \end{bmatrix} + k \det \begin{bmatrix} \vdots \\ - \ \vec{v}_i \ - \\ \vdots \\ - \ \vec{v}_i \ - \\ \vdots \end{bmatrix} = \det(A),$$

by Example 4.

Fact 6.2.4   **Elementary row operations and determinants**
a. If B is obtained from A by dividing a row of A by a scalar k, then

$$\det(B) = (1/k)\det(A).$$

b. If B is obtained from A by a row swap, then

$$\det(B) = -\det(A).$$

c. If B is obtained from A by adding a multiple of a row of A to another row, then

$$\det(B) = \det(A).$$

Analogous results hold for elementary column operations.

Now that we understand how elementary row operations affect determinants, we can describe the relationship between the determinant of a matrix and that of its reduced row-echelon form. Suppose that in the course of Gauss–Jordan elimination, we swap rows s times and divide various rows by the scalars $k_1, k_2, \ldots, k_r$. Then,

$$\det(\text{rref } A) = (-1)^s \frac{1}{k_1 k_2 \ldots k_r} \det(A),$$

or

$$\det(A) = (-1)^s k_1 k_2 \ldots k_r \det(\text{rref } A),$$

by Fact 6.2.4.

Let us examine the cases when A is invertible and when it is not.

If A is invertible, then $\text{rref}(A) = I_n$, so that $\det(\text{rref } A) = 1$, and

$$\det(A) = (-1)^s k_1 k_2 \ldots k_r.$$

Note that this quantity is not zero, because all the scalars k_i are nonzero.

If A is not invertible, then $\text{rref}(A)$ is an upper triangular matrix with some zeros on the diagonal, so that $\det(\text{rref } A) = 0$ and $\det(A) = 0$.

We have established the following fundamental result:

Fact 6.2.5   A square matrix A is invertible if and only if $\det(A) \neq 0$.

If A is invertible, the foregoing discussion also produces a convenient method to compute the determinant, using Gauss–Jordan elimination.

Algorithm 6.2.6   Consider an invertible matrix A. Suppose you swap rows s times as you row-reduce[5] A and you divide various rows by the scalars $k_1, k_2, \ldots, k_r$. Then,

$$\det(A) = (-1)^s k_1 k_2 \ldots k_r.$$

---

[5]Here, it is not necessary to reduce A all the way to rref. It suffices to bring A into upper triangular form with 1's on the diagonal. (Why?)

Suppose you have to find the determinant of a 20×20 matrix. Gauss–Jordan elimination of an $n \times n$ matrix requires fewer than n^3 operations (i.e., fewer than 8,000 for a 20×20 matrix). If a computer performed 1 billion operations a second, this would take less than $\frac{1}{100,000}$ of a second (compared with the more than 1,000 years it takes when we use Definition 6.1.2).

Next we present a much briefer example.

EXAMPLE 5   Find

$$\det \begin{bmatrix} 0 & 2 & 4 & 6 \\ 1 & 1 & 2 & 1 \\ 1 & 1 & 2 & -1 \\ 1 & 1 & 1 & 2 \end{bmatrix}.$$

Solution

We perform Gauss–Jordan elimination, keeping a note of all row swaps and row divisions we do:

$$\begin{bmatrix} 0 & 2 & 4 & 6 \\ 1 & 1 & 2 & 1 \\ 1 & 1 & 2 & -1 \\ 1 & 1 & 1 & 2 \end{bmatrix} \rightarrow \begin{bmatrix} 1 & 1 & 2 & 1 \\ 0 & 2 & 4 & 6 \\ 1 & 1 & 2 & -1 \\ 1 & 1 & 1 & 2 \end{bmatrix} \begin{matrix} \\ \div 2 \\ \rightarrow \\ \end{matrix} \begin{bmatrix} 1 & 1 & 2 & 1 \\ 0 & 1 & 2 & 3 \\ 1 & 1 & 2 & -1 \\ 1 & 1 & 1 & 2 \end{bmatrix} \begin{matrix} \\ \\ -(I) \\ -(I) \end{matrix}$$

$$\begin{bmatrix} 1 & 1 & 2 & 1 \\ 0 & 1 & 2 & 3 \\ 0 & 0 & 0 & -2 \\ 0 & 0 & -1 & 1 \end{bmatrix} \rightarrow \begin{bmatrix} 1 & 1 & 2 & 1 \\ 0 & 1 & 2 & 3 \\ 0 & 0 & -1 & 1 \\ 0 & 0 & 0 & -2 \end{bmatrix} \begin{matrix} \\ \\ \div(-1) \\ \div(-2) \end{matrix} \begin{bmatrix} 1 & 1 & 2 & 1 \\ 0 & 1 & 2 & 3 \\ 0 & 0 & 1 & -1 \\ 0 & 0 & 0 & 1 \end{bmatrix}$$

We made two swaps and performed row divisions by 2, -1, and -2, so that

$$\det(A) = (-1)^2 \cdot 2 \cdot (-1) \cdot (-2) = 4. \qquad \blacksquare$$

The Determinant of a Product

Consider two $n \times n$ matrices A and B. What is the relationship between $\det(AB)$, $\det(A)$, and $\det(B)$? First, suppose A is invertible. Our work applies the following auxiliary result (the proof of which is left for Exercise 20):

Consider an invertible $n \times n$ matrix A and an arbitrary $n \times n$ matrix B. Then,

$$\text{rref}[A \vdots AB] = [I_n \vdots B].$$

Suppose we swap rows s times and divide rows by $k_1, k_2, \ldots, k_r$ as we perform this elimination.

Considering the left and right halves of the matrices separately, we conclude that

$$\det(A) = (-1)^s k_1 k_2 \ldots k_r \det(I_n) = (-1)^s k_1 k_2 \ldots k_r$$

and

$$\det(AB) = (-1)^s k_1 k_2 \ldots k_r \det(B) = \det(A) \det(B).$$

Therefore, $\det(AB) = \det(A)\det(B)$ when A is invertible. If A is not invertible, then neither is AB, so that $\det(AB) = \det(A)\det(B) = 0$.

Fact 6.2.7 **Determinant of a product**
If A and B are $n \times n$ matrices, then
$$\det(AB) = \det(A)\det(B).$$

An alternative proof of this result is outlined in Exercise 42.

The Determinant of an Inverse

If A is an invertible $n \times n$ matrix, what is the relationship between $\det(A)$ and $\det(A^{-1})$?

By definition of the inverse matrix, we have
$$AA^{-1} = I_n.$$

By taking the determinants of both sides and using Fact 6.2.7, we find that
$$\det(AA^{-1}) = \det(A)\det(A^{-1}) = \det(I_n) = 1.$$

Fact 6.2.8 **Determinant of an inverse**
If A is an invertible matrix, then
$$\det(A^{-1}) = (\det\ A)^{-1} = \frac{1}{\det(A)}.$$

EXAMPLE 6 If matrix A is similar to B, what is the relationship between $\det(A)$ and $\det(B)$?

Solution

By Definition 3.4.5, there is an invertible matrix S such that $B = S^{-1}AS$. Now,
$$\det(B) = \det(S^{-1}AS) = \det(S^{-1})\det(A)\det(S) \quad \text{(by Fact 6.2.7)}$$
$$= (\det\ S)^{-1}\det(A)\det(S) \quad \text{(by Fact 6.2.8)}$$
$$= \det(A). \quad \text{(The determinants commute, because they are scalars.)}$$
Thus, $\det(B) = \det(A)$. ∎

Minors and Laplace Expansion[6]

Recall the formula
$$\det(A) = a_{11}a_{22}a_{33} + a_{12}a_{23}a_{31} + a_{13}a_{21}a_{32} - a_{13}a_{22}a_{31} - a_{11}a_{23}a_{32} - a_{12}a_{21}a_{33}$$

[6]Named after the French mathematician Pierre-Simon Marquis de Laplace (1749–1827). Laplace is perhaps best known for his investigation into the stability of the solar system. He was also a prominent member of the committee that aided in the organization of the metric system.

for the determinant of a 3×3 matrix. (See Fact 6.1.1.) Collecting the two terms involving a_{11} and then those involving a_{21} and a_{31}, we can write

$$\det(A) = a_{11}(a_{22}a_{33} - a_{32}a_{23}) \\ + a_{21}(a_{32}a_{13} - a_{12}a_{33}) \\ + a_{31}(a_{12}a_{23} - a_{22}a_{13}).$$

(Where have we seen this formula before?)

Note that computing the determinant this way requires only 9 multiplications, compared with the 12 for Sarrus's formula. Let us analyze the structure of this formula more closely. The terms $(a_{22}a_{33} - a_{32}a_{23})$, $(a_{32}a_{13} - a_{12}a_{33})$, and $(a_{12}a_{23} - a_{22}a_{13})$ are the determinants of submatrices of A. The expression $a_{22}a_{33} - a_{32}a_{23}$ is the determinant of the matrix we get when we omit the first row and the first column of A:

$$\begin{bmatrix} a_{11} & a_{12} & a_{13} \\ a_{21} & a_{22} & a_{23} \\ a_{31} & a_{32} & a_{33} \end{bmatrix}.$$

Likewise for the other summands:

$$\det(A) = a_{11} \det \begin{bmatrix} a_{11} & a_{12} & a_{13} \\ a_{21} & a_{22} & a_{23} \\ a_{31} & a_{32} & a_{33} \end{bmatrix} - a_{21} \det \begin{bmatrix} a_{11} & a_{12} & a_{13} \\ a_{21} & a_{22} & a_{13} \\ a_{31} & a_{32} & a_{33} \end{bmatrix}$$

$$+ a_{31} \det \begin{bmatrix} a_{11} & a_{12} & a_{13} \\ a_{21} & a_{22} & a_{23} \\ a_{31} & a_{32} & a_{33} \end{bmatrix}.$$

To state these observations more succinctly, we introduce some terminology.

Definition 6.2.9

Minors

For an $n \times n$ matrix A, let A_{ij} be the matrix obtained by omitting the ith row and the jth column of A. The $(n - 1) \times (n - 1)$ matrix A_{ij} is called a *minor* of A.

$$A = \begin{bmatrix} a_{11} & a_{12} & \cdots & a_{1j} & \cdots & a_{1n} \\ a_{21} & a_{22} & \cdots & a_{2j} & \cdots & a_{2n} \\ \vdots & \vdots & & \vdots & & \vdots \\ a_{i1} & a_{i2} & \cdots & a_{ij} & \cdots & a_{in} \\ \vdots & \vdots & & \vdots & & \vdots \\ a_{n1} & a_{n2} & \cdots & a_{nj} & \cdots & a_{nn} \end{bmatrix},$$

$$A_{ij} = \begin{bmatrix} a_{11} & a_{12} & \cdots & a_{1j} & \cdots & a_{1n} \\ a_{21} & a_{22} & \cdots & a_{2j} & \cdots & a_{2n} \\ \vdots & \vdots & & \vdots & & \vdots \\ a_{i1} & a_{i2} & \cdots & a_{ij} & \cdots & a_{in} \\ \vdots & \vdots & & \vdots & & \vdots \\ a_{n1} & a_{n2} & \cdots & a_{nj} & \cdots & a_{nn} \end{bmatrix}$$

We can now represent the determinant of a 3×3 matrix more succinctly:

$$\det(A) = a_{11} \det(A_{11}) - a_{21} \det(A_{21}) + a_{31} \det(A_{31}).$$

This representation of the determinant is called the *Laplace expansion of* $\det(A)$ *down the first column*. Likewise, we can expand along the first *row* $\left(\text{since } \det(A^T) = \det(A)\right)$:

$$\det(A) = a_{11} \det(A_{11}) - a_{12} \det(A_{12}) + a_{13} \det(A_{13})$$

In fact, we can expand along any row or down any column. (We can verify this directly or argue in terms of row or column swaps.) For example, the Laplace expansion down the second column is

$$\det(A) = -a_{12} \det(A_{12}) + a_{22} \det(A_{22}) - a_{32} \det(A_{32}),$$

and the Laplace expansion along the third row is

$$\det(A) = a_{31} \det(A_{31}) - a_{32} \det(A_{32}) + a_{33} \det(A_{33}).$$

The rule for the signs is as follows: The summand $a_{ij} \det(A_{ij})$ has a negative sign if the sum of the two indices, $i + j$, is odd. The signs follow a checkerboard pattern:

$$\begin{bmatrix} + & - & + \\ - & + & - \\ + & - & + \end{bmatrix}$$

The following example will allow us to generalize Laplace expansion to larger matrices.

EXAMPLE 7   Consider an $n \times n$ matrix A whose jth column is $\vec{e}_i$:

$$A = \begin{bmatrix} a_{11} & a_{12} & \cdots & 0 & \cdots & a_{1n} \\ a_{21} & a_{22} & \cdots & 0 & \cdots & a_{2n} \\ \vdots & \vdots & & \vdots & & \vdots \\ a_{i1} & a_{i2} & \cdots & 1 & \cdots & a_{in} \\ \vdots & \vdots & & \vdots & & \vdots \\ a_{n1} & a_{n2} & \cdots & 0 & \cdots & a_{nn} \end{bmatrix}$$

$$\nearrow$$
$$j\text{th column}$$

What is the relationship between $\det(A)$ and $\det(A_{ij})$?

Solution

As we compute $\det(A)$, we need to consider only those patterns in A that involve the 1 in the jth column, since all other patterns will contain a zero. Each such pattern corresponds to a pattern of A_{ij}, involving the same numbers, except for the 1 we

have omitted. For example,

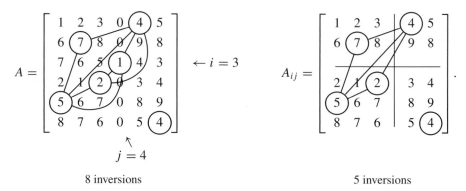

8 inversions   5 inversions

The products associated with the two patterns are the same, but what happens to the number of inversions? We lose all the inversions involving the 1 in the jth column. We leave it as Exercise 36 to show that the number of such inversions is even if $i + j$ is even and odd if $i + j$ is odd. (In our example, $i + j = 7$ is odd, and we are losing 3 inversions.) Since these remarks apply to all the patterns of A involving the 1 in the jth column, we can conclude that $\det(A) = \pm\det(A_{ij})$, taking the plus sign if $i + j$ is even and the negative sign if $i + j$ is odd. We can write this more succinctly as

$$\det(A) = (-1)^{i+j}\det(A_{ij}). \qquad \blacksquare$$

Now we are ready to show how to compute the determinant of any $n \times n$ matrix by Laplace expansion.

Fact 6.2.10

Laplace expansion

We can compute the determinant of an $n \times n$ matrix A by Laplace expansion along any row or down any column.

Expansion along the ith row:

$$\det(A) = \sum_{j=1}^{n}(-1)^{i+j}a_{ij}\det(A_{ij}).$$

Expansion down the jth column:

$$\det(A) = \sum_{i=1}^{n}(-1)^{i+j}a_{ij}\det(A_{ij}).$$

Proof   Since $\det(A^T) = \det(A)$, it suffices to demonstrate Fact 6.2.10 for the columns. Consider an $n \times n$ matrix

$$A = \begin{bmatrix} | & | & & | & & | \\ \vec{v}_1 & \vec{v}_2 & \cdots & \vec{v}_j & \cdots & \vec{v}_n \\ | & | & & | & & | \end{bmatrix},$$

with jth column

$$\vec{v}_j = \begin{bmatrix} a_{1j} \\ a_{2j} \\ \vdots \\ a_{nj} \end{bmatrix} = a_{1j}\vec{e}_1 + a_{2j}\vec{e}_2 + \cdots + a_{nj}\vec{e}_j.$$

By linearity of the determinant in the jth column (Fact 6.2.3), we can write

$$\det(A) = a_{1j} \det \begin{bmatrix} | & & | & & | \\ \vec{v}_1 & \cdots & \vec{e}_1 & \cdots & \vec{v}_n \\ | & & | & & | \end{bmatrix} + a_{2j} \det \begin{bmatrix} | & & | & & | \\ \vec{v}_1 & \cdots & \vec{e}_2 & \cdots & \vec{v}_n \\ | & & | & & | \end{bmatrix}$$

$$+ \cdots + a_{nj} \det \begin{bmatrix} | & & | & & | \\ \vec{v}_1 & \cdots & \vec{e}_n & \cdots & \vec{v}_n \\ | & & | & & | \end{bmatrix}$$

$$= a_{1j}(-1)^{1+j}\det(A_{1j}) + a_{2j}(-1)^{2+j}\det(A_{2j}) + \cdots + a_{nj}(-1)^{n+j}\det(A_{nj})$$

$$= \sum_{i=1}^{n} (-1)^{i+j} a_{ij} \det(A_{ij})$$

(by Example 7). ▲

Again, the signs follow a checkerboard pattern:

$$\begin{bmatrix} + & - & + & - & \cdots \\ - & + & - & + & \cdots \\ + & - & + & - & \cdots \\ - & + & - & + & \cdots \\ \vdots & \vdots & \vdots & \vdots & \ddots \end{bmatrix}.$$

EXAMPLE 8   Use Laplace expansion to compute $\det(A)$ for

$$A = \begin{bmatrix} 1 & 0 & 1 & 2 \\ 9 & 1 & 3 & 0 \\ 9 & 2 & 2 & 0 \\ 5 & 0 & 0 & 3 \end{bmatrix}.$$

Solution

Looking for rows or columns with as many zeros as possible, we can choose the second column:

$$\det(A) = -a_{12}\det(A_{12}) + a_{22}\det(A_{22}) - a_{32}\det(A_{32}) + a_{42}\det(A_{42})$$

$$= 1 \det \begin{bmatrix} 1 & 0 & 1 & 2 \\ 9 & 1 & 3 & 0 \\ 9 & 2 & 2 & 0 \\ 5 & 0 & 0 & 3 \end{bmatrix} - 2 \det \begin{bmatrix} 1 & 0 & 1 & 2 \\ 9 & 1 & 3 & 0 \\ 9 & 2 & 2 & 0 \\ 5 & 0 & 0 & 3 \end{bmatrix}$$

$$= \det \begin{bmatrix} 1 & 1 & 2 \\ 9 & 2 & 0 \\ 5 & 0 & 3 \end{bmatrix} - 2 \det \begin{bmatrix} 1 & 1 & 2 \\ 9 & 3 & 0 \\ 5 & 0 & 3 \end{bmatrix}$$

$$= 2 \det \begin{bmatrix} 9 & 2 \\ 5 & 0 \end{bmatrix} + 3 \det \begin{bmatrix} 1 & 1 \\ 9 & 2 \end{bmatrix} - 2 \left(2 \det \begin{bmatrix} 9 & 3 \\ 5 & 0 \end{bmatrix} + 3 \det \begin{bmatrix} 1 & 1 \\ 9 & 3 \end{bmatrix} \right)$$

↗

Expand down
the last
column

$$= -20 - 21 - 2(-30 - 18) = 55. \qquad \blacksquare$$

Computing the determinant using Laplace expansion is a bit more efficient than using the definition of the determinant, but a lot less efficient than Gauss–Jordan elimination.

The Determinant of a Linear Transformation

(for those who have studied Chapter 4)

If $T(\vec{x}) = A\vec{x}$ is a linear transformation from $\mathbb{R}^n$ to $\mathbb{R}^n$, then it is natural to define the determinant of T as the determinant of matrix A:

$$\det(T) = \det(A).$$

This definition makes sense in view of the fact that an $n \times n$ matrix is essentially the same thing as a linear transformation from $\mathbb{R}^n$ to $\mathbb{R}^n$.

If T is a linear transformation from V to V, where V is a finite-dimensional linear space, then we can introduce coordinates to define the determinant of T; we can choose a basis to transform V into $\mathbb{R}^n$. If $\mathfrak{B}$ is a basis of V and B is the $\mathfrak{B}$-matrix of T, then we define

$$\det(T) = \det(B).$$

We need to think about one issue though. If you pick another basis, $\mathfrak{A}$, of V and consider the $\mathfrak{A}$-matrix A of T, will you end up with the same determinant; that is, will $\det(A)$ equal $\det(B)$? If this isn't the case, then our attempt to define the determinant of a linear transformation has failed.

Fortunately, there is no reason to worry. We know that matrix A is similar to B (by Fact 4.3.4), so that determinants $\det(A)$ and $\det(B)$ are indeed equal, by Example 6 of this section.

Definition 6.2.11

> **The determinant of a linear transformation**
> Consider a linear transformation T from V to V, where V is a finite-dimensional linear space. If $\mathfrak{B}$ is a basis of V and B is the $\mathfrak{B}$-matrix of T, then we define
> $$\det(T) = \det(B).$$
> This determinant is independent of the basis $\mathfrak{B}$ we choose.

EXAMPLE 9   Let V be the space spanned by functions $\cos(2x)$ and $\sin(2x)$. Find the determinant of the linear transformation $D(f) = f'$ from V to V.

Solution

The matrix B of D with respect to the basis $\cos(2x)$, $\sin(2x)$ is

$$B = \begin{bmatrix} 0 & 2 \\ -2 & 0 \end{bmatrix},$$

so that

$$\det(D) = \det(B) = 4.$$ ∎

EXERCISES

GOALS Compute the determinant by Gauss–Jordan elimination or by Laplace expansion. Find the determinants of products, inverses, and transposes. Apply the linearity of the determinant in the rows and columns. Use the determinant to check whether a matrix is invertible.

Find the determinants of the matrices in Exercises 1 through 5 using either Gauss–Jordan elimination or Laplace expansion, but no technology.

1. $\begin{bmatrix} 1 & 2 & 3 \\ 4 & 5 & 6 \\ 7 & 8 & 10 \end{bmatrix}$.

2. $\begin{bmatrix} 1 & 1 & 0 & 0 \\ 2 & 9 & 1 & 0 \\ 0 & 9 & 0 & 0 \\ 1 & 9 & 9 & 5 \end{bmatrix}$.

3. $\begin{bmatrix} 1 & -1 & 2 & -2 \\ -1 & 2 & 1 & 6 \\ 2 & 1 & 14 & 10 \\ -2 & 6 & 10 & 33 \end{bmatrix}$.

4. $\begin{bmatrix} 0 & 2 & 1 & 0 & 1 \\ 0 & 0 & 2 & 0 & 2 \\ 0 & 5 & 3 & 9 & 9 \\ 0 & 7 & 4 & 0 & 1 \\ 3 & 9 & 5 & 4 & 8 \end{bmatrix}$.

5. $\begin{bmatrix} 1 & 1 & 1 & 1 & 1 \\ 1 & 2 & 2 & 2 & 2 \\ 1 & 1 & 3 & 3 & 3 \\ 1 & 1 & 1 & 4 & 4 \\ 1 & 1 & 1 & 1 & 5 \end{bmatrix}$.

6. Find the determinant of the matrix

$$M_n = \begin{bmatrix} 1 & 1 & 1 & \cdots & 1 \\ 1 & 2 & 2 & \cdots & 2 \\ 1 & 2 & 3 & \cdots & 3 \\ \vdots & \vdots & \vdots & & \vdots \\ 1 & 2 & 3 & \cdots & n \end{bmatrix}$$

for arbitrary n. (The ijth entry of M_n is the minimum of i and j.)

7. Find the determinant of the following matrix using paper and pencil:

$$A = \begin{bmatrix} 1 & 0 & 0 & 9 & 8 \\ 2 & 1 & 0 & 7 & 6 \\ 3 & 2 & 1 & 5 & 3 \\ 0 & 0 & 0 & 7 & 4 \\ 0 & 0 & 0 & 5 & 3 \end{bmatrix}$$

8. Consider the linear transformation

$$T(\vec{x}) = \det \begin{bmatrix} | & | & | & & | \\ \vec{x} & \vec{v}_2 & \vec{v}_3 & \cdots & \vec{v}_n \\ | & | & | & & | \end{bmatrix}$$

from $\mathbb{R}^n$ to $\mathbb{R}$, where $\vec{v}_2, \vec{v}_3, \ldots, \vec{v}_n$ are linearly independent vectors in $\mathbb{R}^n$. Describe image and kernel of this transformation, and determine their dimensions.

Consider a 4×4 matrix A with rows $\vec{v}_1$, $\vec{v}_2$, $\vec{v}_3$, $\vec{v}_4$. If $\det(A) = 8$, find the determinants in Exercises 9 through 14.

9. $\det \begin{bmatrix} \vec{v}_1 \\ \vec{v}_2 \\ -9\vec{v}_3 \\ \vec{v}_4 \end{bmatrix}$.

10. $\det \begin{bmatrix} \vec{v}_4 \\ \vec{v}_2 \\ \vec{v}_3 \\ \vec{v}_1 \end{bmatrix}$.

11. $\det \begin{bmatrix} \vec{v}_2 \\ \vec{v}_3 \\ \vec{v}_1 \\ \vec{v}_4 \end{bmatrix}$.

12. $\det \begin{bmatrix} \vec{v}_1 \\ \vec{v}_2 + 9\vec{v}_4 \\ \vec{v}_3 \\ \vec{v}_4 \end{bmatrix}$.

13. det $\begin{bmatrix} \vec{v}_1 \\ \vec{v}_1 + \vec{v}_2 \\ \vec{v}_1 + \vec{v}_2 + \vec{v}_3 \\ \vec{v}_1 + \vec{v}_2 + \vec{v}_3 + \vec{v}_4 \end{bmatrix}$.   **14.** det $\begin{bmatrix} 6\vec{v}_1 + 2\vec{v}_4 \\ \vec{v}_2 \\ \vec{v}_3 \\ 3\vec{v}_1 + \vec{v}_4 \end{bmatrix}$.

15. Justify the following statement: If a square matrix A has two identical columns, then $\det(A) = 0$.

16. Consider two distinct numbers, a and b. We define the function

$$f(t) = \det \begin{bmatrix} 1 & 1 & 1 \\ a & b & t \\ a^2 & b^2 & t^2 \end{bmatrix}.$$

a. Show that $f(t)$ is a quadratic function. What is the coefficient of t^2?

b. Explain why $f(a) = f(b) = 0$. Conclude that $f(t) = k(t - a)(t - b)$, for some constant k. Find k, using your work in part (a).

c. For which values of t is the matrix invertible?

17. *Vandermonde determinants* [introduced by Alexandre Théophile Vandermonde]. Consider distinct scalars a_0, $a_1, \ldots, a_n$. We define the $(n + 1) \times (n + 1)$ matrix

$$A = \begin{bmatrix} 1 & 1 & \cdots & 1 \\ a_0 & a_1 & \cdots & a_n \\ a_0^2 & a_1^2 & \cdots & a_n^2 \\ \vdots & \vdots & & \vdots \\ a_0^n & a_1^n & \cdots & a_n^n \end{bmatrix}.$$

Vandermonde showed that

$$\det(A) = \prod_{i>j}(a_i - a_j),$$

the product of all differences $(a_i - a_j)$, where i exceeds j.

a. Verify this formula in the case of $n = 1$.

b. Suppose the Vandermonde formula holds for $n - 1$. You are asked to demonstrate it for n. Consider the function

$$f(t) = \det \begin{bmatrix} 1 & 1 & \cdots & 1 & 1 \\ a_0 & a_1 & \cdots & a_{n-1} & t \\ a_0^2 & a_1^2 & \cdots & a_{n-1}^2 & t^2 \\ \vdots & \vdots & & \vdots & \vdots \\ a_0^n & a_1^n & \cdots & a_{n-1}^n & t^n \end{bmatrix}.$$

Explain why $f(t)$ is a polynomial of nth degree. Find the coefficient k of t^n using Vandermonde's formula for $a_0, \ldots, a_{n-1}$. Explain why

$$f(a_0) = f(a_1) = \cdots = f(a_{n-1}) = 0.$$

Conclude that

$$f(t) = k(t - a_0)(t - a_1)\ldots(t - a_{n-1})$$

for the scalar k you found above. Substitute $t = a_n$ to demonstrate Vandermonde's formula.

18. Use Exercise 17 to find

$$\det \begin{bmatrix} 1 & 1 & 1 & 1 & 1 \\ 1 & 2 & 3 & 4 & 5 \\ 1 & 4 & 9 & 16 & 25 \\ 1 & 8 & 27 & 64 & 125 \\ 1 & 16 & 81 & 256 & 625 \end{bmatrix}.$$

Do not use technology.

19. For n distinct scalars $a_1, a_2, \ldots, a_n$, find

$$\det \begin{bmatrix} a_1 & a_2 & \cdots & a_n \\ a_1^2 & a_2^2 & \cdots & a_n^2 \\ \vdots & \vdots & & \vdots \\ a_1^n & a_2^n & \cdots & a_n^n \end{bmatrix}.$$

20. a. For an invertible $n \times n$ matrix A and an arbitrary $n \times n$ matrix B, show that

$$\text{rref}[A \;\vdots\; AB] = [I_n \;\vdots\; B].$$

Hint: The left part of $\text{rref}[A \;\vdots\; AB]$ is $\text{rref}(A) = I_n$. Write $\text{rref}[A \;\vdots\; AB] = [I_n \;\vdots\; M]$; we have to show that $M = B$. To demonstrate this, note that the columns of matrix

$$\begin{bmatrix} B \\ -I_n \end{bmatrix}$$

are in the kernel of $[A \;\vdots\; AB]$ and therefore in the kernel of $[I_n \;\vdots\; M]$.

b. What does the formula

$$\text{rref}[A \;\vdots\; AB] = [I_n \;\vdots\; B]$$

tell you if $B = A^{-1}$?

21. Consider two distinct points $\begin{bmatrix} a_1 \\ a_2 \end{bmatrix}$ and $\begin{bmatrix} b_1 \\ b_2 \end{bmatrix}$ in the plane. Explain why the solutions $\begin{bmatrix} x_1 \\ x_2 \end{bmatrix}$ of the equation

$$\det \begin{bmatrix} 1 & 1 & 1 \\ x_1 & a_1 & b_1 \\ x_2 & a_2 & b_2 \end{bmatrix} = 0$$

form a line and why this line goes through the two points $\begin{bmatrix} a_1 \\ a_2 \end{bmatrix}$ and $\begin{bmatrix} b_1 \\ b_2 \end{bmatrix}$.

22. Consider three distinct points $\begin{bmatrix} a_1 \\ a_2 \end{bmatrix}$, $\begin{bmatrix} b_1 \\ b_2 \end{bmatrix}$, $\begin{bmatrix} c_1 \\ c_2 \end{bmatrix}$ in the plane. Describe the set of all points $\begin{bmatrix} x_1 \\ x_2 \end{bmatrix}$ satisfying the equation

$$\det \begin{bmatrix} 1 & 1 & 1 & 1 \\ x_1 & a_1 & b_1 & c_1 \\ x_2 & a_2 & b_2 & c_2 \\ x_1^2 + x_2^2 & a_1^2 + a_2^2 & b_1^2 + b_2^2 & c_1^2 + c_2^2 \end{bmatrix} = 0.$$

23. Consider an $n \times n$ matrix A such that both A and A^{-1} have integer entries. What are the possible values of $\det(A)$?

24. If $\det(A) = 3$ for some $n \times n$ matrix, what is $\det(A^T A)$?

25. If A is an invertible matrix, what can you say about the sign of $\det(A^T A)$?

26. If A is an orthogonal matrix, what are the possible values of $\det(A)$?

27. Consider a skew-symmetric $n \times n$ matrix A, where n is odd. Show that A is noninvertible, by showing that $\det(A) = 0$.

28. Consider an $m \times n$ matrix

$$A = QR,$$

where Q is an $m \times n$ matrix with orthonormal columns and R is an upper triangular $n \times n$ matrix with positive diagonal entries $r_{11}, \ldots, r_{nn}$. Express $\det(A^T A)$ in terms of the scalars r_{ii}. What can you say about the sign of $\det(A^T A)$?

29. Consider two vectors $\vec{v}$ and $\vec{w}$ in $\mathbb{R}^n$. Form the matrix $A = [\vec{v} \ \ \vec{w}]$. Express $\det(A^T A)$ in terms of $\|\vec{v}\|$, $\|\vec{w}\|$, and $\vec{v} \cdot \vec{w}$. What can you say about the sign of the result?

30. *The cross product in $\mathbb{R}^n$.* Consider the vectors $\vec{v}_2, \vec{v}_3, \ldots, \vec{v}_n$ in $\mathbb{R}^n$. The transformation

$$T(\vec{x}) = \det \begin{bmatrix} | & | & | & & | \\ \vec{x} & \vec{v}_2 & \vec{v}_3 & \cdots & \vec{v}_n \\ | & | & | & & | \end{bmatrix}$$

is linear. Therefore, there is a unique vector $\vec{u}$ in $\mathbb{R}^n$ such that

$$T(\vec{x}) = \vec{x} \cdot \vec{u}$$

for all $\vec{x}$ in $\mathbb{R}^n$. (Compare this with Exercise 2.1.43(c).) This vector $\vec{u}$ is called the *cross product* of $\vec{v}_2, \vec{v}_3, \ldots, \vec{v}_n$, written as

$$\vec{u} = \vec{v}_2 \times \vec{v}_3 \times \cdots \times \vec{v}_n.$$

In other words, the cross product is defined by the fact that

$$\vec{x} \cdot (\vec{v}_2 \times \vec{v}_3 \times \cdots \times \vec{v}_n) = \det \begin{bmatrix} | & | & | & & | \\ \vec{x} & \vec{v}_2 & \vec{v}_3 & \cdots & \vec{v}_n \\ | & | & | & & | \end{bmatrix},$$

for all $\vec{x}$ in $\mathbb{R}^n$. Note that the cross product in $\mathbb{R}^n$ is defined for $n - 1$ vectors only. (For example, you cannot form the cross product of just 2 vectors in $\mathbb{R}^4$.) Since the ith component of a vector $\vec{w}$ is $\vec{e}_i \cdot \vec{w}$, we can find the cross product by components as follows:

ith component of $\vec{v}_2 \times \vec{v}_3 \times \cdots \times \vec{v}_n = \vec{e}_i \cdot (\vec{v}_2 \times \cdots \times \vec{v}_n)$

$$= \det \begin{bmatrix} | & | & | & & | \\ \vec{e}_i & \vec{v}_2 & \vec{v}_3 & \cdots & \vec{v}_n \\ | & | & | & & | \end{bmatrix}.$$

a. When is $\vec{v}_2 \times \vec{v}_3 \times \cdots \times \vec{v}_n = \vec{0}$? Give your answer in terms of linear independence.

b. Find $\vec{e}_2 \times \vec{e}_3 \times \cdots \times \vec{e}_n$.

c. Show that $\vec{v}_2 \times \vec{v}_3 \times \cdots \times \vec{v}_n$ is orthogonal to all the vectors $\vec{v}_i$, for $i = 2, \ldots, n$.

d. What is the relationship between $\vec{v}_2 \times \vec{v}_3 \times \cdots \times \vec{v}_n$ and $\vec{v}_3 \times \vec{v}_2 \times \cdots \times \vec{v}_n$? (We swap the first two factors.)

e. Express $\det[\vec{v}_2 \times \vec{v}_3 \times \cdots \times \vec{v}_n \ \ \vec{v}_2 \ \ \vec{v}_3 \ \ \cdots \ \ \vec{v}_n]$ in terms of $\|\vec{v}_2 \times \vec{v}_3 \times \cdots \times \vec{v}_n\|$.

f. How do we know that the cross product of two vectors in $\mathbb{R}^3$, as defined here, is the same as the standard cross product in $\mathbb{R}^3$? (See Definition A.9 of the Appendix.)

31. Find the derivative of the function

$$f(x) = \det \begin{bmatrix} 1 & 1 & 2 & 3 & 4 \\ 9 & 0 & 2 & 3 & 4 \\ 9 & 0 & 0 & 3 & 4 \\ x & 1 & 2 & 9 & 1 \\ 7 & 0 & 0 & 0 & 4 \end{bmatrix}.$$

32. Given some numbers $a, b, c, d, e,$ and f such that

$$\det \begin{bmatrix} a & 1 & d \\ b & 1 & e \\ c & 1 & f \end{bmatrix} = 7 \quad \text{and} \quad \det \begin{bmatrix} a & 1 & d \\ b & 2 & e \\ c & 3 & f \end{bmatrix} = 11,$$

a. Find

$$\det \begin{bmatrix} a & 3 & d \\ b & 3 & e \\ c & 3 & f \end{bmatrix}.$$

b. Find

$$\det \begin{bmatrix} a & 3 & d \\ b & 5 & e \\ c & 7 & f \end{bmatrix}.$$

33. Is the function

$$T \begin{bmatrix} a & b \\ c & d \end{bmatrix} = ad + bc$$

linear in the rows and columns of the matrix?

34. Let M_n be the $n \times n$ matrix with 2's on the main diagonal, 1's directly above and below the main diagonal, and 0's elsewhere. For example,

$$M_5 = \begin{bmatrix} 2 & 1 & 0 & 0 & 0 \\ 1 & 2 & 1 & 0 & 0 \\ 0 & 1 & 2 & 1 & 0 \\ 0 & 0 & 1 & 2 & 1 \\ 0 & 0 & 0 & 1 & 2 \end{bmatrix}.$$

We define $D_n = \det(M_n)$.

a. Find D_1, D_2, D_3, and D_4.

b. If n exceeds 2, find a formula for D_n in terms of D_{n-1} and D_{n-2}. Justify your answer.

c. Find a formula for D_n in terms of n.

35. Consider the $n \times n$ matrix Q_n defined the same way as M_n in Exercise 34 except that the nnth entry is 1 (instead of 2). For example,

$$Q_3 = \begin{bmatrix} 2 & 1 & 0 \\ 1 & 2 & 1 \\ 0 & 1 & 1 \end{bmatrix}.$$

Find a formula for $\det(Q_n)$, first in terms of $\det(Q_{n-1})$ and $\det(Q_{n-2})$, and then in terms of n.

36. Consider a pattern in an $n \times n$ matrix, and choose an entry a_{ij} in this pattern. Show that the number of inversions involving a_{ij} is even if $(i + j)$ is even and odd if $(i + j)$ is odd. *Hint:* Suppose there are k numbers in the pattern to the left and above a_{ij}. Express the number of inversions involving a_{ij} in terms of k.

37. Let P_n be the $n \times n$ matrix whose entries are all ones, except for zeros directly below the main diagonal; for example,

$$P_5 = \begin{bmatrix} 1 & 1 & 1 & 1 & 1 \\ 0 & 1 & 1 & 1 & 1 \\ 1 & 0 & 1 & 1 & 1 \\ 1 & 1 & 0 & 1 & 1 \\ 1 & 1 & 1 & 0 & 1 \end{bmatrix}.$$

Find the determinant of P_n.

38. A zero–one matrix is a matrix whose entries are all zeros and ones. What is the smallest possible number of zeros you can have in an invertible zero–one matrix of size $n \times n$? Exercise 37 is helpful.

39. Consider an invertible 2×2 matrix A with integer entries.

a. Show that if the entries of A^{-1} are integers, then $\det(A) = 1$ or $\det(A) = -1$.

b. Show the converse: If $\det(A) = 1$ or $\det(A) = -1$, then the entries of A^{-1} are integers.

40. Let A and B be 2×2 matrices with integer entries such that A, $A + B$, $A + 2B$, $A + 3B$, and $A + 4B$ are all invertible matrices whose inverses have integer entries. Show that $A + 5B$ is invertible and that its inverse has integer entries. This question was in the William Lowell Putnam Mathematical Competition in 1994. *Hint:* Consider the function $f(t) = \big(\det(A + tB)\big)^2 - 1$. Show that this is a polynomial; what can you say about its degree? Find the values $f(0)$, $f(1)$, $f(2)$, $f(3)$, $f(4)$, using Exercise 39. Now you can determine $f(t)$ by using a familiar result: If a polynomial $f(t)$ of degree $\leq m$ has more than m zeros, then $f(t) = 0$ for all t.

41. For a fixed positive integer n, let D be a function which assigns to any $n \times n$ matrix A a number $D(A)$ such that

a. D is linear in the rows (see Facts 6.2.2 and 6.2.3),

b. $D(B) = -D(A)$ if B is obtained from A by a row swap, and

c. $D(I_n) = 1$.

Show that $D(A) = \det(A)$ for all $n \times n$ matrices A. *Hint:* Consider $E = \text{rref}(A)$. Think about the relationship between $D(A)$ and $D(E)$, mimicking Algorithm 6.2.6.

The point of this exercise is that $\det(A)$ can be characterized by the three properties a, b, and c; the determinant can in fact be *defined* in terms of these properties. Ever since this approach was first presented in the 1880s by the German mathematician Karl Weierstrass (1817–1897), this definition has been generally used in advanced linear algebra courses because it allows a more elegant presentation of the theory of determinants.

42. Use the characterization of the determinant given in Exercise 41 to show that

$$\det(AM) = \det(A)\det(M).$$

Hint: For a fixed invertible matrix M consider the function

$$D(A) = \frac{\det(AM)}{\det(M)}.$$

Show that this function has the three properties a, b, and c listed in Exercise 42, and therefore $D(A) = \det(A)$.

43. Consider a linear transformation T from $\mathbb{R}^{m+n}$ to $\mathbb{R}^m$. The matrix A of T can be written in partitioned form as $A = [A_1 \quad A_2]$, where A_1 is $m \times m$ and A_2 is $m \times n$. Suppose that $\det(A_1) \neq 0$. Show that for every vector $\vec{x}$ in $\mathbb{R}^n$ there is a unique $\vec{y}$ in $\mathbb{R}^m$ such that

$$T \begin{bmatrix} \vec{y} \\ \vec{x} \end{bmatrix} = \vec{0}.$$

Show that the transformation

$$\vec{x} \to \vec{y}$$

from $\mathbb{R}^n$ to $\mathbb{R}^m$ is linear, and find its matrix M (in terms of A_1 and A_2). (This is the linear version of the *implicit function theorem* of multivariable calculus.)

44. Find the matrix M introduced in Exercise 43 for the linear transformation

$$T(\vec{v}) = \begin{bmatrix} 1 & 2 & 1 & 2 \\ 3 & 7 & 4 & 3 \end{bmatrix} \vec{v}.$$

You can either follow the approach outlined in Exercise 43 or use Gaussian elimination, expressing the leading variables y_1, y_2 in terms of the nonleading variables x_1, x_2, where

$$\vec{v} = \begin{bmatrix} y_1 \\ y_2 \\ x_1 \\ x_2 \end{bmatrix}.$$

Note that this procedure amounts to finding the kernel of T, in the familiar way; we just interpret the result somewhat differently.

Find the determinants of the linear transformations in Exercises 45 through 56.

45. $T(f) = 2f + 3f'$ from P_2 to P_2.

46. $T(f(t)) = f(3t - 2)$ from P_2 to P_2.

47. $T(f(t)) = f(-t)$ from P_2 to P_2.

48. $L(A) = A^T$ from $\mathbb{R}^{2 \times 2}$ to $\mathbb{R}^{2 \times 2}$.

49. $T(f(t)) = f(-t)$ from P_3 to P_3.

50. $T(f(t)) = f(-t)$ from P_n to P_n.

51. $L(A) = A^T$ from $\mathbb{R}^{n \times n}$ to $\mathbb{R}^{n \times n}$.

52. $T(z) = (2 + 3i)z$ from $\mathbb{C}$ to $\mathbb{C}$.

53. $T(M) = \begin{bmatrix} 2 & 3 \\ 0 & 4 \end{bmatrix} M$ from the space V of upper triangular 2×2 matrices to V.

54. $T(M) = \begin{bmatrix} 1 & 2 \\ 2 & 3 \end{bmatrix} M + M \begin{bmatrix} 1 & 2 \\ 2 & 3 \end{bmatrix}$ from the space V of symmetric 2×2 matrices into V.

55. $T(f) = af' + bf''$, where a and b are arbitrary constants, from the space V spanned by $\cos(x)$ and $\sin(x)$ to V.

56. $T(\vec{v}) = \begin{bmatrix} 1 \\ 2 \\ 3 \end{bmatrix} \times \vec{v}$ from the plane E given by $x_1 + 2x_2 + 3x_3 = 0$ to E.

6.3 GEOMETRICAL INTERPRETATIONS OF THE DETERMINANT; CRAMER'S RULE

We now present several ways to think about the determinant in geometric terms. Here is a preliminary exercise.

EXAMPLE 1   What are the possible values of the determinant of an orthogonal matrix A?

Solution

We know that

$$A^T A = I_n$$

(by Fact 5.3.7). Taking the determinants of both sides and using Facts 6.2.1 and 6.2.7, we find that

$$\det(A^T A) = \det(A^T) \det(A) = (\det A)^2 = 1.$$

Therefore, $\det(A)$ is either 1 or -1.   ∎

Fact 6.3.1 The determinant of an orthogonal matrix is either 1 or −1.

For example,

$$\det \begin{bmatrix} \cos\alpha & -\sin\alpha \\ \sin\alpha & \cos\alpha \end{bmatrix} = 1$$

for rotation matrices, and

$$\det \begin{bmatrix} \cos\alpha & \sin\alpha \\ \sin\alpha & -\cos\alpha \end{bmatrix} = -1.$$

For matrices of larger sizes, we define:

Definition 6.3.2 An orthogonal $n \times n$ matrix A with $\det(A) = 1$ is called a *rotation matrix,* and the linear transformation $T(\vec{x}) = A\vec{x}$ is called a *rotation.*

The Determinant as Area and Volume

We will now examine what the Gram–Schmidt process (or, equivalently, the QR factorization) tells us about the determinant of a matrix.

EXAMPLE 2 Below we illustrate the Gram–Schmidt process for two linearly independent vectors $\vec{v}_1$ and $\vec{v}_2$ in $\mathbb{R}^2$. As usual, we define $V_1 = \operatorname{span}(\vec{v}_1)$.

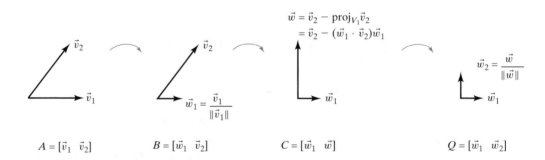

$$A = [\vec{v}_1 \ \ \vec{v}_2] \qquad B = [\vec{w}_1 \ \ \vec{v}_2] \qquad C = [\vec{w}_1 \ \ \vec{w}] \qquad Q = [\vec{w}_1 \ \ \vec{w}_2]$$

What is the relationship between the determinants of the matrices A, B, C, and Q?

Solution

Since we are dividing the first column by $\|\vec{v}_1\|$,

$$\det(B) = \frac{\det(A)}{\|\vec{v}_1\|}.$$

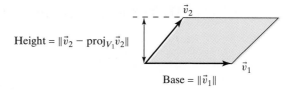

Height = $\|\vec{v}_2 - \text{proj}_{V_1}\vec{v}_2\|$

$\vec{v}_2$

$\vec{v}_1$

Base = $\|\vec{v}_1\|$

Figure 1

Since we are subtracting a multiple of the first column from the second,

$$\det(C) = \det(B).$$

Since we are dividing the second column by $\|\vec{w}\| = \|\vec{v}_2 - \text{proj}_{V_1}\vec{v}_2\|$,

$$\det(Q) = \frac{\det(C)}{\|\vec{w}\|}.$$

We can conclude that

$$\det(A) = \|\vec{v}_1\|\,\|\vec{v}_2 - \text{proj}_{V_1}\vec{v}_2\|\,\det(Q).$$

Since $\det(Q) = 1$ or $\det(Q) = -1$, by Fact 6.3.1, we find that

$$\det(A) = \pm\|\vec{v}_1\|\,\|\vec{v}_2 - \text{proj}_{V_1}\vec{v}_2\|,$$

or

$$|\det(A)| = \|\vec{v}_1\|\,\|\vec{v}_2 - \text{proj}_{V_1}\vec{v}_2\|. \qquad \blacksquare$$

This formula provides us with a simple geometric interpretation of the determinant:

The area of the shaded parallelogram in Figure 1 is

$$(\text{base})(\text{height}) = \|\vec{v}_1\|\,\|\vec{v}_2 - \text{proj}_{V_1}\vec{v}_2\| = |\det(A)|.$$

Fact 6.3.3   Consider a 2×2 matrix $A = [\vec{v}_1 \quad \vec{v}_2]$. Then, the area of the parallelogram defined by $\vec{v}_1$ and $\vec{v}_2$ is $|\det(A)|$.

There are easier ways to derive this fact, but they do not generalize to higher dimensional cases as well as the approach taken in Example 2.

What about the *sign* of $\det(A) = \det[\vec{v}_1 \quad \vec{v}_2]$? As we saw before, $\det(A)$ will be positive if $\det(Q) = 1$ for the orthogonal matrix Q in the QR factorization of A; if $\det(Q) = -1$, then $\det(A)$ will be negative.

Consider the following two examples:

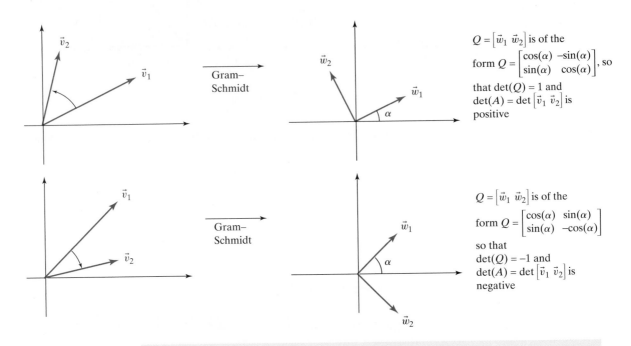

$Q = \begin{bmatrix} \vec{w}_1 & \vec{w}_2 \end{bmatrix}$ is of the form $Q = \begin{bmatrix} \cos(\alpha) & -\sin(\alpha) \\ \sin(\alpha) & \cos(\alpha) \end{bmatrix}$, so that $\det(Q) = 1$ and $\det(A) = \det\begin{bmatrix} \vec{v}_1 & \vec{v}_2 \end{bmatrix}$ is positive

$Q = \begin{bmatrix} \vec{w}_1 & \vec{w}_2 \end{bmatrix}$ is of the form $Q = \begin{bmatrix} \cos(\alpha) & \sin(\alpha) \\ \sin(\alpha) & -\cos(\alpha) \end{bmatrix}$ so that $\det(Q) = -1$ and $\det(A) = \det\begin{bmatrix} \vec{v}_1 & \vec{v}_2 \end{bmatrix}$ is negative

We see that if the direction of $\vec{v}_2$ is obtained by rotating $\vec{v}_1$ through a positive (i.e., counterclockwise) angle between 0 and π, then $\det(A) = \det[\vec{v}_1 \quad \vec{v}_2]$ will be positive. If we rotate through a negative (clockwise) angle between 0 and $-\pi$, then $\det(A)$ will be negative.

Now consider an invertible $n \times n$ matrix

$$A = \begin{bmatrix} | & | & & | \\ \vec{v}_1 & \vec{v}_2 & \cdots & \vec{v}_n \\ | & | & & | \end{bmatrix}.$$

By Fact 5.2.2, we can write

$$A = QR,$$

where Q is an orthogonal matrix, and R is an upper triangular matrix whose diagonal entries are

$$r_{11} = \|\vec{v}_1\| \qquad \text{and} \qquad r_{jj} = \|\vec{v}_j - \text{proj}_{V_{j-1}} \vec{v}_j\|, \qquad \text{for } j \geq 2.$$

We conclude that

$$\det(A) = \det(Q)\det(R)$$
$$= \pm\|\vec{v}_1\| \, \|\vec{v}_2 - \text{proj}_{V_1} \vec{v}_2\| \ldots \|\vec{v}_n - \text{proj}_{V_{n-1}} \vec{v}_n\|,$$

because $\det(Q) = \pm 1$ by Fact 6.3.1 (since Q is orthogonal). Since R is triangular, its determinant is the product of the diagonal entries (by Fact 6.1.3).

Fact 6.3.4

If A is an $n \times n$ matrix with columns $\vec{v}_1, \ldots, \vec{v}_n$, then

$$|\det(A)| = \|\vec{v}_1\| \, \|\vec{v}_2 - \operatorname{proj}_{V_1} \vec{v}_2\| \ldots \|\vec{v}_n - \operatorname{proj}_{V_{n-1}} \vec{v}_n\|,$$

where $V_j = \operatorname{span}(\vec{v}_1, \vec{v}_2, \ldots, \vec{v}_j)$.

Above, we verified this result for an invertible $n \times n$ matrix A; we leave the non-invertible case as Exercise 8.

As an example, consider the 3×3 matrix

$$A = \begin{bmatrix} | & | & | \\ \vec{v}_1 & \vec{v}_2 & \vec{v}_3 \\ | & | & | \end{bmatrix},$$

with

$$|\det(A)| = \|\vec{v}_1\| \, \|\vec{v}_2 - \operatorname{proj}_{V_1} \vec{v}_2\| \, \|\vec{v}_3 - \operatorname{proj}_{V_2} \vec{v}_3\|.$$

As we observed, $\|\vec{v}_1\| \, \|\vec{v}_2 - \operatorname{proj}_{V_1} \vec{v}_2\|$ is the area of the parallelogram defined by $\vec{v}_1$ and $\vec{v}_2$. (See Figure 1.) Now consider the parallelepiped defined by $\vec{v}_1$, $\vec{v}_2$, and $\vec{v}_3$ (i.e., the set of all vectors of the form $c_1 \vec{v}_1 + c_2 \vec{v}_2 + c_3 \vec{v}_3$, where the c_i are between 0 and 1, as shown in Figure 2).

The volume of this parallelepiped is

$$\begin{aligned} \text{volume} &= (\text{base area})(\text{height}) \\ &= \|\vec{v}_1\| \, \|\vec{v}_2 - \operatorname{proj}_{V_1} \vec{v}_2\| \, \|\vec{v}_3 - \operatorname{proj}_{V_2} \vec{v}_3\| \\ &= |\det(A)| \end{aligned}$$

(by Fact 6.3.4).

Fact 6.3.5

Consider a 3×3 matrix $A = [\vec{v}_1 \quad \vec{v}_2 \quad \vec{v}_3]$. Then, the volume of the parallelepiped defined by $\vec{v}_1$, $\vec{v}_2$, and $\vec{v}_3$ is $|\det(A)|$.

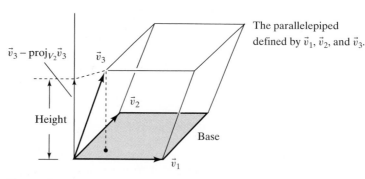

Figure 2

For a geometrical interpretation of the *sign* of det(A) see Exercises 19 through 21.
Let us generalize these observations to higher dimensions.

Definition 6.3.6

> **k-parallelepipeds and k-volume**
>
> Consider the vectors $\vec{v}_1, \vec{v}_2, \ldots, \vec{v}_k$ in $\mathbb{R}^n$. The k-parallelepiped defined by the vectors $\vec{v}_1, \ldots, \vec{v}_k$ is the set of all vectors in $\mathbb{R}^n$ of the form $c_1\vec{v}_1 + c_2\vec{v}_2 + \cdots + c_k\vec{v}_k$, where $0 \leq c_i \leq 1$. The k-volume $V(\vec{v}_1, \ldots, \vec{v}_k)$ of this k-parallelepiped is defined recursively by $V(\vec{v}_1) = \|\vec{v}_1\|$ and
>
> $$V(\vec{v}_1, \ldots, \vec{v}_k) = V(\vec{v}_1, \ldots, \vec{v}_{k-1})\|\vec{v}_k - \text{proj}_{V_{k-1}}\vec{v}_k\|.$$

Note that this formula for the k-volume mimics the formula

$$(\text{base})(\text{height})$$

we used to compute the area of a parallelogram (i.e., a 2-parallelepiped) and the volume of a 3-parallelepiped in $\mathbb{R}^3$.

Alternatively, we can write the formula for the k-volume as

$$V(\vec{v}_1, \ldots, \vec{v}_k) = \|\vec{v}_1\| \, \|\vec{v}_2 - \text{proj}_{V_1}\vec{v}_2\| \ldots \|\vec{v}_k - \text{proj}_{V_{k-1}}\vec{v}_k\|.$$

Let A be the $n \times k$ matrix whose columns are $\vec{v}_1, \ldots, \vec{v}_k$. We will show that the k-volume $V(\vec{v}_1, \ldots, \vec{v}_k)$ is closely related to the quantity $\det(A^T A)$. If the columns of A are linearly independent, we can consider the QR factorization $A = QR$. Then, $A^T A = R^T Q^T Q R = R^T R$, because $Q^T Q = I_k$ (since the columns of Q are orthonormal). Therefore,

$$\begin{aligned}
\det(A^T A) &= \det(R^T R) = (\det R)^2 \\
&= (r_{11}r_{22} \ldots r_{kk})^2 \\
&= \left(\|\vec{v}_1\|\|\vec{v}_2 - \text{proj}_{V_1}\vec{v}_2\| \ldots \|\vec{v}_k - \text{proj}_{V_{k-1}}\vec{v}_k\|\right)^2 \\
&= \left(V(\vec{v}_1, \ldots, \vec{v}_k)\right)^2.
\end{aligned}$$

We can conclude:

Fact 6.3.7

> Consider the vectors $\vec{v}_1, \vec{v}_2, \ldots, \vec{v}_k$ in $\mathbb{R}^n$. Then the k-volume of the k-parallelepiped defined by the vectors $\vec{v}_1, \ldots, \vec{v}_k$ is
>
> $$\sqrt{\det(A^T A)},$$
>
> where A is the $n \times k$ matrix with columns $\vec{v}_1, \vec{v}_2, \ldots, \vec{v}_k$.
> In particular, if $k = n$, this volume is
>
> $$|\det(A)|.$$

(Compare this with Fact 6.3.4.)

We leave it to the reader to verify Fact 6.3.7 for linearly dependent vectors $\vec{v}_1, \ldots, \vec{v}_k$. (See Exercise 15.)

As a simple example, consider the 2-volume (i.e., area) of the 2-parallelepiped (i.e., parallelogram) defined by the vectors

$$\vec{v}_1 = \begin{bmatrix} 1 \\ 1 \\ 1 \end{bmatrix} \quad \text{and} \quad \vec{v}_2 = \begin{bmatrix} 1 \\ 2 \\ 3 \end{bmatrix}$$

in $\mathbb{R}^3$. By Fact 6.3.7, this area is

$$\sqrt{\det\left(\begin{bmatrix} 1 & 1 & 1 \\ 1 & 2 & 3 \end{bmatrix} \begin{bmatrix} 1 & 1 \\ 1 & 2 \\ 1 & 3 \end{bmatrix}\right)} = \sqrt{\det\begin{bmatrix} 3 & 6 \\ 6 & 14 \end{bmatrix}} = \sqrt{6}.$$

In this special case, we can also determine the area as the norm $\| \vec{v}_1 \times \vec{v}_2 \|$ of the cross product of the two vectors.

The Determinant as Expansion Factor

Consider a linear transformation T from $\mathbb{R}^2$ to $\mathbb{R}^2$. In Chapter 5, we examined how a linear transformation T affects various geometric quantities such as lengths and angles. For example, we observed that a rotation preserves both the length of vectors and the angle between vectors. Similarly, we can ask how a linear transformation affects the area of a region Ω in the plane. (See Figure 3.)

We might be interested in finding the *expansion factor*, the ratio

$$\frac{\text{area of } T(\Omega)}{\text{area of } \Omega}.$$

The simplest example is the case when Ω is the unit square. (See Figure 4.)

Since the area of Ω is 1 here, the expansion factor is simply the area of the parallelogram $T(\Omega)$, which is $|\det(A)|$, by Fact 6.3.3.

More generally, let Ω be the parallelogram defined by $\vec{v}_1$ and $\vec{v}_2$, as shown in Figure 5.

Figure 3

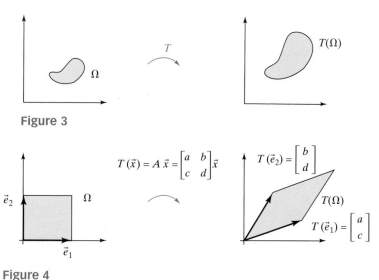

$$T(\vec{x}) = A\,\vec{x} = \begin{bmatrix} a & b \\ c & d \end{bmatrix}\vec{x}$$

$$T(\vec{e}_2) = \begin{bmatrix} b \\ d \end{bmatrix}$$

$$T(\Omega)$$

$$T(\vec{e}_1) = \begin{bmatrix} a \\ c \end{bmatrix}$$

Figure 4

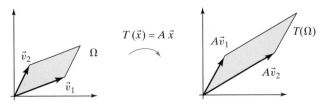

Figure 5

Let $B = [\vec{v}_1 \quad \vec{v}_2]$. Then,

$$\text{area of } \Omega = |\det(B)|,$$

and

$$\text{area of } T(\Omega) = |\det[A\vec{v}_1 \quad A\vec{v}_2]| = |\det(AB)| = |\det(A)|\,|\det(B)|,$$

and the expansion factor is

$$\frac{\text{area of } T(\Omega)}{\text{area of } \Omega} = \frac{|\det(A)|\,|\det(B)|}{|\det(B)|} = |\det(A)|.$$

It is remarkable that the linear transformation $T(\vec{x}) = A\vec{x}$ expands the area of *all* parallelograms by the same factor (namely, $|\det(A)|$).

Fact 6.3.8

Expansion factor
Consider a linear transformation $T(\vec{x}) = A\vec{x}$ from $\mathbb{R}^2$ to $\mathbb{R}^2$. Then, $|\det(A)|$ is the *expansion factor*

$$\frac{\text{area of } T(\Omega)}{\text{area of } \Omega}$$

of T on parallelograms Ω.

Likewise, for a linear transformation $T(\vec{x}) = A\vec{x}$ from $\mathbb{R}^n$ to $\mathbb{R}^n$, $|\det(A)|$ is the expansion factor of T on n-parallelepipeds:

$$V(A\vec{v}_1, \ldots, A\vec{v}_n) = |\det(A)|\,V(\vec{v}_1, \ldots, \vec{v}_n),$$

for all vectors $\vec{v}_1, \ldots, \vec{v}_n$ in $\mathbb{R}^n$.

This interpretation allows us to think about the formulas $\det(A^{-1}) = 1/\det(A)$ and $\det(AB) = \det(A)\det(B)$ from a geometric point of view. See Figures 6 and 7.

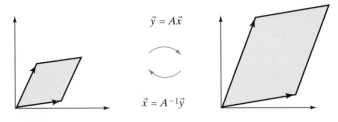

Figure 6

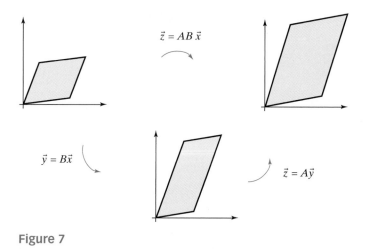

Figure 7

The expansion factor $|\det(A^{-1})|$ is the reciprocal of the expansion factor $|\det(A)|$:

$$|\det(A^{-1})| = \frac{1}{|\det(A)|}$$

The expansion factor $|\det(AB)|$ of the composite transformation is the product of the expansion factors $|\det(A)|$ and $|\det(B)|$:

$$|\det(AB)| = |\det(A)|\,|\det(B)|$$

Using techniques of calculus, you can verify that $|\det(A)|$ gives us the expansion factor of the transformation $T(\vec{x}) = A\vec{x}$ on any region Ω in the plane. The approach uses inscribed parallelograms (or even squares) to approximate the area of the region, as shown in Figure 8. Note that the expansion factor of T on each of these squares is $|\det(A)|$. Choosing smaller and smaller squares and applying calculus, you can conclude that the expansion factor of T on Ω itself is $|\det(A)|$.

We will conclude this chapter with the discussion of a *closed-form solution* for the linear system $A\vec{x} = \vec{b}$ in the case when the coefficient matrix A is invertible.

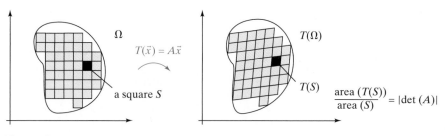

Figure 8

Cramer's Rule

If a matrix

$$A = \begin{bmatrix} a_{11} & a_{12} \\ a_{21} & a_{22} \end{bmatrix}$$

is invertible, we can express its inverse in terms of its determinant:

$$A^{-1} = \frac{1}{\det(A)} \begin{bmatrix} a_{22} & -a_{12} \\ -a_{21} & a_{11} \end{bmatrix}.$$

This formula can be used to find a *closed-formula solution* for a linear system

$$\begin{vmatrix} a_{11}x_1 + a_{12}x_2 = b_1 \\ a_{21}x_1 + a_{22}x_2 = b_2 \end{vmatrix}$$

when the coefficient matrix is invertible. We write the system as $A\vec{x} = \vec{b}$, where

$$A = \begin{bmatrix} a_{11} & a_{12} \\ a_{21} & a_{22} \end{bmatrix}, \qquad \vec{x} = \begin{bmatrix} x_1 \\ x_2 \end{bmatrix}, \qquad \vec{b} = \begin{bmatrix} b_1 \\ b_2 \end{bmatrix}.$$

Then,

$$\begin{bmatrix} x_1 \\ x_2 \end{bmatrix} = \vec{x} = A^{-1}\vec{b} = \frac{1}{\det(A)} \begin{bmatrix} a_{22} & -a_{12} \\ -a_{21} & a_{11} \end{bmatrix} \begin{bmatrix} b_1 \\ b_2 \end{bmatrix} = \frac{1}{\det(A)} \begin{bmatrix} a_{22}b_1 - a_{12}b_2 \\ a_{11}b_2 - a_{21}b_1 \end{bmatrix}.$$

To write this formula more succinctly, we observe that

$$a_{22}b_1 - a_{12}b_2 = \det \begin{bmatrix} b_1 & a_{12} \\ b_2 & a_{22} \end{bmatrix} \quad \longleftarrow \quad \text{replace the first column of } A \text{ by } \vec{b}.$$

$$a_{11}b_2 - a_{21}b_1 = \det \begin{bmatrix} a_{11} & b_1 \\ a_{21} & b_2 \end{bmatrix} \quad \longleftarrow \quad \text{replace the second column of } A \text{ by } \vec{b}.$$

Let A_i be the matrix obtained by replacing the ith column of A by $\vec{b}$:

$$A_1 = \begin{bmatrix} b_1 & a_{12} \\ b_2 & a_{22} \end{bmatrix}, \qquad A_2 = \begin{bmatrix} a_{11} & b_1 \\ a_{21} & b_2 \end{bmatrix}.$$

The solution of the system $A\vec{x} = \vec{b}$ can now be written as

$$x_1 = \frac{\det(A_1)}{\det(A)}, \qquad x_2 = \frac{\det(A_2)}{\det(A)}.$$

EXAMPLE 3   Use the preceding formula to solve the system

$$\begin{vmatrix} 2x_1 + 3x_2 = 7 \\ 4x_1 + 5x_2 = 13 \end{vmatrix}.$$

Solution

$$x_1 = \frac{\det \begin{bmatrix} 7 & 3 \\ 13 & 5 \end{bmatrix}}{\det \begin{bmatrix} 2 & 3 \\ 4 & 5 \end{bmatrix}} = 2, \qquad x_2 = \frac{\det \begin{bmatrix} 2 & 7 \\ 4 & 13 \end{bmatrix}}{\det \begin{bmatrix} 2 & 3 \\ 4 & 5 \end{bmatrix}} = 1$$

■

This method is not particularly helpful for solving numerically given linear systems; Gauss–Jordan elimination is preferable in this case. However, in many applications, we have to deal with systems whose coefficients contain parameters. Often we want to know how the solution changes as we change the parameters. The closed-formula solution given before is well suited to deal with such questions.

EXAMPLE 4   Solve the system

$$\begin{vmatrix} (b-1)x_1 + & ax_2 & = 0 \\ -ax_1 & + (b-1)x_2 = C \end{vmatrix},$$

where a, b, C are arbitrary positive constants.

Solution

$$x_1 = \frac{\det \begin{bmatrix} 0 & a \\ C & b-1 \end{bmatrix}}{\det \begin{bmatrix} b-1 & a \\ -a & b-1 \end{bmatrix}} = \frac{-aC}{(b-1)^2 + a^2}$$

$$x_2 = \frac{\det \begin{bmatrix} b-1 & 0 \\ -a & C \end{bmatrix}}{\det \begin{bmatrix} b-1 & a \\ -a & b-1 \end{bmatrix}} = \frac{(b-1)C}{(b-1)^2 + a^2} \qquad\blacksquare$$

EXAMPLE 5   Consider the linear system

$$\begin{vmatrix} ax + by = 1 \\ cx + dy = 1 \end{vmatrix}, \qquad \text{where} \quad d > b > 0 \quad \text{and} \quad a > c > 0.$$

How does the solution x change as we change the parameters a and c? More precisely, find $\partial x / \partial a$ and $\partial x / \partial c$, and determine the signs of these quantities.

Solution

$$x = \frac{\det \begin{bmatrix} 1 & b \\ 1 & d \end{bmatrix}}{\det \begin{bmatrix} a & b \\ c & d \end{bmatrix}} = \frac{d-b}{ad-bc} > 0, \qquad \frac{\partial x}{\partial a} = \frac{-d(d-b)}{(ad-bc)^2} < 0,$$

$$\frac{\partial x}{\partial c} = \frac{b(d-b)}{(ad-bc)^2} > 0.$$

See Figure 9. $\qquad\blacksquare$

An interesting application of these simple results in biology is to the study of castes. [See E. O. Wilson, "The Ergonomics of Caste in the Social Insects," *American Naturalist,* **102** (923): 41–66 (1968).]

The closed formula for solving linear systems of two equations with two unknowns generalizes easily to larger systems.

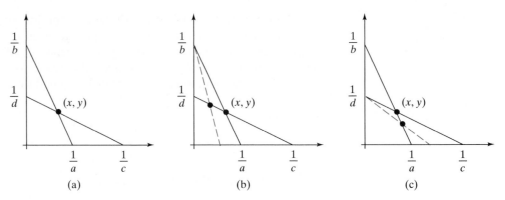

Figure 9 (a) Both components x and y of the solution are positive. (b) $\partial x/\partial a < 0$: as a increases, the component x of the solution decreases. (c) $\partial x/\partial c > 0$: as c increases, the component x of the solution increases.

Fact 6.3.9

Cramer's rule
Consider the linear system

$$A\vec{x} = \vec{b},$$

where A is an invertible $n \times n$ matrix. The components x_i of the solution vector $\vec{x}$ are

$$x_i = \frac{\det(A_i)}{\det(A)},$$

where A_i is the matrix obtained by replacing the ith column of A by $\vec{b}$.

This result is due to the Swiss mathematician Gabriel Cramer (1704–1752). The rule appeared in an appendix to his 1750 book, *Introduction à l'analyse des lignes courbes algébriques.*

Proof Write $A = [\vec{w}_1 \ \vec{w}_2 \ \ldots \ \vec{w}_i \ \ldots \ \vec{w}_n]$. If $\vec{x}$ is the solution of the system $A\vec{x} = \vec{b}$, then

$$\det(A_i) = \det[\vec{w}_1 \ \vec{w}_2 \ \ldots \ \vec{b} \ \ldots \ \vec{w}_n] = \det[\vec{w}_1 \ \vec{w}_2 \ \ldots \ A\vec{x} \ \ldots \ \vec{w}_n]$$

$$= \det[\vec{w}_1 \ \vec{w}_2 \ \ldots \ (x_1\vec{w}_1 + x_2\vec{w}_2 + \cdots + x_i\vec{w}_i + \cdots + x_n\vec{w}_n) \ \ldots \ \vec{w}_n]$$

$$= \det[\vec{w}_1 \ \vec{w}_2 \ \ldots \ x_i\vec{w}_i \ \ldots \ \vec{w}_n] = x_i \det[\vec{w}_1 \ \vec{w}_2 \ \ldots \ \vec{w}_i \ \ldots \ \vec{w}_n]$$

$$= x_i \det(A).$$

Note that we have used the linearity of the determinant in the ith column (Fact 6.2.3). Therefore,

$$x_i = \frac{\det(A_i)}{\det(A)}. \qquad \blacktriangle$$

Cramer's rule allows us to find a closed formula for A^{-1}, generalizing the result

$$\begin{bmatrix} a & b \\ c & d \end{bmatrix}^{-1} = \frac{1}{\det(A)} \begin{bmatrix} d & -b \\ -c & a \end{bmatrix}$$

for 2×2 matrices.

Consider an invertible $n \times n$ matrix A and write

$$A^{-1} = \begin{bmatrix} m_{11} & m_{12} & \cdots & m_{1j} & \cdots & m_{1n} \\ m_{21} & m_{22} & \cdots & m_{2j} & \cdots & m_{2n} \\ \vdots & \vdots & & \vdots & & \vdots \\ m_{n1} & m_{n2} & \cdots & m_{nj} & \cdots & m_{nn} \end{bmatrix}.$$

We know that $AA^{-1} = I_n$. Picking out the jth column of A^{-1}, we find that

$$A \begin{bmatrix} m_{1j} \\ m_{2j} \\ \vdots \\ m_{nj} \end{bmatrix} = \vec{e}_j.$$

By Cramer's rule, $m_{ij} = \det(A_i)/\det(A)$, where the ith column of A is replaced by $\vec{e}_j$.

$$A_i = \begin{bmatrix} a_{11} & a_{12} & \cdots & 0 & \cdots & a_{1n} \\ a_{21} & a_{22} & \cdots & 0 & \cdots & a_{2n} \\ \vdots & \vdots & & \vdots & & \vdots \\ a_{j1} & a_{j2} & \cdots & 1 & \cdots & a_{jn} \\ \vdots & \vdots & & \vdots & & \vdots \\ a_{n1} & a_{n2} & \cdots & 0 & \cdots & a_{nn} \end{bmatrix} \quad \leftarrow \quad j\text{th row}$$

$$\underset{\substack{\uparrow \\ i\text{th} \\ \text{column}}}{}$$

By Example 7 of Section 6.2, $\det(A_i) = (-1)^{i+j}\det(A_{ji})$, so that

$$m_{ij} = (-1)^{i+j}\frac{\det(A_{ji})}{\det(A)}.$$

We have shown the following result:

Fact 6.3.10

> **Corollary to Cramer's rule**
> Consider an invertible $n \times n$ matrix A. The *classical adjoint* $\operatorname{adj}(A)$ is the $n \times n$ matrix whose ijth entry is $(-1)^{i+j}\det(A_{ji})$. Then,
> $$A^{-1} = \frac{1}{\det(A)}\operatorname{adj}(A).$$

For an invertible 2×2 matrix $A = \begin{bmatrix} a & b \\ c & d \end{bmatrix}$, we find

$$\operatorname{adj}(A) = \begin{bmatrix} d & -b \\ -c & a \end{bmatrix} \quad \text{and} \quad A^{-1} = \frac{1}{ad-bc}\begin{bmatrix} d & -b \\ -c & a \end{bmatrix}.$$

(Compare this with Fact 2.3.6.)

For an invertible 3×3 matrix

$$A = \begin{bmatrix} a & b & c \\ d & e & f \\ g & h & k \end{bmatrix},$$

the formula is

$$A^{-1} = \frac{1}{\det(A)} \begin{bmatrix} ek - fh & ch - bk & bf - ce \\ fg - dk & ak - cg & cd - af \\ dh - eg & bg - ah & ae - bd \end{bmatrix}.$$

We can interpret Cramer's rule geometrically.

EXAMPLE 6 For the vectors $\vec{w}_1$, $\vec{w}_2$, and $\vec{b}$ shown in Figure 10, consider the linear system $A\vec{x} = \vec{b}$, where $A = [\vec{w}_1 \quad \vec{w}_2]$.

Using the terminology introduced in Cramer's rule, let $A_2 = [\vec{w}_1 \quad \vec{b}]$. Note that $\det(A)$ and $\det(A_2)$ are both positive, according to the criteria we discussed after Fact 6.3.3. Cramer's rule tells us that

$$x_2 = \frac{\det(A_2)}{\det(A)} \qquad \text{or} \qquad \det(A_2) = x_2 \det(A).$$

Explain this last equation geometrically, in terms of areas of parallelograms (Fact 6.3.3).

Solution

We can write the system $A\vec{x} = \vec{b}$ as $x_1\vec{w}_1 + x_2\vec{w}_2 = \vec{b}$. The geometrical solution is shown in Figure 11.

Now,

$$\det(A_2) = \det[\vec{w}_1 \quad \vec{b}] = \text{area of the parallelogram defined by } \vec{w}_1 \text{ and } \vec{b}$$
$$= \text{area of the parallelogram}^7 \text{ defined by } \vec{w}_1 \text{ and } x_2\vec{w}_2$$
$$= x_2 \left(\text{area of the parallelogram}^8 \text{ defined by } \vec{w}_1 \text{ and } \vec{w}_2 \right) = x_2 \det(A),$$

as claimed. Note that this geometrical proof mimics the algebraic proof of Cramer's rule on page 280. ∎

----
[7] The two parallelograms have the same base and the same height.
[8] Again, think about base and height

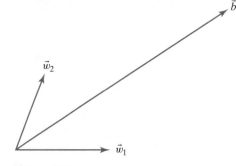

Figure 10

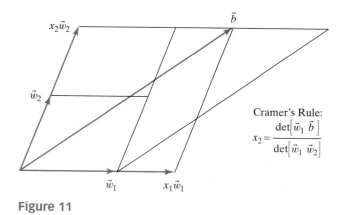

Cramer's Rule:
$$x_2 = \frac{\det\begin{bmatrix} \vec{w}_1 & \vec{b} \end{bmatrix}}{\det\begin{bmatrix} \vec{w}_1 & \vec{w}_2 \end{bmatrix}}$$

Figure 11

The ambitious and artistically inclined reader is encouraged to draw an analogous figure illustrating Cramer's rule for a system of three linear equations with three unknowns.

EXERCISES

GOALS Interpret the determinant as an area or volume and as an expansion factor. Use Cramer's rule.

1. Find the area of the parallelogram defined by $\begin{bmatrix} 3 \\ 7 \end{bmatrix}$ and $\begin{bmatrix} 8 \\ 2 \end{bmatrix}$.

2. Find the area of the triangle defined by $\begin{bmatrix} 3 \\ 7 \end{bmatrix}$ and $\begin{bmatrix} 8 \\ 2 \end{bmatrix}$.

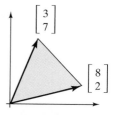

3. Find the area of the following triangle:

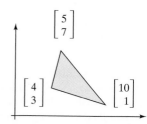

4. Consider the area A of the triangle with vertices $\begin{bmatrix} a_1 \\ a_2 \end{bmatrix}$, $\begin{bmatrix} b_1 \\ b_2 \end{bmatrix}$, $\begin{bmatrix} c_1 \\ c_2 \end{bmatrix}$. Express A in terms of

$$\det \begin{bmatrix} a_1 & b_1 & c_1 \\ a_2 & b_2 & c_2 \\ 1 & 1 & 1 \end{bmatrix}.$$

5. The tetrahedron defined by three vectors $\vec{v}_1, \vec{v}_2, \vec{v}_3$ in $\mathbb{R}^3$ is the set of all vectors of the form $c_1\vec{v}_1 + c_2\vec{v}_2 + c_3\vec{v}_3$, where $c_i \geq 0$ and $c_1 + c_2 + c_3 \leq 1$. Draw a sketch to explain why the volume of this tetrahedron is one-sixth of the volume of the parallelepiped defined by $\vec{v}_1, \vec{v}_2, \vec{v}_3$.

6. What is the relationship between the volume of the tetrahedron defined by the vectors

$$\begin{bmatrix} a_1 \\ a_2 \\ 1 \end{bmatrix}, \quad \begin{bmatrix} b_1 \\ b_2 \\ 1 \end{bmatrix}, \quad \begin{bmatrix} c_1 \\ c_2 \\ 1 \end{bmatrix}$$

and the area of the triangle with vertices

$$\begin{bmatrix} a_1 \\ a_2 \end{bmatrix}, \quad \begin{bmatrix} b_1 \\ b_2 \end{bmatrix}, \quad \begin{bmatrix} c_1 \\ c_2 \end{bmatrix}?$$

(See Exercises 4 and 5.) Explain this relationship geometrically. *Hint:* Consider the top face of the tetrahedron.

7. Find the area of the following region:

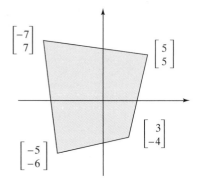

$$\begin{bmatrix} -7 \\ 7 \end{bmatrix} \quad \begin{bmatrix} 5 \\ 5 \end{bmatrix} \quad \begin{bmatrix} 3 \\ -4 \end{bmatrix} \quad \begin{bmatrix} -5 \\ -6 \end{bmatrix}$$

8. Demonstrate the equation

$$|\det(A)| = \|\vec{v}_1\| \, \|\vec{v}_2 - \mathrm{proj}_{V_1} \vec{v}_2\| \, \ldots \, \|\vec{v}_n - \mathrm{proj}_{V_{n-1}} \vec{v}_n\|$$

for a noninvertible $n \times n$ matrix $A = [\vec{v}_1 \quad \vec{v}_2 \quad \ldots \quad \vec{v}_n]$ (Fact 6.3.4).

9. Suppose two nonzero vectors $\vec{v}_1$ and $\vec{v}_2$ in $\mathbb{R}^2$ enclose an angle α (with $0 \le \alpha \le \pi$). Explain why

$$|\det[\vec{v}_1 \quad \vec{v}_2]| = \|\vec{v}_1\| \, \|\vec{v}_2\| \sin(\alpha).$$

10. Consider an $n \times n$ matrix $A = [\vec{v}_1 \quad \vec{v}_2 \quad \ldots \quad \vec{v}_n]$. What is the relationship between the product $\|\vec{v}_1\| \, \|\vec{v}_2\| \, \ldots \, \|\vec{v}_n\|$ and $|\det(A)|$? When is $|\det(A)| = \|\vec{v}_1\| \, \|\vec{v}_2\| \, \ldots \, \|\vec{v}_n\|$?

11. Consider a linear transformation $T(\vec{x}) = A\vec{x}$ from $\mathbb{R}^2$ to $\mathbb{R}^2$. Suppose for two vectors $\vec{v}_1$ and $\vec{v}_2$ in $\mathbb{R}^2$ we have $T(\vec{v}_1) = 3\vec{v}_1$ and $T(\vec{v}_2) = 4\vec{v}_2$. What can you say about $\det(A)$? Justify your answer carefully.

12. Consider those 4×4 matrices whose entries are all 1, -1, or 0. What is the maximal value of the determinant of a matrix of this type? Give an example of a matrix whose determinant has this maximal value.

13. Find the area (or 2-volume) of the parallelogram (or 2-parallelepiped) defined by the vectors

$$\begin{bmatrix} 1 \\ 1 \\ 1 \\ 1 \end{bmatrix} \quad \text{and} \quad \begin{bmatrix} 1 \\ 2 \\ 3 \\ 4 \end{bmatrix}.$$

14. Find the 3-volume of the 3-parallelepiped defined by the vectors

$$\begin{bmatrix} 1 \\ 0 \\ 0 \\ 0 \end{bmatrix}, \quad \begin{bmatrix} 1 \\ 1 \\ 1 \\ 1 \end{bmatrix}, \quad \begin{bmatrix} 1 \\ 2 \\ 3 \\ 4 \end{bmatrix}.$$

15. Demonstrate Fact 6.3.7 for linearly dependent vectors $\vec{v}_1, \ldots, \vec{v}_k$.

16. *True or false?* If Ω is a parallelogram in $\mathbb{R}^3$ and $T(\vec{x}) = A\vec{x}$ is a linear transformation from $\mathbb{R}^3$ to $\mathbb{R}^3$, then

$$\text{area of } T(\Omega) = |\det(A)| \, (\text{area of } \Omega).$$

17. (For some background on the cross product in $\mathbb{R}^n$, see Exercise 6.2.30.) Consider three linearly independent vectors $\vec{v}_1, \vec{v}_2, \vec{v}_3$ in $\mathbb{R}^4$.

a. What is the relationship between $V(\vec{v}_1, \vec{v}_2, \vec{v}_3)$ and $V(\vec{v}_1, \vec{v}_2, \vec{v}_3, \vec{v}_1 \times \vec{v}_2 \times \vec{v}_3)$? See Definition 6.3.6. Exercise 6.2.30(c) is helpful.

b. Express $V(\vec{v}_1, \vec{v}_2, \vec{v}_3, \vec{v}_1 \times \vec{v}_2 \times \vec{v}_3)$ in terms of $\|\vec{v}_1 \times \vec{v}_2 \times \vec{v}_3\|$.

c. Use parts (a) and (b) to express $V(\vec{v}_1, \vec{v}_2, \vec{v}_3)$ in terms of $\|\vec{v}_1 \times \vec{v}_2 \times \vec{v}_3\|$. Is your result still true when the $\vec{v}_i$ are linearly dependent?

(Note the analogy to the fact that for two vectors $\vec{v}_1$ and $\vec{v}_2$ in $\mathbb{R}^3$, $\|\vec{v}_1 \times \vec{v}_2\|$ is the area of the parallelogram defined by $\vec{v}_1$ and $\vec{v}_2$.)

18. If $T(\vec{x}) = A\vec{x}$ is an invertible linear transformation from $\mathbb{R}^2$ to $\mathbb{R}^2$, then the image $T(\Omega)$ of the unit circle Ω is an ellipse. (See Exercise 2.2.50.)

a. Sketch this ellipse when $A = \begin{bmatrix} p & 0 \\ 0 & q \end{bmatrix}$, where p and q are positive. What is its area?

b. For an arbitrary invertible transformation $T(\vec{x}) = A\vec{x}$, denote the lengths of the semi-major and the semi-minor axes of $T(\Omega)$ by a and b, respectively. What is the relationship between a, b, and $\det(A)$?

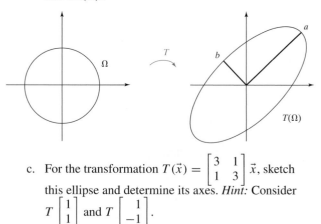

c. For the transformation $T(\vec{x}) = \begin{bmatrix} 3 & 1 \\ 1 & 3 \end{bmatrix} \vec{x}$, sketch this ellipse and determine its axes. *Hint:* Consider $T\begin{bmatrix} 1 \\ 1 \end{bmatrix}$ and $T\begin{bmatrix} 1 \\ -1 \end{bmatrix}$.

19. A basis $\vec{v}_1, \vec{v}_2, \vec{v}_3$ of $\mathbb{R}^3$ is called *positively oriented* if $\vec{v}_1$ encloses an acute angle with $\vec{v}_2 \times \vec{v}_3$. Illustrate this definition with a sketch. Show that the basis is positively oriented if (and only if) $\det[\vec{v}_1 \quad \vec{v}_2 \quad \vec{v}_3]$ is positive.

20. We say that a linear transformation T from $\mathbb{R}^3$ to $\mathbb{R}^3$ *preserves orientation* if it transforms any positively oriented basis into another positively oriented basis. (See Exercise 19.) Explain why a linear transformation $T(\vec{x}) = A\vec{x}$ preserves orientation if (and only if) $\det(A)$ is positive.

21. Arguing geometrically, determine whether the following orthogonal transformations from $\mathbb{R}^3$ to $\mathbb{R}^3$ preserve or reverse orientation. (See Exercise 20.)

  a. Reflection in a plane.

  b. Reflection in a line.

  c. Reflection in the origin.

Use Cramer's rule to solve the systems in Exercises 22 through 24.

22. $\begin{vmatrix} 3x + 7y = 1 \\ 4x + 11y = 3 \end{vmatrix}.$

23. $\begin{vmatrix} 5x_1 - 3x_2 = 1 \\ -6x_1 + 7x_2 = 0 \end{vmatrix}.$

24. $\begin{vmatrix} 2x + 3y \quad\;\; = \;\; 8 \\ 4y + 5z = \;\; 3 \\ 6x \quad\;\; + 7z = -1 \end{vmatrix}.$

25. Find the classical adjoint of the matrix

$$A = \begin{bmatrix} 1 & 0 & 1 \\ 0 & 1 & 0 \\ 2 & 0 & 1 \end{bmatrix},$$

and use the result to find A^{-1}.

26. Consider an $n \times n$ matrix A with integer entries such that $\det(A) = 1$. Are the entries of A^{-1} necessarily integers? Explain.

27. Consider two positive numbers a and b. Solve the following system:

$$\begin{vmatrix} ax - by = 1 \\ bx + ay = 0 \end{vmatrix}$$

What are the signs of the solutions x and y? How does x change as b increases?

28. In an economics text,[9] we find the following system:

$$sY + ar = I° + G$$
$$mY - hr = M_s - M°.$$

Solve for Y and r.

29. In an economics text[10] we find the following system:

$$\begin{bmatrix} -R_1 & R_1 & -(1-\alpha) \\ \alpha & 1-\alpha & -(1-\alpha)^2 \\ R_2 & -R_2 & \dfrac{-(1-\alpha)^2}{\alpha} \end{bmatrix} \begin{bmatrix} dx_1 \\ dy_1 \\ dp \end{bmatrix} = \begin{bmatrix} 0 \\ 0 \\ -R_2\, de_2 \end{bmatrix}.$$

Solve for dx_1, dy_1, and dp. In your answer, you may refer to the determinant of the coefficient matrix as D. (You need not compute D.) The quantities R_1, R_2, and D are positive, and α is between zero and one. If de_2 is positive, what can you say about the signs of dy_1 and dp?

30. (For those who have studied multivariable calculus.) Let T be an invertible linear transformation from $\mathbb{R}^2$ to $\mathbb{R}^2$, represented by the matrix M. Let Ω_1 be the unit square in $\mathbb{R}^2$ and Ω_2 its image under T. Consider a continuous function $f(x, y)$ from $\mathbb{R}^2$ to $\mathbb{R}$, and define the function $g(u, v) = f\big(T(u, v)\big)$. What is the relationship between the following two double integrals?

$$\iint\limits_{\Omega_2} f(x, y)\, dA \qquad \text{and} \qquad \iint\limits_{\Omega_1} g(u, v)\, dA$$

Your answer will involve the matrix M. *Hint:* What happens when $f(x, y) = 1$, for all x, y?

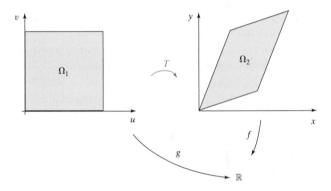

31. What is the area of the largest ellipse you can inscribe into a triangle with side lengths 3, 4, and 5? *Hint:* The largest ellipse you can inscribe into an equilateral triangle is a circle.

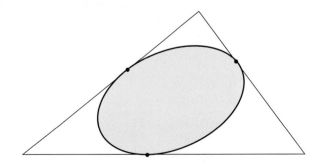

32. What are the lengths of the semiaxes of the largest ellipse you can inscribe into a triangle with sides 3, 4, and 5? See Exercise 31.

---
[9]Simon and Blume, *Mathematics for Economists,* Norton, 1994.
[10]Simon and Blume, *op. cit.*

Chapter 6

TRUE OR FALSE?

1. The determinant of any diagonal $n \times n$ matrix is the product of its diagonal entries.

2. If matrix B is obtained by swapping two rows of an $n \times n$ matrix A, then the equation $\det(B) = -\det(A)$ must hold.

3. If $A = [\vec{u} \ \ \vec{v} \ \ \vec{w}]$ is any 3×3 matrix, then $\det(A) = \vec{u} \cdot (\vec{v} \times \vec{w})$.

4. $\det(4A) = 4 \det(A)$ for all 4×4 matrices A.

5. $\det(A + B) = \det(A) + \det(B)$ for all 5×5 matrices A and B.

6. The equation $\det(-A) = \det(A)$ holds for all 6×6 matrices.

7. If all the entries of a 7×7 matrix A are 7, then $\det(A)$ must be 7^7.

8. An 8×8 matrix fails to be invertible if (and only if) its determinant is nonzero.

9. If B is obtained be multiplying a column of A by 9, then the equation $\det(B) = 9 \det(A)$ must hold.

10. $\det(A^{10}) = (\det A)^{10}$ for all 10×10 matrices A.

11. If two $n \times n$ matrices A and B are similar, then the equation $\det(A) = \det(B)$ must hold.

12. The determinant of all orthogonal matrices is 1.

13. If A is any $n \times n$ matrix, then $\det(AA^T) = \det(A^T A)$.

14. There is an invertible matrix of the form
$$\begin{bmatrix} a & e & f & j \\ b & 0 & g & 0 \\ c & 0 & h & 0 \\ d & 0 & i & 0 \end{bmatrix}.$$

15. The matrix $\begin{bmatrix} k^2 & 1 & 4 \\ k & -1 & -2 \\ 1 & 1 & 1 \end{bmatrix}$ is invertible for all positive constants k.

16. $\det \begin{bmatrix} 0 & 1 & 0 & 0 \\ 0 & 0 & 1 & 0 \\ 0 & 0 & 0 & 1 \\ 1 & 0 & 0 & 0 \end{bmatrix} = 1.$

17. Matrix $\begin{bmatrix} 9 & 100 & 3 & 7 \\ 5 & 4 & 100 & 8 \\ 100 & 9 & 8 & 7 \\ 6 & 5 & 4 & 100 \end{bmatrix}$ is invertible.

18. If A is an invertible $n \times n$ matrix, then $\det(A^T)$ must equal $\det(A^{-1})$.

19. If the determinant of a 4×4 matrix A is 4, then its rank must be 4.

20. There is a nonzero 4×4 matrix A such that $\det(A) = \det(4A)$.

21. If all the columns of a square matrix A are unit vectors, then the determinant of A must be less than or equal to 1.

22. If A is any noninvertible square matrix, then $\det(A) = \det(\text{rref } A)$.

23. If the determinant of a square matrix is -1, then A must be an orthogonal matrix.

24. If all the entries of an invertible matrix A are integers, then the entries of A^{-1} must be integers as well.

25. There is a 4×4 matrix A whose entries are all 1 or -1, and such that $\det(A) = 16$.

26. If the determinant of a 2×2 matrix A is 4, then the inequality $\|A\vec{v}\| \le 4\|\vec{v}\|$ must hold for all vectors $\vec{v}$ in $\mathbb{R}^2$.

27. If $A = [\vec{u} \ \ \vec{v} \ \ \vec{w}]$ is a 3×3 matrix, then the formula $\det(A) = \vec{v} \cdot (\vec{u} \times \vec{w})$ must hold.

28. There are invertible 2×2 matrices A and B such that $\det(A + B) = \det(A) + \det(B)$.

29. If all the entries of a square matrix are 1 or 0, then $\det(A)$ must be 1, 0, or -1.

30. If all the entries of a square matrix A are integers and $\det(A) = 1$, then the entries of matrix A^{-1} must be integers as well.

31. If A is any symmetric matrix, then $\det(A) = 1$ or $\det(A) = -1$.

32. If A is any skew-symmetric 4×4 matrix, then $\det(A) = 0$.

33. If $\det(A) = \det(B)$ for two $n \times n$ matrices A and B, then matrices A and B must be similar.

34. Suppose A is an $n \times n$ matrix and B is obtained from A by swapping two rows of A. If $\det(B) < \det(A)$, then A must be invertible.

35. If an $n \times n$ matrix A is invertible, then there must be an $(n-1) \times (n-1)$ submatrix of A (obtained by deleting a row and a column of A) that is invertible as well.

36. If all the entries of matrices A and A^{-1} are integers, then the equation $\det(A) = \det(A^{-1})$ must hold.

37. If a square matrix A is invertible, then its classical adjoint $\text{adj}(A)$ is invertible as well.

38. There is a 3×3 matrix A such that $A^2 = -I_3$.

39. There are invertible 3×3 matrices A and S such that $S^{-1}AS = 2A$.

40. There are invertible 3×3 matrices A and S such that $S^T AS = -A$.

41. If all the diagonal entries of an $n \times n$ matrix A are odd integers and all the other entries are even integers, then A must be an invertible matrix.

42. If all the diagonal entries of an $n \times n$ matrix A are even integers and all the other entries are odd integers, then A must be an invertible matrix.

43. For every nonzero 2×2 matrix A there exists a 2×2 matrix B such that $\det(A + B) \neq \det(A) + \det(B)$.

44. If A is a 4×4 matrix whose entries are all 1 or -1, then $\det(A)$ must be divisible by 8 (i.e., $\det(A) = 8k$ for some integer k).

7

Eigenvalues and Eigenvectors

7.1 DYNAMICAL SYSTEMS AND EIGENVECTORS: AN INTRODUCTORY EXAMPLE

A stretch of desert in northwestern Mexico is populated mainly by two species of animals: coyotes and roadrunners. We wish to model the populations $c(t)$ and $r(t)$ of coyotes and roadrunners t years from now if the current populations c_0 and r_0 are known.[1]

For this habitat, the following equations model the transformation of this system from one year to the next, from time t to time $(t + 1)$:

$$\begin{vmatrix} c(t + 1) = & 0.86c(t) + 0.08r(t) \\ r(t + 1) = & -0.12c(t) + 1.14r(t) \end{vmatrix}.$$

Why is the coefficient of $c(t)$ in the first equation less than 1, while the coefficient of $r(t)$ in the second equation exceeds 1? What is the practical significance of the signs of the other two coefficients, 0.08 and -0.12?

The two equations can be written in matrix form, as

$$\begin{bmatrix} c(t + 1) \\ r(t + 1) \end{bmatrix} = \begin{bmatrix} 0.86c(t) + 0.08r(t) \\ -0.12c(t) + 1.14r(t) \end{bmatrix} = \begin{bmatrix} 0.86 & 0.08 \\ -0.12 & 1.14 \end{bmatrix} \begin{bmatrix} c(t) \\ r(t) \end{bmatrix}.$$

The vector

$$\vec{x}(t) = \begin{bmatrix} c(t) \\ r(t) \end{bmatrix}$$

---

[1] The point of this lighthearted story is to present an introductory example where neither messy data nor a complicated scenario distracts us from the mathematical ideas we wish to develop.

is called the *state vector* of the system at time t, because it completely describes this system at time t. If we let

$$A = \begin{bmatrix} 0.86 & 0.08 \\ -0.12 & 1.14 \end{bmatrix},$$

we can write the matrix equation above more succinctly as

$$\vec{x}(t+1) = A\vec{x}(t).$$

The transformation the system undergoes over the period of one year is linear, represented by the matrix A.

$$\vec{x}(t) \xrightarrow{A} \vec{x}(t+1)$$

Suppose we know the initial state

$$\vec{x}(0) = \vec{x}_0 = \begin{bmatrix} c_0 \\ r_0 \end{bmatrix}.$$

We wish to find $\vec{x}(t)$, for an arbitrary positive integer t:

$$\vec{x}(0) \xrightarrow{A} \vec{x}(1) \xrightarrow{A} \vec{x}(2) \xrightarrow{A} \vec{x}(3) \xrightarrow{A} \cdots \xrightarrow{A} \vec{x}(t) \xrightarrow{A} \cdots$$

We can find $\vec{x}(t)$ by applying the transformation t times to $\vec{x}(0)$:

$$\vec{x}(t) = A^t \vec{x}(0) = A^t \vec{x}_0.$$

Although it is extremely tedious to find $\vec{x}(t)$ with paper and pencil for large t, we can easily compute $\vec{x}(t)$ using technology. For example, given

$$\vec{x}_0 = \begin{bmatrix} 100 \\ 100 \end{bmatrix},$$

we find that

$$\vec{x}(10) = A^{10}\vec{x}_0 \approx \begin{bmatrix} 80 \\ 170 \end{bmatrix}.$$

To understand the long-term behavior of this system and how it depends on the initial values, we must go beyond numerical experimentation. It would be useful to have *closed formulas* for $c(t)$ and $r(t)$, expressing these quantities as functions of t. We will first do this for certain (carefully chosen) initial state vectors.

Case 1 ◆ Suppose we have $c_0 = 100$ and $r_0 = 300$. Initially, there are 100 coyotes and 300 roadrunners, so that $\vec{x}_0 = \begin{bmatrix} 100 \\ 300 \end{bmatrix}$. Then

$$\vec{x}(1) = A\vec{x}_0 = \begin{bmatrix} 0.86 & 0.08 \\ -0.12 & 1.14 \end{bmatrix} \begin{bmatrix} 100 \\ 300 \end{bmatrix} = \begin{bmatrix} 110 \\ 330 \end{bmatrix}.$$

Note that each population has grown by 10%. This means that the state vector $\vec{x}(1)$ is a scalar multiple of $\vec{x}_0$ (see Figure 1):

$$\vec{x}(1) = A\vec{x}_0 = 1.1\vec{x}_0$$

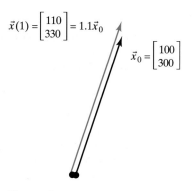

$$\vec{x}(1) = \begin{bmatrix} 110 \\ 330 \end{bmatrix} = 1.1\vec{x}_0$$

$$\vec{x}_0 = \begin{bmatrix} 100 \\ 300 \end{bmatrix}$$

Figure 1

It is now easy to compute $\vec{x}(t)$ for arbitrary t, using linearity:

$$\vec{x}(2) = A\vec{x}(1) = A(1.1\vec{x}_0) = 1.1A\vec{x}_0 = 1.1^2\vec{x}_0$$

$$\vec{x}(3) = A\vec{x}(2) = A(1.1^2\vec{x}_0) = 1.1^2 A\vec{x}_0 = 1.1^3\vec{x}_0$$

$$\vdots$$

$$\vec{x}(t) = 1.1^t\vec{x}_0$$

We keep multiplying the state vector by 1.1 each time we apply the transformation A.

Recall that our goal is to find closed formulas for $c(t)$ and $r(t)$. We have

$$\vec{x}(t) = \begin{bmatrix} c(t) \\ r(t) \end{bmatrix} = 1.1^t\vec{x}_0 = 1.1^t \begin{bmatrix} 100 \\ 300 \end{bmatrix},$$

so that

$$c(t) = 100(1.1)^t$$

and

$$r(t) = 300(1.1)^t.$$

Both populations will grow exponentially, by 10% each year.

Case 2 ◆ Suppose we have $c_0 = 200$ and $r_0 = 100$. Then

$$\vec{x}(1) = A\vec{x}_0 = \begin{bmatrix} 0.86 & 0.08 \\ -0.12 & 1.14 \end{bmatrix} \begin{bmatrix} 200 \\ 100 \end{bmatrix} = \begin{bmatrix} 180 \\ 90 \end{bmatrix} = 0.9\vec{x}_0.$$

Both populations decline by 10% in the first year and will therefore decline another 10% each subsequent year. Thus

$$\vec{x}(t) = 0.9^t\vec{x}_0,$$

so that

$$c(t) = 200(0.9)^t$$

and

$$r(t) = 100(0.9)^t.$$

The initial populations are mismatched: too many coyotes are chasing too few roadrunners, a bad state of affairs for both species.

Case 3 ◆ Suppose we have $c_0 = r_0 = 1000$. Then

$$\vec{x}(1) = A\vec{x}_0 = \begin{bmatrix} 0.86 & 0.08 \\ -0.12 & 1.14 \end{bmatrix}\begin{bmatrix} 1000 \\ 1000 \end{bmatrix} = \begin{bmatrix} 940 \\ 1020 \end{bmatrix}.$$

Things are not working out as nicely as in the first two cases we considered: The state vector $\vec{x}(1)$ is not a scalar multiple of the initial state $\vec{x}_0$. Just by computing $\vec{x}(2), \vec{x}(3), \ldots$, we could not easily detect a trend that would allow us to generate a closed formula for $c(t)$ and $r(t)$. We have to look for another approach.

The idea is to work with the two vectors

$$\vec{v}_1 = \begin{bmatrix} 100 \\ 300 \end{bmatrix} \quad \text{and} \quad \vec{v}_2 = \begin{bmatrix} 200 \\ 100 \end{bmatrix}$$

considered in the first two cases, for which $A^t\vec{v}_i$ was easy to compute. Since $\vec{v}_1$ and $\vec{v}_2$ form a basis of $\mathbb{R}^2$, any vector in $\mathbb{R}^2$ can be written uniquely as a linear combination of $\vec{v}_1$ and $\vec{v}_2$. This holds in particular for the initial state vector

$$\vec{x}_0 = \begin{bmatrix} 1000 \\ 1000 \end{bmatrix}$$

of the coyote–roadrunner system:

$$\vec{x}_0 = c_1\vec{v}_1 + c_2\vec{v}_2$$

A straightforward computation shows that the coordinates are $c_1 = 2$ and $c_2 = 4$:

$$\vec{x}_0 = 2\vec{v}_1 + 4\vec{v}_2$$

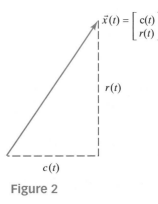

Recall that $A^t\vec{v}_1 = (1.1)^t\vec{v}_1$ and $A^t\vec{v}_2 = (0.9)^t\vec{v}_2$. Therefore,

$$\vec{x}(t) = A^t\vec{x}_0 = A^t(2\vec{v}_1 + 4\vec{v}_2) = 2A^t\vec{v}_1 + 4A^t\vec{v}_2$$
$$= 2(1.1)^t\vec{v}_1 + 4(0.9)^t\vec{v}_2$$
$$= 2(1.1)^t\begin{bmatrix} 100 \\ 300 \end{bmatrix} + 4(0.9)^t\begin{bmatrix} 200 \\ 100 \end{bmatrix}.$$

Figure 2

Considering the components of this equation, we can find formulas for $c(t)$ and $r(t)$:

$$c(t) = 200(1.1)^t + 800(0.9)^t$$
$$r(t) = 600(1.1)^t + 400(0.9)^t$$

Since the terms involving 0.9^t approach zero as t increases, both populations eventually grow by about 10% a year, and their ratio $r(t)/c(t)$ approaches $600/200 = 3$.

Note that the ratio $r(t)/c(t)$ can be interpreted as the slope of the state vector $\vec{x}(t)$, as shown in Figure 2.

How can we represent the computations performed above graphically?

Figure 3 shows the representation $\vec{x}_0 = 2\vec{v}_1 + 4\vec{v}_2$ of $\vec{x}_0$ as the sum of a vector on $L_1 = \text{span}(\vec{v}_1)$ and a vector on $L_2 = \text{span}(\vec{v}_2)$. The formula

$$\vec{x}(t) = (1.1)^t2\vec{v}_1 + (0.9)^t4\vec{v}_2$$

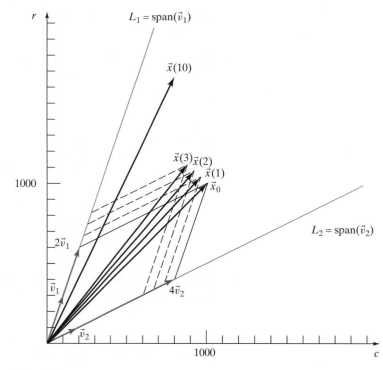

Figure 3

now tells us that the component in L_1 grows by 10% each year, while the component in L_2 shrinks by 10%. The component $(0.9)^t 4\vec{v}_2$ in L_2 approaches $\vec{0}$, which means that the tip of the state vector $\vec{x}(t)$ approaches the line L_1, so that the slope of the state vector will approach 3, the slope of L_1.

To show the evolution of the system more clearly, we can sketch just the endpoints of the state vectors $\vec{x}(t)$. Then, the changing state of the system will be traced out as a sequence of points in the c–r plane.

It is natural to connect the dots to create the illusion of a continuous *trajectory*. (Although, of course, we do not know what really happens between times t and $t + 1$.)

Sometimes, we are interested in the state of the system in the past, at times -1, -2, ... Note that $\vec{x}(0) = A\vec{x}(-1)$, so that $\vec{x}(-1) = A^{-1}\vec{x}_0$ if A is invertible (as in our example). Likewise, $\vec{x}(-t) = (A^t)^{-1}\vec{x}_0$, for $t = 2, 3, \ldots$. The trajectory (future and past) for our coyote–roadrunner system is shown in Figure 4.

To get a feeling for the long-term behavior of this system and how it depends on the initial state, we can draw a rough sketch that shows a number of different trajectories, representing the various qualitative types of behavior. Such a sketch is called a *phase portrait* of the system. In our example, a phase portrait might show the foregoing three cases, as well as a trajectory that starts above L_1 and one that starts below L_2. See Figure 5.

To sketch these trajectories, express the initial state vector $\vec{x}_0$ as the sum of a vector $\vec{w}_1$ on L_1 and a vector $\vec{w}_2$ on L_2. Then see how these two vectors change over time. If $\vec{x}_0 = \vec{w}_1 + \vec{w}_2$, then

$$\vec{x}(t) = (1.1)^t \vec{w}_1 + (0.9)^t \vec{w}_2.$$

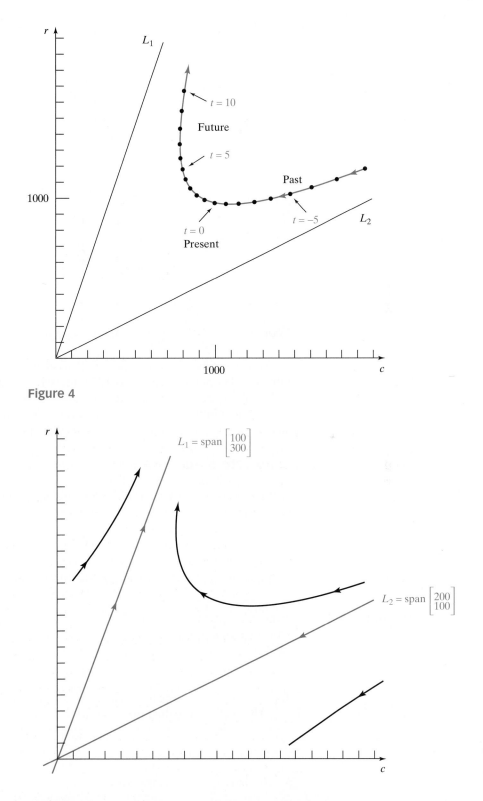

Figure 4

Figure 5

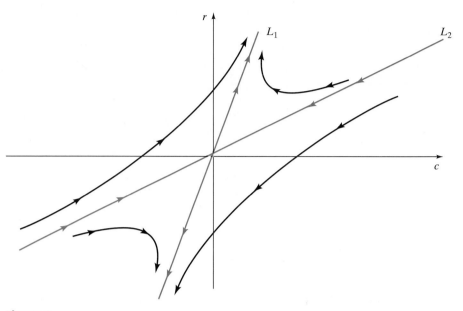

Figure 6

We see that the two populations will prosper over the long term if the ratio r_0/c_0 of the initial populations exceeds $1/2$; otherwise, both populations will die out.

From a mathematical point of view, it is informative to sketch a phase portrait of this system in the whole c–r-plane, even though the trajectories outside the first quadrant are meaningless in terms of our population study. (See Figure 6.)

Eigenvectors and Eigenvalues

In the example just discussed, it turned out to be very useful to have some nonzero vectors $\vec{v}$ such that $A\vec{v}$ is a scalar multiple of $\vec{v}$, or

$$A\vec{v} = \lambda\vec{v},$$

for some scalar λ.

Such vectors are important in many other contexts as well.

Definition 7.1.1

Eigenvectors[2] and eigenvalues
Consider an $n \times n$ matrix A. A nonzero vector $\vec{v}$ in $\mathbb{R}^n$ is called an *eigenvector* of A if $A\vec{v}$ is a scalar multiple of $\vec{v}$, that is, if

$$A\vec{v} = \lambda\vec{v},$$

for some scalar λ. Note that this scalar λ may be zero.

The scalar λ is called the *eigenvalue* associated with the eigenvector $\vec{v}$.

---

[2]From German *eigen:* proper, characteristic.

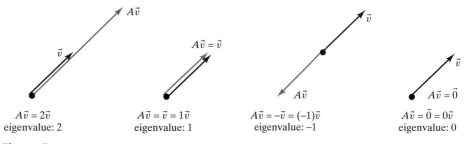

Figure 7

A nonzero vector $\vec{v}$ is an eigenvector of A if the vectors $\vec{v}$ and $A\vec{v}$ are parallel, as shown in Figure 7. (See Definition A.3 in the Appendix.)

If $\vec{v}$ is an eigenvector of matrix A, then $\vec{v}$ is an eigenvector of matrices A^2, A^3, ..., as well, with

$$A^2\vec{v} = \lambda^2\vec{v}, \quad A^3\vec{v} = \lambda^3\vec{v}, \ldots, \quad A^t\vec{v} = \lambda^t\vec{v},$$

for all positive integers t.

EXAMPLE 1   Find all eigenvectors and eigenvalues of the identity matrix I_n.

Solution

Since $I_n\vec{v} = \vec{v} = 1\vec{v}$ for all $\vec{v}$ in $\mathbb{R}^n$, all nonzero vectors in $\mathbb{R}^n$ are eigenvectors, with eigenvalue 1. ∎

EXAMPLE 2   Let T be the orthogonal projection onto a line L in $\mathbb{R}^2$. Describe the eigenvectors of T geometrically and find all eigenvalues of T.

Solution

We find the eigenvectors by inspection: can you think of any nonzero vectors $\vec{v}$ in the plane such that $T(\vec{v})$ is a scalar multiple of $\vec{v}$? Clearly, any vector $\vec{v}$ on L will do (with $T(\vec{v}) = 1\vec{v}$), as well as any vector $\vec{w}$ perpendicular to L (with $T(\vec{w}) = \vec{0} = 0\vec{w}$). See Figure 8.

The eigenvalues are 1 and 0. ∎

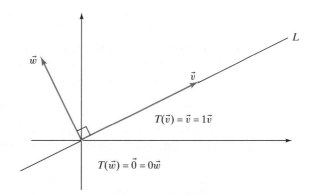

Figure 8

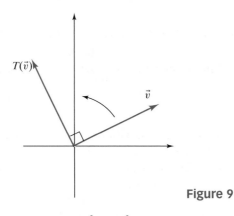

Figure 9

EXAMPLE 3 Let T from $\mathbb{R}^2$ to $\mathbb{R}^2$ be the rotation in the plane through an angle of $90°$ in the counterclockwise direction. Find all eigenvalues and eigenvectors of T.

Solution

If $\vec{v}$ is any nonzero vector in $\mathbb{R}^2$, then $T(\vec{v})$ is not parallel to $\vec{v}$ (it's perpendicular). See Figure 9. There are no eigenvectors and eigenvalues here. ∎

EXAMPLE 4 What are the possible real[3] eigenvalues of an orthogonal[4] matrix A?

Solution

Recall that the linear transformation $T(\vec{x}) = A\vec{x}$ preserves length: $\|T(\vec{x})\| = \|A\vec{x}\| = \|\vec{x}\|$, for all vectors $\vec{x}$. (See Definition 5.3.1.) Consider an eigenvector $\vec{v}$ of A, with eigenvalue λ:

$$A\vec{v} = \lambda\vec{v}.$$

Then

$$\|\vec{v}\| = \|A\vec{v}\| = \|\lambda\vec{v}\| = |\lambda|\|\vec{v}\|,$$

so that $\lambda = 1$ or $\lambda = -1$. The two possibilities are illustrated in Figure 10. ∎

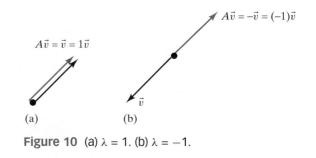

Figure 10 (a) $\lambda = 1$. (b) $\lambda = -1$.

---

[3]In Section 7.5, we will consider *complex* eigenvalues.

[4]Example 4 and Fact 7.1.2 are for those who have studied Chapter 5.

Fact 7.1.2   The possible real eigenvalues of an orthogonal[4] matrix are 1 and -1.

As an example, consider the reflection in a line in $\mathbb{R}^2$. (See Exercise 15.)

Dynamical Systems and Eigenvectors

Consider a physical system whose state at any given time t is described by some quantities $x_1(t)$, $x_2(t)$, $\ldots$, $x_n(t)$. (In our introductory example, there were two such quantities, the populations $c(t)$ and $r(t)$.) We can represent the quantities $x_1(t)$, $x_2(t)$, $\ldots$, $x_n(t)$ by the state vector

$$\vec{x}(t) = \begin{bmatrix} x_1(t) \\ x_2(t) \\ \vdots \\ x_n(t) \end{bmatrix}.$$

Suppose that the state of the system at time $t + 1$ is determined by the state at time t and that the transformation of the system from time t to time $t + 1$ is linear, represented by an $n \times n$ matrix A:

$$\vec{x}(t + 1) = A\vec{x}(t).$$

Then

$$\vec{x}(t) = A^t \vec{x}_0.$$

Such a system is called a *discrete linear dynamical system*. ("Discrete" indicates that we model the change of the system from time t to time $t + 1$, rather than modeling the *continuous* rate of change, which is described by differential equations.)

For an initial state $\vec{x}_0$, it is often our goal to find *closed formulas* for $x_1(t)$, $x_2(t)$, $\ldots$, $x_n(t)$ (i.e., formulas expressing $x_i(t)$ as a function of t alone, as opposed to a recursive formula, for example, which would merely express $x_i(t + 1)$ in terms of $x_1(t)$, $x_2(t)$, $\ldots$, $x_n(t)$).

Fact 7.1.3   **Discrete dynamical systems**
Consider the dynamical system

$$\vec{x}(t + 1) = A\vec{x}(t) \qquad \text{with} \qquad \vec{x}(0) = \vec{x}_0.$$

Then

$$\vec{x}(t) = A^t \vec{x}_0.$$

Suppose we can find a basis $\vec{v}_1, \vec{v}_2, \ldots, \vec{v}_n$ of $\mathbb{R}^n$ consisting of eigenvectors of A, with

$$A\vec{v}_1 = \lambda_1 \vec{v}_1, \, A\vec{v}_2 = \lambda_2 \vec{v}_2, \ldots, A\vec{v}_n = \lambda_n \vec{v}_n.$$

(Continued)

Fact 7.1.3

Discrete dynamical systems (*Continued*)

Find the coordinates $c_1, c_2, \ldots, c_n$ of vector $\vec{x}_0$ with respect to basis $\vec{v}_1, \vec{v}_2, \ldots, \vec{v}_n$:

$$\vec{x}_0 = c_1\vec{v}_1 + c_2\vec{v}_2 + \cdots + c_n\vec{v}_n.$$

Then

$$\vec{x}(t) = c_1\lambda_1^t\vec{v}_1 + c_2\lambda_2^t\vec{v}_2 + \cdots + c_n\lambda_n^t\vec{v}_n.$$

We can write this equation in matrix form as

$$\vec{x}(t) = \begin{bmatrix} | & | & & | \\ \vec{v}_1 & \vec{v}_2 & \cdots & \vec{v}_n \\ | & | & & | \end{bmatrix} \begin{bmatrix} \lambda_1^t & 0 & . & 0 \\ 0 & \lambda_2^t & . & 0 \\ . & . & . & . \\ 0 & 0 & . & \lambda_n^t \end{bmatrix} \begin{bmatrix} c_1 \\ c_2 \\ . \\ c_n \end{bmatrix}$$

$$= S \begin{bmatrix} \lambda_1 & 0 & . & 0 \\ 0 & \lambda_2 & . & 0 \\ . & . & . & . \\ 0 & 0 & . & \lambda_n \end{bmatrix}^t S^{-1}\vec{x}_0,$$

where $S = \begin{bmatrix} | & | & & | \\ \vec{v}_1 & \vec{v}_2 & \cdots & \vec{v}_n \\ | & | & & | \end{bmatrix}$. Note that $\vec{x}_0 = S \begin{bmatrix} c_1 \\ c_2 \\ . \\ c_n \end{bmatrix}$, so that $\begin{bmatrix} c_1 \\ c_2 \\ . \\ c_n \end{bmatrix} = $

$S^{-1}\vec{x}_0.$

We are left with two questions: How can we find the eigenvalues and eigenvectors of an $n \times n$ matrix A? When is there a basis of $\mathbb{R}^n$ consisting of eigenvectors of A? These issues are central to linear algebra; they will keep us busy for the rest of this long chapter.

Definition 7.1.4

Discrete trajectories and phase portraits

Consider a discrete dynamical system

$$\vec{x}(t + 1) = A\vec{x}(t) \quad \text{with initial value } \vec{x}(0) = \vec{x}_0,$$

where A is a 2×2 matrix. In this case, the state vector $\vec{x}(t) = \begin{bmatrix} x_1(t) \\ x_2(t) \end{bmatrix}$ can be represented geometrically in the x_1–x_2-plane.

The endpoints of state vectors $\vec{x}(0) = \vec{x}_0, \vec{x}(1) = A\vec{x}_0, \vec{x}(2) = A^2\vec{x}_0, \ldots$ form the (discrete) *trajectory* of this system, representing its evolution in the future. Sometimes we are interested in the past states $\vec{x}(-1) = A^{-1}\vec{x}_0, \vec{x}(-2) = (A^2)^{-1}\vec{x}_0, \ldots$ as well. It is suggestive to "connect the dots" to create the illusion of a continuous trajectory. Take another look at Figure 4.

A (discrete) *phase portrait* of the system $\vec{x}(t + 1) = A\vec{x}(t)$ shows discrete trajectories for various initial states, capturing all the qualitatively different scenarios (as in Figure 6).

In Figure 11, we sketch phase portraits for the case when A has two eigenvalues $\lambda_1 > \lambda_2 > 0$ with associated eigenvectors $\vec{v}_1$ and $\vec{v}_2$. We leave out the special case when one of the eigenvalues is 1. Start by sketching the trajectories along the lines

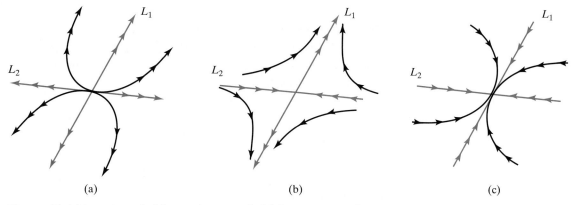

Figure 11 (a) $\lambda_1 > \lambda_2 > 1$. (b) $\lambda_1 > 1 > \lambda_2 > 0$. (c) $1 > \lambda_1 > \lambda_2 > 0$.

$L_1 = \text{span}(\vec{v}_1)$ and $L_2 = \text{span}(\vec{v}_2)$. As you sketch the other trajectories

$$\vec{x}(t) = c_1\lambda_1^t\vec{v}_1 + c_2\lambda_2^t\vec{v}_2,$$

think about the summands $c_1\lambda_1^t\vec{v}_1$ and $c_2\lambda_2^t\vec{v}_2$. Note that for a large positive t the vector $\vec{x}(t)$ will be almost parallel to L_1, since λ_1^t will be much larger than λ_2^t. Likewise, for large negative t the vector $\vec{x}(t)$ will be almost parallel to L_2.

We conclude this section with a third (and final) version of the summing-up theorem. (Compare this with Summary 3.3.11.)

Summary 7.1.4   Consider an $n \times n$ matrix

$$A = \begin{bmatrix} | & | & & | \\ \vec{v}_1 & \vec{v}_2 & \dots & \vec{v}_n \\ | & | & & | \end{bmatrix}.$$

Then the following statements are equivalent:

  i. A is invertible.
 ii. The linear system $A\vec{x} = \vec{b}$ has a unique solution $\vec{x}$, for all $\vec{b}$ in $\mathbb{R}^n$.
iii. $\text{rref}(A) = I_n$.
 iv. $\text{rank}(A) = n$.
  v. $\text{im}(A) = \mathbb{R}^n$.
 vi. $\text{ker}(A) = \{\vec{0}\}$.
vii. The $\vec{v}_i$ are a basis of $\mathbb{R}^n$.
viii. The $\vec{v}_i$ span $\mathbb{R}^n$.
 ix. The $\vec{v}_i$ are linearly independent.
  x. $\det(A) \neq 0$.
 xi. 0 fails to be an eigenvalue of A.

Characterization (x) was given in Fact 6.2.5. The equivalence of (vi) and (xi) follows from the definition of an eigenvalue. (Note that an eigenvector of A with eigenvalue 0 is a nonzero vector in $\text{ker}(A)$.)

EXERCISES

GOALS   Apply the concept of eigenvalues and eigenvectors. Use eigenvectors to analyze discrete dynamical systems.

In Exercises 1 through 4, let A be an invertible n × n matrix and $\vec{v}$ an eigenvector of A with associated eigenvalue λ.

1. Is $\vec{v}$ an eigenvector of A^3? If so, what is the eigenvalue?

2. Is $\vec{v}$ an eigenvector of A^{-1}? If so, what is the eigenvalue?

3. Is $\vec{v}$ an eigenvector of $A + 2I_n$? If so, what is the eigenvalue?

4. Is $\vec{v}$ an eigenvector of $7A$? If so, what is the eigenvalue?

5. If a vector $\vec{v}$ is an eigenvector of both A and B, is $\vec{v}$ necessarily an eigenvector of $A + B$?

6. If a vector $\vec{v}$ is an eigenvector of both A and B, is $\vec{v}$ necessarily an eigenvector of AB?

7. If $\vec{v}$ is an eigenvector of the n × n matrix A with associated eigenvalue λ, what can you say about

$$\ker(\lambda I_n - A)?$$

Is the matrix $\lambda I_n - A$ invertible?

8. Find all 2 × 2 matrices for which $\vec{e}_1 = \begin{bmatrix} 1 \\ 0 \end{bmatrix}$ is an eigenvector with associated eigenvalue 5.

9. Find all 2 × 2 matrices for which $\vec{e}_1$ is an eigenvector.

10. Find all 2 × 2 matrices for which $\begin{bmatrix} 1 \\ 2 \end{bmatrix}$ is an eigenvector with associated eigenvalue 5.

11. Find all 2 × 2 matrices for which $\begin{bmatrix} 1 \\ 2 \end{bmatrix}$ is an eigenvector.

12. Consider the matrix $A = \begin{bmatrix} 2 & 0 \\ 3 & 4 \end{bmatrix}$. Show that 2 and 4 are eigenvalues of A and find all corresponding eigenvectors.

13. Show that 4 is an eigenvalue of $A = \begin{bmatrix} -6 & 6 \\ -15 & 13 \end{bmatrix}$ and find all corresponding eigenvectors.

14. Find all 4 × 4 matrices for which $\vec{e}_2$ is an eigenvector.

Arguing geometrically, find all eigenvectors and eigenvalues of the linear transformations in Exercises 15 through 22. Find a basis consisting of eigenvectors if possible.

15. Reflection in a line L in $\mathbb{R}^2$.

16. Rotation through an angle of 180° in $\mathbb{R}^2$.

17. Counterclockwise rotation through an angle of 45° followed by dilation by 2 in $\mathbb{R}^2$.

18. Reflection in a plane E in $\mathbb{R}^3$.

19. Orthogonal projection onto a line L in $\mathbb{R}^3$.

20. Rotation about the $\vec{e}_3$-axis through an angle of 90°, counterclockwise as viewed from the positive $\vec{e}_3$-axis in $\mathbb{R}^3$.

21. Dilation by 5 in $\mathbb{R}^3$.

22. The shear with $T(\vec{v}) = \vec{v}$ and $T(\vec{w}) = \vec{v} + \vec{w}$ for the vectors $\vec{v}$ and $\vec{w}$ in $\mathbb{R}^2$ sketched below.

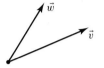

23. a. Consider an invertible matrix
$S = [\vec{v}_1 \quad \vec{v}_2 \quad \dots \quad \vec{v}_n]$. Find $S^{-1}\vec{v}_i$.
Hint: $S\vec{e}_i = \vec{v}_i$.

   b. Consider an n × n matrix A and an invertible n × n matrix $S = [\vec{v}_1 \quad \vec{v}_2 \quad \dots \quad \vec{v}_n]$ whose columns are eigenvectors of A with $A\vec{v}_i = \lambda_i \vec{v}_i$. Find $S^{-1}AS$. *Hint:* Think about the product $S^{-1}AS$ column by column.

In Exercises 24 through 29, consider a dynamical system

$$\vec{x}(t+1) = A\vec{x}(t)$$

with two components. The accompanying sketch shows the initial state vector $\vec{x}_0$ and two eigenvectors, $\vec{v}_1$ and $\vec{v}_2$, of A (with eigenvalues λ_1 and λ_2, respectively). For the given values of λ_1 and λ_2, sketch a rough trajectory. Think about the future and the past of the system.

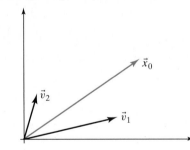

24. $\lambda_1 = 1.1, \lambda_2 = 0.9.$ **25.** $\lambda_1 = 1, \lambda_2 = 0.9.$

26. $\lambda_1 = 1.1, \lambda_2 = 1.$ **27.** $\lambda_1 = 0.9, \lambda_2 = 0.8.$

28. $\lambda_1 = 1.2, \lambda_2 = 1.1.$ **29.** $\lambda_1 = 0.9, \lambda_2 = 0.9.$

In Exercises 30 through 32, consider the dynamical system

$$\vec{x}(t+1) = \begin{bmatrix} 1.1 & 0 \\ 0 & \lambda \end{bmatrix} \vec{x}(t).$$

Sketch a phase portrait of this system for the given values of λ:

30. $\lambda = 1.2$.   **31.** $\lambda = 1$.   **32.** $\lambda = 0.9$.

33. Consider the coyotes–roadrunner system discussed in this section. Find closed formulas for $c(t)$ and $r(t)$, for the initial populations $c_0 = 100$, $r_0 = 800$.

34. Find a 2×2 matrix A such that $\begin{bmatrix} 3 \\ 1 \end{bmatrix}$ and $\begin{bmatrix} 1 \\ 2 \end{bmatrix}$ are eigenvectors of A, with eigenvalues 5 and 10, respectively.

35. Find a 2×2 matrix A such that

$$\vec{x}(t) = \begin{bmatrix} 2^t - 6^t \\ 2^t + 6^t \end{bmatrix}$$

is a trajectory of the dynamical system

$$\vec{x}(t + 1) = A\vec{x}(t).$$

36. Imagine that you are diabetic and have to pay close attention to how your body metabolizes glucose. Let $g(t)$ represent the *excess glucose concentration* in your blood, usually measured in milligrams of glucose per 100 milliliters of blood. (Excess means that we measure how much the glucose concentration deviates from your fasting level, i.e., the level your system approaches after many hours of fasting.) A negative value of $g(t)$ indicates that the glucose concentration is below fasting level at time t. Shortly after you eat a heavy meal, the function $g(t)$ will reach a peak, and then it will slowly return to 0. Certain hormones help regulate glucose, especially the hormone *insulin*. Let $h(t)$ represent the excess hormone concentration in your blood. Researchers have developed mathematical models for the glucose regulatory system. The following is one such model, in slightly simplified form (these formulas apply between meals; obviously, the system is disturbed during and right after a meal):

$$\begin{vmatrix} g(t + 1) = ag(t) - bh(t) \\ h(t + 1) = cg(t) + dh(t) \end{vmatrix},$$

here time t is measured in minutes; a and d are constants slightly less than 1; and b and c are small positive constants. For your system, the equations might be

$$\begin{vmatrix} g(t + 1) = 0.978\, g(t) - 0.006\, h(t) \\ h(t + 1) = 0.004\, g(t) + 0.992\, h(t) \end{vmatrix}.$$

The term $-0.006h(t)$ in the first equation is negative, because insulin helps your body absorb glucose. The term $0.004g(t)$ is positive, because glucose in your blood stimulates the cells of the pancreas to secrete insulin (for a more thorough discussion of this model, read E. Ackerman et al., "Blood glucose regulation and diabetes," Chapter 4 in *Concepts and Models of Biomathematics,* Marcel Dekker, 1969).

Consider the coefficient matrix

$$A = \begin{bmatrix} 0.978 & -0.006 \\ 0.004 & 0.992 \end{bmatrix}$$

of this dynamical system.

a. We are told that $\begin{bmatrix} -1 \\ 2 \end{bmatrix}$ and $\begin{bmatrix} 3 \\ -1 \end{bmatrix}$ are eigenvectors of A. Find the associated eigenvalues.

b. After you have consumed a heavy meal, the concentrations in your blood are $g_0 = 100$ and $h_0 = 0$. Find closed formulas for $g(t)$ and $h(t)$. Sketch the trajectory. Briefly describe the evolution of this system in practical terms.

c. For the case discussed in part (b), how long does it take for the glucose concentration to fall below fasting level? (This quantity is useful in diagnosing diabetes: a period of more than four hours may indicate mild diabetes.)

37. Consider the matrix

$$A = \begin{bmatrix} 3 & 4 \\ 4 & -3 \end{bmatrix}.$$

a. Use the geometric interpretation of this transformation as a reflection–dilation to find the eigenvalues of A.

b. Find two linearly independent eigenvectors for A.

38. Consider the growth of a lilac bush. The state of this lilac bush for several years (at year's end) is shown in the accompanying sketch. Let $n(t)$ be the number of new branches (grown in the year t) and $a(t)$ the number of old branches. In the sketch, the new branches are represented by shorter lines. Each old branch will grow two new branches in the following year. We assume that no branches ever die.

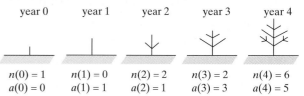

year 0	year 1	year 2	year 3	year 4
$n(0) = 1$	$n(1) = 0$	$n(2) = 2$	$n(3) = 2$	$n(4) = 6$
$a(0) = 0$	$a(1) = 1$	$a(2) = 1$	$a(3) = 3$	$a(4) = 5$

a. Find the matrix A such that

$$\begin{bmatrix} n(t + 1) \\ a(t + 1) \end{bmatrix} = A \begin{bmatrix} n(t) \\ a(t) \end{bmatrix}.$$

b. Verify that $\begin{bmatrix} 1 \\ 1 \end{bmatrix}$ and $\begin{bmatrix} 2 \\ -1 \end{bmatrix}$ are eigenvectors of A. Find the associated eigenvalues.

c. Find closed formulas for $n(t)$ and $a(t)$.

39. Show that similar matrices have the same eigenvalues. *Hint:* If $\vec{v}$ is an eigenvector of $S^{-1}AS$, then $S\vec{v}$ is an eigenvector of A.

40. Suppose $\vec{v}$ is an eigenvector of the $n \times n$ matrix A, with eigenvalue 4. Explain why $\vec{v}$ is an eigenvector of $A^2 + 2A + 3I_n$. What is the associated eigenvalue?

7.2 FINDING THE EIGENVALUES OF A MATRIX

In the previous section, we used eigenvalues to analyze a dynamical system

$$\vec{x}(t+1) = A\vec{x}(t).$$

Now we will see how we can actually find those eigenvalues.

Consider an $n \times n$ matrix A and a scalar λ. By definition, λ is an eigenvalue of A if there is a nonzero vector $\vec{v}$ in $\mathbb{R}^n$ such that

$$A\vec{v} = \lambda\vec{v},$$

or

$$\lambda\vec{v} - A\vec{v} = \vec{0},$$

or

$$(\lambda I_n - A)\vec{v} = \vec{0}.$$

This means, by definition of the kernel, that

$$\ker(\lambda I_n - A) \neq \{\vec{0}\}.$$

(That is, there are other vectors in the kernel besides the zero vector.)

This is the case if (and only if) the matrix $\lambda I_n - A$ is not invertible (by Fact 3.1.7), that is, if $\det(\lambda I_n - A) = 0$ (by Fact 6.2.5).

> **Fact 7.2.1**
>
> Consider an $n \times n$ matrix A and a scalar λ. Then λ is an eigenvalue of A if (and only if) $\det(\lambda I_n - A) = 0$.

To clarify things, we represent the observations made above as a series of equivalent statements:

λ is an eigenvalue of A.

$\updownarrow$

There is a nonzero vector $\vec{v}$ such that $A\vec{v} = \lambda\vec{v}$ or $(\lambda I_n - A)\vec{v} = \vec{0}$.

$\updownarrow$

$\ker(\lambda I_n - A) \neq \{\vec{0}\}$.

$\updownarrow$

$\lambda I_n - A$ is not invertible.

$\updownarrow$

$\det(\lambda I_n - A) = 0$.

EXAMPLE 1 Find the eigenvalues of the matrix

$$A = \begin{bmatrix} 1 & 2 \\ 4 & 3 \end{bmatrix}.$$

Solution

By Fact 7.2.1, we have to look for numbers λ such that $\det(\lambda I_2 - A) = 0$:

$$\det(\lambda I_2 - A) = \det\left(\begin{bmatrix} \lambda & 0 \\ 0 & \lambda \end{bmatrix} - \begin{bmatrix} 1 & 2 \\ 4 & 3 \end{bmatrix}\right)$$

$$= \det\begin{bmatrix} \lambda - 1 & -2 \\ -4 & \lambda - 3 \end{bmatrix}$$

$$= (\lambda - 1)(\lambda - 3) - 8$$

$$= \lambda^2 - 4\lambda - 5$$

$$= (\lambda - 5)(\lambda + 1)$$

The equation $\det(\lambda I_2 - A) = (\lambda - 5)(\lambda + 1) = 0$ holds for $\lambda_1 = 5$ and $\lambda_2 = -1$. These two scalars are the eigenvalues of A. ∎

EXAMPLE 2 Find the eigenvalues of

$$A = \begin{bmatrix} 1 & 2 & 3 & 4 & 5 \\ 0 & 2 & 3 & 4 & 5 \\ 0 & 0 & 3 & 4 & 5 \\ 0 & 0 & 0 & 4 & 5 \\ 0 & 0 & 0 & 0 & 5 \end{bmatrix}.$$

Solution

Again, we have to solve the equation $\det(\lambda I_5 - A) = 0$:

$$\det(\lambda I_5 - A) = \det\begin{bmatrix} \lambda - 1 & -2 & -3 & -4 & -5 \\ 0 & \lambda - 2 & -3 & -4 & -5 \\ 0 & 0 & \lambda - 3 & -4 & -5 \\ 0 & 0 & 0 & \lambda - 4 & -5 \\ 0 & 0 & 0 & 0 & \lambda - 5 \end{bmatrix}$$

$$= (\lambda - 1)(\lambda - 2)(\lambda - 3)(\lambda - 4)(\lambda - 5) = 0$$

Recall that the determinant of a triangular matrix is just the product of the diagonal entries (Fact 6.1.3). The solutions are 1, 2, 3, 4, 5; these are the eigenvalues of A. ∎

Fact 7.2.2 The eigenvalues of a triangular matrix are its diagonal entries.

By Fact 7.2.1, we can think of the eigenvalues of an $n \times n$ matrix A as the zeros of the function

$$f_A(\lambda) = \det(\lambda I_n - A).$$

EXAMPLE 3   Find $f_A(\lambda)$ for the 2×2 matrix

$$A = \begin{bmatrix} a & b \\ c & d \end{bmatrix}.$$

Solution

A straightforward computation shows that

$$f_A(\lambda) = \det(\lambda I_2 - A) = \det \begin{bmatrix} \lambda - a & -b \\ -c & \lambda - d \end{bmatrix} = \lambda^2 - (a + d)\lambda + (ad - bc).$$

This is a *quadratic polynomial* with constant term $\det(A) = ad - bc$. This makes sense, because the constant term is $f_A(0) = \det(0I_2 - A) = \det(-A) = \det(A)$.

The coefficient of λ is the negative of the sum of the diagonal entries of A. Since this sum is important in many other contexts as well, we introduce a name for it. ∎

Definition 7.2.3

Trace

The sum of the diagonal entries of an $n \times n$ matrix A is called the *trace* of A, denoted by $\operatorname{tr}(A)$.

We can now write the function $f_A(\lambda)$ associated with a 2×2 matrix A succinctly as follows:

Fact 7.2.4

If A is a 2×2 matrix, then

$$f_A(\lambda) = \det(\lambda I_2 - A) = \lambda^2 - \operatorname{tr}(A)\lambda + \det(A).$$

For the matrix $A = \begin{bmatrix} 1 & 2 \\ 4 & 3 \end{bmatrix}$, we have $\operatorname{tr}(A) = 4$ and $\det(A) = -5$, so that

$$f_A(\lambda) = \lambda^2 - 4\lambda - 5,$$

as we found in Example 1.

What is the format of $f_A(\lambda)$ for an $n \times n$ matrix A?

$$f_A(\lambda) = \det(\lambda I_n - A)$$

$$= \det \begin{bmatrix} \lambda - a_{11} & -a_{12} & \cdots & -a_{1n} \\ -a_{21} & \lambda - a_{22} & \cdots & -a_{2n} \\ \vdots & \vdots & \ddots & \vdots \\ -a_{n1} & -a_{n2} & \cdots & \lambda - a_{nn} \end{bmatrix}.$$

The contribution a pattern makes to this determinant is a polynomial in λ of degree less than or equal to n. This implies that the determinant itself, being the sum of all these contributions, is a polynomial of degree less than or equal to n.

We can be more precise: The diagonal pattern of the matrix $\lambda I_n - A$ makes the contribution

$$(\lambda - a_{11})(\lambda - a_{22}) \ldots (\lambda - a_{nn})$$

$$= \lambda^n - (a_{11} + a_{22} + \cdots + a_{nn})\lambda^{n-1} + \big(\text{a polynomial of degree} \leq (n-2)\big)$$

$$= \lambda^n - \operatorname{tr}(A)\lambda^{n-1} + \big(\text{a polynomial of degree} \leq (n-2)\big).$$

Any other pattern involves at least two scalars off the diagonal (see Exercise 6.1.39), and its contribution toward the determinant is therefore a polynomial of degree less than or equal to $n - 2$. This implies that

$$f_A(\lambda) = \lambda^n - \text{tr}(A)\lambda^{n-1} + \left(\text{a polynomial of degree} \leq (n-2)\right).$$

The constant term of this polynomial is $f_A(0) = \det(-A) = (-1)^n \det(A)$. It is possible to describe the coefficients of $\lambda, \lambda^2, \ldots, \lambda^{n-2}$ as well, but these formulas are complicated, and we do not need them here.

Fact 7.2.5

Characteristic polynomial

Consider an $n \times n$ matrix A. Then $f_A(\lambda) = \det(\lambda I_n - A)$ is a polynomial of degree n of the form

$$f_A(\lambda) = \lambda^n - \text{tr}(A)\lambda^{n-1} + \cdots + (-1)^n \det(A).$$

$f_A(\lambda)$ is called the *characteristic polynomial* of A.

Note that Fact 7.2.4 represents a special case of Fact 7.2.5.

What does Fact 7.2.5 tell us about the number of eigenvalues of an $n \times n$ matrix A? We know from elementary algebra that a polynomial of degree n has at most n zeros. Therefore, an $n \times n$ matrix has at most n eigenvalues. If the characteristic polynomial $f_A(\lambda)$ is of odd degree, then there is at least one zero, by the intermediate value theorem, since

$$\lim_{\lambda \to \infty} f_A(\lambda) = \infty \qquad \text{and} \qquad \lim_{\lambda \to -\infty} f_A(\lambda) = -\infty.$$

See Figure 1.

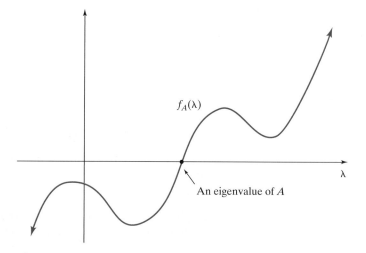

Figure 1

EXAMPLE 4   Find all eigenvalues of

$$A = \begin{bmatrix} 1 & 2 & 3 & 4 & 5 \\ 0 & 2 & 3 & 4 & 5 \\ 0 & 0 & 1 & 2 & 3 \\ 0 & 0 & 0 & 2 & 3 \\ 0 & 0 & 0 & 0 & 1 \end{bmatrix}.$$

Solution

The characteristic polynomial is $f_A(\lambda) = (\lambda - 1)^3 (\lambda - 2)^2$. The eigenvalues are 1 and 2. Since 1 is a root of multiplicity 3 of the characteristic polynomial, we say that the eigenvalue 1 has *algebraic multiplicity* 3. Likewise, the eigenvalue 2 has algebraic multiplicity 2.   ∎

Definition 7.2.6

> **Algebraic multiplicity of an eigenvalue**
> We say that an eigenvalue λ_0 of a square matrix A has *algebraic multiplicity k* if
>
> $$f_A(\lambda) = (\lambda - \lambda_0)^k g(\lambda)$$
>
> for some polynomial $g(\lambda)$ with $g(\lambda_0) \neq 0$ (i.e., if λ_0 is a root of multiplicity k of $f_A(\lambda)$).

In Example 4, we had

$$f_A(\lambda) = (\lambda - 1)^3 \underbrace{(\lambda - 2)^2}_{g(\lambda)}.$$

with k over the exponent 3 of $(\lambda-1)^3$ and λ_0 pointing to the base $(\lambda-1)$.

The algebraic multiplicity of the eigenvalue $\lambda_0 = 1$ is $k = 3$.

EXAMPLE 5   Find the eigenvalues of

$$A = \begin{bmatrix} 2 & -1 & -1 \\ -1 & 2 & -1 \\ -1 & -1 & 2 \end{bmatrix}$$

with their algebraic multiplicities.

Solution

$$f_A(\lambda) = \det \begin{bmatrix} \lambda - 2 & 1 & 1 \\ 1 & \lambda - 2 & 1 \\ 1 & 1 & \lambda - 2 \end{bmatrix} = (\lambda - 2)^3 + 2 - 3(\lambda - 2) = (\lambda - 3)^2 \lambda$$

The details of the computation are left to the reader. We find two distinct eigenvalues, 3 and 0, with algebraic multiplicities 2 and 1, respectively.   ∎

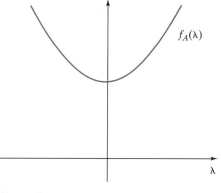

Figure 2

Let us summarize:

Fact 7.2.7

> An $n \times n$ matrix has at most n eigenvalues, even if they are counted with their algebraic multiplicities.
>
> If n is odd, then an $n \times n$ matrix has at least one eigenvalue.

If n is even, an $n \times n$ matrix A need not have any real eigenvalues. Perhaps the simplest example is

$$A = \begin{bmatrix} 0 & -1 \\ 1 & 0 \end{bmatrix},$$

with $f_A(\lambda) = \lambda^2 + 1$. See Figure 2.

Geometrically, it makes sense that A has no real eigenvalues: The transformation $T(\vec{x}) = A\vec{x}$ is a counterclockwise rotation through an angle of $90°$. Compare with Example 7.1.3.

EXAMPLE 6 Describe all possible cases for the number of real eigenvalues of a 3×3 matrix and their algebraic multiplicities. Give an example in each case and graph the characteristic polynomial.

Solution

The characteristic polynomial either factors completely,

$$f_A(\lambda) = (\lambda - \lambda_1)(\lambda - \lambda_2)(\lambda - \lambda_3),$$

or it has a quadratic factor without real zeros:

$$f_A(\lambda) = (\lambda - \lambda_1)p(\lambda).$$

In the first case, the λ_i could all be distinct, two of them could be equal, or they could all be equal. This leads to the following possibilities:

Case	No. of Distinct Eigenvalues	Algebraic Multiplicities
1	3	1 each
2	2	2 and 1
3	1	3
4	1	1

Examples for each case follow:

Case 1 ♦

$$A = \begin{bmatrix} 1 & 0 & 0 \\ 0 & 2 & 0 \\ 0 & 0 & 3 \end{bmatrix}, \qquad f_A(\lambda) = (\lambda - 1)(\lambda - 2)(\lambda - 3).$$

See Figure 3.

Case 2 ♦

$$A = \begin{bmatrix} 1 & 0 & 0 \\ 0 & 1 & 0 \\ 0 & 0 & 2 \end{bmatrix}, \qquad f_A(\lambda) = (\lambda - 1)^2(\lambda - 2).$$

See Figure 4.

Case 3 ♦

$$A = I_3, \qquad f_A(\lambda) = (\lambda - 1)^3.$$

See Figure 5.

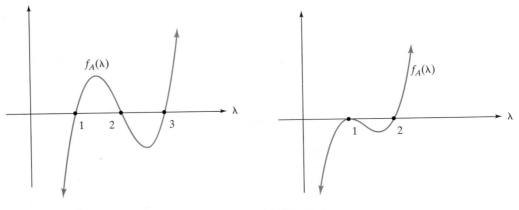

Figure 3

Figure 4

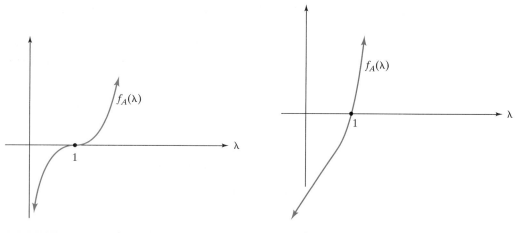

Figure 5

Figure 6

Case 4 ◆

$$A = \begin{bmatrix} 1 & 0 & 0 \\ 0 & 0 & -1 \\ 0 & 1 & 0 \end{bmatrix}, \qquad f_A(\lambda) = (\lambda - 1)(\lambda^2 + 1).$$

See Figure 6.

You can recognize an eigenvalue λ_0 whose algebraic multiplicity exceeds 1 on the graph of $f_A(\lambda)$ by the fact that $f_A(\lambda_0) = f'_A(\lambda_0) = 0$. (The derivative is zero.) The verification of this observation is left as Exercise 37. ∎

Finding the Eigenvalues of a Matrix in Practice

To find the eigenvalues of an $n \times n$ matrix with the method developed in this section, we have to find the zeros of the characteristic function, a polynomial of degree n. For $n = 2$, this is a trivial matter: We can either factor the polynomial by inspection or use the quadratic formula. The problem of finding the zeros of a polynomial of higher degree is nontrivial: It has been of considerable interest throughout the history of mathematics. In the early 1500s, Italian mathematicians found formulas in the cases $n = 3$ and $n = 4$, published in the *Ars Magna* by Gerolamo Cardano.[5] (See Exercise 38 for the case $n = 3$.) During the next 300 years, people tried to find a general formula to solve the quintic (a polynomial equation of fifth order). In 1824, the Norwegian mathematician Niels Henrik Abel (1802–1829) showed that no such general solution is possible, putting an end to the long search. The first numerical example of a quintic that cannot be solved by

---

[5]Cardano (1501–1576) was a humanist with a wide range of interests. In his book *Liber de ludo aleae* he presents the first systematic computations of probabilities. Trained as a physician, he gave the first clinical description of typhus fever. In his book *Somniorum Synesiorum* (Basel, 1562) he explores the meaning of dreams (sample: "To dream of living in a new and unknown city means imminent death"). Still, he is best known today as the most outstanding mathematician of his time and the author of the *Ars Magna*. In 1570, he was arrested on accusation of heresy; he lost his academic position and the right to publish.

For an English translation of parts of Chapter XI of the *Ars Magna* (dealing with cubic equations) see D. J. Struik (editor), *A Source Book in Mathematics* 1200–1800, Princeton University Press, 1986.

radicals was given by the French mathematician Évariste Galois (1811–1832). (Note the short life spans of these two brilliant mathematicians. Abel died from tuberculosis; Galois died in a duel.)

It is usually impossible to find the exact eigenvalues of a matrix. To find approximations for the eigenvalues, you could graph the characteristic polynomial (using technology). The graph may give you an idea of the number of eigenvalues and their approximate values. Numerical analysts tell us that this is not a very efficient way to go about finding the eigenvalues; other techniques are used in practice. (See Exercise 7.5.33 for an example; another approach uses QR factorization.[6])

If you have to find the eigenvalues of a matrix, it may be worth trying out a few small integers, such as ± 1 and ± 2. (Those matrices considered in introductory linear algebra texts often happen to have such eigenvalues.)

EXERCISES

GOALS Use the characteristic polynomial $f_A(\lambda) = \det(\lambda I_n - A)$ to find the eigenvalues of a matrix A, with their algebraic multiplicities.

For each of the matrices in Exercises 1 through 13, find all real eigenvalues, with their algebraic multiplicities. Show your work. Do not use technology.

1. $\begin{bmatrix} 1 & 2 \\ 0 & 3 \end{bmatrix}$

2. $\begin{bmatrix} 2 & 0 & 0 & 0 \\ 2 & 1 & 0 & 0 \\ 2 & 1 & 2 & 0 \\ 2 & 1 & 2 & 1 \end{bmatrix}$

3. $\begin{bmatrix} 5 & -4 \\ 2 & -1 \end{bmatrix}$

4. $\begin{bmatrix} 0 & 4 \\ -1 & 4 \end{bmatrix}$

5. $\begin{bmatrix} 11 & -15 \\ 6 & -7 \end{bmatrix}$

6. $\begin{bmatrix} 1 & 2 \\ 3 & 4 \end{bmatrix}$

7. I_3

8. $\begin{bmatrix} -1 & -1 & -1 \\ -1 & -1 & -1 \\ -1 & -1 & -1 \end{bmatrix}$

9. $\begin{bmatrix} 3 & -2 & 5 \\ 1 & 0 & 7 \\ 0 & 0 & 2 \end{bmatrix}$

10. $\begin{bmatrix} -3 & 0 & 4 \\ 0 & -1 & 0 \\ -2 & 7 & 3 \end{bmatrix}$

11. $\begin{bmatrix} 5 & 1 & -5 \\ 2 & 1 & 0 \\ 8 & 2 & -7 \end{bmatrix}$

12. $\begin{bmatrix} 2 & -2 & 0 & 0 \\ 1 & -1 & 0 & 0 \\ 0 & 0 & 3 & -4 \\ 0 & 0 & 2 & -3 \end{bmatrix}$

13. $\begin{bmatrix} 0 & 1 & 0 \\ 0 & 0 & 1 \\ 1 & 0 & 0 \end{bmatrix}$

14. Consider a 4×4 matrix $A = \begin{bmatrix} B & C \\ 0 & D \end{bmatrix}$, where B, C, and D are 2×2 matrices. What is the relationship between the eigenvalues of A, B, C, and D?

15. Consider the matrix $A = \begin{bmatrix} 1 & k \\ 1 & 1 \end{bmatrix}$, where k is an arbitrary constant. For which values of k does the matrix have two distinct real eigenvalues? When is there no real eigenvalue?

16. Consider the matrix $A = \begin{bmatrix} a & b \\ b & c \end{bmatrix}$, where a, b, and c are nonzero constants. For which choices of a, b, and c does A have two distinct eigenvalues?

17. Consider the matrix $A = \begin{bmatrix} a & b \\ b & -a \end{bmatrix}$, where a and b are arbitrary constants. Find all eigenvalues of A. Explain in terms of the geometric interpretation of the linear transformation $T(\vec{x}) = A\vec{x}$.

18. Consider the matrix $A = \begin{bmatrix} a & b \\ b & a \end{bmatrix}$, where a and b are arbitrary constants. Find all eigenvalues of A.

19. *True or false?* If the determinant of a 2×2 matrix A is negative, then A has two distinct real eigenvalues.

20. Consider a 2×2 matrix A with two distinct real eigenvalues, λ_1 and λ_2. Express $\det(A)$ in terms of λ_1 and λ_2. Do the same for the trace of A. *Hint:* The characteristic polynomial is $f_A(\lambda) = (\lambda - \lambda_1)(\lambda - \lambda_2)$.

---

[6]See, for example, J. Stoer and R. Bulisch, *Introduction to Numerical Analysis*, Springer Verlag, 1980.

Verify your answers in the case of the matrix

$$A = \begin{bmatrix} 1 & 2 \\ 4 & 3 \end{bmatrix}.$$

21. Consider an $n \times n$ matrix A with n distinct real eigenvalues, $\lambda_1, \lambda_2, \ldots, \lambda_n$. Express the determinant of A in terms of the λ_i. Do the same for the trace of A.

22. Consider an arbitrary $n \times n$ matrix A. What is the relationship between the characteristic polynomials of A and A^T? What does your answer tell you about the eigenvalues of A and A^T?

23. Suppose matrix A is similar to B. What is the relationship between the characteristic polynomials of A and B? What does your answer tell you about the eigenvalues of A and B?

24. Find all eigenvalues of the matrix

$$A = \begin{bmatrix} 0.5 & 0.25 \\ 0.5 & 0.75 \end{bmatrix}.$$

25. Consider a 2×2 matrix of the form

$$A = \begin{bmatrix} a & c \\ b & d \end{bmatrix},$$

where a, b, c, d are positive numbers such that $a + b = c + d = 1$. (The matrix in Exercise 24 has this form.) Such a matrix is called a *regular transition matrix*. Verify that $\begin{bmatrix} c \\ b \end{bmatrix}$ and $\begin{bmatrix} 1 \\ -1 \end{bmatrix}$ are eigenvectors of A. What are the associated eigenvalues? Is the absolute value of these eigenvalues more or less than 1? Sketch a phase portrait.

26. Based on your answer to Exercise 25, sketch a phase portrait of the dynamical system

$$\vec{x}(t + 1) = \begin{bmatrix} 0.5 & 0.25 \\ 0.5 & 0.75 \end{bmatrix} \vec{x}(t).$$

27. a. Based on your answer to Exercise 25, find closed formulas for the components of the dynamical system

$$\vec{x}(t + 1) = \begin{bmatrix} 0.5 & 0.25 \\ 0.5 & 0.75 \end{bmatrix} \vec{x}(t),$$

with initial value $\vec{x}_0 = \vec{e}_1$. Then do the same for the initial value $\vec{x}_0 = \vec{e}_2$. Sketch the two trajectories.

b. Consider the matrix

$$A = \begin{bmatrix} 0.5 & 0.25 \\ 0.5 & 0.75 \end{bmatrix}.$$

Using technology, compute some powers of the matrix A, say, A^2, A^5, A^{10}, ... What do you observe? Explain your answer carefully.

c. If $A = \begin{bmatrix} a & c \\ b & d \end{bmatrix}$ is an arbitrary regular transition matrix, what can you say about the powers A^t as t goes to infinity?

28. Consider the isolated Swiss town of Andelfingen, inhabited by 1,200 families. Each family takes a weekly shopping trip to the only grocery store in town, run by Mr. and Mrs. Wipf, until the day when a new, fancier (and cheaper) chain store, Migros, opens its doors. It is not expected that everybody will immediately run to the new store, but we do anticipate that 20% of those shopping at Wipf's one week switch to Migros the following week. Some people who do switch miss the personal service (and the gossip) and switch back: We expect that 10% of those shopping at Migros one week go to Wipf's the following week. The state of this town (as far as grocery shopping is concerned) can be represented by the vector

$$\vec{x}(t) = \begin{bmatrix} w(t) \\ m(t) \end{bmatrix},$$

where $w(t)$ and $m(t)$ are the numbers of families shopping at Wipf's and at Migros, respectively, t weeks after Migros opens. Suppose $w(0) = 1200$ and $m(0) = 0$.

a. Find a 2×2 matrix A such that $\vec{x}(t + 1) = A\vec{x}(t)$. Verify that A is a regular transition matrix.

b. How many families will shop at each store after t weeks? Give closed formulas.

c. The Wipfs expect that they must close down when they have less than 250 customers a week. When does that happen?

29. Consider an $n \times n$ matrix A such that the sum of the entries in each *row* is 1. Show that the vector

$$\vec{e} = \begin{bmatrix} 1 \\ 1 \\ \vdots \\ 1 \end{bmatrix}$$

in $\mathbb{R}^n$ is an eigenvector of A. What is the corresponding eigenvalue?

30. a. Consider an $n \times n$ matrix A such that the sum of the entries in each row is 1 and that all entries are positive. Consider an eigenvector $\vec{v}$ of A with positive components. Show that the associated eigenvalue is less than or equal to 1. *Hint:* Consider the largest entry of $\vec{v}$. What can you say about the corresponding entry of $A\vec{v}$?

b. If we drop the requirement that the components of the eigenvector $\vec{v}$ be positive, is it still true that the associated eigenvalue is less than or equal to 1 in absolute value? Justify your answer.

31. Consider a matrix A with positive entries such that the entries in each column add up to 1. Explain why 1 is an eigenvalue of A. What can you say about the other eigenvalues? Is

$$\vec{e} = \begin{bmatrix} 1 \\ 1 \\ \vdots \\ 1 \end{bmatrix}$$

necessarily an eigenvector? *Hint:* Consider Exercises 22, 29, and 30.

32. Consider the matrix

$$A = \begin{bmatrix} 0 & 1 & 0 \\ 0 & 0 & 1 \\ k & 3 & 0 \end{bmatrix},$$

where k is an arbitrary constant. For which values of k does A have three distinct real eigenvalues? What happens in the other cases? *Hint:* Graph the function $g(\lambda) = \lambda^3 - 3\lambda$. Find its local maxima and minima.

33. a. Find the characteristic polynomial of the matrix

$$A = \begin{bmatrix} 0 & 1 & 0 \\ 0 & 0 & 1 \\ a & b & c \end{bmatrix}.$$

b. Can you find a 3×3 matrix M whose characteristic polynomial is

$$\lambda^3 - 17\lambda^2 + 5\lambda - \pi \text{ ?}$$

34. Suppose a certain 4×4 matrix A has two distinct real eigenvalues. What could the algebraic multiplicities of these eigenvalues be? Give an example for each possible case and sketch the characteristic polynomial.

35. Give an example of a 4×4 matrix A without real eigenvalues.

36. For an arbitrary positive integer n, give a $2n \times 2n$ matrix A without real eigenvalues.

37. Consider an eigenvalue λ_0 of an $n \times n$ matrix A. We are told that the algebraic multiplicity of λ_0 exceeds 1. Show that $f_A'(\lambda_0) = 0$ (i.e., the *derivative* of the characteristic polynomial of A vanishes at λ_0).

38. In his groundbreaking text *Ars Magna* (Nuremberg, 1545), the Italian mathematician Gerolamo Cardano explains how to solve cubic equations. In Chapter XI, he considers the following example:

$$x^3 + 6x = 20.$$

a. Explain why this equation has exactly one (real) solution. Here, this solution is easy to find by inspection. The point of the exercise is to show a systematic way to find it.

b. Cardano explains his method as follows (we are using modern notation for the variables): "I take two cubes v^3 and u^3 whose difference shall be 20, so that the product vu shall be 2, that is, a third of the coefficient of the unknown x. Then, I say that $v - u$ is the value of the unknown x." Show that if v and u are chosen as stated by Cardano, then $x = v - u$ is indeed the solution of the equation $x^3 + 6x = 20$.

c. Solve the system

$$\begin{vmatrix} v^3 - u^3 = 20 \\ vu = 2 \end{vmatrix}$$

to find u and v.

d. Consider the equation

$$x^3 + px = q,$$

where p is positive. Using your work in parts (a), (b), and (c) as a guide, show that the unique solution of this equation is

$$x = \sqrt[3]{\frac{q}{2} + \sqrt{\left(\frac{q}{2}\right)^2 + \left(\frac{p}{3}\right)^3}}$$
$$- \sqrt[3]{-\frac{q}{2} + \sqrt{\left(\frac{q}{2}\right)^2 + \left(\frac{p}{3}\right)^3}}.$$

This solution can also be written as

$$x = \sqrt[3]{\frac{q}{2} + \sqrt{\left(\frac{q}{2}\right)^2 + \left(\frac{p}{3}\right)^3}}$$
$$+ \sqrt[3]{\frac{q}{2} - \sqrt{\left(\frac{q}{2}\right)^2 + \left(\frac{p}{3}\right)^3}}.$$

What can go wrong when p is negative?

e. Consider an arbitrary cubic equation

$$x^3 + ax^2 + bx + c = 0.$$

Show that the substitution $x = t - (a/3)$ allows you to write this equation as

$$t^3 + pt = q.$$

7.3 FINDING THE EIGENVECTORS OF A MATRIX

After we have found an eigenvalue λ of an $n \times n$ matrix A, we may be interested in the corresponding eigenvectors. We have to find the vectors $\vec{v}$ in $\mathbb{R}^n$ such that

$$A\vec{v} = \lambda\vec{v}, \quad \text{or} \quad (\lambda I_n - A)\vec{v} = \vec{0}.$$

In other words, we have to find the *kernel* of the matrix $\lambda I_n - A$. The following notation is useful:

Definition 7.3.1

> **Eigenspace**
>
> Consider an eigenvalue λ of an $n \times n$ matrix A. Then the kernel of the matrix $\lambda I_n - A$ is called the *eigenspace* associated with λ, denoted by E_λ:
>
> $$E_\lambda = \ker(\lambda I_n - A).$$
>
> Note that E_λ consists of all solutions $\vec{v}$ of the linear system
>
> $$A\vec{v} = \lambda\vec{v}.$$

In other words, the eigenspace E_λ consists of all eigenvectors with eigenvalue λ, together with the zero vector.

EXAMPLE 1 Let $T(\vec{x}) = A\vec{x}$ be the orthogonal projection onto a plane E in $\mathbb{R}^3$. Describe the eigenspaces geometrically.

Solution

We can find the eigenvectors and eigenvalues by inspection: The nonzero vectors $\vec{v}$ in E are eigenvectors with eigenvalue 1, because $A\vec{v} = \vec{v} = 1\vec{v}$. The eigenspace E_1 consists of all vectors $\vec{v}$ in $\mathbb{R}^3$ such that $A\vec{v} = 1\vec{v} = \vec{v}$; that is, E_1 is just the plane E. Likewise, E_0 is the line $E^\perp$ perpendicular to E. Note that E_0 is simply the kernel of A; that is, E_0 consists of all solutions of the equation $A\vec{v} = \vec{0}$. See Figure 1. ∎

To find the eigenvectors associated with a known eigenvalue λ algebraically, we seek a basis of the eigenspace E_λ. Since $E_\lambda = \ker(\lambda I_n - A)$, this amounts to finding a basis of a kernel, a problem we can handle. (See Section 3.3.)

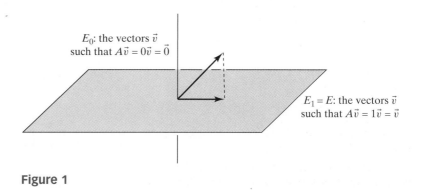

E_0: the vectors $\vec{v}$ such that $A\vec{v} = 0\vec{v} = \vec{0}$

$E_1 = E$: the vectors $\vec{v}$ such that $A\vec{v} = 1\vec{v} = \vec{v}$

Figure 1

EXAMPLE 2 Find the eigenvectors of the matrix $A = \begin{bmatrix} 1 & 2 \\ 4 & 3 \end{bmatrix}$.

Solution

In Section 7.2, Example 1, we saw that the eigenvalues are 5 and -1. Then

$$E_5 = \ker(5I_2 - A) = \ker \begin{bmatrix} 4 & -2 \\ -4 & 2 \end{bmatrix}.$$

It is easy enough to find this kernel algebraically, using the reduced row-echelon form:

$$E_5 = \ker \begin{bmatrix} 4 & -2 \\ -4 & 2 \end{bmatrix} = \ker \begin{bmatrix} 1 & -1/2 \\ 0 & 0 \end{bmatrix} = \text{span} \begin{bmatrix} 1/2 \\ 1 \end{bmatrix} = \text{span} \begin{bmatrix} 1 \\ 2 \end{bmatrix}.$$

Alternatively, we can think about this problem geometrically: We are looking for a vector that is perpendicular to the two rows (those are parallel, of course). A vector perpendicular to $[\,4 \quad -2\,]$ is $\begin{bmatrix} 2 \\ 4 \end{bmatrix}$. (Swap the two components and change one of the signs.) Therefore,

$$E_5 = \ker \begin{bmatrix} 4 & -2 \\ -4 & 2 \end{bmatrix} = \text{span} \begin{bmatrix} 2 \\ 4 \end{bmatrix} = \text{span} \begin{bmatrix} 1 \\ 2 \end{bmatrix},$$

and

$$E_{-1} = \ker \begin{bmatrix} -2 & -2 \\ -4 & -4 \end{bmatrix} = \text{span} \begin{bmatrix} 2 \\ -2 \end{bmatrix} = \text{span} \begin{bmatrix} 1 \\ -1 \end{bmatrix}.$$

We can (and should) check that the vectors we found are indeed eigenvectors of A, with the eigenvalues we claim:

$$\begin{bmatrix} 1 & 2 \\ 4 & 3 \end{bmatrix} \begin{bmatrix} 1 \\ 2 \end{bmatrix} = \begin{bmatrix} 5 \\ 10 \end{bmatrix} = 5 \begin{bmatrix} 1 \\ 2 \end{bmatrix} \quad \text{and} \quad \begin{bmatrix} 1 & 2 \\ 4 & 3 \end{bmatrix} \begin{bmatrix} 1 \\ -1 \end{bmatrix} = \begin{bmatrix} -1 \\ 1 \end{bmatrix} = (-1) \begin{bmatrix} 1 \\ -1 \end{bmatrix}$$

Both eigenspaces are lines, as shown in Figure 2. ∎

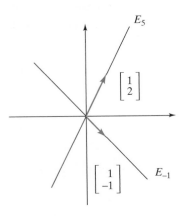

Figure 2

EXAMPLE 3   Find the eigenvectors of

$$A = \begin{bmatrix} 1 & 1 & 1 \\ 0 & 0 & 1 \\ 0 & 0 & 1 \end{bmatrix}.$$

Solution

The eigenvalues are 1 and 0 (the diagonal entries of A), with algebraic multiplicities 2 and 1, respectively. Then

$$E_1 = \ker \begin{bmatrix} 0 & -1 & -1 \\ 0 & 1 & -1 \\ 0 & 0 & 0 \end{bmatrix}.$$

To find this kernel, bring the matrix into reduced row-echelon form and solve the corresponding system:

$$\begin{bmatrix} 0 & -1 & -1 \\ 0 & 1 & -1 \\ 0 & 0 & 0 \end{bmatrix} \overset{\text{rref}}{\rightarrow} \begin{bmatrix} 0 & 1 & 0 \\ 0 & 0 & 1 \\ 0 & 0 & 0 \end{bmatrix}$$

The general solution of the system

$$\begin{vmatrix} x_2 & & = 0 \\ & x_3 = 0 \end{vmatrix}$$

is

$$\begin{bmatrix} x_1 \\ 0 \\ 0 \end{bmatrix} = x_1 \begin{bmatrix} 1 \\ 0 \\ 0 \end{bmatrix}.$$

Therefore,

$$E_1 = \operatorname{span} \begin{bmatrix} 1 \\ 0 \\ 0 \end{bmatrix}.$$

Now let us think about the eigenspace E_0:

$$E_0 = \ker \begin{bmatrix} 1 & 1 & 1 \\ 0 & 0 & 1 \\ 0 & 0 & 1 \end{bmatrix} = \ker \begin{bmatrix} 1 & 1 & 0 \\ 0 & 0 & 1 \\ 0 & 0 & 0 \end{bmatrix}$$

$$\begin{vmatrix} \textcircled{x_1} + x_2 & & = 0 \\ & \textcircled{x_3} & = 0 \end{vmatrix}$$

The general solution is

$$\begin{bmatrix} -x_2 \\ x_2 \\ 0 \end{bmatrix} = x_2 \begin{bmatrix} -1 \\ 1 \\ 0 \end{bmatrix}$$

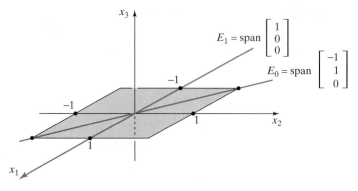

Figure 3

Therefore,

$$E_0 = \operatorname{span} \begin{bmatrix} -1 \\ 1 \\ 0 \end{bmatrix}.$$

Both eigenspaces are lines in the x_1–x_2-plane, as shown in Figure 3. ∎

Note that Example 3 is qualitatively different from Example 1, where we studied the orthogonal projection onto a plane in $\mathbb{R}^3$. There, too, we have two eigenvalues, 1 and 0, but one of the eigenspaces, E_1, is a plane.

To discuss these different cases, the following terminology is useful:

Definition 7.3.2

Geometric multiplicity
Consider an eigenvalue λ of a matrix A. Then the dimension of eigenspace $E_\lambda = \ker(\lambda I_n - A)$ is called the *geometric multiplicity* of eigenvalue λ. Thus, the geometric multiplicity of λ is the *nullity* of matrix $\lambda I_n - A$.

Example 3 shows that the geometric multiplicity of an eigenvalue may be different from the algebraic multiplicity: We have

(algebraic multiplicity of eigenvalue 1) $= 2$,

but

(geometric multiplicity of eigenvalue 1) $= 1$.

However, the following inequality always holds:

Fact 7.3.3

Consider an eigenvalue λ of a matrix A. Then

(geometric multiplicity of λ) $\leq$ (algebraic multiplicity of λ).

To give an elegant proof, we need some additional machinery, which we will develop in Section 7.4; the proof will then be left as Exercise 7.4.60.

The following example illustrates the relationship between the algebraic and geometric multiplicity of an eigenvalue.

EXAMPLE 4   Consider an upper triangular matrix of the form

$$
A = \begin{bmatrix} 1 & \bullet & \bullet & \bullet & \bullet \\ 0 & 2 & \bullet & \bullet & \bullet \\ 0 & 0 & 4 & \bullet & \bullet \\ 0 & 0 & 0 & 4 & \bullet \\ 0 & 0 & 0 & 0 & 4 \end{bmatrix},
$$

where the bullets are placeholders for arbitrary entries. Without referring to Fact 7.3.3, what can you say about the geometric multiplicity of the eigenvalue 4?

Solution

$$
E_4 = \ker \begin{bmatrix} 3 & \bullet & \bullet & \bullet & \bullet \\ 0 & 2 & \bullet & \bullet & \bullet \\ 0 & 0 & 0 & \bullet & \bullet \\ 0 & 0 & 0 & 0 & \bullet \\ 0 & 0 & 0 & 0 & 0 \end{bmatrix}.
$$

The reduced row-echelon form is

$$
\begin{bmatrix} \boxed{1} & 0 & \bullet & \bullet & \bullet \\ 0 & \boxed{1} & \bullet & \bullet & \bullet \\ 0 & 0 & 0 & \bullet & \bullet \\ 0 & 0 & 0 & 0 & \bullet \\ 0 & 0 & 0 & 0 & 0 \end{bmatrix},
$$

where any of the bullets in rows 3 and 4 could be leading 1's.

The rank of this matrix will be between 2 and 4, and its nullity will be between $3 (= 5 - 2)$ and $1 (= 5 - 4)$. We can conclude that the geometric multiplicity of the eigenvalue 4 is between 1 and 3. This result agrees with Fact 7.3.3, since the algebraic multiplicity of the eigenvalue 4 is 3.   ∎

Using Example 4 as a guide, can you give a proof of Fact 7.3.3 for triangular matrices? See Exercise 30.

When we analyze a dynamical system

$$
\vec{x}(t + 1) = A\vec{x}(t),
$$

where A is an $n \times n$ matrix, we may be interested in finding a basis $\vec{v}_1, \vec{v}_2, \ldots, \vec{v}_n$ of $\mathbb{R}^n$ that consists of *eigenvectors* of A. (Recall the introductory example in Section 7.1, and Fact 7.1.3.) Such a basis deserves a name.

Definition 7.3.4

Eigenbasis

Consider an $n \times n$ matrix A. A basis of $\mathbb{R}^n$ consisting of eigenvectors of A is called an *eigenbasis* for A.

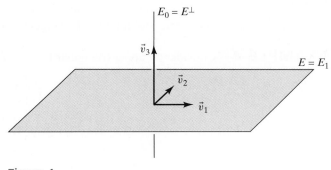

Figure 4

Recall Examples 1 through 3.

Example 1 Revisited ■ *Projection onto a plane E in* $\mathbb{R}^3$. Pick a basis $\vec{v}_1$, $\vec{v}_2$ of E and a nonzero $\vec{v}_3$ in $E^\perp$. The vectors $\vec{v}_1$, $\vec{v}_2$, $\vec{v}_3$ form an eigenbasis. See Figure 4.

Example 2 Revisited ■ $A = \begin{bmatrix} 1 & 2 \\ 4 & 3 \end{bmatrix}$.

The vectors $\begin{bmatrix} 1 \\ 2 \end{bmatrix}$ and $\begin{bmatrix} 1 \\ -1 \end{bmatrix}$ form an eigenbasis for A, as shown in Figure 5.

Example 3 Revisited ■ $A = \begin{bmatrix} 1 & 1 & 1 \\ 0 & 0 & 1 \\ 0 & 0 & 1 \end{bmatrix}$.

There are not enough eigenvectors to form an eigenbasis. For example, we are unable to express $\vec{e}_3$ as a linear combination of eigenvectors. See Figure 6.

Consider an $n \times n$ matrix A. If the sum of the dimensions of the eigenspaces of A is less than n, then there are not enough linearly independent eigenvectors to form an eigenbasis, as illustrated in Example 3.

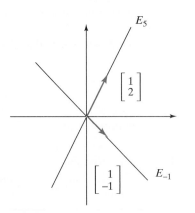

Figure 5

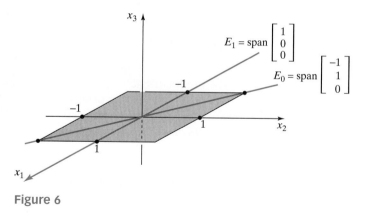

Figure 6

Conversely, suppose the dimensions of the eigenspaces do add up to n, as in Examples 1 and 2. Can we construct an eigenbasis for A simply by picking a basis of each eigenspace and combining these vectors? This approach certainly works in the two preceding examples. However, in higher-dimensional cases, we must worry about the linear independence of the n vectors we find by combining bases of the various eigenspaces. Let us first think about a simple case.

EXAMPLE 5   Consider a 3×3 matrix A with three eigenvalues, 1, 2, and 3. Let $\vec{v}_1$, $\vec{v}_2$, and $\vec{v}_3$ be corresponding eigenvectors. Are vectors $\vec{v}_1$, $\vec{v}_2$, and $\vec{v}_3$ necessarily linearly independent?

Solution

Consider the plane E spanned by $\vec{v}_1$ and $\vec{v}_2$. We have to examine whether $\vec{v}_3$ could be contained in this plane. More generally, we examine whether the plane E could contain any eigenvectors besides the multiples of $\vec{v}_1$ and $\vec{v}_2$. Consider a vector $\vec{x} = c_1\vec{v}_1 + c_2\vec{v}_2$ in E (with $c_1 \neq 0$ and $c_2 \neq 0$). Then $A\vec{x} = c_1 A\vec{v}_1 + c_2 A\vec{v}_2 = c_1\vec{v}_1 + 2c_2\vec{v}_2$. This vector is not a scalar multiple of $\vec{x}$; that is, $\vec{x}$ is not an eigenvector of A. (See Figure 7.)

We have shown that the plane E does not contain any eigenvectors besides the multiples of $\vec{v}_1$ and $\vec{v}_2$; in particular, $\vec{v}_3$ is not contained in E. We conclude that the vectors $\vec{v}_1$, $\vec{v}_2$, and $\vec{v}_3$ are linearly independent.   ■

The observation we made in Example 5 generalizes to higher dimensions.

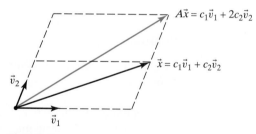

Figure 7

Fact 7.3.5   Consider the eigenvectors $\vec{v}_1, \vec{v}_2, \ldots, \vec{v}_m$ of an $n \times n$ matrix A, with distinct eigenvalues $\lambda_1, \lambda_2, \ldots, \lambda_m$. Then the $\vec{v}_i$ are linearly independent.

Proof   The proofs of Facts 7.3.5 and 7.3.7 are somewhat technical. You may wish to skip them in a first reading of this text.

We argue by induction on m. We leave the case $m = 1$ as an exercise, and assume that the claim holds for $m - 1$. Consider a relation

$$c_1 \vec{v}_1 + \cdots + c_{m-1} \vec{v}_{m-1} + c_m \vec{v}_m = \vec{0}. \qquad \textbf{(I)}$$

First, we apply the transformation $T(\vec{x}) = A\vec{x}$ to both sides of equation (I), keeping in mind that $A\vec{v}_i = \lambda_i \vec{v}_i$:

$$c_1 \lambda_1 \vec{v}_1 + \cdots + c_{m-1} \lambda_{m-1} \vec{v}_{m-1} + c_m \lambda_m \vec{v}_m = \vec{0} \qquad \textbf{(II)}$$

Next, we multiply both sides of equation (I) by λ_m:

$$c_1 \lambda_m \vec{v}_1 + \cdots + c_{m-1} \lambda_m \vec{v}_{m-1} + c_m \lambda_m \vec{v}_m = \vec{0} \qquad \textbf{(III)}$$

Now subtract equation (III) from equation (II):

$$c_1 (\lambda_1 - \lambda_m) \vec{v}_1 + \cdots + c_{m-1} (\lambda_{m-1} - \lambda_m) \vec{v}_{m-1} = \vec{0} \qquad \textbf{(IV)}$$

By induction, the $\vec{v}_i$ in equation (IV) are linearly independent. Therefore, equation (IV) must represent the trivial relation; that is, $c_i (\lambda_i - \lambda_m) = 0$, for $i = 1, \ldots, m-1$. The eigenvalues were assumed to be distinct; therefore $\lambda_i - \lambda_m \neq 0$, for $i = 1, \ldots, m-1$. We conclude that $c_i = 0$, for $i = 1, \ldots, m-1$. Equation (I) now tells us that $c_m \vec{v}_m = \vec{0}$, so that $c_m = 0$ as well.   ▲

Let's apply Fact 7.3.5 to the special case when $m = n$.

Fact 7.3.6   If an $n \times n$ matrix A has n distinct eigenvalues, then there is an eigenbasis for A. We can construct an eigenbasis by choosing an eigenvector for each eigenvalue.

EXAMPLE 6   Is there an eigenbasis for the following matrix?

$$A = \begin{bmatrix} 1 & 2 & 3 & 4 & 5 & 6 \\ 0 & 2 & 3 & 4 & 5 & 6 \\ 0 & 0 & 3 & 4 & 5 & 6 \\ 0 & 0 & 0 & 4 & 5 & 6 \\ 0 & 0 & 0 & 0 & 5 & 6 \\ 0 & 0 & 0 & 0 & 0 & 6 \end{bmatrix}$$

Solution

Yes, because there are 6 distinct eigenvalues, the diagonal entries of A: 1, 2, 3, 4, 5, 6.   ■

What if an $n \times n$ matrix A has fewer than n distinct eigenvalues? Then we have to consider the geometric multiplicities of the eigenvalues.

Fact 7.3.7   Consider an $n \times n$ matrix A. If the geometric multiplicities of the eigenvalues of A add up to n, then there is an eigenbasis for A: We can construct an eigenbasis by choosing a basis of each eigenspace and combining these vectors.

Proof   Suppose the eigenvalues are $\lambda_1, \lambda_2, \ldots, \lambda_m$, with $\dim(E_{\lambda_i}) = d_i$. We first choose a basis $\vec{v}_1, \vec{v}_2, \ldots, \vec{v}_{d_1}$ of E_{λ_1} and then a basis $\vec{v}_{d_1+1}, \ldots, \vec{v}_{d_1+d_2}$ of E_{λ_2}, and so on.

We have to show that the vectors $\vec{v}_1, \ldots, \vec{v}_n$ found in this way are linearly independent. Consider a relation

$$\underbrace{c_1\vec{v}_1 + \cdots + c_{d_1}\vec{v}_{d_1}}_{\vec{w}_1 \text{ in } E_{\lambda_1}} + \underbrace{\cdots + c_{d_1+d_2}\vec{v}_{d_1+d_2}}_{\vec{w}_2 \text{ in } E_{\lambda_2}} + \cdots + \underbrace{\cdots + c_n\vec{v}_n}_{\vec{w}_m \text{ in } E_{\lambda_m}} = \vec{0}.$$

Each of the vectors $\vec{w}_i$ either is an eigenvector with eigenvalue λ_i or else is $\vec{0}$. We wish to show that the $\vec{w}_i$ are in fact all $\vec{0}$. We argue indirectly, assuming that some of the $\vec{w}_i$ are nonzero and showing that this assumption leads to a contradiction. Those $\vec{w}_i$ that are nonzero are eigenvectors with distinct eigenvalues; they are linearly independent, by Fact 7.3.5. But here is our contradiction: The sum of the $\vec{w}_i$ is $\vec{0}$, giving us a nontrivial relation among the $\vec{w}_i$.

Because $\vec{w}_1 = \vec{0}$, it follows that $c_1 = c_2 = \cdots = c_{d_1} = 0$, since $\vec{v}_1, \vec{v}_2, \ldots, \vec{v}_{d_1}$ are linearly independent. Likewise, all the other c_j are zero.   ▲

We conclude this section by working another example of a dynamical system:

EXAMPLE 7   Consider an Albanian mountain farmer who raises goats. This particular breed of goats has a life span of three years. At the end of each year t, the farmer conducts a census of his goats. He counts the number of young goats $j(t)$ (those born in the year t), the middle-aged ones $m(t)$ (born the year before), and the old ones $a(t)$ (born in the year $t - 2$). The state of the herd can be represented by the vector

$$\vec{x}(t) = \begin{bmatrix} j(t) \\ m(t) \\ a(t) \end{bmatrix}.$$

How do we expect the population to change from year to year? Suppose that for this breed and this environment the evolution of the system can be modeled by

$$\vec{x}(t + 1) = A\vec{x}(t), \qquad \text{where} \qquad A = \begin{bmatrix} 0 & 0.95 & 0.6 \\ 0.8 & 0 & 0 \\ 0 & 0.5 & 0 \end{bmatrix}.$$

We leave it as an exercise to interpret the entries of A in terms of reproduction rates and survival rates.

Suppose the initial populations are $j_0 = 750$ and $m_0 = a_0 = 200$. What will the populations be after t years, according to this model? What will happen in the long term?

Solution

To answer these questions, we have to find an eigenbasis for A. (See Fact 7.1.3.)

Step 1   Find the *eigenvalues* of A. The characteristic polynomial is

$$f_A(\lambda) = \det \begin{bmatrix} \lambda & -0.95 & -0.6 \\ -0.8 & \lambda & 0 \\ 0 & -0.5 & \lambda \end{bmatrix} = \lambda^3 - 0.76\lambda - 0.24.$$

Note that $\lambda_1 = 1$ is an eigenvalue. To find the others, we factor:

$$f_A(\lambda) \div (\lambda - \lambda_1) = (\lambda^3 - 0.76\lambda - 0.24) \div (\lambda - 1)$$

$$= \lambda^2 + \lambda + 0.24$$

$$= (\lambda + 0.6)(\lambda + 0.4).$$

The characteristic polynomial is

$$f_A(\lambda) = (\lambda - 1)(\lambda + 0.6)(\lambda + 0.4),$$

and the eigenvalues of A are

$$\lambda_1 = 1, \qquad \lambda_2 = -0.6, \qquad \lambda_3 = -0.4.$$

Now we know that an eigenbasis exists for A, by Fact 7.3.6 (or Example 5).

Step 2   Construct an *eigenbasis* by finding a basis of each eigenspace:

$$E_1 = \ker \begin{bmatrix} 1 & -0.95 & -0.6 \\ -0.8 & 1 & 0 \\ 0 & -0.5 & 1 \end{bmatrix}$$

$$= \ker \begin{bmatrix} 1 & 0 & -2.5 \\ 0 & 1 & -2 \\ 0 & 0 & 0 \end{bmatrix} \qquad \text{(use technology to find rref)}$$

$$\begin{vmatrix} x_1 & & -2.5x_3 = 0 \\ & x_2 - & 2x_3 = 0 \end{vmatrix}$$

The general solution is

$$\begin{bmatrix} 2.5x_3 \\ 2x_3 \\ x_3 \end{bmatrix} = x_3 \begin{bmatrix} 2.5 \\ 2 \\ 1 \end{bmatrix}$$

$$E_1 = \text{span} \begin{bmatrix} 2.5 \\ 2 \\ 1 \end{bmatrix} = \text{span} \begin{bmatrix} 5 \\ 4 \\ 2 \end{bmatrix}$$

Similarly, we find that

$$E_{-0.6} = \text{span} \begin{bmatrix} 9 \\ -12 \\ 10 \end{bmatrix}, \qquad E_{-0.4} = \text{span} \begin{bmatrix} -2 \\ 4 \\ -5 \end{bmatrix}.$$

We have constructed the eigenbasis

$$\vec{v}_1 = \begin{bmatrix} 5 \\ 4 \\ 2 \end{bmatrix}, \qquad \vec{v}_2 = \begin{bmatrix} 9 \\ -12 \\ 10 \end{bmatrix}, \qquad \vec{v}_3 = \begin{bmatrix} -2 \\ 4 \\ -5 \end{bmatrix},$$

with associated eigenvalues

$$\lambda_1 = 1, \qquad \lambda_2 = -0.6, \qquad \lambda_3 = -0.4.$$

Step 3 Express the initial state vector as a linear combination of eigenvectors. We have to solve the linear system $\vec{x}_0 = c_1\vec{v}_1 + c_2\vec{v}_2 + c_3\vec{v}_3$. Using technology, we find that $c_1 = 100$, $c_2 = 50$, and $c_3 = 100$.

Step 4 Write the closed formula and discuss the long-term behavior (see Fact 7.1.3):

$$\vec{x}_0 = c_1\vec{v}_1 + c_2\vec{v}_2 + c_3\vec{v}_3.$$

Then

$$\vec{x}(t) = A^t\vec{x}_0 = c_1 A^t\vec{v}_1 + c_2 A^t\vec{v}_2 + c_3 A^t\vec{v}_3$$

$$= c_1\lambda_1^t\vec{v}_1 + c_2\lambda_2^t\vec{v}_2 + c_3\lambda_3^t\vec{v}_3$$

$$= 100\begin{bmatrix} 5 \\ 4 \\ 2 \end{bmatrix} + 50(-0.6)^t\begin{bmatrix} 9 \\ -12 \\ 10 \end{bmatrix} + 100(-0.4)^t\begin{bmatrix} -2 \\ 4 \\ -5 \end{bmatrix}.$$

The individual populations are given by

$$\begin{aligned} j(t) &= 500 + 450(-0.6)^t - 200(-0.4)^t, \\ m(t) &= 400 - 600(-0.6)^t + 400(-0.4)^t, \\ a(t) &= 200 + 500(-0.6)^t - 500(-0.4)^t. \end{aligned}$$

In the long term, the populations approach the equilibrium values

$$j = 500, \qquad m = 400, \qquad a = 200. \qquad \blacksquare$$

Eigenvalues and Similarity

If matrix A is similar to B, what is the relationship between the eigenvalues of A and B? The following theorem shows that this relationship is very close indeed.

Fact 7.3.8

The eigenvalues of similar matrices
Suppose matrix A is similar to B. Then

a. Matrices A and B have the same characteristic polynomial; that is, $f_A(\lambda) = f_B(\lambda)$

b. $\text{rank}(A) = \text{rank}(B)$ and $\text{nullity}(A) = \text{nullity}(B)$.

c. Matrices A and B have the same eigenvalues, with the same algebraic and geometric multiplicities. (However, the eigenvectors need not be the same.)

d. $\det(A) = \det(B)$ and $\text{tr}(A) = \text{tr}(B)$

Proof   a. If $B = S^{-1}AS$, then

$$f_B(\lambda) = \det(\lambda I_n - B) = \det(\lambda I_n - S^{-1}AS) = \det\left(S^{-1}(\lambda I_n - A)S\right)$$

$$= (\det S)^{-1}\det(\lambda I_n - A)\det(S) = \det(\lambda I_n - A) = f_A(\lambda) \quad \text{for all scalars } \lambda.$$

b. See Exercises 45 and 46 of Section 3.4. An alternative proof is suggested in Exercise 34 of this section.

c. It follows from part (a) that matrices A and B have the same eigenvalues, with the same algebraic multiplicities. (See Fact 7.2.1 and Definition 7.2.6.) As for the geometric multiplicity, note that $\lambda I_n - A$ is similar to $\lambda I_n - B$ for all λ (see Exercise 33), so that nullity$(\lambda I_n - A) =$ nullity$(\lambda I_n - B)$ for all eigenvalues λ, by part (b). (See Definition 7.3.2.)

d. These equations follow from part (a) and Fact 7.2.5: Trace and determinant are coefficients of the characteristic polynomial.   ▲

EXERCISES

GOALS For a given eigenvalue, find a basis of the associated eigenspace. Use the geometric multiplicities of the eigenvalues to determine whether there is an eigenbasis for a matrix.

For each of the matrices in Exercises 1 through 18, find all (real) eigenvalues. Then find a basis of each eigenspace, and find an eigenbasis, if you can. Do not use technology.

1. $\begin{bmatrix} 7 & 8 \\ 0 & 9 \end{bmatrix}$

2. $\begin{bmatrix} 1 & 1 \\ 1 & 1 \end{bmatrix}$

3. $\begin{bmatrix} 6 & 3 \\ 2 & 7 \end{bmatrix}$

4. $\begin{bmatrix} 0 & -1 \\ 1 & 2 \end{bmatrix}$

5. $\begin{bmatrix} 4 & 5 \\ -2 & -2 \end{bmatrix}$

6. $\begin{bmatrix} 2 & 3 \\ 4 & 5 \end{bmatrix}$

7. $\begin{bmatrix} 1 & 0 & 0 \\ 0 & 2 & 0 \\ 0 & 0 & 3 \end{bmatrix}$

8. $\begin{bmatrix} 1 & 1 & 0 \\ 0 & 2 & 2 \\ 0 & 0 & 3 \end{bmatrix}$

9. $\begin{bmatrix} 1 & 0 & 1 \\ 0 & 1 & 0 \\ 0 & 0 & 0 \end{bmatrix}$

10. $\begin{bmatrix} 1 & 1 & 0 \\ 0 & 1 & 0 \\ 0 & 0 & 0 \end{bmatrix}$

11. $\begin{bmatrix} 1 & 1 & 1 \\ 1 & 1 & 1 \\ 1 & 1 & 1 \end{bmatrix}$

12. $\begin{bmatrix} 1 & 1 & 0 \\ 0 & 1 & 1 \\ 0 & 0 & 1 \end{bmatrix}$

13. $\begin{bmatrix} 3 & 0 & -2 \\ -7 & 0 & 4 \\ 4 & 0 & -3 \end{bmatrix}$

14. $\begin{bmatrix} 1 & 0 & 0 \\ -5 & 0 & 2 \\ 0 & 0 & 1 \end{bmatrix}$

15. $\begin{bmatrix} -1 & 0 & 1 \\ -3 & 0 & 1 \\ -4 & 0 & 3 \end{bmatrix}$

16. $\begin{bmatrix} 1 & 1 & 0 \\ 0 & -1 & -1 \\ 2 & 2 & 0 \end{bmatrix}$

17. $\begin{bmatrix} 0 & 0 & 0 & 0 \\ 0 & 1 & 1 & 0 \\ 0 & 0 & 0 & 0 \\ 0 & 0 & 0 & 1 \end{bmatrix}$

18. $\begin{bmatrix} 0 & 0 & 0 & 0 \\ 0 & 1 & 0 & 1 \\ 0 & 0 & 0 & 0 \\ 0 & 0 & 0 & 1 \end{bmatrix}$

19. Consider the matrix

$$A = \begin{bmatrix} 1 & a & b \\ 0 & 1 & c \\ 0 & 0 & 1 \end{bmatrix},$$

where a, b, c are arbitrary constants. How does the geometric multiplicity of the eigenvalue 1 depend on the constants a, b, c? When is there an eigenbasis for A?

20. Consider the matrix

$$A = \begin{bmatrix} 1 & a & b \\ 0 & 1 & c \\ 0 & 0 & 2 \end{bmatrix},$$

where a, b, c are arbitrary constants. How do the geometric multiplicities of the eigenvalues 1 and 2 depend on the constants a, b, c? When is there an eigenbasis for A?

21. Find a 2×2 matrix A for which

$$E_1 = \text{span}\begin{bmatrix} 1 \\ 2 \end{bmatrix} \quad \text{and} \quad E_2 = \text{span}\begin{bmatrix} 2 \\ 3 \end{bmatrix}.$$

How many such matrices are there?

22. Find all 2×2 matrices A for which

$$E_7 = \mathbb{R}^2.$$

23. Find all eigenvalues and eigenvectors of $A = \begin{bmatrix} 1 & 1 \\ 0 & 1 \end{bmatrix}$. Is there an eigenbasis? Interpret your result geometrically.

24. Find a 2×2 matrix A for which

$$E_1 = \text{span} \begin{bmatrix} 2 \\ 1 \end{bmatrix}$$

is the only eigenspace.

25. What can you say about the geometric multiplicity of the eigenvalues of a matrix of the form

$$A = \begin{bmatrix} 0 & 1 & 0 \\ 0 & 0 & 1 \\ a & b & c \end{bmatrix},$$

where a, b, c are arbitrary constants?

26. Show that if a 6×6 matrix A has a negative determinant, then A has at least one positive eigenvalue. *Hint:* Sketch the graph of the characteristic polynomial.

27. Consider a 2×2 matrix A. Suppose that $\text{tr}(A) = 5$ and $\det(A) = 6$. Find the eigenvalues of A.

28. Consider the matrix

$$J_n(\lambda) = \begin{bmatrix} \lambda & 1 & 0 & \cdots & 0 \\ 0 & \lambda & 1 & \cdots & 0 \\ \vdots & \vdots & \ddots & \ddots & \vdots \\ 0 & 0 & 0 & \ddots & 1 \\ 0 & 0 & 0 & \cdots & \lambda \end{bmatrix}$$

(with all λ's on the diagonal and 1's directly above), where λ is an arbitrary constant. Find the eigenvalue(s) of $J_n(\lambda)$, and determine their algebraic and geometric multiplicities.

29. Consider a diagonal $n \times n$ matrix A with $\text{rank}(A) = r < n$. Find the algebraic and the geometric multiplicity of the eigenvalue 0 of A in terms of r and n.

30. Consider an upper triangular $n \times n$ matrix A with $a_{ii} \neq 0$ for $i = 1, 2, \ldots, m$ and $a_{ii} = 0$ for $i = m + 1$, $\ldots, n$. Find the algebraic multiplicity of the eigenvalue 0 of A. Without using Fact 7.3.3, what can you say about the geometric multiplicity?

31. Suppose there is an eigenbasis for a matrix A. What is the relationship between the algebraic and geometric multiplicities of its eigenvalues?

32. Consider an eigenvalue λ of an $n \times n$ matrix A. We know that λ is an eigenvalue of A^T as well (since A and A^T have the same characteristic polynomial). Compare the geometric multiplicities of λ as an eigenvalue of A and A^T.

33. Show that if matrix A is similar to B, then $\lambda I_n - A$ is similar to $\lambda I_n - B$, for all scalars λ.

34. Suppose that $B = S^{-1} A S$ for some $n \times n$ matrices A, B, and S.

a. Show that if $\vec{x}$ is in $\ker(B)$, then $S\vec{x}$ is in $\ker(A)$.

b. Show that the linear transformation $T(\vec{x}) = S\vec{x}$ from $\ker(B)$ to $\ker(A)$ is an isomorphism.

c. Show that $\text{nullity}(A) = \text{nullity}(B)$ and $\text{rank}(A) = \text{rank}(B)$.

35. Is matrix $\begin{bmatrix} 1 & 2 \\ 0 & 3 \end{bmatrix}$ similar to $\begin{bmatrix} 3 & 0 \\ 1 & 2 \end{bmatrix}$?

36. Is matrix $\begin{bmatrix} 0 & 1 \\ 5 & 3 \end{bmatrix}$ similar to $\begin{bmatrix} 1 & 2 \\ 4 & 3 \end{bmatrix}$?

37. Consider a symmetric $n \times n$ matrix A.

a. Show that if $\vec{v}$ and $\vec{w}$ are two vectors in $\mathbb{R}^n$, then

$$A\vec{v} \cdot \vec{w} = \vec{v} \cdot A\vec{w}.$$

b. Show that if $\vec{v}$ and $\vec{w}$ are two eigenvectors of A, with distinct eigenvalues, then $\vec{w}$ is orthogonal to $\vec{v}$.

38. Consider a rotation $T(\vec{x}) = A\vec{x}$ in $\mathbb{R}^3$. (That is, A is an orthogonal 3×3 matrix with determinant 1.) Show that T has a nonzero fixed point (i.e., a vector $\vec{v}$ with $T(\vec{v}) = \vec{v}$). This result is known as *Euler's theorem,* after the great Swiss mathematician Leonhard Euler (1707–1783). *Hint:* Consider the characteristic polynomial f_A. Pay attention to the intercepts with both axes. Use Fact 7.1.2.

39. Consider a subspace V of $\mathbb{R}^n$ with $\dim(V) = m$.

a. Suppose the $n \times n$ matrix A represents the orthogonal projection onto V. What can you say about the eigenvalues of A and their algebraic and geometric multiplicities?

b. Suppose the $n \times n$ matrix B represents the reflection in V. What can you say about the eigenvalues of B and their algebraic and geometric multiplicities?

40. Let $x(t)$ and $y(t)$ be the annual defense budgets of two antagonistic nations (expressed in billions of US dollars). The change of these budgets is modeled by the equations

$$x(t + 1) = a\,x(t) + b\,y(t),$$
$$y(t + 1) = b\,x(t) + a\,y(t),$$

where a is a constant slightly less than 1, expressing the fact that defense budgets tend to decline when there is no perceived threat. The constant b is a small positive number. You may assume that a exceeds b.

Suppose $x(0) = 3$ and $y(0) = 0.5$. What will happen in the long term? There are three possible cases, depending on the numerical values of a and b. Sketch a trajectory for each case, and discuss the outcome in practical terms. Include the eigenspaces in all your sketches.

41. Consider a modification of Example 7: Suppose the transformation matrix is

$$A = \begin{bmatrix} 0 & 1.4 & 1.2 \\ 0.8 & 0 & 0 \\ 0 & 0.4 & 0 \end{bmatrix}.$$

The initial populations are $j(0) = 600$, $m(0) = 100$, and $a(0) = 250$. Find closed formulas for $j(t)$, $m(t)$, and $a(t)$. Describe the long-term behavior. What can you say about the proportion $j(t) : m(t) : a(t)$ in the long term?

42. A street magician at Montmartre begins to perform at 11:00 p.m. on Saturday night. He starts out with no onlookers, but he attracts passersby at a rate of 10 per minute. Some get bored and wander off: Of the people present t minutes after 11:00 p.m., 20% will have left a minute later (but everybody stays for at least a minute). Let $C(t)$ be the size of the crowd t minutes after 11:00 p.m. Find a 2×2 matrix A such that

$$\begin{bmatrix} C(t+1) \\ 1 \end{bmatrix} = A \begin{bmatrix} C(t) \\ 1 \end{bmatrix}.$$

Find a closed formula for $C(t)$, and graph this function. What is the long-term behavior of $C(t)$?

43. Three friends, Alberich, Brunnhilde, and Carl, play a number game together: Each thinks of a (real) number and announces it to the others. In the first round, each player finds the average of the numbers chosen by the two others; that is his or her new score. In the second round, the corresponding averages of the scores in the first round are taken, and so on. Here is an example:

	A	**B**	**C**
Initial choice	7	11	5
After 1st round	8	6	9
After 2nd round	7.5	8.5	7

Whoever is ahead after 1,001 rounds wins.

a. The state of the game after t rounds can be represented as a vector:

$$\vec{x}(t) = \begin{bmatrix} a(t) \\ b(t) \\ c(t) \end{bmatrix} \quad \begin{matrix} \text{Alberich's score} \\ \text{Brunnhilde's score} \\ \text{Carl's score} \end{matrix}$$

Find the matrix A such that $\vec{x}(t+1) = A\vec{x}(t)$.

b. With the initial values mentioned above ($a_0 = 7$, $b_0 = 11$, $c_0 = 5$), what is the score after 10 rounds? After 50 rounds? Use technology.

c. Now suppose that Alberich and Brunnhilde initially pick the numbers 1 and 2, respectively. If Carl picks

the number c_0, what is the state of the game after t rounds? (Find closed formulas for $a(t)$, $b(t)$, $c(t)$, in terms of c_0.) For which choices of c_0 does Carl win the game?

44. In an unfortunate accident involving an Austrian truck, 100 kg of a highly toxic substance are spilled into Lake Silvaplana, in the Swiss Engadine Valley. The river Inn carries the pollutant down to Lake Sils and later to Lake St. Moritz. This sorry state, t weeks after the accident, can be described by the vector

$$\vec{x}(t) = \begin{bmatrix} x_1(t) \\ x_2(t) \\ x_3(t) \end{bmatrix} \quad \left. \begin{matrix} \text{pollutant in Lake Silvaplana} \\ \text{pollutant in Lake Sils} \\ \text{pollutant in Lake St. Moritz} \end{matrix} \right\} \text{ (in kg).}$$

Suppose that

$$\vec{x}(t+1) = \begin{bmatrix} 0.7 & 0 & 0 \\ 0.1 & 0.6 & 0 \\ 0 & 0.2 & 0.8 \end{bmatrix} \vec{x}(t).$$

a. Explain the significance of the entries of the transformation matrix in practical terms.

b. Find closed formulas for the amount of pollutant in each of the three lakes t weeks after the accident. Graph the three functions against time (on the same axes). When does the pollution in Lake Sils reach a maximum?

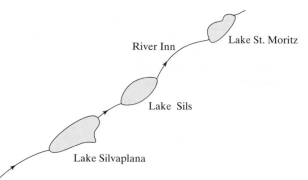

45. Consider a dynamical system

$$\vec{x}(t) = \begin{bmatrix} x_1(t) \\ x_2(t) \end{bmatrix}$$

whose transformation from time t to time $t + 1$ is given by the following equations:

$$x_1(t+1) = 0.1x_1(t) + 0.2x_2(t) + 1,$$
$$x_2(t+1) = 0.4x_1(t) + 0.3x_2(t) + 2.$$

Such a system, with constant terms in the equations, is not linear, but *affine*.

a. Find a 2×2 matrix A and a vector $\vec{b}$ in $\mathbb{R}^2$ such that

$$\vec{x}(t + 1) = A\vec{x}(t) + \vec{b}.$$

b. Introduce a new state vector

$$\vec{y}(t) = \begin{bmatrix} x_1(t) \\ x_2(t) \\ 1 \end{bmatrix},$$

with a "dummy" 1 in the last component. Find a 3×3 matrix B such that

$$\vec{y}(t + 1) = B\vec{y}(t).$$

How is B related to the matrix A and the vector $\vec{b}$ in part (a)? Can you write B as a partitioned matrix involving A and $\vec{b}$?

c. What is the relationship between the eigenvalues of A and B? What about eigenvectors?

d. For arbitrary values of $x_1(0)$ and $x_2(0)$, what can you say about the long-term behavior of $x_1(t)$ and $x_2(t)$?

46. A machine contains the grid of wires shown in the accompanying sketch. At the seven indicated points, the temperature is kept fixed at the given values (in °C). Consider the temperatures $T_1(t)$, $T_2(t)$, and $T_3(t)$ at the other three mesh points. Because of heat flow along the wires, the temperatures $T_i(t)$ changes according to the formula

$$T_i(t + 1) = T_i(t) - \frac{1}{10} \sum \left(T_i(t) - T_{\text{adj}}(t) \right),$$

where the sum is taken over the four adjacent points in the grid and time is measured in minutes. For example,

$$T_2(t + 1) = T_2(t) - \frac{1}{10} \left(T_2(t) - T_1(t) \right) - \frac{1}{10} \left(T_2(t) - 200 \right)$$

$$- \frac{1}{10} \left(T_2(t) - 0 \right) - \frac{1}{10} \left(T_2(t) - T_3(t) \right).$$

Note that each of the four terms we subtract represents the cooling caused by heat flowing along one of the wires. Let

$$\vec{x}(t) = \begin{bmatrix} T_1(t) \\ T_2(t) \\ T_3(t) \end{bmatrix}.$$

a. Find a 3×3 matrix A and a vector $\vec{b}$ in $\mathbb{R}^3$ such that

$$\vec{x}(t + 1) = A\vec{x}(t) + \vec{b}.$$

b. Introduce the state vector

$$\vec{y}(t) = \begin{bmatrix} T_1(t) \\ T_2(t) \\ T_3(t) \\ 1 \end{bmatrix},$$

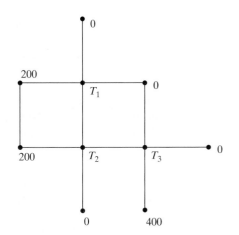

with a "dummy" 1 as the last component. Find a 4×4 matrix B such that

$$\vec{y}(t + 1) = B\vec{y}(t).$$

(This technique for converting an affine system into a linear system is introduced in Exercise 45; see also Exercise 42.)

c. Suppose the initial temperatures are $T_1(0) = T_2(0) = T_3(0) = 0$. Using technology, find the temperatures at the three points at $t = 10$ and $t = 30$. What long-term behavior do you expect?

d. Using technology, find numerical approximations for the eigenvalues of the matrix B. Find an eigenvector for the largest eigenvalue. Use the results to confirm your conjecture in part (c).

47. The color of snapdragons is determined by a pair of genes, which we designate by the letters A and a. The pair of genes is called the flower's *genotype*. Genotype AA produces red flowers, genotype Aa pink ones, and genotype aa white ones. A biologist undertakes a breeding program, starting with a large population of flowers of genotype AA. Each flower is fertilized with pollen from a plant of genotype Aa (taken from another population), and one offspring is produced. Since it is a matter of chance which of the genes a parent passes on, we expect half of the flowers in the next generation to be red (genotype AA) and the other half pink (genotype Aa). All the flowers in this generation are now fertilized with pollen from plants of genotype Aa (taken from another population), and so on.

a. Find closed formulas for the fractions of red, pink, and white flowers in the tth generation. We know that $r(0) = 1$ and $p(0) = w(0) = 0$, and we found that $r(1) = p(1) = \frac{1}{2}$ and $w(1) = 0$.

b. What is the proportion $r(t) : p(t) : w(t)$ in the long run?

48. *Leonardo of Pisa: The rabbit problem.* Leonardo of Pisa (c. 1170–1240), also known as Fibonacci, was the first outstanding European mathematician after the ancient Greeks. He traveled widely in the Islamic world and studied Arabic mathematical writing. His work is in the spirit of the Arabic mathematics of his day. Fibonacci brought the decimal-position system to Europe. In his book *Liber abaci* (1202), Fibonacci discusses the following problem:

> How many pairs of rabbits can be bred from one pair in one year? A man has one pair of rabbits at a certain place entirely surrounded by a wall. We wish to know how many pairs can be bred from it in one year, if the nature of these rabbits is such that they breed every month one other pair and begin to breed in the second month after their birth. Let the first pair breed a pair in the first month, then duplicate it and there will be 2 pairs in a month. From these pairs one, namely, the first, breeds a pair in the second month, and thus there are 3 pairs in the second month. From these, in one month, two will become pregnant, so that in the third month 2 pairs of rabbits will be born. Thus, there are 5 pairs in this month. From these, in the same month, 3 will be pregnant, so that in the fourth month there will be 8 pairs. From these pairs, 5 will breed 5 other pairs, which, added to the 8 pairs, gives 13 pairs in the fifth month, from which 5 pairs (which were bred in that same month) will not conceive in that month, but the other 8 will be pregnant. Thus, there will be 21 pairs in the sixth month. When we add to these the 13 pairs that are bred in the seventh month, then there will be in that month 34 pairs [and so on, 55, 89, 144, 233, 377, ...]. Finally, there will be 377, and this number of pairs has been born from the first-mentioned pair at the given place in one year.

Let $j(t)$ be the number of juvenile pairs and $a(t)$ the number of adult pairs after t months. Fibonacci starts his thought experiment in rabbit breeding with one adult pair, so $j(0) = 0$ and $a(0) = 1$. At $t = 1$, the adult pair will have bred a (juvenile) pair, so $a(1) = 1$ and $j(1) = 1$. At $t = 2$, the initial adult pair will have bred another (juvenile) pair, and last month's juvenile pair will have grown up, so $a(2) = 2$ and $j(2) = 1$.

a. Find formulas expressing $a(t + 1)$ and $j(t + 1)$ in terms of $a(t)$ and $j(t)$. Find the matrix A

such that

$$\vec{x}(t + 1) = A\vec{x}(t),$$

where

$$\vec{x}(t) = \begin{bmatrix} a(t) \\ j(t) \end{bmatrix}.$$

b. Find closed formulas for $a(t)$ and $j(t)$. (*Note:* You will have to deal with irrational quantities here.)

c. Find the limit of the ratio $a(t)/j(t)$ as t approaches infinity. The result is known as the *golden section*. The golden section of a line segment AB is given by the point P such that

$$\frac{\overline{AB}}{\overline{AP}} = \frac{\overline{AP}}{\overline{PB}}$$

$$A \qquad\qquad\qquad P \qquad\quad B$$

49. Consider an $n \times n$ matrix A with zeros on the diagonal and below the diagonal, and "random" entries above the diagonal. What is the geometric multiplicity of the eigenvalue 0 likely to be?

50. a. Sketch a phase portrait for the dynamical system $\vec{x}(t + 1) = A\vec{x}(t)$, where

$$A = \begin{bmatrix} 2 & 1 \\ 3 & 2 \end{bmatrix}.$$

b. In his paper "On the Measurement of the Circle," the great Greek mathematician Archimedes (c. 280–210 BC) uses the approximation

$$\frac{265}{153} < \sqrt{3} < \frac{1351}{780}$$

to estimate $\cos(30°)$. He does not explain how he arrived at these estimates. Explain how we can obtain these approximations from the dynamical system in part (a). *Hint:*

$$A^4 = \begin{bmatrix} 97 & 56 \\ 168 & 97 \end{bmatrix}, \qquad A^6 = \begin{bmatrix} 1351 & 780 \\ 2340 & 1351 \end{bmatrix}.$$

c. Without using technology, explain why

$$\frac{1351}{780} - \sqrt{3} < 10^{-6}.$$

Hint: Consider $\det(A^6)$.

d. Based on the data in part (b), give an underestimate of the form p/q of $\sqrt{3}$ that is better than the one given by Archimedes.

7.4 DIAGONALIZATION

EXAMPLE 1 In Example 2 of Section 7.3, we found that vectors $\vec{v}_1 = \begin{bmatrix} 1 \\ -1 \end{bmatrix}$ and $\vec{v}_2 = \begin{bmatrix} 1 \\ 2 \end{bmatrix}$ form an eigenbasis for matrix $A = \begin{bmatrix} 1 & 2 \\ 4 & 3 \end{bmatrix}$, with associated eigenvalues $\lambda_1 = -1$ and $\lambda_2 = 5$. Find the matrix B of the linear transformation $T(\vec{x}) = A\vec{x}$ with respect to this eigenbasis $\mathcal{B}$.

Solution

In Section 3.4, we learned three methods to do problems of this kind: Using Definition 3.4.2, constructing B column by column (using Fact 3.4.3), or by means of the formula $B = S^{-1}AS$. (See Fact 3.4.4.)

Let us construct B column by column:

$$B = \begin{bmatrix} [T(\vec{v}_1)]_{\mathcal{B}} & [T(\vec{v}_2)]_{\mathcal{B}} \end{bmatrix} = \begin{bmatrix} [\lambda_1\vec{v}_1]_{\mathcal{B}} & [\lambda_2\vec{v}_2]_{\mathcal{B}} \end{bmatrix} = \begin{bmatrix} \lambda_1 & 0 \\ 0 & \lambda_2 \end{bmatrix} = \begin{bmatrix} -1 & 0 \\ 0 & 5 \end{bmatrix}.$$

It is remarkable that B turns out to be a diagonal matrix, with the eigenvalues of A as its diagonal entries.

If we use the c_1–c_2 coordinate system defined by eigenvectors $\vec{v}_1 = \begin{bmatrix} 1 \\ -1 \end{bmatrix}$ and $\vec{v}_2 = \begin{bmatrix} 1 \\ 2 \end{bmatrix}$, then transformation T maps $\begin{bmatrix} c_1 \\ c_2 \end{bmatrix}$ into $B\begin{bmatrix} c_1 \\ c_2 \end{bmatrix} = \begin{bmatrix} -1 & 0 \\ 0 & 5 \end{bmatrix}\begin{bmatrix} c_1 \\ c_2 \end{bmatrix} = \begin{bmatrix} -c_1 \\ 5c_2 \end{bmatrix}$. The c_1-component is reversed and the c_2-component is magnified by the factor 5, as shown in Figure 1. ∎

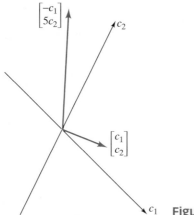

Figure 1

We can generalize Example 1.

Fact 7.4.1

The matrix of a linear transformation with respect to an eigenbasis is diagonal.

More specifically, consider a linear transformation $T(\vec{x}) = A\vec{x}$, where A is an $n \times n$ matrix. Suppose $\mathfrak{D}$ is an eigenbasis for T consisting of vectors $\vec{v}_1, \vec{v}_2, \ldots, \vec{v}_n$, with $A\vec{v}_i = \lambda_i \vec{v}_i$. Then the $\mathfrak{D}$-matrix D of T is

$$D = S^{-1} A S = \begin{bmatrix} \lambda_1 & 0 & \cdot & 0 \\ 0 & \lambda_2 & \cdot & 0 \\ \cdot & \cdot & \cdot & \cdot \\ 0 & 0 & \cdot & \lambda_n \end{bmatrix},$$

the diagonal matrix with eigenvalues $\lambda_1, \lambda_2, \ldots, \lambda_n$ as its diagonal entries. Here

$$S = \begin{bmatrix} | & | & & | \\ \vec{v}_1 & \vec{v}_2 & \ldots & \vec{v}_n \\ | & | & & | \end{bmatrix}.$$

Compare this with Fact 3.4.4.

To justify Fact 7.4.1, consider the columns of D (see Fact 3.4.3), just as we did in Example 1:

$$(i\text{th column of } D) = [T(\vec{v}_i)]_{\mathfrak{D}} = [\lambda_i \vec{v}_i]_{\mathfrak{D}} = \begin{bmatrix} 0 \\ 0 \\ \cdots \\ \lambda_i \\ \cdots \\ 0 \end{bmatrix} \swarrow i\text{th component}$$

Applying this observation to all the columns (for $i = 1, 2, \ldots, n$), we prove Fact 7.4.1.

The converse of Fact 7.4.1 holds as well: If the matrix of a linear transformation T with respect to a basis $\mathfrak{D}$ is diagonal, then $\mathfrak{D}$ is an eigenbasis for T (Exercise 53).

Fact 7.4.1 motivates the following definition.

Definition 7.4.2

Diagonalizable matrices

An $n \times n$ matrix A is called *diagonalizable* if A is similar to a diagonal matrix D, that is, if there is an invertible $n \times n$ matrix S such that $D = S^{-1} A S$ is diagonal.

As we just observed, matrix $S^{-1} A S$ is diagonal if (and only if) the columns of S form an eigenbasis for A. This implies the following result.

Fact 7.4.3

Matrix A is diagonalizable if (and only if) there is an eigenbasis for A.

In particular, if an $n \times n$ matrix A has n distinct eigenvalues, then A is diagonalizable (by Fact 7.3.6).

If an $n \times n$ matrix A has fewer than n distinct eigenvalues, then it may or may not be diagonalizable; consider the diagonalizable matrix $\begin{bmatrix} 1 & 0 \\ 0 & 1 \end{bmatrix}$ and the nondiagonalizable $\begin{bmatrix} 1 & 1 \\ 0 & 1 \end{bmatrix}$.

Asking whether a matrix A is diagonalizable amounts to asking whether there is an eigenbasis for A. In the past three sections, we outlined a method for answering this question. Let us summarize this process.

Algorithm 7.4.4   **Diagonalization**

Suppose we are asked to decide whether a given $n \times n$ matrix A is diagonalizable, and, if so, to find an invertible matrix S such that $S^{-1}AS$ is diagonal. We can proceed as follows.

a. Find the eigenvalues of A, that is, solve the equation $f_A(\lambda) = \det(\lambda I_n - A) = 0$.

b. For each eigenvalue λ, find a basis of the eigenspace $E_\lambda = \ker(\lambda I_n - A)$.

c. Matrix A is diagonalizable if (and only if) the dimensions of the eigenspaces add up to n. In this case, we find an eigenbasis $\vec{v}_1, \vec{v}_2, \ldots, \vec{v}_n$ for A by combining the bases of the eigenspaces we found in step (b). Let $S = [\vec{v}_1 \quad \vec{v}_2 \quad \ldots \quad \vec{v}_n]$. Then matrix $S^{-1}AS$ is diagonal, with the corresponding eigenvalues on the diagonal.

EXAMPLE 2   Diagonalize the matrix

$$A = \begin{bmatrix} 1 & 1 & 1 \\ 0 & 0 & 0 \\ 0 & 0 & 0 \end{bmatrix}$$

if you can.

Solution

We will follow the steps outlined in Algorithm 7.4.4:

a. The eigenvalues are 0 and 1, the diagonal entries. (Eigenvalue 0 has algebraic multiplicity 2.) At this point, we cannot tell whether A is diagonalizable; the answer will depend on the dimension of eigenspace E_0.

b. A routine computation shows that

$$E_0 = \ker(A) = \text{span}\left(\begin{bmatrix} -1 \\ 1 \\ 0 \end{bmatrix}, \begin{bmatrix} -1 \\ 0 \\ 1 \end{bmatrix} \right) \quad \text{and} \quad E_1 = \ker(I_3 - A) = \text{span}\begin{bmatrix} 1 \\ 0 \\ 0 \end{bmatrix}.$$

c. Matrix A is indeed diagonalizable, since the sum of the dimensions of the two eigenspaces is 3. We have the eigenbasis

$$\begin{bmatrix} -1 \\ 1 \\ 0 \end{bmatrix}, \quad \begin{bmatrix} -1 \\ 0 \\ 1 \end{bmatrix}, \quad \begin{bmatrix} 1 \\ 0 \\ 0 \end{bmatrix}$$

$$\updownarrow \qquad\qquad \updownarrow \qquad\qquad \updownarrow$$
$$0 \qquad\qquad 0 \qquad\qquad 1$$

If we let

$$S = \begin{bmatrix} -1 & -1 & 1 \\ 1 & 0 & 0 \\ 0 & 1 & 0 \end{bmatrix},$$

then

$$D = S^{-1}AS = \begin{bmatrix} 0 & 0 & 0 \\ 0 & 0 & 0 \\ 0 & 0 & 1 \end{bmatrix}$$

Note that you need not actually compute $S^{-1}AS$. Fact 7.4.1 guarantees that $S^{-1}AS$ is diagonal, with the eigenvalues on the diagonal. To check your work, you may wish to verify that $S^{-1}AS = D$, or, equivalently, that $AS = SD$ (this way you need not find the inverse of S):

$$AS = \begin{bmatrix} 1 & 1 & 1 \\ 0 & 0 & 0 \\ 0 & 0 & 0 \end{bmatrix} \begin{bmatrix} -1 & -1 & 1 \\ 1 & 0 & 0 \\ 0 & 1 & 0 \end{bmatrix} = \begin{bmatrix} 0 & 0 & 1 \\ 0 & 0 & 0 \\ 0 & 0 & 0 \end{bmatrix}$$

$$SD = \begin{bmatrix} -1 & -1 & 1 \\ 1 & 0 & 0 \\ 0 & 1 & 0 \end{bmatrix} \begin{bmatrix} 0 & 0 & 0 \\ 0 & 0 & 0 \\ 0 & 0 & 1 \end{bmatrix} = \begin{bmatrix} 0 & 0 & 1 \\ 0 & 0 & 0 \\ 0 & 0 & 0 \end{bmatrix} \quad \checkmark \qquad \blacksquare$$

Powers of a Matrix

EXAMPLE 3 Consider the matrix $A = \begin{bmatrix} 1/2 & 3/4 \\ 1/2 & 1/4 \end{bmatrix}$.

a. Find a formula for A^t, for any positive integer t.
b. Find $\lim_{t \to \infty} A^t$

(*Hint:* Diagonalize A.)

Solution

To get some sense for this problem, it is useful to compute A^t for a few values of t first, using a calculator or computer. For example,

$$A^{10} \approx \begin{bmatrix} 0.6000003815 & 0.5999994278 \\ 0.399996183 & 0.4000005722 \end{bmatrix}, \quad \text{close to} \quad \begin{bmatrix} 0.6 & 0.6 \\ 0.4 & 0.4 \end{bmatrix}.$$

Further numerical experimentation suggests that $\lim_{t \to \infty} A^t = \begin{bmatrix} 0.6 & 0.6 \\ 0.4 & 0.4 \end{bmatrix}$. We cannot be certain of this result, of course, for (at least) two reasons: We can only check finitely many values of t, and the results the calculator produces are afflicted with round-off errors.

To show that our conjecture, $\lim\limits_{t\to\infty} A^t = \begin{bmatrix} 0.6 & 0.6 \\ 0.4 & 0.4 \end{bmatrix}$, does indeed hold, we follow the hint and diagonalize matrix A:

- $f_A(\lambda) = \det \begin{bmatrix} \lambda - \frac{1}{2} & -\frac{3}{4} \\ -\frac{1}{2} & \lambda - \frac{1}{4} \end{bmatrix} = \lambda^2 - \frac{3}{4}\lambda + \frac{1}{8} - \frac{3}{8} = \lambda^2 - \frac{3}{4}\lambda - \frac{1}{4} =$
  $(\lambda - 1)\left(\lambda + \frac{1}{4}\right) = 0$. The eigenvalues are 1 and $-\frac{1}{4}$.

- $E_1 = \ker \begin{bmatrix} 1/2 & -3/4 \\ -1/2 & 3/4 \end{bmatrix} = \text{span} \begin{bmatrix} 3 \\ 2 \end{bmatrix}$ and $E_{-\frac{1}{4}} = \ker \begin{bmatrix} -3/4 & -3/4 \\ -1/2 & -1/2 \end{bmatrix} =$
  $\text{span} \begin{bmatrix} 1 \\ -1 \end{bmatrix}$.

- Thus $S^{-1}AS = D = \begin{bmatrix} 1 & 0 \\ 0 & -1/4 \end{bmatrix}$, where $S = \begin{bmatrix} 3 & 1 \\ 2 & -1 \end{bmatrix}$.

To compute the powers of matrix A, we solve the equation $S^{-1}AS = D$ for A and find that $A = SDS^{-1}$. Now

$$A^t = \overbrace{(SDS^{-1})(SDS^{-1})\ldots(SDS^{-1})}^{t\ times} = SD^tS^{-1};$$

note the cancellation of the terms of the form $S^{-1}S$. (Compare this with Example 7 of Section 3.4.)

Matrices S^{-1} and D^t are easy to compute: To find D^t, just raise the diagonal entries of D to the tth power. Thus, the problem of finding A^t is essentially solved.

$$S^{-1} = \frac{1}{5}\begin{bmatrix} 1 & 1 \\ 2 & -3 \end{bmatrix}, \qquad D^t = \begin{bmatrix} 1 & 0 \\ 0 & \left(-\frac{1}{4}\right)^t \end{bmatrix},$$

so that

$$A^t = SD^tS^{-1} = \frac{1}{5}\begin{bmatrix} 3 & 1 \\ 2 & -1 \end{bmatrix}\begin{bmatrix} 1 & 0 \\ 0 & \left(-\frac{1}{4}\right)^t \end{bmatrix}\begin{bmatrix} 1 & 1 \\ 2 & -3 \end{bmatrix}$$

$$= \frac{1}{5}\begin{bmatrix} 3 + 2\left(-\frac{1}{4}\right)^t & 3 - 3\left(-\frac{1}{4}\right)^t \\ 2 - 2\left(-\frac{1}{4}\right)^t & 2 + 3\left(-\frac{1}{4}\right)^t \end{bmatrix}.$$

Since the term $\left(-\frac{1}{4}\right)^t$ approaches 0 as t goes to infinity, we have

$$\lim_{t\to\infty} A^t = \frac{1}{5}\begin{bmatrix} 3 & 3 \\ 2 & 2 \end{bmatrix} = \begin{bmatrix} 0.6 & 0.6 \\ 0.4 & 0.4 \end{bmatrix},$$

confirming our earlier conjecture. ∎

Algorithm 7.4.5

Powers of a diagonalizable matrix
To compute the powers A^t of a diagonalizable matrix A (where t is a positive integer), proceed as follows:

Use Algorithm 7.4.4 to diagonalize A, that is, find an invertible S and a diagonal D such that $S^{-1}AS = D$.

Then

$$A = SDS^{-1} \quad \text{and} \quad A^t = SD^tS^{-1}.$$

To compute D^t, raise the diagonal entries of D to the tth power.

Algorithm 7.4.5 provides us with an alternative method for analyzing discrete dynamical systems. (See Fact 7.1.3.)

EXAMPLE 4 Consider the dynamical system

$$\vec{x}(t+1) = \begin{bmatrix} 1/2 & 3/4 \\ 1/2 & 1/4 \end{bmatrix} \vec{x}(t) \quad \text{with initial value} \quad \vec{x}_0 = \begin{bmatrix} 100 \\ 0 \end{bmatrix}.$$

Use your work in Example 3 to give a closed formula for $\vec{x}(t)$, and find $\lim_{t \to \infty} \vec{x}(t)$.
We know that

$$\vec{x}(t) = A^t \vec{x}_0, \quad \text{where} \quad A = \begin{bmatrix} 1/2 & 3/4 \\ 1/2 & 1/4 \end{bmatrix}.$$

In Section 7.1, we solved problems of this kind by finding the coordinates of the initial state vector $\vec{x}_0$ with respect to an eigenbasis of A. Alternatively, we can use the formula for A^t from Example 3.

$$\vec{x}(t) = A^t \vec{x}_0 = SD^tS^{-1}\vec{x}_0 = \frac{1}{5} \begin{bmatrix} 3 + 2\left(-\frac{1}{4}\right)^t & 3 - 3\left(-\frac{1}{4}\right)^t \\ 2 - 2\left(-\frac{1}{4}\right)^t & 2 + 3\left(-\frac{1}{4}\right)^t \end{bmatrix} \begin{bmatrix} 100 \\ 0 \end{bmatrix}$$

$$= \begin{bmatrix} 60 + 40\left(-\frac{1}{4}\right)^t \\ 40 - 40\left(-\frac{1}{4}\right)^t \end{bmatrix}$$

and

$$\lim_{t \to \infty} \vec{x}(t) = \begin{bmatrix} 60 \\ 40 \end{bmatrix}.$$

Note that we have seen the formula $\vec{x}(t) = SD^tS^{-1}\vec{x}_0$ in Fact 7.1.3 already, although it was derived in a different way. ∎

The Eigenvalues of a Linear Transformation

(for those who have studied Chapter 4)
In the preceding three sections, we developed the theory of eigenvalues and eigenvectors for $n \times n$ matrices, or, equivalently, for linear transformations $T(\vec{x}) = A\vec{x}$ from $\mathbb{R}^n$ to $\mathbb{R}^n$. These concepts can be generalized easily to a linear transformation from V to V, where V is any linear space. In the case of a finite-dimensional space V, we can generalize the idea of diagonalization as well.

Definition 7.4.6

> **The eigenvalues of a linear transformation**
> Consider a linear transformation T from V to V, where V is a linear space. A scalar λ is called an *eigenvalue* of T if there is a nonzero element f of V such that
>
> $$T(f) = \lambda f.$$
>
> Such an f is called an *eigenfunction* if V consists of functions, an *eigenmatrix* if V consists of matrices, etc. In theoretical work, the inclusive term *eigenvector* is often used for f.
>
> Now suppose that V is finite-dimensional. Then transformation T is called *diagonalizable* if there is a basis $\mathfrak{D}$ of V such that the $\mathfrak{D}$-matrix D of T is diagonal. Matrix D is diagonal if (and only if) basis $\mathfrak{D}$ consists of eigenvectors of T. (Compare this with Fact 7.4.1.)

EXAMPLE 5 Consider the linear transformation $T(f) = f'$ (the derivative) from C^∞ to C^∞. Show that all real numbers are eigenvalues of T. (*Hint:* Apply T to exponential functions.)

Solution

Following the hint, we observe that $T(e^x) = (e^x)' = e^x = 1 \cdot e^x$; this shows that e^x is an eigenfunction of T, with associated eigenvalue 1.

More generally,

$$T(e^{kx}) = (e^{kx})' = k(e^{kx}) \quad \text{(use the chain rule)},$$

showing that e^{kx} is an eigenfunction with associated eigenvalue k. Here k can be any real number, proving our claim. ∎

EXAMPLE 6 Consider the linear transformation $L(A) = A^T$ from $\mathbb{R}^{2\times 2}$ to $\mathbb{R}^{2\times 2}$ (the transpose[7]). Is transformation L diagonalizable? If so, find an eigenbasis for L. (*Hint:* Think about eigenvalues 1 and -1.)

Solution

A matrix A in $\mathbb{R}^{2\times 2}$ is an eigenmatrix with eigenvalue 1 if $L(A) = A^T = 1 \cdot A = A$, that is, if A is symmetric. The symmetric matrices of the form $\begin{bmatrix} a & b \\ b & c \end{bmatrix}$ form a three-dimensional space, with basis $\begin{bmatrix} 1 & 0 \\ 0 & 0 \end{bmatrix}, \begin{bmatrix} 0 & 1 \\ 1 & 0 \end{bmatrix}, \begin{bmatrix} 0 & 0 \\ 0 & 1 \end{bmatrix}$. We need only one more matrix to form an eigenbasis for L, since $\mathbb{R}^{2\times 2}$ is four-dimensional.

Following the hint, let's think about the eigenvalue -1. A matrix A in $\mathbb{R}^{2\times 2}$ is an eigenmatrix with eigenvalue -1 if $L(A) = A^T = (-1)A = -A$; those are the skew-symmetric matrices, of the form $\begin{bmatrix} 0 & b \\ -b & 0 \end{bmatrix}$. An example is $\begin{bmatrix} 0 & 1 \\ -1 & 0 \end{bmatrix}$.

---

[7]If you have skipped Chapter 5, read Definition 5.3.5 and Examples 5 and 6 following that definition.

We have found enough eigenmatrices to form an eigenbasis for L:

$$\begin{bmatrix} 1 & 0 \\ 0 & 0 \end{bmatrix}, \quad \begin{bmatrix} 0 & 1 \\ 1 & 0 \end{bmatrix}, \quad \begin{bmatrix} 0 & 0 \\ 0 & 1 \end{bmatrix}, \quad \begin{bmatrix} 0 & 1 \\ -1 & 0 \end{bmatrix}.$$

Thus, L is diagonalizable. ∎

EXAMPLE 7 Consider the linear transformation $T(a + bx + cx^2) = a + b(2x - 1) + c(2x - 1)^2$ from P_2 to P_2. (We replace x by $2x - 1$ wherever it appears.) Is this transformation T diagonalizable? If so, find an eigenbasis $\mathfrak{D}$ and the $\mathfrak{D}$-matrix D. What is the determinant of T?

Solution

Here it would be hard to find eigenvalues and eigenfunctions by inspection; we need a systematic approach. The idea is to find the matrix B of T with respect to some (convenient) basis $\mathfrak{B}$. Then we can use Algorithm 7.4.4 to determine whether B is diagonalizable, and, if so, to find an eigenbasis for B. Finally, we can transform this eigenbasis back into P_2 to find an eigenbasis for T.

The matrix B of T with respect to the standard basis $1, x, x^2$ is

$$B = \begin{bmatrix} 1 & -1 & 1 \\ 0 & 2 & -4 \\ 0 & 0 & 4 \end{bmatrix}.$$

(Verify this!)

The upper triangular matrix B has the three distinct eigenvalues 1, 2, 4, so that B will be diagonalizable (by Fact 7.4.3). A straightforward computation produces the eigenbasis

$$\begin{bmatrix} 1 \\ 0 \\ 0 \end{bmatrix}, \quad \begin{bmatrix} -1 \\ 1 \\ 0 \end{bmatrix}, \quad \begin{bmatrix} 1 \\ -2 \\ 1 \end{bmatrix}$$

for B. Transforming these vectors back into P_2, we find the eigenbasis

$$1, \quad x - 1, \quad x^2 - 2x + 1 = (x - 1)^2$$

for T. To check our work, we can verify that these are indeed eigenfunctions of T:

$$T(1) = 1,$$
$$T(x - 1) = (2x - 1) - 1 = 2x - 2 = 2(x - 1),$$
$$T((x - 1)^2) = ((2x - 1) - 1)^2 = (2x - 2)^2 = 4(x - 1)^2.$$

The matrix of T with respect to this basis is

$$D = \begin{bmatrix} 1 & 0 & 0 \\ 0 & 2 & 0 \\ 0 & 0 & 4 \end{bmatrix},$$

and $\det(T) = \det(D) = \det(B) = 1 \cdot 2 \cdot 4 = 8$. ∎

EXERCISES

GOALS Use the concept of a diagonalizable matrix. Find the eigenvalues of a linear transformation.

Decide which of the matrices A in Exercises 1 through 20 are diagonalizable. If possible, find an invertible S and a diagonal D such that $S^{-1}AS = D$. Do not use technology.

1. $A = \begin{bmatrix} 2 & 0 \\ 0 & 3 \end{bmatrix}$

2. $A = \begin{bmatrix} 2 & 1 \\ 0 & 3 \end{bmatrix}$

3. $A = \begin{bmatrix} 1 & 1 \\ 2 & 2 \end{bmatrix}$

4. $A = \begin{bmatrix} 1 & 2 \\ 3 & 6 \end{bmatrix}$

5. $A = \begin{bmatrix} 1 & 1 \\ 0 & 1 \end{bmatrix}$

6. $A = \begin{bmatrix} 2 & 0 \\ -1 & 2 \end{bmatrix}$

7. $\begin{bmatrix} 1 & 4 \\ 1 & -2 \end{bmatrix}$

8. $A = \begin{bmatrix} 1 & 3 \\ 3 & 1 \end{bmatrix}$

9. $A = \begin{bmatrix} 4 & 9 \\ -1 & -2 \end{bmatrix}$

10. $A = \begin{bmatrix} 3 & 4 \\ -1 & -1 \end{bmatrix}$

11. $A = \begin{bmatrix} 1 & 0 & 1 \\ 0 & 2 & 0 \\ 0 & 0 & 1 \end{bmatrix}$

12. $A = \begin{bmatrix} 2 & 0 & 1 \\ 0 & 1 & 0 \\ 0 & 0 & 1 \end{bmatrix}$

13. $\begin{bmatrix} 1 & 1 & 0 \\ 0 & 2 & 2 \\ 0 & 0 & 3 \end{bmatrix}$

14. $A = \begin{bmatrix} 3 & 1 & 1 \\ 0 & 2 & 1 \\ 0 & 0 & 1 \end{bmatrix}$

15. $A = \begin{bmatrix} 3 & -4 & 0 \\ 2 & -3 & 0 \\ 0 & 0 & 1 \end{bmatrix}$

16. $A = \begin{bmatrix} 4 & 0 & -2 \\ 0 & 1 & 0 \\ 1 & 0 & 1 \end{bmatrix}$

17. $A = \begin{bmatrix} 1 & 1 & 1 \\ 1 & 1 & 1 \\ 1 & 1 & 1 \end{bmatrix}$

18. $A = \begin{bmatrix} 1 & 0 & 1 \\ 0 & 1 & 0 \\ 1 & 0 & 1 \end{bmatrix}$

19. $A = \begin{bmatrix} 1 & 1 & 1 \\ 0 & 1 & 0 \\ 0 & 1 & 0 \end{bmatrix}$

20. $A = \begin{bmatrix} 1 & 0 & 1 \\ 1 & 1 & 1 \\ 1 & 0 & 1 \end{bmatrix}$

For which values of constants a, b, and c are the matrices in Exercises 21 through 30 diagonalizable?

21. $\begin{bmatrix} 1 & a \\ 0 & 2 \end{bmatrix}$

22. $\begin{bmatrix} 1 & a \\ 0 & b \end{bmatrix}$

23. $\begin{bmatrix} 1 & 1 \\ a & 1 \end{bmatrix}$

24. $\begin{bmatrix} a & b \\ b & c \end{bmatrix}$

25. $\begin{bmatrix} 1 & a & b \\ 0 & 2 & c \\ 0 & 0 & 3 \end{bmatrix}$

26. $\begin{bmatrix} 1 & a & b \\ 0 & 2 & c \\ 0 & 0 & 1 \end{bmatrix}$

27. $\begin{bmatrix} 1 & a & b \\ 0 & 1 & c \\ 0 & 0 & 1 \end{bmatrix}$

28. $\begin{bmatrix} 0 & 0 & 0 \\ 1 & 0 & a \\ 0 & 1 & 0 \end{bmatrix}$

29. $\begin{bmatrix} 0 & 0 & a \\ 1 & 0 & 0 \\ 0 & 1 & 0 \end{bmatrix}$

30. $\begin{bmatrix} 0 & 0 & a \\ 1 & 0 & 3 \\ 0 & 1 & 0 \end{bmatrix}$

For the matrices A in Exercises 31 through 34, find formulas for the entries of A^t, where t is a positive integer. Also, find the vector $A^t \begin{bmatrix} 1 \\ 2 \end{bmatrix}$.

31. $A = \begin{bmatrix} 1 & 2 \\ 4 & 3 \end{bmatrix}$

32. $A = \begin{bmatrix} 4 & -2 \\ 1 & 1 \end{bmatrix}$

33. $A = \begin{bmatrix} 1 & 2 \\ 3 & 6 \end{bmatrix}$

34. $A = \begin{bmatrix} 1/2 & 1/4 \\ 1/2 & 3/4 \end{bmatrix}$

35. Is matrix $\begin{bmatrix} -1 & 6 \\ -2 & 6 \end{bmatrix}$ similar to $\begin{bmatrix} 3 & 0 \\ 0 & 2 \end{bmatrix}$?

36. Is matrix $\begin{bmatrix} -1 & 6 \\ -2 & 6 \end{bmatrix}$ similar to $\begin{bmatrix} 1 & 2 \\ -1 & 4 \end{bmatrix}$?

37. If A and B are two 2×2 matrices with $\det(A) = \det(B) = \operatorname{tr}(A) = \operatorname{tr}(B) = 7$, is A necessarily similar to B?

38. If A and B are two 2×2 matrices with $\det(A) = \det(B) = \operatorname{tr}(A) = \operatorname{tr}(B) = 4$, is A necessarily similar to B?

Find all the eigenvalues and "eigenvectors" of the linear transformations in Exercises 39 through 52.

39. $T(f) = f' - f$ from C^∞ to C^∞.

40. $T(f) = 5f' - 3f$ from C^∞ to C^∞.

41. $L(A) = A + A^T$ from $\mathbb{R}^{2 \times 2}$ to $\mathbb{R}^{2 \times 2}$. Is L diagonalizable?

42. $L(A) = A - A^T$ from $\mathbb{R}^{2 \times 2}$ to $\mathbb{R}^{2 \times 2}$. Is L diagonalizable?

43. $T(x + iy) = x - iy$ from $\mathbb{C}$ to $\mathbb{C}$. Is T diagonalizable?

44. $T(x_0, x_1, x_2, \ldots) = (x_1, x_2, \ldots)$ from the space V of infinite sequences into V. (We drop the first term of the sequence.)

45. $T(x_0, x_1, x_2, \ldots) = (0, x_0, x_1, x_2, \ldots)$ from the space V of infinite sequences into V. (We insert a zero at the beginning.)

46. $T(x_0, x_1, x_2, x_3, x_4, \ldots) = (x_0, x_2, x_4, \ldots)$ from the space V of infinite sequences into V. (We drop every other term.)

47. $T\big(f(x)\big) = f(-x)$ from P_2 to P_2. Is T diagonalizable?

48. $T\big(f(x)\big) = f(2x)$ from P_2 to P_2. Is T diagonalizable?

49. $T\big(f(x)\big) = f(3x - 1)$ from P_2 to P_2. Is T diagonalizable?

50. $T\big(f(x)\big) = f(x - 3)$ from P_2 to P_2. Is T diagonalizable?

51. $T(f) = f'$ from P to P.

52. $T\big(f(x)\big) = x\big(f'(x)\big)$ from P to P.

53. Show that if the matrix B of a linear transformation T with respect to a basis $\mathfrak{D}$ is diagonal, then $\mathfrak{D}$ is an eigenbasis for T.

54. Are the following matrices similar?

$$A = \begin{bmatrix} 0 & 1 & 0 & 0 \\ 0 & 0 & 0 & 0 \\ 0 & 0 & 0 & 1 \\ 0 & 0 & 0 & 0 \end{bmatrix}, \quad B = \begin{bmatrix} 0 & 1 & 0 & 0 \\ 0 & 0 & 1 & 0 \\ 0 & 0 & 0 & 0 \\ 0 & 0 & 0 & 0 \end{bmatrix}$$

55. Find two 2×2 matrices A and B such that AB fails to be similar to BA. *Hint:* It can be arranged that AB is zero, but BA isn't.

56. Show that if A and B are two $n \times n$ matrices, then the matrices AB and BA have the same characteristic polynomial, and thus the same eigenvalues (matrices AB and BA need not be similar though; see Exercise 55). *Hint:*

$$\begin{bmatrix} AB & 0 \\ B & 0 \end{bmatrix} \begin{bmatrix} I_n & A \\ 0 & I_n \end{bmatrix} = \begin{bmatrix} I_n & A \\ 0 & I_n \end{bmatrix} \begin{bmatrix} 0 & 0 \\ B & BA \end{bmatrix}.$$

57. Consider an $m \times n$ matrix A and an $n \times m$ matrix B. Using Exercise 56 as a guide, show that matrices AB and BA have the same *nonzero* eigenvalues, with the same algebraic multiplicities. What about eigenvalue 0?

58. Consider a diagonalizable $n \times n$ matrix A with m distinct eigenvalues $\lambda_1, \ldots, \lambda_m$. Show that

$$(A - \lambda_1 I_n)(A - \lambda_2 I_n)\ldots(A - \lambda_m I_n) = 0.$$

59. Consider a diagonalizable $n \times n$ matrix A with characteristic polynomial

$$f_A(\lambda) = \lambda^n + a_{n-1}\lambda^{n-1} + \cdots + a_1\lambda + a_0.$$

Show that

$$f_A(A) = A^n + a_{n-1}A^{n-1} + \cdots + a_1 A + a_0 I_n = 0.$$

60. In this exercise, we will show that the geometric multiplicity of an eigenvalue is less than or equal to the algebraic multiplicity. Suppose λ_0 is an eigenvalue of an $n \times n$ matrix A, with geometric multiplicity d.

a. Explain why there is a basis $\mathfrak{B}$ of $\mathbb{R}^n$ whose first d vectors are eigenvectors of A, with eigenvalue λ_0.

b. Let B be the matrix of the linear transformation $T(\vec{x}) = A\vec{x}$ with respect to $\mathfrak{B}$. What do the first d columns of B look like?

c. Explain why the characteristic polynomial of B is of the form

$$f_B(\lambda) = (\lambda - \lambda_0)^d g(\lambda),$$

for some polynomial $g(\lambda)$. Conclude that the algebraic multiplicity of λ_0 as an eigenvalue of B (and A) is at least d.

61. We say that two $n \times n$ matrices A and B are *simultaneously diagonalizable* if there is an invertible $n \times n$ matrix S such that $S^{-1}AS$ and $S^{-1}BS$ are both diagonal.

a. Are the matrices

$$A = \begin{bmatrix} 1 & 0 & 0 \\ 0 & 1 & 0 \\ 0 & 0 & 1 \end{bmatrix} \quad \text{and} \quad B = \begin{bmatrix} 1 & 2 & 3 \\ 0 & 2 & 3 \\ 0 & 0 & 3 \end{bmatrix}$$

simultaneously diagonalizable? Explain.

b. Show that if A and B are simultaneously diagonalizable then $AB = BA$.

c. Give an example of two $n \times n$ matrices such that $AB = BA$, but A and B are not simultaneously diagonalizable.

d. Let D be a diagonal $n \times n$ matrix with n distinct entries on the diagonal. Find all $n \times n$ matrices B that commute with D.

e. Show that if $AB = BA$ and A has n distinct eigenvalues, then A and B are simultaneously diagonalizable. *Hint:* Part (d) is useful.

62. Consider the linear transformation $T(f) = f'' + af' + bf$ from C^∞ to C^∞, where a and b are arbitrary constants. What does Fact 4.1.5 tell you about the eigenvalues of T? What about the dimension of the eigenspaces of T?

63. Consider the linear transformation $T(f) = f''$ from C^∞ to C^∞. For each of the following eigenvalues, find a basis of the associated eigenspace. See Exercise 62. (a) $\lambda = 1$, (b) $\lambda = 0$, (c) $\lambda = -1$, (d) $\lambda = -4$.

7.5 COMPLEX EIGENVALUES

Imagine that you are diabetic and have to pay close attention to how your body metabolizes glucose. After you eat a heavy meal, the glucose concentration will reach a peak, and then it will slowly return to the fasting level. Certain hormones help regulate the glucose metabolism, especially the hormone insulin. (Compare with Exercise 7.1.36.) Let $g(t)$ be the excess glucose concentration in your blood, usually measured in milligrams of glucose per 100 milliliters of blood. (Excess means that we measure how much the glucose concentration deviates from the fasting level.) A negative value of $g(t)$ indicates that the glucose concentration is below fasting level at time t. Let $h(t)$ be the excess insulin concentration in your blood. Researchers have developed mathematical models for the glucose regulatory system. The following is one such model, in slightly simplified (linearized) form.

$$g(t+1) = a\,g(t) - b\,h(t)$$
$$h(t+1) = c\,g(t) + d\,h(t)$$

(These formulas apply between meals; obviously, the system is disturbed during and right after a meal.)

In these formulas, a, b, c, and d are positive constants; constants a and d will be less than 1. The term $-bh(t)$ expresses the fact that insulin helps your body absorb glucose, and the term $cg(t)$ represents the fact that the glucose in your blood stimulates the pancreas to secrete insulin.

For your system, the equations might be

$$g(t+1) = 0.9g(t) - 0.4h(t)$$
$$h(t+1) = 0.1g(t) + 0.9h(t),$$

with initial values $g(0) = 100$ and $h(0) = 0$, right after a heavy meal. Here, time t is measured in hours.

After one hour, the values will be $g(1) = 90$ and $h(1) = 10$. Some of the glucose has been absorbed, and the excess glucose has stimulated the pancreas to produce 10 extra units of insulin.

The rounded values of $g(t)$ and $h(t)$ in the following table give you some sense for the evolution of this dynamical system.

t	0	1	2	3	4	5	6	7	8	15	22	29
$g(t)$	100	90	77	62.1	46.3	30.6	15.7	2.3	−9.3	−29	1.6	9.5
$h(t)$	0	10	18	23.9	27.7	29.6	29.7	28.3	25.7	−2	−8.3	0.3

We can "connect the dots" to sketch a rough trajectory, visualizing the long-term behavior.

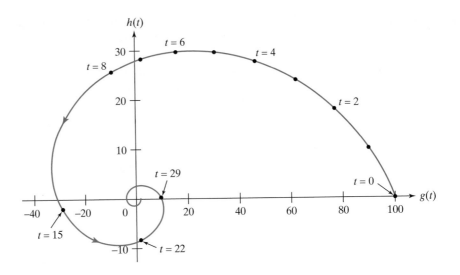

We see that after 7 hours the excess glucose is almost gone, but now there are about 30 units of excess insulin in the system. Since this excess insulin helps to reduce glucose further, the glucose concentration will now fall below fasting level, reaching about -30 after 15 hours. (You will feel awfully hungry by now.) Under normal circumstances, you would have taken another meal in the meantime, of course, but let's consider the case of (voluntary or involuntary) fasting.

We leave it to the reader to explain the concentrations after 22 and 29 hours, in terms of how glucose and insulin concentrations influence each other, according to our model. The *spiraling trajectory* indicates an *oscillatory behavior* of the system: Both glucose and insulin levels will swing back and forth around the fasting level, like a damped pendulum. Both concentrations will approach the fasting level (thus the name).

Another way to visualize this oscillatory behavior is to graph the functions $g(t)$ and $h(t)$ against time, using the values from our table.

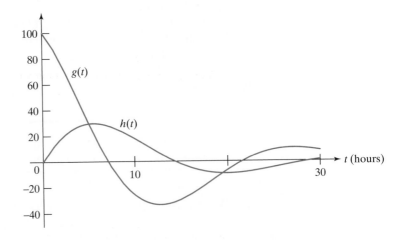

Next, we try to use the tools developed in the last three sections to analyze this system. We can introduce the transformation matrix

$$A = \begin{bmatrix} 0.9 & -0.4 \\ 0.1 & 0.9 \end{bmatrix}$$

and the state vector

$$\vec{x}(t) = \begin{bmatrix} g(t) \\ h(t) \end{bmatrix}.$$

Then

$$\vec{x}(t+1) = A\vec{x}(t) \quad \text{and thus} \quad \vec{x}(t) = A^t\vec{x}(0) = A^t \begin{bmatrix} 100 \\ 0 \end{bmatrix}.$$

To find formulas for $g(t)$ and $h(t)$, we need to know the eigenvalues and eigenvectors of matrix A. The characteristic polynomial of A is

$$f_A(\lambda) = \lambda^2 - 1.8\lambda + 0.85,$$

and the quadratic equation gives

$$\lambda_{1,2} = \frac{1.8 \pm \sqrt{3.24 - 3.4}}{2} = \frac{1.8 \pm \sqrt{-0.16}}{2}.$$

Since the square of a real number cannot be negative, there are no *real* eigenvalues here. However, if we allow complex solutions, then we have the eigenvalues

$$\lambda_{1,2} = \frac{1.8 \pm \sqrt{-0.16}}{2} = \frac{1.8 \pm i\sqrt{0.16}}{2} = 0.9 \pm 0.2i.$$

In this section, we will first review some basic facts on complex numbers. Then we will examine how the theory of eigenvalues and eigenvectors developed in Sections 7.1 through 7.4 can be adapted to the complex case. In Section 7.6 we will apply this work to dynamical systems. A great many dynamical systems, in physics, chemistry, biology, and economics, show oscillatory behavior; we will see that we can expect complex eigenvalues in this case.

These tools will enable you to find formulas for $g(t)$ and $h(t)$. (See Exercise 7.6.32.)

Complex Numbers: A Brief Review

Let us review some basic facts about complex numbers. We trust that you have at least a fleeting acquaintance with complex numbers. Without attempting a definition, we recall that a complex number can be expressed as

$$z = a + ib,$$

where a and b are real numbers.[8] Addition of complex numbers is defined in a natural way, by the rule

$$(a + ib) + (c + id) = (a + c) + i(b + d),$$

---

[8]The letter i for the imaginary unit was introduced by Leonhard Euler, the most prolific mathematician in history. For a fascinating glimpse at the history of the complex numbers see Tobias Dantzig, *Number: The Language of Science,* Macmillan, 1954.

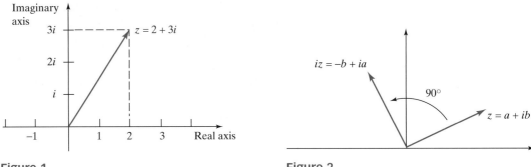

Figure 1   **Figure 2**

and multiplication is defined by the rule

$$(a + ib)(c + id) = (ac - bd) + i(ad + bc);$$

that is, we let $i \cdot i = -1$ and distribute.

If $z = a + ib$ is a complex number, we call a its *real part* (denoted by $\text{Re}(z)$) and b its *imaginary part* (denoted by $\text{Im}(z)$). A complex number of the form ib (with $a = 0$) is called *imaginary*.

The set of all complex numbers is denoted by $\mathbb{C}$. The real numbers, $\mathbb{R}$, form a subset of $\mathbb{C}$ (namely, those complex numbers with imaginary part 0).

Complex numbers can be represented as vectors (or points) in the complex plane,[9] as shown in Figure 1. This is a graphical representation of the isomorphism $T \begin{bmatrix} a \\ b \end{bmatrix} = a + ib$ from $\mathbb{R}^2$ to $\mathbb{C}$.

EXAMPLE 1   Consider a nonzero complex number z. What is the geometric relationship between z and iz in the complex plane?

Solution

If $z = a + ib$, then $iz = -b + ia$. We obtain the vector $\begin{bmatrix} -b \\ a \end{bmatrix}$ (representing iz) by rotating the vector $\begin{bmatrix} a \\ b \end{bmatrix}$ (representing z) through an angle of 90° in the counterclockwise direction. (See Figure 2.)   ∎

The *conjugate* of a complex number $z = a + ib$ is defined by

$$\bar{z} = a - ib.$$

(The sign of the imaginary part is reversed.) We say that z and $\bar{z}$ form a *conjugate pair* of complex numbers. Geometrically, the conjugate $\bar{z}$ is the reflection of z in the real axis, as shown in Figure 3.

Sometimes it is useful to describe a complex number in *polar coordinates,* as shown in Figure 4.

---

[9]Also called "Argand plane," after the Swiss mathematician Jean Robert Argand (1768–1822). The representation of complex numbers in the plane was introduced independently by Argand, by Gauss, and by the Norwegian mathematician Caspar Wessel (1745–1818).

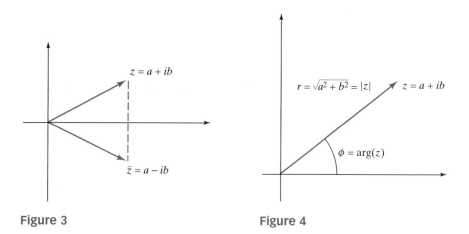

Figure 3   **Figure 4**

The length r of the vector is called the *modulus* of z, denoted by $|z|$. The angle ϕ is called an *argument* of z; note that the argument is determined only up to a multiple of 2π. (Mathematicians say "modulo 2π.") For example, for $z = -1$, we can choose the argument π, $-\pi$, or 3π.

EXAMPLE 2   Find the modulus and argument of $z = -2 + 2i$.

Solution

$|z| = \sqrt{2^2 + 2^2} = \sqrt{8}$. Representing z in the complex plane, we see that $\frac{3}{4}\pi$ is an argument of z. (See Figure 5.)   ∎

If z is a complex number with modulus r and argument ϕ, we can write z as

$$z = r(\cos\phi) + ir(\sin\phi) = r(\cos\phi + i\sin\phi),$$

as shown in Figure 6.
    The representation

$$z = r(\cos\phi + i\sin\phi)$$

is called the *polar form* of the complex number z.

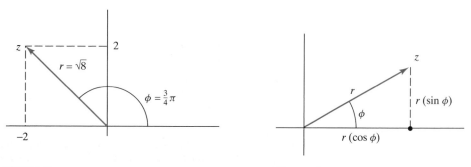

Figure 5   **Figure 6**

EXAMPLE 3 Consider the complex numbers $z = \cos\alpha + i\sin\alpha$ and $w = \cos\beta + i\sin\beta$. Find the polar form of the product zw.

Solution

Apply the addition formulas from trigonometry (see Exercise 2.2.32):

$$zw = (\cos\alpha + i\sin\alpha)(\cos\beta + i\sin\beta)$$
$$= (\cos\alpha\cos\beta - \sin\alpha\sin\beta) + i(\sin\alpha\cos\beta + \cos\alpha\sin\beta)$$
$$= \cos(\alpha + \beta) + i\sin(\alpha + \beta)$$

We conclude that the modulus of zw is 1, and $\alpha + \beta$ is an argument of zw. (See Figure 7.) ∎

In general, if $z = r(\cos\alpha + i\sin\alpha)$ and $w = s(\cos\beta + i\sin\beta)$, then

$$zw = rs\big(\cos(\alpha + \beta) + i\sin(\alpha + \beta)\big).$$

When we multiply two complex numbers, we multiply the moduli, and we add the arguments:

$$|zw| = |z|\,|w|$$
$$\arg(zw) = \arg z + \arg w \qquad (\text{modulo } 2\pi)$$

EXAMPLE 4 Describe the transformation $T(z) = (3 + 4i)z$ from $\mathbb{C}$ to $\mathbb{C}$ geometrically.

Solution

$$|T(z)| = |3 + 4i|\,|z| = 5|z|,$$
$$\arg\big(T(z)\big) = \arg(3 + 4i) + \arg(z) = \arctan\left(\frac{4}{3}\right) + \arg(z) \approx 53° + \arg(z).$$

The transformation T is a *rotation–dilation* in the complex plane. (See Figure 8.)

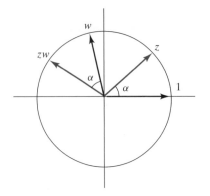

Figure 7

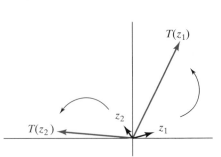

Figure 8 Rotate through about 53° and stretch the vector by a factor of 5.

Alternatively, we observe that the matrix of the linear transformation T with respect to the basis $1, i$ is the rotation–dilation matrix $\begin{bmatrix} 3 & -4 \\ 4 & 3 \end{bmatrix}$. ∎

The polar form is convenient for finding *powers* of a complex number z: if

$$z = r(\cos\phi + i\sin\phi),$$

then

$$z^2 = r^2\big(\cos(2\phi) + i\sin(2\phi)\big),$$

$$\vdots$$

$$z^n = r^n\big(\cos(n\phi) + i\sin(n\phi)\big),$$

for any positive integer n. Each time we multiply by z, the modulus is multiplied by r and the argument increases by ϕ. The preceding formula was found by the French mathematician Abraham de Moivre (1667–1754).

Fact 7.5.1   **De Moivre's formula**

$$(\cos\phi + i\sin\phi)^n = \cos(n\phi) + i\sin(n\phi).$$

EXAMPLE 5   Consider the complex number $z = 0.5 + 0.8i$. Represent the powers $z^2, z^3, \ldots$ in the complex plane. What is $\lim_{n\to\infty} z^n$?

Solution

To study the powers, write z in polar form:

$$z = r(\cos\phi + i\sin\phi).$$

Here

$$r = \sqrt{0.5^2 + 0.8^2} \approx 0.943$$

and

$$\phi = \arctan\frac{0.8}{0.5} \approx 58°.$$

We have

$$z^n = r^n\big(\cos(n\phi) + i\sin(n\phi)\big).$$

The vector representation of z^{n+1} is a little shorter than that of z^n (by about 5.7%), and z^{n+1} makes an angle $\phi \approx 58°$ with z^n. If we connect the tips of consecutive vectors, we see a trajectory that spirals in toward the origin, as shown in Figure 9. Note that $\lim_{n\to\infty} z^n = 0$, since $r = |z| < 1$. ∎

Perhaps the most remarkable property of the complex numbers is expressed in the fundamental theorem of algebra, first demonstrated by Carl Friedrich Gauss (in his thesis, at age 22).

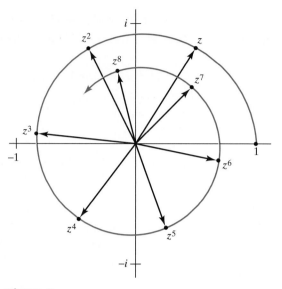

Figure 9

Fact 7.5.2

Fundamental theorem of algebra
Any polynomial $p(\lambda)$ with complex coefficients can be written as a product of linear factors

$$p(\lambda) = k(\lambda - \lambda_1)(\lambda - \lambda_2) \ldots (\lambda - \lambda_n),$$

for some complex numbers $\lambda_1, \lambda_2, \ldots, \lambda_n$, and k. (The λ_i need not be distinct.)

Therefore, a polynomial $p(\lambda)$ of degree n has precisely n complex roots if they are properly counted with their multiplicities.

For example, the polynomial

$$p(\lambda) = \lambda^2 + 1,$$

which does not have any real zeros, can be factored over $\mathbb{C}$:

$$p(\lambda) = (\lambda + i)(\lambda - i).$$

More generally, for a quadratic polynomial

$$q(\lambda) = \lambda^2 + b\lambda + c,$$

where b and c are real, we can find the complex roots

$$\lambda_{1,2} = \frac{-b \pm \sqrt{b^2 - 4c}}{2}$$

and

$$q(\lambda) = (\lambda - \lambda_1)(\lambda - \lambda_2).$$

Proving the fundamental theorem of algebra would lead us too far afield. Read any introduction to complex analysis or check Gauss's original proof.[10]

Complex Eigenvalues and Eigenvectors

The complex numbers share some basic algebraic properties with the real numbers.[11] Mathematicians summarize these properties by saying that both the real numbers $\mathbb{R}$ and the complex numbers $\mathbb{C}$ form a *field*. The rational numbers $\mathbb{Q}$ are another important example of a field; the integers $\mathbb{Z}$ on the other hand don't form a field. (Which of the 10 properties listed in the footnote fail to hold in this case?)

Which of the results and techniques derived in this text thus far still apply when we work with complex numbers throughout, that is, when we consider complex scalars, vectors with complex components, and matrices with complex entries? We observe that everything works the same way except for those geometrical concepts that are defined in terms of the dot product (length, angles, orthogonality, and so on, discussed in Chapter 5 and Section 6.3). The dot product in $\mathbb{C}^n$ is defined in a way that we will not discuss in this introductory text. The whole body of "core linear algebra" can be generalized without difficulty, however: echelon form, linear transformation, kernel, image, linear independence, basis, dimension, coordinates, linear spaces, determinant, eigenvalues, eigenvectors, and diagonalization.

EXAMPLE 6   Diagonalize the rotation–dilation matrix $A = \begin{bmatrix} a & -b \\ b & a \end{bmatrix}$ "over $\mathbb{C}$." Here, a and b are real numbers, and b is nonzero.

Solution

Following Algorithm 7.4.4, we will find the eigenvalues of A first:

$$f_A(\lambda) = \det \begin{bmatrix} \lambda - a & b \\ -b & \lambda - a \end{bmatrix} = (\lambda - a)^2 + b^2 = 0$$

when

$$(\lambda - a)^2 = -b^2 \quad \text{or} \quad \lambda_{1,2} - a = \pm ib \quad \text{or} \quad \lambda_{1,2} = a \pm ib.$$

---

[10]C. F. Gauss: *Werke*, III, 3–56. For an English translation see D. J. Struik (editor): *A Source Book in Mathematics* 1200–1800, Princeton University Press, 1986.

[11]Here is a list of these properties:
1. Addition is commutative.
2. Addition is associative.
3. There is a unique number n such that $a + n = a$, for all numbers a. This number n is denoted by 0.
4. For each number a there is a unique number b such that $a + b = 0$. This number b is denoted by $-a$. (*Comment:* This property says that we can subtract in this number system.)
5. Multiplication is commutative.
6. Multiplication is associative.
7. There is a unique number e such that $ea = a$, for all numbers a. This number e is denoted by 1.
8. For each nonzero number a there is a unique number b such that $ab = 1$. This number b is denoted by a^{-1}. (*Comment:* This property says that we can divide by a nonzero number.)
9. Multiplication distributes over addition: $a(b + c) = ab + ac$.
10. The numbers 0 and 1 are not equal.

Now we find the eigenvectors:

$$E_{a+ib} = \ker \begin{bmatrix} ib & b \\ -b & ib \end{bmatrix} = \text{span} \begin{bmatrix} i \\ 1 \end{bmatrix}, \quad E_{a-ib} = \ker \begin{bmatrix} -ib & b \\ -b & -ib \end{bmatrix} = \text{span} \begin{bmatrix} -i \\ 1 \end{bmatrix}.$$

Thus,

$$R^{-1} \begin{bmatrix} a & -b \\ b & a \end{bmatrix} R = \begin{bmatrix} a+ib & 0 \\ 0 & a-ib \end{bmatrix}, \quad \text{where} \quad R = \begin{bmatrix} i & -i \\ 1 & 1 \end{bmatrix}. \qquad \blacksquare$$

EXAMPLE 7 Let A be a real 2×2 matrix with eigenvalues $a \pm ib$. Show that A is similar (over $\mathbb{R}$) to the rotation–dilation matrix $\begin{bmatrix} a & -b \\ b & a \end{bmatrix}$.

Solution

Let $\vec{v} \pm i\vec{w}$ be eigenvectors of A with eigenvalues $a \pm ib$. (See Exercise 42.) By Fact 7.4.1, matrix A is similar to $\begin{bmatrix} a+ib & 0 \\ 0 & a-ib \end{bmatrix}$; more precisely, $\begin{bmatrix} a+ib & 0 \\ 0 & a-ib \end{bmatrix} = P^{-1}AP$, where $P = \begin{bmatrix} \vec{v}+i\vec{w} & \vec{v}-i\vec{w} \end{bmatrix}$. By Example 6, matrix $\begin{bmatrix} a & -b \\ b & a \end{bmatrix}$ is similar to $\begin{bmatrix} a+ib & 0 \\ 0 & a-ib \end{bmatrix}$ as well, with $\begin{bmatrix} a+ib & 0 \\ 0 & a-ib \end{bmatrix} = R^{-1} \begin{bmatrix} a & -b \\ b & a \end{bmatrix} R$, where $R = \begin{bmatrix} i & -i \\ 1 & 1 \end{bmatrix}$.

Thus,

$$P^{-1}AP = R^{-1} \begin{bmatrix} a & -b \\ b & a \end{bmatrix} R,$$

and

$$\begin{bmatrix} a & -b \\ b & a \end{bmatrix} = RP^{-1}APR^{-1} = S^{-1}AS,$$

where $S = PR^{-1}$ and $S^{-1} = (PR^{-1})^{-1} = RP^{-1}$.

A straightforward computation shows that

$$S = PR^{-1} = \frac{1}{2i} \begin{bmatrix} \vec{v}+i\vec{w} & \vec{v}-i\vec{w} \end{bmatrix} \begin{bmatrix} 1 & i \\ -1 & i \end{bmatrix} = \begin{bmatrix} \vec{w} & \vec{v} \end{bmatrix};$$

note that S has real entries, as claimed. $\qquad \blacksquare$

Fact 7.5.3 **Complex eigenvalues and rotation–dilation matrices**

If A is a real 2×2 matrix with eigenvalues $a \pm ib$, and if $\vec{v} + i\vec{w}$ is an eigenvector of A with eigenvalue $a + ib$, then

$$S^{-1}AS = \begin{bmatrix} a & -b \\ b & a \end{bmatrix}, \quad \text{where} \quad S = \begin{bmatrix} \vec{w} & \vec{v} \end{bmatrix}.$$

We see that matrix A is similar to a rotation–dilation matrix. Those who have studied Section 5.5 can go a step further: If we introduce the inner product $\langle \vec{x}, \vec{y} \rangle = S^{-1}\vec{x} \cdot S^{-1}\vec{y}$ in $\mathbb{R}^2$ and define the length of vectors and the angle between vectors with respect to this inner product, then the transformation $T(\vec{x}) = A\vec{x}$ is a rotation–dilation in that inner product space. (Think about it!)

EXAMPLE 8   For $A = \begin{bmatrix} 3 & -5 \\ 1 & -1 \end{bmatrix}$, find an invertible 2×2 matrix S such that $S^{-1}AS$ is a rotation–dilation matrix.

Solution

We will use the method outlined in Fact 7.5.3:

$$f_A(\lambda) = \lambda^2 - 2\lambda + 2, \quad \text{so that} \quad \lambda_{1,2} = \frac{2 \pm \sqrt{4-8}}{2} = 1 \pm i.$$

Now

$$E_{1+i} = \ker \begin{bmatrix} -2+i & 5 \\ -5 & 2+i \end{bmatrix} = \operatorname{span} \begin{bmatrix} -5 \\ -2+i \end{bmatrix},$$

and

$$\begin{bmatrix} -5 \\ -2+i \end{bmatrix} = \begin{bmatrix} -5 \\ -2 \end{bmatrix} + i \begin{bmatrix} 0 \\ 1 \end{bmatrix}, \quad \text{so that} \quad \vec{w} = \begin{bmatrix} 0 \\ 1 \end{bmatrix}, \quad \vec{v} = \begin{bmatrix} -5 \\ -2 \end{bmatrix}.$$

Therefore,

$$S^{-1}AS = \begin{bmatrix} 1 & -1 \\ 1 & 1 \end{bmatrix}, \quad \text{where} \quad S = \begin{bmatrix} 0 & -5 \\ 1 & -2 \end{bmatrix}. \qquad \blacksquare$$

The great advantage of complex eigenvalues is that there are so many of them. By the fundamental theorem of algebra, Fact 7.5.2, the characteristic polynomial always factors completely:

$$f_A(\lambda) = (\lambda - \lambda_1)(\lambda - \lambda_2) \ldots (\lambda - \lambda_n).$$

Fact 7.5.4   A complex $n \times n$ matrix has n complex eigenvalues if they are counted with their algebraic multiplicities.

Although a complex $n \times n$ matrix may have fewer than n *distinct* complex eigenvalues (examples are I_n or $\begin{bmatrix} 0 & 1 \\ 0 & 0 \end{bmatrix}$), this is literally a coincidence. (Some of the λ_i in the factorization of the characteristic polynomial $f_A(\lambda)$ coincide.) "Most" complex $n \times n$ matrices do have n distinct eigenvalues, so that **most complex $n \times n$ matrices are diagonalizable** (by Fact 7.4.3). An example of a matrix that fails to be diagonalizable over $\mathbb{C}$ is $\begin{bmatrix} 0 & 1 \\ 0 & 0 \end{bmatrix}$.

In this text, we focus on diagonalizable matrices and often dismiss others as rare aberrations. Some theorems will be proven in the diagonalizable case only, with the nondiagonalizable case being left as an exercise. Much attention is given to nondiagonalizable matrices in more advanced linear algebra courses.

EXAMPLE 9   Consider an $n \times n$ matrix A with complex eigenvalues $\lambda_1, \lambda_2, \ldots, \lambda_n$, listed with their algebraic multiplicities. What is the relationship between the λ_i and the determinant of A? (*Hint:* Evaluate the characteristic polynomial at $\lambda = 0$.)

Solution

$$f_A(\lambda) = \det(\lambda I_n - A) = (\lambda - \lambda_1)(\lambda - \lambda_2) \ldots (\lambda - \lambda_n),$$

$$f_A(0) = \det(-A) = (-1)^n \lambda_1 \lambda_2 \ldots \lambda_n.$$

Recall that $\det(-A) = (-1)^n \det(A)$, by linearity in the rows. We can conclude that

$$\det(A) = \lambda_1 \lambda_2 \ldots \lambda_n. \qquad \blacksquare$$

Can you interpret this result geometrically when A is a 3×3 matrix with a real eigenbasis? (*Hint:* Think about the expansion factor. See Exercise 18.)

In Example 9, we found that the determinant of a matrix is the product of its complex eigenvalues. Likewise, the trace is the *sum* of the eigenvalues. The verification is left as Exercise 35.

Fact 7.5.5

> **Trace, determinant, and eigenvalues**
> Consider an $n \times n$ matrix A with complex eigenvalues $\lambda_1, \lambda_2, \ldots, \lambda_n$, listed with their algebraic multiplicities. Then
>
> $$\text{tr}(A) = \lambda_1 + \lambda_2 + \cdots + \lambda_n,$$
>
> and
>
> $$\det(A) = \lambda_1 \lambda_2 \ldots \lambda_n.$$

Note that this result is obvious for a triangular matrix: In this case, the eigenvalues are the diagonal entries.

EXERCISES

GOALS   Use the basic properties of complex numbers. Write products and powers of complex numbers in polar form. Apply the fundamental theorem of algebra.

1. Write the complex number $z = 3 - 3i$ in polar form.

2. Find all complex numbers z such that $z^4 = 1$. Represent your answers graphically in the complex plane.

3. For an arbitrary positive integer n, find all complex numbers z such that $z^n = 1$ (in polar form). Represent your answers graphically.

4. Show that if z is a nonzero complex number, then there are exactly two complex numbers w such that $w^2 = z$. If z is in polar form, describe w in polar form.

5. Show that if z is a nonzero complex number, then there are exactly n complex numbers w such that $w^n = z$. If z is in polar form, write w in polar form. Represent the vectors w in the complex plane.

6. If z is a nonzero complex number in polar form, describe $1/z$ in polar form. What is the relationship between the complex conjugate $\bar{z}$ and $1/z$? Represent the numbers z, $\bar{z}$, and $1/z$ in the complex plane.

7. Describe the transformation $T(z) = (1 - i)z$ from $\mathbb{C}$ to $\mathbb{C}$ geometrically.

8. Use de Moivre's formula to express $\cos(3\phi)$ and $\sin(3\phi)$ in terms of $\cos\phi$ and $\sin\phi$.

9. Consider the complex number $z = 0.8 - 0.7i$. Represent the powers $z^2, z^3, \ldots$ in the complex plane and explain their long-term behavior.

10. Prove the fundamental theorem of algebra for cubic polynomials with real coefficients.

11. Express the polynomial $f(\lambda) = \lambda^3 - 3\lambda^2 + 7\lambda - 5$ as a product of linear factors over $\mathbb{C}$.

12. Consider a polynomial $f(\lambda)$ with real coefficients. Show that if a complex number λ_0 is a root of $f(\lambda)$, then so is its complex conjugate, $\bar{\lambda}_0$.

For Exercises 13 through 17 state whether the given set is a field (with the customary addition and multiplication).

13. The rational numbers $\mathbb{Q}$.

14. The integers $\mathbb{Z}$.

15. The binary digits (introduced in Exercises 3.1.53 and 3.1.54).

16. The rotation–dilation matrices of the form $\begin{bmatrix} p & -q \\ q & p \end{bmatrix}$, where p and q are real numbers.

17. The set H considered in Exercise 5.3.44.

18. Consider a real 2×2 matrix A with two distinct real eigenvalues, λ_1 and λ_2. Explain the formula $\det(A) = \lambda_1 \lambda_2$ geometrically, thinking of $|\det(A)|$ as an expansion factor. Illustrate your explanation with a sketch. Is there a similar geometric interpretation for a 3×3 matrix?

19. Consider a subspace V of $\mathbb{R}^n$, with $\dim(V) = m < n$.
    a. If the $n \times n$ matrix A represents the orthogonal projection onto V, what is $\text{tr}(A)$? What is $\det(A)$?
    b. If the $n \times n$ matrix B represents the reflection in V, what is $\text{tr}(B)$? What is $\det(B)$?

Find all complex eigenvalues of the matrices in Exercises 20 through 26 (including the real ones, of course). Do not use technology. Show all your work.

20. $\begin{bmatrix} 3 & -5 \\ 2 & -3 \end{bmatrix}$

21. $\begin{bmatrix} 11 & -15 \\ 6 & -7 \end{bmatrix}$

22. $\begin{bmatrix} 1 & 3 \\ -4 & 10 \end{bmatrix}$

23. $\begin{bmatrix} 0 & 0 & 1 \\ 1 & 0 & 0 \\ 0 & 1 & 0 \end{bmatrix}$

24. $\begin{bmatrix} 0 & 1 & 0 \\ 0 & 0 & 1 \\ 5 & -7 & 3 \end{bmatrix}$

25. $\begin{bmatrix} 0 & 0 & 0 & 1 \\ 1 & 0 & 0 & 0 \\ 0 & 1 & 0 & 0 \\ 0 & 0 & 1 & 0 \end{bmatrix}$

26. $\begin{bmatrix} 1 & -1 & 1 & -1 \\ 1 & 1 & 1 & 1 \\ 0 & 0 & 1 & 1 \\ 0 & 0 & 1 & 1 \end{bmatrix}$

27. Suppose a real 3×3 matrix A has only two distinct eigenvalues. Suppose that $\text{tr}(A) = 1$ and $\det(A) = 3$. Find the eigenvalues of A with their algebraic multiplicities.

28. Suppose a 3×3 matrix A has the real eigenvalue 2 and two complex conjugate eigenvalues. Also, suppose that $\det(A) = 50$ and $\text{tr}(A) = 8$. Find the complex eigenvalues.

29. Consider a matrix of the form
$$A = \begin{bmatrix} 0 & a & b \\ c & 0 & 0 \\ 0 & d & 0 \end{bmatrix},$$
where a, b, c, and d are positive real numbers. Suppose the matrix A has three distinct real eigenvalues. What can you say about the signs of the eigenvalues? (How many of them are positive, negative, zero?) Is the eigenvalue with the largest absolute value positive or negative?

30. A real $n \times n$ matrix A is called a regular transition matrix if all entries of A are positive, and the entries in each column add up to 1. (See Exercises 24 through 31 of Section 7.2.) An example is
$$A = \begin{bmatrix} 0.4 & 0.3 & 0.1 \\ 0.5 & 0.1 & 0.2 \\ 0.1 & 0.6 & 0.7 \end{bmatrix}.$$
You may take the following properties of a regular transition matrix for granted (a partial proof is outlined in Exercise 7.2.31):
- 1 is an eigenvalue of A, with $\dim(E_1) = 1$.
- If λ is a complex eigenvalue of A other than 1, then $|\lambda| < 1$.

    a. Consider a regular $n \times n$ transition matrix A and a vector $\vec{x}$ in $\mathbb{R}^n$ whose entries add up to 1. Show that the entries of $A\vec{x}$ will also add up to 1.
    b. Pick a regular transition matrix A, and compute some powers of A (using technology): $A^2, \ldots, A^{10}, \ldots, A^{100}, \ldots$. What do you observe? Explain your observation. Here, you may assume that there is a complex eigenbasis for A.

31. Form a 5×5 matrix by writing the integers 1, 2, 3, 4, 5 into each column in any order you like. An

example is

$$A = \begin{bmatrix} 5 & 1 & 2 & 2 & 1 \\ 1 & 4 & 3 & 3 & 2 \\ 3 & 3 & 5 & 4 & 3 \\ 2 & 5 & 4 & 1 & 4 \\ 4 & 2 & 1 & 5 & 5 \end{bmatrix}.$$

(Optional question for combinatorics aficionados: How many such matrices are there?) Take higher and higher powers of the matrix you have chosen (using technology), and compare the columns of the matrices you get. What do you observe? Explain the result. (Exercise 30 is helpful.)

32. Most long-distance telephone service in the United States is provided by three companies: AT&T, MCI, and Sprint. The three companies are in fierce competition, offering discounts or even cash to those who switch. If the figures advertised by the companies are to be believed, people are switching their long distance provider from one month to the next according to the following diagram:

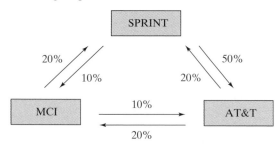

For example, 20% of the people who use AT&T go to Sprint one month later.

a.   We introduce the state vector

$$\vec{x}(t) = \begin{bmatrix} a(t) \\ m(t) \\ s(t) \end{bmatrix} \quad \begin{matrix} \text{fraction using AT\&T} \\ \text{fraction using MCI} \\ \text{fraction using Sprint} \end{matrix}$$

Find the matrix A such that $\vec{x}(t+1) = A\vec{x}(t)$, assuming that the customer base remains unchanged. Note that A is a regular transition matrix.

b.   Which fraction of the customers will be with each company in the long term? Do you have to know the current market shares to answer this question? Use the power method introduced in Exercise 30.

33. *The power method for finding eigenvalues.* Consider Exercises 30 and 31 for some background. Using technology, generate a random 5×5 matrix A with nonnegative entries. (Depending on the technology you are using, the entries could be integers between zero and nine, or numbers between zero and one.) Using

technology, compute $B = A^{20}$ (or another high power of A). We wish to compare the columns of B. This is hard to do by inspection, particularly because the entries of B may get rather large.

To get a better hold on B, form the diagonal 5×5 matrix D whose ith diagonal element is b_{1i}, the ith entry of the first row of B. Compute $C = BD^{-1}$.

a.   How is C obtained from B? Give your answer in terms of elementary row or column operations.

b.   Take a look at the columns of the matrix C you get. What do you observe? What does your answer tell you about the columns of $B = A^{20}$?

c.   Explain the observations you made in part (b). You may assume that A has five distinct (complex) eigenvalues and that the eigenvalue with maximal modulus is real and positive. (We cannot explain here why this will usually be the case.)

d.   Compute AC. What is the significance of the entries in the top row of this matrix in terms of the eigenvalues of A? What is the significance of the columns of C (or B) in terms of the eigenvectors of A?

34. Exercise 33 illustrates how you can use the powers of a matrix to find its dominant eigenvalue (i.e., the eigenvalue with maximal modulus), at least when this eigenvalue is real. But what about the other eigenvalues?

a.   Consider an $n \times n$ matrix A with n distinct complex eigenvalues $\lambda_1, \lambda_2, \ldots, \lambda_n$, where λ_1 is real. Suppose you have a good (real) approximation λ of λ_1 (good in that $|\lambda - \lambda_1| < |\lambda - \lambda_i|$, for $i = 2, \ldots, n$). Consider the matrix $\lambda I_n - A$. What are its eigenvalues? Which has the smallest modulus? Now consider the matrix $(\lambda I_n - A)^{-1}$. What are its eigenvalues? Which has the largest modulus? What is the relationship between the eigenvectors of A and $(\lambda I_n - A)^{-1}$? Consider higher and higher powers of $(\lambda I_n - A)^{-1}$. How does this help you to find an eigenvector of A with eigenvalue λ_1, and λ_1 itself? Use the results of Exercise 33.

b.   As an example of part (a), consider the matrix

$$A = \begin{bmatrix} 1 & 2 & 3 \\ 4 & 5 & 6 \\ 7 & 8 & 10 \end{bmatrix}.$$

We wish to find the eigenvectors and eigenvalues of A without using the corresponding commands on the computer (which is, after all, a "black box"). First, we find approximations for the eigenvalues by graphing the characteristic polynomial (use technology). Approximate the three real eigenvalues of A to the nearest integer. One of the three eigenvalues of A is negative. Find a good approximation for this eigenvalue and a corresponding eigenvector by using the procedure outlined in part (a). You are not asked to do the same for the two other eigenvalues.

35. Demonstrate the formula

$$\operatorname{tr}(A) = \lambda_1 + \lambda_2 + \cdots + \lambda_n,$$

where the λ_i are the complex eigenvalues of the matrix A, counted with their algebraic multiplicities. *Hint:* Consider the coefficient of λ^{n-1} in $f_A(\lambda) = (\lambda - \lambda_1)(\lambda - \lambda_2) \cdots (\lambda - \lambda_n)$, and compare the result with Fact 7.2.5.

36. In 1990, the population of the African country Benin was about 4.6 million people. Its composition by age was as follows:

Age Bracket	0–15	15–30	30–45	45–60	60–75	75–90
Percent of Population	46.6	25.7	14.7	8.4	3.8	0.8

We represent these data in a state vector whose components are the populations in the various age brackets, in millions:

$$\vec{x}(0) = 4.6 \begin{bmatrix} 0.466 \\ 0.257 \\ 0.147 \\ 0.084 \\ 0.038 \\ 0.008 \end{bmatrix} \approx \begin{bmatrix} 2.14 \\ 1.18 \\ 0.68 \\ 0.39 \\ 0.17 \\ 0.04 \end{bmatrix}.$$

We measure time in increments of 15 years, with $t = 0$ in 1990. For example, $\vec{x}(3)$ gives the age composition in the year 2035 ($1990 + 3 \cdot 15$). If current age-dependent birth and death rates are extrapolated, we have the following model:

$$\vec{x}(t+1) = \begin{bmatrix} 1.1 & 1.6 & 0.6 & 0 & 0 & 0 \\ 0.82 & 0 & 0 & 0 & 0 & 0 \\ 0 & 0.89 & 0 & 0 & 0 & 0 \\ 0 & 0 & 0.81 & 0 & 0 & 0 \\ 0 & 0 & 0 & 0.53 & 0 & 0 \\ 0 & 0 & 0 & 0 & 0.29 & 0 \end{bmatrix} \vec{x}(t) = A\vec{x}(t).$$

a. Explain the significance of all the entries in the matrix A in terms of population dynamics.

b. Find the eigenvalue of A with largest modulus and an associated eigenvector. (Use technology.) What is the significance of these quantities in terms of population dynamics? (For a summary on matrix techniques used in the study of age-structured populations, see Dmitrii O. Logofet, *Matrices and*

Graphs: Stability Problems in Mathematical Ecology, Chapters 2 and 3, CRC Press, 1993.)

37. Consider the set $\mathbb{H}$ of all complex 2×2 matrices of the form

$$A = \begin{bmatrix} w & -\bar{z} \\ z & \bar{w} \end{bmatrix},$$

where w and z are arbitrary complex numbers.

a. Show that $\mathbb{H}$ is closed under addition and multiplication. (That is, show that the sum and the product of two matrices in $\mathbb{H}$ are again in $\mathbb{H}$.)

b. Which matrices in $\mathbb{H}$ are invertible?

c. If a matrix in $\mathbb{H}$ is invertible, is the inverse in $\mathbb{H}$ as well?

d. Find two matrices A and B in $\mathbb{H}$ such that $AB \neq BA$.
   $\mathbb{H}$ is an example of a *skew field:* It satisfies all axioms for a field, except for the commutativity of multiplication. [The skew field $\mathbb{H}$ was introduced by the Irish mathematician Sir William Hamilton (1805–1865); its elements are called the *quaternions*. Another way to define the quaternions is discussed in Exercise 5.3.44.]

38. Consider the matrix

$$C_4 = \begin{bmatrix} 0 & 0 & 0 & 1 \\ 1 & 0 & 0 & 0 \\ 0 & 1 & 0 & 0 \\ 0 & 0 & 1 & 0 \end{bmatrix}.$$

a. Find the powers $C_4^2, C_4^3, C_4^4, \ldots$.

b. Find all complex eigenvalues of C_4, and construct a complex eigenbasis.

c. A 4×4 matrix M is called *circulant* if it is of the form

$$M = \begin{bmatrix} a & d & c & b \\ b & a & d & c \\ c & b & a & d \\ d & c & b & a \end{bmatrix}.$$

Circulant matrices play an important role in statistics. Show that any circulant 4×4 matrix M can be expressed as a linear combination of I_4, C_4, C_4^2, C_4^3. Use this representation to find an eigenbasis for M. What are the eigenvalues (in terms of a, b, c, d)?

39. Consider the $n \times n$ matrix C_n which has ones directly below the main diagonal and in the right upper corner, and zeros everywhere else. (See Exercise 38 for a discussion of C_4.)

a. Describe the powers of C_n.

b. Find all complex eigenvalues of C_n, and construct a complex eigenbasis.

c. Generalize part (c) of Exercise 38.

40. Consider a cubic equation

$$x^3 + px = q,$$

where $(p/3)^3 + (q/2)^2$ is negative. Show that this equation has three real solutions; write the solutions in the form $x_j = A\cos(\phi_j)$ for $j = 1, 2, 3$, expressing A and ϕ_j in terms of p and q. How many of the solutions are in the interval $(\sqrt{-p/3}, 2\sqrt{-p/3})$? Can there be solutions larger than $2\sqrt{-p/3}$? *Hint:* Cardano's formula derived in Exercise 7.2.38 is useful.

41. In his high school final examination (Aarau, Switzerland, 1896), young Albert Einstein (1879–1955) was given the following problem: In a triangle ABC, let P be the center of the inscribed circle. We are told that $\overline{AP} = 1$, $\overline{BP} = \frac{1}{2}$, and $\overline{CP} = \frac{1}{3}$. Find the radius ρ of the inscribed circle. Einstein worked through this problem as follows:

$$\sin\left(\frac{\alpha}{2}\right) = \rho,$$

$$\sin\left(\frac{\beta}{2}\right) = 2\rho,$$

$$\sin\left(\frac{\gamma}{2}\right) = 3\rho.$$

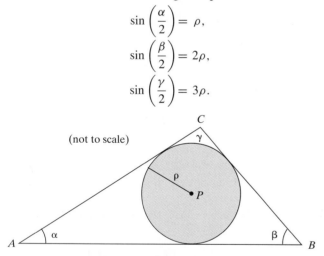

(not to scale)

For every triangle the following equation holds:

$$\sin^2\left(\frac{\alpha}{2}\right) + \sin^2\left(\frac{\beta}{2}\right) + \sin^2\left(\frac{\gamma}{2}\right)$$

$$+ 2\sin\left(\frac{\alpha}{2}\right)\sin\left(\frac{\beta}{2}\right)\sin\left(\frac{\gamma}{2}\right) = 1.$$

In our case:

$$14\rho^2 + 12\rho^3 - 1 = 0.$$

Now let

$$\rho = \frac{1}{x}.$$

At this point we interrupt Einstein's work and ask you to finish the job. *Hint:* Exercise 40 is helpful. Find the exact solution (in terms of trigonometric and inverse trigonometric functions), and give a numerical approximation as well. (By the way, Einstein, who was allowed to use a logarithm table, solved the problem correctly.) *Source: The Collected Papers of Albert Einstein,* Vol. 1, Princeton University Press, 1987.

42. Consider a complex $m \times n$ matrix A. The conjugate $\overline{A}$ is defined by taking the conjugate of each entry of A. For example, if

$$A = \begin{bmatrix} 2+3i & 5 \\ 2i & 9 \end{bmatrix}, \quad \text{then} \quad \overline{A} = \begin{bmatrix} 2-3i & 5 \\ -2i & 9 \end{bmatrix}.$$

a. Show that if A and B are complex $m \times n$ and $n \times p$ matrices, respectively, then

$$\overline{AB} = \overline{A}\,\overline{B}.$$

b. Let A be a real $n \times n$ matrix and $\vec{v} + i\vec{w}$ an eigenvector of A with eigenvalue $p + iq$. Show that the vector $\vec{v} - i\vec{w}$ is an eigenvector of A with eigenvalue $p - iq$.

43. Consider two real $n \times n$ matrices A and B that are "similar over $\mathbb{C}$": That is, there is a complex invertible $n \times n$ matrix S such that $B = S^{-1}AS$. Show that A and B are in fact "similar over $\mathbb{R}$": That is, there is a real R such that $B = R^{-1}AR$. *Hint:* Write $S = S_1 + iS_2$, where S_1 and S_2 are real. Consider the function $f(x) = \det(S_1 + xS_2)$, where x is a complex variable. Show that $f(x)$ is a nonzero polynomial. Conclude that there is a real number x_0 such that $f(x_0) \neq 0$. Show that $R = S_1 + x_0 S_2$ does the job.

44. Show that every complex 2×2 matrix is similar to an upper triangular 2×2 matrix. Can you generalize this result to square matrices of larger size? *Hint:* Argue by induction.

For which values of the real constant a are the matrices in Exercises 45 through 50 diagonalizable over $\mathbb{C}$?

45. $\begin{bmatrix} 1 & 1 \\ a & 1 \end{bmatrix}$ 46. $\begin{bmatrix} 0 & -a \\ a & 0 \end{bmatrix}$ 47. $\begin{bmatrix} 0 & 0 & 0 \\ 1 & 0 & a \\ 0 & 1 & 0 \end{bmatrix}$

48. $\begin{bmatrix} 0 & 0 & a \\ 1 & 0 & 3 \\ 0 & 1 & 0 \end{bmatrix}$ 49. $\begin{bmatrix} 0 & 1 & 0 \\ 0 & 0 & 1 \\ 0 & 1-a & a \end{bmatrix}$

50. $\begin{bmatrix} -a & a & -a \\ -a-1 & a+1 & -a-1 \\ 0 & 0 & 0 \end{bmatrix}$

7.6 STABILITY

In applications, the long-term behavior is often the most important qualitative feature of a dynamical system. We are frequently faced with the following situation: The state $\vec{0}$ represents an equilibrium of the system (in physics, ecology, or economics, for example). If the system is disturbed (moved into another state, away from the equilibrium $\vec{0}$) and then left to its own devices, will it always return to the equilibrium state $\vec{0}$?

EXAMPLE 1   Consider a dynamical system $\vec{x}(t+1) = A\vec{x}(t)$ where A is an $n \times n$ matrix. Suppose an initial state vector $\vec{x}_0$ is given. We are told that A has n distinct complex eigenvalues and that the modulus of each eigenvalue is less than 1. What can you say about the long-term behavior of the system, that is, about $\lim_{t \to \infty} \vec{x}(t)$?

Solution

For each complex eigenvalue λ_i, we can choose a complex eigenvector $\vec{v}_i$. Then the $\vec{v}_i$ form a complex eigenbasis for A (by Fact 7.3.6). We can write $\vec{x}_0$ as a complex linear combination of the $\vec{v}_i$:

$$\vec{x}_0 = c_1\vec{v}_1 + \cdots + c_n\vec{v}_n$$

Then

$$\vec{x}(t) = A^t\vec{x}_0 = c_1\lambda_1^t\vec{v}_1 + \cdots + c_n\lambda_n^t\vec{v}_n.$$

By Example 5 of Section 7.5,

$$\lim_{t \to \infty} \lambda_i^t = 0, \qquad \text{since} \quad |\lambda_i| < 1.$$

Therefore,

$$\lim_{t \to \infty} \vec{x}(t) = \vec{0}. \qquad \blacksquare$$

For the discussion of the long-term behavior of a dynamical system, the following definition is useful:

Definition 7.6.1

> **Stable equilibrium**
> Consider a dynamical system
>
> $$\vec{x}(t+1) = A\vec{x}(t).$$
>
> We say that $\vec{0}$ is an (asymptotically) *stable equilibrium* for this system if
>
> $$\lim_{t \to \infty} \vec{x}(t) = \vec{0}$$
>
> for all its trajectories.[12]

---

[12] In this text, "stable" will always mean asymptotically stable. Several other notions of stability are used in applied mathematics.

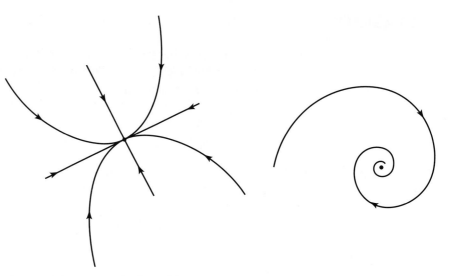

Figure 1(a) Asymptotically stable.

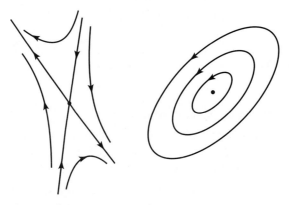

Figure 1(b) Not asymptotically stable.

Note that the zero state is stable if (and only if)

$$\lim_{t \to \infty} A^t = 0$$

(meaning that all entries of A^t approach zero). See Exercise 36.

Consider the examples shown in Figure 1.

Generalizing Example 1, we have the following result:

Fact 7.6.2   **Stability and eigenvalues**
Consider a dynamical system $\vec{x}(t + 1) = A\vec{x}(t)$. The zero state is asymptotically stable if (and only if) the modulus of all the complex eigenvalues of A is less than 1.

Example 1 illustrates this fact only when A is diagonalizable (i.e., when there is a complex eigenbasis for A); recall that this is the case for most matrices A.

In Exercises 37 through 42 of Section 8.1, we will discuss the nondiagonalizable case.

For an illustration of Fact 7.6.2, see Figure 11 of Section 7.1, where we sketched the phase portraits of 2×2 matrices with two distinct positive eigenvalues.

We will now turn our attention to the phase portraits of 2×2 matrices with complex eigenvalues $p \pm iq$ (where $q \neq 0$).

EXAMPLE 2   Consider the dynamical system

$$\vec{x}(t+1) = \begin{bmatrix} p & -q \\ q & p \end{bmatrix} \vec{x}(t),$$

where p and q are real, and q is nonzero. Examine the stability of this system. Sketch phase portraits. Discuss your results in terms of Fact 7.6.2.

Solution

As in Fact 2.2.3, we can write

$$\begin{bmatrix} p & -q \\ q & p \end{bmatrix} = r \begin{bmatrix} \cos(\phi) & -\sin(\phi) \\ \sin(\phi) & \cos(\phi) \end{bmatrix},$$

representing the transformation as a rotation through an angle ϕ followed by a dilation by $r = \sqrt{p^2 + q^2}$. Then

$$\vec{x}(t) = \begin{bmatrix} p & -q \\ q & p \end{bmatrix}^t \vec{x}_0 = r^t \begin{bmatrix} \cos(\phi t) & -\sin(\phi t) \\ \sin(\phi t) & \cos(\phi t) \end{bmatrix} \vec{x}_0,$$

representing a rotation through an angle ϕt followed be a dilation by r^t.

Figure 2 illustrates that the zero state is stable if $r = \sqrt{p^2 + q^2} < 1$.

Alternatively, we can use the Fact 7.6.2 to examine the stability of the system. From Example 6 of Section 7.5, we know that the eigenvalues of $\begin{bmatrix} p & -q \\ q & p \end{bmatrix}$ are $\lambda_{1,2} = p \pm iq$, with $|\lambda_1| = |\lambda_2| = \sqrt{p^2 + q^2}$. By Fact 7.6.2, the zero state is stable if $\sqrt{p^2 + q^2} < 1$. ■

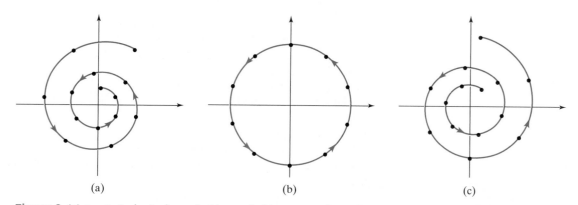

(a)   (b)   (c)

Figure 2 (a) $r < 1$: trajectories spiral inward. (b) $r = 1$: trajectories are circles. (c) $r > 1$: trajectories spiral outward.

Let us generalize Example 2. If A is any 2×2 matrix with eigenvalues $\lambda_{1,2} = p \pm iq$, what does the phase portrait of the dynamical system $\vec{x}(t+1) = A\vec{x}(t)$ look like? Let $\vec{v} + i\vec{w}$ be an eigenvector of A with eigenvalue $p + iq$. From Fact 7.5.3, we know that A is similar to the rotation–dilation matrix $\begin{bmatrix} p & -q \\ q & p \end{bmatrix}$, with

$$S^{-1}AS = \begin{bmatrix} p & -q \\ q & p \end{bmatrix} \quad \text{or} \quad A = S \begin{bmatrix} p & -q \\ q & p \end{bmatrix} S^{-1}, \quad \text{where} \quad S = \begin{bmatrix} \vec{w} & \vec{v} \end{bmatrix}.$$

Using the terminology introduced in Example 2, we find that

$$\vec{x}(t) = A^t\vec{x}_0 = S \begin{bmatrix} p & -q \\ q & p \end{bmatrix}^t S^{-1}\vec{x}_0 = r^t S \begin{bmatrix} \cos(\phi t) & -\sin(\phi t) \\ \sin(\phi t) & \cos(\phi t) \end{bmatrix} S^{-1}\vec{x}_0.$$

Fact 7.6.3

Dynamical systems with complex eigenvalues

Consider the dynamical system $\vec{x}(t+1) = A\vec{x}(t)$, where A is a real 2×2 matrix with eigenvalues

$$\lambda_{1,2} = p \pm iq = r\big(\cos(\phi) \pm i\sin(\phi)\big), \quad \text{where} \quad q \neq 0.$$

Let $\vec{v} + i\vec{w}$ be an eigenvector of A with eigenvalue $p + iq$.
Then

$$\vec{x}(t) = r^t S \begin{bmatrix} \cos(\phi t) & -\sin(\phi t) \\ \sin(\phi t) & \cos(\phi t) \end{bmatrix} S^{-1}\vec{x}_0, \quad \text{where} \quad S = \begin{bmatrix} \vec{w} & \vec{v} \end{bmatrix}.$$

Note that $S^{-1}\vec{x}_0$ is the coordinate vector of $\vec{x}_0$ with respect to basis $\vec{w}, \vec{v}$.

EXAMPLE 3   Consider the dynamical system

$$\vec{x}(t+1) = \begin{bmatrix} 3 & -5 \\ 1 & -1 \end{bmatrix} \vec{x}(t) \quad \text{with initial state} \quad \vec{x}_0 = \begin{bmatrix} 0 \\ 1 \end{bmatrix}.$$

Use Fact 7.6.3 to find a closed formula for $\vec{x}(t)$, and sketch the trajectory.

Solution

In Example 8 of Section 7.5, we found the eigenvalues

$$\lambda_{1,2} = 1 \pm i.$$

The polar coordinates of eigenvalue $1 + i$ are $r = \sqrt{2}$ and $\phi = \frac{\pi}{4}$. Furthermore, we found that

$$S = \begin{bmatrix} \vec{w} & \vec{v} \end{bmatrix} = \begin{bmatrix} 0 & -5 \\ 1 & -2 \end{bmatrix}.$$

Since

$$S^{-1}\vec{x}_0 = \begin{bmatrix} 1 \\ 0 \end{bmatrix},$$

Fact 7.6.3 gives

$$\vec{x}(t) = (\sqrt{2})^t \begin{bmatrix} 0 & -5 \\ 1 & -2 \end{bmatrix} \begin{bmatrix} \cos(\frac{\pi}{4}t) & -\sin(\frac{\pi}{4}t) \\ \sin(\frac{\pi}{4}t) & \cos(\frac{\pi}{4}t) \end{bmatrix} \begin{bmatrix} 1 \\ 0 \end{bmatrix}$$

$$= (\sqrt{2})^t \begin{bmatrix} -5\sin(\frac{\pi}{4}t) \\ \cos(\frac{\pi}{4}t) - 2\sin(\frac{\pi}{4}t) \end{bmatrix}.$$

We leave it to the reader to work out the details of this computation.

Next, let's think about the trajectory. We will develop the trajectory step by step:

- The points

$$\begin{bmatrix} \cos(\frac{\pi}{4}t) & -\sin(\frac{\pi}{4}t) \\ \sin(\frac{\pi}{4}t) & \cos(\frac{\pi}{4}t) \end{bmatrix} \begin{bmatrix} 1 \\ 0 \end{bmatrix} \quad \text{(for } t = 0, 1, 2, \ldots)$$

are located on the unit circle, as shown in Figure 3(a). Note that at $t = 8$ the system returns to its initial position, $\begin{bmatrix} 1 \\ 0 \end{bmatrix}$; the period of this system is 8.

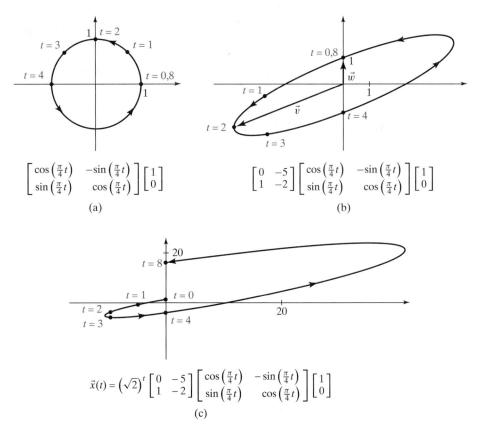

Figure 3

- In Exercise 2.2.50, we saw that an invertible linear transformation maps the unit circle into an ellipse. Thus, the points

$$\begin{bmatrix} 0 & -5 \\ 1 & -2 \end{bmatrix} \begin{bmatrix} \cos(\frac{\pi}{4}t) & -\sin(\frac{\pi}{4}t) \\ \sin(\frac{\pi}{4}t) & \cos(\frac{\pi}{4}t) \end{bmatrix} \begin{bmatrix} 1 \\ 0 \end{bmatrix}$$

are located on an ellipse, as shown in Figure 3(b). The two column vectors of

$$S = \begin{bmatrix} 0 & -5 \\ 1 & -2 \end{bmatrix} = \begin{bmatrix} \vec{w} & \vec{v} \end{bmatrix}$$

are shown in that figure as well. Again, the period of this system is 8.
- The exponential growth factor $(\sqrt{2})^t$ will produce longer and longer vectors

$$\vec{x}(t) = (\sqrt{2})^t \begin{bmatrix} 0 & -5 \\ 1 & -2 \end{bmatrix} \begin{bmatrix} \cos(\frac{\pi}{4}t) & -\sin(\frac{\pi}{4}t) \\ \sin(\frac{\pi}{4}t) & \cos(\frac{\pi}{4}t) \end{bmatrix} \begin{bmatrix} 1 \\ 0 \end{bmatrix}.$$

Thus, the trajectory *spirals outward,* as shown in Figure 3(c). (We are using different scales in Figures 3(a), (b), and (c).) Note that $\vec{x}(8) = (\sqrt{2})^8 \vec{x}(0) = 16\vec{x}(0)$. ∎

We can generalize our findings in Example 3.

Fact 7.6.4

Phase portrait of a system with complex eigenvalues
Consider a dynamical system

$$\vec{x}(t+1) = A\vec{x}(t),$$

where A is a real 2×2 matrix with eigenvalues $\lambda_{1,2} = p \pm iq$. Let

$$r = |\lambda_1| = |\lambda_2| = \sqrt{p^2 + q^2}.$$

If $r = 1$, then the points $\vec{x}(t)$ are located on an ellipse; if r exceeds 1, then the trajectory spirals outward; and if r is less than 1, then the trajectory spirals inward, approaching the origin.

Fact 7.6.4 provides another illustration of Fact 7.6.2: The zero state is stable if (and only if) $r = |\lambda_1| = |\lambda_2| < 1$.

If you have to sketch a trajectory of a system with complex eigenvalues without the aid of technology, it helps to compute and plot the first few points $\vec{x}(0)$, $\vec{x}(1)$, $\vec{x}(2), \ldots$, until you see a trend.

EXERCISES

GOALS Use eigenvalues to determine the stability of a dynamical system. Analyze the dynamical system $\vec{x}(t+1) = A\vec{x}(t)$, where A is a real 2×2 matrix with eigenvalues $p \pm iq$.

For the matrices A in Exercises 1 through 10, determine whether the zero state is a stable equilibrium of the dynamical

system $\vec{x}(t+1) = A\vec{x}(t)$.

1. $A = \begin{bmatrix} 0.9 & 0 \\ 0 & 0.8 \end{bmatrix}$

2. $A = \begin{bmatrix} -1.1 & 0 \\ 0 & 0.9 \end{bmatrix}$

3. $A = \begin{bmatrix} 0.8 & 0.7 \\ -0.7 & 0.8 \end{bmatrix}$ **4.** $A = \begin{bmatrix} -0.9 & -0.4 \\ 0.4 & -0.9 \end{bmatrix}$

5. $A = \begin{bmatrix} 0.5 & 0.6 \\ -0.3 & 1.4 \end{bmatrix}$ **6.** $A = \begin{bmatrix} -1 & 3 \\ -1.2 & 2.6 \end{bmatrix}$

7. $A = \begin{bmatrix} 2.4 & -2.5 \\ 1 & -0.6 \end{bmatrix}$ **8.** $A = \begin{bmatrix} 1 & -0.2 \\ 0.1 & 0.7 \end{bmatrix}$

9. $A = \begin{bmatrix} 0.8 & 0 & -0.6 \\ 0 & 0.7 & 0 \\ 0.6 & 0 & 0.8 \end{bmatrix}$

10. $A = \begin{bmatrix} 0.3 & 0.3 & 0.3 \\ 0.3 & 0.3 & 0.3 \\ 0.3 & 0.3 & 0.3 \end{bmatrix}$

Consider the matrices A in Exercises 11 through 16. For which real numbers k is the zero state a stable equilibrium of the dynamical system $\vec{x}(t + 1) = A\vec{x}(t)$?

11. $A = \begin{bmatrix} k & 0 \\ 0 & 0.9 \end{bmatrix}$ **12.** $A = \begin{bmatrix} 0.6 & k \\ -k & 0.6 \end{bmatrix}$

13. $A = \begin{bmatrix} 0.7 & k \\ 0 & -0.9 \end{bmatrix}$ **14.** $A = \begin{bmatrix} k & k \\ k & k \end{bmatrix}$

15. $A = \begin{bmatrix} 1 & k \\ 0.01 & 1 \end{bmatrix}$ **16.** $A = \begin{bmatrix} 0.1 & k \\ 0.3 & 0.3 \end{bmatrix}$

For the matrices A in Exercises 17 through 24, find real closed formulas for the trajectory $\vec{x}(t + 1) = A\vec{x}(t)$, where $\vec{x}(0) = \begin{bmatrix} 0 \\ 1 \end{bmatrix}$. Draw a rough sketch.

17. $A = \begin{bmatrix} 0.6 & -0.8 \\ 0.8 & 0.6 \end{bmatrix}$ **18.** $A = \begin{bmatrix} -0.8 & 0.6 \\ -0.8 & -0.8 \end{bmatrix}$

19. $A = \begin{bmatrix} 2 & -3 \\ 3 & 2 \end{bmatrix}$ **20.** $A = \begin{bmatrix} 4 & 3 \\ -3 & 4 \end{bmatrix}$

21. $A = \begin{bmatrix} 1 & 5 \\ -2 & 7 \end{bmatrix}$ **22.** $A = \begin{bmatrix} 7 & -15 \\ 6 & -11 \end{bmatrix}$

23. $A = \begin{bmatrix} -0.5 & 1.5 \\ -0.6 & 1.3 \end{bmatrix}$ **24.** $A = \begin{bmatrix} 1 & -3 \\ 1.2 & -2.6 \end{bmatrix}$

Consider an invertible $n \times n$ matrix A such that the zero state is a stable equilibrium of the dynamical system $\vec{x}(t + 1) = A\vec{x}(t)$. What can you say about the stability of the systems listed in Exercises 25 through 30?

25. $\vec{x}(t + 1) = A^{-1}\vec{x}(t)$ **26.** $\vec{x}(t + 1) = A^T\vec{x}(t)$

27. $\vec{x}(t + 1) = -A\vec{x}(t)$ **28.** $\vec{x}(t + 1) = (A - 2I_n)\vec{x}(t)$

29. $\vec{x}(t + 1) = (A + I_n)\vec{x}(t)$ **30.** $\vec{x}(t + 1) = A^2\vec{x}(t)$

31. Let A be a real 2×2 matrix. Show that the zero state is a stable equilibrium of the dynamical system $\vec{x}(t + 1) = A\vec{x}(t)$ if (and only if)

$$|\mathrm{tr}(A)| - 1 < \det(A) < 1.$$

32. Let's revisit the introductory example of Section 7.5: The glucose regulatory system of a certain patient can be modeled by the equations

$$g(t + 1) = 0.9g(t) - 0.4h(t)$$
$$h(t + 1) = 0.1g(t) + 0.9h(t).$$

Find closed formulas for $g(t)$ and $h(t)$, and draw the trajectory. Does your trajectory look like the one on page 340?

33. Consider a real 2×2 matrix A with eigenvalues $p \pm iq$ and corresponding eigenvectors $\vec{v} \pm i\vec{w}$. Show that if a real vector $\vec{x}_0$ is written as $\vec{x}_0 = c_1(\vec{v} + i\vec{w}) + c_2(\vec{v} - i\vec{w})$, then $c_2 = \bar{c}_1$.

34. Consider a dynamical system $\vec{x}(t + 1) = A\vec{x}(t)$, where A is a real $n \times n$ matrix.

a. If $|\det(A)| \geq 1$, what can you say about the stability of the zero state?

b. If $|\det(A)| < 1$, what can you say about the stability of the zero state?

35. a. Consider a real $n \times n$ matrix with n distinct real eigenvalues $\lambda_1, \ldots, \lambda_n$, where $|\lambda_i| \leq 1$ for all $i = 1, \ldots, n$. Let $\vec{x}(t)$ be a trajectory of the dynamical system $\vec{x}(t + 1) = A\vec{x}(t)$. Show that this trajectory is *bounded;* that is, there is a positive number M such that $\|\vec{x}(t)\| \leq M$ for all positive integers t.

b. Are all trajectories of the dynamical system

$$\vec{x}(t + 1) = \begin{bmatrix} 1 & 1 \\ 0 & 1 \end{bmatrix} \vec{x}(t)$$

bounded? Explain.

36. Show that the zero state is a stable equilibrium of the dynamical system $\vec{x}(t + 1) = A\vec{x}(t)$ if (and only if)

$$\lim_{t \to \infty} A^t = 0$$

(meaning that all entries of A^t approach zero).

37. Consider the national income of a country, which consists of consumption, investment, and government expenditures. Here we assume the government expenditure to be constant, at G_0, while the national income $Y(t)$, consumption $C(t)$, and investment $I(t)$ change over time. According to a simple model, we have

$$\begin{vmatrix} Y(t) & = C(t) + I(t) + G_0 \\ C(t+1) & = \gamma Y(t) \\ I(t+1) & = \alpha\big(C(t+1) - C(t)\big) \end{vmatrix} \quad \begin{matrix} (0 < \gamma < 1), \\ (\alpha > 0) \end{matrix}$$

where γ is the marginal propensity to consume and α is the acceleration coefficient. (See Paul E. Samuelson, "Interactions between the Multiplier Analysis and the Principle of Acceleration," *Review of Economic Statistics,* May 1939, pp. 75–78.)

a. Find the equilibrium solution of these equations, when $Y(t+1) = Y(t)$, $C(t+1) = C(t)$, and $I(t+1) = I(t)$.

b. Let $y(t)$, $c(t)$, and $i(t)$ be the deviations of $Y(t)$, $C(t)$, and $I(t)$, respectively, from the equilibrium state you found in part (a). These quantities are related by the equations

$$\begin{vmatrix} y(t) & = c(t) + i(t) \\ c(t+1) & = \gamma y(t) \\ i(t+1) & = \alpha\big(c(t+1) - c(t)\big) \end{vmatrix}.$$

(Verify this!) By substituting $y(t)$ into the second equation, set up equations of the form

$$\begin{vmatrix} c(t+1) & = p\,c(t) + q\,i(t) \\ i(t+1) & = r\,c(t) + s\,i(t) \end{vmatrix}.$$

c. When $\alpha = 5$ and $\gamma = 0.2$, determine the stability of the zero state of this system.

d. When $\alpha = 1$ (and γ is arbitrary, $0 < \gamma < 1$), determine the stability of the zero state.

e. For each of the four sectors in the α–γ-plane, determine the stability of the zero state.

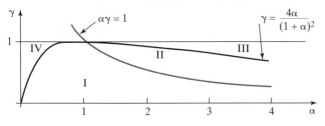

Discuss the various cases, in practical terms.

38. Consider an affine transformation

$$T(\vec{x}) = A\vec{x} + \vec{b},$$

where A is an $n \times n$ matrix and $\vec{b}$ is a vector in $\mathbb{R}^n$. (Compare this with Exercise 7.3.45.) Suppose that 1 is

not an eigenvalue of A.

a. Find the vector $\vec{v}$ in $\mathbb{R}^n$ such that $T(\vec{v}) = \vec{v}$; this vector is called the *equilibrium state* of the dynamical system $\vec{x}(t+1) = T\big(\vec{x}(t)\big)$.

b. When is the equilibrium $\vec{v}$ in part (a) stable (meaning that $\lim_{t \to \infty} \vec{x}(t) = \vec{v}$ for all trajectories)?

39. Consider the dynamical system

$$x_1(t+1) = 0.1x_1(t) + 0.2x_2(t) + 1,$$

$$x_2(t+1) = 0.4x_1(t) + 0.3x_2(t) + 2.$$

(See Exercise 7.3.45.) Find the equilibrium state of this system and determine its stability. (See Exercise 38.) Sketch a phase portrait.

40. Consider the matrix

$$A = \begin{bmatrix} p & -q & -r & -s \\ q & p & s & -r \\ r & -s & p & q \\ s & r & -q & p \end{bmatrix},$$

where p, q, r, s are arbitrary real numbers. (Compare this with Exercise 5.3.44.)

a. Compute $A^T A$.

b. For which choices of p, q, r, s is A invertible? Find the inverse if it exists.

c. Find the determinant of A.

d. Find the complex eigenvalues of A.

e. If $\vec{x}$ is a vector in $\mathbb{R}^4$, what is the relationship between $\|\vec{x}\|$ and $\|A\vec{x}\|$?

f. Consider the numbers

$$59 = 3^2 + 3^2 + 4^2 + 5^2$$

and

$$37 = 1^2 + 2^2 + 4^2 + 4^2.$$

Express the number

$$2183$$

as the sum of the squares of four integers:

$$2183 = a^2 + b^2 + c^2 + d^2.$$

Hint: Part (e) is useful. Note that $2183 = 59 \cdot 37$.

g. The French mathematician Joseph-Louis Lagrange (1736–1813) showed that any prime number can be expressed as the sum of the squares of four integers. Using this fact and your work in part (f) as a guide, show that any positive integer can be expressed in this way.

41. Find a 2×2 matrix A without real eigenvalues and a vector $\vec{x}_0$ in $\mathbb{R}^2$ such that for all positive integers t the point $A^t \vec{x}_0$ is located on the ellipse in the accompanying sketch.

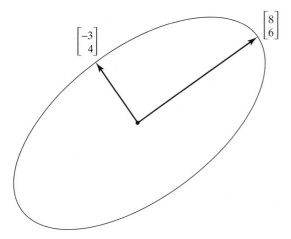

42. We quote from a text on computer graphics (M. Beeler et al., "HAKMEM," MIT Artificial Intelligence Report AIM-239, 1972):

Here is an elegant way to draw almost circles on a point-plotting display.

```
CIRCLE ALGORITHM:
NEW X = OLD X − K * OLD Y;
NEW Y = OLD Y + K * NEW X.
```

This makes a very round ellipse centered at the origin with its size determined by the initial point. The circle algorithm was invented by mistake when I tried to save a register in a display hack!

(In the preceding formula, k is a small number.) Here, a dynamical system is defined in "computer lingo." In our terminology, the formulas are

$$x(t + 1) = x(t) - k \, y(t),$$

$$y(t + 1) = y(t) + k \, x(t + 1).$$

a. Find the matrix of this transformation. (Note the entry $x(t + 1)$ in the second formula.)

b. Explain why the trajectories are ellipses, as claimed.

Chapter 7

TRUE OR FALSE?

1. The eigenvalues of any triangular matrix are its diagonal entries.

2. The trace of any square matrix is the sum of its diagonal entries.

3. The algebraic multiplicity of an eigenvalue cannot exceed its geometric multiplicity.

4. If an $n \times n$ matrix A is diagonalizable (over $\mathbb{R}$), then there must be a basis of $\mathbb{R}^n$ consisting of eigenvectors of A.

5. If the standard vectors $\vec{e}_1, \vec{e}_2, \ldots, \vec{e}_n$ are eigenvectors of an $n \times n$ matrix A, then A must be diagonal.

6. If $\vec{v}$ is an eigenvector of A, then $\vec{v}$ must be an eigenvector of A^3 as well.

7. There is a diagonalizable 5×5 matrix with only two distinct eigenvalues (over $\mathbb{C}$).

8. There is a real 5×5 matrix without any real eigenvalues.

9. If 0 is an eigenvalue of a matrix A, then $\det(A) = 0$.

10. The eigenvalues of a 2×2 matrix A are the solutions of the equation $\lambda^2 - (\text{tr } A)\lambda + (\det A) = 0$.

11. If 1 is the only eigenvalue of an $n \times n$ matrix A, then A must be I_n.

12. If A and B are $n \times n$ matrices, if α is an eigenvalue of A, and if β is an eigenvalue of B, then $\alpha\beta$ must be an eigenvalue of AB.

13. If 3 is an eigenvalue of an $n \times n$ matrix A, then 9 must be an eigenvalue of A^2.

14. The matrix of any orthogonal projection onto a subspace V of $\mathbb{R}^n$ is diagonalizable.

15. If matrices A and B have the same eigenvalues (over $\mathbb{C}$), with the same algebraic multiplicities, then matrices A and B must have the same trace.

16. If a real matrix A has only the eigenvalues 1 and -1, then A must be orthogonal.

17. Any real rotation–dilation matrix is diagonalizable over $\mathbb{C}$.

18. If A is a noninvertible $n \times n$ matrix, then the geometric multiplicity of eigenvalue 0 is $n - \text{rank}(A)$.

19. If matrix A is diagonalizable, then its transpose A^T must be diagonalizable as well.

20. If A and B are two 3×3 matrices such that $\text{tr}(A) = \text{tr}(B)$ and $\det(A) = \det(B)$, then A and B must have the same eigenvalues.

21. If matrix A^2 is diagonalizable, then matrix A must be diagonalizable as well.

22. The determinant of a matrix is the product of its eigenvalues (over $\mathbb{C}$), counted with their algebraic multiplicities.

23. All lower triangular matrices are diagonalizable (over $\mathbb{C}$).

24. If two $n \times n$ matrices A and B are diagonalizable, then AB must be diagonalizable as well.

25. If an invertible matrix A is diagonalizable, then A^{-1} must be diagonalizable as well.

26. If $\det(A) = \det(A^T)$, then matrix A must be symmetric.

27. If matrix $A = \begin{bmatrix} 7 & a & b \\ 0 & 7 & c \\ 0 & 0 & 7 \end{bmatrix}$ is diagonalizable, then $a, b,$ and c must all be zero.

28. If two $n \times n$ matrices A and B are diagonalizable, then $A + B$ must be diagonalizable as well.

29. All diagonalizable matrices are invertible.

30. If vector $\vec{v}$ is an eigenvector of both A and B, then $\vec{v}$ must be an eigenvector of $A + B$.

31. If an $n \times n$ matrix A is diagonalizable, then A must have n distinct eigenvalues.

32. If two 3×3 matrices A and B both have the eigenvalues 1, 2, and 3, then A must be similar to B.

33. If $\vec{v}$ is an eigenvector of A, then $\vec{v}$ must be an eigenvector of A^T as well.

34. All invertible matrices are diagonalizable.

35. If $\vec{v}$ and $\vec{w}$ are linearly independent eigenvectors of matrix A, then $\vec{v} + \vec{w}$ must be an eigenvector of A as well.

36. If a 2×2 matrix R represents a reflection in a line L, then R must be diagonalizable.

37. If A is a 2×2 matrix such that $\operatorname{tr}(A) = 1$ and $\det(A) = -6$, then A must be diagonalizable.

38. If a matrix is diagonalizable, then the algebraic multiplicity of each of its eigenvalues λ must equal the geometric multiplicity of λ.

39. If $\vec{u}, \vec{v}, \vec{w}$ are eigenvectors of a 4×4 matrix A, with associated eigenvalues 3, 7, and 11, respectively, then vectors $\vec{u}, \vec{v}, \vec{w}$ must be linearly independent.

40. If a 4×4 matrix A is diagonalizable, then the matrix $A + 4I_4$ must be diagonalizable as well.

41. All orthogonal matrices are diagonalizable (over $\mathbb{R}$).

42. If A is an $n \times n$ matrix and λ is an eigenvalue of the partitioned matrix $M = \begin{bmatrix} A & A \\ 0 & A \end{bmatrix}$, then λ must be an eigenvalue of matrix A.

43. If two matrices A and B have the same characteristic polynomials, then they must be similar.

44. If A is a diagonalizable 4×4 matrix with $A^4 = 0$, then A must be the zero matrix.

45. If an $n \times n$ matrix A is diagonalizable (over $\mathbb{R}$), then every vector $\vec{v}$ in $\mathbb{R}^n$ can be expressed as a sum of eigenvectors of A.

46. If vector $\vec{v}$ is an eigenvector of both A and B, then $\vec{v}$ is an eigenvector of AB.

47. Similar matrices have the same characteristic polynomials.

48. If a matrix A has k distinct eigenvalues, then $\operatorname{rank}(A) \geq k$.

49. If the rank of a square matrix A is 1, then all the nonzero vectors in the image of A are eigenvectors of A.

50. If the rank of an $n \times n$ matrix A is 1, then A must be diagonalizable.

51. If A is a 4×4 matrix with $A^4 = 0$, then 0 is the only eigenvalue of A.

52. If two $n \times n$ matrices A and B are both diagonalizable, then they must commute.

53. If $\vec{v}$ is an eigenvector of A, then $\vec{v}$ must be in the kernel of A or in the image of A.

54. All symmetric 2×2 matrices are diagonalizable (over $\mathbb{R}$).

55. If A is a 2×2 matrix with eigenvalues 3 and 4 and if $\vec{u}$ is a unit eigenvector in $\mathbb{R}^2$, then the length of vector $A\vec{u}$ cannot exceed 4.

56. If $\vec{u}$ is a nonzero vector in $\mathbb{R}^n$, then $\vec{u}$ must be an eigenvector of matrix $\vec{u}\vec{u}^T$.

57. If $\vec{v}_1, \vec{v}_2, \ldots, \vec{v}_n$ is an eigenbasis of both A and B, then matrices A and B must commute.

58. If $\vec{v}$ is an eigenvector of a 2×2 matrix $A = \begin{bmatrix} a & b \\ c & d \end{bmatrix}$, then $\vec{v}$ must be an eigenvector of its classical adjoint $adj(A) = \begin{bmatrix} d & -b \\ -c & a \end{bmatrix}$ as well.

C H A P T E R

8

Symmetric Matrices and Quadratic Forms

8.1 SYMMETRIC MATRICES

In this chapter we will work with real numbers throughout, except for a brief digression into $\mathbb{C}$ in the discussion of Fact 8.1.3.

Our work in the last chapter dealt with the following central question:

When is a given square matrix A *diagonalizable?* That is, when is there an *eigenbasis* for A?

In geometry, we prefer to work with *orthonormal* bases, which raises the question:

For which matrices is there an *orthonormal* eigenbasis? Or, equivalently, for which matrices A is there an *orthogonal* matrix S such that $S^{-1}AS = S^T AS$ is diagonal?

(Recall that $S^{-1} = S^T$ for orthogonal matrices, by Fact 5.3.7.) We say that A is *orthogonally diagonalizable* if there is an orthogonal S such that $S^{-1}AS = S^T AS$ is diagonal. Then, the question is:

Which matrices are orthogonally diagonalizable?

Simple examples of orthogonally diagonalizable matrices are diagonal matrices (we can let $S = I_n$) and the matrices of orthogonal projections and reflections.

365

EXAMPLE 1   If A is orthogonally diagonalizable, what is the relationship between A^T and A?

Solution

We have

$$S^{-1}AS = D \quad \text{or} \quad A = SDS^{-1} = SDS^T,$$

for an orthogonal S and a diagonal D. Then

$$A^T = (SDS^T)^T = SD^T S^T = SDS^T = A.$$

We find that A is symmetric:

$$A^T = A. \qquad \blacksquare$$

Surprisingly, the converse is true as well:

Fact 8.1.1   **Spectral theorem**
A matrix A is *orthogonally diagonalizable* (i.e., there is an orthogonal S such that $S^{-1}AS = S^T AS$ is diagonal) if and only if A is *symmetric* (i.e., $A^T = A$).

We will prove this theorem later in this section, based on two preliminary results, Facts 8.1.2 and 8.1.3. First, we will illustrate the spectral theorem with an example.

EXAMPLE 2   For the symmetric matrix $A = \begin{bmatrix} 4 & 2 \\ 2 & 7 \end{bmatrix}$, find an orthogonal S such that $S^{-1}AS$ is diagonal.

Solution

We will first find an eigenbasis. The eigenvalues of A are 3 and 8, with corresponding eigenvectors $\begin{bmatrix} 2 \\ -1 \end{bmatrix}$ and $\begin{bmatrix} 1 \\ 2 \end{bmatrix}$, respectively. (See Figure 1.)

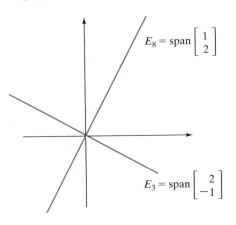

Figure 1

Note that the two eigenspaces, E_3 and E_8, are perpendicular. (This is no coincidence, as we will see in Fact 8.1.2.) Therefore, we can find an orthonormal eigenbasis simply by dividing the given eigenvectors by their lengths:

$$\vec{v}_1 = \frac{1}{\sqrt{5}} \begin{bmatrix} 2 \\ -1 \end{bmatrix}, \qquad \vec{v}_2 = \frac{1}{\sqrt{5}} \begin{bmatrix} 1 \\ 2 \end{bmatrix}.$$

If we define the orthogonal matrix

$$S = \begin{bmatrix} | & | \\ \vec{v}_1 & \vec{v}_2 \\ | & | \end{bmatrix} = \frac{1}{\sqrt{5}} \begin{bmatrix} 2 & 1 \\ -1 & 2 \end{bmatrix},$$

then $S^{-1}AS$ will be diagonal, namely, $S^{-1}AS = \begin{bmatrix} 3 & 0 \\ 0 & 8 \end{bmatrix}$. ■

The key observation we made in Example 2 generalizes as follows:

Fact 8.1.2   Consider a symmetric matrix A. If $\vec{v}_1$ and $\vec{v}_2$ are eigenvectors of A with *distinct* eigenvalues λ_1 and λ_2, then $\vec{v}_1 \cdot \vec{v}_2 = 0$; that is, $\vec{v}_2$ is orthogonal to $\vec{v}_1$.

Proof   We compute the product

$$\vec{v}_1^T A \vec{v}_2$$

in two different ways:

$$\vec{v}_1^T A \vec{v}_2 = \vec{v}_1^T (\lambda_2 \vec{v}_2) = \lambda_2 (\vec{v}_1 \cdot \vec{v}_2)$$
$$\vec{v}_1^T A \vec{v}_2 = \vec{v}_1^T A^T \vec{v}_2 = (A\vec{v}_1)^T \vec{v}_2 = (\lambda_1 \vec{v}_1)^T \vec{v}_2 = \lambda_1 (\vec{v}_1 \cdot \vec{v}_2)$$

Comparing the results, we find

$$\lambda_1 (\vec{v}_1 \cdot \vec{v}_2) = \lambda_2 (\vec{v}_1 \cdot \vec{v}_2),$$

or

$$(\lambda_1 - \lambda_2)(\vec{v}_1 \cdot \vec{v}_2) = 0.$$

Since the first factor in this product, $\lambda_1 - \lambda_2$, is nonzero, the second factor, $\vec{v}_1 \cdot \vec{v}_2$, must be zero, as claimed. ▲

Fact 8.1.2 tells us that the eigenspaces of a symmetric matrix are perpendicular to one another. Here is another illustration of this property:

EXAMPLE 3   For the symmetric matrix

$$A = \begin{bmatrix} 1 & 1 & 1 \\ 1 & 1 & 1 \\ 1 & 1 & 1 \end{bmatrix},$$

find an orthogonal S such that $S^{-1}AS$ is diagonal.

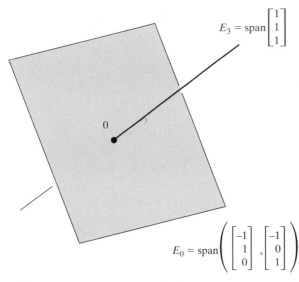

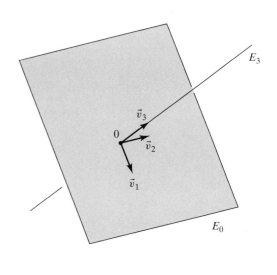

Figure 2 The eigenspaces E_0 and E_3 are orthogonal complements.

Figure 3

Solution

The eigenvalues are 0 and 3, with

$$E_0 = \text{span} \left(\begin{bmatrix} -1 \\ 1 \\ 0 \end{bmatrix}, \begin{bmatrix} -1 \\ 0 \\ 1 \end{bmatrix} \right) \qquad \text{and} \qquad E_3 = \text{span} \begin{bmatrix} 1 \\ 1 \\ 1 \end{bmatrix}.$$

Note that the two eigenspaces are indeed perpendicular to one another, in accordance with Fact 8.1.2. (See Figure 2.)

We can construct an orthonormal eigenbasis for A by picking an orthonormal basis of each eigenspace (using the Gram–Schmidt process in the case of E_0). See Figure 3.

In Figure 3, the vectors $\vec{v}_1$, $\vec{v}_2$ form an orthonormal basis of E_0, and $\vec{v}_3$ is a unit vector in E_3. Then, $\vec{v}_1$, $\vec{v}_2$, $\vec{v}_3$ is an orthonormal eigenbasis for A. We can let $S = [\vec{v}_1 \quad \vec{v}_2 \quad \vec{v}_3]$ to diagonalize A orthogonally.

If we apply the Gram–Schmidt[1] process to the vectors

$$\begin{bmatrix} -1 \\ 1 \\ 0 \end{bmatrix}, \qquad \begin{bmatrix} -1 \\ 0 \\ 1 \end{bmatrix}$$

spanning E_0, we find

$$\vec{v}_1 = \frac{1}{\sqrt{2}} \begin{bmatrix} -1 \\ 1 \\ 0 \end{bmatrix} \qquad \text{and} \qquad \vec{v}_2 = \frac{1}{\sqrt{6}} \begin{bmatrix} -1 \\ -1 \\ 2 \end{bmatrix}.$$

---

[1] Alternatively, we could find a unit vector $\vec{v}_1$ in E_0 and a unit vector $\vec{v}_3$ in E_3, and then let $\vec{v}_2 = \vec{v}_3 \times \vec{v}_1$.

The computations are left as an exercise. For E_3, we get

$$\vec{v}_3 = \frac{1}{\sqrt{3}} \begin{bmatrix} 1 \\ 1 \\ 1 \end{bmatrix}.$$

Therefore, the orthogonal matrix

$$S = \begin{bmatrix} | & | & | \\ \vec{v}_1 & \vec{v}_2 & \vec{v}_3 \\ | & | & | \end{bmatrix} = \begin{bmatrix} -1/\sqrt{2} & -1/\sqrt{6} & 1/\sqrt{3} \\ 1/\sqrt{2} & -1/\sqrt{6} & 1/\sqrt{3} \\ 0 & 2/\sqrt{6} & 1/\sqrt{3} \end{bmatrix}$$

diagonalizes the matrix A:

$$S^{-1}AS = \begin{bmatrix} 0 & 0 & 0 \\ 0 & 0 & 0 \\ 0 & 0 & 3 \end{bmatrix} \qquad \blacksquare$$

By Fact 8.1.2, if a symmetric matrix is diagonalizable, then it is orthogonally diagonalizable. We still have to show that symmetric matrices are diagonalizable in the first place (over $\mathbb{R}$). The key point is the following observation:

Fact 8.1.3   A symmetric $n \times n$ matrix A has n real eigenvalues if they are counted with their algebraic multiplicities.

Proof   (This proof is for those who have studied Section 7.5.) By Fact 7.5.4, we need to show that all the complex eigenvalues of matrix A are in fact real. Consider two complex conjugate eigenvalues $p \pm iq$ of A with corresponding eigenvectors $\vec{v} \pm i\vec{w}$. (Compare this with Exercise 7.5.42b.) We wish to show that these eigenvalues are real; that is, $q = 0$. Note first that

$$(\vec{v} + i\vec{w})^T (\vec{v} - i\vec{w}) = \|\vec{v}\|^2 + \|\vec{w}\|^2.$$

(Verify this.) Now we compute the product

$$(\vec{v} + i\vec{w})^T A(\vec{v} - i\vec{w})$$

in two different ways:

$$(\vec{v} + i\vec{w})^T A(\vec{v} - i\vec{w}) = (\vec{v} + i\vec{w})^T (p - iq)(\vec{v} - i\vec{w})$$
$$= (p - iq)(\|\vec{v}\|^2 + \|\vec{w}\|^2)$$
$$(\vec{v} + i\vec{w})^T A(\vec{v} - i\vec{w}) = \big(A(\vec{v} + i\vec{w})\big)^T (\vec{v} - i\vec{w}) = (p + iq)(\vec{v} + i\vec{w})^T (\vec{v} - i\vec{w})$$
$$= (p + iq)(\|\vec{v}\|^2 + \|\vec{w}\|^2)$$

Comparing the results, we find that $p + iq = p - iq$, so that $q = 0$, as claimed. $\blacktriangle$

The foregoing proof is not very enlightening. A more transparent proof would follow if we were to define the dot product for complex vectors, but to do so would lead us too far afield.

We are now ready to prove Fact 8.1.1: Symmetric matrices are orthogonally diagonalizable.

Even though this is not logically necessary, let us first examine the case of a symmetric $n \times n$ matrix A with n distinct real eigenvalues. For each eigenvalue, we can choose an eigenvector of length 1. By Fact 8.1.2, these eigenvectors will form an orthonormal eigenbasis, that is, the matrix A will be orthogonally diagonalizable, as claimed.

Proof   (of Fact 8.1.1): This proof is somewhat technical; it may be skipped in a first reading of this text without harm.

We prove by induction on n that a symmetric $n \times n$ matrix A is orthogonally diagonalizable.

For a 1×1 matrix A, we can let $S = [1]$.

Now assume that the claim is true for $n - 1$; we show that it holds for n. Pick a real eigenvalue λ of A (this is possible by Fact 8.1.3), and choose an eigenvector $\vec{v}_1$ of length 1 for λ. We can find an orthonormal basis $\vec{v}_1, \vec{v}_2, \ldots, \vec{v}_n$ of $\mathbb{R}^n$. (Think about how you could construct such a basis.) Form the orthogonal matrix

$$P = \begin{bmatrix} | & | & & | \\ \vec{v}_1 & \vec{v}_2 & \cdots & \vec{v}_n \\ | & | & & | \end{bmatrix},$$

and compute

$$P^{-1}AP.$$

The first column of $P^{-1}AP$ is $\lambda \vec{e}_1$. (Why?) Also note that $P^{-1}AP = P^T AP$ is symmetric: $(P^T AP)^T = P^T A^T P = P^T AP$, because A is symmetric. Combining these two statements, we conclude that $P^{-1}AP$ is of the form

$$P^{-1}AP = \begin{bmatrix} \lambda & 0 \\ 0 & B \end{bmatrix}, \tag{I}$$

where B is a symmetric $(n - 1) \times (n - 1)$ matrix. By induction, B is orthogonally diagonalizable; that is, there is an orthogonal $(n - 1) \times (n - 1)$ matrix Q such that

$$Q^{-1}BQ = D$$

is a diagonal $(n - 1) \times (n - 1)$ matrix. Now introduce the orthogonal $n \times n$ matrix

$$R = \begin{bmatrix} 1 & 0 \\ 0 & Q \end{bmatrix}.$$

Then

$$R^{-1}\begin{bmatrix} \lambda & 0 \\ 0 & B \end{bmatrix}R = \begin{bmatrix} 1 & 0 \\ 0 & Q^{-1} \end{bmatrix}\begin{bmatrix} \lambda & 0 \\ 0 & B \end{bmatrix}\begin{bmatrix} 1 & 0 \\ 0 & Q \end{bmatrix} = \begin{bmatrix} \lambda & 0 \\ 0 & D \end{bmatrix} \tag{II}$$

is diagonal.

Combining equations (I) and (II), we find that

$$R^{-1}P^{-1}APR = \begin{bmatrix} \lambda & 0 \\ 0 & D \end{bmatrix} \tag{III}$$

is diagonal. Consider the orthogonal matrix $S = PR$. (Recall Fact 5.3.4a: The product of orthogonal matrices is orthogonal.) Note that $S^{-1} = (PR)^{-1} = R^{-1}P^{-1}$.

Therefore, equation (III) can be written

$$S^{-1}AS = \begin{bmatrix} \lambda & 0 \\ 0 & D \end{bmatrix},$$

proving our claim. ▲

The method outlined in the proof of Fact 8.1.1 is not a sensible way to find the matrix S in a numerical example. Rather, we can proceed as in Example 3:

Algorithm 8.1.4

Orthogonal diagonalization of a symmetric matrix A

a. Find the eigenvalues of A, and find a basis of each eigenspace.
b. Using the Gram–Schmidt process, find an *orthonormal* basis of each eigenspace.
c. Form an orthonormal eigenbasis $\vec{v}_1, \vec{v}_2, \ldots, \vec{v}_n$ for A by combining the vectors you found in part (b), and let

$$S = \begin{bmatrix} | & | & & | \\ \vec{v}_1 & \vec{v}_2 & \cdots & \vec{v}_n \\ | & | & & | \end{bmatrix}.$$

S is orthogonal (by Fact 8.1.2), and $S^{-1}AS$ will be diagonal.

We conclude this section with an example of a geometric nature:

EXAMPLE 4   Consider an invertible symmetric 2×2 matrix A. Show that the linear transformation $T(\vec{x}) = A\vec{x}$ maps the unit circle into an ellipse, and find the lengths of the semimajor and the semiminor axes of this ellipse in terms of the eigenvalues of A. Compare this with Exercise 2.2.50.

Solution

The spectral theorem tells us that there is an orthonormal eigenbasis $\vec{v}_1, \vec{v}_2$ for T, with associated real eigenvalues λ_1 and λ_2. Suppose that $|\lambda_1| \geq |\lambda_2|$. These eigenvalues will be nonzero, since A is invertible. The unit circle in $\mathbb{R}^2$ consists of all vectors of the form

$$\vec{v} = \cos(t)\vec{v}_1 + \sin(t)\vec{v}_2.$$

The image of the unit circle consists of the vectors

$$\begin{aligned} T(\vec{v}) &= \cos(t)\, T(\vec{v}_1) + \sin(t)\, T(\vec{v}_2) \\ &= \cos(t)\, \lambda_1\vec{v}_1 + \sin(t)\, \lambda_2\vec{v}_2, \end{aligned}$$

an ellipse whose semimajor axis $\lambda_1\vec{v}_1$ has the length $\|\lambda_1\vec{v}_1\| = |\lambda_1|$, while the length of the semiminor axis is $\|\lambda_2\vec{v}_2\| = |\lambda_2|$. (See Figure 4.)

In the example illustrated in Figure 4, the eigenvalue λ_1 is positive, and λ_2 is negative. ■

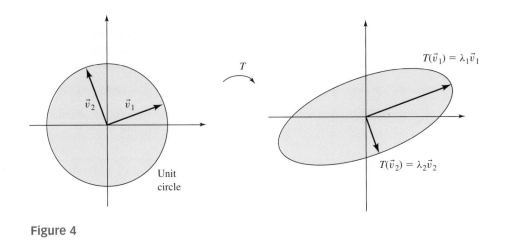

Figure 4

EXERCISES

GOALS   Find orthonormal eigenbases for symmetric matrices. Apply the spectral theorem.

For each of the matrices in Exercises 1 through 6, find an orthonormal eigenbasis. Do not use technology.

1. $\begin{bmatrix} 1 & 0 \\ 0 & 2 \end{bmatrix}$
2. $\begin{bmatrix} 1 & 1 \\ 1 & 1 \end{bmatrix}$
3. $\begin{bmatrix} 6 & 2 \\ 2 & 3 \end{bmatrix}$

4. $\begin{bmatrix} 0 & 0 & 1 \\ 0 & 0 & 1 \\ 1 & 1 & 1 \end{bmatrix}$
5. $\begin{bmatrix} 0 & 1 & 1 \\ 1 & 0 & 1 \\ 1 & 1 & 0 \end{bmatrix}$
6. $\begin{bmatrix} 0 & 2 & 2 \\ 2 & 1 & 0 \\ 2 & 0 & -1 \end{bmatrix}$

For each of the matrices A in the Exercises 7 through 11 find an orthogonal matrix S and a diagonal matrix D such that $S^{-1}AS = D$. Do not use technology.

7. $A = \begin{bmatrix} 3 & 2 \\ 2 & 3 \end{bmatrix}$
8. $A = \begin{bmatrix} 3 & 3 \\ 3 & -5 \end{bmatrix}$

9. $A = \begin{bmatrix} 0 & 0 & 3 \\ 0 & 2 & 0 \\ 3 & 0 & 0 \end{bmatrix}$
10. $A = \begin{bmatrix} 1 & -2 & 2 \\ -2 & 4 & -4 \\ 2 & -4 & 4 \end{bmatrix}$

11. $A = \begin{bmatrix} 1 & 0 & 1 \\ 0 & 1 & 0 \\ 1 & 0 & 1 \end{bmatrix}$

12. Let L from $\mathbb{R}^3$ to $\mathbb{R}^3$ be a reflection in the line spanned by

$$\vec{v} = \begin{bmatrix} 1 \\ 0 \\ 2 \end{bmatrix}.$$

   a.   Find an orthonormal eigenbasis $\mathcal{B}$ for L.

   b.   Find the matrix B of L with respect to $\mathcal{B}$.

   c.   Find the matrix A of L with respect to the standard basis of $\mathbb{R}^3$.

13. Consider a symmetric 3×3 matrix A with $A^2 = I_3$. Is the linear transformation $T(\vec{x}) = A\vec{x}$ necessarily the reflection in a subspace of $\mathbb{R}^3$?

14. In Example 3 of this section, we diagonalized the matrix

$$A = \begin{bmatrix} 1 & 1 & 1 \\ 1 & 1 & 1 \\ 1 & 1 & 1 \end{bmatrix}$$

by means of an orthogonal matrix S. Use this result to diagonalize the following matrices orthogonally (find S and D in each case):

   a. $\begin{bmatrix} 2 & 2 & 2 \\ 2 & 2 & 2 \\ 2 & 2 & 2 \end{bmatrix}$
   b. $\begin{bmatrix} -2 & 1 & 1 \\ 1 & -2 & 1 \\ 1 & 1 & -2 \end{bmatrix}$

   c. $\begin{bmatrix} 0 & \frac{1}{2} & \frac{1}{2} \\ \frac{1}{2} & 0 & \frac{1}{2} \\ \frac{1}{2} & \frac{1}{2} & 0 \end{bmatrix}$

15. If A is invertible and orthogonally diagonalizable, is A^{-1} orthogonally diagonalizable as well?

16. a.   Find the eigenvalues of the matrix

$$A = \begin{bmatrix} 1 & 1 & 1 & 1 & 1 \\ 1 & 1 & 1 & 1 & 1 \\ 1 & 1 & 1 & 1 & 1 \\ 1 & 1 & 1 & 1 & 1 \\ 1 & 1 & 1 & 1 & 1 \end{bmatrix}$$

with their multiplicities. Note that the algebraic multiplicity agrees with the geometric multiplicity. (Why?) *Hint:* What is the kernel of A?

b. Find the eigenvalues of the matrix

$$B = \begin{bmatrix} 3 & 1 & 1 & 1 & 1 \\ 1 & 3 & 1 & 1 & 1 \\ 1 & 1 & 3 & 1 & 1 \\ 1 & 1 & 1 & 3 & 1 \\ 1 & 1 & 1 & 1 & 3 \end{bmatrix}$$

with their multiplicities. Do not use technology.

c. Use your result in part (b) to find $\det(B)$.

17. Use the approach of Exercise 16 to find the determinant of the $n \times n$ matrix B that has p's on the diagonal and q's elsewhere:

$$B = \begin{bmatrix} p & q & \cdots & q \\ q & p & \cdots & q \\ \vdots & \vdots & \ddots & \vdots \\ q & q & \cdots & p \end{bmatrix}.$$

18. Consider unit vectors $\vec{v}_1, \ldots, \vec{v}_n$ in $\mathbb{R}^n$ such that the angle between $\vec{v}_i$ and $\vec{v}_j$ is $60°$ for all $i \neq j$. Find the n-volume of the n-parallelepiped spanned by $\vec{v}_1, \ldots, \vec{v}_n$. *Hint:* Let $A = [\vec{v}_1 \ \cdots \ \vec{v}_n]$, and think about the matrix $A^T A$ and its determinant. Exercise 17 is useful.

19. Consider a linear transformation L from $\mathbb{R}^n$ to $\mathbb{R}^m$. Show that there is an orthonormal basis $\vec{v}_1, \vec{v}_2, \ldots, \vec{v}_n$ of $\mathbb{R}^n$ such that the vectors $L(\vec{v}_1), L(\vec{v}_2), \ldots, L(\vec{v}_n)$ are orthogonal. Note that some of the vectors $L(\vec{v}_i)$ may be zero. *Hint:* Consider an orthonormal eigenbasis $\vec{v}_1, \vec{v}_2, \ldots, \vec{v}_n$ for the symmetric matrix $A^T A$.

20. Consider a linear transformation T from $\mathbb{R}^n$ to $\mathbb{R}^m$, where n does not exceed m. Show that there is an orthonormal basis $\vec{v}_1, \ldots, \vec{v}_n$ of $\mathbb{R}^n$ and an orthonormal basis $\vec{w}_1, \ldots, \vec{w}_m$ of $\mathbb{R}^m$ such that $T(\vec{v}_i)$ is a scalar multiple of $\vec{w}_i$, for $i = 1, \ldots, n$. *Hint:* Exercise 19 is helpful.

21. Consider a symmetric 3×3 matrix A with eigenvalues 1, 2, and 3. How many different orthogonal matrices S are there such that $S^{-1}AS$ is diagonal?

22. Consider the matrix

$$A = \begin{bmatrix} 0 & 2 & 0 & 0 \\ k & 0 & 2 & 0 \\ 0 & k & 0 & 2 \\ 0 & 0 & k & 0 \end{bmatrix},$$

where k is a constant.

a. Find a value of k such that the matrix A is diagonalizable.

b. Find a value of k such that A fails to be diagonalizable.

23. If an $n \times n$ matrix A is both symmetric and orthogonal, what can you say about the eigenvalues of A? What about the eigenspaces? Interpret the linear transformation $T(\vec{x}) = A\vec{x}$ geometrically in the cases $n = 2$ and $n = 3$.

24. Consider the matrix

$$A = \begin{bmatrix} 0 & 0 & 0 & 1 \\ 0 & 0 & 1 & 0 \\ 0 & 1 & 0 & 0 \\ 1 & 0 & 0 & 0 \end{bmatrix}.$$

Find an orthonormal eigenbasis for A.

25. Consider the matrix

$$\begin{bmatrix} 0 & 0 & 0 & 0 & 1 \\ 0 & 0 & 0 & 1 & 0 \\ 0 & 0 & 1 & 0 & 0 \\ 0 & 1 & 0 & 0 & 0 \\ 1 & 0 & 0 & 0 & 0 \end{bmatrix}.$$

Find an orthogonal 5×5 matrix S such that $S^{-1}AS$ is diagonal.

26. Let J_n be the $n \times n$ matrix with all ones on the "other diagonal" and zeros elsewhere. (In Exercises 24 and 25, we studied J_4 and J_5, respectively.) Find the eigenvalues of J_n, with their multiplicities.

27. Diagonalize the $n \times n$ matrix

$$\begin{bmatrix} 1 & 0 & 0 & \cdots & 0 & 1 \\ 0 & 1 & 0 & \cdots & 1 & 0 \\ & & \ddots & & & \\ 0 & 0 & 1 & \cdots & 0 & 0 \\ 0 & 1 & 0 & \cdots & 1 & 0 \\ 1 & 0 & 0 & \cdots & 0 & 1 \end{bmatrix}.$$

(All ones along both diagonals, and zeros elsewhere.)

28. Diagonalize the 13×13 matrix

$$\begin{bmatrix} 0 & 0 & 0 & \cdots & 0 & 1 \\ 0 & 0 & 0 & \cdots & 0 & 1 \\ & & & \ddots & & \\ 0 & 0 & 0 & \cdots & 0 & 1 \\ 1 & 1 & 1 & \cdots & 1 & 1 \end{bmatrix}.$$

(All ones in the last row and the last column, and zeros elsewhere.)

29. Consider a symmetric matrix A. If the vector $\vec{v}$ is in the image of A and $\vec{w}$ is in the kernel of A, is $\vec{v}$ necessarily orthogonal to $\vec{w}$? Justify your answer.

30. Consider an orthogonal matrix R whose first column is $\vec{v}$. Form the symmetric matrix $A = \vec{v}\vec{v}^T$. Find an orthogonal matrix S and a diagonal matrix D such that $S^{-1}AS = D$. Describe S in terms of R.

31. *True or false?* If A is a symmetric matrix, then rank$(A) = $ rank(A^2).

32. Consider the $n \times n$ matrix with all ones on the main diagonal and all q's elsewhere. For which choices of q is this matrix invertible? *Hint:* Exercise 17 is helpful.

33. For which angle(s) α can you find three distinct unit vectors in $\mathbb{R}^2$ such that the angle between any two of them is α? Draw a sketch.

34. For which angle(s) α can you find four distinct unit vectors in $\mathbb{R}^3$ such that the angle between any two of them is α? Draw a sketch.

35. Consider $n + 1$ distinct unit vectors in $\mathbb{R}^n$ such that the angle between any two of them is α. Find α.

36. Consider a symmetric $n \times n$ matrix A with $A^2 = A$. Is the linear transformation $T(\vec{x}) = A\vec{x}$ necessarily the orthogonal projection onto a subspace of $\mathbb{R}^n$?

37. We say that an $n \times n$ matrix A is *triangulizable* if A is similar to an upper triangular $n \times n$ matrix B.

    a. Give an example of a matrix with real entries that is not triangulizable over $\mathbb{R}$.

    b. Show that any $n \times n$ matrix with complex entries is triangulizable over $\mathbb{C}$. *Hint:* Give a proof by induction analogous to the proof of Fact 8.1.1.

38. a. Consider a complex upper triangular $n \times n$ matrix U with zeros on the diagonal. Show that U is *nilpotent* (i.e., that $U^n = 0$). Compare with Exercises 56 and 57 of Section 3.3.

    b. Consider a complex $n \times n$ matrix A that has zero as its only eigenvalue (with algebraic multiplicity n). Use Exercise 37 to show that A is nilpotent.

39. Let us first introduce two notations.

For a complex $n \times n$ matrix A, let $|A|$ be the matrix whose ijth entry is $|a_{ij}|$.

For two real $n \times n$ matrices A and B, we write $A \leq B$ if $a_{ij} \leq b_{ij}$ for all entries. Show that

    a. $|AB| \leq |A||B|$, for all complex $n \times n$ matrices A and B, and

    b. $|A^t| \leq |A|^t$, for all complex $n \times n$ matrices A and all positive integers t.

40. Let $U \geq 0$ be a real upper triangular $n \times n$ matrix with zeros on the diagonal. Show that

$$(I_n + U)^t \leq t^n (I_n + U + U^2 + \cdots + U^{n-1})$$

for all positive integers t. See Exercises 38 and 39.

41. Let R be a complex upper triangular $n \times n$ matrix with $|r_{ii}| < 1$ for $i = 1, \ldots, n$. Show that

$$\lim_{t \to \infty} R^t = 0,$$

meaning that the modulus of all entries of R^t approaches zero. *Hint:* We can write $|R| \leq \lambda(I_n + U)$, for some positive real number $\lambda < 1$ and an upper triangular matrix $U \geq 0$ with zeros on the diagonal. Exercises 39 and 40 are helpful.

42. a. Let A be a complex $n \times n$ matrix such that $|\lambda| < 1$ for all eigenvalues λ of A. Show that

$$\lim_{t \to \infty} A^t = 0,$$

meaning that the modulus of all entries of A^t approaches zero.

    b. Prove Fact 7.6.2.

8.2 QUADRATIC FORMS

In this section, we will present an important application of the spectral theorem (Fact 8.1.1).

In a multivariable calculus text, we found the following problem:

EXAMPLE 1   Consider the function

$$q(x_1, x_2) = 8x_1^2 - 4x_1 x_2 + 5x_2^2$$

from $\mathbb{R}^2$ to $\mathbb{R}$.

Determine whether $q(0, 0) = 0$ is the global maximum, the global minimum, or neither.

Recall that $q(0, 0)$ is called the global (or absolute) minimum if $q(0, 0) \leq q(x_1, x_2)$ for all real numbers x_1, x_2; the global maximum is defined analogously.

Solution

There are a number of ways to do this problem, some of which you may have seen in a previous course. Here we present an approach based on matrix techniques. We will first develop some theory, and then do the example.

Note that we can write

$$q \begin{bmatrix} x_1 \\ x_2 \end{bmatrix} = 8x_1^2 - 4x_1 x_2 + 5x_2^2$$

$$= \begin{bmatrix} x_1 \\ x_2 \end{bmatrix} \cdot \begin{bmatrix} 8x_1 - 2x_2 \\ -2x_1 + 5x_2 \end{bmatrix} \qquad \text{We "split" the contribution } -4x_1 x_2 \text{ equally among the two components.}$$

More succinctly, we can write

$$q(\vec{x}) = \vec{x} \cdot A\vec{x}, \quad \text{where} \quad A = \begin{bmatrix} 8 & -2 \\ -2 & 5 \end{bmatrix},$$

or

$$q(\vec{x}) = \vec{x}^T A \vec{x}.$$

The matrix A is symmetric by construction. By the spectral theorem (Fact 8.1.1), there is an orthonormal eigenbasis $\vec{v}_1$, $\vec{v}_2$ for A. We find

$$\vec{v}_1 = \frac{1}{\sqrt{5}} \begin{bmatrix} 2 \\ -1 \end{bmatrix}, \qquad \vec{v}_2 = \frac{1}{\sqrt{5}} \begin{bmatrix} 1 \\ 2 \end{bmatrix},$$

with associated eigenvalues $\lambda_1 = 9$ and $\lambda_2 = 4$. (Verify this.)

If we write $\vec{x} = c_1 \vec{v}_1 + c_2 \vec{v}_2$, we can express the value of the function as follows:

$$q(\vec{x}) = \vec{x} \cdot A\vec{x} = (c_1 \vec{v}_1 + c_2 \vec{v}_2) \cdot (c_1 \lambda_1 \vec{v}_1 + c_2 \lambda_2 \vec{v}_2) = \lambda_1 c_1^2 + \lambda_2 c_2^2 = 9c_1^2 + 4c_2^2.$$

(Recall that $\vec{v}_1 \cdot \vec{v}_1 = 1$, $\vec{v}_1 \cdot \vec{v}_2 = 0$, and $\vec{v}_2 \cdot \vec{v}_2 = 1$, since $\vec{v}_1$, $\vec{v}_2$ is an orthonormal basis of $\mathbb{R}^2$.)

The formula $q(\vec{x}) = 9c_1^2 + 4c_2^2$ shows that $q(\vec{x}) > 0$ for all nonzero $\vec{x}$, because at least one of the terms $9c_1^2$ and $4c_2^2$ is positive.

Thus $q(0, 0) = 0$ is the global minimum of the function.

The preceding work shows that the c_1–c_2 coordinate system defined by an orthonormal eigenbasis for A is "well adjusted" to the function q. The formula

$$9c_1^2 + 4c_2^2$$

is easier to work with than the original formula

$$8x_1^2 - 4x_1 x_2 + 5x_2^2,$$

because no term involves $c_1 c_2$:

$$q(x_1, x_2) = 8x_1^2 - 4x_1 x_2 + 5x_2^2$$
$$= 9c_1^2 + 4c_2^2$$

The two coordinate systems are shown in Figure 1. ∎

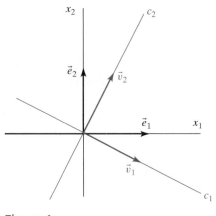

Figure 1

Let us present these ideas in greater generality:

Definition 8.2.1

> **Quadratic forms**
> A function $q(x_1, x_2, \ldots, x_n)$ from $\mathbb{R}^n$ to $\mathbb{R}$ is called a *quadratic form* if it is a linear combination of functions of the form $x_i x_j$ (where i and j may be equal). A quadratic form can be written as
> $$q(\vec{x}) = \vec{x} \cdot A\vec{x} = \vec{x}^T A\vec{x},$$
> for a symmetric $n \times n$ matrix A.

The set Q_n of quadratic forms $q(x_1, x_2, \ldots, x_n)$ is a *subspace* of the linear space of all functions from $\mathbb{R}^n$ to $\mathbb{R}$. In Exercise 42 you will think about the dimension of this space.

EXAMPLE 2 Consider the quadratic form

$$q(x_1, x_2, x_3) = 9x_1^2 + 7x_2^2 + 3x_3^2 - 2x_1x_2 + 4x_1x_3 - 6x_2x_3.$$

Find a symmetric matrix A such that $q(\vec{x}) = \vec{x} \cdot A\vec{x}$ for all $\vec{x}$ in $\mathbb{R}^3$.

Solution

As in Example 1, we let

$$a_{ii} = \left(\text{coefficient of } x_i^2\right),$$

$$a_{ij} = a_{ji} = \tfrac{1}{2}(\text{coefficient of } x_i x_j), \quad \text{if } i \neq j.$$

Therefore,

$$A = \begin{bmatrix} 9 & -1 & 2 \\ -1 & 7 & -3 \\ 2 & -3 & 3 \end{bmatrix}. \qquad \blacksquare$$

The observation we made in Example 1 can now be generalized as follows:

Fact 8.2.2

Consider a quadratic form $q(\vec{x}) = \vec{x} \cdot A\vec{x}$ from $\mathbb{R}^n$ to $\mathbb{R}$. Let $\mathfrak{B}$ be an orthonormal eigenbasis for A, with associated eigenvalues $\lambda_1, \ldots, \lambda_n$. Then

$$q(\vec{x}) = \lambda_1 c_1^2 + \lambda_2 c_2^2 + \cdots + \lambda_n c_n^2,$$

where the c_i are the coordinates of $\vec{x}$ with respect to $\mathfrak{B}$.[2]

Again, note that we have been able to get rid of the mixed terms: no summand involves $c_i c_j$ (with $i \neq j$) in the formula above. To justify the formula stated in Fact 8.2.2, we can proceed as in Example 1. We leave the details as an exercise.

When we study a quadratic form q, we are often interested in finding out whether $q(\vec{x}) > 0$ for all nonzero $\vec{x}$ (as in Example 1). In this context, it is useful to introduce the following terminology:

Definition 8.2.3

Positive definite quadratic forms

Consider a quadratic form $q(\vec{x}) = \vec{x} \cdot A\vec{x}$, where A is a symmetric $n \times n$ matrix.

We say that A is *positive definite* if $q(\vec{x})$ is positive for all nonzero $\vec{x}$ in $\mathbb{R}^n$, and we call A *positive semidefinite* if $q(\vec{x}) \geq 0$, for all $\vec{x}$ in $\mathbb{R}^n$.

Negative definite and negative semidefinite symmetric matrices are defined analogously.

Finally, we call A *indefinite* if q takes positive as well as negative values.

EXAMPLE 3

Consider an $m \times n$ matrix A. Show that the function $q(\vec{x}) = \|A\vec{x}\|^2$ is a quadratic form, find its matrix, and determine its definiteness.

Solution

We can write $q(\vec{x}) = (A\vec{x}) \cdot (A\vec{x}) = (A\vec{x})^T (A\vec{x}) = \vec{x}^T A^T A\vec{x} = \vec{x} \cdot (A^T A\vec{x})$. This shows that q is a quadratic form, with matrix $A^T A$. This quadratic form is positive semidefinite, because $q(\vec{x}) = \|A\vec{x}\|^2 \geq 0$ for all vectors $\vec{x}$ in $\mathbb{R}^n$. Note that $q(\vec{x}) = 0$ if and only if $\vec{x}$ is in the kernel of A. Therefore, the quadratic form is positive definite if and only if $\ker(A) = \{\vec{0}\}$. ∎

By Fact 8.2.2, the definiteness of a symmetric matrix A is easy to determine from its eigenvalues:

Fact 8.2.4

Eigenvalues and definiteness

A symmetric matrix A is *positive definite* if (and only if) all of its eigenvalues are positive. The matrix A is *positive semidefinite* if (and only if) all of its eigenvalues are positive or zero.

---

[2]The basic properties of quadratic forms were first derived by the Dutchman Johan de Witt (1625–1672) in his *Elementa curvarum linearum*. De Witt was one of the leading statesmen of his time, guiding his country through two wars against England. He consolidated his nation's commercial and naval power. De Witt met an unfortunate end when he was literally torn to pieces by an angry mob. (He should have stayed with math!)

These facts follow immediately from the formula

$$q(\vec{x}) = \lambda_1 c_1^2 + \cdots + \lambda_n c_n^2. \quad \text{(See Fact 8.2.2.)}$$

The determinant of a positive definite matrix is positive, since the determinant is the product of the eigenvalues. The converse is not true, however: Consider a symmetric 3×3 matrix A with one positive and two negative eigenvalues. Then $\det(A)$ is positive, but $q(\vec{x}) = \vec{x} \cdot A\vec{x}$ is indefinite. In practice, the following criterion for positive definiteness is often used (a proof is outlined in Exercise 34):

Fact 8.2.5

> **Principal submatrices and definiteness**
> Consider a symmetric $n \times n$ matrix A. For $m = 1, \ldots, n$, let $A^{(m)}$ be the $m \times m$ matrix obtained by omitting all rows and columns of A past the mth. These matrices $A^{(m)}$ are called the *principal submatrices* of A.
>   The matrix A is positive definite if (and only if) $\det(A^{(m)}) > 0$, for all $m = 1, \ldots, n$.

As an example, consider the matrix

$$A = \begin{bmatrix} 9 & -1 & 2 \\ -1 & 7 & -3 \\ 2 & -3 & 3 \end{bmatrix}$$

from Example 2:

$$\det\left(A^{(1)}\right) = \det[9] = 9 > 0$$

$$\det\left(A^{(2)}\right) = \det\begin{bmatrix} 9 & -1 \\ -1 & 7 \end{bmatrix} = 62 > 0$$

$$\det\left(A^{(3)}\right) = \det(A) = 89 > 0$$

We can conclude that A is positive definite.

Alternatively, we could find the eigenvalues of A and use Fact 8.2.4. Using technology, we find that $\lambda_1 \approx 10.7$, $\lambda_2 \approx 7.1$, and $\lambda_3 \approx 1.2$, confirming our result.

Principal Axes

When we study a function $f(x_1, x_2, \ldots, x_n)$ from $\mathbb{R}^n$ to $\mathbb{R}$, we are often interested in the solutions of the equations

$$f(x_1, x_2, \ldots, x_n) = k,$$

for a fixed k in $\mathbb{R}$, called the *level sets* of f (*level curves* for $n = 2$, *level surfaces* for $n = 3$).

Here we will think about the level curves of a quadratic form $q(x_1, x_2)$ of two variables. For simplicity, we focus on the level curve $q(x_1, x_2) = 1$.

Let us first think about the case when there is no mixed term in the formula. We trust that you had at least a brief encounter with those level curves in a previous course. Let us discuss the two major cases:

Case 1 ◆ $q(x_1, x_2) = ax_1^2 + bx_2^2 = 1$, where $b > a > 0$. This curve is an *ellipse*, as shown in Figure 2. The lengths of the semimajor and the semiminor axes are $1/\sqrt{a}$

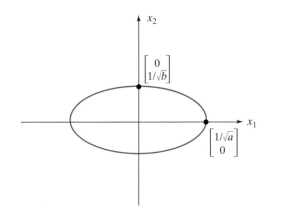

Figure 2

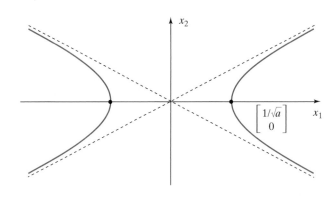

Figure 3

and $1/\sqrt{b}$, respectively. This ellipse can be parameterized by

$$\begin{bmatrix} x_1 \\ x_2 \end{bmatrix} = \cos(t) \begin{bmatrix} 1/\sqrt{a} \\ 0 \end{bmatrix} + \sin(t) \begin{bmatrix} 0 \\ 1/\sqrt{b} \end{bmatrix}.$$

Case 2 ♦ $q(x_1, x_2) = ax_1^2 + bx_2^2 = 1$, where a is positive and b negative. This is a hyperbola, with x_1-intercepts $\begin{bmatrix} \pm 1/\sqrt{a} \\ 0 \end{bmatrix}$, as shown in Figure 3. What are the slopes of the asymptotes, in terms of a and b?

Now consider the level curve

$$q(\vec{x}) = \vec{x} \cdot A\vec{x} = 1,$$

where A is an invertible symmetric 2×2 matrix. By Fact 8.2.2, we can write this equation as

$$\lambda_1 c_1^2 + \lambda_2 c_2^2 = 1,$$

where c_1, c_2 are the coordinates of $\vec{x}$ with respect to an orthonormal eigenbasis for A, and λ_1, λ_2 are the associated eigenvalues.

This curve is an ellipse if both eigenvalues are positive and a hyperbola if one eigenvalue is positive and one negative. (What happens when both eigenvalues are negative?)

EXAMPLE 4   Sketch the curve

$$8x_1^2 - 4x_1x_2 + 5x_2^2 = 1.$$

(See Example 1.)

Solution

In Example 1, we found that we can write this equation as

$$9c_1^2 + 4c_2^2 = 1,$$

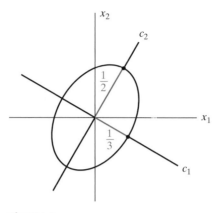

Figure 4

where c_1, c_2 are the coordinates of $\vec{x}$ with respect to the orthonormal eigenbasis

$$\vec{v}_1 = \frac{1}{\sqrt{5}}\begin{bmatrix} 2 \\ -1 \end{bmatrix}, \qquad \vec{v}_2 = \frac{1}{\sqrt{5}}\begin{bmatrix} 1 \\ 2 \end{bmatrix},$$

for $A = \begin{bmatrix} 8 & -2 \\ -2 & 5 \end{bmatrix}$. We sketch this ellipse in Figure 4.

The c_1- and the c_2-axes are called the *principal axes* of the quadratic form $q(x_1, x_2) = 8x_1^2 - 4x_1x_2 + 5x_2^2$. Note that these are the eigenspaces of the matrix

$$A = \begin{bmatrix} 8 & -2 \\ -2 & 5 \end{bmatrix}$$

of the quadratic form. ∎

Definition 8.2.6

Principal axes
Consider a quadratic form $q(\vec{x}) = \vec{x} \cdot A\vec{x}$, where A is a symmetric $n \times n$ matrix with n distinct eigenvalues. Then the eigenspaces of A are called the *principal axes* of q. (Note that these eigenspaces will be one-dimensional.)

Let's return to the case of a quadratic form of two variables. We can summarize our findings as follows:

Fact 8.2.7

Ellipses and hyperbolas
Consider the curve C in $\mathbb{R}^2$ defined by

$$q(x_1, x_2) = ax_1^2 + bx_1x_2 + cx_2^2 = 1.$$

Let λ_1 and λ_2 be the eigenvalues of the matrix $\begin{bmatrix} a & b/2 \\ b/2 & c \end{bmatrix}$ of q.

If both λ_1 and λ_2 are positive, then C is an *ellipse*. If there is a positive and a negative eigenvalue, then C is a *hyperbola*.

EXERCISES

GOALS Apply the concept of a quadratic form. Use an orthonormal eigenbasis for A to analyze the quadratic form $q(\vec{x}) = \vec{x} \cdot A\vec{x}$.

For each of the quadratic forms q listed in Exercises 1 through 3, find a symmetric matrix A such that $q(\vec{x}) = \vec{x} \cdot A\vec{x}$.

1. $q(x_1, x_2) = 6x_1^2 - 7x_1x_2 + 8x_2^2$.

2. $q(x_1, x_2) = x_1x_2$.

3. $q(x_1, x_2, x_3) = 3x_1^2 + 4x_2^2 + 5x_3^2 + 6x_1x_3 + 7x_2x_3$.

Determine the definiteness of the quadratic forms in Exercises 4 through 7.

4. $q(x_1, x_2) = 6x_1^2 + 4x_1x_2 + 3x_2^2$.

5. $q(x_1, x_2) = x_1^2 + 4x_1x_2 + x_2^2$.

6. $q(x_1, x_2) = 2x_1^2 + 6x_1x_2 + 4x_2^2$.

7. $q(x_1, x_2, x_3) = 3x_2^2 + 4x_1x_3$.

8. If A is a symmetric matrix, what can you say about the definiteness of A^2? When is A^2 positive definite?

9. A real square matrix A is called *skew-symmetric* if $A^T = -A$.

   a. If A is skew-symmetric, is A^2 skew-symmetric as well? Or is A^2 symmetric?

   b. If A is skew-symmetric, what can you say about the definiteness of A^2? What about the eigenvalues of A^2?

   c. What can you say about the complex eigenvalues of a skew-symmetric matrix? Which skew-symmetric matrices are diagonalizable over $\mathbb{R}$?

10. Consider a quadratic form $q(\vec{x}) = \vec{x} \cdot A\vec{x}$ on $\mathbb{R}^n$ and a fixed vector $\vec{v}$ in $\mathbb{R}^n$. Is the transformation

$$L(\vec{x}) = q(\vec{x} + \vec{v}) - q(\vec{x}) - q(\vec{v})$$

linear? If so, what is its matrix?

11. If A is an invertible symmetric matrix, what is the relationship between the definiteness of A and A^{-1}?

12. Show that a quadratic form $q(\vec{x}) = \vec{x} \cdot A\vec{x}$ of two variables is indefinite if (and only if) $\det(A) < 0$. Here, A is a symmetric 2×2 matrix.

13. Show that the diagonal elements of a positive definite matrix A are positive.

14. Consider a 2×2 matrix $A = \begin{bmatrix} a & b \\ b & c \end{bmatrix}$, where a and $\det(A)$ are both positive. Without using Fact 8.2.5, show that A is positive definite. *Hint:* Show first that c is

positive, and thus $\operatorname{tr}(A)$ is positive. Then think about the signs of the eigenvalues.

Sketch the curves defined in Exercises 15 through 20. In each case, draw and label the principal axes, label the intercepts of the curve with the principal axes, and give the formula of the curve in the coordinate system defined by the principal axes.

15. $6x_1^2 + 4x_1x_2 + 3x_2^2 = 1$.   16. $x_1x_2 = 1$.

17. $3x_1^2 + 4x_1x_2 = 1$.   18. $9x_1^2 - 4x_1x_2 + 6x_2^2 = 1$.

19. $x_1^2 + 4x_1x_2 + 4x_2^2 = 1$.   20. $-3x_1^2 + 6x_1x_2 + 5x_2^2 = 1$.

21. a. Sketch the following three surfaces:

$$x_1^2 + 4x_2^2 + 9x_3^2 = 1,$$
$$x_1^2 + 4x_2^2 - 9x_3^2 = 1,$$
$$-x_1^2 - 4x_2^2 + 9x_3^2 = 1.$$

Which of these are bounded? Which are connected? Label the points closest to and farthest from the origin (if there are any).

   b. Consider the surface

$$x_1^2 + 2x_2^2 + 3x_3^2 + x_1x_2 + 2x_1x_3 + 3x_2x_3 = 1.$$

Which of the three surfaces in part (a) does this surface qualitatively resemble most? Which points on this surface are closest to the origin? Give a rough approximation; use technology.

22. On the surface

$$-x_1^2 + x_2^2 - x_3^2 + 10x_1x_3 = 1,$$

find the two points closest to the origin.

23. Consider an $n \times n$ matrix M that is not symmetric, and define the function $g(\vec{x}) = \vec{x} \cdot M\vec{x}$ from $\mathbb{R}^n$ to $\mathbb{R}$. Is g necessarily a quadratic form? If so, give a symmetric matrix A (in terms of M) such that

$$g(\vec{x}) = \vec{x} \cdot A\vec{x}.$$

24. Consider a quadratic form

$$q(\vec{x}) = \vec{x} \cdot A\vec{x},$$

where A is a symmetric $n \times n$ matrix. Find $q(\vec{e}_1)$. Give your answer in terms of the entries of the matrix A.

25. Consider a quadratic form

$$q(\vec{x}) = \vec{x} \cdot A\vec{x},$$

where A is a symmetric $n \times n$ matrix. Let $\vec{v}$ be a unit eigenvector of A, with associated eigenvalue λ. Find $q(\vec{v})$.

26. Consider a quadratic form

$$q(\vec{x}) = \vec{x} \cdot A\vec{x},$$

where A is a symmetric $n \times n$ matrix. *True or false?* If there is a nonzero vector $\vec{v}$ in $\mathbb{R}^n$ such that $q(\vec{v}) = 0$, then A fails to be invertible.

27. Consider a quadratic form $q(\vec{x}) = \vec{x} \cdot A\vec{x}$, where A is a symmetric $n \times n$ matrix with the positive eigenvalues $\lambda_1 \geq \lambda_2 \geq \cdots \geq \lambda_n$. Let S^{n-1} be the set of all unit vectors in $\mathbb{R}^n$. Describe the image of S^{n-1} under q, in terms of the eigenvalues of A.

28. Show that any positive definite $n \times n$ matrix A can be written as $A = BB^T$, where B is an $n \times n$ matrix with orthogonal columns. *Hint:* There is an orthogonal matrix S such that $S^{-1}AS = S^T AS = D$ is a diagonal matrix with positive diagonal entries. Then $A = SDS^T$. Now write D as the square of a diagonal matrix.

29. For the matrix $A = \begin{bmatrix} 8 & -2 \\ -2 & 5 \end{bmatrix}$ write $A = BB^T$ as discussed in Exercise 28. See Example 1.

30. Show that any positive definite matrix A can be written as $A = B^2$, where B is a positive definite matrix.

31. For the matrix $A = \begin{bmatrix} 8 & -2 \\ -2 & 5 \end{bmatrix}$ write $A = B^2$ as discussed in Exercise 30. See Example 1.

32. *Cholesky factorization* for 2×2 matrices. Show that any positive definite 2×2 matrix A can be written uniquely as $A = LL^T$, where L is a lower triangular 2×2 matrix with positive entries on the diagonal. *Hint:* Solve the equation

$$\begin{bmatrix} a & b \\ b & c \end{bmatrix} = \begin{bmatrix} x & 0 \\ y & z \end{bmatrix} \begin{bmatrix} x & y \\ 0 & z \end{bmatrix}.$$

33. Find the Cholesky factorization (discussed in Exercise 32) for

$$A = \begin{bmatrix} 8 & -2 \\ -2 & 5 \end{bmatrix}.$$

34. A *Cholesky factorization* of a symmetric matrix A is a factorization of the form $A = LL^T$, where L is lower triangular with positive diagonal entries.

Show that for a symmetric $n \times n$ matrix A the following are equivalent:

   i. A is positive definite.

   ii. All principal submatrices $A^{(m)}$ of A are positive definite. (See Fact 8.2.5.)

   iii. $\det(A^{(m)}) > 0$ for $m = 1, \ldots, n$.

   iv. A has a Cholesky factorization $A = LL^T$.

Hint: Show that (i) implies (ii), (ii) implies (iii), (iii) implies (iv), and (iv) implies (i). The hardest step is the implication from (iii) to (iv): arguing by induction on n, you may assume that $A^{(n-1)}$ has a Cholesky factorization $A^{(n-1)} = BB^T$. Now show that there is a vector $\vec{x}$ in $\mathbb{R}^{n-1}$ and a scalar t such that

$$A = \begin{bmatrix} A^{(n-1)} & \vec{v} \\ \vec{v}^T & k \end{bmatrix} = \begin{bmatrix} B & 0 \\ \vec{x}^T & 1 \end{bmatrix} \begin{bmatrix} B^T & \vec{x} \\ 0 & t \end{bmatrix}.$$

Explain why the scalar t is positive. Therefore, we have the Cholesky factorization

$$A = \begin{bmatrix} B & 0 \\ \vec{x}^T & \sqrt{t} \end{bmatrix} \begin{bmatrix} B^T & \vec{x} \\ 0 & \sqrt{t} \end{bmatrix}.$$

This reasoning also shows that the Cholesky factorization of A is unique. Alternatively, you can use the LDL^T factorization of A to show that (iii) implies (iv). (See Exercise 5.3.43.)

To show that (i) implies (ii), consider a nonzero vector $\vec{x}$ in $\mathbb{R}^m$, and define

$$\vec{y} = \begin{bmatrix} \vec{x} \\ 0 \\ \vdots \\ 0 \end{bmatrix}$$

in $\mathbb{R}^n$ (fill in $n - m$ zeros). Then

$$\vec{x}^T A^{(m)} \vec{x} = \vec{y}^T A\vec{y} > 0.$$

35. Find the Cholesky factorization of the matrix

$$A = \begin{bmatrix} 4 & -4 & 8 \\ -4 & 13 & 1 \\ 8 & 1 & 26 \end{bmatrix}.$$

36. Consider an invertible $n \times n$ matrix A. What is the relationship between the matrix R in the QR factorization of A and the matrix L in the Cholesky factorization of $A^T A$?

37. Consider the quadratic form

$$q(x_1, x_2) = ax_1^2 + bx_1x_2 + cx_2^2.$$

We define

$$q_{11} = \frac{\partial^2 q}{\partial x_1^2}, \qquad q_{12} = q_{21} = \frac{\partial^2 q}{\partial x_1 \, \partial x_2}, \qquad q_{22} = \frac{\partial^2 q}{\partial x_2^2}.$$

The discriminant D of q is defined as

$$D = \det \begin{bmatrix} q_{11} & q_{12} \\ q_{21} & q_{22} \end{bmatrix} = q_{11}q_{22} - (q_{12})^2.$$

The *second derivative test* tells us that if D and q_{11} are both positive, then $q(x_1, x_2)$ has a minimum at $(0, 0)$. Justify this fact, using the theory developed in this section.

38. For which choices of the constants p and q is the $n \times n$ matrix

$$B = \begin{bmatrix} p & q & \cdots & q \\ q & p & \cdots & q \\ \vdots & \vdots & \ddots & \vdots \\ q & q & \cdots & p \end{bmatrix}$$

positive definite? (B has p's on the diagonal and q's elsewhere.) *Hint:* Exercise 8.1.17 is helpful.

39. For which angles α can you find a basis of $\mathbb{R}^n$ such that the angle between any two vectors in this basis is α?

40. Show that for every symmetric $n \times n$ matrix A there is a constant k such that matrix $A + kI_n$ is positive definite.

41. Find the dimension of the space Q_2 of all quadratic forms in two variables.

42. Find the dimension of the space Q_n of all quadratic forms in n variables.

43. Consider the transformation $T\big(q(x_1, x_2)\big) = q(x_1, 0)$ from Q_2 to P_2. Is T a linear transformation? If so, find image, rank, kernel, and nullity of T.

44. Consider the transformation $T\big(q(x_1, x_2)\big) = q(x_1, 1)$ from Q_2 to P_2. Is T a linear transformation? Is it an isomorphism?

45. Consider the transformation $T\big(q(x_1, x_2, x_3)\big) = q(x_1, 1, 1)$ from Q_3 to P_2. Is T a linear transformation? If so, find image, rank, kernel, and nullity of T.

46. Consider the linear transformation $T\big(q(x_1, x_2, x_3)\big) = q(x_1, x_2, x_1)$ from Q_3 to Q_2. Find image, kernel, rank, and nullity of this transformation.

47. Consider the function $T(A)(\vec{x}) = \vec{x}^T A \vec{x}$ from $\mathbb{R}^{n \times n}$ to Q_n. Show that T is a linear transformation. Find image, kernel, rank, and nullity of T.

48. Consider the linear transformation $T\big(q(x_1, x_2)\big) = q(x_2, x_1)$ from Q_2 to Q_2. Find all the eigenvalues and eigenfunctions of T. Is transformation T diagonalizable?

49. Consider the linear transformation $T\big(q(x_1, x_2)\big) = q(x_1, 2x_2)$ from Q_2 to Q_2. Find all the eigenvalues and eigenfunctions of T. Is transformation T diagonalizable?

50. Consider the linear transformation

$$T\big(q(x_1, x_2)\big) = x_1 \frac{\partial q}{\partial x_2} + x_2 \frac{\partial q}{\partial x_1}$$

from Q_2 to Q_2. Find all the eigenvalues and eigenfunctions of T. Is transformation T diagonalizable?

51. What are the signs of the determinants of the principal submatrices of a negative definite matrix? (See Fact 8.2.5.)

8.3 SINGULAR VALUES

In Exercise 47 of Section 2.2, we stated the following remarkable fact.

EXAMPLE 1   Show that if $L(\vec{x}) = A\vec{x}$ is a linear transformation from $\mathbb{R}^2$ to $\mathbb{R}^2$, then there are two orthogonal unit vectors $\vec{v}_1$ and $\vec{v}_2$ in $\mathbb{R}^2$ such that vectors $L(\vec{v}_1)$ and $L(\vec{v}_2)$ are orthogonal as well (although not necessarily unit vectors). See Figure 1. (*Hint:* Consider an orthonormal eigenbasis $\vec{v}_1, \vec{v}_2$ of the symmetric matrix $A^T A$.)

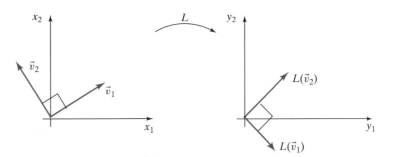

Figure 1

Solution

This statement is clear for some classes of transformations, for example,

- If L is an orthogonal transformation, then any two orthogonal unit vectors $\vec{v}_1$ and $\vec{v}_2$ will do, by Fact 5.3.2.
- If $L(\vec{x}) = A\vec{x}$, where A is symmetric, then we can choose two orthogonal unit eigenvectors, by the spectral theorem, Fact 8.1.1. See also Example 4 of Section 8.1.

However, for an arbitrary linear transformation L, the statement isn't that obvious; think about the case of a shear, for example.

In Exercise 47 of Section 2.2, we suggested a proof based on the intermediate value theorem for continuous functions. Here we will present a proof in the spirit of linear algebra that generalizes more easily to higher-dimensional spaces.

Following the hint, we first note that matrix $A^T A$ is symmetric, since $(A^T A)^T = A^T (A^T)^T = A^T A$. The spectral theorem (Fact 8.1.1) tells us that there is an orthonormal eigenbasis $\vec{v}_1, \vec{v}_2$ for $A^T A$, with associated eigenvalues λ_1, λ_2. We can verify that vectors $L(\vec{v}_1) = A\vec{v}_1$ and $L(\vec{v}_2) = A\vec{v}_2$ are orthogonal, as claimed:

$$(A\vec{v}_1) \cdot (A\vec{v}_2) = (A\vec{v}_1)^T A\vec{v}_2 = \vec{v}_1^T A^T A\vec{v}_2 = \vec{v}_1^T (\lambda_2 \vec{v}_2) = \lambda_2 (\vec{v}_1 \cdot \vec{v}_2) = 0.$$

It is worth mentioning that $\vec{v}_1$ and $\vec{v}_2$ need not be eigenvectors of matrix A. ∎

EXAMPLE 2   Consider the linear transformation $L(\vec{x}) = A\vec{x}$, where $A = \begin{bmatrix} 6 & 2 \\ -7 & 6 \end{bmatrix}$.

a. Find an orthonormal basis $\vec{v}_1, \vec{v}_2$ of $\mathbb{R}^2$ such that vectors $L(\vec{v}_1)$ and $L(\vec{v}_2)$ are orthogonal.

b. Show that the image of the unit circle under transformation L is an ellipse. Find the lengths of the two semiaxes of this ellipse, in terms of the eigenvalues of matrix $A^T A$.

Solution

a. Using the ideas of Example 1, we will find an orthonormal eigenbasis for matrix $A^T A$:

$$A^T A = \begin{bmatrix} 6 & -7 \\ 2 & 6 \end{bmatrix} \begin{bmatrix} 6 & 2 \\ -7 & 6 \end{bmatrix} = \begin{bmatrix} 85 & -30 \\ -30 & 40 \end{bmatrix}$$

The characteristic polynomial of $A^T A$ is

$$\lambda^2 - 125\lambda + 2500 = (\lambda - 100)(\lambda - 25),$$

so that the eigenvalues of $A^T A$ are $\lambda_1 = 100$ and $\lambda_2 = 25$. Now we can find the eigenspaces of $A^T A$.

$$E_{100} = \ker \begin{bmatrix} 15 & 30 \\ 30 & 60 \end{bmatrix} = \operatorname{span} \begin{bmatrix} 2 \\ -1 \end{bmatrix} \quad \text{and} \quad E_{25} = \ker \begin{bmatrix} -60 & 30 \\ 30 & -15 \end{bmatrix} = \operatorname{span} \begin{bmatrix} 1 \\ 2 \end{bmatrix}.$$

To find an ortho*normal* basis, we need to multiply these vectors by the reciprocals of their lengths:

$$\vec{v}_1 = \frac{1}{\sqrt{5}} \begin{bmatrix} 2 \\ -1 \end{bmatrix}, \qquad \vec{v}_2 = \frac{1}{\sqrt{5}} \begin{bmatrix} 1 \\ 2 \end{bmatrix}$$

b. The unit circle consists of the vectors of the form $\vec{x} = \cos(t)\vec{v}_1 + \sin(t)\vec{v}_2$, and the image of the unit circle consists of the vectors $L(\vec{x}) = \cos(t)L(\vec{v}_1) + \sin(t)L(\vec{v}_2)$. This image is the ellipse whose semimajor and semiminor axes are $L(\vec{v}_1)$ and $L(\vec{v}_2)$. What are the lengths of these axes?

$$\|L(\vec{v}_1)\|^2 = (A\vec{v}_1) \cdot (A\vec{v}_1) = \vec{v}_1^T A^T A\vec{v}_1 = \vec{v}_1^T(\lambda_1\vec{v}_1) = \lambda_1(\vec{v}_1 \cdot \vec{v}_1) = \lambda_1$$

Likewise,

$$\|L(\vec{v}_2)\|^2 = \lambda_2.$$

Thus,

$$\|L(\vec{v}_1)\| = \sqrt{\lambda_1} = \sqrt{100} = 10 \qquad \text{and} \qquad \|L(\vec{v}_2)\| = \sqrt{\lambda_2} = \sqrt{25} = 5.$$

See Figure 2. We can compute the lengths of vectors $L(\vec{v}_1)$ and $L(\vec{v}_2)$ directly, of course, but the way we did it before is more informative. For example,

$$L(\vec{v}_1) = A\vec{v}_1 = \frac{1}{\sqrt{5}} \begin{bmatrix} 6 & 2 \\ -7 & 6 \end{bmatrix} \begin{bmatrix} 2 \\ -1 \end{bmatrix} = \frac{1}{\sqrt{5}} \begin{bmatrix} 10 \\ -20 \end{bmatrix}, \qquad \text{so that}$$

$$\|L(\vec{v}_1)\| = \left\| \frac{1}{\sqrt{5}} \begin{bmatrix} 10 \\ -20 \end{bmatrix} \right\| = 10. \quad \blacksquare$$

Part (b) of Example 2 shows that the square roots of the eigenvalues of matrix $A^T A$ play an important role in the geometrical interpretation of the transformation $L(\vec{x}) = A\vec{x}$. This observation motivates the following definition.

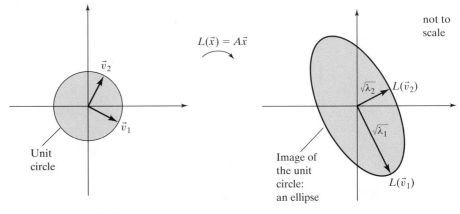

Figure 2

Definition 8.3.1

> **Singular values**
>
> The *singular values* of an $m \times n$ matrix A are the square roots of the eigenvalues of the symmetric $n \times n$ matrix $A^T A$, listed with their algebraic multiplicities. It is customary to denote the singular values by $\sigma_1, \sigma_2, \ldots, \sigma_n$, and to list them in decreasing order:
>
> $$\sigma_1 \geq \sigma_2 \geq \cdots \geq \sigma_n$$

The singular values of the matrix $A = \begin{bmatrix} 6 & 2 \\ -7 & 6 \end{bmatrix}$ considered in Example 2 are $\sigma_1 = \sqrt{\lambda_1} = 10$ and $\sigma_2 = \sqrt{\lambda_2} = 5$, since the eigenvalues of $A^T A$ are $\lambda_1 = 100$ and $\lambda_2 = 25$.

We can now generalize our work in Example 2.

Fact 8.3.2

The image of the unit circle

Let $L(\vec{x}) = A\vec{x}$ be an invertible linear transformation from $\mathbb{R}^2$ to $\mathbb{R}^2$. The image of the unit circle under L is an ellipse E. The lengths of the semimajor and the semiminor axes of E are the singular values σ_1 and σ_2 of A, respectively.

Take another look at Figure 2.

We can generalize our findings in Examples 1 and 2 to matrices of arbitrary size.

Fact 8.3.3

Let $L(\vec{x}) = A\vec{x}$ be a linear transformation from $\mathbb{R}^n$ to $\mathbb{R}^m$. Then there is an orthonormal basis $\vec{v}_1, \vec{v}_2, \ldots, \vec{v}_n$ of $\mathbb{R}^n$ such that

a. Vectors $L(\vec{v}_1), L(\vec{v}_2), \ldots, L(\vec{v}_n)$ are orthogonal, and
b. The lengths of vectors $L(\vec{v}_1), L(\vec{v}_2), \ldots, L(\vec{v}_n)$ are the singular values $\sigma_1, \sigma_2, \ldots, \sigma_n$ of matrix A.

To construct $\vec{v}_1, \vec{v}_2, \ldots, \vec{v}_n$, find an orthonormal eigenbasis for matrix $A^T A$. Make sure that the corresponding eigenvalues $\lambda_1, \lambda_2, \ldots, \lambda_n$ appear in descending order:

$$\lambda_1 \geq \lambda_2 \geq \cdots \geq \lambda_n$$

The proof is analogous to the special case $n = m = 2$ considered in Examples 1 and 2:

a. $L(\vec{v}_i) \cdot L(\vec{v}_j) = (A\vec{v}_i) \cdot (A\vec{v}_j) = (A\vec{v}_i)^T A\vec{v}_j = \vec{v}_i^T A^T A\vec{v}_j = \vec{v}_i^T (\lambda_j \vec{v}_j) = \lambda_j(\vec{v}_i \cdot \vec{v}_j) = 0$ when $i \neq j$, and
b. $\|L(\vec{v}_i)\|^2 = (A\vec{v}_i) \cdot (A\vec{v}_i) = \vec{v}_i^T A^T A\vec{v}_i = \vec{v}_i^T (\lambda_i \vec{v}_i) = \lambda_i(\vec{v}_i \cdot \vec{v}_i) = \lambda_i = \sigma_i^2$, so that $\|L(\vec{v}_i)\| = \sigma_i$.

EXAMPLE 3 Consider the linear transformation

$$L(\vec{x}) = A\vec{x}, \quad \text{where} \quad A = \begin{bmatrix} 0 & 1 & 1 \\ 1 & 1 & 0 \end{bmatrix}.$$

a. Find the singular values of A.

b. Find orthonormal vectors $\vec{v}_1, \vec{v}_2, \vec{v}_3$ in $\mathbb{R}^3$ such that $L(\vec{v}_1), L(\vec{v}_2)$, and $L(\vec{v}_3)$ are orthogonal.

c. Sketch and describe the image of the unit sphere under the transformation L.

Solution

a.

$$A^T A = \begin{bmatrix} 0 & 1 \\ 1 & 1 \\ 1 & 0 \end{bmatrix} \begin{bmatrix} 0 & 1 & 1 \\ 1 & 1 & 0 \end{bmatrix} = \begin{bmatrix} 1 & 1 & 0 \\ 1 & 2 & 1 \\ 0 & 1 & 1 \end{bmatrix}.$$

The eigenvalues are $\lambda_1 = 3, \lambda_2 = 1, \lambda_3 = 0$. The singular values of A are

$$\sigma_1 = \sqrt{\lambda_1} = \sqrt{3}, \qquad \sigma_2 = \sqrt{\lambda_2} = 1, \qquad \sigma_3 = \sqrt{\lambda_3} = 0.$$

b. Find an orthonormal eigenbasis v_1, v_2, v_3 for $A^T A$ (we omit the details):

$$E_3 = \text{span} \begin{bmatrix} 1 \\ 2 \\ 1 \end{bmatrix}, \qquad E_1 = \text{span} \begin{bmatrix} 1 \\ 0 \\ -1 \end{bmatrix}, \qquad E_0 = \ker(A) = \text{span} \begin{bmatrix} 1 \\ -1 \\ 1 \end{bmatrix}$$

$$\vec{v}_1 = \frac{1}{\sqrt{6}} \begin{bmatrix} 1 \\ 2 \\ 1 \end{bmatrix}, \qquad \vec{v}_2 = \frac{1}{\sqrt{2}} \begin{bmatrix} 1 \\ 0 \\ -1 \end{bmatrix}, \qquad \vec{v}_3 = \frac{1}{\sqrt{3}} \begin{bmatrix} 1 \\ -1 \\ 1 \end{bmatrix}.$$

We compute $A\vec{v}_1, A\vec{v}_2, A\vec{v}_3$ and check orthogonality:

$$A\vec{v}_1 = \frac{1}{\sqrt{6}} \begin{bmatrix} 3 \\ 3 \end{bmatrix}, \qquad A\vec{v}_2 = \frac{1}{\sqrt{2}} \begin{bmatrix} -1 \\ 1 \end{bmatrix}, \qquad A\vec{v}_3 = \begin{bmatrix} 0 \\ 0 \end{bmatrix}.$$

We can also check that the length of $A\vec{v}_i$ is σ_i:

$$\|A\vec{v}_1\| = \sqrt{3} = \sigma_1, \qquad \|A\vec{v}_2\| = 1 = \sigma_2, \qquad \|A\vec{v}_3\| = 0 = \sigma_3.$$

c. The unit sphere in $\mathbb{R}^3$ consists of all vectors of the form

$$\vec{x} = c_1 \vec{v}_1 + c_2 \vec{v}_2 + c_3 \vec{v}_3, \quad \text{where} \quad c_1^2 + c_2^2 + c_3^2 = 1.$$

The image of the unit sphere consists of the vectors

$$L(\vec{x}) = c_1 L(\vec{v}_1) + c_2 L(\vec{v}_2),$$

where $c_1^2 + c_2^2 \leq 1$. (Recall that $L(\vec{v}_3) = \vec{0}$.)
The image is the full ellipse shaded in Figure 3. ■

Example 3 shows that some of the singular values of a matrix may be zero. Suppose the singular values $\sigma_1, \ldots, \sigma_s$ of an $m \times n$ matrix A are nonzero, while $\sigma_{s+1}, \ldots, \sigma_n$ are zero. Choose vectors $\vec{v}_1, \ldots, \vec{v}_s, \vec{v}_{s+1}, \ldots, \vec{v}_n$ for A as introduced in Fact 8.3.3. Note that $\|A\vec{v}_i\| = \sigma_i = 0$ and therefore $A\vec{v}_i = \vec{0}$ for $i = s + 1$, $\ldots, n$. We claim that the vectors $A\vec{v}_1, \ldots, A\vec{v}_s$ form a basis of the image of A.

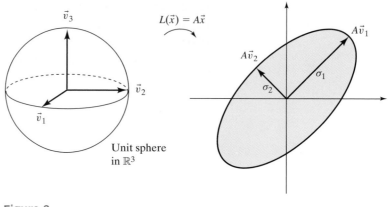

Figure 3

Indeed, these vectors are linearly independent (because they are orthogonal and nonzero), and they span the image, since any vector in the image of A can be written as

$$A\vec{x} = A(c_1\vec{v}_1 + \cdots + c_s\vec{v}_s + \cdots + c_n\vec{v}_n)$$
$$= c_1 A\vec{v}_1 + \cdots + c_s A\vec{v}_s.$$

This shows that $s = \dim(\text{im } A) = \text{rank}(A)$.

Fact 8.3.4   If A is an $m \times n$ matrix of rank r, then the singular values $\sigma_1, \ldots, \sigma_r$ are nonzero, while $\sigma_{r+1}, \ldots, \sigma_n$ are zero.

The Singular Value Decomposition

Just as we expressed the Gram–Schmidt process in terms of a matrix decomposition (the QR factorization), we will now express Fact 8.3.3 in terms of a matrix decomposition.

Consider a linear transformation $L(\vec{x}) = A\vec{x}$ from $\mathbb{R}^n$ to $\mathbb{R}^m$, and choose an orthonormal basis $\vec{v}_1, \ldots, \vec{v}_n$ as in Fact 8.3.3. Let $r = \text{rank}(A)$. We know that the vectors $A\vec{v}_1, \ldots, A\vec{v}_r$ are orthogonal and nonzero, with $\|A\vec{v}_i\| = \sigma_i$. We introduce the unit vectors

$$\vec{u}_1 = \frac{1}{\sigma_1} A\vec{v}_1, \ldots, \vec{u}_r = \frac{1}{\sigma_r} A\vec{v}_r.$$

We can expand the sequence $\vec{u}_1, \ldots, \vec{u}_r$ to an orthonormal basis $\vec{u}_1, \ldots, \vec{u}_m$ of $\mathbb{R}^m$. Then we can write

$$A\vec{v}_i = \sigma_i \vec{u}_i \quad \text{for } i = 1, \ldots, r$$

and

$$A\vec{v}_i = \vec{0} \quad \text{for } i = r+1, \ldots, n.$$

We can express these equations in matrix form as follows:

$$A \underbrace{\begin{bmatrix} | & & | & | & & | \\ \vec{v}_1 & \cdots & \vec{v}_r & \vec{v}_{r+1} & \cdots & \vec{v}_n \\ | & & | & | & & | \end{bmatrix}}_{V} = \begin{bmatrix} | & & | & | & & | \\ \sigma_1\vec{u}_1 & \cdots & \sigma_r\vec{u}_r & \vec{0} & \cdots & \vec{0} \\ | & & | & | & & | \end{bmatrix}$$

$$= \begin{bmatrix} | & & | & | & & | \\ \vec{u}_1 & \cdots & \vec{u}_r & \vec{0} & \cdots & \vec{0} \\ | & & | & | & & | \end{bmatrix} \begin{bmatrix} \sigma_1 & & & \\ & \ddots & & 0 \\ & & \sigma_r & \\ 0 & & & 0 \end{bmatrix}$$

$$= \underbrace{\begin{bmatrix} | & & | & | & & | \\ \vec{u}_1 & \cdots & \vec{u}_r & \vec{u}_{r+1} & \cdots & \vec{u}_m \\ | & & | & | & & | \end{bmatrix}}_{U} \underbrace{\begin{bmatrix} \sigma_1 & & & \\ & \ddots & & 0 \\ & & \sigma_r & \\ 0 & & & 0 \end{bmatrix}}_{\Sigma},$$

or, more succinctly,

$$AV = U\Sigma.$$

Note that V is an orthogonal $n \times n$ matrix, U is an orthogonal $m \times m$ matrix, and Σ is an $m \times n$ matrix whose first r diagonal entries are $\sigma_1, \ldots, \sigma_r$, and all other entries are zero.

Multiplying the equation $AV = U\Sigma$ with V^T from the right, we find that $A = U\Sigma V^T$.

Fact 8.3.5

Singular-value decomposition (SVD)

Any $m \times n$ matrix A can be written as

$$A = U\Sigma V^T,$$

where U is an orthogonal $m \times m$ matrix; V is an orthogonal $n \times n$ matrix; and Σ is an $m \times n$ matrix whose first r diagonal entries are the nonzero singular values $\sigma_1, \ldots, \sigma_r$ of A, and all other entries are zero (where $r = \mathrm{rank}(A)$).

Alternatively, this singular value decomposition can be written as

$$A = \sigma_1\vec{u}_1\vec{v}_1^T + \cdots + \sigma_r\vec{u}_r\vec{v}_r^T,$$

where the $\vec{u}_i$ and the $\vec{v}_i$ are the columns of U and V, respectively. (See Exercise 29.)

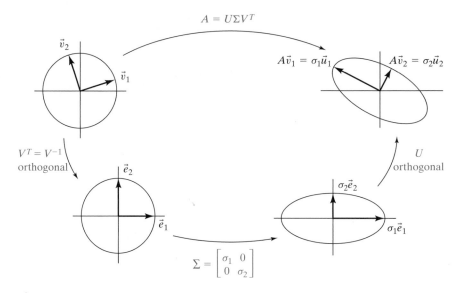

Figure 4

A singular value decomposition of a 2×2 matrix A is presented in Figure 4.

Here are two numerical examples.

EXAMPLE 4   Find an SVD for $A = \begin{bmatrix} 6 & 2 \\ -7 & 6 \end{bmatrix}$. (Compare with Example 2.)

Solution

In Example 2, we found $\vec{v}_1 = \dfrac{1}{\sqrt{5}} \begin{bmatrix} 2 \\ -1 \end{bmatrix}$ and $\vec{v}_2 = \dfrac{1}{\sqrt{5}} \begin{bmatrix} 1 \\ 2 \end{bmatrix}$, so that

$$V = \frac{1}{\sqrt{5}} \begin{bmatrix} 2 & 1 \\ -1 & 2 \end{bmatrix}.$$

The columns $\vec{u}_1$ and $\vec{u}_2$ of U are defined by

$$\vec{u}_1 = \frac{1}{\sigma_1} A\vec{v}_1 = \frac{1}{\sqrt{5}} \begin{bmatrix} 1 \\ -2 \end{bmatrix},$$

$$\vec{u}_2 = \frac{1}{\sigma_2} A\vec{v}_2 = \frac{1}{\sqrt{5}} \begin{bmatrix} 2 \\ 1 \end{bmatrix},$$

and therefore

$$U = \frac{1}{\sqrt{5}} \begin{bmatrix} 1 & 2 \\ -2 & 1 \end{bmatrix}.$$

Finally,

$$\Sigma = \begin{bmatrix} \sigma_1 & 0 \\ 0 & \sigma_2 \end{bmatrix} = \begin{bmatrix} 10 & 0 \\ 0 & 5 \end{bmatrix}.$$

You can check that

$$A = U\Sigma V^T.$$

EXAMPLE 5 Find an SVD for $A = \begin{bmatrix} 0 & 1 & 1 \\ 1 & 1 & 0 \end{bmatrix}$. (Compare with Example 3.)

Solution

Using our work in Example 3, we find that

$$V = \begin{bmatrix} 1/\sqrt{6} & 1/\sqrt{2} & 1/\sqrt{3} \\ 2/\sqrt{6} & 0 & -1/\sqrt{3} \\ 1/\sqrt{6} & -1/\sqrt{2} & 1/\sqrt{3} \end{bmatrix},$$

$$U = \begin{bmatrix} 1/\sqrt{2} & -1/\sqrt{2} \\ 1/\sqrt{2} & 1/\sqrt{2} \end{bmatrix},$$

and

$$\Sigma = \begin{bmatrix} \sqrt{3} & 0 & 0 \\ 0 & 1 & 0 \end{bmatrix}.$$

Check that $A = U\Sigma V^T$.

Consider a singular value decomposition

$$A = U\Sigma V^T,$$

where

$$V = \begin{bmatrix} | & & | \\ \vec{v}_1 & \cdots & \vec{v}_n \\ | & & | \end{bmatrix} \quad \text{and} \quad U = \begin{bmatrix} | & & | \\ \vec{u}_1 & \cdots & \vec{u}_m \\ | & & | \end{bmatrix}.$$

We know that

$$A\vec{v}_i = \sigma_i \vec{u}_i \quad \text{for } i = 1, \dots, r$$

and

$$A\vec{v}_i = \vec{0} \quad \text{for } i = r+1, \dots, n.$$

These equations tell us that

$$\ker(A) = \text{span}(\vec{v}_{r+1}, \dots, \vec{v}_n)$$

and

$$\text{im}(A) = \text{span}(\vec{u}_1, \dots, \vec{u}_r).$$

(Fill in the details.) We see that an SVD provides us with orthonormal bases for the kernel and image of A.

Likewise, we have

$$A^T = V\Sigma^T U^T \quad \text{or} \quad A^T U = V\Sigma^T.$$

Reading the last equation column by column, we find that

$$A^T \vec{u}_i = \sigma_i \vec{v}_i, \quad \text{for } i = 1, \ldots, r,$$

and

$$A^T \vec{u}_i = \vec{0}, \quad \text{for } i = r + 1, \ldots, m.$$

(Observe that the roles of vectors $\vec{u}_i$ and the $\vec{v}_i$ are reversed.)

As before, we have

$$\text{im}(A^T) = \text{span}(\vec{v}_1, \ldots, \vec{v}_r)$$

and

$$\ker(A^T) = \text{span}(\vec{u}_{r+1}, \ldots, \vec{u}_m).$$

In Figure 5, we make an attempt to visualize these observations. We represent each of the kernels and images simply as a line.

Note that $\text{im}(A)$ and $\ker(A^T)$ are orthogonal complements, as observed in Fact 5.4.1.

We conclude this section with a brief discussion of one of the many applications of the SVD—an application to data compression. We follow the exposition of Gilbert Strang (*Linear Algebra and Its Applications,* 3d ed., Harcourt, 1986).

Suppose a satellite transmits a picture containing 1000×1000 pixels. If the color of each pixel is digitized, this information can be represented in a 1000×1000 matrix A. How can we transmit the essential information contained in this picture without sending all 1,000,000 numbers?

Suppose we know an SVD

$$A = \sigma_1 \vec{u}_1 \vec{v}_1^T + \cdots + \sigma_r \vec{u}_r \vec{v}_r^T.$$

Even if the rank r of the matrix A is large, most of the singular values will typically be very small (relative to σ_1). If we neglect those, we get a good approximation $A \approx \sigma_1 \vec{u}_1 \vec{v}_1^T + \cdots + \sigma_s \vec{u}_s \vec{v}_s^T$, where s is much smaller than r. For example, if we choose $s = 10$, we need to transmit only the 20 vectors $\sigma_1 \vec{u}_1, \ldots, \sigma_{10} \vec{u}_{10}$ and $\vec{v}_1, \ldots, \vec{v}_{10}$ in $\mathbb{R}^{1000}$, that is, 20,000 numbers.

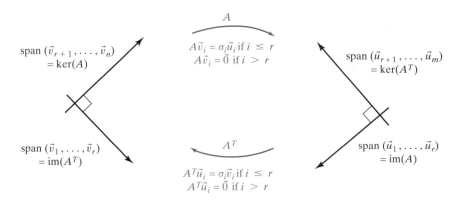

Figure 5

EXERCISES

GOALS Find the singular values and a singular value decomposition of a matrix. Interpret the singular values of a 2×2 matrix in terms of the image of the unit circle.

1. Find the singular values of $A = \begin{bmatrix} 1 & 0 \\ 0 & -2 \end{bmatrix}$.

2. Let A be an orthogonal 2×2 matrix. Use the image of the unit circle to find the singular values of A.

3. Let A be an orthogonal $n \times n$ matrix. Find the singular values of A algebraically.

4. Find the singular values of $A = \begin{bmatrix} 1 & 1 \\ 0 & 1 \end{bmatrix}$.

5. Find the singular values of $A = \begin{bmatrix} p & -q \\ q & p \end{bmatrix}$. Explain your answer geometrically.

6. Find the singular values of $A = \begin{bmatrix} 1 & 2 \\ 2 & 4 \end{bmatrix}$. Find a unit vector $\vec{v}_1$ such that $\|A\vec{v}_1\| = \sigma_1$. Sketch the image of the unit circle.

Find singular value decompositions for the matrices listed in Exercises 7 through 14. Work with paper and pencil.

7. $\begin{bmatrix} 1 & 0 \\ 0 & -2 \end{bmatrix}$ **8.** $\begin{bmatrix} p & -q \\ q & p \end{bmatrix}$ **9.** $\begin{bmatrix} 1 & 2 \\ 2 & 4 \end{bmatrix}$

10. $\begin{bmatrix} 6 & -7 \\ 2 & 6 \end{bmatrix}$ (See Example 4.)

11. $\begin{bmatrix} 1 & 0 \\ 0 & 2 \\ 0 & 0 \end{bmatrix}$

12. $\begin{bmatrix} 0 & 1 \\ 1 & 1 \\ 1 & 0 \end{bmatrix}$ (See Example 5.)

13. $\begin{bmatrix} 6 & 3 \\ -1 & 2 \end{bmatrix}$ **14.** $\begin{bmatrix} 2 & 3 \\ 0 & 2 \end{bmatrix}$

15. If A is an invertible 2×2 matrix, what is the relationship between the singular values of A and A^{-1}? Justify your answer in terms of the image of the unit circle.

16. If A is an invertible $n \times n$ matrix, what is the relationship between the singular values of A and A^{-1}?

17. Consider an $m \times n$ matrix A with $\text{rank}(A) = n$, and a singular value decomposition $A = U\Sigma V^T$. Show that the least-squares solution of a linear system $A\vec{x} = \vec{b}$ can be written as

$$\vec{x}^* = \frac{\vec{b} \cdot \vec{u}_1}{\sigma_1} \vec{v}_1 + \cdots + \frac{\vec{b} \cdot \vec{u}_n}{\sigma_n} \vec{v}_n.$$

18. Consider the 4×2 matrix

$$A = \frac{1}{10} \begin{bmatrix} 1 & 1 & 1 & 1 \\ 1 & 1 & -1 & -1 \\ 1 & -1 & 1 & -1 \\ 1 & -1 & -1 & 1 \end{bmatrix} \begin{bmatrix} 2 & 0 \\ 0 & 1 \\ 0 & 0 \\ 0 & 0 \end{bmatrix} \begin{bmatrix} 3 & -4 \\ 4 & 3 \end{bmatrix}.$$

Use the result of Exercise 17 to find the least-squares solution of the linear system

$$A\vec{x} = \vec{b}, \quad \text{where} \quad \vec{b} = \begin{bmatrix} 1 \\ 2 \\ 3 \\ 4 \end{bmatrix}.$$

Work with paper and pencil.

19. Consider an $m \times n$ matrix A of rank r, and a singular value decomposition $A = U\Sigma V^T$. Explain how you can express the least-squares solutions of a system $A\vec{x} = \vec{b}$ as linear combinations of the columns $\vec{v}_1, \ldots, \vec{v}_n$ of V.

20. a. Explain how any square matrix A can be written as

$$A = QS,$$

where Q is orthogonal and S is symmetric positive semidefinite. *Hint:* Write $A = U\Sigma V^T = U V^T V \Sigma V^T$.

b. Is it possible to write $A = S_1 Q_1$, where Q_1 is orthogonal and S_1 is symmetric positive semidefinite?

21. Find a decomposition $A = QS$ as discussed in Exercise 20 for $A = \begin{bmatrix} 6 & 2 \\ -7 & 6 \end{bmatrix}$. (Compare this with Examples 2 and 4.)

22. Consider an arbitrary 2×2 matrix A and an orthogonal 2×2 matrix S.

a. Explain in terms of the image of the unit circle why A and SA have the same singular values.

b. Explain algebraically why A and SA have the same singular values.

23. Consider an SVD

$$A = U\Sigma V^T$$

of an $m \times n$ matrix A. Show that the columns of U form an orthonormal eigenbasis for AA^T. What are the associated eigenvalues? What does your answer tell you about the relationship between the eigenvalues of $A^T A$ and AA^T? (Compare this with Exercise 7.4.57.)

24. If A is a symmetric $n \times n$ matrix, what is the relationship between the eigenvalues and the singular values of A?

25. Let A be a 2×2 matrix and $\vec{u}$ a unit vector in $\mathbb{R}^2$. Show that

$$\sigma_2 \leq \|A\vec{u}\| \leq \sigma_1,$$

where σ_1, σ_2 are the singular values of A. Illustrate this inequality with a sketch, and justify it algebraically.

26. Let A be an $m \times n$ matrix and $\vec{v}$ a vector in $\mathbb{R}^n$. Show that

$$\sigma_n\|\vec{v}\| \leq \|A\vec{v}\| \leq \sigma_1\|\vec{v}\|,$$

where σ_1 and σ_n are the largest and the smallest singular values of A, respectively. (Compare this with Exercise 25.)

27. Let λ be a real eigenvalue of an $n \times n$ matrix A. Show that

$$\sigma_n \leq |\lambda| \leq \sigma_1,$$

where σ_1 and σ_n are the largest and the smallest singular values of A, respectively.

28. If A is an $n \times n$ matrix, what is the product of its singular values $\sigma_1, \ldots, \sigma_n$? State the product in terms of the determinant of A. For a 2×2 matrix A, explain this result in terms of the image of the unit circle.

29. Show that an SVD

$$A = U\Sigma V^T$$

can be written as

$$A = \sigma_1\vec{u}_1\vec{v}_1^T + \cdots + \sigma_r\vec{u}_r\vec{v}_r^T.$$

30. Find a decomposition

$$A = \sigma_1\vec{u}_1\vec{v}_1^T + \sigma_2\vec{u}_2\vec{v}_2^T$$

for $A = \begin{bmatrix} 6 & 2 \\ -7 & 6 \end{bmatrix}$. (See Exercise 29 and Example 2.)

31. Show that any matrix of rank r can be written as the sum of r matrices of rank 1.

32. Consider an $m \times n$ matrix A, an orthogonal $m \times m$ matrix S, and an orthogonal $n \times n$ matrix R. Compare the singular values of A and SAR.

33. If the singular values of an $n \times n$ matrix A are all 1, is A necessarily orthogonal?

34. For which square matrices A is there a singular value decomposition $A = U\Sigma V^T$ with $U = V$?

35. Consider a singular value decomposition $A = U\Sigma V^T$ of an $m \times n$ matrix A with $\text{rank}(A) = n$. Let $\vec{v}_1, \ldots, \vec{v}_n$ be the columns of V and $\vec{u}_1, \ldots, \vec{u}_m$ the columns of U. Without using the results of Chapter 5, compute $(A^T A)^{-1}A^T\vec{u}_i$. Explain the result in terms of least-squares approximations.

36. Consider a singular value decomposition $A = U\Sigma V^T$ of an $m \times n$ matrix A with $\text{rank}(A) = n$. Let $\vec{u}_1, \ldots, \vec{u}_m$ be the columns of U. Without using the results of Chapter 5, compute $A(A^T A)^{-1}A^T\vec{u}_i$. Explain your result in terms of Fact 5.4.8.

C h a p t e r 8

TRUE OR FALSE?

(Work with real numbers throughout.)

1. All symmetric matrices are diagonalizable.

2. If the matrix $\begin{bmatrix} a & b \\ b & c \end{bmatrix}$ is positive definite, then a must be positive.

3. If A is an orthogonal matrix, then there must be a symmetric invertible matrix S such that $S^{-1}AS$ is diagonal.

4. The singular value of the 2×1 matrix $\begin{bmatrix} 3 \\ 4 \end{bmatrix}$ is 5.

5. The function $q(x_1, x_2) = 3x_1^2 + 4x_1x_2 + 5x_2^2$ is a quadratic form.

6. The singular values of any matrix A are the eigenvalues of matrix $A^T A$.

7. If matrix A is positive definite, then all the eigenvalues of A must be positive.

8. The function $q(\vec{x}) = \vec{x}^T \begin{bmatrix} 1 & 2 \\ 2 & 4 \end{bmatrix} \vec{x}$ is a quadratic form.

9. The singular values of any diagonal matrix D are the absolute values of the diagonal entries of D.

10. The equation $2x^2 + 5xy + 3y^2 = 1$ defines an ellipse.

11. All skew-symmetric matrices are diagonalizable (over $\mathbb{R}$).

12. If A is any matrix, then matrix AA^T is diagonalizable.

13. All positive definite matrices are invertible.

14. Matrix $\begin{bmatrix} 3 & 2 & 1 \\ 2 & 3 & 2 \\ 1 & 2 & 3 \end{bmatrix}$ is diagonalizable.

15. The singular values of any triangular matrix are the absolute values of its diagonal entries.

16. If A is any matrix, then matrix $A^T A$ is the transpose of $A A^T$.

17. If the singular values of a 2×2 matrix are 3 and 4, then there must be a unit vector $\vec{u}$ in $\mathbb{R}^2$ such that $\|A\vec{u}\| = 4$.

18. The determinant of a negative definite 4×4 matrix must be positive.

19. If A is a symmetric matrix such that $A\vec{v} = 3\vec{v}$ and $A\vec{w} = 4\vec{w}$, then the equation $\vec{v} \cdot \vec{w} = 0$ must hold.

20. Matrix $\begin{bmatrix} -2 & 1 & 1 \\ 1 & -2 & 1 \\ 1 & 1 & -2 \end{bmatrix}$ is negative definite.

21. If A and S are invertible $n \times n$ matrices, then matrices A and $S^T A S$ must be similar.

22. If A is negative definite, then all the diagonal entries of A must be negative.

23. If the positive definite matrix A is similar to the symmetric matrix B, then B must be positive definite as well.

24. If A is a symmetric matrix, then there must be an orthogonal matrix S such that $S A S^T$ is diagonal.

25. If $\vec{v}$ and $\vec{w}$ are linearly independent eigenvectors of a symmetric matrix A, then $\vec{w}$ must be orthogonal to $\vec{v}$.

26. For any $m \times n$ matrix A there exists an orthogonal $n \times n$ matrix S such that the columns of matrix $A S$ are orthogonal.

27. If A is a symmetric $n \times n$ matrix such that $A^n = 0$, then A must be the zero matrix.

28. If $q(\vec{x})$ is a positive definite quadratic form, then so is $kq(\vec{x})$, for any scalar k.

29. If A is an invertible symmetric matrix, then A^2 must be positive definite.

30. If the two columns $\vec{v}$ and $\vec{w}$ of a 2×2 matrix A are orthogonal, then the singular values of A must be $\|\vec{v}\|$ and $\|\vec{w}\|$.

31. The product of two quadratic forms in 3 variables must be a quadratic form as well.

32. The function $q(\vec{x}) = \vec{x}^T \begin{bmatrix} 1 & 2 \\ 3 & 4 \end{bmatrix} \vec{x}$ is a quadratic form.

33. If the determinants of all the principal submatrices of a symmetric 3×3 matrix A are negative, then A must be negative definite.

34. If A and B are positive definite $n \times n$ matrices, then matrix $A + B$ must be positive definite as well.

35. If A is a positive definite $n \times n$ matrix and $\vec{x}$ is a nonzero vector in $\mathbb{R}^n$, then the angle between $\vec{x}$ and $A\vec{x}$ must be acute.

36. If the 2×2 matrix A has the singular values 2 and 3 and the 2×2 matrix B has the singular values 4 and 5, then both singular values of matrix AB must be ≤ 15.

37. The equation $A^T A = A A^T$ holds for all square matrices A.

38. For every symmetric $n \times n$ matrix A there is a constant k such that $A + kI_n$ is positive definite.

39. If A and B are 2×2 matrices, then the singular values of matrices AB and BA must be the same.

40. If A is any orthogonal matrix, then matrix $A + A^{-1}$ is diagonalizable (over $\mathbb{R}$).

41. If matrix $\begin{bmatrix} a & b & c \\ b & d & e \\ c & e & f \end{bmatrix}$ is positive definite, then af must exceed c^2.

42. If A is positive definite, then all the entries of A must be positive or zero.

43. If A is indefinite, then 0 must be an eigenvalue of A.

44. If A is a 2×2 matrix with singular values 3 and 5, then there must be a unit vector $\vec{u}$ in $\mathbb{R}^2$ such that $\|A\vec{u}\| = 4$.

45. If A is skew-symmetric, then A^2 must be negative semidefinite.

46. The product of the n singular values of an $n \times n$ matrix A must be $|\det A|$.

47. If $A = \begin{bmatrix} 1 & 2 \\ 2 & 3 \end{bmatrix}$, then there are exactly 4 orthogonal 2×2 matrices S such that $S^{-1} A S$ is diagonal.

48. The sum of two quadratic forms in 3 variables must be a quadratic form as well.

49. The eigenvalues of a symmetric matrix A must be equal to the singular values of A.

50. Similar matrices must have the same singular values.

51. If A is a symmetric 2×2 matrix with eigenvalues 1 and 2, then the angle between $\vec{x}$ and $A\vec{x}$ must be less than $\pi/6$, for all nonzero vectors $\vec{x}$ in $\mathbb{R}^2$.

52. If both singular values of a 2×2 matrix A are less than 5, then all the entries of A must be less than 5.

53. If A is a positive definite matrix, then the largest entry of A must be on the diagonal.

C H A P T E R

9

Linear Differential Equations

9.1 AN INTRODUCTION TO CONTINUOUS DYNAMICAL SYSTEMS

There are two fundamentally different ways to model the evolution of a dynamical system over time: the *discrete* approach and the *continuous* approach. As a simple example, consider a dynamical system with only one component.

EXAMPLE 1 You want to open a savings account and you shop around for the best available interest rate. You learn that *DiscreetBank* pays 7%, compounded annually. Its competitor, the *Bank of Continuity,* offers 6% annual interest, compounded continuously. Everything else being equal, where should you open the account?

Solution

Let us examine what will happen to your investment at the two banks. At Discreet-Bank, the balance grows by 7% each year if no deposits or withdrawals are made.

$$
\begin{array}{ccccc}
\text{new} & = & \text{old} & + & \text{interest} \\
\text{balance} & & \text{balance} & & \\
\downarrow & & \downarrow & & \downarrow \\
x(t+1) & = & x(t) & + & 0.07x(t) \\
x(t+1) & = & 1.07x(t) & &
\end{array}
$$

This equation describes a discrete linear dynamical system with one component. The balance after t years is

$$ x(t) = (1.07)^t x_0. $$

396

The balance grows exponentially with time.

At the Bank of Continuity, by definition of continuous compounding, the balance $x(t)$ grows at an *instantaneous rate* of 6% of the current balance:

$$\frac{dx}{dt} = 6\% \text{ of balance } x(t),$$

or

$$\frac{dx}{dt} = 0.06x.$$

Here, we use a *differential equation* to model a continuous linear dynamical system with one component. We will solve the differential equation in two ways, by separating variables and by making an educated guess.

Let us try to guess the solution. We think about an easier problem first. Do we know a function $x(t)$ that is its own derivative: $dx/dt = x$? You may recall from calculus that $x(t) = e^t$ is such a function. (Some people *define* $x(t) = e^t$ by this property.) More generally, the function $x(t) = Ce^t$ is its own derivative, for any constant C. How can we modify $x(t) = Ce^t$ to get a function whose derivative is 0.06 times itself? By the chain rule, $x(t) = Ce^{0.06t}$ will do:

$$\frac{dx}{dt} = \frac{d}{dt}(Ce^{0.06t}) = 0.06Ce^{0.06t} = 0.06x(t).$$

Note that $x(0) = Ce^0 = C$; that is, C is the initial value, x_0. We conclude that the balance after t years is

$$x(t) = e^{0.06t}x_0.$$

Again, the balance $x(t)$ grows exponentially.

Alternatively, we can solve the differential equation $dx/dt = 0.06x$ by separating variables. Write

$$\frac{dx}{x} = 0.06dt$$

and integrate both sides to get

$$\ln(x) = 0.06t + k,$$

for some constant k. Exponentiating gives

$$x = e^{\ln(x)} = e^{0.06t+k} = e^{0.06t}C,$$

where $C = e^k$.

Which bank offers the better deal? We have to compare the exponential functions $(1.07)^t$ and $e^{0.06t}$. Using a calculator, we compute

$$e^{0.06t} = (e^{0.06})^t \approx (1.0618)^t$$

to see that *DiscreetBank* offers the better deal. The extra interest from continuous compounding does not make up for the one-point difference in the nominal interest rate. ∎

We can generalize.

Fact 9.1.1   **Exponential growth and decay**
Consider the linear differential equation

$$\frac{dx}{dt} = kx,$$

with initial value x_0 (k is an arbitrary constant). The solution is

$$x(t) = e^{kt} x_0.$$

The quantity x will grow or decay exponentially (depending on the sign of k).
See Figure 1.

Now consider a dynamical system with state vector $\vec{x}(t)$ and components
$x_1(t), \ldots, x_n(t)$. In Chapter 7, we use the *discrete* approach to model this dynamical
system: we take a snapshot of the system at times $t = 1, 2, 3, \ldots$, and we describe the transformation the system undergoes between these snapshots. If $\vec{x}(t + 1)$
depends linearly on $\vec{x}(t)$, we can write

$$\vec{x}(t + 1) = A\vec{x}(t),$$

or

$$\vec{x}(t) = A^t \vec{x}_0,$$

for some $n \times n$ matrix A.

In the *continuous* approach, we model the gradual change the system undergoes
as time goes by. Mathematically speaking, we model the (instantaneous) *rates of
change* of the components of the state vector $\vec{x}(t)$, or their derivatives

$$\frac{dx_1}{dt}, \quad \frac{dx_2}{dt}, \quad \ldots, \quad \frac{dx_n}{dt}.$$

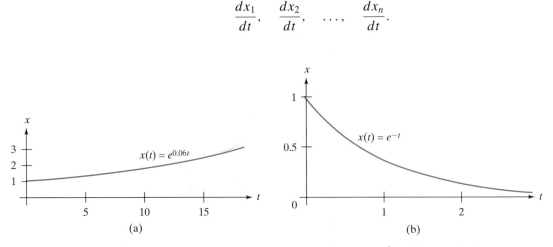

Figure 1 (a) $x(t) = e^{kt}$ with *positive k*. Exponential *growth*. (b) $x(t) = e^{kt}$ with *negative k*.
Exponential *decay*.

If these rates depend linearly on $x_1, x_2, \ldots, x_n$, then we can write

$$
\left|
\begin{aligned}
\frac{dx_1}{dt} &= a_{11}x_1 + a_{12}x_2 + \cdots + a_{1n}x_n \\
\frac{dx_2}{dt} &= a_{21}x_1 + a_{22}x_2 + \cdots + a_{2n}x_n \\
&\vdots \qquad \vdots \qquad \vdots \qquad\qquad \vdots \\
\frac{dx_n}{dt} &= a_{n1}x_1 + a_{n2}x_2 + \cdots + a_{nn}x_n
\end{aligned}
\right|,
$$

or, in matrix form,

$$
\frac{d\vec{x}}{dt} = A\vec{x},
$$

where

$$
A = \begin{bmatrix}
a_{11} & a_{12} & \cdots & a_{1n} \\
a_{21} & a_{22} & \cdots & a_{2n} \\
\vdots & \vdots & & \vdots \\
a_{n1} & a_{n2} & \cdots & a_{nn}
\end{bmatrix}.
$$

The derivative of the parameterized curve $\vec{x}(t)$ is defined component-wise:

$$
\frac{d\vec{x}}{dt} = \begin{bmatrix}
\dfrac{dx_1}{dt} \\[2mm]
\dfrac{dx_2}{dt} \\[2mm]
\vdots \\[2mm]
\dfrac{dx_n}{dt}
\end{bmatrix}
$$

We summarize these observations:

A *linear dynamical system* can be modeled by

$$
\vec{x}(t+1) = B\vec{x}(t) \qquad \text{(discrete model),}
$$

or

$$
\frac{d\vec{x}}{dt} = A\vec{x} \qquad \text{(continuous model).}
$$

A and B are $n \times n$ matrices, where n is the number of components of the system.

We will first think about the equation

$$
\frac{d\vec{x}}{dt} = A\vec{x}
$$

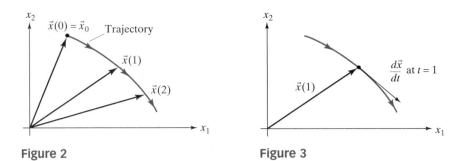

Figure 2   **Figure 3**

from a graphical point of view when A is a 2×2 matrix. We are looking for the parameterized curve

$$\vec{x}(t) = \begin{bmatrix} x_1(t) \\ x_2(t) \end{bmatrix}$$

that represents the evolution of the system from a given initial value $\vec{x}_0$. Each point on the curve $\vec{x}(t)$ will represent the state of the system at a certain moment in time, as shown in Figure 2.

It is natural to think of the trajectory $\vec{x}(t)$ in Figure 2 as the path of a moving particle in the x_1–x_2-plane. As you may have seen in a previous course, the *velocity vector* $d\vec{x}/dt$ of this moving particle is tangent to the trajectory at each point.[1] See Figure 3.

In other words, to solve the system

$$\frac{d\vec{x}}{dt} = A\vec{x}$$

for a given initial state $\vec{x}_0$, we have to find the trajectory in the x_1–x_2-plane that starts at $\vec{x}_0$ and whose velocity vector at each point $\vec{x}$ is the vector $A\vec{x}$. The existence and uniqueness of such a trajectory seems intuitively obvious. Our intuition can be misleading in such matters, however, and it is comforting to know that we can establish the existence and uniqueness of the trajectory later. See Facts 9.1.2 and 9.2.3 and Exercise 9.3.48.

We can represent $A\vec{x}$ graphically as a *vector field* in the x_1–x_2-plane: At the endpoint of each vector $\vec{x}$, we attach the vector $A\vec{x}$. To get a clearer picture, we often sketch merely a *direction field* for $A\vec{x}$, which means that we will not necessarily sketch the vectors $A\vec{x}$ to scale. (We care only about their direction.)

To find the trajectory $\vec{x}(t)$, we simply follow the vector field (or direction field); that is, we follow the arrows of the field, starting at the point representing the initial state $\vec{x}_0$. The trajectories are also called the *flow lines* of the vector field $A\vec{x}$.

To put it differently, imagine a traffic officer standing at each point of the plane, showing us in which direction to go and how fast to move (in other words, defining our velocity). As we follow his directions, we trace out a trajectory.

---

[1] It is sensible to attach the velocity vector $d\vec{x}/dt$ at the endpoint of the state vector $\vec{x}(t)$, indicating the path the particle would take if it were to maintain its direction at time t.

EXAMPLE 2 Consider the linear system $d\vec{x}/dt = A\vec{x}$, where $A = \begin{bmatrix} 1 & 2 \\ 4 & 3 \end{bmatrix}$. In Figure 4, we sketch a direction field for $A\vec{x}$. Draw rough trajectories for the three given initial values.

Solution

Sketch the flow lines for the three given points by following the arrows, as shown in Figure 5.

This picture does not tell the whole story about a trajectory $\vec{x}(t)$. We don't know the position $\vec{x}(t)$ of the moving particle at a specific time t. In other words, we know roughly which path the particle takes, but we don't know how fast it moves along that path. ∎

As we look at Figure 5, our eye's attention is attracted to two special lines, along which the vectors $A\vec{x}$ point either radially away from the origin or directly toward the origin. In either case, the vector $A\vec{x}$ is *parallel* to $\vec{x}$:

$$A\vec{x} = \lambda\vec{x},$$

for some scalar λ. This means that the nonzero vectors along these two special lines are just the eigenvectors of A, and the special lines themselves are the eigenspaces. See Figure 6.

We have seen earlier that the eigenvalues of $A = \begin{bmatrix} 1 & 2 \\ 4 & 3 \end{bmatrix}$ are 5 and -1, with corresponding eigenvectors $\begin{bmatrix} 1 \\ 2 \end{bmatrix}$ and $\begin{bmatrix} 1 \\ -1 \end{bmatrix}$. These results agree with our graphical work in Figures 4 and 5. See Figure 7.

As in the case of a discrete dynamical system, we can sketch a phase portrait for the system $d\vec{x}/dt = A\vec{x}$ that shows some representative trajectories. See Figure 8.

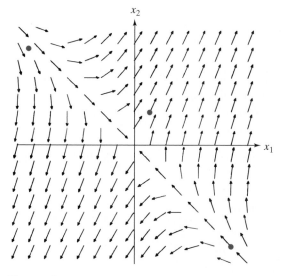

Figure 4

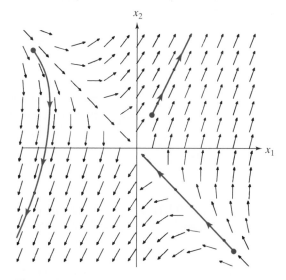

Figure 5

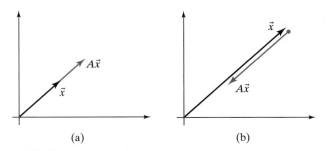

Figure 6 (a) $A\vec{x} = \lambda\vec{x}$, for a *positive* λ. (b) $A\vec{x} = \lambda\vec{x}$, for a *negative* λ.

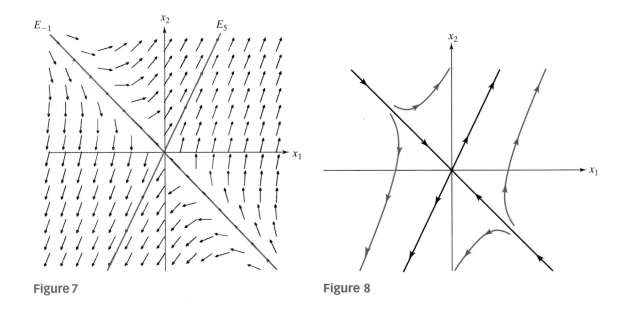

Figure 7   Figure 8

In summary, if the initial state vector $\vec{x}_0$ is an eigenvector, then the trajectory moves along the corresponding eigenspace, away from the origin if the eigenvalue is positive and toward the origin if the eigenvalue is negative. If the eigenvalue is zero, then $\vec{x}_0$ is an equilibrium solution: $\vec{x}(t) = \vec{x}_0$, for all times t.

How can we solve the system $d\vec{x}/dt = A\vec{x}$ analytically? We start with a simple case.

EXAMPLE 3   Find all solutions of the system

$$\frac{d\vec{x}}{dt} = \begin{bmatrix} 2 & 0 \\ 0 & 3 \end{bmatrix} \vec{x}.$$

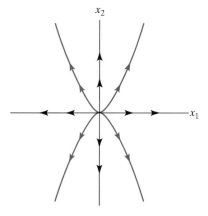

Figure 9

Solution

The two differential equations

$$\frac{dx_1}{dt} = 2x_1$$

$$\frac{dx_2}{dt} = 3x_2$$

are unrelated, or *uncoupled;* we can solve them separately, using Fact 9.1.1:

$$x_1(t) = e^{2t}x_1(0),$$
$$x_2(t) = e^{3t}x_2(0).$$

Thus,

$$\vec{x}(t) = \begin{bmatrix} e^{2t}x_1(0) \\ e^{3t}x_2(0) \end{bmatrix}.$$

Both components of $\vec{x}(t)$ grow exponentially, and the second one will grow faster than the first. In particular, if one of the components is initially 0, it remains 0 for all future times. In Figure 9, we sketch a rough phase portrait for this system.   ■

Now let's do a slightly harder example:

EXAMPLE 4   Find all solutions of the system

$$\frac{d\vec{x}}{dt} = \begin{bmatrix} 1 & 2 \\ -1 & 4 \end{bmatrix} \vec{x}.$$

Solution

We have seen that the eigenvalues and eigenvectors of A tell us a lot about the behavior of the solutions of the system $d\vec{x}/dt = A\vec{x}$. The eigenvalues of A are

$\lambda_1 = 2$ and $\lambda_2 = 3$ with corresponding eigenvectors $\vec{v}_1 = \begin{bmatrix} 2 \\ 1 \end{bmatrix}$ and $\vec{v}_2 = \begin{bmatrix} 1 \\ 1 \end{bmatrix}$. This means that $S^{-1}AS = B$, where $S = \begin{bmatrix} 2 & 1 \\ 1 & 1 \end{bmatrix}$ and $B = \begin{bmatrix} 2 & 0 \\ 0 & 3 \end{bmatrix}$, the matrix considered in Example 3.

We can write the system

$$\frac{d\vec{x}}{dt} = A\vec{x}$$

as

$$\frac{d\vec{x}}{dt} = SBS^{-1}\vec{x},$$

or

$$S^{-1}\frac{d\vec{x}}{dt} = BS^{-1}\vec{x},$$

or (see Exercise 51)

$$\frac{d}{dt}(S^{-1}\vec{x}) = B(S^{-1}\vec{x}).$$

Let us introduce the notation $\vec{c}(t) = S^{-1}\vec{x}(t)$; note that $\vec{c}(t)$ is the coordinate vector of $\vec{x}(t)$ with respect to the eigenbasis $\vec{v}_1, \vec{v}_2$. Then the system takes the form

$$\frac{d\vec{c}}{dt} = B\vec{c},$$

which is just the equation we solved in Example 3. We found that the solutions are of the form

$$\vec{c}(t) = \begin{bmatrix} e^{2t}c_1 \\ e^{3t}c_2 \end{bmatrix},$$

where c_1 and c_2 are arbitrary constants. Therefore, the solutions of the original system

$$\frac{d\vec{x}}{dt} = A\vec{x}$$

are

$$\vec{x}(t) = S\vec{c}(t) = \begin{bmatrix} 2 & 1 \\ 1 & 1 \end{bmatrix} \begin{bmatrix} e^{2t}c_1 \\ e^{3t}c_2 \end{bmatrix} = c_1 e^{2t} \begin{bmatrix} 2 \\ 1 \end{bmatrix} + c_2 e^{3t} \begin{bmatrix} 1 \\ 1 \end{bmatrix}.$$

We can write this formula in more general terms as

$$\vec{x}(t) = c_1 e^{\lambda_1 t} \vec{v}_1 + c_2 e^{\lambda_2 t} \vec{v}_2.$$

Note that c_1 and c_2 are the coordinates of $\vec{x}(0)$ with respect to the basis $\vec{v}_1, \vec{v}_2$, since

$$\vec{x}(0) = c_1 \vec{v}_1 + c_2 \vec{v}_2.$$

It is informative to consider a few special trajectories: If $c_1 = 1$ and $c_2 = 0$, the trajectory

$$\vec{x}(t) = e^{2t} \begin{bmatrix} 2 \\ 1 \end{bmatrix}$$

moves along the eigenspace E_2 spanned by $\begin{bmatrix} 2 \\ 1 \end{bmatrix}$, as expected. Likewise, if $c_1 = 0$ and $c_2 = 1$, we have the trajectory

$$\vec{x}(t) = e^{3t} \begin{bmatrix} 1 \\ 1 \end{bmatrix}$$

moving along the eigenspace E_3.

If $c_2 \neq 0$, then the entries of $c_2 e^{3t} \begin{bmatrix} 1 \\ 1 \end{bmatrix}$ will become much larger (in absolute value) than the entries of $c_1 e^{2t} \begin{bmatrix} 2 \\ 1 \end{bmatrix}$ as t goes to infinity. The *dominant term* $c_2 e^{3t} \begin{bmatrix} 1 \\ 1 \end{bmatrix}$, associated with the larger eigenvalue, determines the behavior of the system in the distant future. The state vector $\vec{x}(t)$ is almost parallel to E_3 for large t. For large negative t, on the other hand, the state vector is very small and almost parallel to E_2.

In Figure 10, we sketch a rough phase portrait for the system $d\vec{x}/dt = A\vec{x}$.

This is a linear distortion of the phase portrait we sketched in Figure 9. More precisely, the matrix $S = \begin{bmatrix} 2 & 1 \\ 1 & 1 \end{bmatrix}$ transforms the phase portraits in Figure 9 into the phase portrait sketched in Figure 10 (transforming $\vec{e}_1$ into the eigenvector $\begin{bmatrix} 2 \\ 1 \end{bmatrix}$ and $\vec{e}_2$ into $\begin{bmatrix} 1 \\ 1 \end{bmatrix}$). ∎

Our work in Examples 3 and 4 generalizes readily to any $n \times n$ matrix A that is diagonalizable over $\mathbb{R}$ (i.e., for which there is an eigenbasis in $\mathbb{R}^n$):

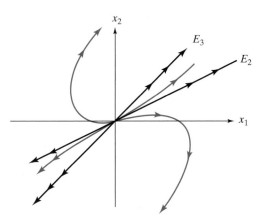

Figure 10

Fact 9.1.2   **Continuous dynamical systems**

Consider the system $d\vec{x}/dt = A\vec{x}$. Suppose there is a real eigenbasis $\vec{v}_1, \ldots, \vec{v}_n$ for A, with associated eigenvalues $\lambda_1, \ldots, \lambda_n$. Then the general solution of the system is

$$\vec{x}(t) = c_1 e^{\lambda_1 t} \vec{v}_1 + \cdots + c_n e^{\lambda_n t} \vec{v}_n.$$

Scalars $c_1, c_2, \ldots, c_n$ are the coordinates of $\vec{x}_0$ with respect to the basis $\vec{v}_1, \vec{v}_2, \ldots, \vec{v}_n$.

We can write the preceding equation in matrix form as

$$\vec{x}(t) = \begin{bmatrix} | & | & & | \\ \vec{v}_1 & \vec{v}_2 & \cdots & \vec{v}_n \\ | & | & & | \end{bmatrix} \begin{bmatrix} e^{\lambda_1 t} & 0 & \cdot & 0 \\ 0 & e^{\lambda_2 t} & \cdot & 0 \\ \cdot & \cdot & \cdot & \cdot \\ 0 & 0 & \cdot & e^{\lambda_n t} \end{bmatrix} \begin{bmatrix} c_1 \\ c_2 \\ \cdot \\ c_n \end{bmatrix}$$

$$= S \begin{bmatrix} e^{\lambda_1 t} & 0 & \cdot & 0 \\ 0 & e^{\lambda_2 t} & \cdot & 0 \\ \cdot & & \cdot & \cdot \\ 0 & 0 & \cdot & e^{\lambda_n t} \end{bmatrix} S^{-1} \vec{x}_0,$$

where $S = \begin{bmatrix} | & | & & | \\ \vec{v}_1 & \vec{v}_2 & \cdots & \vec{v}_n \\ | & | & & | \end{bmatrix}$.

We can think of the general solution as a linear combination of the solutions $e^{\lambda_i t} \vec{v}_i$ associated with the eigenvectors $\vec{v}_i$. Note the similarity between this solution and the general solution of the *discrete* dynamical system $\vec{x}(t + 1) = A\vec{x}(t)$,

$$\vec{x}(t) = c_1 \lambda_1^t \vec{v}_1 + \cdots + c_n \lambda_n^t \vec{v}_n.$$

See Fact 7.1.3.

The terms λ_i^t are replaced by $e^{\lambda_i t}$. We have already observed this fact in a dynamical system with only one component. (See Example 1.)

We can state Fact 9.1.2 in the language of linear spaces. The solutions of the system $d\vec{x}/dt = A\vec{x}$ form a subspace of the space $F(\mathbb{R}, \mathbb{R}^n)$ of all functions from $\mathbb{R}$ to $\mathbb{R}^n$. (See Exercises 22 and 23.) This space is n-dimensional, with basis $e^{\lambda_1 t} \vec{v}_1, e^{\lambda_2 t} \vec{v}_2, \ldots, e^{\lambda_n t} \vec{v}_n$.

EXAMPLE 5   Consider a system $d\vec{x}/dt = A\vec{x}$, where A is diagonalizable over $\mathbb{R}$. When is the zero state a stable equilibrium solution? Give your answer in terms of the eigenvalues of A.

Solution

Note that $\lim_{t \to \infty} e^{\lambda t} = 0$ if (and only if) λ is negative. Therefore, we observe stability if (and only if) all eigenvalues of A are negative.   ∎

Consider an invertible 2×2 matrix A with two distinct eigenvalues $\lambda_1 > \lambda_2$. Then the phase portrait of $d\vec{x}/dt = A\vec{x}$ looks qualitatively like one of the three sketches in Figure 11. We observe stability only in Figure 11(c).

Consider a trajectory that does not run along one of the eigenspaces. In all three cases, the state vector $\vec{x}(t)$ is almost parallel to the dominant eigenspace E_{λ_1} for large t. For large negative t, on the other hand, the state vector is almost parallel to E_{λ_2}. Compare with Figure 7.1.11.

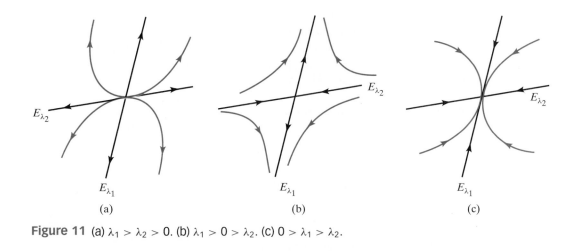

Figure 11 (a) $\lambda_1 > \lambda_2 > 0$. (b) $\lambda_1 > 0 > \lambda_2$. (c) $0 > \lambda_1 > \lambda_2$.

EXERCISES

GOALS Use the concept of a continuous dynamical system. Solve the differential equation $dx/dt = kx$. Solve the system $d\vec{x}/dt = A\vec{x}$ when A is diagonalizable over $\mathbb{R}$, and sketch the phase portrait for 2×2 matrices A.

Solve the initial value problems posed in Exercises 1 through 5. Graph the solution.

1. $\dfrac{dx}{dt} = 5x$ with $x(0) = 7$.

2. $\dfrac{dx}{dt} = -0.71x$ with $x(0) = -e$.

3. $\dfrac{dP}{dt} = 0.03P$ with $P(0) = 7$.

4. $\dfrac{dy}{dt} = 0.8t$ with $y(0) = -0.8$.

5. $\dfrac{dy}{dt} = 0.8y$ with $y(0) = -0.8$.

Solve the nonlinear differential equations in Exercises 6 through 11 using the method of separation of variables: Write the differential equation $dx/dt = f(x)$ as $dx/f(x) = dt$ and integrate both sides.

6. $\dfrac{dx}{dt} = \dfrac{1}{x}$, $x(0) = 1$.

7. $\dfrac{dx}{dt} = x^2$, $x(0) = 1$. Describe the behavior of your solution as t increases.

8. $\dfrac{dx}{dt} = \sqrt{x}$, $x(0) = 4$.

9. $\dfrac{dx}{dt} = x^k$ (with $k \neq 1$), $x(0) = 1$.

10. $\dfrac{dx}{dt} = \dfrac{1}{\cos(x)}$, $x(0) = 0$.

11. $\dfrac{dx}{dt} = 1 + x^2$, $x(0) = 0$.

12. Find a differential equation of the form $dx/dt = kx$ for which $x(t) = 3^t$ is a solution.

13. In 1778, a wealthy Pennsylvanian merchant named Jacob DeHaven lent \$450,000 to the Continental Congress to support the troops at Valley Forge. The loan was never repaid. Mr. DeHaven's descendants are taking the U.S. Government to court to collect what they believe they are owed. The going interest rate at the time was 6%. How much were the DeHavens owed in 1990

    a. if interest is compounded yearly?

    b. if interest is compounded continuously?

    (Adapted from *The New York Times,* May 27, 1990.)

14. The carbon in living matter contains a minute proportion of the radioactive isotope C-14. This radiocarbon arises from cosmic-ray bombardment in the upper atmosphere and enters living systems by exchange processes. After the death of an organism, exchange stops, and the carbon decays. Therefore, carbon dating enables us to calculate the time at which an organism died. Let $x(t)$ be the proportion of the original C-14 still present t years after death. By definition, $x(0) = 1 = 100\%$. We are told that $x(t)$ satisfies the differential equation

$$\frac{dx}{dt} = -\frac{1}{8270}x.$$

a. Find a formula for $x(t)$. Determine the half-life of C-14 (that is, the time it takes for half of the C-14 to decay).

b. *The Iceman.* In 1991, the body of a man was found in melting snow in the Alps of Northern Italy. A well-known historian in Innsbruck, Austria, determined that the man had lived in the Bronze Age, which started about 2000 B.C. in that region. Examination of tissue samples performed independently at Zürich and Oxford revealed that 47% of the C-14 present in the body at the time of his death had decayed. When did this man die? Is the result of the carbon dating compatible with the estimate of the Austrian historian?

15. Justify the "rule of 69": If a quantity grows at an instantaneous rate of $k\%$, then its doubling time is about $69/k$. *Example:* In 1995 the population of India grew at a rate of about 1.9%, with a doubling time of about $69/1.9 \approx 36$ years.
Consider the system

$$\frac{d\vec{x}}{dt} = \begin{bmatrix} \lambda_1 & 0 \\ 0 & \lambda_2 \end{bmatrix} \vec{x}.$$

For the values of λ_1 and λ_2 given in Exercises 16 through 19, sketch the trajectories for all nine initial values shown below. For each of the points, trace out both the future and the past of the system.

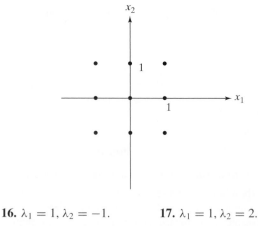

16. $\lambda_1 = 1, \lambda_2 = -1.$
17. $\lambda_1 = 1, \lambda_2 = 2.$
18. $\lambda_1 = -1, \lambda_2 = -2.$
19. $\lambda_1 = 0, \lambda_2 = 1.$

20. Consider the system $d\vec{x}/dt = A\vec{x}$ with $A = \begin{bmatrix} 0 & -1 \\ 1 & 0 \end{bmatrix}$.
Sketch a direction field for $A\vec{x}$. Based on your sketch, describe the trajectories geometrically. From your sketch, can you guess a formula for the solution with $\vec{x}_0 = \begin{bmatrix} 1 \\ 0 \end{bmatrix}$? Verify your guess by substituting into the equations.

21. Consider the system $d\vec{x}/dt = A\vec{x}$ with $A = \begin{bmatrix} 0 & 1 \\ 0 & 0 \end{bmatrix}$.
Sketch a direction field of $A\vec{x}$. Based on your sketch, describe the trajectories geometrically. Can you find the solutions analytically?

22. Consider a linear system $d\vec{x}/dt = A\vec{x}$ of arbitrary size. Suppose $\vec{x}_1(t)$ and $\vec{x}_2(t)$ are solutions of the system. Is the sum $\vec{x}(t) = \vec{x}_1(t) + \vec{x}_2(t)$ a solution as well? How do you know?

23. Consider a linear system $d\vec{x}/dt = A\vec{x}$ of arbitrary size. Suppose $\vec{x}_1(t)$ is a solution of the system and k is an arbitrary constant. Is $\vec{x}(t) = k\vec{x}_1(t)$ a solution as well? How do you know?

24. Let A be an $n \times n$ matrix and k a scalar. Consider the following two systems:

$$(\text{I}) \quad \frac{d\vec{x}}{dt} = A\vec{x},$$

$$(\text{II}) \quad \frac{d\vec{c}}{dt} = (A + kI_n)\vec{c}.$$

Show that if $\vec{x}(t)$ is a solution of system (I), then $\vec{c}(t) = e^{kt}\vec{x}(t)$ is a solution of system (II).

25. Let A be an $n \times n$ matrix and k a scalar. Consider the following two systems:

$$(\text{I}) \quad \frac{d\vec{x}}{dt} = A\vec{x},$$

$$(\text{II}) \quad \frac{d\vec{c}}{dt} = kA\vec{c}.$$

Show that if $\vec{x}(t)$ is a solution of system (I), then $\vec{c}(t) = \vec{x}(kt)$ is a solution of system (II). Compare the vector fields of the two systems.

In Exercises 26 through 31, solve the system with the given initial value.

26. $\dfrac{d\vec{x}}{dt} = \begin{bmatrix} 1 & 2 \\ 3 & 0 \end{bmatrix} \vec{x}$ with $\vec{x}(0) = \begin{bmatrix} 7 \\ 2 \end{bmatrix}$.

27. $\dfrac{d\vec{x}}{dt} = \begin{bmatrix} -4 & 3 \\ 2 & -3 \end{bmatrix} \vec{x}$ with $\vec{x}(0) = \begin{bmatrix} 1 \\ 0 \end{bmatrix}$.

28. $\dfrac{d\vec{x}}{dt} = \begin{bmatrix} 4 & 3 \\ 4 & 8 \end{bmatrix} \vec{x}$ with $\vec{x}(0) = \begin{bmatrix} 1 \\ 1 \end{bmatrix}$.

29. $\dfrac{d\vec{x}}{dt} = \begin{bmatrix} 1 & 2 \\ 2 & 4 \end{bmatrix} \vec{x}$ with $\vec{x}(0) = \begin{bmatrix} 5 \\ 0 \end{bmatrix}$.

30. $\dfrac{d\vec{x}}{dt} = \begin{bmatrix} 1 & 2 \\ 2 & 4 \end{bmatrix} \vec{x}$ with $\vec{x}(0) = \begin{bmatrix} 2 \\ -1 \end{bmatrix}$.

31. $\dfrac{d\vec{x}}{dt} = \begin{bmatrix} 2 & 1 & 1 \\ 1 & 3 & 3 \\ 3 & 2 & 2 \end{bmatrix} \vec{x}$ with $\vec{x}(0) = \begin{bmatrix} 1 \\ -2 \\ 1 \end{bmatrix}$.

Sketch rough phase portraits for the dynamical systems given in Exercises 32 through 39.

32. $\dfrac{d\vec{x}}{dt} = \begin{bmatrix} 1 & 2 \\ 3 & 0 \end{bmatrix} \vec{x}.$

33. $\dfrac{d\vec{x}}{dt} = \begin{bmatrix} -4 & 3 \\ 2 & -3 \end{bmatrix} \vec{x}.$

34. $\dfrac{d\vec{x}}{dt} = \begin{bmatrix} 4 & 3 \\ 4 & 8 \end{bmatrix} \vec{x}.$

35. $\dfrac{d\vec{x}}{dt} = \begin{bmatrix} 1 & 2 \\ 2 & 4 \end{bmatrix} \vec{x}.$

36. $\vec{x}(t+1) = \begin{bmatrix} 0.9 & 0.2 \\ 0.2 & 1.2 \end{bmatrix} \vec{x}(t).$

37. $\vec{x}(t+1) = \begin{bmatrix} 1 & 0.3 \\ -0.2 & 1.7 \end{bmatrix} \vec{x}(t).$

38. $\vec{x}(t+1) = \begin{bmatrix} 1.1 & 0.2 \\ -0.4 & 0.5 \end{bmatrix} \vec{x}(t).$

39. $\vec{x}(t+1) = \begin{bmatrix} 0.8 & -0.4 \\ 0.3 & 1.6 \end{bmatrix} \vec{x}(t).$

40. Find a 2×2 matrix A such that the system $d\vec{x}/dt = A\vec{x}$ has

$$\vec{x}(t) = \begin{bmatrix} 2e^{2t} + 3e^{3t} \\ 3e^{2t} + 4e^{3t} \end{bmatrix}$$

as one of its solutions.

41. Consider a noninvertible 2×2 matrix A with two distinct eigenvalues. (Note that one of the eigenvalues must be 0.) Choose two eigenvectors $\vec{v}_1$ and $\vec{v}_2$ with eigenvalues $\lambda_1 = 0$ and λ_2. Suppose λ_2 is *negative*. Sketch a phase portrait for the system $d\vec{x}/dt = A\vec{x}$, clearly indicating the shape and long-term behavior of the trajectories.

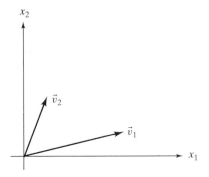

42. Consider the interaction of two species of animals in a habitat. We are told that the change of the populations $x(t)$ and $y(t)$ can be modeled by the equations

$$\left| \begin{aligned} \frac{dx}{dt} &= 1.4x - 1.2y \\ \frac{dy}{dt} &= 0.8x - 1.4y \end{aligned} \right|,$$

where time t is measured in years.

a. What kind of interaction do we observe (symbiosis, competition, or predator-prey)?

b. Sketch a phase portrait for this system. From the nature of the problem, we are interested only in the first quadrant.

c. What will happen in the long term? Does the outcome depend on the initial populations? If so, how?

43. Answer the questions posed in Exercise 42 for the following system:

$$\left| \begin{aligned} \frac{dx}{dt} &= 5x - y \\ \frac{dy}{dt} &= -2x + 4y \end{aligned} \right|.$$

44. Answer the questions posed in Exercise 42 for the following system:

$$\left| \begin{aligned} \frac{dx}{dt} &= x + 4y \\ \frac{dy}{dt} &= 2x - y \end{aligned} \right|.$$

45. Two herds of vicious animals are fighting each other to the death. During the fight, the populations $x(t)$ and $y(t)$ of the two species can be modeled by the following system:[2]

$$\left| \begin{aligned} \frac{dx}{dt} &= -4y \\ \frac{dy}{dt} &= -x \end{aligned} \right|.$$

a. What is the significance of the constants -4 and -1 in these equations? Which species has the more vicious (or more efficient) fighters?

b. Sketch a phase portrait for this system.

c. Who wins the fight (in the sense that some individuals of that species are left while the other herd is eradicated)? How does your answer depend on the initial populations?

---

[2]This is the simplest in a series of combat models developed by F. W. Lanchester during World War I (F. W. Lanchester, *Aircraft in Warfare, the Dawn of the Fourth Arm*, Tiptree, Constable and Co., Ltd., 1916).

46. Repeat Exercise 45 for the system

$$\left| \begin{aligned} \frac{dx}{dt} &= \qquad -py \\ \frac{dy}{dt} &= -qx \end{aligned} \right|,$$

where p and q are two positive constants.[3]

47. The interaction of two populations of animals is modeled by the differential equations

$$\left| \begin{aligned} \frac{dx}{dt} &= -x + ky \\ \frac{dy}{dt} &= kx - 4y \end{aligned} \right|,$$

for some positive constant k.

a. What kind of interaction do we observe? What is the practical significance of the constant k?

b. Find the eigenvalues of the coefficient matrix of the system. What can you say about the signs of these eigenvalues? How does your answer depend on the value of the constant k?

c. For each case you discussed in part (b), sketch a rough phase portrait. What does each phase portrait tell you about the future of the two populations?

48. Repeat Exercise 47 for the system

$$\left| \begin{aligned} \frac{dx}{dt} &= -x + ky \\ \frac{dy}{dt} &= x - 4y \end{aligned} \right|,$$

where k is a positive constant.

49. Here is a continuous model of a person's glucose regulatory system. (Compare this with Exercise 7.1.36.) Let $g(t)$ and $h(t)$ be the excess glucose and insulin concentrations in a person's blood. We are told that

$$\left| \begin{aligned} \frac{dg}{dt} &= -g - 0.2h \\ \frac{dh}{dt} &= 0.6g - 0.2h \end{aligned} \right|,$$

where time t is measured in hours. After a heavy holiday dinner, we measure $g(0) = 30$ and $h(0) = 0$. Find closed formulas for $g(t)$ and $h(t)$. Sketch the trajectory.

50. Consider a linear system $d\vec{x}/dt = A\vec{x}$, where A is a 2×2 matrix that is diagonalizable over $\mathbb{R}$. When is

the zero state a stable equilibrium solution? Give your answer in terms of the determinant and the trace of A.

51. Let $\vec{x}(t)$ be a differentiable curve in $\mathbb{R}^n$ and S an $n \times n$ matrix. Show that

$$\frac{d}{dt}(S\vec{x}) = S\frac{d\vec{x}}{dt}.$$

52. Find all solutions of the system

$$\frac{d\vec{x}}{dt} = \begin{bmatrix} \lambda & 1 \\ 0 & \lambda \end{bmatrix} \vec{x},$$

where λ is an arbitrary constant. *Hint:* Exercises 21 and 24 are helpful. Sketch a phase portrait. For which choices of λ is the zero state a stable equilibrium solution?

53. Solve the initial value problem

$$\frac{d\vec{x}}{dt} = \begin{bmatrix} p & -q \\ q & p \end{bmatrix} \vec{x} \quad \text{with} \quad \vec{x}_0 = \begin{bmatrix} 1 \\ 0 \end{bmatrix}.$$

Sketch the trajectory for the cases when p is positive, negative, or 0. In which cases does the trajectory approach the origin? *Hint:* Exercises 20, 24, and 25 are helpful.

54. Consider a door that opens to only one side (as most doors do). A spring mechanism closes the door automatically. The state of the door at a given time t (measured in seconds) is determined by the *angular displacement* $\alpha(t)$ (measured in radians) and the *angular velocity* $\omega(t) = d\alpha/dt$. Note that α is always positive or zero (since the door opens to only one side), but ω can be positive or negative (depending on whether the door is opening or closing).

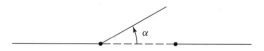

When the door is moving freely (nobody is pushing or pulling), its movement is subject to the following differential equations:

$$\left| \begin{aligned} \frac{d\alpha}{dt} &= \qquad \omega \\ \frac{d\omega}{dt} &= -2\alpha - 3\omega \end{aligned} \right|$$

(the definition of ω)

(-2α reflects the force of the spring, and -3ω models friction).

---

[3]The result is known as Lanchester's square law.

a. Sketch a phase portrait for this system.

b. Discuss the movement of the door represented by the qualitatively different trajectories. For which initial states does the door slam (i.e., reach $\alpha = 0$ with velocity $\omega < 0$)?

55. Answer the questions posed in Exercise 54 for the system

$$\left| \begin{array}{rcl} \dfrac{d\alpha}{dt} & = & \omega \\[2mm] \dfrac{d\omega}{dt} & = & -p\alpha - q\omega \end{array} \right|,$$

where p and q are positive, and $q^2 > 4p$.

9.2 THE COMPLEX CASE: EULER'S FORMULA

Consider a linear system

$$\frac{d\vec{x}}{dt} = A\vec{x},$$

where the $n \times n$ matrix A is diagonalizable over $\mathbb{C}$: There is a complex eigenbasis $\vec{v}_1, \ldots, \vec{v}_n$ for A, with associated complex eigenvalues $\lambda_1, \ldots, \lambda_n$. You may wonder whether the formula

$$\vec{x}(t) = c_1 e^{\lambda_1 t} \vec{v}_1 + \cdots + c_n e^{\lambda_n t} \vec{v}_n$$

(with complex c_i) produces the general complex solution of the system, just as in the real case (Fact 9.1.2).

Before we can make sense out of the formula above, we have to think about the idea of a complex-valued function and in particular about the exponential function $e^{\lambda t}$ for complex λ.

Complex-Valued Functions

A complex-valued function $z = f(t)$ is a function from $\mathbb{R}$ to $\mathbb{C}$ (with domain $\mathbb{R}$ and codomain $\mathbb{C}$): The input t is real, and the output z is complex. Here are two examples:

$$z = t + it^2,$$
$$z = \cos(t) + i\sin(t).$$

For each t, the output z can be represented as a point in the complex plane. As we let t vary, we trace out a *trajectory* in the complex plane. In Figure 1, we sketch the trajectories of the two complex-valued functions just defined.

We can write a complex-valued function $z(t)$ in terms of its real and imaginary parts:

$$z(t) = x(t) + iy(t).$$

(Consider the two preceding examples.) If $x(t)$ and $y(t)$ are differentiable real-valued functions, then the *derivative* of the complex-valued function $z(t)$ is defined by

$$\frac{dz}{dt} = \frac{dx}{dt} + i\frac{dy}{dt}.$$

For example, if

$$z(t) = t + it^2,$$

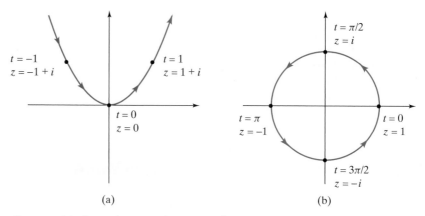

Figure 1 (a) The trajectory of $z = t + it^2$. (b) The trajectory of $z = \cos(t) + i \sin(t)$.

then

$$\frac{dz}{dt} = 1 + 2it.$$

If

$$z(t) = \cos(t) + i \sin(t),$$

then

$$\frac{dz}{dt} = -\sin(t) + i \cos(t).$$

Please verify that the basic rules of differential calculus (the sum, product, and quotient rules) apply to complex-valued functions. The chain rule holds in the following form: If $z = f(t)$ is a differentiable complex-valued function and $t = g(s)$ is a differentiable function from $\mathbb{R}$ to $\mathbb{R}$, then

$$\frac{dz}{ds} = \frac{dz}{dt}\frac{dt}{ds}.$$

The derivative dz/dt of a complex-valued function $z(t)$, for a given t, can be visualized as a tangent vector to the trajectory at $z(t)$, as shown in Figure 2.

Next let's think about the complex-valued exponential function $z = e^{\lambda t}$, where λ is complex and t real. How should the function $z = e^{\lambda t}$ be defined? We can get some inspiration from the real case: The exponential function $x = e^{kt}$ (for real k) is the unique function such that $dx/dt = kx$ and $x(0) = 1$. (Compare this with Fact 9.1.1.)

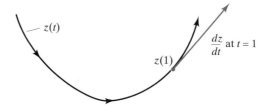

Figure 2

We can use this fundamental property of real exponential functions to define the complex exponential functions:

Definition 9.2.1

> **Complex exponential functions**
> If λ is a complex number, then $z = e^{\lambda t}$ is the unique complex-valued function such that
>
> $$\frac{dz}{dt} = \lambda z \qquad \text{and} \qquad z(0) = 1.$$

(The existence of such a function, for any λ, will be established below; the proof of uniqueness is left as Exercise 38.)

It follows that the unique complex-valued function $z(t)$ with

$$\frac{dz}{dt} = \lambda z \qquad \text{and} \qquad z(0) = z_0$$

is

$$z(t) = e^{\lambda t} z_0,$$

for an arbitrary complex initial value z_0.

Let us first consider the simplest case, $z = e^{it}$, where $\lambda = i$. We are looking for a complex-valued function $z(t)$ such that $dz/dt = iz$ and $z(0) = 1$.

From a graphical point of view, we are looking for the trajectory $z(t)$ in the complex plane that starts at $z = 1$ and whose tangent vector $dz/dt = iz$ is perpendicular to z at each point. (See Example 1 of Section 7.5.) In other words, we are looking for the flow line of the vector field in Figure 3 starting at $z = 1$.

The *unit circle,* with parametrization $z(t) = \cos(t) + i \sin(t)$, satisfies

$$\frac{dz}{dt} = -\sin(t) + i \cos(t) = iz(t),$$

and $z(0) = 1$. See Figure 4.

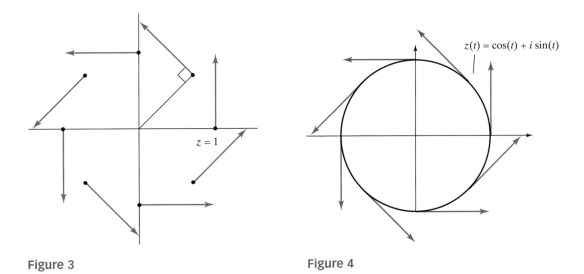

Figure 3   **Figure 4**

Figure 5 Euler's likeness and his celebrated formula are shown on a Swiss postage stamp.

We have shown the following fundamental result:

Fact 9.2.2   **Euler's formula**

$$e^{it} = \cos(t) + i\sin(t).$$

The case $t = \pi$ leads to the intriguing formula $e^{i\pi} = -1$; this has been called the most beautiful formula in all of mathematics.[4]

Euler's formula can be used to write the polar form of a complex number more succinctly:

$$z = r(\cos\phi + i\sin\phi) = re^{i\phi}$$

Now consider $z = e^{\lambda t}$, where λ is an arbitrary complex number, $\lambda = p + iq$. By manipulating exponentials as if they were real, we find that

$$e^{\lambda t} = e^{(p+iq)t} = e^{pt}e^{iqt} = e^{pt}\big(\cos(qt) + i\sin(qt)\big).$$

We can validate this result by checking that the complex-valued function

$$z(t) = e^{pt}\big(\cos(qt) + i\sin(qt)\big)$$

does indeed satisfy the definition of $e^{\lambda t}$, namely, $dz/dt = \lambda z$ and $z(0) = 1$:

$$\frac{dz}{dt} = pe^{pt}\big(\cos(qt) + i\sin(qt)\big) + e^{pt}\big(-q\sin(qt) + iq\cos(qt)\big)$$

$$= (p + iq)e^{pt}\big(\cos(qt) + i\sin(qt)\big) = \lambda z.$$

EXAMPLE 2   Sketch the trajectory of the complex-valued function $z(t) = e^{(0.1+i)t}$ in the complex plane.

Solution

$$z(t) = e^{0.1t}e^{it} = e^{0.1t}\big(\cos(t) + i\sin(t)\big)$$

The trajectory spirals outward as shown in Figure 6, since the function $e^{0.1t}$ grows exponentially. ∎

---

[4]Benjamin Peirce (1809–1880), a Harvard mathematician, after observing that $e^{i\pi} = -1$, used to turn to his students and say, "Gentlemen, that is surely true, it is absolutely paradoxical, we cannot understand it, and we don't know what it means, but we have proved it, and therefore we know it must be the truth." Do you not now think that we understand not only that the formula is true but also what it means?

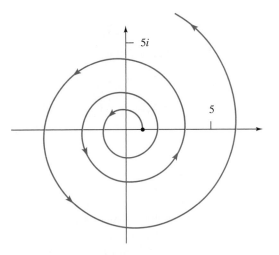

Figure 6

EXAMPLE 3   For which complex numbers λ is $\lim\limits_{t\to\infty} e^{\lambda t} = 0$?

Solution

Recall that

$$e^{\lambda t} = e^{(p+iq)t} = e^{pt}\big(\cos(qt) + i\sin(qt)\big),$$

so that $|e^{\lambda t}| = e^{pt}$. This quantity approaches zero if (and only if) p is negative (i.e., if e^{pt} decays exponentially).

   We summarize: $\lim\limits_{t\to\infty} e^{\lambda t} = 0$ if (and only if) the real part of λ is negative. ∎

   We are now ready to tackle the problem posed at the beginning of this section: Consider a system $d\vec{x}/dt = A\vec{x}$, where the $n \times n$ matrix A has a complex eigenbasis $\vec{v}_1, \ldots, \vec{v}_n$, with eigenvalues $\lambda_1, \ldots, \lambda_n$. Find all complex solutions $\vec{x}(t)$ of this system. By a complex solution we mean a function from $\mathbb{R}$ to $\mathbb{C}^n$ (that is, t is real and $\vec{x}$ is in $\mathbb{C}^n$). In other words, the component functions $x_1(t), \ldots, x_n(t)$ of $\vec{x}(t)$ are complex-valued functions.

   As you review our work in the last section, you will find that the approach we took to the *real* case applies to the *complex* case as well, without modifications:

Fact 9.2.3   Consider a linear system

$$\frac{d\vec{x}}{dt} = A\vec{x}.$$

Suppose there is a complex eigenbasis $\vec{v}_1, \ldots, \vec{v}_n$ for A, with associated complex eigenvalues $\lambda_1, \ldots, \lambda_n$. Then the general complex solution of the system is

$$\vec{x}(t) = c_1 e^{\lambda_1 t} \vec{v}_1 + \cdots + c_n e^{\lambda_n t} \vec{v}_n,$$

where the c_i are arbitrary complex numbers.

   We can write this solution in matrix form, as in Fact 9.1.2.

We can check that the given curve $\vec{x}(t)$ satisfies the equation $d\vec{x}/dt = A\vec{x}$: We have

$$\frac{d\vec{x}}{dt} = c_1 \lambda_1 e^{\lambda_1 t} \vec{v}_1 + \cdots + c_n \lambda_n e^{\lambda_n t} \vec{v}_n$$

(by Definition 9.2.1), and

$$A\vec{x} = c_1 e^{\lambda_1 t} \lambda_1 \vec{v}_1 + \cdots + c_n e^{\lambda_n t} \lambda_n \vec{v}_n,$$

because the $\vec{v}_i$ are eigenvectors. The two answers match.

When is the zero state a stable equilibrium solution for the system $d\vec{x}/dt = A\vec{x}$? Considering Example 3 and the form of the solution given in Fact 9.2.3, we can conclude that this is the case if (and only if) the real parts of all eigenvalues are negative (at least when A is diagonalizable over $\mathbb{C}$). The nondiagonalizable case is left as Exercise 9.3.48.

Fact 9.2.4

Stability of a continuous dynamical system
For a system

$$\frac{d\vec{x}}{dt} = A\vec{x},$$

the zero state is an asymptotically stable equilibrium solution if (and only if) the real parts of all eigenvalues of A are negative.

EXAMPLE 4   Consider the system $d\vec{x}/dt = A\vec{x}$, where A is a (real) 2×2 matrix. When is the zero state a stable equilibrium solution for this system? Give your answer in terms of $\operatorname{tr}(A)$ and $\det(A)$.

Solution

We observe stability either if A has two negative eigenvalues or if A has two conjugate eigenvalues $p \pm iq$, where p is negative. In both cases, $\operatorname{tr}(A)$ is negative and $\det(A)$ is positive. Check that in all other cases $\operatorname{tr}(A) \geq 0$ or $\det(A) \leq 0$.   ∎

Fact 9.2.5

Determinant, trace, and stability
Consider the system

$$\frac{d\vec{x}}{dt} = A\vec{x},$$

where A is a real 2×2 matrix. Then the zero state is an asymptotically stable equilibrium solution if (and only if) $\operatorname{tr}(A) < 0$ and $\det(A) > 0$.

As a special case of Fact 9.2.3, let's consider the system

$$\frac{d\vec{x}}{dt} = A\vec{x},$$

where A is a real 2×2 matrix with eigenvalues $\lambda_{1,2} = p \pm iq$ and corresponding eigenvectors $\vec{v}_{1,2} = \vec{v} \pm i\vec{w}$.

Facts 9.1.2 and 9.2.3 tell us that

$$\vec{x}(t) = P \begin{bmatrix} e^{\lambda_1 t} & 0 \\ 0 & e^{\lambda_2 t} \end{bmatrix} P^{-1} \vec{x}_0 = P \begin{bmatrix} e^{(p+iq)t} & 0 \\ 0 & e^{(p-iq)t} \end{bmatrix} P^{-1} \vec{x}_0$$

$$= e^{pt} P \begin{bmatrix} \cos(qt) + i\sin(qt) & 0 \\ 0 & \cos(qt) - i\sin(qt) \end{bmatrix} P^{-1} \vec{x}_0,$$

where $P = \begin{bmatrix} \vec{v} + i\vec{w} & \vec{v} - i\vec{w} \end{bmatrix}$. Note that we have used Euler's formula (Fact 9.2.2).

We can write this formula in terms of real quantities. By Example 6 of Section 7.5,

$$\begin{bmatrix} \cos(qt) + i\sin(qt) & 0 \\ 0 & \cos(qt) - i\sin(qt) \end{bmatrix} = R^{-1} \begin{bmatrix} \cos(qt) & -\sin(qt) \\ \sin(qt) & \cos(qt) \end{bmatrix} R,$$

$$\text{where} \quad R = \begin{bmatrix} i & -i \\ 1 & 1 \end{bmatrix}.$$

Thus,

$$\vec{x}(t) = e^{pt} P R^{-1} \begin{bmatrix} \cos(qt) & -\sin(qt) \\ \sin(qt) & \cos(qt) \end{bmatrix} R P^{-1} \vec{x}_0 = e^{pt} S \begin{bmatrix} \cos(qt) & -\sin(qt) \\ \sin(qt) & \cos(qt) \end{bmatrix} S^{-1} \vec{x}_0,$$

where

$$S = P R^{-1} = \frac{1}{2i} \begin{bmatrix} \vec{v} + i\vec{w} & \vec{v} - i\vec{w} \end{bmatrix} \begin{bmatrix} 1 & i \\ -1 & i \end{bmatrix} = \begin{bmatrix} \vec{w} & \vec{v} \end{bmatrix}, \qquad \text{and}$$

$$S^{-1} = (P R^{-1})^{-1} = R P^{-1}.$$

Recall that we have performed the same computations in Example 7 of Section 7.5.

Fact 9.2.6

> **Continuous dynamical systems with complex eigenvalues**
> Consider the linear system
>
> $$\frac{d\vec{x}}{dt} = A\vec{x},$$
>
> where A is a real 2×2 matrix with complex eigenvalues $p \pm iq$ (and $q \neq 0$). Consider an eigenvector $\vec{v} + i\vec{w}$ with eigenvalue $p + iq$. Then
>
> $$\vec{x}(t) = e^{pt} S \begin{bmatrix} \cos(qt) & -\sin(qt) \\ \sin(qt) & \cos(qt) \end{bmatrix} S^{-1} \vec{x}_0,$$
>
> where $S = \begin{bmatrix} \vec{w} & \vec{v} \end{bmatrix}$. Recall that $S^{-1} \vec{x}_0$ is the coordinate vector of $\vec{x}_0$ with respect to basis $\vec{w}, \vec{v}$.
>
> The trajectories are either ellipses (linearly distorted circles), if $p = 0$, or spirals, spiraling outward if p is positive and inward if p is negative. In the case of an ellipse, the trajectories have a period of $2\pi/q$.

Note the analogy between Fact 9.2.6 and the formula

$$\vec{x}(t) = r^t S \begin{bmatrix} \cos(\phi t) & -\sin(\phi t) \\ \sin(\phi t) & \cos(\phi t) \end{bmatrix} S^{-1}\vec{x}_0$$

in the case of the discrete system $\vec{x}(t+1) = A\vec{x}(t)$ (Fact 7.6.3).

EXAMPLE 5   Solve the system

$$\frac{d\vec{x}}{dt} = \begin{bmatrix} 3 & -2 \\ 5 & -3 \end{bmatrix} \vec{x} \qquad \text{with} \qquad \vec{x}_0 = \begin{bmatrix} 0 \\ 1 \end{bmatrix}.$$

Solution

The eigenvalues are $\lambda_{1,2} = \pm i$, so that $p = 0$ and $q = 1$. This tells us that the trajectory is an ellipse. To determine the direction of the trajectory (clockwise or counterclockwise) and its rough shape, we can draw the direction field $A\vec{x}$ for a few simple points $\vec{x}$, say, $\vec{x} = \pm\vec{e}_1$ and $\vec{x} = \pm\vec{e}_2$, and sketch the flow line starting at $\begin{bmatrix} 0 \\ 1 \end{bmatrix}$. See Figure 7.

Now let us find a formula for the trajectory.

$$E_i = \ker \begin{bmatrix} i-3 & 2 \\ -5 & i+3 \end{bmatrix} = \operatorname{span} \begin{bmatrix} -2 \\ i-3 \end{bmatrix}$$

$$\begin{bmatrix} -2 \\ i-3 \end{bmatrix} = \underbrace{\begin{bmatrix} -2 \\ -3 \end{bmatrix}}_{\vec{v}} + i \underbrace{\begin{bmatrix} 0 \\ 1 \end{bmatrix}}_{\vec{w}}.$$

Therefore,

$$\vec{x}(t) = e^{pt} S \begin{bmatrix} \cos(qt) & -\sin(qt) \\ \sin(qt) & \cos(qt) \end{bmatrix} S^{-1}\vec{x}_0 = \begin{bmatrix} 0 & -2 \\ 1 & -3 \end{bmatrix} \begin{bmatrix} \cos(t) & -\sin(t) \\ \sin(t) & \cos(t) \end{bmatrix} \begin{bmatrix} 1 \\ 0 \end{bmatrix}$$

$$= \begin{bmatrix} -2\sin(t) \\ \cos(t) - 3\sin(t) \end{bmatrix} = \cos(t) \begin{bmatrix} 0 \\ 1 \end{bmatrix} + \sin(t) \begin{bmatrix} -2 \\ -3 \end{bmatrix}.$$

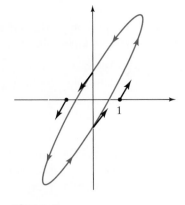

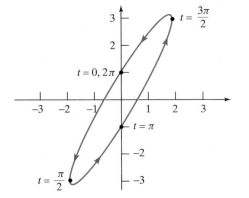

Figure 7

Figure 8

You can check that

$$\frac{d\vec{x}}{dt} = A\vec{x} \qquad \text{and} \qquad \vec{x}(0) = \begin{bmatrix} 0 \\ 1 \end{bmatrix}.$$

The trajectory is the ellipse shown in Figure 8. ∎

Consider a 2×2 matrix A. The various scenarios for the system $d\vec{x}/dt = A\vec{x}$ can be conveniently represented in the $\text{tr}(A)$–$\det(A)$ plane, where a 2×2 matrix A is represented by the point $\big(\text{tr}(A), \det(A)\big)$. Recall that the characteristic equation is

$$\lambda^2 - \text{tr}(A)\lambda + \det(A) = 0$$

and the eigenvalues are

$$\lambda_{1,2} = \frac{1}{2}\left(\text{tr}(A) \pm \sqrt{(\text{tr } A)^2 - 4\det(A)}\right).$$

Therefore, the eigenvalues of A are real if (and only if) the point $\big(\text{tr}(A), \det(A)\big)$ is located below or on the parabola

$$\det(A) = \left(\frac{\text{tr}(A)}{2}\right)^2$$

in the $\text{tr}(A)$–$\det(A)$ plane. See Figure 9.

Note that there are five major cases, corresponding to the regions in Figure 9, and some exceptional cases, corresponding to the dividing lines.

What does the phase portrait look like when $\det(A) = 0$ and $\text{tr}(A) \neq 0$?

In Figure 10 we take another look at the five major types of phase portraits. Both in the discrete and in the continuous case, we sketch the phase portraits produced by various eigenvalues. We include the case of an ellipse, since it is important in applications.

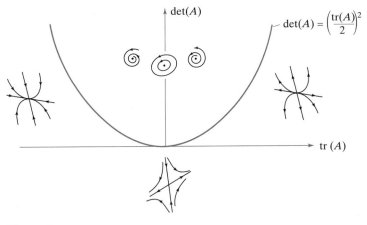

Figure 9

Discrete	Continuous	Phase Portrait
$\lambda_1 > \lambda_2 > 1$	$\lambda_1 > \lambda_2 > 0$	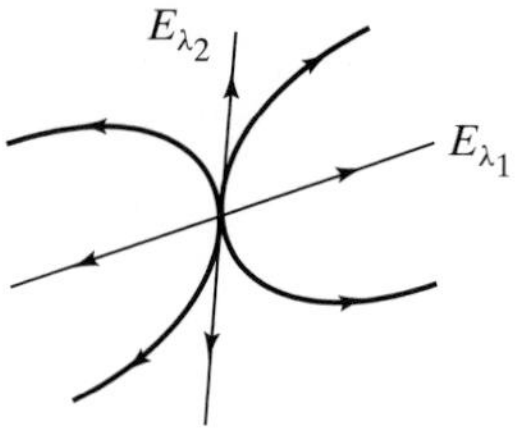
$\lambda_1 > 1 > \lambda_2 > 0$	$\lambda_1 > 0 > \lambda_2$	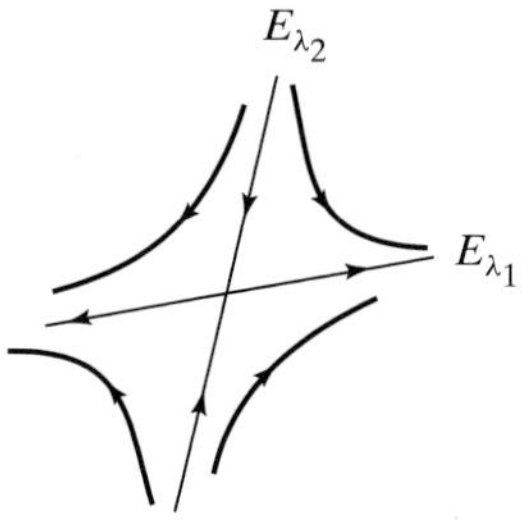
$1 > \lambda_1 > \lambda_2 > 0$	$0 > \lambda_1 > \lambda_2$	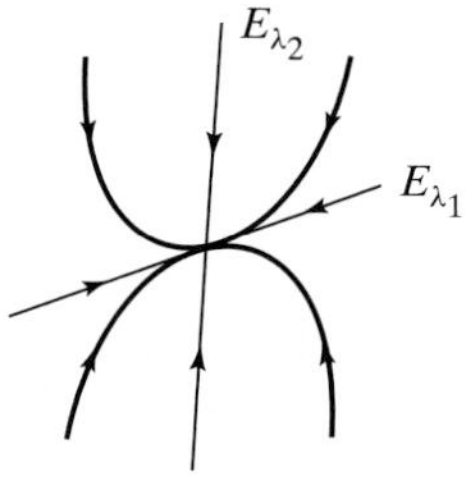
$\lambda_{1,2} = p \pm iq$ $p^2 + q^2 > 1$	$\lambda_{1,2} = p \pm iq$ $p > 0$	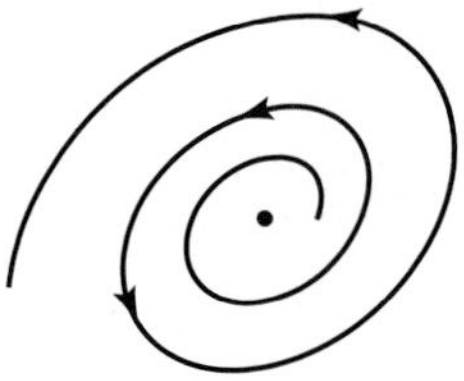
$\lambda_{1,2} = p \pm iq$ $p^2 + q^2 < 1$	$\lambda_{1,2} = p \pm iq$ $p < 0$	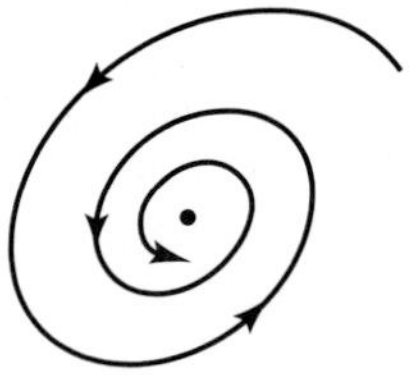
$\lambda_{1,2} = p \pm iq$ $p^2 + q^2 = 1$	$\lambda_{1,2} = \pm iq$	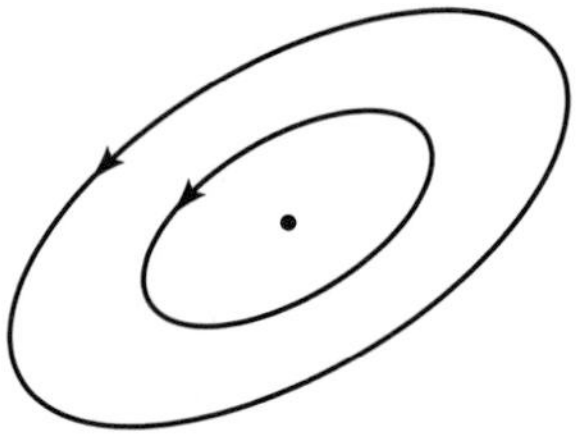

Figure 10 The major types of phase portraits.

EXERCISES

GOALS Use the definition of the complex-valued exponential function $z = e^{\lambda t}$. Solve the system

$$\frac{d\vec{x}}{dt} = A\vec{x}$$

for a 2×2 matrix A with complex eigenvalues $p \pm iq$.

1. Find $e^{2\pi i}$.

2. Find $e^{(1/2)\pi i}$.

3. Write $z = -1 + i$ in polar form as $z = re^{i\phi}$.

4. Sketch the trajectory of the complex-valued function

$$z = e^{3it}.$$

What is the period?

5. Sketch the trajectory of the complex-valued function

$$z = e^{(-0.1-2i)t}.$$

6. Find all complex solutions of the system

$$\frac{d\vec{x}}{dt} = \begin{bmatrix} 3 & -2 \\ 5 & -3 \end{bmatrix} \vec{x}$$

in the form given in Fact 9.2.3. What solution do you get if you let $c_1 = c_2 = 1$?

7. Determine the stability of the system

$$\frac{d\vec{x}}{dt} = \begin{bmatrix} -1 & 2 \\ 3 & -4 \end{bmatrix} \vec{x}.$$

8. Consider a system

$$\frac{d\vec{x}}{dt} = A\vec{x},$$

where A is a symmetric matrix. When is the zero state a stable equilibrium solution? Give your answer in terms of the definiteness of the matrix A.

9. Consider a system

$$\frac{d\vec{x}}{dt} = A\vec{x},$$

where A is a 2×2 matrix with $\mathrm{tr}(A) < 0$. We are told that A has no real eigenvalues. What can you say about the stability of the system?

10. Consider a quadratic form $q(\vec{x}) = \vec{x} \cdot A\vec{x}$ of two variables, x_1 and x_2. Consider the system of differential equations

$$\begin{vmatrix} \dfrac{dx_1}{dt} = \dfrac{\partial q}{\partial x_1} \\[2mm] \dfrac{dx_2}{dt} = \dfrac{\partial q}{\partial x_2} \end{vmatrix},$$

or, more succinctly,

$$\frac{d\vec{x}}{dt} = \mathrm{grad}(q).$$

a. Show that the system $d\vec{x}/dt = \mathrm{grad}(q)$ is linear by finding a matrix B (in terms of the symmetric matrix A) such that $\mathrm{grad}(q) = B\vec{x}$.

b. When q is negative definite, draw a sketch showing possible level curves of q. On the same sketch, draw a few trajectories of the system $d\vec{x}/dt = \mathrm{grad}(q)$. What does your sketch suggest about the stability of the system $d\vec{x}/dt = \mathrm{grad}(q)$?

c. Do the same as in part (b) for an indefinite quadratic form.

d. Explain the relationship between the definiteness of the form q and the stability of the system $d\vec{x}/dt = \mathrm{grad}(q)$.

11. Do parts (a) and (d) of Exercise 10 for a quadratic form of n variables.

12. Determine the stability of the system

$$\frac{d\vec{x}}{dt} = \begin{bmatrix} 0 & 1 & 0 \\ 0 & 0 & 1 \\ -1 & -1 & -2 \end{bmatrix} \vec{x}.$$

13. If the system $d\vec{x}/dt = A\vec{x}$ is stable, is $d\vec{x}/dt = A^{-1}\vec{x}$ stable as well? How can you tell?

14. *Negative feedback loops.* Suppose some quantities $x_1(t), x_2(t), \ldots, x_n(t)$ can be modeled by differential equations of the form

$$\begin{vmatrix} \dfrac{dx_1}{dt} = -k_1 x_1 & & & -bx_n \\[2mm] \dfrac{dx_2}{dt} = & x_1 - k_2 x_2 & & \\[2mm] \vdots & & \ddots & \\[2mm] \dfrac{dx_n}{dt} = & & x_{n-1} - k_n x_n & \end{vmatrix},$$

where b is positive and the k_i are positive. (The matrix of this system has negative numbers on the diagonal, 1's directly below the diagonal, and a negative number in the top right corner.) We say that the quantities $x_1, \ldots, x_n$ describe a (linear) negative feedback loop.

a. Describe the significance of the entries in the preceding system, in practical terms.

b. Is a negative feedback loop with two components ($n = 2$) necessarily stable? *(continued)*

c. Is a negative feedback loop with three components necessarily stable?

15. Consider a noninvertible 2×2 matrix A with a positive trace. What does the phase portrait of the system $d\vec{x}/dt = A\vec{x}$ look like?

16. Consider the system

$$\frac{d\vec{x}}{dt} = \begin{bmatrix} 0 & 1 \\ a & b \end{bmatrix} \vec{x},$$

where a and b are arbitrary constants. For which choices of a and b is the zero state a stable equilibrium solution?

17. Consider the system

$$\frac{d\vec{x}}{dt} = \begin{bmatrix} -1 & k \\ k & -1 \end{bmatrix} \vec{x},$$

where k is an arbitrary constant. For which choices of k is the zero state a stable equilibrium solution?

18. Consider a diagonalizable 3×3 matrix A such that the zero state is a stable equilibrium solution of the system $d\vec{x}/dt = A\vec{x}$. What can you say about the determinant and the trace of A?

19. *True or false?* If the trace and the determinant of a 3×3 matrix A are both negative, then the origin is a stable equilibrium solution of the system $d\vec{x}/dt = A\vec{x}$. Justify your answer.

20. Consider a 2×2 matrix A with eigenvalues $\pm \pi i$. Let $\vec{v} + i\vec{w}$ be an eigenvector of A with eigenvalue πi. Solve the initial-value problem

$$\frac{d\vec{x}}{dt} = A\vec{x}, \quad \text{with} \quad \vec{x}_0 = \vec{w}.$$

Draw the solution in the accompanying figure. Mark the vectors $\vec{x}(0)$, $\vec{x}(\frac{1}{2})$, $\vec{x}(1)$, and $\vec{x}(2)$.

21. Ngozi opens a bank account with an initial balance of 1,000 Nigerian naira. Let $b(t)$ be the balance in the account at time t; we are told that $b(0) = 1,000$. The bank is paying interest at a continuous rate of 5% per year. Ngozi makes deposits into the account at a continuous rate of $s(t)$ (measured in naira per year). We are told that $s(0) = 1,000$ and that $s(t)$ is increasing at a continuous rate of 7% per year. (Ngozi can save more as her income goes up over time.)

a. Set up a linear system of the form

$$\begin{vmatrix} \dfrac{db}{dt} = ?b + ?s \\[2mm] \dfrac{ds}{dt} = ?b + ?s \end{vmatrix}.$$

(Time is measured in years.)

b. Find $b(t)$ and $s(t)$.

For each of the linear systems in Exercises 22 through 26, find the matching phase portrait on the next page.

22. $\vec{x}(t+1) = \begin{bmatrix} 3 & 0 \\ -2.5 & 0.5 \end{bmatrix} \vec{x}(t)$

23. $\vec{x}(t+1) = \begin{bmatrix} -1.5 & -1 \\ 2 & 0.5 \end{bmatrix} \vec{x}(t)$

24. $\dfrac{d\vec{x}}{dt} = \begin{bmatrix} 3 & 0 \\ -2.5 & 0.5 \end{bmatrix} \vec{x}$

25. $\dfrac{d\vec{x}}{dt} = \begin{bmatrix} -1.5 & -1 \\ 2 & 0.5 \end{bmatrix} \vec{x}$

26. $\dfrac{d\vec{x}}{dt} = \begin{bmatrix} -2 & 0 \\ 3 & 1 \end{bmatrix} \vec{x}$

Find all real solutions of the systems in Exercises 27 through 30.

27. $\dfrac{d\vec{x}}{dt} = \begin{bmatrix} 0 & -3 \\ 3 & 0 \end{bmatrix} \vec{x}$

28. $\dfrac{d\vec{x}}{dt} = \begin{bmatrix} 0 & 4 \\ -9 & 0 \end{bmatrix} \vec{x}$

29. $\dfrac{d\vec{x}}{dt} = \begin{bmatrix} 2 & 4 \\ -4 & 2 \end{bmatrix} \vec{x}$

30. $\dfrac{d\vec{x}}{dt} = \begin{bmatrix} -11 & 15 \\ -6 & 7 \end{bmatrix} \vec{x}$

Solve the systems in Exercises 31 through 34. Give the solution in real form. Sketch the solution.

31. $\dfrac{d\vec{x}}{dt} = \begin{bmatrix} -1 & -2 \\ 2 & -1 \end{bmatrix} \vec{x}$ with $\vec{x}(0) = \begin{bmatrix} 1 \\ -1 \end{bmatrix}$.

32. $\dfrac{d\vec{x}}{dt} = \begin{bmatrix} 0 & 1 \\ -4 & 0 \end{bmatrix} \vec{x}$ with $\vec{x}(0) = \begin{bmatrix} 1 \\ 0 \end{bmatrix}$.

33. $\dfrac{d\vec{x}}{dt} = \begin{bmatrix} -1 & 1 \\ -2 & 1 \end{bmatrix} \vec{x}$ with $\vec{x}(0) = \begin{bmatrix} 0 \\ 1 \end{bmatrix}$.

34. $\dfrac{d\vec{x}}{dt} = \begin{bmatrix} 7 & 10 \\ -4 & -5 \end{bmatrix} \vec{x}$ with $\vec{x}(0) = \begin{bmatrix} 1 \\ 0 \end{bmatrix}$.

35. Consider the following mass–spring system:

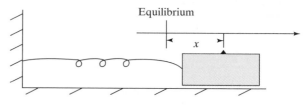

Equilibrium

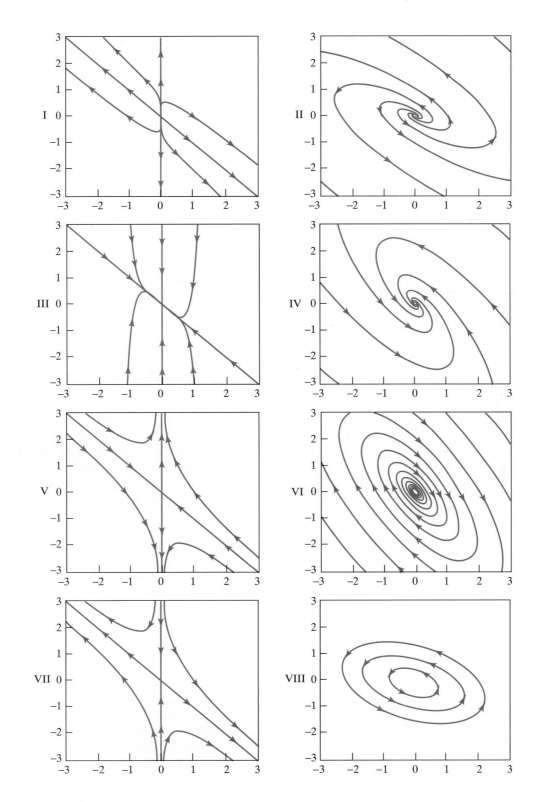

Let $x(t)$ be the deviation of the block from the equilibrium position at time t. Consider the velocity $v(t) = dx/dt$ of the block. There are two forces acting on the mass: The spring force F_s, which is assumed to be proportional to the displacement x, and the force F_f of friction, which is assumed to be proportional to the velocity

$$F_s = -px, \qquad F_f = -qv,$$

where $p > 0$ and $q \geq 0$. (q is 0 if the oscillation is frictionless.) Therefore, the total force acting on the mass is

$$F = F_s + F_f = -px - qv.$$

By Newton's second law of motion, we have

$$F = ma = m\frac{dv}{dt},$$

where a represents acceleration and m the mass of the block. Combining the last two equations, we find that

$$m\frac{dv}{dt} = -px - qv,$$

or

$$\frac{dv}{dt} = -\frac{p}{m}x - \frac{q}{m}v.$$

Let $b = p/m$ and $c = q/m$ for simplicity. Then the dynamics of this mass–spring system are described by the system

$$\left| \begin{aligned} \frac{dx}{dt} &= v \\ \frac{dv}{dt} &= -bx - cv \end{aligned} \right. \qquad (b > 0, \ c \geq 0).$$

Sketch a phase portrait for this system in each of the following cases, and describe briefly the significance of your trajectories in terms of the movement of the block. Comment on the stability in each case.

a. $c = 0$ (frictionless). Find the period.

b. $c^2 < 4b$ (underdamped).

c. $c^2 > 4b$ (overdamped).

36. Prove the product rule for derivatives of complex-valued functions.

37. a. For a differentiable complex-valued function $z(t)$, find the derivative of

$$\frac{1}{z(t)}.$$

b. Prove the quotient rule for derivatives of complex-valued functions.

In both parts of this exercise, you may use the product rule. (See Exercise 36.)

38. Let $z_1(t)$ and $z_2(t)$ be two complex-valued solutions of the initial-value problem

$$\frac{dz}{dt} = \lambda z, \quad \text{with} \quad z(0) = 1,$$

where λ is a complex number. Suppose that $z_2(t) \neq 0$ for all t.

a. Using the quotient rule (Exercise 37), show that the derivative of

$$\frac{z_1(t)}{z_2(t)}$$

is zero. Conclude that $z_1(t) = z_2(t)$.

b. Show that the initial-value problem

$$\frac{dz}{dt} = \lambda z, \quad \text{with} \quad z(0) = 1,$$

has a unique complex-valued solution $z(t)$. *Hint:* One solution is given in the text.

39. Solve the system

$$\frac{d\vec{x}}{dt} = \begin{bmatrix} \lambda & 1 & 0 \\ 0 & \lambda & 1 \\ 0 & 0 & \lambda \end{bmatrix} \vec{x}.$$

Compare this with Exercise 9.1.24. When is the zero state a stable equilibrium solution?

40. An eccentric mathematician is able to gain autocratic power in a small Alpine country. In her first decree, she announces the introduction of a new currency, the Euler, which is measured in complex units. Banks are ordered to pay only imaginary interest on deposits.

a. If you invest 1,000 Euler at $5i\%$ interest, compounded annually, how much money do you have after 1 year, after 2 years, after t years? Describe the effect of compounding in this case. Sketch a trajectory showing the evolution of the balance in the complex plane.

b. Do part (a) in the case when the $5i\%$ interest is compounded continuously.

c. Suppose people's social standing is determined by the modulus of the balance of their bank account. Under these circumstances, would you choose an account with annual compounding or with continuous compounding of interest?

(This problem is based on an idea of Prof. D. Mumford, Brown University.)

9.3 LINEAR DIFFERENTIAL OPERATORS AND LINEAR DIFFERENTIAL EQUATIONS

In this final section, we will study an important class of linear transformations from C^∞ to C^∞. Here, C^∞ denotes the linear space of complex-values smooth functions (i.e., functions from $\mathbb{R}$ to $\mathbb{C}$), which we consider as a linear space over $\mathbb{C}$.

Definition 9.3.1

> **Linear differential operators and linear differential equations**
> A transformation T from C^∞ to C^∞ of the form
>
> $$T(f) = f^{(n)} + a_{n-1}f^{(n-1)} + \cdots + a_1 f' + a_0 f$$
>
> is called an nth-order *linear differential operator*.[5,6] Here $f^{(k)}$ denotes the kth derivative of function f, and the coefficients a_k are complex scalars.
>   If T is an nth-order linear differential operator and g is a smooth function, then the equation
>
> $$T(f) = g \qquad \text{or} \qquad f^{(n)} + a_{n-1}f^{(n-1)} + \cdots + a_1 f' + a_0 f = g$$
>
> is called an nth-order *linear differential equation* (DE). The DE is called *homogeneous* if $g = 0$ and inhomogeneous otherwise.

Verify that a linear differential operator is indeed a linear transformation. Examples of linear differential operators are

$$D(f) = f',$$
$$T(f) = f'' - 5f' + 6f, \qquad \text{and}$$
$$L(f) = f''' - 6f'' + 5f,$$

of first, second, and third order, respectively.
   Examples of linear DEs are

$$f'' - f' - 6f = 0 \quad \text{(second order, homogeneous)}$$

and

$$f'(t) - 5f(t) = \sin(t) \quad \text{(first order, inhomogeneous)}.$$

Note that solving a homogeneous DE $T(f) = 0$ amounts to finding the kernel of T.
   We will first think about the relationship between the solutions of the DEs $T(f) = 0$ and $T(f) = g$.
   More generally, consider a linear transformation T from V to W, where V and W are arbitrary linear spaces. What is the relationship between the kernel of T and the solutions f of the equation $T(f) = g$, provided that this equation has solutions at all? (Compare this with Exercise 1.3.48.)

---

[5]More precisely, this is a linear differential operator *with constant coefficients*. More advanced texts consider the case when the a_k are functions.

[6]The term "operator" is often used for a transformation whose domain and codomain consist of functions.

Here is a simple example:

EXAMPLE 1 Consider the linear transformation $T(\vec{x}) = \begin{bmatrix} 1 & 2 & 3 \\ 2 & 4 & 6 \end{bmatrix} \vec{x}$ from $\mathbb{R}^3$ to $\mathbb{R}^2$. Describe the relationship between the kernel of T and the solutions of the linear system $T(\vec{x}) = \begin{bmatrix} 6 \\ 12 \end{bmatrix}$, both algebraically and geometrically.

Solution

Using Gauss–Jordan elimination, we find that the kernel of T consists of all vectors of the form

$$\begin{bmatrix} -2x_2 - 3x_3 \\ x_2 \\ x_3 \end{bmatrix} = x_2 \begin{bmatrix} -2 \\ 1 \\ 0 \end{bmatrix} + x_3 \begin{bmatrix} -3 \\ 0 \\ 1 \end{bmatrix},$$

with basis

$$\begin{bmatrix} -2 \\ 1 \\ 0 \end{bmatrix}, \quad \begin{bmatrix} -3 \\ 0 \\ 1 \end{bmatrix}.$$

The solution set of the system $T(\vec{x}) = \begin{bmatrix} 6 \\ 12 \end{bmatrix}$ consists of all vectors of the form

$$\begin{bmatrix} 6 - 2x_2 - 3x_3 \\ x_2 \\ x_3 \end{bmatrix} = \underbrace{x_2 \begin{bmatrix} -2 \\ 1 \\ 0 \end{bmatrix} + x_3 \begin{bmatrix} -3 \\ 0 \\ 1 \end{bmatrix}}_{\substack{\text{a vector in the} \\ \text{kernel of } T}} + \underbrace{\begin{bmatrix} 6 \\ 0 \\ 0 \end{bmatrix}}_{\substack{\text{a particular solution} \\ \text{of the system} \\ T(\vec{x}) = \begin{bmatrix} 6 \\ 12 \end{bmatrix}.}}$$

The kernel of T and the solution set of $T(\vec{x}) = \begin{bmatrix} 6 \\ 12 \end{bmatrix}$ form two parallel planes in $\mathbb{R}^3$, as shown in Figure 1.

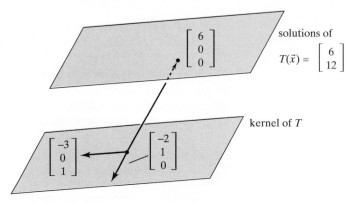

Figure 1

These observations generalize as follows:

Fact 9.3.2

Consider a linear transformation T from V to W, where V and W are arbitrary linear spaces. Suppose we have a basis $f_1, f_2, \ldots, f_n$ of the kernel of T. Consider an equation $T(f) = g$ with a particular solution f_p. Then the solutions f of the equation $T(f) = g$ are of the form

$$f = c_1 f_1 + c_2 f_2 + \cdots + c_n f_n + f_p,$$

where the c_i are arbitrary constants.

Note that $T(f) = T(c_1 f_1 + \cdots + c_n f_n) + T(f_p) = 0 + g = g$, so that f is indeed a solution. Verify that all solutions are of this form.

What is the significance of Fact 9.3.2 for linear differential equations? At the end of this section, we will demonstrate the following fundamental result:

Fact 9.3.3

The kernel of an nth-order linear differential operator is n-dimensional.

Fact 9.3.2 now provides us with the following strategy for solving linear differential equations:

Fact 9.3.4

To solve an nth-order linear DE

$$T(f) = g,$$

we have to find

a. a basis $f_1, \ldots, f_n$ of kernel(T), and
b. a particular solution f_p of the DE.

Then the solutions f are of the form

$$f = c_1 f_1 + \cdots + c_n f_n + f_p,$$

where the c_i are arbitrary constants.

EXAMPLE 2   Find all solutions of the DE

$$f''(t) + f(t) = e^t.$$

We are told that $f_p(t) = \frac{1}{2} e^t$ is a particular solution (verify this).

Solution

Consider the linear differential operator $T(f) = f'' + f$. A basis of the kernel of T is $f_1(t) = \cos(t)$ and $f_2(t) = \sin(t)$ (compare with Example 1 of Section 4.1). Therefore, the solutions f of the DE $f'' + f = e^t$ are of the form

$$f(t) = c_1 \cos(t) + c_2 \sin(t) + \frac{1}{2} e^t,$$

where c_1 and c_2 are arbitrary constants.   ∎

We now present an approach that allows us to find solutions to homogeneous linear DE's more systematically.

The Eigenfunction Approach to Solving Linear DEs

Definition 9.3.5

> **Eigenfunctions**
> Consider a linear differential operator T from C^∞ to C^∞. A smooth function f is called an *eigenfunction* of T if $T(f) = \lambda f$ for some complex scalar λ; this scalar λ is called the eigenvalue associated with the eigenfunction f.

EXAMPLE 3   Find all eigenfunctions and eigenvalues of the operator $D(f) = f'$.

Solution

We have to solve the differential equation

$$f' = \lambda f.$$

We know that for a given λ, the solutions are all exponential functions of the form $f(t) = Ce^{\lambda t}$. This means that all complex numbers are eigenvalues of D, and the eigenspace associated with the eigenvalue λ is one-dimensional, spanned by $e^{\lambda t}$. (Compare this with Definition 9.2.1.)   ∎

It follows that the exponential functions are eigenfunctions for all linear differential operators: if

$$T(f) = f^{(n)} + a_{n-1}f^{(n-1)} + \cdots + a_1 f' + a_0 f,$$

then

$$T(e^{\lambda t}) = (\lambda^n + a_{n-1}\lambda^{n-1} + \cdots + a_1\lambda + a_0)e^{\lambda t}.$$

This observation motivates the following definition:

Definition 9.3.6

> **Characteristic polynomial**
> Consider the linear differential operator
>
> $$T(f) = f^{(n)} + a_{n-1}f^{(n-1)} + \cdots + a_1 f' + a_0 f.$$
>
> The *characteristic polynomial* of T is defined as
>
> $$p_T(\lambda) = \lambda^n + a_{n-1}\lambda^{n-1} + \cdots + a_1\lambda + a_0.$$

Fact 9.3.7   If T is a linear differential operator, then $e^{\lambda t}$ is an eigenfunction of T, with associated eigenvalue $p_T(\lambda)$, for all λ:

$$T(e^{\lambda t}) = p_T(\lambda)e^{\lambda t}$$

In particular, if $p_T(\lambda) = 0$, then $e^{\lambda t}$ is in the kernel of T.

EXAMPLE 4   Find all exponential functions $e^{\lambda t}$ in the kernel of the linear differential operator

$$T(f) = f'' + f' - 6f.$$

Solution

The characteristic polynomial is $p_T(\lambda) = \lambda^2 + \lambda - 6 = (\lambda + 3)(\lambda - 2)$, with roots 2 and -3. Therefore, the functions e^{2t} and e^{-3t} are in the kernel of T. We can check this:

$$T(e^{2t}) = 4e^{2t} + 2e^{2t} - 6e^{2t} = 0,$$
$$T(e^{-3t}) = 9e^{-3t} - 3e^{-3t} - 6e^{-3t} = 0. \qquad \blacksquare$$

Since most polynomials of degree n have n distinct complex roots, we can find n distinct exponential functions $e^{\lambda_1 t}, \ldots, e^{\lambda_n t}$ in the kernel of most nth-order linear differential operators. Note that these functions are linearly independent. (They are eigenfunctions of D with distinct eigenvalues; the proof of Fact 7.3.5 applies.)

Now we can use Fact 9.3.3.

Fact 9.3.8

The kernel of a linear differential operator

Consider an nth-order linear differential operator T whose characteristic polynomial $p_T(\lambda)$ has n distinct roots $\lambda_1, \ldots, \lambda_n$. Then the exponential functions

$$e^{\lambda_1 t}, e^{\lambda_2 t}, \ldots, e^{\lambda_n t}$$

form a basis of the kernel of T; that is, they form a basis of the solution space of the homogeneous DE

$$T(f) = 0.$$

See Exercise 38 for the case of an nth-order linear differential operator whose characteristic polynomial has fewer than n distinct roots.

EXAMPLE 5   Find all solutions f of the differential equation

$$f'' + 2f' - 3f = 0.$$

Solution

The characteristic polynomial of the operator $T(f) = f'' + 2f' - 3f$ is $p_T(\lambda) = \lambda^2 + 2\lambda - 3 = (\lambda + 3)(\lambda - 1)$, with roots 1 and -3. The exponential functions e^t and e^{-3t} form a basis of the solution space, i.e., the solutions are of the form

$$f(t) = c_1 e^t + c_2 e^{-3t}. \qquad \blacksquare$$

EXAMPLE 6   Find all solutions f of the differential equation

$$f'' - 6f' + 13f = 0.$$

Solution

The characteristic polynomial is $p_T(\lambda) = \lambda^2 - 6\lambda + 13$, with complex roots $3 \pm 2i$. The exponential functions

$$e^{(3+2i)t} = e^{3t}\big(\cos(2t) + i\sin(2t)\big)$$

and

$$e^{(3-2i)t} = e^{3t}\big(\cos(2t) - i\sin(2t)\big)$$

form a basis of the solution space. We may wish to find a basis of the solution space consisting of real-valued functions. The following observation is helpful: if $f(t) = g(t) + ih(t)$ is a solution of the DE $T(f) = 0$, then $T(f) = T(g) + iT(h) = 0$, so that g and h are solutions as well. We can apply this remark to the real and the imaginary parts of the solution $e^{(3+2i)t}$: The functions

$$e^{3t}\cos(2t) \qquad \text{and} \qquad e^{3t}\sin(2t)$$

are a basis of the solution space (they are clearly linearly independent), and the general solution is

$$f(t) = c_1 e^{3t}\cos(2t) + c_2 e^{3t}\sin(2t) = e^{3t}\big(c_1\cos(2t) + c_2\sin(2t)\big). \qquad \blacksquare$$

Fact 9.3.9   Consider a differential equation

$$T(f) = f'' + af' + bf = 0,$$

where the coefficients a and b are real. Suppose the zeros of $p_T(\lambda)$ are $p \pm iq$, with $q \neq 0$. Then the solutions of the given DE are

$$f(t) = e^{pt}\big(c_1\cos(qt) + c_2\sin(qt)\big),$$

where c_1 and c_2 are arbitrary constants.

    The special case when $a = 0$ and $b > 0$ is important in many applications. Then $p = 0$ and $q = \sqrt{b}$, so that the solutions of the DE

$$f'' + bf = 0$$

are

$$f(t) = c_1\cos(\sqrt{b}\,t) + c_2\sin(\sqrt{b}\,t).$$

Note that the function

$$f(t) = e^{pt}\big(c_1\cos(qt) + c_2\sin(qt)\big)$$

is the product of an exponential and a sinusoidal function. The case when p is negative comes up frequently in physics, when we model a *damped oscillator*. See Figure 2.

What about nonhomogeneous differential equations? Let us discuss an example that is particularly important in applications.

EXAMPLE 7   Consider the differential equation

$$f''(t) + f'(t) - 6f(t) = 8\cos(2t).$$

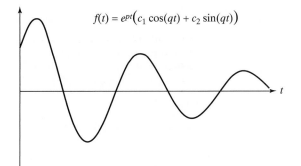

$$f(t) = e^{pt}\big(c_1 \cos(qt) + c_2 \sin(qt)\big)$$

Figure 2

a. Let V be the linear space consisting of all functions of the form $c_1 \cos(2t) + c_2 \sin(2t)$. Show that the linear differential operator $T(f) = f'' + f' - 6f$ defines an isomorphism from V to V.

b. Part (a) implies that the DE $T(f) = 8 \cos(2t)$ has a unique particular solution $f_p(t)$ in V. Find this solution.

c. Find all solutions of the DE $T(f) = 8 \cos(2t)$.

Solution

a. Consider the matrix A of T with respect to the basis $\cos(2t), \sin(2t)$. A straightforward computation shows that

$$A = \begin{bmatrix} -10 & 2 \\ -2 & -10 \end{bmatrix},$$

a rotation–dilation matrix. Since A is invertible, T defines an isomorphism from V to V.

b. If we work in coordinates with respect to the basis $\cos(2t), \sin(2t)$, the DE $T(f) = 8 \cos(2t)$ takes the form $A\vec{x} = \begin{bmatrix} 8 \\ 0 \end{bmatrix}$, with the solution

$$\vec{x} = A^{-1}\begin{bmatrix} 8 \\ 0 \end{bmatrix} = \frac{1}{104}\begin{bmatrix} -10 & -2 \\ 2 & -10 \end{bmatrix}\begin{bmatrix} 8 \\ 0 \end{bmatrix} = \begin{bmatrix} -10/13 \\ 2/13 \end{bmatrix}.$$

The particular solution in V is

$$f_p(t) = -\frac{10}{13}\cos(2t) + \frac{2}{13}\sin(2t).$$

A more straightforward way to find $f_p(t)$ is to set $f_p(t) = P\cos(2t) + Q\sin(2t)$ and substitute this trial solution into the DE to determine P and Q.

c. In Example 4, we have seen that the functions $f_1(t) = e^{2t}$ and $f_2(t) = e^{-3t}$ form a basis of the kernel of T. By Fact 9.3.4, the solutions of the DE are of the form

$$f(t) = c_1 f_1(t) + c_2 f_2(t) + f_p(t)$$

$$= c_1 e^{2t} + c_2 e^{-3t} - \frac{10}{13}\cos(2t) + \frac{2}{13}\sin(2t). \qquad \blacksquare$$

Let us summarize the methods developed in Example 7:

Fact 9.3.10

Consider the linear differential equation

$$f''(t) + af'(t) + bf(t) = C\cos(\omega t),$$

where a, b, C, and ω are real numbers. Suppose that $a \neq 0$ or $b \neq \omega^2$. This DE has a particular solution of the form

$$f_p(t) = P\cos(\omega t) + Q\sin(\omega t).$$

Now use Facts 9.3.4 and 9.3.8 to find all solutions f of the DE.

What goes wrong when $a = 0$ and $b = \omega^2$?

The Operator Approach to Solving Linear DE's

We will now present an alternative, deeper approach to DE's, which allows us to solve any linear DE (at least if we can find the zeros of the characteristic polynomial). This approach will lead us to a better understanding of the kernel and image of a linear differential operator; in particular, it will enable us to prove Fact 9.3.3.

Let us first introduce a more succinct notation for linear differential operators. Recall the notation $Df = f'$ for the derivative operator. We let

$$D^m = \underbrace{D \circ D \circ \cdots \circ D}_{m \text{ times}};$$

that is,

$$D^m f = f^{(m)}.$$

Then the operator

$$T(f) = f^{(n)} + a_{n-1}f^{(n-1)} + \cdots + a_1 f' + a_0 f$$

can be written more succinctly as

$$T = D^n + a_{n-1}D^{n-1} + \cdots + a_1 D + a_0,$$

the characteristic polynomial $p_T(\lambda)$ "evaluated at D."

For example, the operator

$$T(f) = f'' + f' - 6f$$

can be written as

$$T = D^2 + D - 6.$$

Treating T formally as a polynomial in D, we can write

$$T = (D+3) \circ (D-2).$$

We can verify that this formula gives us a decomposition of the operator T:

$$\big((D+3)\circ(D-2)\big)f = (D+3)(f'-2f) = f''-2f'+3f'-6f = (D^2+D-6)f.$$

This works because D is linear: We have $D(f' - 2f) = f'' - 2f'$.

The fundamental theorem of algebra (Fact 7.5.2) now tells us the following:

Fact 9.3.11

An nth-order linear differential operator T can be expressed as the composite of n first-order linear differential operators:

$$T = D^n + a_{n-1}D^{n-1} + \cdots + a_1 D + a_0$$
$$= (D - \lambda_1)(D - \lambda_2)\ldots(D - \lambda_n),$$

where the λ_i are complex numbers.

We can therefore hope to understand all linear differential operators by studying first-order operators.

EXAMPLE 8   Find the kernel of the operator $T = D - a$, where a is a complex number. Do not use Fact 9.3.3.

Solution

We have to solve the homogeneous differential equation $T(f) = 0$ or $f'(t) - af(t) = 0$ or $f'(t) = af(t)$. By definition of an exponential function, the solutions are the functions of the form $f(t) = Ce^{at}$, where C is an arbitrary constant. (See Definition 9.2.1.) ∎

Fact 9.3.12

The kernel of the operator

$$T = D - a$$

is one-dimensional, spanned by

$$f(t) = e^{at}.$$

Next we think about the nonhomogeneous equation

$$(D - a)f = g,$$

or

$$f'(t) - af(t) = g(t),$$

where $g(t)$ is a smooth function. It will turn out to be useful to multiply both sides of this equation with the function e^{-at}:

$$e^{-at}f'(t) - ae^{-at}f(t) = e^{-at}g(t).$$

We recognize the left-hand side of this equation as the derivative of the function $e^{-at}f(t)$, so that we can write

$$\left(e^{-at}f(t)\right)' = e^{-at}g(t).$$

Integrating, we get

$$e^{-at} f(t) = \int e^{-at} g(t)\, dt$$

and

$$f(t) = e^{at} \int e^{-at} g(t)\, dt,$$

where $\int e^{-at} g(t)\, dt$ denotes the indefinite integral, that is, the family of all anti-derivatives of the functions $e^{-at} g(t)$, involving a parameter C.

Fact 9.3.13

First-order linear differential equations
Consider the differential equation

$$f'(t) - a f(t) = g(t),$$

where $g(t)$ is a smooth function and a a constant.
Then

$$f(t) = e^{at} \int e^{-at} g(t)\, dt.$$

Fact 9.3.13 shows that the differential equation $(D - a)f = g$ has solutions f, for any smooth function g; this means that

$$\operatorname{im}(D - a) = C^{\infty}.$$

EXAMPLE 9 Find the solutions f of the DE

$$f' - af = ce^{at},$$

where c is an arbitrary constant.

Solution

Using Fact 9.3.13 we find that

$$f(t) = e^{at} \int e^{-at} ce^{at}\, dt = e^{at} \int c\, dt = e^{at}(ct + C),$$

where C is another arbitrary constant. ∎

Now consider an nth-order DE $T(f) = g$, where

$$T = D^n + a_{n-1} D^{n-1} + \cdots + a_1 D + a_0,$$
$$= (D - \lambda_1)(D - \lambda_2) \ldots (D - \lambda_{n-1})(D - \lambda_n).$$

We can break this DE down into n first-order DEs:

$$f \xrightarrow{D - \lambda_n} f_{n-1} \xrightarrow{D - \lambda_{n-1}} f_{n-2} \quad \cdots \quad f_2 \xrightarrow{D - \lambda_2} f_1 \xrightarrow{D - \lambda_1} g$$

We can successively solve the first-order DEs:

$$(D - \lambda_1)f_1 = g,$$
$$(D - \lambda_2)f_2 = f_1,$$
$$\vdots$$
$$(D - \lambda_{n-1})f_{n-1} = f_{n-2},$$
$$(D - \lambda_n)f = f_{n-1}.$$

In particular, the DE $T(f) = g$ does have solutions f.

Fact 9.3.14

The image of a linear differential operator
The image of all linear differential operators (from C^∞ to C^∞) is C^∞; that is, any linear DE $T(f) = g$ has solutions f.

EXAMPLE 10   Find all solutions of the DE

$$T(f) = f'' - 2f' + f = 0.$$

Note that $p_{_T}(\lambda) = \lambda^2 - 2\lambda + 1 = (\lambda - 1)^2$ has only one root, 1, so that we cannot use Fact 9.3.8.

Solution

We break the DE down into two first-order DE's, as discussed above:

$$f \xrightarrow{\;\;D-1\;\;} f_1 \xrightarrow{\;\;D-1\;\;} 0$$

The DE $(D - 1)f_1 = 0$ has the general solution $f_1(t) = c_1 e^t$, where c_1 is an arbitrary constant.

Then the DE $(D-1)f = f_1 = c_1 e^t$ has the general solution $f(t) = e^t(c_1 t + c_2)$, where c_2 is another arbitrary constant. (See Example 9.)

The functions e^t and te^t form a basis of the solution space (i.e., of the kernel of T). Note that the kernel is two-dimensional, since we pick up an arbitrary constant each time we solve a first-order DE.   ∎

Now we can explain why the kernel of an nth-order linear differential operator T is n-dimensional. Roughly speaking, this is true because the general solution of the DE $T(f) = 0$ contains n arbitrary constants. (We pick up one each time we solve a first-order linear DE.)

Here is a formal proof of Fact 9.3.3. We will argue by induction on n. Fact 9.3.12 takes care of the case $n = 1$. By Fact 9.3.11, we can write an nth-order linear differential operator T as $T = (D - \lambda) \circ L$, where L is of order $n - 1$. Arguing by induction, we can assume that the kernel of L is $(n - 1)$-dimensional, with basis $f_1, f_2, \ldots, f_{n-1}$. The function $e^{\lambda t}$ is in the kernel of operator $(D - \lambda)$, and by Fact 9.3.14 there is a function f_n such that $L(f_n) = e^{\lambda t}$. Note that this function f_n will be in the kernel of operator T, since $T(f_n) = (D - \lambda)(L(f_n)) = (D - \lambda)e^{\lambda t} = 0$.

We claim that the n functions $f_1, f_2, \ldots, f_{n-1}, f_n$ form a basis of the kernel of T. To show linear independence, consider a relation $c_1 f_1 + c_2 f_2 + \cdots + c_{n-1}f_{n-1} + c_n f_n = 0$. Applying transformation L to both sides gives $c_n L(f_n) = c_n e^{\lambda t} = 0$, so that $c_n = 0$. Now $c_1 = c_2 = \cdots = c_{n-1} = 0$, since functions $f_1, f_2, \ldots, f_{n-1}$ are linearly independent, by construction.

To show that functions $f_1, f_2, \ldots, f_{n-1}, f_n$ span the kernel of T, consider any function f in the kernel of T. Then $L(f)$ will be in the kernel of $(D - \lambda)$, so that $L(f) = ce^{\lambda t}$ for some constant c, by Fact 9.3.12. Then $f - cf_n$ will be in the kernel of L, since $L(f - cf_n) = L(f) - cL(f_n) = ce^{\lambda t} - ce^{\lambda t} = 0$. Thus, $f - cf_n = c_1 f_1 + c_2 f_2 + \cdots + c_{n-1} f_{n-1}$ for some scalars $c_1, c_2, \ldots, c_{n-1}$. It follows that $f = c_1 f_1 + c_2 f_2 + \cdots + c_{n-1} f_{n-1} + cf_n$, as claimed.   ▲

Let's summarize the main techniques we discussed in this section.

Strategy for linear differential equations

Suppose you have to solve an nth-order linear differential equation $T(f) = g$

Step 1   Find n linearly independent solutions of the DE $T(f) = 0$

- Write the characteristic polynomial $p_T(\lambda)$ of T (replacing $f^{(k)}$ by λ^k).
- Find the solutions $\lambda_1, \lambda_2, \ldots, \lambda_n$ of the equation $p_T(\lambda) = 0$.
- If λ is a solution of the equation $p_T(\lambda) = 0$, then $e^{\lambda t}$ is a solution of $T(f) = 0$.
- If λ is a solution of $p_T(\lambda) = 0$ with multiplicity m, then $e^{\lambda t}, te^{\lambda t}, t^2 e^{\lambda t}, \ldots, t^{m-1} e^{\lambda t}$ are solutions of the DE $T(f) = 0$. (See Exercise 38.)
- If $p \pm iq$ are complex solutions of $p_T(\lambda) = 0$, then $e^{pt} \cos(qt)$ and $e^{pt} \sin(qt)$ are real solutions of the DE $T(f) = 0$.

Step 2   If the DE is inhomogeneous (i.e., if $g \neq 0$), find one particular solution f_p of the DE $T(f) = g$.

- If g is of the form $g(t) = A \cos(\omega t) + B \sin(\omega t)$, look for a particular solution of the same form, $f_p(t) = P \cos(\omega t) + Q \sin(\omega t)$.
- If g is constant, look for a constant particular solution $f_p(t) = c$.[7]
- If the DE is of first order, of the form $f'(t) - af(t) = g(t)$, use the formula $f(t) = e^{at} \int e^{-at} g(t) \, dt$.
- If none of the preceding techniques applies, write $T = (D - \lambda_1)(D - \lambda_2) \cdots (D - \lambda_n)$, and solve the corresponding first order DEs.

Step 3   The solutions of the DE $T(f) = g$ are of the form

$$f(t) = c_1 f_1(t) + c_2 f_2(t) + \cdots + c_n f_n(t) + f_p(t),$$

where $f_1, f_2, \ldots, f_n$ are the solutions from step 1 and f_p is the solution from step 2.

---

[7]More generally, it is often helpful to look for a particular solution of the same form as $g(t)$, for example, a polynomial of a certain degree, or an exponential function Ce^{kt}. This technique is explored more fully in a course on differential equations.

Take another look at Examples 2, 5, 6, 7, 9, and 10.

EXAMPLE 11 Find all solutions of the DE

$$f'''(t) + f''(t) - f'(t) - f(t) = 10.$$

Solution

We will follow the approach just outlined.

Step 1

- $p_T(\lambda) = \lambda^3 + \lambda^2 - \lambda - 1$
- We recognize $\lambda = 1$ as a root, and we can use long division to factor:

$$p_T(\lambda) = \lambda^3 + \lambda^2 - \lambda - 1 = (\lambda - 1)(\lambda^2 + 2\lambda + 1) = (\lambda - 1)(\lambda + 1)^2.$$

- Since $\lambda = 1$ is a solution of the equation $p_T(\lambda) = 0$, we let $f_1(t) = e^t$.
- Since $\lambda = -1$ is a solution of $p_T(\lambda) = 0$ with multiplicity 2, we let $f_2(t) = e^{-t}$ and $f_3(t) = te^{-t}$.

Step 2 Since $g(t) = 10$ is a constant, we look for a constant solution, $f_p(t) = c$. Plugging into the DE, we find that $c = -10$. Thus $f_p(t) = -10$.

Step 3 The solutions are of the form

$$f(t) = c_1 e^t + c_2 e^{-t} + c_3 te^{-t} - 10,$$

where $c_1, c_2,$ and c_3 are arbitrary constants. ∎

EXERCISES

GOAL Solve linear differential equations.

Find all real solutions of the differential equations in Exercises 1 through 22.

1. $f'(t) - 5f(t) = 0.$

2. $\dfrac{dx}{dt} + 3x = 7.$

3. $f'(t) + 2f(t) = e^{3t}.$

4. $\dfrac{dx}{dt} - 2x = \cos(3t).$

5. $f'(t) - f(t) = t.$

6. $f'(t) - 2f(t) = e^{2t}.$

7. $f''(t) + f'(t) - 12f(t) = 0.$

8. $\dfrac{d^2x}{dt^2} + 3\dfrac{dx}{dt} - 10x = 0.$

9. $f''(t) - 9f(t) = 0.$

10. $f''(t) + f(t) = 0.$

11. $\dfrac{d^2x}{dt^2} - 2\dfrac{dx}{dt} + 2x = 0.$

12. $f''(t) - 4f'(t) + 13f(t) = 0.$

13. $f''(t) + 2f'(t) + f(t) = 0.$

14. $f''(t) + 3f'(t) = 0.$

15. $f''(t) = 0.$

16. $f''(t) + 4f'(t) + 13f(t) = \cos(t).$

17. $f''(t) + 2f'(t) + f(t) = \sin(t).$

18. $f''(t) + 3f'(t) + 2f(t) = \cos(t)$.

19. $\dfrac{d^2x}{dt^2} + 2x = \cos(t)$.

20. $f'''(t) - 3f''(t) + 2f'(t) = 0$.

21. $f'''(t) + 2f''(t) - f'(t) - 2f(t) = 0$.

22. $f'''(t) - f''(t) - 4f'(t) + 4f(t) = 0$.

Solve the initial value problems in Exercises 23 through 29.

23. $f'(t) - 5f(t) = 0$, $f(0) = 3$.

24. $\dfrac{dx}{dt} + 3x = 7$, $x(0) = 0$.

25. $f'(t) + 2f(t) = 0$, $f(1) = 1$.

26. $f''(t) - 9f(t) = 0$, $f(0) = 0$, $f'(0) = 1$.

27. $f''(t) + 9f(t) = 0$, $f(0) = 0$, $f\left(\frac{\pi}{2}\right) = 1$.

28. $f''(t) + f'(t) - 12f(t) = 0$, $f(0) = f'(0) = 0$.

29. $f''(t) + 4f(t) = \sin(t)$, $f(0) = f'(0) = 0$.

30. The temperature of a hot cup of coffee can be modeled by the DE

$$T'(t) = -k(T(t) - A).$$

a. What is the significance of the constants k and A?

b. Solve the DE for $T(t)$, in terms of k, A, and the initial temperature T_0. *Hint:* There is a constant particular solution.

31. The speed $v(t)$ of a falling object can sometimes be modeled by

$$m\frac{dv}{dt} = mg - kv,$$

or

$$\frac{dv}{dt} + \frac{k}{m}v = g,$$

where m is the mass of the body, g the gravitational acceleration, and k a constant related to the air resistance. Solve this DE when $v(0) = 0$. Describe the long term behavior of $v(t)$. Sketch a graph.

32. Consider the balance $B(t)$ of a bank account, with initial balance $B(0) = B_0$. We are withdrawing money at a continuous rate r (in Euro/year). The interest rate is k (%/year), compounded continuously. Set up a differential equation for $B(t)$, and solve it in terms of B_0, r, and k. What will happen in the long run? Describe all possible scenarios. Sketch a graph for $B(t)$ in each case.

33. Consider a pendulum of length L. Let $x(t)$ be the angle the pendulum makes with the vertical (measured in radians). For small angles, the motion is well

approximated by the DE

$$\frac{d^2x}{dt^2} = -\frac{g}{L}x,$$

where g is the acceleration due to gravity ($g \approx 9.81$ m/sec^2). How long does the pendulum have to be so that it swings from one extreme position to the other in exactly one second?

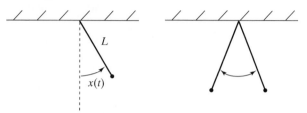

Note: $x(t)$ is negative when the pendulum is on the left.

The two extreme positions of the pendulum.

Historical note: The result of this exercise was considered as a possible *definition* of the meter. The French committee reforming the measures in the 1790's finally adopted another definition: a meter is the 10,000,000th part of the distance from the North Pole to the Equator, measured along the meridian through Paris.

34. Consider a wooden block in the shape of a cube whose edges are 10 cm long. The density of the wood is 0.8 g/cm^3. The block is submersed in water; a guiding mechanism guarantees that the top and the bottom surfaces of the block are parallel to the surface of the water at all times. Let $x(t)$ be the depth of the block in the water at time t. Assume that x is between 0 and 10 at all times.

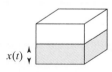

a. Two forces are acting on the block: its weight and the buoyancy (the weight of the displaced water). Recall that the density of water is 1 g/cm^3. Find formulas for these two forces.

b. Set up a differential equation for $x(t)$. Find the solution, assuming that the block is initially completely submersed ($x(0) = 10$) and at rest.

c. How does the period of the oscillation change if you change the dimensions of the block? (Consider a larger or smaller cube.) What if the wood has a different density, or if the initial state is different? What if you conduct the experiment on the moon?

35. The displacement $x(t)$ of a certain oscillator can be modeled by the DE

$$\frac{d^2x}{dt^2} + 3\frac{dx}{dt} + 2x = 0.$$

a. Find all solutions of this DE.

b. Find the solution with initial values $x(0) = 1$, $x'(0) = 0$. Graph the solution.

c. Find the solution with initial values $x(0) = 1$, $x'(0) = -3$. Graph the solution.

d. Describe the qualitative difference of the solutions in parts (b) and parts (c), in terms of the motion of the oscillator. How many times will the oscillator go through the equilibrium state $x = 0$ in each case?

36. The displacement $x(t)$ of a certain oscillator can be modeled by the DE

$$\frac{d^2x}{dt^2} + 2\frac{dx}{dt} + 101x = 0.$$

Find all solutions of this DE, and graph a typical solution. How many times will the oscillator go through the equilibrium state $x = 0$?

37. The displacement $x(t)$ of a certain oscillator can be modeled by the DE

$$\frac{d^2x}{dt^2} + 6\frac{dx}{dt} + 9x = 0.$$

Find the solution $x(t)$ for the initial values $x(0) = 0$, $x'(0) = 1$. Sketch the graph of the solution. How many times will the oscillator go through the equilibrium state $x = 0$ in this case?

38. a. If $p(t)$ is a polynomial and λ a scalar, show that

$$(D - \lambda)\left(p(t)e^{\lambda t}\right) = p'(t)e^{\lambda t}.$$

b. If $p(t)$ is a polynomial of degree less than m, what is

$$(D - \lambda)^m \left(p(t)e^{\lambda t}\right) ?$$

c. Find a basis of the kernel of the linear differential operator $(D - \lambda)^m$.

d. If $\lambda_1, \ldots, \lambda_r$ are distinct scalars and $m_1, \ldots, m_r$ are positive integers, find a basis of the kernel of the linear differential operator

$$(D - \lambda_1)^{m_1} \ldots (D - \lambda_r)^{m_r}.$$

39. Find all solutions of the linear DE

$$f'''(t) + 3f''(t) + 3f'(t) + f(t) = 0.$$

Hint: Use Exercise 38.

40. Find all solutions of the linear DE

$$\frac{d^3x}{dt^3} + \frac{d^2x}{dt^2} - \frac{dx}{dt} - x = 0.$$

Hint: Use Exercise 38.

41. If T is an nth-order linear differential operator and λ is an arbitrary scalar, is λ necessarily an eigenvalue of T? If so, what is the dimension of the eigenspace associated with λ?

42. Let C^∞ be the space of all *real-valued* smooth functions.

a. Consider the linear differential operator $T = D^2$ from C^∞ to C^∞. Find all (real) eigenvalues of T. For each eigenvalue, find a basis of the associated eigenspace.

b. Let P be the subspace of C^∞ consisting of all *periodic* functions $f(t)$ with period one (i.e., $f(t + 1) = f(t)$, for all t). Consider the linear differential operator $L = D^2$ from P to P. Find all (real) eigenvalues and eigenfunctions of L.

43. The displacement of a certain forced oscillator can be modeled by the DE

$$\frac{d^2x}{dt^2} + 5\frac{dx}{dt} + 6x = \cos(t).$$

a. Find all solutions of this DE.

b. Describe the long-term behavior of this oscillator.

44. The displacement of a certain forced oscillator can be modeled by the DE

$$\frac{d^2x}{dt^2} + 4\frac{dx}{dt} + 5x = \cos(3t).$$

a. Find all solutions of this DE.

b. Describe the long-term behavior of this oscillator.

45. Use Fact 9.3.13 to solve the initial value problem

$$\frac{d\vec{x}}{dt} = \begin{bmatrix} 1 & 2 \\ 0 & 1 \end{bmatrix}\vec{x}, \quad \text{with} \quad \vec{x}(0) = \begin{bmatrix} 1 \\ -1 \end{bmatrix}.$$

Hint: Find first $x_2(t)$ and then $x_1(t)$.

46. Use Fact 9.3.13 to solve the initial value problem

$$\frac{d\vec{x}}{dt} = \begin{bmatrix} 2 & 3 & 1 \\ 0 & 1 & 2 \\ 0 & 0 & 1 \end{bmatrix}\vec{x}, \quad \text{with} \quad \vec{x}(0) = \begin{bmatrix} 2 \\ 1 \\ -1 \end{bmatrix}.$$

Hint: Find first $x_3(t)$, then $x_2(t)$, and then $x_1(t)$.

47. Consider the initial-value problem

$$\frac{d\vec{x}}{dt} = A\vec{x}, \quad \text{with} \quad \vec{x}(0) = \vec{x}_0,$$

where A is an upper triangular $n \times n$ matrix with m distinct diagonal entries $\lambda_1, \ldots, \lambda_m$. See the examples in Exercises 45 and 46.

a. Show that this problem has a unique solution $\vec{x}(t)$, whose components $x_i(t)$ are of the form

$$x_i(t) = p_1(t)e^{\lambda_1 t} + \cdots + p_m(t)e^{\lambda_m t},$$

for some polynomials $p_j(t)$.

Hint: Find first $x_n(t)$, then $x_{n-1}(t)$, and so on.

b. Show that the zero state is a stable equilibrium solution of this system if (and only if) the real part of all the λ_i is negative.

48. Consider an $n \times n$ matrix A with m distinct eigenvalues $\lambda_1, \ldots, \lambda_m$.

a. Show that the initial-value problem

$$\frac{d\vec{x}}{dt} = A\vec{x}, \quad \text{with} \quad \vec{x}(0) = \vec{x}_0,$$

has a unique solution $\vec{x}(t)$.

b. Show that the zero state is a stable equilibrium solution of the system

$$\frac{d\vec{x}}{dt} = A\vec{x}$$

if and only if the real part of all the λ_i is negative.

Hint: Exercises 8.1.37 and Exercise 47 of this section are helpful.

A

Vectors

Here we will provide a concise summary of basic facts on vectors. In Section 1.2, vectors are defined as matrices with only one column: $\vec{v} = \begin{bmatrix} v_1 \\ v_2 \\ \vdots \\ v_n \end{bmatrix}$. The scalars v_i are called the *components* of the vector.[1] The set of all vectors with n components is denoted by $\mathbb{R}^n$.

You may be accustomed to a different notation for vectors. Writing the components in a column is the most convenient notation for linear algebra.

Vector Algebra

Definition A.1

> **Vector addition**
>
> a. The sum of two vectors $\vec{v}$ and $\vec{w}$ in $\mathbb{R}^n$ is defined "componentwise":
>
> $$\vec{v} + \vec{w} = \begin{bmatrix} v_1 \\ v_2 \\ \vdots \\ v_n \end{bmatrix} + \begin{bmatrix} w_1 \\ w_2 \\ \vdots \\ w_n \end{bmatrix} = \begin{bmatrix} v_1 + w_1 \\ v_2 + w_2 \\ \vdots \\ v_n + w_n \end{bmatrix}$$
>
> *(Continued)*

---

[1] In vector and matrix algebra, the term "scalar" is synonymous with (real) number.

Vector addition (*Continued*)

b. *Scalar multiplication*
The product of a scalar k and a vector $\vec{v}$ is defined componentwise as well:

$$k\vec{v} = k \begin{bmatrix} v_1 \\ v_2 \\ \vdots \\ v_n \end{bmatrix} = \begin{bmatrix} kv_1 \\ kv_2 \\ \vdots \\ kv_n \end{bmatrix}$$

EXAMPLE 1

$$\begin{bmatrix} 1 \\ 2 \\ 3 \\ 4 \end{bmatrix} + \begin{bmatrix} 4 \\ 2 \\ 0 \\ -1 \end{bmatrix} = \begin{bmatrix} 5 \\ 4 \\ 3 \\ 3 \end{bmatrix}$$

∎

EXAMPLE 2

$$3 \begin{bmatrix} 1 \\ 2 \\ 0 \\ -1 \end{bmatrix} = \begin{bmatrix} 3 \\ 6 \\ 0 \\ -3 \end{bmatrix}$$

∎

The *negative* or *opposite* of a vector $\vec{v}$ in $\mathbb{R}^n$ is defined as

$$-\vec{v} = (-1)\vec{v} \ .$$

The *difference* $\vec{v} - \vec{w}$ of two vectors $\vec{v}$ and $\vec{w}$ in $\mathbb{R}^n$ is defined componentwise. Alternatively, we can express the difference of two vectors as

$$\vec{v} - \vec{w} = \vec{v} + (-\vec{w}) \ .$$

The vector in $\mathbb{R}^n$ that consists of n zeros is called the *zero vector* in $\mathbb{R}^n$:

$$\vec{0} = \begin{bmatrix} 0 \\ 0 \\ \vdots \\ 0 \end{bmatrix}$$

Fact A.2

Rules of vector algebra
The following formulas hold for all vectors $\vec{u}$, $\vec{v}$, $\vec{w}$ in $\mathbb{R}^n$ and for all scalars c and k:

1. $(\vec{u} + \vec{v}) + \vec{w} = \vec{u} + (\vec{v} + \vec{w})$. (Addition is *associative*.)
2. $\vec{v} + \vec{w} = \vec{w} + \vec{v}$. (Addition is *commutative*.)
3. $\vec{v} + \vec{0} = \vec{v}$.

4. For each $\vec{v}$ in $\mathbb{R}^n$, there is a unique $\vec{x}$ in $\mathbb{R}^n$ such that $\vec{v} + \vec{x} = 0$, namely, $\vec{x} = -\vec{v}$.

5. $k(\vec{v} + \vec{w}) = k\vec{v} + k\vec{w}$

6. $(c + k)\vec{v} = c\vec{v} + k\vec{v}$

7. $c(k\vec{v}) = (ck)\vec{v}$

8. $1\vec{v} = \vec{v}$

These rules follow from the corresponding rules for scalars (commutativity, associativity, distributivity); for example:

$$\vec{v} + \vec{w} = \begin{bmatrix} v_1 \\ v_2 \\ \vdots \\ v_n \end{bmatrix} + \begin{bmatrix} w_1 \\ w_2 \\ \vdots \\ w_n \end{bmatrix} = \begin{bmatrix} v_1 + w_1 \\ v_2 + w_2 \\ \vdots \\ v_n + w_n \end{bmatrix} = \begin{bmatrix} w_1 + v_1 \\ w_2 + v_2 \\ \vdots \\ w_n + v_n \end{bmatrix} = \begin{bmatrix} w_1 \\ w_2 \\ \vdots \\ w_n \end{bmatrix} + \begin{bmatrix} v_1 \\ v_2 \\ \vdots \\ v_n \end{bmatrix} = \vec{w} + \vec{v}.$$

Geometrical Representation of Vectors

The *standard representation* of a vector

$$\vec{x} = \begin{bmatrix} x_1 \\ x_2 \end{bmatrix}$$

in the Cartesian coordinate plane is as an *arrow* (a directed line segment) connecting the origin to the point (x_1, x_2), as shown in Figure 1.

Occasionally, it is helpful to translate (or shift) the vector in the plane (preserving its direction and length), so that it will connect some point (a_1, a_2) to the point $(a_1 + x_1, a_2 + x_2)$. See Figure 2.

In this text, we consider the *standard representation* of vectors, unless we explicitly state that the vector has been translated.

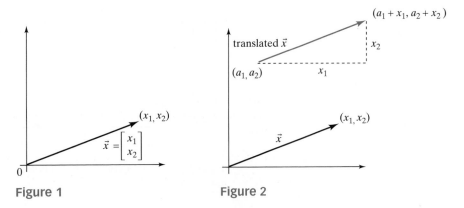

Figure 1   Figure 2

A vector in $\mathbb{R}^2$ (in standard representation) is uniquely determined by its endpoint. Conversely, with each point in the plane we can associate its *position vector,* which connects the origin to the given point. See Figure 3.

We need not clearly distinguish between a vector and its endpoint; we can identify them as long as we consistently use the standard representation of vectors.

For example, we will talk about "the vectors on a line L" when we really mean the vectors whose endpoints are on the line L (in standard representation). Likewise, we can talk about "the vectors in a region R" in the plane. See Figure 4.

Adding vectors in $\mathbb{R}^2$ can be represented by means of a parallelogram, as shown in Figure 5.

If k is a positive scalar, then $k\vec{v}$ is obtained by stretching the vector $\vec{v}$ by a factor of k, leaving its direction unchanged. If k is negative, then the direction is reversed. See Figure 6.

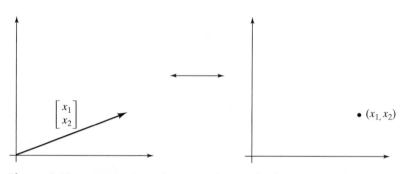

Figure 3 The components of a vector in standard representation are the coordinates of its endpoint.

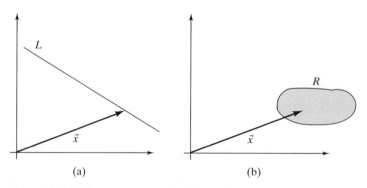

Figure 4 (a) $\vec{x}$ is a vector on the line L. (b) $\vec{x}$ is a vector in the region R.

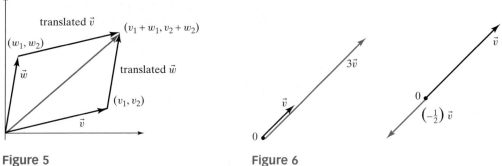

Figure 5

Figure 6

Definition A.3

We say that two vectors $\vec{v}$ and $\vec{w}$ in $\mathbb{R}^n$ are *parallel* if one of them is a scalar multiple of the other.

EXAMPLE 3   The vectors

$$\begin{bmatrix} 1 \\ 3 \\ 2 \\ -2 \end{bmatrix} \quad \text{and} \quad \begin{bmatrix} 3 \\ 9 \\ 6 \\ -6 \end{bmatrix}$$

are parallel, since

$$\begin{bmatrix} 3 \\ 9 \\ 6 \\ -6 \end{bmatrix} = 3 \begin{bmatrix} 1 \\ 3 \\ 2 \\ -2 \end{bmatrix}.$$

■

EXAMPLE 4   The vectors

$$\begin{bmatrix} 1 \\ 2 \\ 3 \\ 4 \end{bmatrix} \quad \text{and} \quad \begin{bmatrix} 0 \\ 0 \\ 0 \\ 0 \end{bmatrix}$$

are parallel, since

$$\begin{bmatrix} 0 \\ 0 \\ 0 \\ 0 \end{bmatrix} = 0 \begin{bmatrix} 1 \\ 2 \\ 3 \\ 4 \end{bmatrix}.$$

■

Let us briefly review Cartesian coordinates in *space:* If we choose an origin 0 and three mutually perpendicular coordinate axes through 0, we can describe any point in space by a triple of numbers, (x_1, x_2, x_3). See Figure 7.

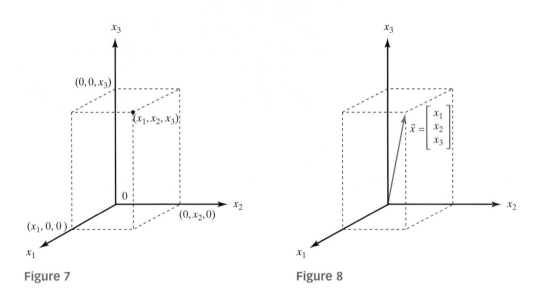

Figure 7 Figure 8

The standard representation of the vector

$$\vec{x} = \begin{bmatrix} x_1 \\ x_2 \\ x_3 \end{bmatrix}$$

is the arrow connecting the origin to the point (x_1, x_2, x_3), as shown in Figure 8.

Dot Product, Length, Orthogonality

Definition A.4 Consider two vectors $\vec{v}$ and $\vec{w}$ with components $v_1, v_2, \ldots, v_n$ and $w_1, w_2, \ldots, w_n$, respectively. Here $\vec{v}$ and $\vec{w}$ may be column or row vectors, and they need not be of the same type (these conventions are convenient in linear algebra). The *dot product* of $\vec{v}$ and $\vec{w}$ is defined as

$$\vec{v} \cdot \vec{w} = v_1 w_1 + v_2 w_2 + \cdots + v_n w_n.$$

Note that the dot product of two vectors is a *scalar*.

EXAMPLE 5

$$\begin{bmatrix} 1 \\ 2 \\ 1 \end{bmatrix} \cdot \begin{bmatrix} 3 \\ -1 \\ -1 \end{bmatrix} = 1 \cdot 3 + 2 \cdot (-1) + 1 \cdot (-1) = 0.$$

∎

EXAMPLE 6

$$[1 \quad 2 \quad 3 \quad 4] \cdot \begin{bmatrix} 3 \\ 1 \\ 0 \\ -1 \end{bmatrix} = 3 + 2 + 0 - 4 = 1.$$

∎

Fact A.5   **Rules for dot products**
The following equations hold for all column or row vectors $\vec{u}, \vec{v}, \vec{w}$ with n components, and for all scalars k:

a. $\vec{v} \cdot \vec{w} = \vec{w} \cdot \vec{v}$.

b. $(\vec{u} + \vec{v}) \cdot \vec{w} = \vec{u} \cdot \vec{w} + \vec{v} \cdot \vec{w}$.

c. $(k\vec{v}) \cdot \vec{w} = k(\vec{v} \cdot \vec{w})$.

d. $\vec{v} \cdot \vec{v} > 0$ for all nonzero $\vec{v}$.

The verification of these rules is straightforward. Let us justify rule (d): since $\vec{v}$ is nonzero, at least one of the components v_i is nonzero, so that v_i^2 is positive. Then

$$\vec{v} \cdot \vec{v} = v_1^2 + v_2^2 + \cdots + v_i^2 + \cdots + v_n^2$$

is positive as well.

Let us think about the *length* of a vector. The length of a vector

$$\vec{x} = \begin{bmatrix} x_1 \\ x_2 \end{bmatrix}$$

in $\mathbb{R}^2$ is $\sqrt{x_1^2 + x_2^2}$ by the Pythagorean theorem. See Figure 9.

This length is often denoted by $\|\vec{x}\|$. Note that we have

$$\vec{x} \cdot \vec{x} = \begin{bmatrix} x_1 \\ x_2 \end{bmatrix} \cdot \begin{bmatrix} x_1 \\ x_2 \end{bmatrix} = x_1^2 + x_2^2 = \|\vec{x}\|^2;$$

therefore,

$$\|\vec{x}\| = \sqrt{\vec{x} \cdot \vec{x}}.$$

Verify that this formula holds for vectors $\vec{x}$ in $\mathbb{R}^3$ as well.

We can use this formula to *define* the length of a vector in $\mathbb{R}^n$:

Definition A.6   The *length* (or *norm*) $\|\vec{x}\|$ of a vector $\vec{x}$ in $\mathbb{R}^n$ is

$$\|\vec{x}\| = \sqrt{\vec{x} \cdot \vec{x}} = \sqrt{x_1^2 + x_2^2 + \cdots + x_n^2}.$$

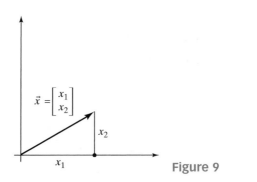

Figure 9

EXAMPLE 7   Find $\|\vec{x}\|$ for

$$\vec{x} = \begin{bmatrix} 7 \\ 1 \\ 7 \\ -1 \end{bmatrix}.$$

Solution

$$\|\vec{x}\| = \sqrt{\vec{x} \cdot \vec{x}} = \sqrt{49 + 1 + 49 + 1} = 10. \qquad \blacksquare$$

Definition A.7

> A vector $\vec{u}$ in $\mathbb{R}^n$ is called a *unit vector* if $\|\vec{u}\| = 1$; that is, the length of the vector $\vec{u}$ is 1.

Consider two perpendicular vectors $\vec{x}$ and $\vec{y}$ in $\mathbb{R}^2$, as shown in Figure 10. By the theorem of Pythagoras,

$$\|\vec{x} + \vec{y}\|^2 = \|\vec{x}\|^2 + \|\vec{y}\|^2,$$

or

$$(\vec{x} + \vec{y}) \cdot (\vec{x} + \vec{y}) = \vec{x} \cdot \vec{x} + \vec{y} \cdot \vec{y}.$$

By Fact A.5,

$$\vec{x} \cdot \vec{x} + 2(\vec{x} \cdot \vec{y}) + \vec{y} \cdot \vec{y} = \vec{x} \cdot \vec{x} + \vec{y} \cdot \vec{y},$$

or

$$\vec{x} \cdot \vec{y} = 0.$$

You can read these equations backward to show that $\vec{x} \cdot \vec{y} = 0$ if and only if $\vec{x}$ and $\vec{y}$ are perpendicular. This reasoning applies to vectors in $\mathbb{R}^3$ as well.

We can use this characterization to *define* perpendicular vectors in $\mathbb{R}^n$:

Definition A.8

> Two (row or column) vectors $\vec{v}$ and $\vec{w}$ are called *perpendicular* (or *orthogonal*) if $\vec{v} \cdot \vec{w} = 0$.

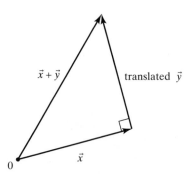

Figure 10

Cross Product

Here we present the cross product for vectors in $\mathbb{R}^3$ only; for a generalization to $\mathbb{R}^n$, see Exercises 6.2.30 and 6.3.17.

In Chapter 6, we discuss the cross product in the context of linear algebra.

Definition A.9

The cross product of two vectors $\vec{v}$ and $\vec{w}$ in $\mathbb{R}^3$ is

$$\vec{v} \times \vec{w} = \begin{bmatrix} v_1 \\ v_2 \\ v_3 \end{bmatrix} \times \begin{bmatrix} w_1 \\ w_2 \\ w_3 \end{bmatrix} = \begin{bmatrix} v_2 w_3 - v_3 w_2 \\ v_3 w_1 - v_1 w_3 \\ v_1 w_2 - v_2 w_1 \end{bmatrix}.$$

Unlike the dot product, the cross product $\vec{v} \times \vec{w}$ is a *vector* in $\mathbb{R}^3$.

EXAMPLE 8

$$\begin{bmatrix} 1 \\ 2 \\ -1 \end{bmatrix} \times \begin{bmatrix} -1 \\ 1 \\ 1 \end{bmatrix} = \begin{bmatrix} 3 \\ 0 \\ 3 \end{bmatrix}$$ ∎

Fact A.10

Geometric interpretation of the cross product

Let $\vec{v} \times \vec{w}$ be the cross product of two vectors $\vec{v}$ and $\vec{w}$ in $\mathbb{R}^3$. Then

a. $\vec{v} \times \vec{w}$ is orthogonal to both $\vec{v}$ and $\vec{w}$.

b. The length of the vector $\vec{v} \times \vec{w}$ is numerically equal to the area of the parallelogram defined by $\vec{v}$ and $\vec{w}$. (See Figure 11(a).)

c. If $\vec{v}$ and $\vec{w}$ are not parallel, then the vectors

$$\vec{v}, \ \vec{w}, \ \vec{v} \times \vec{w}$$

form a right-handed system. (See Figure 11(b).)

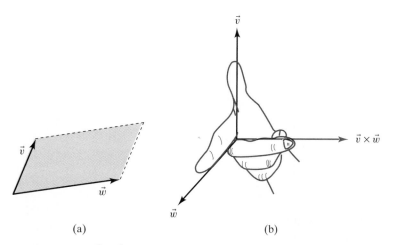

(a)   (b)

Figure 11 (a) $\|\vec{v} \times \vec{w}\|$ is numerically equal to the shaded area.
(b) A right-handed system.

Answers to Odd-Numbered Exercises

Chapter 1

1.1

1. $(x, y) = (-1, 1)$
3. No solutions
5. $(x, y) = (0, 0)$
7. No solutions
9. $(x, y, z) = (t, \frac{1}{2} - 2t, t)$, where t is arbitrary
11. $(x, y) = (4, 1)$
13. No solutions
15. $(x, y, z) = (0, 0, 0)$
17. $(x, y) = (-5a + 2b, 3a - b)$
19. a. Products are competing.
    b. $P_1 = 26, P_2 = 46$
21. $a = 400, b = 300$
23. a. $(x, y) = (t, 2t)$;   b. $(x, y) = (t, -3t)$;
    c. $(x, y) = (0, 0)$
25. a. If $k = 7$
    b. If $k = 7$, there are infinitely many solutions.
    c. If $k = 7$, the solutions are $(x, y, z) =$
       $(1 - t, 2t - 3, t)$.
27. 7 children (3 boys and 4 girls)
29. $f(t) = 1 - 5t + 3t^2$
31. If $a - 2b + c = 0$
33. a. The intercepts of the line $x + y = 1$ are $(1, 0)$
       and $(0, 1)$. The intercepts of the line
       $x + \frac{1}{2}y = t$ are $(t, 0)$ and $(0, 2)$. The lines
       intersect if $t \neq 2$.
    b. $x = -\dfrac{t}{t - 2}, y = \dfrac{2t - 2}{t - 2}$
35. There are many correct answers. Example:
$$\begin{vmatrix} x & -5z = -4 \\ y & -3z = -2 \end{vmatrix}$$

1.2

1. $\begin{bmatrix} x \\ y \\ z \end{bmatrix} = \begin{bmatrix} 10t + 13 \\ -8t - 8 \\ t \end{bmatrix}$

3. $\begin{bmatrix} x \\ y \\ z \end{bmatrix} = \begin{bmatrix} 4 - 2s - 3t \\ s \\ t \end{bmatrix}$

5. $\begin{bmatrix} x_1 \\ x_2 \\ x_3 \\ x_4 \end{bmatrix} = \begin{bmatrix} -t \\ t \\ -t \\ t \end{bmatrix}$

7. $\begin{bmatrix} x_1 \\ x_2 \\ x_3 \\ x_4 \\ x_5 \end{bmatrix} = \begin{bmatrix} -2t \\ t \\ 0 \\ 0 \\ 0 \end{bmatrix}$

9. $\begin{bmatrix} x_1 \\ x_2 \\ x_3 \\ x_4 \\ x_5 \\ x_6 \end{bmatrix} = \begin{bmatrix} t - s - 2r \\ r \\ -t + s & +1 \\ t - 2s & +2 \\ s \\ t \end{bmatrix}$

11. $\begin{bmatrix} x_1 \\ x_2 \\ x_3 \\ x_4 \end{bmatrix} = \begin{bmatrix} -2t \\ 3t + 4 \\ t \\ -2 \end{bmatrix}$

13. No solutions

15. $\begin{bmatrix} x \\ y \\ z \end{bmatrix} = \begin{bmatrix} 4 \\ 2 \\ 1 \end{bmatrix}$

17. $\begin{bmatrix} x_1 \\ x_2 \\ x_3 \\ x_4 \\ x_5 \end{bmatrix} = \begin{bmatrix} -8221/4340 \\ 8591/8680 \\ 4695/434 \\ -459/434 \\ 699/434 \end{bmatrix}$

19. $\begin{bmatrix} 0 \\ 0 \\ 0 \\ 0 \end{bmatrix}$ and $\begin{bmatrix} 1 \\ 0 \\ 0 \\ 0 \end{bmatrix}$

21. 4 types
25. Yes; perform the operations backwards.
27. No; you cannot make the last column zero by
    elementary row operations.
29. $a = 2, b = c = d = 1$
31. $f(t) = 1 - 5t + 4t^2 + 3t^3 - 2t^4$
33. $f(t) = -5 + 13t - 10t^2 + 3t^3$

35. $\begin{bmatrix} -t \\ 6t \\ -9t \\ 4t \end{bmatrix}$, where t is arbitrary.

37. $\begin{bmatrix} x_1 \\ x_2 \\ x_3 \end{bmatrix} = \begin{bmatrix} 500 \\ 300 \\ 400 \end{bmatrix}$

39. a. Neither the manufacturing nor the energy
       sector makes demands on agriculture.
    b. $x_1 \approx 18.67, x_2 \approx 22.60, x_3 \approx 3.63$
41. $m_1 = \dfrac{2}{3}m_2$
43. $a \approx 12.17, b \approx -1.15, c \approx 0.18$. The longest
    day is about 13.3 hours.

1.3   1. a. No solutions   b. One solution
c. Infinitely many solutions
3. rank is 1
5. a. $x \begin{bmatrix} 1 \\ 3 \end{bmatrix} + y \begin{bmatrix} 2 \\ 1 \end{bmatrix} = \begin{bmatrix} 7 \\ 11 \end{bmatrix}$
b. $x = 3, y = 2$
7. One solution
9. $\begin{bmatrix} 1 & 2 & 3 \\ 4 & 5 & 6 \\ 7 & 8 & 9 \end{bmatrix} \begin{bmatrix} x \\ y \\ z \end{bmatrix} = \begin{bmatrix} 1 \\ 4 \\ 9 \end{bmatrix}$
11. Undefined
13. $\begin{bmatrix} 29 \\ 65 \end{bmatrix}$
15. 70
17. Undefined
19. $\begin{bmatrix} 0 \\ 0 \\ 0 \end{bmatrix}$
21. $\begin{bmatrix} 158 \\ 70 \\ 81 \\ 123 \end{bmatrix}$
23. $\begin{bmatrix} 1 & 0 & 0 \\ 0 & 1 & 0 \\ 0 & 0 & 1 \\ 0 & 0 & 0 \end{bmatrix}$
25. The system $A\vec{x} = \vec{c}$ has infinitely many solutions or none.
27. a. True   b. False
29. True
31. $\begin{bmatrix} 1 & 0 & 0 & 0 \\ 0 & 1 & 0 & 0 \\ 0 & 0 & 1 & 0 \\ 0 & 0 & 0 & 1 \end{bmatrix}$
33. $A\vec{x} = \vec{x}$
35. $A\vec{e}_i$ is the ith column of A.
37. $\vec{x} = \begin{bmatrix} 2 - 2t \\ t \\ 1 \end{bmatrix}$, where t is arbitrary
39. $\begin{bmatrix} 1 & 0 & 0 & * \\ 0 & 1 & 0 & * \\ 0 & 0 & 1 & * \end{bmatrix}$
41. One solution
43. No solutions
47. a. $\vec{x} = 0$ is a solution.
b. By part (a) and Fact 1.3.3
c. $A(\vec{x}_1 + \vec{x}_2) = A\vec{x}_1 + A\vec{x}_2 = \vec{0} + \vec{0} = \vec{0}$
d. $A(k\vec{x}) = k(A\vec{x}) = k\vec{0} = \vec{0}$

49. a. Infinitely many solutions or none
b. One solution or none
c. No solutions
d. Infinitely many solutions
51. If $n = r$ and $s = p$
53. Yes
55. Yes; $\begin{bmatrix} 7 \\ 8 \\ 9 \end{bmatrix} = -1 \begin{bmatrix} 1 \\ 2 \\ 3 \end{bmatrix} + 2 \begin{bmatrix} 4 \\ 5 \\ 6 \end{bmatrix}$
57. $\begin{bmatrix} 7 \\ 11 \end{bmatrix} = \begin{bmatrix} 3 \\ 9 \end{bmatrix} + \begin{bmatrix} 4 \\ 2 \end{bmatrix}$

Chapter 2
2.1   1. Not linear
3. Not linear
5. $A = \begin{bmatrix} 7 & 6 & -13 \\ 11 & 9 & 17 \end{bmatrix}$
7. T is linear; $A = \begin{bmatrix} \vec{v}_1 & \vec{v}_2 & \dots & \vec{v}_n \end{bmatrix}$
9. Not invertible
11. The inverse is $\begin{bmatrix} 3 & -2/3 \\ -1 & 1/3 \end{bmatrix}$.
15. A is invertible if $a \neq 0$ or $b \neq 0$. In this case,
$$A^{-1} = \frac{1}{a^2 + b^2} \begin{bmatrix} a & b \\ -b & a \end{bmatrix}$$
17. Reflection in the origin; this transformation is its own inverse.
19. Projection onto the $\vec{e}_1$ axis; not invertible.
21. Rotation through an angle of 90° in the clockwise direction; invertible.
23. Clockwise rotation through an angle of 90°, followed by a dilation by a factor of 2; invertible.
25. Dilation by a factor of 2
27. Reflection in the $\vec{e}_1$ axis
29. Reflection in the origin
31. Reflection in the $\vec{e}_2$ axis, represented by the matrix $\begin{bmatrix} -1 & 0 \\ 0 & 1 \end{bmatrix}$
33. $\frac{\sqrt{2}}{2} \begin{bmatrix} 1 & -1 \\ 1 & 1 \end{bmatrix}$
35. $\begin{bmatrix} 1 & 2 \\ 2 & 1 \end{bmatrix}$
37. $T(\vec{x}) = T(\vec{v}) + k\big(T(\vec{w}) - T(\vec{v})\big)$ is on the line segment.
41. $y = c_1 x_1 + c_2 x_2$; the graph is a plane through the origin in $\mathbb{R}^3$.
43. a. T is linear, represented by the matrix $\begin{bmatrix} 2 & 3 & 4 \end{bmatrix}$.

b. If $\vec{v} = \begin{bmatrix} v_1 \\ v_2 \\ v_3 \end{bmatrix}$, then the matrix of T is

$[v_1 \quad v_2 \quad v_3]$.

c. $y = c_1 x_1 + c_2 x_2 + c_3 x_3 = \begin{bmatrix} c_1 \\ c_2 \\ c_3 \end{bmatrix} \cdot \begin{bmatrix} x_1 \\ x_2 \\ x_3 \end{bmatrix}$

2.2 1. The parallelogram in $\mathbb{R}^2$ defined by the vectors
$\begin{bmatrix} 3 \\ 1 \end{bmatrix}$ and $\begin{bmatrix} 1 \\ 2 \end{bmatrix}$

3. The parallelogram in $\mathbb{R}^3$ defined by the vectors $T(\vec{e}_1)$ and $T(\vec{e}_2)$

5. About 2.5 radians

7. $\frac{1}{9} \begin{bmatrix} 11 \\ 1 \\ 11 \end{bmatrix}$

9. A shear parallel to the $\vec{e}_2$ axis

11. $\frac{1}{25} \begin{bmatrix} 7 & 24 \\ 24 & -7 \end{bmatrix}$

13. $\begin{bmatrix} 2u_1^2 - 1 & 2u_1 u_2 \\ 2u_1 u_2 & 2u_2^2 - 1 \end{bmatrix}$

15. $a_{ii} = 2u_i^2 - 1$, and $a_{ij} = 2u_i u_j$ when $i \neq j$

17. Projection onto the line spanned by $\begin{bmatrix} 1 \\ 1 \end{bmatrix}$,

followed by a dilation by a factor of 2

19. $\begin{bmatrix} 1 & 0 & 0 \\ 0 & 1 & 0 \\ 0 & 0 & 0 \end{bmatrix}$

21. $\begin{bmatrix} 0 & -1 & 0 \\ 1 & 0 & 0 \\ 0 & 0 & 1 \end{bmatrix}$

23. $\begin{bmatrix} 1 & 0 & 0 \\ 0 & 0 & 1 \\ 0 & 1 & 0 \end{bmatrix}$

25. $\begin{bmatrix} 1 & -k \\ 0 & 1 \end{bmatrix}$; you shear back.

27. Yes; use Fact 2.2.1.

29. Yes

31. $A = \begin{bmatrix} -2 & 0 & 0 \\ 1 & 0 & 0 \\ 0 & 0 & 0 \end{bmatrix}$

33. Use a parallelogram

35. Yes; mimic Example 2.2.2.

37. Write $\vec{w} = c_1 \vec{v}_1 + c_2 \vec{v}_2$; then
$T(\vec{w}) = c_1 T(\vec{v}_1) + c_2 T(\vec{v}_2)$

39. $A = \begin{bmatrix} 1 - ab & -b^2 \\ a^2 & 1 + ab \end{bmatrix}$

41. $\text{ref}_Q \vec{x} = -\text{ref}_P \vec{x}$

43. $\begin{bmatrix} \cos \alpha & \sin \alpha \\ -\sin \alpha & \cos \alpha \end{bmatrix}$, a clockwise rotation through the angle α.

45. Write $\vec{x} = c_1 \vec{v}_1 + c_2 \vec{v}_2$ and use Fact 2.2.1 to compute $T(\vec{x})$ and $L(\vec{x})$.

47. Write $T(\vec{x}) = \begin{bmatrix} a & b \\ c & d \end{bmatrix} \vec{x}$. Express $f(t)$ in terms of a, b, c, d.

49. The image is an ellipse with semimajor axes $\pm 5\vec{e}_1$ and semiminor axes $\pm 2\vec{e}_2$.

51. The curve C is the image of the unit circle under the transformation with matrix $[\vec{w}_1 \quad \vec{w}_2]$.

2.3 1. $\begin{bmatrix} 8 & -3 \\ -5 & 2 \end{bmatrix}$

3. $\begin{bmatrix} -\frac{1}{2} & 1 \\ \frac{1}{2} & 0 \end{bmatrix}$

5. Not invertible

7. Not invertible

9. Not invertible

11. $\begin{bmatrix} 1 & 0 & -1 \\ 0 & 1 & 0 \\ 0 & 0 & 1 \end{bmatrix}$

13. $\begin{bmatrix} 1 & 0 & 0 & 0 \\ -2 & 1 & 0 & 0 \\ 1 & -2 & 1 & 0 \\ 0 & 1 & -2 & 1 \end{bmatrix}$

15. $\begin{bmatrix} -6 & 9 & -5 & 1 \\ 9 & -1 & -5 & 2 \\ -5 & -5 & 9 & -3 \\ 1 & 2 & -3 & 1 \end{bmatrix}$

17. Not invertible

19. $x_1 = 3y_1 - 2.5y_2 + 0.5y_3$
$x_2 = -3y_1 + 4y_2 - y_3$
$x_3 = y_1 - 1.5y_2 + 0.5y_3$

21. Not invertible

23. Invertible

25. Invertible

27. Not invertible

29. For all k except $k = 1$ and $k = 2$

31. It's never invertible

33. If $a^2 + b^2 = 1$

35. a. Invertible if a, d, f are all nonzero
b. Invertible if all diagonal entries are nonzero
c. Yes; use Fact 2.3.5.
d. Invertible if all diagonal entries are nonzero

37. $(cA)^{-1} = \frac{1}{c} A^{-1}$

39. M is invertible; if $m_{ij} = k$ (where $i \neq j$), then the ijth entry of M^{-1} is $-k$; all other entries are the same.

41. The transformations in parts (a), (c), (d) are invertible, while the projection in part (b) is not.

43. Yes; $\vec{x} = B^{-1}(A^{-1}\vec{y})$

45. a. $3^3 = 27$ b. n^3 c. $\dfrac{12^3}{3^3} = 64$ (seconds)

47. $f(x) = x^2$ is not invertible, but the equation $f(x) = 0$ has the unique solution $x = 0$.

2.4

1. $\begin{bmatrix} 4 & 6 \\ 3 & 4 \end{bmatrix}$

3. Undefined

5. $\begin{bmatrix} a & b \\ c & d \\ 0 & 0 \end{bmatrix}$

7. $\begin{bmatrix} -1 & 1 & 0 \\ 5 & 3 & 4 \\ -6 & -2 & -4 \end{bmatrix}$

9. $\begin{bmatrix} 0 & 0 \\ 0 & 0 \end{bmatrix}$

11. $[10]$

13. $[h]$

15. I_3; Fact 2.4.9 applies to *square* matrices only.

17. False

19. False

21. True

23. True

25. True

27. $\left[\begin{array}{c|c} 1 & 0 \\ 2 & 0 \\ \hline 19 & 16 \end{array}\right]$

29. For example, $B = \begin{bmatrix} 3 & 0 \\ -1 & 0 \end{bmatrix}$

31. $\begin{bmatrix} t+2 & s-1 \\ -2t-1 & -2s+1 \\ t & s \end{bmatrix}$, for arbitrary t and s

33. No

35. a. $\vec{x} = \vec{0}$ b. $\vec{x} = A\vec{b}$
    c. $\operatorname{rank}(A) = \operatorname{rank}(B) = m$
    d. By Example 1.3.3a

37. All diagonal 2×2 matrices

39. The matrices $\begin{bmatrix} k & 0 \\ 0 & k \end{bmatrix}$, for an arbitrary k

41. a. The matrices $D_\alpha D_\beta$ and $D_\beta D_\alpha$ both represent the counterclockwise rotation through the angle $\alpha + \beta$.
    b. $D_\alpha D_\beta = D_\beta D_\alpha = \begin{bmatrix} \cos(\alpha + \beta) & -\sin(\alpha + \beta) \\ \sin(\alpha + \beta) & \cos(\alpha + \beta) \end{bmatrix}$

43. $A = \begin{bmatrix} \cos(2\pi/3) & -\sin(2\pi/3) \\ \sin(2\pi/3) & \cos(2\pi/3) \end{bmatrix} = \dfrac{1}{2}\begin{bmatrix} -1 & -\sqrt{3} \\ \sqrt{3} & -1 \end{bmatrix}$

45. $A = BS^{-1}$

47. $A = \dfrac{1}{5}\begin{bmatrix} 9 & 3 \\ -2 & 16 \end{bmatrix}$

49. $\begin{cases} \text{matrix of } T: \begin{bmatrix} 0 & 0 & 1 \\ -1 & 0 & 0 \\ 0 & -1 & 0 \end{bmatrix} \\ \\ \text{matrix of } L: \begin{bmatrix} 0 & 1 & 0 \\ 1 & 0 & 0 \\ 0 & 0 & 1 \end{bmatrix} \end{cases}$

51. Yes; yes; each elementary row operation can be "undone" by an elementary row operation.

53. a. Use Exercise 52; let $S = E_1 E_2 \ldots E_p$
    b. $S = \begin{bmatrix} 1 & 0 \\ -4 & 1 \end{bmatrix}\begin{bmatrix} \frac{1}{2} & 0 \\ 0 & 1 \end{bmatrix} = \begin{bmatrix} \frac{1}{2} & 0 \\ -2 & 1 \end{bmatrix}$

55. $\begin{bmatrix} k & 0 \\ 0 & 1 \end{bmatrix}, \begin{bmatrix} 1 & 0 \\ 0 & k \end{bmatrix}, \begin{bmatrix} 1 & c \\ 0 & 1 \end{bmatrix}, \begin{bmatrix} 1 & 0 \\ c & 1 \end{bmatrix}, \begin{bmatrix} 0 & 1 \\ 1 & 0 \end{bmatrix},$
    where k is nonzero and c is arbitrary. Cases 3 and 4 represent shears, and case 5 is a reflection.

57. Yes; use Fact 2.4.4.

59. a. $\vec{y} = \begin{bmatrix} -3 \\ 5 \\ 2 \\ 0 \end{bmatrix}$ b. $\vec{x} = \begin{bmatrix} 1 \\ -1 \\ 2 \\ 0 \end{bmatrix}$

61. a. Write $A = \begin{bmatrix} A^{(m)} & A_2 \\ A_3 & A_4 \end{bmatrix}$,
    $L = \begin{bmatrix} L^{(m)} & 0 \\ L_3 & L_4 \end{bmatrix}, U = \begin{bmatrix} U^{(m)} & U_2 \\ 0 & U_4 \end{bmatrix}$.
    Use Fact 2.4.12.
    c. Solve the equation
    $A = \begin{bmatrix} A^{(n-1)} & \vec{v} \\ \vec{w} & k \end{bmatrix} = \begin{bmatrix} L' & 0 \\ \vec{x} & t \end{bmatrix}\begin{bmatrix} U' & \vec{y} \\ 0 & 1 \end{bmatrix}$
    for $\vec{x}, \vec{y}$, and t.

65. A is invertible if both A_{11} and A_{22} are invertible. In this case,
    $A^{-1} = \begin{bmatrix} A_{11}^{-1} & 0 \\ 0 & A_{22}^{-1} \end{bmatrix}$

67. (ith row of AB) = (ith row of A)B

69. $\operatorname{rank}(A) = \operatorname{rank}(A_{11}) + \operatorname{rank}(A_{23})$

71. Only $A = I_n$

73. (ijth entry of AB) = $\sum_{k=1}^{n} a_{ik}b_{kj} \le s\sum_{k=1}^{n} b_{kj} \le sr$

79. $g(f(x)) = x$, for all x.
    $f(g(x)) = \begin{cases} x & \text{if } x \text{ is even} \\ x+1 & \text{if } x \text{ is odd} \end{cases}$
    The functions f and g are not invertible.

Chapter 3

3.1   1. $\ker(A) = \{\vec{0}\}$

3. $\vec{e}_1, \vec{e}_2$

5. $\begin{bmatrix} 1 \\ -2 \\ 1 \end{bmatrix}$

7. $\ker(A) = \{\vec{0}\}$

9. $\ker(A) = \{\vec{0}\}$

11. $\begin{bmatrix} -2 \\ 3 \\ 1 \\ 0 \end{bmatrix}$

13. $\begin{bmatrix} -2 \\ 1 \\ 0 \\ 0 \\ 0 \\ 0 \end{bmatrix}, \begin{bmatrix} -3 \\ 0 \\ -2 \\ -1 \\ 1 \\ 0 \end{bmatrix}, \begin{bmatrix} 0 \\ 0 \\ 0 \\ 0 \\ 0 \\ 1 \end{bmatrix}$

15. $\begin{bmatrix} 1 \\ 1 \end{bmatrix}, \begin{bmatrix} 1 \\ 2 \end{bmatrix}$

17. All of $\mathbb{R}^2$

19. The line spanned by $\begin{bmatrix} 1 \\ -2 \end{bmatrix}$

21. All of $\mathbb{R}^3$

23. kernel is $\{\vec{0}\}$, image is all of $\mathbb{R}^2$

25. Same as Exercise 23

27. $f(x) = x^3 - x$

29. $f\begin{bmatrix} \varphi \\ \vartheta \end{bmatrix} = \begin{bmatrix} \sin(\varphi)\cos(\vartheta) \\ \sin(\varphi)\sin(\vartheta) \\ \cos(\varphi) \end{bmatrix}$ (compare with spherical coordinates)

31. $A = \begin{bmatrix} -2 & -3 \\ 0 & 1 \\ 1 & 0 \end{bmatrix}$

33. $T\begin{bmatrix} x \\ y \\ z \end{bmatrix} = x + 2y + 3z$

35. $\ker(T)$ is the plane with normal vector $\vec{v}$; $\text{im}(T) = \mathbb{R}$.

37. $\text{im}(A) = \text{span}(\vec{e}_1, \vec{e}_2)$; $\ker(A) = \text{span}(\vec{e}_1)$; $\text{im}(A^2) = \text{span}(\vec{e}_1)$; $\ker(A^2) = \text{span}(\vec{e}_1, \vec{e}_2)$; $A^3 = 0$ so $\ker(A^3) = \mathbb{R}^3$ and $\text{im}(A^3) = \{\vec{0}\}$

39. a. $\ker(B)$ is contained in $\ker(AB)$, but they need not be equal.
   b. $\text{im}(AB)$ is contained in $\text{im}(A)$, but they need not be equal.

41. a. $\text{im}(A)$ is the line spanned by $\begin{bmatrix} 3 \\ 4 \end{bmatrix}$, and $\ker(A)$ is the perpendicular line, spanned by $\begin{bmatrix} -4 \\ 3 \end{bmatrix}$.
   b. $A^2 = A$; if $\vec{v}$ is in $\text{im}(A)$, then $A\vec{v} = \vec{v}$.

   c. Orthogonal projection onto the line spanned by $\begin{bmatrix} 3 \\ 4 \end{bmatrix}$.

43. Suppose A is an $m \times n$ matrix of rank r. Let B be the matrix you get when you omit the first r rows and the first n columns of $\text{rref}[A : I_m]$. (What can you do when $r = m$?)

45. There are $n - r$ nonleading variables, which can be chosen freely. The general vector in the kernel can be written as a linear combination of $n - r$ vectors, with the nonleading variables as coefficients.

47. $\text{im}(T) = L_2$ and $\ker(T) = L_1$

51. $\ker(AB) = \{\vec{0}\}$

53. a. $\begin{bmatrix} 1 \\ 1 \\ 1 \\ 1 \\ 0 \\ 0 \\ 0 \end{bmatrix}, \begin{bmatrix} 0 \\ 1 \\ 1 \\ 0 \\ 1 \\ 0 \\ 0 \end{bmatrix}, \begin{bmatrix} 1 \\ 0 \\ 1 \\ 0 \\ 0 \\ 1 \\ 0 \end{bmatrix}, \begin{bmatrix} 1 \\ 1 \\ 0 \\ 0 \\ 0 \\ 0 \\ 1 \end{bmatrix}$

   b. $\ker(H) = \text{span}(\vec{v}_1, \vec{v}_2, \vec{v}_3, \vec{v}_4)$, by part (a), and $\text{im}(M) = \text{span}(\vec{v}_1, \vec{v}_2, \vec{v}_3, \vec{v}_4)$, by Fact 3.1.3. Thus $\ker(H) = \text{im}(M)$. $H(M\vec{x}) = \vec{0}$, since $M\vec{x}$ is in $\text{im}(M) = \ker(H)$.

3.2   1. Not a subspace

3. W is a subspace

7. Yes

9. Dependent

11. Independent

13. Dependent

15. Dependent

17. Independent

19. Dependent

21. Dependent

25. $\begin{bmatrix} -2 \\ 1 \\ 0 \\ 0 \\ 0 \end{bmatrix}, \begin{bmatrix} -3 \\ 0 \\ 1 \\ 0 \\ 0 \end{bmatrix}, \begin{bmatrix} -4 \\ 0 \\ 0 \\ 1 \\ 0 \end{bmatrix}, \begin{bmatrix} -5 \\ 0 \\ 0 \\ 0 \\ 1 \end{bmatrix}$

27. $\begin{bmatrix} 1 \\ 1 \\ 1 \end{bmatrix}, \begin{bmatrix} 1 \\ 2 \\ 3 \end{bmatrix}$

29. $\begin{bmatrix} 1 \\ 4 \end{bmatrix}, \begin{bmatrix} 2 \\ 5 \end{bmatrix}$

31. $\begin{bmatrix} 1 \\ 2 \\ 3 \\ 5 \end{bmatrix}, \begin{bmatrix} 5 \\ 6 \\ 7 \\ 8 \end{bmatrix}$

33. $\begin{bmatrix} 1 \\ 0 \\ 0 \\ 0 \end{bmatrix}, \begin{bmatrix} 0 \\ 1 \\ 0 \\ 0 \end{bmatrix}, \begin{bmatrix} 0 \\ 0 \\ 1 \\ 0 \end{bmatrix}$

35. Consider a relation
$c_1 \vec{v}_1 + \cdots + c_i \vec{v}_i + \cdots + c_m \vec{v}_m = \vec{0}$ where $c_i \neq 0$
but $c_j = 0$ for all $j > i$. Solve for $\vec{v}_i$.

37. The vectors $T(\vec{v}_1), \ldots, T(\vec{v}_m)$ are not necessarily independent.

39. The vectors $\vec{v}_1, \ldots, \vec{v}_m, \vec{v}$ are linearly independent.

41. The columns of B are linearly independent, while the columns of A are dependent.

43. The vectors are linearly independent.

45. Yes

47. $\begin{bmatrix} 1 & 0 & 0 \\ 0 & 1 & 0 \\ 0 & 0 & 1 \\ 0 & 0 & 0 \end{bmatrix}$

49. $L = \text{im} \begin{bmatrix} 1 \\ 1 \\ 1 \end{bmatrix} = \ker \begin{bmatrix} 1 & 0 & -1 \\ 0 & 1 & -1 \end{bmatrix}$

51. a. Consider a relation $c_1 \vec{v}_1 + \cdots + c_p \vec{v}_p + d_1 \vec{w}_1 + \cdots + d_q \vec{w}_q = \vec{0}$. Then $c_1 \vec{v}_1 + \cdots + c_p \vec{v}_p = -d_1 \vec{w}_1 - \cdots - d_q \vec{w}_q$ is $\vec{0}$, because this vector is both in V and in W. The claim follows.

b. From part (a) we know that the vectors $\vec{v}_1, \ldots, \vec{v}_p, \vec{w}_1, \ldots, \vec{w}_q$ are linearly independent. Consider a vector $\vec{x}$ in $V + W$. By definition of $V + W$ we can write $\vec{x} = \vec{v} + \vec{w}$ for a $\vec{v}$ in V and a $\vec{w}$ in W. The $\vec{v}$ is a linear combination of the $\vec{v}_i$, and $\vec{w}$ is a linear combination of the $\vec{w}_j$. This shows that the vectors $\vec{v}_1, \ldots, \vec{v}_p, \vec{w}_1, \ldots, \vec{w}_q$ span $V + W$.

3.3 1. $\begin{bmatrix} -2 \\ 1 \end{bmatrix}$

3. $\begin{bmatrix} -5 \\ 1 \\ 1 \end{bmatrix}$

5. $\begin{bmatrix} 4 \\ 5 \\ 0 \end{bmatrix}, \begin{bmatrix} -3 \\ 0 \\ 5 \end{bmatrix}$

7. $\begin{bmatrix} 1 \\ 0 \\ 0 \\ 0 \\ 0 \end{bmatrix}, \begin{bmatrix} 0 \\ -2 \\ 1 \\ 0 \\ 0 \end{bmatrix}, \begin{bmatrix} 0 \\ -3 \\ 0 \\ -4 \\ 1 \end{bmatrix}$

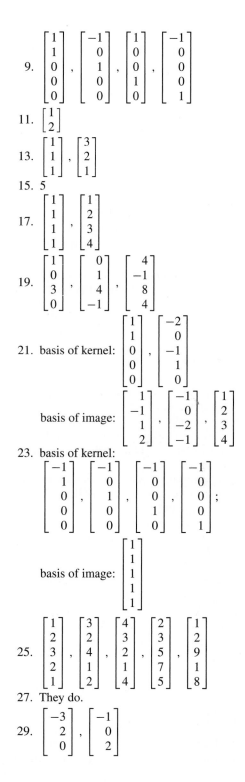

9. $\begin{bmatrix} 1 \\ 1 \\ 0 \\ 0 \\ 0 \end{bmatrix}, \begin{bmatrix} -1 \\ 0 \\ 1 \\ 0 \\ 0 \end{bmatrix}, \begin{bmatrix} 1 \\ 0 \\ 0 \\ 1 \\ 0 \end{bmatrix}, \begin{bmatrix} -1 \\ 0 \\ 0 \\ 0 \\ 1 \end{bmatrix}$

11. $\begin{bmatrix} 1 \\ 2 \end{bmatrix}$

13. $\begin{bmatrix} 1 \\ 1 \\ 1 \end{bmatrix}, \begin{bmatrix} 3 \\ 2 \\ 1 \end{bmatrix}$

15. 5

17. $\begin{bmatrix} 1 \\ 1 \\ 1 \\ 1 \end{bmatrix}, \begin{bmatrix} 1 \\ 2 \\ 3 \\ 4 \end{bmatrix}$

19. $\begin{bmatrix} 1 \\ 0 \\ 3 \\ 0 \end{bmatrix}, \begin{bmatrix} 0 \\ 1 \\ 4 \\ -1 \end{bmatrix}, \begin{bmatrix} 4 \\ -1 \\ 8 \\ 4 \end{bmatrix}$

21. basis of kernel: $\begin{bmatrix} 1 \\ 1 \\ 0 \\ 0 \\ 0 \end{bmatrix}, \begin{bmatrix} -2 \\ 0 \\ -1 \\ 1 \\ 0 \end{bmatrix}$

basis of image: $\begin{bmatrix} 1 \\ -1 \\ 1 \\ 2 \end{bmatrix}, \begin{bmatrix} -1 \\ 0 \\ -2 \\ -1 \end{bmatrix}, \begin{bmatrix} 1 \\ 2 \\ 3 \\ 4 \end{bmatrix}$

23. basis of kernel:
$\begin{bmatrix} -1 \\ 1 \\ 0 \\ 0 \\ 0 \end{bmatrix}, \begin{bmatrix} -1 \\ 0 \\ 1 \\ 0 \\ 0 \end{bmatrix}, \begin{bmatrix} -1 \\ 0 \\ 0 \\ 1 \\ 0 \end{bmatrix}, \begin{bmatrix} -1 \\ 0 \\ 0 \\ 0 \\ 1 \end{bmatrix};$

basis of image: $\begin{bmatrix} 1 \\ 1 \\ 1 \\ 1 \\ 1 \end{bmatrix}$

25. $\begin{bmatrix} 1 \\ 2 \\ 3 \\ 2 \\ 1 \end{bmatrix}, \begin{bmatrix} 3 \\ 2 \\ 4 \\ 1 \\ 2 \end{bmatrix}, \begin{bmatrix} 4 \\ 3 \\ 2 \\ 1 \\ 4 \end{bmatrix}, \begin{bmatrix} 2 \\ 3 \\ 5 \\ 7 \\ 5 \end{bmatrix}, \begin{bmatrix} 1 \\ 2 \\ 9 \\ 1 \\ 8 \end{bmatrix}$

27. They do.

29. $\begin{bmatrix} -3 \\ 2 \\ 0 \end{bmatrix}, \begin{bmatrix} -1 \\ 0 \\ 2 \end{bmatrix}$

31. $A = \begin{bmatrix} 1 & -2 & -4 \\ 1 & 0 & 0 \\ 0 & 1 & 0 \\ 0 & 0 & 1 \end{bmatrix}$

33. The dimension of a hyperplane in $\mathbb{R}^n$ is $n - 1$.

35. The dimension is $n - 1$.

37. $A = \begin{bmatrix} 1 & 0 & 0 & 0 & 0 \\ 0 & 1 & 0 & 0 & 0 \\ 0 & 0 & 0 & 0 & 0 \\ 0 & 0 & 0 & 0 & 0 \end{bmatrix}$

39. $\ker(C)$ is at least 1-dimensional, and $\ker(C)$ is contained in $\ker(A)$.

41. A basis of V is also a basis of W, by Fact 3.3.4c.

43. $\dim(V + W) = \dim(V) + \dim(W)$, by Exercise 3.2.51.

45. The first p columns of $\mathrm{rref}(A)$ contain leading 1's because the $\vec{v}_i$ are linearly independent. Now apply Fact 3.3.7.

49. $[0 \quad 1 \quad 0 \quad 2 \quad 0], [0 \quad 0 \quad 1 \quad 3 \quad 0],$ $[0 \quad 0 \quad 0 \quad 0 \quad 1]$

51. a. A and E have the same row space, since elementary row operations leave the row space unchanged.
   b. $\mathrm{rank}(A) = \dim(\mathrm{rowspace}(A))$, by part (a) and Exercise 50.

55. Suppose $\mathrm{rank}(A) = m$. The submatrix of A consisting of the m pivot columns of A is invertible, since the pivot columns are linearly independent.
   Conversely, if A has an invertible $m \times m$ submatrix, then the columns of that submatrix span $\mathbb{R}^m$, so $\mathrm{im}(A) = \mathbb{R}^m$ and $\mathrm{rank}(A) = m$.

57. Let m be the smallest number such that $A^m = 0$. By Exercise 56 there are m linearly independent vectors in $\mathbb{R}^n$; therefore, $m \leq n$, and $A^n = A^m A^{n-m} = 0$.

61. a. 3, 4, or 5   b. 0, 1, or 2

63. a. $\mathrm{rank}(AB) \leq \mathrm{rank}(A)$
   b. $\mathrm{rank}(AB) \leq \mathrm{rank}(B)$

3.4 1. $\begin{bmatrix} 3 \\ 4 \end{bmatrix}$

3. $\begin{bmatrix} 1/2 \\ 1/2 \end{bmatrix}$

5. $\begin{bmatrix} 4 \\ -3 \\ 2 \end{bmatrix}$

7. If $\vec{v}$ is any vector in the plane that is not parallel to $\vec{x}$, then $\vec{v}, \frac{1}{3}(\vec{x} - 2\vec{v})$ is a basis with the desired

property. For example, $\vec{v} = \begin{bmatrix} 3 \\ 2 \\ 0 \end{bmatrix}$ gives the basis $\begin{bmatrix} 3 \\ 2 \\ 0 \end{bmatrix}, \frac{1}{3}\begin{bmatrix} -4 \\ -4 \\ -1 \end{bmatrix}$.

9. $\begin{bmatrix} -4 \\ 3 \end{bmatrix}$

11. $\begin{bmatrix} 1 \\ -1 \\ 0 \end{bmatrix}$

13. $\begin{bmatrix} -1 & -1 \\ 4 & 6 \end{bmatrix}$

15a. $\begin{bmatrix} 1 & 0 \\ 0 & 0 \end{bmatrix}$

15b. $\begin{bmatrix} 0.1 & 0.3 \\ 0.3 & 0.9 \end{bmatrix}$

17. $\begin{bmatrix} -149 & -231 \\ 99 & 154 \end{bmatrix}$

19. $\begin{bmatrix} d & c \\ b & a \end{bmatrix}$

21. $\begin{bmatrix} 1/2 \\ -1/2 \end{bmatrix}$

23. $\begin{bmatrix} -1 \\ -1 \end{bmatrix}$

27. $\begin{bmatrix} 40 \\ 58 \end{bmatrix}$

29. $\begin{bmatrix} -1/2 & 1 \\ 1/2 & 0 \end{bmatrix}$

31. $\begin{bmatrix} 0 & 0 & -1 \\ 0 & 0 & 0 \\ 1 & 0 & 0 \end{bmatrix}$

33. Yes

35. $\begin{bmatrix} -9 \\ 6 \end{bmatrix}, \begin{bmatrix} 0 \\ 1 \end{bmatrix}$, for example

37. Yes

41. $\begin{bmatrix} 0 & bc - ad \\ 1 & a + d \end{bmatrix}$

43. $S = \begin{bmatrix} 1 & 2 \\ 2 & 1 \end{bmatrix}$, for example, and $D = \begin{bmatrix} 3 & 0 \\ 0 & -1 \end{bmatrix}$

45. a. $A(S\vec{x}) = SB\vec{x} = \vec{0}$, so that $S\vec{x}$ is in $\ker(A)$.
   b. Since we have the p linearly independent vectors $S\vec{v}_1, S\vec{v}_2, \ldots, S\vec{v}_p$ in $\ker(A)$, we know that $\dim(\ker B) = p \leq \dim(\ker A)$, by Fact 3.3.4a. Reversing the roles of A and B, we find that $\dim(\ker A) \leq \dim(\ker B)$. Thus, $\mathrm{nullity}(A) = \dim(\ker A) = \dim(\ker B) = \mathrm{nullity}(B)$.

47. $\begin{bmatrix} 0.36 & 0.48 & 0.8 \\ 0.48 & 0.64 & -0.6 \\ -0.8 & 0.6 & 0 \end{bmatrix}$

49. $\begin{bmatrix} 0 & -1 \\ 1 & 0 \end{bmatrix}$

51. $b_{ij} = a_{n+1-i,\,n+1-j}$

Chapter 4

4.1
1. Not a subspace
3. Subspace with basis $1 - t, 2 - t^2$
5. Subspace with basis t
7. Subspace
9. Not a subspace
11. Not a subspace
13. Not a subspace
15. Subspace
17. Matrices with one entry equal to 1 and all other entries equal to 0. The dimension is mn.
19. A basis is $\begin{bmatrix} 1 \\ 0 \end{bmatrix}, \begin{bmatrix} i \\ 0 \end{bmatrix}, \begin{bmatrix} 0 \\ 1 \end{bmatrix}, \begin{bmatrix} 0 \\ i \end{bmatrix}$, so that the dimension is 4.
21. A basis is $\begin{bmatrix} 1 & 0 \\ 0 & 0 \end{bmatrix}, \begin{bmatrix} 0 & 0 \\ 0 & 1 \end{bmatrix}$, so that the dimension is 2.
23. A basis is $\begin{bmatrix} 1 & 0 \\ 0 & 0 \end{bmatrix}, \begin{bmatrix} 0 & 0 \\ 1 & 0 \end{bmatrix}, \begin{bmatrix} 0 & 0 \\ 0 & 1 \end{bmatrix}$, so that the dimension is 3.
25. A basis is $1 - t, 1 - t^2$, so that the dimension is 2.
27. A basis is $\begin{bmatrix} 1 & 0 \\ 0 & 0 \end{bmatrix}, \begin{bmatrix} 0 & 0 \\ 0 & 1 \end{bmatrix}$, so that the dimension is 2.
29. A basis is $\begin{bmatrix} -1 & 1 \\ 0 & 0 \end{bmatrix}, \begin{bmatrix} 0 & 0 \\ -1 & 1 \end{bmatrix}$, so that the dimension is 2.
31. A basis is $\begin{bmatrix} 1 & 0 \\ 1 & 0 \end{bmatrix}, \begin{bmatrix} 0 & -1 \\ 0 & 1 \end{bmatrix}$, so that the dimension is 2.
33. Yes, both even and odd functions form a subspace of $F(\mathbb{R}^n, \mathbb{R}^n)$.
35. Yes
37. $f(x) = ae^{3t} + be^{4t}$

4.2
1. Not linear
3. Linear; not an isomorphism
5. Not linear
7. Linear; an isomorphism
9. Linear; not an isomorphism
11. Linear; an isomorphism
13. Linear; not an isomorphism
15. Linear; an isomorphism
17. Linear; an isomorphism
19. Linear; not an isomorphism
21. Linear; not an isomorphism
23. Linear; not an isomorphism
25. $\ker(T)$ consists of all matrices of the form $\begin{bmatrix} a & b \\ 0 & a \end{bmatrix}$, so that the nullity is 2.
27. The image consists of all linear functions, of the form $mt + b$, so that the rank is 2. The kernel consists of the constant functions, so that the nullity is 1.
29. The image consists of all infinite sequences, and the kernel consists of all sequences of the form $(0, x_1, 0, x_3, 0, x_5, \ldots)$.
31. The kernel consists of all functions of the form $ae^{2t} + be^{3t}$, so that the nullity is 2.
33. The kernel has the basis $t - 7, (t - 7)^2$, so that the nullity is 2. The image is all of $\mathbb{R}$, so that the rank is 1.
35. The kernel consists of the zero function alone, and the image consists of all polynomials $g(t)$ whose constant term is zero (that is, $g(0) = 0$).
37. Impossible, since $\dim(P_3) \neq \dim(\mathbb{R}^3)$.
39. b. $\ker(T)$ consists of the zero matrix alone. d. This dimension is mn.
43. Yes, the composite of linear transformations is linear. Yes, the composite of isomorphisms is an isomorphism.
45. Surprisingly, yes.

4.3
1. Yes
3. Yes
5. $\begin{bmatrix} 1 & 0 & 0 \\ 0 & 1 & 2 \\ 0 & 0 & 3 \end{bmatrix}$
7. $\begin{bmatrix} 0 & 0 & 0 \\ 0 & 0 & 4 \\ 0 & 0 & 0 \end{bmatrix}$
9. $\begin{bmatrix} 1 & 0 & 0 \\ 0 & 2 & 0 \\ 0 & 0 & 1 \end{bmatrix}$
11. $\begin{bmatrix} 1 & 0 & 0 \\ 0 & 1 & 0 \\ 0 & 0 & 3 \end{bmatrix}$
13. $\begin{bmatrix} 1 & 0 & 1 & 0 \\ 0 & 1 & 0 & 1 \\ 2 & 0 & 2 & 0 \\ 0 & 2 & 0 & 2 \end{bmatrix}$
15. $\begin{bmatrix} 1 & 0 \\ 0 & -1 \end{bmatrix}$
17. $\begin{bmatrix} 0 & -1 \\ 1 & 0 \end{bmatrix}$

19. $\begin{bmatrix} p & -q \\ q & p \end{bmatrix}$

21. $\begin{bmatrix} -3 & 1 & 0 \\ 0 & -3 & 2 \\ 0 & 0 & -3 \end{bmatrix}$

23. $\begin{bmatrix} 1 & 3 & 9 \\ 0 & 0 & 0 \\ 0 & 0 & 0 \end{bmatrix}$

25. $\begin{bmatrix} 1 & 0 & 0 \\ 0 & -1 & 0 \\ 0 & 0 & 1 \end{bmatrix}$

27. $\begin{bmatrix} 1 & -1 & 1 \\ 0 & 2 & -4 \\ 0 & 0 & 4 \end{bmatrix}$

29. $\begin{bmatrix} 2 & 2 & 8/3 \\ 0 & 0 & 0 \\ 0 & 0 & 0 \end{bmatrix}$

31. $\begin{bmatrix} 0 & 1 & 0 \\ 0 & 0 & 2 \\ 0 & 0 & 0 \end{bmatrix}$

33. $\begin{bmatrix} 1 & 0 & 0 \\ 0 & 1 & 0 \\ 0 & 0 & 0 \end{bmatrix}$

35. Image has basis $\begin{bmatrix} 1 & 0 \\ 2 & 0 \end{bmatrix}, \begin{bmatrix} 0 & 1 \\ 0 & 2 \end{bmatrix}$, and kernel has basis $\begin{bmatrix} -1 & 0 \\ 1 & 0 \end{bmatrix}, \begin{bmatrix} 0 & -1 \\ 0 & 1 \end{bmatrix}$.

37. Image is all of P_2; kernel consists of the zero polynomial alone.

39. Kernel has basis $(t-3), (t-3)^2$, and image consists of the constant functions.

41. $\begin{bmatrix} 1 & 0 & 0 \\ 0 & 1 & 1 \\ 0 & 0 & 1 \end{bmatrix}$

43. $\begin{bmatrix} 1 & 0 & 0 \\ -1 & 1 & 1 \\ 0 & 1 & 0 \end{bmatrix}$

45. $\begin{bmatrix} 1 & 1 \\ 1 & -1 \end{bmatrix}$

47. $\begin{bmatrix} 1 & -1 & 1 \\ 0 & 1 & -2 \\ 0 & 0 & 1 \end{bmatrix}$

49. $\begin{bmatrix} 2 & 2 \\ -2 & 2 \end{bmatrix}$

51. $\begin{bmatrix} 0 & -1 \\ 1 & 0 \end{bmatrix}$

53. $\begin{bmatrix} \cos(\delta) & -\sin(\delta) \\ \sin(\delta) & \cos(\delta) \end{bmatrix}$

55. $\begin{bmatrix} 2/9 & -14/9 \\ -1/9 & 7/9 \end{bmatrix}$

57. $\begin{bmatrix} -1 & 3 \\ -1 & 0 \end{bmatrix}$

59. $T(f(t)) = t \cdot f(t)$ from P to P, for example.

61. a. $\begin{bmatrix} 0 & 0 & 0 & 2 \\ 0 & 0 & -2 & 0 \\ 0 & 0 & 0 & 0 \\ 0 & 0 & 0 & 0 \end{bmatrix}$ b. $f(t) = \dfrac{1}{2}t\sin(t)$

Chapter 5

5.1

1. $\sqrt{170}$

3. $\sqrt{54}$

5. $\arccos\left(\dfrac{20}{\sqrt{406}}\right) \approx 0.12$ (radians)

7. obtuse

9. acute

11. $\arccos\left(\dfrac{1}{\sqrt{n}}\right) \to \dfrac{\pi}{2}$ (as $n \to \infty$)

13. $2\arccos(0.8) \approx 74°$

15. $\begin{bmatrix} -2 \\ 1 \\ 0 \\ 0 \end{bmatrix}, \begin{bmatrix} -3 \\ 0 \\ 1 \\ 0 \end{bmatrix}, \begin{bmatrix} -4 \\ 0 \\ 0 \\ 1 \end{bmatrix}$

17. $\begin{bmatrix} 1 \\ -2 \\ 1 \\ 0 \end{bmatrix}, \begin{bmatrix} 2 \\ -3 \\ 0 \\ 1 \end{bmatrix}$

19. a. Orthogonal projection onto $L^\perp$
    b. Reflection in $L^\perp$ c. Reflection in L

21. For example: $b = d = e = g = 0$,
    $a = \dfrac{1}{2}, c = \dfrac{\sqrt{3}}{2}, f = -\dfrac{\sqrt{3}}{2}$

25. a. $\|k\vec{v}\| = \sqrt{(k\vec{v}) \cdot (k\vec{v})} = \sqrt{k^2(\vec{v} \cdot \vec{v})} = \sqrt{k^2}\sqrt{\vec{v} \cdot \vec{v}} = |k| \, \|\vec{v}\|$

    b. By part (a), $\left\| \dfrac{1}{\|\vec{v}\|}\vec{v} \right\| = \dfrac{1}{\|\vec{v}\|}\|\vec{v}\| = 1$.

27. $\begin{bmatrix} 8 \\ 0 \\ 2 \\ -2 \end{bmatrix}$

29. By Pythagoras, $\|\vec{x}\| = \sqrt{49 + 9 + 4 + 1 + 1} = 8$.

31. $p \le \|\vec{x}\|^2$. Equality holds if (and only if) $\vec{x}$ is a linear combination of the vectors $\vec{v}_i$.

33. The vector whose n components are all $1/n$

35. $-\dfrac{1}{\sqrt{14}}\begin{bmatrix} 1 \\ 2 \\ 3 \end{bmatrix}$

37. $R(\vec{x}) = 2(\vec{v}_1 \cdot \vec{x})\vec{v}_1 + 2(\vec{v}_2 \cdot \vec{x})\vec{v}_2 - \vec{x}$

39. No; if $\vec{u}$ is a unit vector in L, then
    $\vec{x} \cdot \text{proj}_L\vec{x} = \vec{x} \cdot (\vec{u} \cdot \vec{x})\vec{u} = (\vec{u} \cdot \vec{x})^2 \ge 0$.

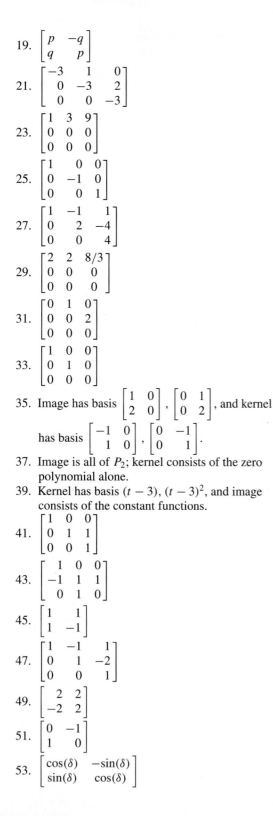

5.2

1. $\begin{bmatrix} 2/3 \\ 1/3 \\ -2/3 \end{bmatrix}$

3. $\begin{bmatrix} 4/5 \\ 0 \\ 3/5 \end{bmatrix}, \begin{bmatrix} 3/5 \\ 0 \\ -4/5 \end{bmatrix}$

5. $\begin{bmatrix} 2/3 \\ 2/3 \\ 1/3 \end{bmatrix}, \dfrac{1}{\sqrt{18}}\begin{bmatrix} -1 \\ -1 \\ 4 \end{bmatrix}$

7. $\begin{bmatrix} 2/3 \\ 2/3 \\ 1/3 \end{bmatrix}, \begin{bmatrix} -2/3 \\ 1/3 \\ 2/3 \end{bmatrix}, \begin{bmatrix} 1/3 \\ -2/3 \\ 2/3 \end{bmatrix}$

9. $\begin{bmatrix} 1/2 \\ 1/2 \\ 1/2 \\ 1/2 \end{bmatrix}, \begin{bmatrix} -1/10 \\ 7/10 \\ -7/10 \\ 1/10 \end{bmatrix}$

11. $\begin{bmatrix} 4/5 \\ 0 \\ 0 \\ 3/5 \end{bmatrix}, \begin{bmatrix} -3/15 \\ 2/15 \\ 14/15 \\ 4/15 \end{bmatrix}$

13. $\begin{bmatrix} 1/2 \\ 1/2 \\ 1/2 \\ 1/2 \end{bmatrix}, \begin{bmatrix} 1/2 \\ -1/2 \\ -1/2 \\ 1/2 \end{bmatrix}, \begin{bmatrix} 1/2 \\ 1/2 \\ -1/2 \\ -1/2 \end{bmatrix}$

15. $\begin{bmatrix} 2/3 \\ 1/3 \\ -2/3 \end{bmatrix} [3]$

17. $\begin{bmatrix} 4/5 & 3/5 \\ 0 & 0 \\ 3/5 & -4/5 \end{bmatrix}\begin{bmatrix} 5 & 5 \\ 0 & 35 \end{bmatrix}$

19. $\begin{bmatrix} 2/3 & -1/\sqrt{18} \\ 2/3 & -1/\sqrt{18} \\ 1/3 & 4/\sqrt{18} \end{bmatrix}\begin{bmatrix} 3 & 3 \\ 0 & \sqrt{18} \end{bmatrix}$

21. $\dfrac{1}{3}\begin{bmatrix} 2 & -2 & 1 \\ 2 & 1 & -2 \\ 1 & 2 & 2 \end{bmatrix}\begin{bmatrix} 3 & 0 & 12 \\ 0 & 3 & -12 \\ 0 & 0 & 6 \end{bmatrix}$

23. $\begin{bmatrix} 1/2 & -1/10 \\ 1/2 & 7/10 \\ 1/2 & -7/10 \\ 1/2 & 1/10 \end{bmatrix}\begin{bmatrix} 2 & 4 \\ 0 & 10 \end{bmatrix}$

25. $\begin{bmatrix} 4/5 & -3/15 \\ 0 & 2/15 \\ 0 & 14/15 \\ 3/5 & 4/15 \end{bmatrix}\begin{bmatrix} 5 & 10 \\ 0 & 15 \end{bmatrix}$

27. $\dfrac{1}{2}\begin{bmatrix} 1 & 1 & 1 \\ 1 & -1 & 1 \\ 1 & -1 & -1 \\ 1 & 1 & -1 \end{bmatrix}\begin{bmatrix} 2 & 1 & 1 \\ 0 & 1 & -2 \\ 0 & 0 & 1 \end{bmatrix}$

29. $\begin{bmatrix} -3/5 \\ 4/5 \end{bmatrix}, \begin{bmatrix} 4/5 \\ 3/5 \end{bmatrix}$

31. $\vec{e}_1, \vec{e}_2, \vec{e}_3$

33. $\dfrac{1}{\sqrt{2}}\begin{bmatrix} 1 \\ 0 \\ 0 \\ -1 \end{bmatrix}, \dfrac{1}{\sqrt{2}}\begin{bmatrix} 0 \\ 1 \\ -1 \\ 0 \end{bmatrix}$

35. $\begin{bmatrix} 1/3 \\ 2/3 \\ 2/3 \end{bmatrix}, \begin{bmatrix} 2/3 \\ 1/3 \\ -2/3 \end{bmatrix}$

37. $\dfrac{1}{2}\begin{bmatrix} 1 & 1 \\ 1 & -1 \\ 1 & -1 \\ 1 & 1 \end{bmatrix}\begin{bmatrix} 3 & 4 \\ 0 & 5 \end{bmatrix}$

39. $\dfrac{1}{\sqrt{14}}\begin{bmatrix} 1 \\ 2 \\ 3 \end{bmatrix}, \dfrac{1}{\sqrt{3}}\begin{bmatrix} 1 \\ 1 \\ -1 \end{bmatrix}, \dfrac{1}{\sqrt{42}}\begin{bmatrix} 5 \\ -4 \\ 1 \end{bmatrix}$

41. Q is diagonal with $q_{ii} = 1$ if $a_{ii} > 0$ and $q_{ii} = -1$ if $a_{ii} < 0$. You can get R from A by multiplying the ith row of A with -1 whenever a_{ii} is negative.

43. Write the QR factorization of A in partitioned form as $A = [\, A_1 \;\; A_2 \,] =$ $[Q_1 \;\; Q_2]\begin{bmatrix} R_1 & R_2 \\ 0 & R_4 \end{bmatrix}$. Then $A_1 = Q_1 R_1$ is the QR factorization of A_1.

45. Yes

5.3

1. $(A\vec{v}) \cdot \vec{w} = (A\vec{v})^T \vec{w} = \vec{v}^T A^T \vec{w} = \vec{v} \cdot (A^T \vec{w})$.

3. $\sphericalangle(L(\vec{v}), L(\vec{w})) = \arccos\dfrac{L(\vec{v}) \cdot L(\vec{w})}{\|L(\vec{v})\| \, \|L(\vec{w})\|} =$ $\arccos\dfrac{\vec{v} \cdot \vec{w}}{\|\vec{v}\| \, \|\vec{w}\|} = \sphericalangle(\vec{v}, \vec{w})$. (The equation $L(\vec{v}) \cdot L(\vec{w}) = \vec{v} \cdot \vec{w}$ is shown in Exercise 2.)

5. Yes, by Fact 5.3.4a.

7. Yes, since $AA^T = I_n$.

9. The first column is a unit vector; we can write it as $\vec{v}_1 = \begin{bmatrix} \cos(\varphi) \\ \sin(\varphi) \end{bmatrix}$ for some φ. The second column is a unit vector orthogonal to $\vec{v}_1$; there are two choices: $\begin{bmatrix} -\sin(\varphi) \\ \cos(\varphi) \end{bmatrix}$ and $\begin{bmatrix} \sin(\varphi) \\ -\cos(\varphi) \end{bmatrix}$. Solution: $\begin{bmatrix} \cos(\varphi) & -\sin(\varphi) \\ \sin(\varphi) & \cos(\varphi) \end{bmatrix}$ and $\begin{bmatrix} \cos(\varphi) & \sin(\varphi) \\ \sin(\varphi) & -\cos(\varphi) \end{bmatrix}$, for arbitrary φ.

11. For example, $T(\vec{x}) = \dfrac{1}{3}\begin{bmatrix} 1 & -2 & 2 \\ -2 & 1 & 2 \\ 2 & 2 & 1 \end{bmatrix}\vec{x}$.

13. No, by Fact 5.3.2.

15. No; consider $A = \begin{bmatrix} 1 & 0 \\ 0 & 0 \end{bmatrix}$ and $B = \begin{bmatrix} 0 & 1 \\ 1 & 0 \end{bmatrix}$.

17. Yes, since $(A^{-1})^T = (A^T)^{-1} = A^{-1}$.

19. (ijth entry of A) $= v_i v_j$

21. All entries of A are $\dfrac{1}{n}$.

23. A represents the reflection in the line spanned by $\vec{u}$ (compare with Example 2), and B represents the reflection in the plane with normal vector $\vec{u}$.

25. $\dim\big(\ker(A)\big) = n - \mathrm{rank}(A)$ (by Fact 3.3.5) and $\dim\big(\ker(A^T)\big) = m - \mathrm{rank}(A^T) = m - \mathrm{rank}(A)$ (by Fact 5.3.9c). Therefore, the dimensions of the two kernels are equal if (and only if) $m = n$, that is, if A is a square matrix.

27. $A^T A = (QR)^T QR = R^T Q^T QR = R^T R$

29. By Exercise 4.2.45, we can write $A^T = QL$ where Q is orthogonal and L is lower triangular. Then $A = (QL)^T = L^T Q^T$ does the job.

31. a. $I_m = Q_1^T Q_1 = S^T Q_2^T Q_2 S = S^T S$, so that S is orthogonal.

   b. $R_2 R_1^{-1}$ is both orthogonal (by part (a)) and upper triangular, with positive diagonal entries. By Exercise 30a, we have $R_2 R_1^{-1} = I_m$ so that $R_2 = R_1$ and $Q_1 = Q_2$, as claimed.

33. $\begin{bmatrix} 0 & 1 & 0 \\ -1 & 0 & 0 \\ 0 & 0 & 0 \end{bmatrix}, \begin{bmatrix} 0 & 0 & 1 \\ 0 & 0 & 0 \\ -1 & 0 & 0 \end{bmatrix}, \begin{bmatrix} 0 & 0 & 0 \\ 0 & 0 & 1 \\ 0 & -1 & 0 \end{bmatrix}$;

   dimension 3

35. $\dfrac{n(n+1)}{2}$

37. Yes, Yes.

39. The kernel consists of the symmetric $n \times n$ matrices, and the image consists of the skew-symmetric matrices.

41. $\begin{bmatrix} 0 & 0 & 0 & 0 \\ 0 & 0 & 0 & 0 \\ 0 & 0 & 0 & 0 \\ 0 & 0 & 0 & 2 \end{bmatrix}$

43. If $A = LDU$, then $A^T = U^T DL^T$ is the LDU factorization of A^T. Since $A = A^T$ the two factorizations are identical, so that $U = L^T$, as claimed.

5.4

1. $\mathrm{im}(A) = \mathrm{span}\begin{bmatrix} 2 \\ 3 \end{bmatrix}$ and $\ker(A^T) = \mathrm{span}\begin{bmatrix} -3 \\ 2 \end{bmatrix}$.

3. The vectors form a basis of $\mathbb{R}^n$.

5. $V^\perp = \big(\ker(A)\big)^\perp = \mathrm{im}(A^T)$, where

   $A = \begin{bmatrix} 1 & 1 & 1 & 1 \\ 1 & 2 & 5 & 4 \end{bmatrix}$.

   Basis of $V^\perp$: $\begin{bmatrix} 1 \\ 1 \\ 1 \\ 1 \end{bmatrix}, \begin{bmatrix} 1 \\ 2 \\ 5 \\ 4 \end{bmatrix}$.

7. $\mathrm{im}(A) = \big(\ker(A)\big)^\perp$

9. $\|\vec{x}_0\| < \|\vec{x}\|$ for all other vectors $\vec{x}$ in S.

11. b. $L\big(L^+(\vec{y})\big) = \vec{y}$

   c. $L^+\big(L(\vec{x})\big) = \mathrm{proj}_V \vec{x}$, where $V = \big(\ker(A)\big)^\perp = \mathrm{im}(A^T)$

   d. $\mathrm{im}(L^+) = \mathrm{im}(A^T)$ and $\ker(L^+) = \{\vec{0}\}$

   e. $L^+(\vec{y}) = \begin{bmatrix} 1 & 0 \\ 0 & 1 \\ 0 & 0 \end{bmatrix} \vec{y}$

13. b. $L^+\big(L(\vec{x})\big) = \mathrm{proj}_V \vec{x}$, where $V = \big(\ker(A)\big)^\perp = \mathrm{im}(A^T)$

   c. $L\big(L^+(\vec{y})\big) = \mathrm{proj}_W \vec{y}$, where $W = \mathrm{im}(A) = \big(\ker(A^T)\big)^\perp$

   d. $\mathrm{im}(L^+) = \mathrm{im}(A^T)$ and $\ker(L^+) = \ker(A^T)$

   e. $L^+(\vec{y}) = \begin{bmatrix} \frac{1}{2} & 0 \\ 0 & 0 \\ 0 & 0 \end{bmatrix} \vec{y}$

15. Let $B = (A^T A)^{-1} A^T$.

17. Yes; note that $\ker(A) = \ker(A^T A)$.

19. $\begin{bmatrix} 1 \\ 1 \end{bmatrix}$

21. $\vec{x}^* = \begin{bmatrix} -1 \\ 2 \end{bmatrix}$, $\|\vec{b} - A\vec{x}^*\| = 42$

23. $\begin{bmatrix} 0 \\ 0 \end{bmatrix}$

25. $\begin{bmatrix} 1 - 3t \\ t \end{bmatrix}$, for arbitrary t

27. $\begin{bmatrix} 7 \\ 11 \end{bmatrix}$

29. $x_1^* = x_2^* \approx \dfrac{1}{2}$

31. $3 + 1.5t$

33. approximately $1.5 + 0.1 \sin(t) - 1.41 \cos(t)$

37. a. Try to solve the system $\begin{vmatrix} c_0 + 35c_1 = \log(35) \\ c_0 + 46c_1 = \log(46) \\ c_0 + 59c_1 = \log(77) \\ c_0 + 69c_1 = \log(133) \end{vmatrix}$.

   Least-squares solution $\begin{bmatrix} c_0^* \\ c_1^* \end{bmatrix} \approx \begin{bmatrix} 0.915 \\ 0.017 \end{bmatrix}$. Use approximation $\log(d) = 0.915 + 0.017t$.

b. Exponentiate the equation in part (a): $d = 10^{\log d} = 10^{0.915+0.017t} \approx 8.221 \cdot 10^{0.017t} \approx 8.221 \cdot 1.04^t$

c. Predicts 259 displays for the A320; there are much fewer

since the A320 is highly computerized.

39. a. Try to solve the system

$$\begin{vmatrix} c_0 + \log(600,000)c_1 = \log(250) \\ c_0 + \log(200,000)c_1 = \log(60) \\ c_0 + \log(60,000)c_1 = \log(25) \\ c_0 + \log(10,000)c_1 = \log(12) \\ c_0 + \log(2,500)c_1 = \log(5) \end{vmatrix}.$$

5.5   3. a. If S is invertible.

b. If S is orthogonal.

5. Yes.

7. For positive k.

9. True.

11. The angle is δ.

13. The two norms are equal, by Fact 5.5.6.

15. If $b = c$ and $b^2 < d$.

17. If $\ker(T) = \{0\}$.

19. The matrices $A = \begin{bmatrix} a & b \\ c & d \end{bmatrix}$ such that

$b = c, a > 0$, and $b^2 < ad$.

21. Yes, $\langle \vec{v}, \vec{w} \rangle = 2(\vec{v} \cdot \vec{w})$.

23. $1, 2t - 1$.

25. $\sqrt{1 + \dfrac{1}{4} + \dfrac{1}{9} + \cdots} = \dfrac{\pi}{\sqrt{6}}.$

27. $a_0 = \dfrac{1}{\sqrt{2}}. c_k = 0$ for all k.

$b_k = \begin{cases} \dfrac{2}{k\pi} & \text{if } k \text{ is odd} \\ 0 & \text{if } k \text{ is even.} \end{cases}$

29. $\displaystyle\sum_{k \text{ odd}} \dfrac{1}{k^2} = \dfrac{\pi^2}{8}$

Chapter 6

6.1   1. -2

3. 6

5. -24

7. 0

9. 0

11. 0

13. 36

15. 24

17. -1

19. $aps - aqr$

21. invertible

23. invertible if a, b, c are all nonzero

25. invertible

27. $3, 8$

29. 1

31. $-1, 0, 2$

33. $\det(A) = 0$

35. $\det(A) = 1$; there are $50 \cdot 99$ inversions.

37. $\det(M) = \det(A) \det(C)$

39. Let a_{ii} be the first diagonal entry that does not belong to the pattern. The pattern must contain a number in the ith row to the right of a_{ii} and also a number in the ith column below a_{ii}.

41. $\det(A) = 2x^2 - x - x^3 = -x(x - 1)^2$. The matrix A is invertible except when $x = 0$ or $x = 1$.

43. $-k$

45. k

47. $\det(-A) = (-1)^n \det(A)$

49. $\det(A^{-1}) = \dfrac{1}{\det(A)}$

51. $\det(A) = 1$ if n is even and $\det(A) = -1$ if n is odd (there are n^2 inversions).

6.2   1. -3

3. 9

5. 24

7. 1

9. -72

11. 8

13. 8

15. Analogous to Example 4

17. a. $\det \begin{bmatrix} 1 & 1 \\ a_0 & a_1 \end{bmatrix} = a_1 - a_0$

b. Use Laplace Expansion down the last column to see that $f(t)$ is a polynomial of degree $\leq n$. The coefficient k of t^n is $\displaystyle\prod_{n-1 \geq i > j} (a_i - a_j)$.

Now $\det(A) = f(a_n) =$

$k(a_n - a_0)(a_n - a_1) \ldots (a_n - a_{n-1}) = \displaystyle\prod_{n \geq i > j} (a_i - a_j)$, as claimed.

19. $\displaystyle\prod_{i=1}^{n} a_i \cdot \prod_{i>j}(a_i - a_j)$ (use linearity in the columns and Exercise 17)

21. $\begin{bmatrix} x_1 \\ x_2 \end{bmatrix} = \begin{bmatrix} a_1 \\ a_2 \end{bmatrix}$ and $\begin{bmatrix} x_1 \\ x_2 \end{bmatrix} = \begin{bmatrix} b_1 \\ b_2 \end{bmatrix}$ are solutions. The equation is of the form $px_1 + qx_2 + b = 0$, that is, it defines a line.

23. ± 1

25. $\det(A^T A) = \left(\det(A)\right)^2 > 0$

27. $\det(A) = \det(A^T) = \det(-A) = (-1)^n \det(A) = -\det(A)$, so $\det(A) = 0$

29. $A^T A = \begin{bmatrix} \|\vec{v}\|^2 & \vec{v} \cdot \vec{w} \\ \vec{v} \cdot \vec{w} & \|\vec{w}\|^2 \end{bmatrix}$, so $\det(A^T A) =$

$\|\vec{v}\|^2 \|\vec{w}\|^2 - (\vec{v} \cdot \vec{w})^2 \geq 0$, by the Cauchy-Schwarz inequality. Equality holds only if $\vec{v}$ and $\vec{w}$ are parallel.

31. Expand down the first column:
$f(x) = -x \det(A_{41}) + $ constant, so
$f'(x) = -\det(A_{41}) = -24$.

33. T is linear in the rows and columns.

35. $\det(Q_1) = \det(Q_2) = 1$ and
$\det(Q_n) = 2\det(Q_{n-1}) - \det(Q_{n-2})$, so
$\det(Q_n) = 1$ for all n.

37. $\det(P_1) = 1$ and $\det(P_n) = \det(P_{n-1})$, by expansion down the first column, so $\det(P_n) = 1$ for all n.

39. a. Note that $\det(A) \det(A^{-1}) = 1$, and both factors are integers.
b. Use the formula for the inverse of a 2×2 matrix (Fact 2.3.6).

45. 8

47. -1

49. 1

51. $(-1)^{n(n-1)/2}$. This is 1 if either n or $(n-1)$ is divisible by 4, and -1 otherwise.

53. 16

55. $a^2 + b^2$

6.3 1. 50

3. 13

7. 110

9. *Geometrically:* $\|\vec{v}_1\|$ is the base and $\|\vec{v}_2\| \sin \alpha$ the height of the parallelogram defined by $\vec{v}_1$ and $\vec{v}_2$.
Algebraically: In Exercise 6.2.29 we learned that
$\det(A^T A) = \|\vec{v}_1\|^2 \|\vec{v}_2\|^2 - (\vec{v}_1 \cdot \vec{v}_2)^2 =$
$\|\vec{v}_1\|^2 \|\vec{v}_2\|^2 - \|\vec{v}_1\| \|\vec{v}_2\|^2 \cos^2 \alpha =$
$\|\vec{v}_1\|^2 \|\vec{v}_2\|^2 \sin^2 \alpha$, so that $|\det(A)| =$
$\sqrt{\det(A^T A)} = \|\vec{v}_1\| \|\vec{v}_2\| \sin \alpha$.

11. $|\det(A)| = 12$, the expansion factor of T on the parallelogram defined by $\vec{v}_1$ and $\vec{v}_2$.

13. $\sqrt{20}$

15. We need to show that both $V(\vec{v}_1, \vec{v}_2, \ldots, \vec{v}_k)$ and $\sqrt{\det(A^T A)}$ are zero. One of the $\vec{v}_i$ is a linear combination of $\vec{v}_1, \ldots, \vec{v}_{i-1}$. Then $\text{proj}_{V_{i-1}} \vec{v}_i = \vec{v}_i$ and $\vec{v}_i - \text{proj}_{V_{i-1}} \vec{v}_i = \vec{0}$. This shows that $V(\vec{v}_1, \ldots, \vec{v}_k) = 0$, by Definition 5.3.6. Fact 4.4.3 implies that $\det(A^T A) = 0$.

17. a. $V(\vec{v}_1, \vec{v}_2, \vec{v}_3, \vec{v}_1 \times \vec{v}_2 \times \vec{v}_3) =$
$V(\vec{v}_1, \vec{v}_2, \vec{v}_3) \|\vec{v}_1 \times \vec{v}_2 \times \vec{v}_3\|$ because $\vec{v}_1 \times \vec{v}_2 \times \vec{v}_3$ is orthogonal to $\vec{v}_1, \vec{v}_2$, and $\vec{v}_3$.
b. $V(\vec{v}_1, \vec{v}_2, \vec{v}_3, \vec{v}_1 \times \vec{v}_2 \times \vec{v}_3) =$
$|\det[\vec{v}_1 \times \vec{v}_2 \times \vec{v}_3 \quad \vec{v}_1 \quad \vec{v}_2 \quad \vec{v}_3]| =$

$\|\vec{v}_1 \times \vec{v}_2 \times \vec{v}_3\|^2$, by definition of the cross product.
c. $V(\vec{v}_1, \vec{v}_2, \vec{v}_3) = \|\vec{v}_1 \times \vec{v}_2 \times \vec{v}_3\|$, by parts (a) and (b).

19. $\det[\vec{v}_1 \quad \vec{v}_2 \quad \vec{v}_3] = \vec{v}_1 \cdot (\vec{v}_2 \times \vec{v}_3)$ is positive if (and only if) $\vec{v}_1$ and $\vec{v}_2 \times \vec{v}_3$ enclose an acute angle.

21. a. reverses   b. preserves   c. reverses

23. $x_1 = \dfrac{\det \begin{bmatrix} 1 & -3 \\ 0 & 7 \end{bmatrix}}{\det \begin{bmatrix} 5 & -3 \\ -6 & 7 \end{bmatrix}} = \dfrac{7}{17}$;

$x_2 = \dfrac{\det \begin{bmatrix} 5 & 1 \\ -6 & 0 \end{bmatrix}}{\det \begin{bmatrix} 5 & -3 \\ -6 & 7 \end{bmatrix}} = \dfrac{6}{17}$

25. $\text{adj}(A) = \begin{bmatrix} 1 & 0 & -1 \\ 0 & -1 & 0 \\ -2 & 0 & 1 \end{bmatrix}$;

$A^{-1} = \dfrac{1}{\det(A)} \text{adj}(A) = -\text{adj}(A) =$
$\begin{bmatrix} -1 & 0 & 1 \\ 0 & 1 & 0 \\ 2 & 0 & -1 \end{bmatrix}$

27. $x = \dfrac{a}{a^2 + b^2} > 0$; $y = \dfrac{-b}{a^2 + b^2} < 0$; x decreases as b increases.

29. $dx_1 = -D^{-1}R_2(1 - R_1)(1 - \alpha)^2 de_2$,
$dy_1 = D^{-1}(1 - \alpha)R_2(R_1(1 - \alpha) + \alpha)de_2 > 0$,
$dp = D^{-1}R_1 R_2 de_2 > 0$.

Chapter 7

7.1 1. Yes; the eigenvalue is λ^3

3. Yes; the eigenvalue is $\lambda + 2$

5. Yes

7. $\ker(\lambda I_n - A) \neq \{\vec{0}\}$ because $(\lambda I_n - A)\vec{v} = \vec{0}$. The matrix $\lambda I_n - A$ is not invertible.

9. $\begin{bmatrix} a & b \\ 0 & d \end{bmatrix}$

11. $\begin{bmatrix} a & b \\ 2a + 4b - 2d & d \end{bmatrix}$

13. All vectors of the form $\begin{bmatrix} 3t \\ 5t \end{bmatrix}$, where $t \neq 0$ (solve the linear system $A\vec{x} = 4\vec{x}$).

15. The nonzero vectors in L are the eigenvectors with eigenvalue 1, and the nonzero vectors in $L^{\perp}$ have eigenvalue -1. Construct an eigenbasis by picking one of each.

17. No eigenvectors and eigenvalues (compare with Example 3).

19. The nonzero vectors in L are the eigenvectors with eigenvalue 1, and the nonzero vectors in the plane $L^{\perp}$ have eigenvalue 0. Construct an

eigenbasis by picking one nonzero vector in L and two linearly independent vectors in $L^\perp$. (Compare with Example 2.)

21. All nonzero vectors in $\mathbb{R}^3$ are eigenvectors with eigenvalue 5. Any basis of $\mathbb{R}^3$ is an eigenbasis.

23. a. $S^{-1}\vec{v}_i = \vec{e}_i$ because $S\vec{e}_i = \vec{v}_i$

   b. $S^{-1}AS = S^{-1}A[\vec{v}_1 \ldots \vec{v}_n] =$
   $S^{-1}[\lambda_1\vec{v}_1 \ldots \lambda_n\vec{v}_n] = [\lambda_1\vec{e}_1 \ldots \lambda_n\vec{e}_n] =$
   $\begin{bmatrix} \lambda_1 & & 0 \\ & \ddots & \\ 0 & & \lambda_n \end{bmatrix}$, the diagonal matrix with
   diagonal entries $\lambda_1, \ldots, \lambda_n$.

25.

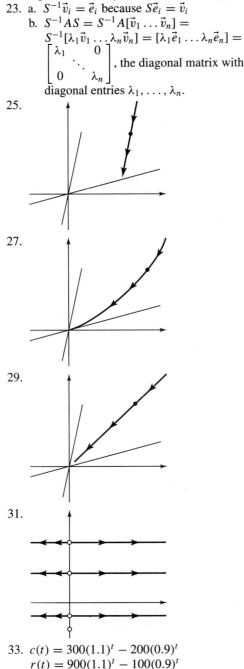

27.

29.

31.

33. $c(t) = 300(1.1)^t - 200(0.9)^t$
   $r(t) = 900(1.1)^t - 100(0.9)^t$

35. $\vec{x}(t) = 2^t \begin{bmatrix} 1 \\ 1 \end{bmatrix} + 6^t \begin{bmatrix} -1 \\ 1 \end{bmatrix}$. We need a matrix A

   with eigenvectors $\begin{bmatrix} 1 \\ 1 \end{bmatrix}, \begin{bmatrix} -1 \\ 1 \end{bmatrix}$, with associated

   eigenvalues 2 and 6, respectively. Let

   $A\begin{bmatrix} 1 & -1 \\ 1 & 1 \end{bmatrix} = \begin{bmatrix} 2 & -6 \\ 2 & 6 \end{bmatrix}$ and solve for A. We

   find $A = \begin{bmatrix} 4 & -2 \\ -2 & 4 \end{bmatrix}$.

37. a. A represents a reflection in a line followed by a dilation by a factor of $\sqrt{3^2 + 4^2} = 5$. The eigenvalues are therefore 5 and -5.

   b. Solving the linear systems $A\vec{x} = 5\vec{x}$ and $A\vec{x} = -5\vec{x}$ we find the eigenbasis $\begin{bmatrix} 2 \\ 1 \end{bmatrix}, \begin{bmatrix} -1 \\ 2 \end{bmatrix}$.

39. If λ is an eigenvalue of $S^{-1}AS$, with corresponding eigenvector $\vec{v}$, then

   $$S^{-1}AS\vec{v} = \lambda\vec{v},$$

   so

   $$AS\vec{v} = S\lambda\vec{v} = \lambda S\vec{v},$$

   and λ is an eigenvalue of A ($S\vec{v}$ is an eigenvector). Likewise, if $\vec{w}$ is an eigenvector of A, then $S^{-1}\vec{w}$ is an eigenvector of $S^{-1}AS$ with the same eigenvalue.

7.2

1. 1, 3

3. 1, 3

5. none

7. 1, 1, 1

9. 1, 2, 2

11. -1

13. 1

15. Eigenvalues $\lambda_{1,2} = 1 \pm \sqrt{k}$. Two distinct real eigenvalues if k is positive; none, if k is negative.

17. A represents a reflection-dilation, with a dilation factor of $\sqrt{a^2 + b^2}$. The eigenvalues are $\pm\sqrt{a^2 + b^2}$.

19. True (the discriminant $\operatorname{tr}(A)^2 - 4\det(A)$ is positive)

21. $\det(A)$ is the product of the eigenvalues, and $\operatorname{tr}(A)$ is their sum. To see this, write
   $f_A(\lambda) = (\lambda - \lambda_1) \cdots (\lambda - \lambda_n) =$
   $\lambda^n - (\lambda_1 + \cdots + \lambda_n)\lambda^{n-1} + \cdots + (-1)^n\lambda_1 \cdots \lambda_n$
   and compare with the formula in Fact 7.2.5.

23. $f_B(\lambda) = \det(\lambda I_n - B) = \det(\lambda I_n - S^{-1}AS)$
   $= \det\left(S^{-1}(\lambda I_n - A)S\right)$
   $= \left(\det(S)\right)^{-1}\det(\lambda I_n - A)\det(S)$
   $= f_A(\lambda)$
   A and B have the same eigenvalues, with the same algebraic multiplicities.

25. $A \begin{bmatrix} c \\ b \end{bmatrix} = \begin{bmatrix} c \\ b \end{bmatrix}$ and $A \begin{bmatrix} 1 \\ -1 \end{bmatrix} = (a - c) \begin{bmatrix} 1 \\ -1 \end{bmatrix}$.

Note that $|a - c| < 1$

Phase portrait when $a > c$:

line spanned by $\begin{bmatrix} c \\ b \end{bmatrix}$

line spanned by $\begin{bmatrix} 1 \\ -1 \end{bmatrix}$

27. a. $\vec{x}(t) = \frac{1}{3} \begin{bmatrix} 1 \\ 2 \end{bmatrix} + \frac{2}{3} \left(\frac{1}{4}\right)^t \begin{bmatrix} 1 \\ -1 \end{bmatrix}$ for $\vec{x}_0 = \vec{e}_1$

$\vec{x}(t) = \frac{1}{3} \begin{bmatrix} 1 \\ 2 \end{bmatrix} - \frac{1}{3} \left(\frac{1}{4}\right)^t \begin{bmatrix} 1 \\ -1 \end{bmatrix}$ for $\vec{x}_0 = \vec{e}_2$

b. $A^t = [A^t \vec{e}_1 \quad A^t \vec{e}_2]$ approaches $\frac{1}{3} \begin{bmatrix} 1 & 1 \\ 2 & 2 \end{bmatrix}$,

   by part (a).

c. $A^t \to \frac{1}{b+c} \begin{bmatrix} c & c \\ b & b \end{bmatrix}$

29. $A\vec{e} = \vec{e}$, so that $\vec{e}$ is an eigenvector with associated eigenvalue 1.

31. A and A^T have the same eigenvalues, by Exercise 22. Since the row sums of A^T are 1, we can use the results of Exercises 29 and 30: 1 is an eigenvalue of A; if λ is an eigenvalue of A, then $|\lambda| \le 1$. $\vec{e}$ need not be an eigenvector of A; consider $A = \begin{bmatrix} 0.9 & 0.9 \\ 0.1 & 0.1 \end{bmatrix}$

33. a. $f_A(\lambda) = \lambda^3 - c\lambda^2 - b\lambda - a$

   b. $M = \begin{bmatrix} 0 & 1 & 0 \\ 0 & 0 & 1 \\ \pi & -5 & 17 \end{bmatrix}$

35. $A = \begin{bmatrix} 0 & -1 & 0 & 0 \\ 1 & 0 & 0 & 0 \\ 0 & 0 & 0 & -1 \\ 0 & 0 & 1 & 0 \end{bmatrix}$

37. We can write $f_A(\lambda) = (\lambda - \lambda_0)^2 g(\lambda)$. By the product rule,

$f'_A(\lambda) = 2(\lambda - \lambda_0)g(\lambda) + (\lambda - \lambda_0)^2 g'(\lambda)$, so that $f'_A(\lambda_0) = 0$.

7.3

1. eigenbasis: $\begin{bmatrix} 1 \\ 0 \end{bmatrix}, \begin{bmatrix} 4 \\ 1 \end{bmatrix}$, with eigenvalues 7, 9.

3. eigenbasis: $\begin{bmatrix} 3 \\ -2 \end{bmatrix}, \begin{bmatrix} 1 \\ 1 \end{bmatrix}$, with eigenvalues 4, 9.

5. No real eigenvalues.

7. eigenbasis: $\vec{e}_1, \vec{e}_2, \vec{e}_3$, with eigenvalues 1, 2, 3.

9. eigenbasis: $\begin{bmatrix} 1 \\ 0 \\ 0 \end{bmatrix}, \begin{bmatrix} 0 \\ 1 \\ 0 \end{bmatrix}, \begin{bmatrix} -1 \\ 0 \\ 1 \end{bmatrix}$, with eigenvalues 1, 1, 0.

11. eigenbasis: $\begin{bmatrix} 1 \\ 1 \\ 1 \end{bmatrix}, \begin{bmatrix} 1 \\ -1 \\ 0 \end{bmatrix}, \begin{bmatrix} 1 \\ 0 \\ -1 \end{bmatrix}$, with eigenvalues 3, 0, 0.

13. eigenbasis: $\begin{bmatrix} 0 \\ 1 \\ 0 \end{bmatrix}, \begin{bmatrix} 1 \\ -3 \\ 1 \end{bmatrix}, \begin{bmatrix} 1 \\ -1 \\ 2 \end{bmatrix}$, with eigenvalues 0, 1, −1.

15. eigenvectors: $\begin{bmatrix} 0 \\ 1 \\ 0 \end{bmatrix}, \begin{bmatrix} 1 \\ -1 \\ 2 \end{bmatrix}$, with eigenvalues 0, 1. No eigenbasis.

17. eigenbasis: $\vec{e}_2, \vec{e}_4, \vec{e}_1, \vec{e}_3 - \vec{e}_2$, with eigenvalues 1, 1, 0, 0.

19. The geometric multiplicity of the eigenvalue 1 is 1 if $a \ne 0$ and $c \ne 0$; 3 if $a = b = c = 0$; and 2 otherwise. Therefore, there is an eigenbasis only if $A = I_3$.

21. We want $A \begin{bmatrix} 1 \\ 2 \end{bmatrix} = \begin{bmatrix} 1 \\ 2 \end{bmatrix}$ and

$A \begin{bmatrix} 2 \\ 3 \end{bmatrix} = 2 \begin{bmatrix} 2 \\ 3 \end{bmatrix} = \begin{bmatrix} 4 \\ 6 \end{bmatrix}$, that is,

$A \begin{bmatrix} 1 & 2 \\ 2 & 3 \end{bmatrix} = \begin{bmatrix} 1 & 4 \\ 2 & 6 \end{bmatrix}$. The unique solution is

$A = \begin{bmatrix} 5 & -2 \\ 6 & -2 \end{bmatrix}$.

23. The only eigenvalue of A is 1, with $E_1 = \text{span}(\vec{e}_1)$. There is no eigenbasis. A represents a shear parallel to the x_1-axis.

25. The geometric multiplicity is always 1.

27. $f_A(\lambda) = \lambda^2 - 5\lambda + 6 = (\lambda - 2)(\lambda - 3)$, so that the eigenvalues are 2, 3.

29. Both multiplicities are $n - r$.

31. They are the same.

33. If $B = S^{-1}AS$, then $B - \lambda I_n = S^{-1}(A - \lambda I_n)S$.

35. No (consider the eigenvalues).

37. a. $A\vec{v} \cdot \vec{w} = (A\vec{v})^T \vec{w} = \vec{v}^T A^T \vec{w} = \vec{v}^T A\vec{w} = \vec{v} \cdot A\vec{w}$

   b. Suppose $A\vec{v} = \lambda\vec{v}$ and $A\vec{w} = \mu\vec{w}$. Then $A\vec{v} \cdot \vec{w} = \lambda(\vec{v} \cdot \vec{w})$ and $\vec{v} \cdot A\vec{w} = \mu(\vec{v} \cdot \vec{w})$. By part (a), $\lambda(\vec{v} \cdot \vec{w}) = \mu(\vec{v} \cdot \vec{w})$, so that

$(\lambda - \mu)(\vec{v} \cdot \vec{w}) = 0$. Since $\lambda \neq \mu$ it follows that $\vec{v} \cdot \vec{w} = 0$, as claimed.

39. a. $E_1 = V$ and $E_0 = V^\perp$, so that the geometric multiplicity of 1 is m and that of 0 is $n - m$. The algebraic multiplicities are the same (see Exercise 31).

   b. $E_1 = V$ and $E_{-1} = V^\perp$, so that the multiplicity of 1 is m and that of -1 is $n - m$.

41. Eigenbasis for A: $\begin{bmatrix} 9 \\ 6 \\ 2 \end{bmatrix}$, $\begin{bmatrix} 2 \\ -2 \\ 1 \end{bmatrix}$, $\begin{bmatrix} 1 \\ -2 \\ 2 \end{bmatrix}$, with

   eigenvalues $1.2, -0.8, -0.4$; $\vec{x}_0 = 50\vec{v}_1 + 50\vec{v}_2 + 50\vec{v}_3$.

   $j(t) = 450(1.2)^t + 100(-0.8)^t + 50(-0.4)^t$
   $m(t) = 300(1.2)^t - 100(-0.8)^t - 100(-0.4)^t$
   $a(t) = 100(1.2)^t + 50(-0.8)^t + 100(-0.4)^t$

   The populations approach the proportion $9:6:2$.

43. a. $A = \dfrac{1}{2} \begin{bmatrix} 0 & 1 & 1 \\ 1 & 0 & 1 \\ 1 & 1 & 0 \end{bmatrix}$

   c. $\vec{x}(t) = \left(1 + \dfrac{c_0}{3} \right) \begin{bmatrix} 1 \\ 1 \\ 1 \end{bmatrix} + \left(-\dfrac{1}{2} \right)^t \begin{bmatrix} 0 \\ 1 \\ -1 \end{bmatrix} +$

   $\left(-\dfrac{1}{2} \right)^t \dfrac{c_0}{3} \begin{bmatrix} -1 \\ -1 \\ 2 \end{bmatrix}$. Carl wins if he chooses

   $c_0 < 1$.

45. a. $A = \begin{bmatrix} 0.1 & 0.2 \\ 0.4 & 0.3 \end{bmatrix}$, $\vec{b} = \begin{bmatrix} 1 \\ 2 \end{bmatrix}$

   b. $B = \begin{bmatrix} A & \vec{b} \\ 0 & 1 \end{bmatrix}$

   c. The eigenvalues of A are 0.5 and -0.1, those of B, $0.5, -0.1, 1$. If $\vec{v}$ is an eigenvector of A, then $\begin{bmatrix} \vec{v} \\ 0 \end{bmatrix}$ is an eigenvector of B. Furthermore,

   $\begin{bmatrix} (I_2 - A)^{-1}\vec{b} \\ 1 \end{bmatrix} = \begin{bmatrix} 2 \\ 4 \\ 1 \end{bmatrix}$ is an eigenvector of B

   with eigenvalue 1.

   d. Will approach $(I_2 - A)^{-1}\vec{b} = \begin{bmatrix} 2 \\ 4 \end{bmatrix}$, for any initial value.

47. Let $\vec{x}(t) = \begin{bmatrix} r(t) \\ p(t) \\ w(t) \end{bmatrix}$. Then $\vec{x}(t+1) = A\vec{x}(t)$

   where $A = \begin{bmatrix} \frac{1}{2} & \frac{1}{4} & 0 \\ \frac{1}{2} & \frac{1}{2} & \frac{1}{2} \\ 0 & \frac{1}{4} & \frac{1}{2} \end{bmatrix}$. Eigenbasis for

   A: $\begin{bmatrix} 1 \\ 2 \\ 1 \end{bmatrix}$, $\begin{bmatrix} 1 \\ 0 \\ -1 \end{bmatrix}$, $\begin{bmatrix} 1 \\ -2 \\ 1 \end{bmatrix}$, with eigenvalues $1, \dfrac{1}{2}, 0$.

$\vec{x}_0 = \vec{e}_1 = \dfrac{1}{4} \begin{bmatrix} 1 \\ 2 \\ 1 \end{bmatrix} + \dfrac{1}{2} \begin{bmatrix} 1 \\ 0 \\ -1 \end{bmatrix} + \dfrac{1}{4} \begin{bmatrix} 1 \\ -2 \\ 1 \end{bmatrix}$,

so $\vec{x}(t) = \dfrac{1}{4} \begin{bmatrix} 1 \\ 2 \\ 1 \end{bmatrix} + \left(\dfrac{1}{2} \right)^{t+1} \begin{bmatrix} 1 \\ 0 \\ -1 \end{bmatrix}$ for $t > 0$.

The proportion in the long run is $1:2:1$.

49. 1 (rref(A) is likely to be the matrix with all 1's directly above the main diagonal and 0's everywhere else)

7.4 1. $S = I_2, D = A$

3. $S = \begin{bmatrix} 1 & 1 \\ -1 & 2 \end{bmatrix}$, $D = \begin{bmatrix} 0 & 0 \\ 0 & 3 \end{bmatrix}$, for example. If you found a different solution, check that $AS = SD$.

5. not diagonalizable

7. $S = \begin{bmatrix} 4 & 1 \\ 1 & -1 \end{bmatrix}$, $D = \begin{bmatrix} 2 & 0 \\ 0 & -3 \end{bmatrix}$

9. not diagonalizable

11. not diagonalizable

13. $S = \begin{bmatrix} 1 & 1 & 1 \\ 0 & 1 & 2 \\ 0 & 0 & 1 \end{bmatrix}$, $D = \begin{bmatrix} 1 & 0 & 0 \\ 0 & 2 & 0 \\ 0 & 0 & 3 \end{bmatrix}$

15. $S = \begin{bmatrix} 2 & 0 & 1 \\ 1 & 0 & 1 \\ 0 & 1 & 0 \end{bmatrix}$, $D = \begin{bmatrix} 1 & 0 & 0 \\ 0 & 1 & 0 \\ 0 & 0 & -1 \end{bmatrix}$

17. $S = \begin{bmatrix} -1 & -1 & 1 \\ 1 & 0 & 1 \\ 0 & 1 & 1 \end{bmatrix}$, $D = \begin{bmatrix} 0 & 0 & 0 \\ 0 & 0 & 0 \\ 0 & 0 & 3 \end{bmatrix}$

19. Not diagonalizable; eigenvalue 1 has algebraic multiplicity 2, but geometric multiplicity 1.

21. diagonalizable for all a

23. diagonalizable for positive a

25. diagonalizable for all a, b, and c.

27. diagonalizable only if $a = b = c = 0$

29. never diagonalizable

31. $A^t = \dfrac{1}{3} \begin{bmatrix} 5^t + 2(-1)^t & 5^t - (-1)^t \\ 2(5^t) - 2(-1)^t & 2(5^t) + (-1)^t \end{bmatrix}$

33. $A^t = 7^{t-1} A$

35. Yes, since $\begin{bmatrix} -1 & 6 \\ -2 & 6 \end{bmatrix}$ is diagonalizable, with eigenvalues 3 and 2.

37. Yes. Both matrices have the characteristic polynomial $\lambda^2 - 7\lambda + 7$, so that they both have

   the eigenvalues $\lambda_{1,2} = \dfrac{7 \pm \sqrt{21}}{2}$. Thus, A and B

   are both similar to $\begin{bmatrix} \lambda_1 & 0 \\ 0 & \lambda_2 \end{bmatrix}$, and therefore, A is

   similar to B.

39. All real numbers λ are eigenvalues, with corresponding eigenfunctions $Ce^{(\lambda+1)t}$.

41. The symmetric matrices are eigenmatrices with eigenvalue 2, and the skew-symmetric matrices have eigenvalue 0. Yes, L is diagonalizable, since the sum of the dimensions of the eigenspaces is 4.

43. 1 and i are "eigenvectors" with eigenvalues 1 and -1, respectively. Yes, T is diagonalizable; 1, i is an eigenbasis.

45. no eigensequences

47. Polynomials of the form $a + cx^2$ are eigenfunctions with eigenvalue 1, and bx has eigenvalue -1. Yes, T is diagonalizable, with eigenbasis 1, x, x^2.

49. 1, $2x - 1$, and $(2x - 1)^2$ are eigenfunctions with eigenvalues 1, 3, and 9, respectively. These functions form an eigenbasis, so that T is indeed diagonalizable.

51. The only eigenfunctions are the nonzero constant functions, with eigenvalue 0.

7.5
1. $\sqrt{18}\left(\cos\left(-\dfrac{\pi}{4} \right) + i \sin\left(-\dfrac{\pi}{4} \right) \right)$

3. $\cos\left(\dfrac{2\pi k}{n} \right) + i \sin\left(\dfrac{2\pi k}{n} \right)$, for $k = 0, \ldots, n-1$

5. If $z = r(\cos(\phi) + i \sin(\phi))$, then $w = $
$$\sqrt[n]{r}\left(\cos\left(\dfrac{\phi + 2\pi k}{n} \right) + i \sin\left(\dfrac{\phi + 2\pi k}{n} \right) \right),$$
for $k = 0, \ldots, n-1$.

7. Clockwise rotation through an angle of $\frac{\pi}{4}$ followed by a dilation by a factor of $\sqrt{2}$.

9. Spirals outward since $|z| > 1$.

11. $f(\lambda) = (\lambda - 1)(\lambda - 1 - 2i)(\lambda - 1 + 2i)$

13. $\mathbb{Q}$ is a field.

15. The binary digits form a field.

17. H is not a field (multiplication is noncommutative).

19. a. $\operatorname{tr}(A) = m$, $\det(A) = 0$
    b. $\operatorname{tr}(B) = 2m - n$, $\det(B) = (-1)^{n-m}$ (compare with Exercise 7.3.39)

21. $2 \pm 3i$

23. $1, -\dfrac{1}{2} \pm \dfrac{\sqrt{3}}{2}i$

25. $\pm 1, \pm i$

27. $-1, -1, 3$

29. $\operatorname{tr}(A) = \lambda_1 + \lambda_2 + \lambda_3 = 0$ and $\det(A) = \lambda_1 \lambda_2 \lambda_3 = bcd > .0$. Therefore, there are one positive and two negative eigenvalues; the positive one is largest in absolute value.

31. $\frac{1}{15}A$ is a regular transition matrix (compare with Exercise 30), so that $\lim\limits_{t \to \infty} \left(\frac{1}{15}A \right)^t$ exists and has

identical columns (see Exercise 30). Therefore, the columns of A^t are nearly identical for large t.

33. c. *Hint:* Let $\lambda_1, \lambda_2, \ldots, \lambda_5$ be the eigenvalues, with $\lambda_1 > |\lambda_j|$, for $j = 2, \ldots, 5$. Let $\vec{v}_1, \vec{v}_2, \ldots, \vec{v}_5$ be corresponding eigenvectors. Write $\vec{e}_i = c_1 \vec{v}_1 + \cdots + c_5 \vec{v}_5$. Then ith column of $A^t = A^t \vec{e}_i$
$$= c_1 \lambda_1^t \vec{v}_1 + \cdots + c_5 \lambda_5^t \vec{v}_5$$
is nearly parallel to $\vec{v}_1$ for large t.

45. If a is nonzero.

47. If a is nonzero.

49. If a is neither 1 nor 2.

7.6
1. stable

3. not stable

5. not stable

7. not stable

9. not stable

11. For $|k| < 1$

13. For all k

15. Never stable

17. $\vec{x}(t) = \begin{bmatrix} -\sin(\phi t) \\ \cos(\phi t) \end{bmatrix}$, where $\phi = \arctan\left(\dfrac{4}{3} \right)$; a circle.

19. $\vec{x}(t) = \sqrt{13}^t \begin{bmatrix} -\sin(\phi t) \\ \cos(\phi t) \end{bmatrix}$, where $\phi = \arctan\left(\dfrac{3}{2} \right)$; spirals outward.

21. $\vec{x}(t) = \sqrt{17}^t \begin{bmatrix} 5\sin(\phi t) \\ \cos(\phi t) + 3\sin(\phi t) \end{bmatrix}$, where $\phi = \arctan\left(\dfrac{1}{4} \right)$; spirals outward.

23. $\vec{x}(t) = \left(\dfrac{1}{2} \right)^t \begin{bmatrix} 5\sin(\phi t) \\ \cos(\phi t) + 3\sin(\phi t) \end{bmatrix}$, where $\phi = \arctan\left(\dfrac{3}{4} \right)$; spirals inward.

25. not stable

27. stable

29. may or may not be stable; consider $A = \pm\frac{1}{2}I_2$

33. The matrix represents a rotation-dilation with a dilation factor of $\sqrt{0.99^2 + 0.01^2} < 1$. Trajectory spirals inward.

35. a. Choose an eigenbasis $\vec{v}_1, \ldots, \vec{v}_n$ and write
$$\vec{x}_0 = c_1 \vec{v}_1 + \cdots + c_n \vec{v}_n.$$
Then
$$\vec{x}(t) = c_1 \lambda_1^t \vec{v}_1 + \cdots + c_n \lambda_n^t \vec{v}_n$$
and
$$\|\vec{x}(t)\| \le |c_1|\|\vec{v}_1\| + \cdots + |c_n|\|\vec{v}_n\| = M$$

(use the triangle inequality $\|\vec{u} + \vec{w}\| \le \|\vec{u}\| + \|\vec{w}\|$, and observe that $|\lambda_i^t| \le 1$)

b. The trajectory $\vec{x}(t) = \begin{bmatrix} 1 & 1 \\ 0 & 1 \end{bmatrix}^t \begin{bmatrix} 0 \\ 1 \end{bmatrix} = \begin{bmatrix} t \\ 1 \end{bmatrix}$

   is not bounded. This does not contradict part (a), since there is no eigenbasis for the matrix $\begin{bmatrix} 1 & 1 \\ 0 & 1 \end{bmatrix}$.

39. $\begin{bmatrix} 2 \\ 4 \end{bmatrix}$ is a stable equilibrium.

Chapter 8

8.1

1. $\begin{bmatrix} 1 \\ 0 \end{bmatrix}, \begin{bmatrix} 0 \\ 1 \end{bmatrix}$

3. $\frac{1}{\sqrt{5}} \begin{bmatrix} 2 \\ 1 \end{bmatrix}, \frac{1}{\sqrt{5}} \begin{bmatrix} -1 \\ 2 \end{bmatrix}$

5. $\frac{1}{\sqrt{2}} \begin{bmatrix} -1 \\ 1 \\ 0 \end{bmatrix}, \frac{1}{\sqrt{6}} \begin{bmatrix} -1 \\ -1 \\ 2 \end{bmatrix}, \frac{1}{\sqrt{3}} \begin{bmatrix} 1 \\ 1 \\ 1 \end{bmatrix}$

7. $S = \frac{1}{\sqrt{2}} \begin{bmatrix} 1 & -1 \\ 1 & 1 \end{bmatrix}, D = \begin{bmatrix} 5 & 0 \\ 0 & 1 \end{bmatrix}$

9. $S = \begin{bmatrix} 1/\sqrt{2} & -1/\sqrt{2} & 0 \\ 0 & 0 & 1 \\ 1/\sqrt{2} & 1/\sqrt{2} & 0 \end{bmatrix}$,

   $D = \begin{bmatrix} 3 & 0 & 0 \\ 0 & -3 & 0 \\ 0 & 0 & 2 \end{bmatrix}$

11. Same S as in 9, $D = \begin{bmatrix} 2 & 0 & 0 \\ 0 & 0 & 0 \\ 0 & 0 & 1 \end{bmatrix}$

13. Yes (reflection in E_1)

15. Yes (can use the same orthonormal eigenbasis)

17. Let A be the $n \times n$ matrix whose entries are 1. The eigenvalues of A are 0 (with multiplicity $n - 1$) and n. Now $B = qA + (p - q)I_n$, so that the eigenvalues of B are $p - q$ (with multiplicity $n - 1$) and $qn + p - q$. Therefore, $\det(B) = (p - q)^{n-1}(qn + p - q)$.

21. $48 = 6 \cdot 4 \cdot 2$ (note that A has 6 unit eigenvectors)

23. The only possible eigenvalues are 1 and -1 (because A is orthogonal), and the eigenspaces E_1 and E_{-1} are orthogonal complements (because A is symmetric). A represents the reflection in a subspace of $\mathbb{R}^n$.

25. $S = \frac{1}{\sqrt{2}} \begin{bmatrix} 1 & 1 & 0 & 0 & 0 \\ 0 & 0 & 1 & 1 & 0 \\ 0 & 0 & 0 & 0 & \sqrt{2} \\ 0 & 0 & 1 & -1 & 0 \\ 1 & -1 & 0 & 0 & 0 \end{bmatrix}$

27. If n is even, we have the eigenbasis $\vec{e}_1 - \vec{e}_n$, $\vec{e}_2 - \vec{e}_{n-1}, \ldots, \vec{e}_{n/2} - \vec{e}_{n/2+1}, \vec{e}_1 + \vec{e}_n$,

$\vec{e}_2 + \vec{e}_{n-1}, \ldots, \vec{e}_{n/2} + \vec{e}_{n/2+1}$, with associated eigenvalues 0 ($n/2$ times) and 2 ($n/2$ times).

29. Yes

31. True

33. $\alpha = \frac{2}{3}\pi = 120°$.

35. $\alpha = \arccos(-\frac{1}{n})$. Hint: If $\vec{v}_0, \ldots, \vec{v}_n$ are such vectors, let $A = [\vec{v}_0 \quad \cdots \quad \vec{v}_n]$. Then the noninvertible matrix $A^T A$ has 1's on the diagonal and $\cos(\alpha)$ everywhere else. Now use Exercise 17.

39. a. ijth entry of $|AB| = \left| \sum_{k=1}^{n} a_{ik}b_{kj} \right|$

   $\le \sum_{k=1}^{n} |a_{ik}||b_{kj}|$

   $= ij$th entry of $|A||B|$

   b. By induction on t, using part (a): $|A^t| = |A^{t-1}A| \le |A^{t-1}||A| \le |A|^{t-1}|A| = |A|^t$

41. Let λ be the maximum of all $|r_{ii}|$, for $i = 1, \ldots, n$. Note that $\lambda < 1$. Then $|R| \le \lambda(I_n + U)$, where U is upper triangular with $u_{ii} = 0$ and $u_{ij} = |r_{ij}|/\lambda$ if $j > i$. Note that $U^n = 0$ (see Exercise 38a). Now $|R^t| \le |R|^t \le \lambda^t(I_n + U)^t \le \lambda^t t^n(I_n + U + \cdots + U^{n-1})$. From calculus we know that $\lim_{t \to \infty} \lambda^t t^n = 0$.

8.2

1. $\begin{bmatrix} 6 & -3.5 \\ -3.5 & 8 \end{bmatrix}$

3. $\begin{bmatrix} 3 & 0 & 3 \\ 0 & 4 & 3.5 \\ 3 & 3.5 & 5 \end{bmatrix}$

5. indefinite

7. indefinite

9. a. A^2 is symmetric

   b. $A^2 = -A^T A$ is negative semidefinite, so that its eigenvalues are ≤ 0.

   c. The eigenvalues of A are imaginary (that is, of the form bi, for a real b). The zero matrix is the only skew-symmetric matrix that is diagonalizable over $\mathbb{R}$.

11. The same (the eigenvalues of A and A^{-1} have the same signs).

13. $a_{ii} = q(\vec{e}_i) > 0$.

15. Ellipse; principal axes spanned by $\begin{bmatrix} 2 \\ 1 \end{bmatrix}$ and $\begin{bmatrix} -1 \\ 2 \end{bmatrix}$; equation $7c_1^2 + 2c_2^2 = 1$

17. Hyperbola; principal axes spanned by $\begin{bmatrix} 2 \\ 1 \end{bmatrix}$ and $\begin{bmatrix} -1 \\ 2 \end{bmatrix}$, equation $4c_1^2 - c_2^2 = 1$.

19. A pair of lines; principal axes spanned by $\begin{bmatrix} 2 \\ -1 \end{bmatrix}$ and $\begin{bmatrix} 1 \\ 2 \end{bmatrix}$; equation $5c_2^2 = 1$.

Note that we can write $x_1^2 + 4x_1x_2 + 4x_2^2 = (x_1 + 2x_2)^2 = 1$, so that $x_1 + 2x_2 = \pm 1$

21. a. The first is an ellipsoid, the second a hyperboloid of one sheet, and the third a hyperboloid of two sheets (see any text in multivariable calculus). Only the ellipsoid is bounded, and the first two surfaces are connected.

b. The matrix A of this quadratic form has positive eigenvalues $\lambda_1 \approx 0.56$, $\lambda_2 \approx 4.44$, and $\lambda_3 = 1$, with corresponding unit eigenvectors

$$\vec{v}_1 \approx \begin{bmatrix} 0.86 \\ 0.19 \\ -0.47 \end{bmatrix}, \vec{v}_2 \approx \begin{bmatrix} 0.31 \\ 0.54 \\ 0.78 \end{bmatrix}, \vec{v}_3 \approx \begin{bmatrix} 0.41 \\ -0.82 \\ 0.41 \end{bmatrix}$$

Since all eigenvalues are positive, the surface is an ellipsoid. The points farthest from the origin are

$$\pm \frac{1}{\sqrt{\lambda_1}} \vec{v}_1 \approx \pm \begin{bmatrix} 1.15 \\ 0.26 \\ -0.63 \end{bmatrix}.$$

and those closest are

$$\pm \frac{1}{\sqrt{\lambda_2}} \vec{v}_2 \approx \pm \begin{bmatrix} 0.15 \\ 0.26 \\ 0.37 \end{bmatrix}.$$

23. Yes; $A = \frac{1}{2}(M + M^T)$

25. $q(\vec{v}) = \vec{v} \cdot \lambda \vec{v} = \lambda$.

27. The closed interval $[\lambda_n, \lambda_1]$

29. $B = \frac{1}{\sqrt{5}} \begin{bmatrix} 6 & 2 \\ -3 & 4 \end{bmatrix}$

31. $B = \frac{1}{5} \begin{bmatrix} 14 & -2 \\ -2 & 11 \end{bmatrix}$

33. $L = \frac{1}{\sqrt{2}} \begin{bmatrix} 4 & 0 \\ -1 & 3 \end{bmatrix}$

35. $L = \begin{bmatrix} 2 & 0 & 0 \\ -2 & 3 & 0 \\ 4 & 3 & 1 \end{bmatrix}$

39. For $0 < \alpha < \arccos\left(-\frac{1}{n-1}\right)$

41. 3

43. $\text{im}(T) = \text{span}(x_1^2)$, $\text{rank}(T) = 1$, $\text{ker}(T) = \text{span}(x_1x_2, x_2^2)$, $\text{nullity}(T) = 2$

45. $\text{im}(T) = P_2$, $\text{rank}(T) = 3$, $\text{ker}(T) = \text{span}(x_3^2 - x_2^2, x_1x_3 - x_1x_2, x_2x_3 - x_2^2)$, $\text{nullity}(T) = 3$

47. The determinant of the mth principal submatrix is positive if m is even, and negative if m is odd.

8.3

1. $\sigma_1 = 2, \sigma_2 = 1$

3. All singular values are 1 (since $A^T A = I_n$)

5. $\sigma_1 = \sigma_2 = \sqrt{p^2 + q^2}$

7. $\begin{bmatrix} 0 & 1 \\ -1 & 0 \end{bmatrix} \begin{bmatrix} 2 & 0 \\ 0 & 1 \end{bmatrix} \begin{bmatrix} 0 & 1 \\ 1 & 0 \end{bmatrix}$

9. $\frac{1}{\sqrt{5}} \begin{bmatrix} 1 & -2 \\ 2 & 1 \end{bmatrix} \begin{bmatrix} 5 & 0 \\ 0 & 0 \end{bmatrix} \frac{1}{\sqrt{5}} \begin{bmatrix} 1 & 2 \\ -2 & 1 \end{bmatrix}$

11. $\begin{bmatrix} 0 & 1 & 0 \\ 1 & 0 & 0 \\ 0 & 0 & 1 \end{bmatrix} \begin{bmatrix} 2 & 0 \\ 0 & 1 \\ 0 & 0 \end{bmatrix} \begin{bmatrix} 0 & 1 \\ 1 & 0 \end{bmatrix}$

13. $I_2 \begin{bmatrix} 3\sqrt{5} & 0 \\ 0 & \sqrt{5} \end{bmatrix} \frac{1}{\sqrt{5}} \begin{bmatrix} 2 & 1 \\ -1 & 2 \end{bmatrix}$

15. Singular values of A^{-1} are the reciprocals of those of A.

21. $\begin{bmatrix} 0.8 & 0.6 \\ -0.6 & 0.8 \end{bmatrix} \begin{bmatrix} 9 & -2 \\ -2 & 6 \end{bmatrix}$

23. $AA^T \vec{u}_i = \begin{cases} \sigma_i^2 \vec{u}_i & \text{for } i = 1, \ldots, r \\ \vec{0} & \text{for } i = r+1, \ldots, m \end{cases}$

The nonzero eigenvalues of $A^T A$ and AA^T are the same.

25. Choose vectors $\vec{v}_1$ and $\vec{v}_2$ as in Fact 8.3.3. Write

$$\vec{u} = c_1\vec{v}_1 + c_2\vec{v}_2.$$

Note that

$$\|\vec{u}\|^2 = c_1^2 + c_2^2 = 1.$$

Now

$$A\vec{u} = c_1 A\vec{v}_1 + c_2 A\vec{v}_2,$$

so that

$$\|A\vec{u}\|^2 = c_1^2 \|A\vec{v}_1\|^2 + c_2^2 \|A\vec{v}_2\|^2$$
$$= c_1^2 \sigma_1^2 + c_2^2 \sigma_2^2$$
$$\leq (c_1^2 + c_2^2)\sigma_1^2$$
$$= \sigma_1^2$$

We conclude that $\|A\vec{u}\| \leq \sigma_1$. The proof of $\sigma_2 \leq \|A\vec{u}\|$ is analogous.

27. Apply Exercise 26 to a unit eigenvector $\vec{v}$ with associated eigenvalue λ.

33. False; consider $A = \begin{bmatrix} 0 & 1 \\ 2 & 0 \end{bmatrix}$.

35. $(A^T A)^{-1} A^T \vec{u}_i = \begin{cases} \frac{1}{\sigma_i} \vec{v}_i & \text{for } i = 1, \ldots, r \\ \vec{0} & \text{for } i = r+1, \ldots, m \end{cases}$

37. Yes, since $A^T A = I_n$

Chapter 9

9.1

1. $x(t) = 7e^{5t}$

3. $P(t) = 7e^{0.03t}$

5. $y(t) = -0.8e^{0.8t}$

7. $x(t) = \frac{1}{1-t}$, has a vertical asymptote at $t = 1$.

9. $x(t) = ((1-k)t + 1)^{1/(1-k)}$

11. $x(t) = \tan(t)$

13. a. about 104 billion dollars
    b. about 150 billion dollars

15. The solution of the equation $e^{kT/100} = 2$ is
$$T = \frac{100 \ln(2)}{k} \approx \frac{69}{k}$$

17.

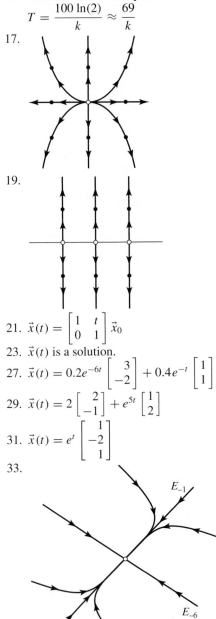

19.

21. $\vec{x}(t) = \begin{bmatrix} 1 & t \\ 0 & 1 \end{bmatrix} \vec{x}_0$

23. $\vec{x}(t)$ is a solution.

27. $\vec{x}(t) = 0.2e^{-6t} \begin{bmatrix} 3 \\ -2 \end{bmatrix} + 0.4e^{-t} \begin{bmatrix} 1 \\ 1 \end{bmatrix}$

29. $\vec{x}(t) = 2 \begin{bmatrix} 2 \\ -1 \end{bmatrix} + e^{5t} \begin{bmatrix} 1 \\ 2 \end{bmatrix}$

31. $\vec{x}(t) = e^t \begin{bmatrix} 1 \\ -2 \\ 1 \end{bmatrix}$

33.

35.

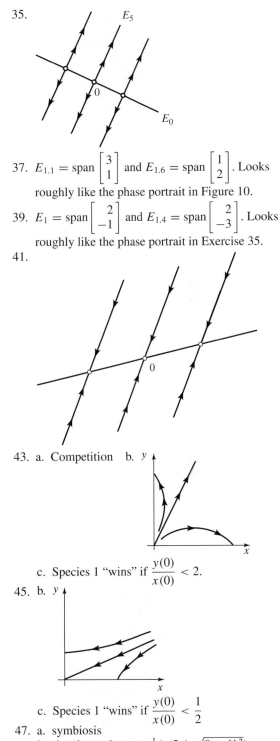

37. $E_{1.1} = \operatorname{span} \begin{bmatrix} 3 \\ 1 \end{bmatrix}$ and $E_{1.6} = \operatorname{span} \begin{bmatrix} 1 \\ 2 \end{bmatrix}$. Looks roughly like the phase portrait in Figure 10.

39. $E_1 = \operatorname{span} \begin{bmatrix} 2 \\ -1 \end{bmatrix}$ and $E_{1.4} = \operatorname{span} \begin{bmatrix} 2 \\ -3 \end{bmatrix}$. Looks roughly like the phase portrait in Exercise 35.

41.

43. a. Competition   b.

c. Species 1 "wins" if $\dfrac{y(0)}{x(0)} < 2$.

45. b.

c. Species 1 "wins" if $\dfrac{y(0)}{x(0)} < \dfrac{1}{2}$

47. a. symbiosis
    b. the eigenvalues are $\frac{1}{2}(-5 \pm \sqrt{9 + 4k^2})$. There are two negative eigenvalues if $k < 2$;

if $k > 2$ there is a negative and a positive eigenvalue.

49. $g(t) = 45e^{-0.8t} - 15e^{-0.4t}$ and
$h(t) = -45e^{-0.8t} + 45e^{-0.4t}$

53. $\vec{x}(t) = e^{pt}\begin{bmatrix}\cos(qt)\\\sin(qt)\end{bmatrix}$, a spiral if $p \neq 0$ and a circle if $p = 0$. Approaches the origin if p is negative.

55. Eigenvalues $\lambda_{1,2} = \frac{1}{2}(-q \pm \sqrt{q^2 - 4p})$; both eigenvalues are negative

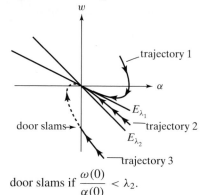

door slams if $\dfrac{\omega(0)}{\alpha(0)} < \lambda_2$.

9.2
1. 1

3. $\sqrt{2}e^{3\pi i/4}$

5. $e^{-0.1t}(\cos(2t) - i\sin(2t))$; spirals inward, in the clockwise direction

7. not stable

9. stable

11. a. $B = 2A$

d. The zero state is a stable equilibrium of the system $\dfrac{d\vec{x}}{dt} = \text{grad}(q)$ if (and only if) q is negative definite (then, the eigenvalues of A and B are all negative).

13. The eigenvalues of A^{-1} are the reciprocals of the eigenvalues of A; the real parts have the same sign.

15.

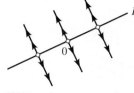

17. If $|k| < 1$

19. False; consider A with eigenvalues $1, 2, -4$.

21. a. $\begin{vmatrix}\dfrac{db}{dt} = 0.5b + & s\\[2mm]\dfrac{ds}{dt} = & 0.07s\end{vmatrix}$

b. $b(t) = 50{,}000e^{0.07t} - 49{,}000e^{0.5t}$
$s(t) = 1{,}000e^{0.07t}$

27. $\vec{x}(t) = \begin{bmatrix}\cos(3t) & -\sin(3t)\\\sin(3t) & \cos(3t)\end{bmatrix}\begin{bmatrix}a\\b\end{bmatrix}$, where a, b are arbitrary constants

29. Eigenvalue $2 + 4i$ with corresponding eigenvector $\begin{bmatrix}i\\-1\end{bmatrix}$. Use Fact 9.2.6, with $p = 2$,
$q = 4, \vec{w} = \begin{bmatrix}1\\0\end{bmatrix}, \vec{v} = \begin{bmatrix}0\\-1\end{bmatrix}$.
$\vec{x}(t) = e^{2t}\begin{bmatrix}1 & 0\\0 & -1\end{bmatrix}\begin{bmatrix}\cos(4t) & -\sin(4t)\\\sin(4t) & \cos(4t)\end{bmatrix}\begin{bmatrix}a\\b\end{bmatrix}$

31. Eigenvalue $-1 + 2i$ with corresponding eigenvector $\begin{bmatrix}i\\1\end{bmatrix}$.
$\vec{x}(t) = e^{-t}\begin{bmatrix}\cos(2t) & -\sin(2t)\\\sin(2t) & \cos(2t)\end{bmatrix}\begin{bmatrix}1\\-1\end{bmatrix} =$
$e^{-t}\begin{bmatrix}\cos(2t) + \sin(2t)\\\sin(2t) - \cos(2t)\end{bmatrix}$. Spirals inward, in the counterclockwise direction.

33. Eigenvalue i with corresponding eigenvector $\begin{bmatrix}1\\1+i\end{bmatrix}$. $\vec{x}(t) = \begin{bmatrix}0 & 1\\1 & 1\end{bmatrix}\begin{bmatrix}\cos(t)\\\sin(t)\end{bmatrix} =$
$\begin{bmatrix}\sin(t)\\\sin(t) + \cos(t)\end{bmatrix}$. An ellipse with clockwise orientation.

39. The system $\dfrac{d\vec{c}}{dt} = \begin{bmatrix}0 & 1 & 0\\0 & 0 & 1\\0 & 0 & 0\end{bmatrix}\vec{c}$ has the solutions
$\vec{c}(t) = \begin{bmatrix}k_1 + k_2 t + k_3 t^2/2\\k_2 + k_3 t\\k_3\end{bmatrix}$,
where k_1, k_2, k_3 are arbitrary constants. The solutions of the given system are $\vec{x}(t) = e^{\lambda t}\vec{c}(t)$, by Exercise 9.1.24. The zero state is a stable equilibrium solution if (and only if) the real part of λ is negative.

9.3
1. Ce^{5t}

3. $\frac{1}{5}e^{3t} + Ce^{-2t}$ (use Fact 9.3.13)

5. $-1 - t + Ce^{t}$

7. $c_1 e^{-4t} + c_2 e^{3t}$

9. $c_1 e^{3t} + c_2 e^{-3t}$

11. $e^{t}(c_1 \cos(t) + c_2 \sin(t))$

13. $e^{-t}(c_1 + c_2 t)$ (compare with Example 10).

15. $c_1 + c_2 t$

17. $e^{-t}(c_1 + c_2 t) - \frac{1}{2}\cos(t)$

19. $\cos(t) + c_1 \cos(\sqrt{2}t) + c_2 \sin(\sqrt{2}t)$

21. $c_1 e^{t} + c_2 e^{-t} + c_3 e^{-2t}$

23. $3e^{5t}$

25. e^{-2t+2}

27. $-\sin(3t)$

29. $\frac{1}{3}\sin(t) - \frac{1}{6}\sin(2t)$

31. $v(t) = \dfrac{mg}{k}(1 - e^{-kt/m})$

$\lim\limits_{t\to\infty} v(t) = \dfrac{mg}{k} = $ terminal velocity.

35. a. $c_1 e^{-t} + c_2 e^{-2t}$

b. $2e^{-t} - e^{-2t}$

c. $-e^{-t} + 2e^{-2t}$

d. In part (c) the oscillator goes through the equilibrium state once; in part (b) it never reaches it.

37. $x(t) = te^{-3t}$

39. $e^{-t}(c_1 + c_2 t + c_3 t^2)$

41. λ is an eigenvalue with $\dim(E_\lambda) = n$, because E_λ is the kernel of the nth-order linear differential operator $T(x) - \lambda x$.

43. $\frac{1}{10}\cos(t) + \frac{1}{10}\sin(t) + c_1 e^{-2t} + c_2 e^{-3t}$

45. $e^t \begin{bmatrix} 1 - 2t \\ -1 \end{bmatrix}$

Subject Index

Name Index